PYRIDINE AND ITS DERIVATIVES

SUPPLEMENT IN FOUR PARTS,
PART TWO

This is the fourteenth volume in the series

THE CHEMISTRY OF HETEROCYCLIC COMPOUNDS

THE CHEMISTRY OF HETEROCYCLIC COMPOUNDS

A SERIES OF MONOGRAPHS

ARNOLD WEISSBERGER and EDWARD C. TAYLOR

Editors

PYRIDINE
AND
ITS DERIVATIVES
SUPPLEMENT
PART TWO

Edited by

R. A. Abramovitch

University of Alabama

AN INTERSCIENCE® PUBLICATION

JOHN WILEY & SONS

NEW YORK · LONDON · SYDNEY · TORONTO

An Interscience® Publication

Copyright © 1974, by John Wiley & Sons, Inc.

Library of Congress Cataloging in Publication Data:

Abramovitch, R. A.   1930–
  Pyridine supplement.

  (The Chemistry of heterocyclic compounds, v. 14)
  "An Interscience publication."
  Supplement to E. Klingsberg's Pyridine and its derivatives.
  Includes bibliographical references.
  1. Pyridine. I. Klingsberg, Erwin, ed. Pyridine and its derivatives. II. Title.

QD401.A22   547′.593   73-9800

ISBN 0-471-37914-X

Printed in the United States of America

10 9 8 7 6 5 4 3 2 1

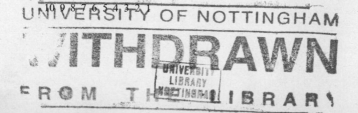

Contributors

R. A. ABRAMOVITCH, *Department of Chemistry, University of Alabama, University, Alabama*

MAX M. BOUDAKIAN, *Olin Corporation, Rochester, New York*

RONALD G. MICETICH, *Raylo Chemicals Limited, Edmonton, Alberta, Canada*

ELIZABETH M. SMITH, *Wayne State University, Detroit, Michigan, and Schering Corporation, Bloomfield, New Jersey*

HARRY L. YALE, *Squibb Institute for Medical Research, New Brunswick, New Jersey*

TO THE MEMORY OF

Michael

The Chemistry of Heterocyclic Compounds

The chemistry of heterocyclic compounds is one of the most complex branches of organic chemistry. It is equally interesting for its theoretical implications, for the diversity of its synthetic procedures, and for the physiological and industrial significance of heterocyclic compounds.

A field of such importance and intrinsic difficulty should be made as readily accessible as possible, and the lack of a modern detailed and comprehensive presentation of heterocyclic chemistry is therefore keenly felt. It is the intention of the present series to fill this gap by expert presentations of the various branches of heterocyclic chemistry. The subdivisions have been designed to cover the field in its entirety by monographs which reflect the importance and the interrelations of the various compounds, and accommodate the specific interests of the authors.

In order to continue to make heterocyclic chemistry as readily accessible as possible new editions are planned for those areas where the respective volumes in the first edition have become obsolete by overwhelming progress. If, however, the changes are not too great so that the first editions can be brought up-to-date by supplementary volumes, supplements to the respective volumes will be published in the first edition.

ARNOLD WEISSBERGER

Research Laboratories
Eastman Kodak Company
Rochester, New York

EDWARD C. TAYLOR

Princeton University
Princeton, New Jersey

Preface

Four volumes covering the pyridines were originally published under the editorship of Dr. Erwin Klingsberg over a period of four years, Part I appearing in 1960 and Part IV in 1964. The large growth of research in this specialty is attested to by the fact that a supplement is needed so soon and that the four supplementary volumes are larger than the original ones. Pyridine chemistry is coming of age. The tremendous variations from the properties of benzene achieved by the replacement of an annular carbon atom by a nitrogen atom are being appreciated, understood, and utilized.

Progress has been made in all aspects of the field. New instrumental methods have been applied to the pyridine system at an accelerating pace, and the mechanisms of many of the substitution reactions of pyridine and its derivatives have been studied extensively. This has led to many new reactions being developed and, in particular, to an emphasis on the direct substitution of hydrogen in the parent ring system. Moreover, many new and important pharmaceutical and agricultural chemicals are pyridine derivatives (these are usually ecologically acceptable, whereas benzene derivatives usually are not). The modifications of the properties of heteroaromatic systems by *N*-oxide formation are being exploited extensively.

For the convenience of practitioners in this area of chemistry and of the users of these volumes, essentially the same format and the same order of the supplementary chapters are maintained as in the original. Only a few changes have been made. Chapter I is now divided into two parts, Part A on pyridine derivatives and Part B on reduced pyridine derivatives. A new chapter has been added on pharmacologically active pyridine derivatives. It had been hoped to have a chapter on complexes of pyridine and its derivatives. This chapter was never received and it was felt that Volume IV could not be held back any longer.

The decision to publish these chapters in the original order has required sacrifices on the part of the authors, for while some submitted their chapters on time, others were less prompt. I thank the authors who finished their chapters early for their forebearance and understanding. Coverage of the literature starts as of 1959, though in many cases earlier references are also given to present sufficient background and make the articles more readable. The literature is covered until 1970 and in many cases includes material up to 1972.

I express my gratitude to my co-workers for their patience during the course of this undertaking, and to my family, who saw and talked to me even less than usual during this time. In particular, I acknowledge the inspiration given me by the strength and smiling courage of my son, Michael, who will never know how much the time spent away from him cost me. I hope he understood.

R. A. ABRAMOVITCH

University, Alabama
June 1973

Contents

Part Two

PYRIDINE AND ITS DERIVATIVES

SUPPLEMENT IN FOUR PARTS,
PART TWO

This is the fourteenth volume in the series
THE CHEMISTRY OF HETEROCYCLIC COMPOUNDS

CHAPTER IV

Pyridine-1-oxides

R. A. ABRAMOVITCH and ELIZABETH M. SMITH*

Department of Chemistry, University of Alabama, University, Alabama
Wayne State University, Detroit, Michigan

* *Present address:* Natural Products Dept., Schering Corporation, Bloomfield, N.J.

Since the early 1940s, Japanese workers (1), as well as den Hertog (2) in Holland and Colonna (3) in Italy opened up the field of pyridine-1-oxides. The chemistry of these compounds has been reviewed by Katritzky (4a, b), Shaw (5), Abramovitch and Saha (6), and Ochiai (7). For the most part, this chapter is a supplement to the one previously written by Shaw (5) in the 1961 edition of this series. Repetition occurs only for continuity.

I. Preparation

Pyridine-1-oxides may be prepared by the direct oxidation of pyridines with per-acids, or by the introduction or the exchange of substituents in the ring with retention of the N-oxide group already present. The latter method permits the synthesis of those N-oxides which are not available by direct oxidation, but this method is discussed separately under the appropriate section on the reactions of pyridine-1-oxides. Only a few pyridine-1-oxides have been prepared from nonpyridine precursors.

1. Direct Oxidation

Pyridine was first oxidized to the N-oxide with perbenzoic acid (8). Perphthalic acid (9) and hydrogen peroxide in glacial acetic acid (10) have also been used. Recently, it has been found that 2,6-dibromopyridine (**IV-1**) can be converted to its N-oxide (**IV-2**) with 30% hydrogen peroxide in trifluoroacetic acid in spite of the steric hindrance and the inductive effects of the bromine atoms which prevented N-oxidation with perbenzoic or peracetic acid (11). This reagent is also effective in the N-oxidation of 2,3,5,6-tetrachloropyridine and 2,3,4,5,6-pentachloropyridine (12), as well as

2-chloro-3-nitropyridine (13), and 2-bromo-5-chloropyridine (14). Using the reagent, 2-(2,4,5-trichlorophenylthio)pyridine (**IV-3**) was converted to the corresponding sulfone-1-oxide (**IV-4**) (15). Oxidation of pentachloropyridine

in trichloroperacetic acid at 50° for 5 hr yields the N-oxide (ca. 20%) and unreacted starting material (16, 17). Higher temperatures or longer heating caused deoxygenation until the concentration of N-oxide fell to ca. 10%.

Loss of oxygen was due to the presence of hydrogen peroxide since prolonged heating of the *N*-oxide in boiling trifluoroacetic acid had no untoward effect. This reaction could involve nucleophilic attack by the *N*-oxide on the peroxy acid, followed by loss of molecular oxygen from the intermediate *N*-peroxide (eq. IV-1) (17a).

$$C_5H_5N^+\!\!-\!\!O^- + O\!\!-\!\!\overset{\displaystyle}{O}COR \longrightarrow C_5H_5N^+\!\!-\!\!O\!\!-\!\!OH + OCOR^- \xrightarrow{-RCO_2H}$$
$$\overset{|}{H}$$

$$C_5H_5N^+\!\!-\!\!O\!\!-\!\!O^- \longrightarrow C_5H_5N + O_2 \qquad\qquad (IV\text{-}1)$$

Pentachloropyridine-1-oxide was obtained in high yield ($>80\%$) from the reaction of pentachloropyridine with a mixture of an organic peracid and concentrated sulfuric acid (17b). For best results, the substrate was dissolved in a mixture of sulfuric acid and acetic acid, and hydrogen peroxide, varying in strength from 30 to 90%, was added slowly with stirring at or below room temperature. The success of a mixture of peracid and sulfuric acid in oxidizing weakly basic nitrogen compounds is probably due to enhancement of the electrophilicity of the peroxidic oxygen through protonation of the acid which must ease the transfer of H^+ to the nucleophile, as indicated in **IV-5** and **IV-6** (17b). Substitution of polyphosphoric acid for sulfuric acid usually gave inferior results with these compounds (17b), but a mixture of polyphosphoric and peracetic acid was effective in the *N*-oxidation of 4-methoxytetrachloropyridine. In the presence of sulfuric acid, this compound suffers hydrolysis and gives the corresponding 4-hydroxytetrachloropyridine-1-oxide (17b).

IV-5   **IV-6**

m-Chloroperbenzoic acid has been used for the *N*-oxidation of heterocyclic compounds (18a) and polyarylpyridines (18b). Monopermaleic acid has been shown to be effective in the oxidation of 2-chloro- (19a), 3-methoxy-, 3-fluoro-, and 3-mesylaminopyridine (**IV-7** (19b)) to their *N*-oxides (**IV-8**).

$$\text{IV-7} \quad \boxed{} -X + \begin{matrix} CHCO_3H \\ \| \\ CHCO_2H \end{matrix} \longrightarrow \quad \boxed{} -X$$

IV-7   IV-8

3-Pyridyltetra-*O*-acetyl-*β*-D-glucopyranoside,   4-pyridyltetra-*O*-acetyl-*β*-D-glucopyranoside, and 4-pyridyl-*β*-D-glucopyranoside were converted to the *N*-oxides with perbenzoic acid (20). Vinylpyridines (21) and polyvinyl-pyridines (22a, b) are also oxidized with peracids to the *N*-oxides.

Recently, a new organic oxidant, (dibenzoyldioxyiodo)benzene, has been used for the preparation of pyridine-1-oxide (85%) in 30 min (eq. IV-2) (23). This reagent is obtained by the reaction of iodosobenzenes with substituted perbenzoic acids in chloroform-methanol (8:1) at 0°. A word of caution: (dibenzoyldioxyiodo)benzenes detonate at 80 to 120° and many spontane-ously ignite or detonate upon manipulation in the solid state at room temperature; also manipulation with metal spatulas is not recommended (23).

$$ArIO + 2ArCO_3H \longrightarrow ArI(O-O-COAr)_2 \xrightarrow{\;C_5H_5N\;} \quad \boxed{} \quad + ArI(OCOAr)_2 \quad (IV-2)$$

The formation and properties of pyridine-1-oxides obtained by direct oxidation are given in Table IV-1.

TABLE IV-1. Preparation of Pyridine-1-oxides by Direct Oxidation

Pyridine (R)	Conditions	Properties	Yield (%)	Ref.
Pyridine	H_2O_2, AcOH, Δ, 6 hr	m.p. 66°	95–100	24
	H_2O_2, AcOH	m.p. 64–6°, b.p. 100–105°/1 mm		25
	$ArI(O_2COAr)_2$, 30 min		85	23
2,6-d_2	H_2O_2, AcOH	b.p. 130–131°/9 mm		26
Pyridine, HCl	(i) 50% H_2O_2, AcOH, tungstic acid (ii) conc. HCl		85	27
	(i) molybdic anhydride, 60°, 2 hr (ii) Ca(OH)$_2$ (iii) HCl			27
Pyridine-$^{14}C_2$, HCl	(i) 40% AcO_2H, <80°, 6 hr (ii) HCl	m.p. 180–181°		28

Table IV-1 (*Continued*)

Pyridine (R)	Conditions	Properties	Yield (%)	Ref.
2-Me	H$_2$O$_2$, AcOH	b.p. 124–125°/15 mm n_D^{20} 1.5910	84	25
	H$_2$O$_2$, AcOH, 1–1.5 hr	b.p. 120–125°/14 mm n_D^{20} 1.591		29
3-Me	H$_2$O$_2$, AcOH	m.p. 33–36° b.p. 150°/15 mm	81	25
	H$_2$O$_2$, AcOH, 1–1.5 hr	b.p. 150–153°/20 mm		29
4-Me	H$_2$O$_2$, AcOH	m.p. 184°	86	25
	H$_2$O$_2$, AcOH, 1–1.5 hr	m.p. 183.5–184°		29
4-Me, HCl	(i) 50% H$_2$O$_2$, 60°, 6 hr (ii) HCl		72.5	27
2,3-Me$_2$	30% H$_2$O$_2$, AcOH	b.p. 118°/4 mm		30
2,4-Me$_2$	H$_2$O$_2$, AcOH, 70°, 16 hr	b.p. 110°/1 mm	79	25
	30% H$_2$O$_2$, AcOH	b.p. 91°/0.3 mm		30
2,5-Me$_2$	H$_2$O$_2$	m.p. 53–58°		31
	30% H$_2$O$_2$, AcOH, 80–90°	m.p. 55–57°; picrate, m.p. 122.5°		32
	30% H$_2$O$_2$, AcOH, 80°, 10 hr	m.p. 55–57°; picrate, m.p. 122.4–124°		33
	30% H$_2$O$_2$, AcOH	b.p. 86°/0.3 mm		30
	30% H$_2$O$_2$, AcOH, 70–80°, 10 hr	m.p. 60°	80.5	34
2,6-Me$_2$	H$_2$O$_2$, AcOH, 1–1.5 hr	b.p. 120–123°/21 mm		29
	30% H$_2$O$_2$, AcOH	b.p. 107–108°/5 mm	76	35
	30% H$_2$O$_2$, phthalic anhydride		35	35
	H$_2$O$_2$, AcOH			36
3,4-Me$_2$	H$_2$O$_2$, AcOH, 70–80°, 18 hr	m.p. 138°	77.4	34
	H$_2$O$_2$, AcOH			37
4-t-Bu-2,5-Me$_2$	30% H$_2$O$_2$, AcOH, 80°, 10 hr	m.p. 77–78°; picrate, m.p. 122.4–124°		33
2,3,6-Me$_3$	30% H$_2$O$_2$, AcOH	m.p. 63°, b.p. 84°/ 0.2 mm; picrate, m.p. 97–98.5°		30
2,4,6-Me$_3$	30% H$_2$O$_2$, AcOH			30
3-t-Bu-2,4,6-Me$_3$	30% H$_2$O$_2$, AcOH, 70–80°	m.p. 120–121°, b.p. 168– 170°/15 mm		38, 39
2-Et	H$_2$O$_2$, AcOH	b.p. 103–105°/3 mm; picrate, m.p. 134–137°		40
5-Et-2-Me	30% H$_2$O$_2$, AcOH	b.p. 120°/2 mm; picrate, m.p. 110°		35
2,6-Et$_2$		b.p. 120–127°/10 mm	57.8	41
2-Pr	H$_2$O$_2$, AcOH, 70°, 3 hr	b.p. 102°/2 mm		42
2,6-Pr$_2$		b.p. 135–139°/10 mm	57.3	41

Table IV-1 (*Continued*)

Pyridine (R)	Conditions	Properties	Yield (%)	Ref.
2,6-Bu$_2$		b.p. 162–164°/10 mm	57	41
2,6-Di-*n*-amyl		b.p. 180–183°/8 mm	59.5	41
2-(CH$_2$)$_7$CH$_3$		b.p. 140°/0.7 mm n_D^{20} 1.5220		43
2-Methyloctyl	30% H$_2$O$_2$, AcOH, cat. H$_2$SO$_4$, R.T.	b.p. 137–138°/0.1 mm n_D^{21} 1.5256; picrate m.p. 88–89°		44
2-(CH$_2$)$_8$CH$_3$	30% H$_2$O$_2$, AcOH, cat. H$_2$SO$_4$, R.T.	m.p. 39–40°, b.p 52°/ 1 mm n_D^{19} 1.5230; HCl, m.p. 85–86°; picrate, m.p. 77–78°		44
2-(CH$_2$)$_9$CH$_3$				43
2-CH[(CH$_2$)$_5$CH$_3$]$_2$				43
2-(CH$_2$)$_{11}$CH$_3$		m.p. 54–56°		43
2-(CH$_2$)$_{16}$CH$_3$	AcO$_2$H	m.p. 67–67.5°		43
	35% H$_2$O$_2$, AcOH, 78–80°, 12 hr	m.p. 70–72°		45
2-Tridecyl	BzO$_2$H in CHCl$_3$, 0.16 hr	m.p. 55–56°		43
3-Tridecyl	BzO$_2$H in CHCl$_3$, 0.16 hr	m.p. 54–55°		43
4-Tridecyl	H$_2$O$_2$, AcOH	m.p. 61–63°		43
3-Trimethylsilyl	H$_2$O$_2$, AcOH	b.p. 146°/6 mm; picrate, m.p. 107–108°		46
4-Et	AcO$_2$H	m.p. 106–109°	84	47
4-Heptadecyl	35% H$_2$O$_2$, AcOH, 75–80°, 12 hr	m.p. 31–32°		45
1,2-Di-(4-pyridyl)ethane		m.p. 236–238°		48
1,2-Di-(6-methyl-2-pyridyl)ethane	30% H$_2$O$_2$, AcOH, 70–80°, 3 hr	m.p. 183–185°		49
1,2-Di(4-ethoxy-3-pyridyl)ethane	AcO$_2$H			50
1,2-Di-(6-methyl-2-pyridyl)ethylene	30% H$_2$O$_2$, AcOH, 70–80°, 3 hr	m.p. 247.5–249°		49
1,3-Di-(2-pyridyl)propane	H$_2$O$_2$, AcOH	m.p. 212–213°; picrate, m.p. 138°	8	51
1,5-Di-(4-pyridyl)pentane		m.p. 218–220°		48
1,6-Di-(2-pyridyl)hexane		m.p. 140–141°		48
1,6-Di-(3-pyridyl)hexane		m.p. 186–188°		48
1,6-Di-(4-pyridyl)hexane		m.p. 241–244°		48
1-(2-Pyridyl)-6-(3-pyridyl)hexane		m.p. 108–110°		48
1-(2-Pyridyl)-6-(4-pyridyl)hexane·$\frac{1}{2}$H$_2$O		m.p. 135–137°		48
1,7-Di-(4-pyridyl)heptane		m.p. 152–154°		48
1,8-Di-(4-pyridyl)octane		m.p. 155.7°		48

Table IV-1 (*Continued*)

Pyridine (R)	Conditions	Properties	Yield (%)	Ref.
2,7-Di-(4-pyridyl)octane		m.p. 130–132°		48
2,2′-Dipyridylmethyl ether	H_2O_2, AcOH	m.p. 155°		52
2-Pyridylmethyl-2-(2-pyridylethyl)ether		m.p. 126°	60	52
Di-(2-pyridyl)glycol	30% H_2O_2, AcOH, 85°, 15 hr	m.p. 213°		53
Di-(6-methyl-2-pyridyl)glycol	30% H_2O_2, AcOH, 85°, 15 hr	m.p. 242°		53
dl-Di-(4-pyridyl)glycol (hydrate)	30% H_2O_2, AcOH, 85°, 15 hr	m.p. 200°		53
meso-Di-(4-pyridyl)glycol	30% H_2O_2, AcOH, 50°, 5 hr		54	53
3-F	30% H_2O_2, AcOH, 70–80°, 2 hr	m.p. 64°		54
	30% H_2O_2, AcOH	m.p. 62.5–63°; picrate, m.p. 107–108°		55
	30% H_2O_2, AcOH, 70–80°, 12 hr	m.p. 63°, b.p. 78–80°/0.3 mm	19	19b
	(i) Permaleic acid, $CHCl_3$, 0° (ii) 25°, 48 hr		38	19b
2-F-5-Me	30% H_2O_2, Ac_2O			56
2-Cl	Maleic anhyd. in CH_2Cl_2 70% H_2O_2, 50°, 1.5 hr		81.6	19a
	succinic anhydride			19a
	30% H_2O_2, AcOH, 70–80°, 12 hr	m.p. 59–60°; picrate, m.p. 165–166°	80	36
	60–90% aq. H_2O_2 in CH_2Cl_2, 45–70°			57
	40% AcO_2H, 70°, 150 min			58
	40% AcO_2H, AcOH, 45°			59
2-Cl·HCl	(i) 35% H_2O_2, 60°, 6 hr (ii) conc. HCl		78	27
2,6-Cl_2	90% H_2O_2, AcOH, CF_3CO_2H, 90°, 18 hr		65	17b
2-Cl-3-NO_2	$(CF_3CO)_2O$, H_2O_2 in CH_2Cl_2	m.p. 99–100°		13, 60
2-Cl-5-NO_2	CF_3CO_3H			60
2,3,5,6-Cl_4	H_2O_2, CF_3CO_2H, 50°	m.p. 212–215°	78	12
	90% H_2O_2, AcOH, CF_3CO_2H, 20°, 48 hr		80	17b
2,3,5,6-Cl_4-4-Me	90% H_2O_2, AcOH, CF_3CO_2H, 60°, 18 hr		70	17b
2,3,5,6-Cl_4-4-OMe	AcO_2H, PPA		80	17b
2,3,5,6-Cl_4-4-NO_2	90% H_2O_2, CF_3CO_2H, H_2SO_4, 20°, 24 hr		67	17b

Table IV-1 (*Continued*)

Pyridine (R)	Conditions	Properties	Yield (%)	Ref.
2,3,4,5,6-Cl$_5$	H$_2$O$_2$, CF$_3$CO$_2$H, 50°	m.p. 177–181°	78	12
	90% H$_2$O$_2$, AcOH, CF$_3$CO$_2$H, 20°, 48 hr		85	17b
3-Cl	30% H$_2$O$_2$, AcOH, 70–80°, 3 hr	m.p. 59–60°, picrate, m.p. 137–139°; b.p. 88–89°/1 mm	99	36
	30% H$_2$O$_2$, AcOH, Δ, 3 hr			55
3-Cl-2,6-Me$_2$	35% H$_2$O$_2$, AcOH	m.p. 38–40°		61
2-Br-5-Cl	H$_2$O$_2$, CF$_3$CO$_2$H, Δ, 3 hr	m.p. 111–112°	69	14
3-Cl-2-(SO$_2$C$_6$H$_4$Cl-*p*)	30% H$_2$O$_2$, CF$_3$CO$_2$H, 3 hr	m.p. 190–192°	79	15
3-Cl-2-(2,4,5-Cl$_3$C$_6$H$_2$SO$_2$)	30% H$_2$O$_2$, CF$_3$CO$_2$H	m.p. 233.5–237°		15
4-Cl-2,6-Me$_2$	35% H$_2$O$_2$, AcOH	m.p. 102–103°, picrate, m.p. 147–148°		61
4-Cl-2-OMe	BzO$_2$H, 8 days			62
2-Br	30% H$_2$O$_2$, AcOH	b.p. 97–99°/0.5 mm		55, 63
	30% H$_2$O$_2$, AcOH, 80–90°, 12 hr	b.p. 118–120°/0.5 mm	40.5	19b
2-Br-3-Et-6-Me	H$_2$O$_2$, AcOH, 60–70°, 4 hr	b.p. 102°/0.7 mm		64
2,6-Br$_2$	30% H$_2$O$_2$, CF$_3$CO$_2$H, Δ, 3 hr	m.p. 186.5–188.5° (decomp.)		11
3-Br-4-Me	BzO$_2$H	m.p. 138–140°	95	31
3-Br-2,5-Me$_2$	BzO$_2$H			31
3-Br-5-OEt				31
3,5-Br$_2$	40% AcO$_2$H, AcOH, 80°, 3 hr	m.p. 143–144°		65
5-Br-2-CONH$_2$	40% AcO$_2$H, 60–70°, 1 hr	m.p. 165–168°		66
4-Cl-5-OMe-2-CH$_2$Cl	30% H$_2$O$_2$, AcOH, 70°, 18 hr	m.p. 163–164°	60	67
3-OH-2,6-Me$_2$	H$_2$O$_2$	m.p. 177°		68
5-CN-3-OH-2-Me	30% H$_2$O$_2$, AcOH	m.p. 278–280° (decomp.)		69
3-OH-2-CO$_2$H		m.p. 94.5°	76	70
3-OH-2-CON(Me)Ph		m.p. 263–265° (decomp.)	95	70
2-OMe-6-Me	40% AcO$_2$H, AcOH			71
2-OEt-6-Me	40% AcO$_2$H, AcOH			71
2-OMe-4-Cl	BzO$_2$H, 8 days			62
2-OMe-6-CO$_2$H	H$_2$O$_2$, AcOH, Δ, 5 hr	m.p. 186–188°		62
2-OMe-4-CONH$_2$	BzO$_2$H	m.p. 198–200°		62
3-OMe	(i) permaleic acid, CHCl$_3$, 0°, 1 hr (ii) 25°, 48 hr	m.p. 100–101°	46	19b
5-Et-3-OMe-2-Me	30% AcO$_2$H, 80°, 5 hr	b.p. 140–143°/5 mm		72

9

Table IV-1 (*Continued*)

Pyridine (R)	Conditions	Properties	Yield (%)	Ref.
3-OAc-2-Me-4,5-(CH$_2$OAc)$_2$	BzO$_2$H, CHCl$_3$	m.p. 114–115°		73
2-CH$_2$OAc	H$_2$O$_2$, AcOH	b.p. 160–175°/5 mm		74
5-Acetoxymethyl-2-(5-nitrofurylvinyl)pyridine				75
3-CH$_2$CO$_2$Et	36% H$_2$O$_2$, AcOH, 60–70°	m.p. 97–98°		76
2-Me-3-CH$_2$CO$_2$Et	H$_2$O$_2$, AcOH, 80–85°, 11 hr	m.p. 54–59°; b.p. 140–150°/4 mm		77
4-CH(Ph)CO$_2$Me		m.p. 127–129°, crude; picrate, m.p. 127–129°	95	78
3-CH$_2$CH$_2$CO$_2$H	H$_2$O$_2$, AcOH, Δ, 3 hr	m.p. 144–153°		79
3-CH$_2$CN	30% H$_2$O$_2$, AcOH	m.p. 135.5–136.5°		80a
2-CH$_2$Cl	AcO$_2$H, H$_2$O$_2$, 70–78°	m.p. 75–77° (benzene) b.p. 90°/0.2 mm (50%)		80b
2-CH$_2$OH	H$_2$O$_2$, AcOH, 80–90°, 10 hr	m.p. 142–143° m.p. 141°		81a 82
6-Me-2-CH$_2$OH		m.p. 113–114°	67	81a
4-Me-2-CH$_2$OH	H$_2$O$_2$, Ac$_2$O	m.p. 86–88°	60	81a
2-CH$_2$CH$_2$OH		m.p. 93–95°; picrate, m.p. 123°		83
2-(CH$_2$)$_3$OH	H$_2$O$_2$, AcOH, 80°, 2 hr	m.p. 52–54°		84
2-CH$_2$CH(OH)CH$_2$CH$_3$	30% H$_2$O$_2$, AcOH, 80°, 2 hr	b.p. 126–127°/0.025 mm		85
2-CH$_2$CH(OH)CHMe$_2$	30% H$_2$O$_2$, AcOH, 80°, 2 hr	b.p. 150–155°/0.025 mm		85
2-CH$_2$CH(OH)C$_9$H$_{19}$	30% H$_2$O$_2$, AcOH, 80°, 2 hr	b.p. 188°/0.02 mm		85
2-CH$_2$C(OH)(Me)Ph	30% H$_2$O$_2$, AcOH, 80°, 2 hr	b.p. 124–127°/0.008 mm		85
2-(1-Hydroxycyclohexyl)-pyridine	30% H$_2$O$_2$, AcOH, 70–80°, 60 hr	m.p. 88°	86.4	34
6-(1-Hydroxycyclohexyl)-3-picoline	30% H$_2$O$_2$, AcOH, 70°, 24 hr	m.p. 123–125°	56.7	34
1-(2-Pyridylmethyl)cyclohexanol	30% H$_2$O$_2$, AcOH, 70°, 18 hr	m.p. 113–134°	89.4	34
3-CH$_2$OH	H$_2$O$_2$, AcOH, 80–90°, 10 hr	m.p. 84–86° m.p. 93–94°	65	82
2-Me-5-CH$_2$OH	H$_2$O$_2$, AcOH, 70–80°, 3 hr	m.p. 87°		86
4-CH$_2$OH	H$_2$O$_2$, AcOH, 80–90°, 10 hr	m.p. 124–125°		81, 82
2-Me-3-CH$_2$OMe	H$_2$O$_2$, AcOH	m.p. 55–60°; b.p. 135–138°/2 mm; picrate, m.p. 92–95°		87
2,3-(CHO)$_2$	Monoperphthalic acid, EtOH, HCl	m.p. 128°		88
2,4-(CHO)$_2$	Monoperphthalic acid, EtOH, HCl	m.p. 148°		88
2,5-(CHO)$_2$	Monoperphthalic acid, EtOH, HCl	m.p. 208°		88

Table IV-1 (*Continued*)

Pyridine (R)	Conditions	Properties	Yield (%)	Ref.
2,6-(CHO)$_2$	Monoperphthalic acid, EtOH, HCl			88
2-COMe		b.p. 82°/0.3 mm; picrolonate, m.p. 157°		89
4-COMe	Phthalic anhydride, H$_2$O$_2$, 50–60°, 24 hr	m.p. 134–136°		90
2-Pyridylaldoxime	30% H$_2$O$_2$, AcOH, 70–80°	m.p. 221–222°; pK_a 9.23	34	91
3-Pyridylaldoxime	30% H$_2$O$_2$, AcOH, 70–80°	m.p. 217–218°; pK_a 9.68	46	91
4-Pyridylaldoxime	30% H$_2$O$_2$, AcOH, 70–80°	m.p. 228–229°; pK_a 9.57	50	91
2-Benzoylpyridineoxime		syn-, m.p. 163–164.5°, anti-, m.p. 219–222°		92
3-Benzoylpyridineoxime		syn-, m.p. 162–163°, anti-, m.p. 178–180°		92
4-Benzoylpyridineoxime		syn-, m.p. 186–188°, anti-, m.p. 222–223°		92
2-CH=NNHPh	BzO$_2$H in C$_6$H$_6$	m.p. 237°		93
4-CH=NNHPh	BzO$_2$H			93
2-CO$_2$H	H$_2$O$_2$, AcOH, 75–80°, 3 hr	m.p. 162°	93	94
	H$_2$O$_2$ (K salt)	m.p. 162.5°		95
	30% H$_2$O$_2$, AcOH	m.p. 166–167°	70	96
2-CO$_2$H-6-Me	H$_2$O$_2$	m.p. 176°		49
2-CO$_2$H-5-Me	80% H$_2$O$_2$, AcOH	m.p. 162–163°	77.6	97
	30% H$_2$O$_2$, AcOH			34
2-CO$_2$H-4-Me			72	98
2-CO$_2$H-4-NO$_2$	H$_2$O$_2$, AcOH			99
2,6-(CO$_2$H)$_2$	NaOH, 45% AcO$_2$H in AcOH, 40°, 11 hr	m.p. 155–157°	73	100, 101, 102
2,5-(CO$_2$H)$_2$	NaOH, 45% AcO$_2$H in AcOH, 40°, 11 hr	m.p. 241–244°		100, 101
2,5-(CO$_2$H)$_2$-4-Ph	30% H$_2$O$_2$, AcOH, 90°, 10 hr	m.p. 167–168° (decomp.)		103
3-CO$_2$H	30% H$_2$O$_2$, 24 hr	m.p. 256°		104
4-CO$_2$H-2-CONH$_2$	H$_2$O$_2$, AcOH, 2 hr	m.p. 268–271°		105
3-CO$_2$Me		m.p. 97–98°		47
	(i) 30% H$_2$O$_2$, AcOH, 65°, 3 hr			47
	(ii) 90°, 8 hr	m.p. 99–100°	46	19b
3-CO$_2$Pr	30% H$_2$O$_2$, AcOH, 70°	m.p. 40–50.2°; picrate, m.p. 91.5–92°		106
3-CO$_2$-isoPr	30% H$_2$O$_2$, AcOH, 70°	m.p. 70°	71	106
3-CO$_2$-s-Bu	30% H$_2$O$_2$, AcOH, 70°	m.p. 62–62.5°; picrate, m.p. 90°	62	106

11

Table IV-1 (*Continued*)

Pyridine (R)	Conditions	Properties	Yield (%)	Ref.
3-CO$_2$Et-2-Me	H$_2$O$_2$, AcOH	m.p. 50–52°, b.p. 152°/4 mm		77, 107
3-CO$_2$Et-2,6-Me$_2$	H$_2$O$_2$, AcOH	m.p. 35–45°, b.p. 112–113°/0.01 mm		107
4-CN-5-CO$_2$Et-2-Me	H$_2$O$_2$, AcOH	m.p. 165–169°		108
3,5-(CO$_2$Et)$_2$-2,6-Me$_2$-4-Ph		m.p. 114°	95	109a
3,5-(CO$_2$Et)$_2$-2,5-Me$_2$-4-β-C$_{10}$H$_7$				109a
4-CO$_2$Me	30% H$_2$O$_2$, AcOH, 70–80°, 24 hr	m.p. 120–122°		109b
4-CO$_2$Et-2-Et	30% H$_2$O$_2$, AcOH, 70–80°, 3 hr	m.p. 140°		110
4-CO$_2$Et-2-n-Pr	30% H$_2$O$_2$, AcOH, 70–80°, 3 hr	m.p. 70°, b.p. 164–166°/7 mm	80	110
4-CO$_2$Me-2-n-Bu	30% H$_2$O$_2$, AcOH, 70–80°, 3 hr	m.p. 81°, b.p. 172–174°/8 mm	70	110
2-CONH$_2$	50% H$_2$O$_2$, Ac$_2$O, Δ, 6 hr	m.p. 161–162°	95	111
	40% AcO$_2$H, 80°, 6 hr	m.p. 165–166°	65	
2-CONHPh	Peracetic acid, THF	m.p. 143–145	84	70
2-CON(Ph)Me-3-OH		m.p. 263–265° (decomp.)	95	70
3-CONH$_2$		m.p. 287–288°		112
3-CONHEt	40% AcO$_2$H, AcOH, 80°, 3 hr	m.p. 123–124°	65	
3-CONHCH$_2$Ph-2-Me	H$_2$O$_2$, AcOH	m.p. 175–176°		77
4-CONH$_2$	34% H$_2$O$_2$, AcOH, 100°, 5 hr	m.p. 306°		110
2-CN	Peracetic acid	m.p. 116–118		113
3-CN	30% H$_2$O$_2$, AcOH, 70°	m.p. 174–175°; picrate, m.p. 91.5–92°	47	106, 113
	30% H$_2$O$_2$, AcOH, 80–90°, 7 hr	m.p. 174–175°	25	19b
4-CN	30% H$_2$O$_2$, AcOH, 80–90°, 4 hr	m.p. 224–226°		114
4-CN-2,6-d_2	H$_2$O$_2$, AcOH	m.p. 228.5–229°		115
4-CN-2-Me	30% H$_2$O$_2$, AcOH, Δ, 4 hr	m.p. 132–133°		99, 116
4-CN-2,6-Me$_2$	30% H$_2$O$_2$, AcOH, 4 hr	m.p. 187–188°		116
3-NO$_2$	(i) 40% AcO$_2$H, R.T., 4 hr (ii) 75°, 5 hr	m.p. 172–173°		117, 118
2,4-Me$_2$-3-NO$_2$	30% H$_2$O$_2$, AcOH, 80–85°, 48 hr	m.p. 118–119°; picrate, m.p. 113–114°		119
2,4-Me$_2$-5-NO$_2$	30% H$_2$O$_2$, AcOH, 80–85°, 48 hr	m.p. 102–103°		119
2,6-Me$_2$-3-NO$_2$	PhCO$_3$H, CHCl$_3$ 30% H$_2$O$_2$, Ac$_2$O	m.p. 100–102°	51	120, 121

Table IV-1 (*Continued*)

Pyridine (R)	Conditions	Properties	Yield (%)	Ref.
2-NH$_2$, HCl	(i) BzO$_2$H in C$_6$H$_6$, 10°, 3 hr (ii) R.T., 2 hr	m.p. 154–156°	76	122
2-NHMe		m.p. 67–68°; picrate, m.p. 156–158°		122
2-NHAc-5-NO$_2$		m.p. 196–197		123
2-N(COPh)$_2$	27.5% H$_2$O, AcOH, 70–80°, 5 hr	m.p. 98–99°		124
2-(NHCO$_2$Et)-5-Me	AcO$_2$H	m.p. 90–93°		125
2-(NHCO$_2$Et)-4,5-Me$_2$	AcO$_2$H	m.p. 111–112°		125
2,6-(NHCO$_2$Et)$_2$	AcO$_2$H	m.p. 114–116°		125
2,2′-Dipyridylamine	9% AcO$_2$H, 50–60°, 3 hr	m.p. 233–234°		126a
2-(NHC$_6$H$_3$Me$_2$-2,3)	30% H$_2$O$_2$, AcOH, 75°, 15 hr	m.p. 108–109°		126b
2-(NHC$_6$H$_4$CF$_3$-*m*)	30% H$_2$O$_2$, AcOH, 75°, 15 hr	m.p. 135–137°		126b
2-(NHC$_6$H$_4$Cl-*p*)-6-Me	30% H$_2$O$_2$, AcOH, 75°, 15 hr	m.p. 126–128°	35	126b
2-(NHC$_6$H$_4$CF$_3$-*m*)-6-Me	30% H$_2$O$_2$, AcOH, 75°, 15 hr	m.p. 101–102°		126b
3-NHCH$_2$Ph	H$_2$O$_2$, AcOH	m.p. 209–210°		127
3-(NHC$_6$H$_4$CF$_3$-*m*)	30% H$_2$O$_2$, AcOH, 80–85°, 19 hr	m.p. 135–137°		128
3-(NHCO$_2$Et)	30% H$_2$O$_2$, AcOH, 9 hr	m.p. 197°		129
3-NHCHO-4-Me	40% AcO$_2$H	m.p. 155–156°; picrate, m.p. 141–143°		130
3-NHAc	AcO$_2$H	m.p. 209–210.5°		130
	AcO$_2$H, 65°, 4 hr	m.p. 216°		117
		m.p. 208–211°		131
3-NHAc-4-Me	40% AcO$_2$H	m.p. 194–196°		130
3-NHAc-2,6-Me$_2$	30% H$_2$O$_2$, AcOH, 55–60°, 6 days	m.p. 165–167°		132
3-NHSO$_2$Me	(i) permaleic acid, CHCl$_3$, 0°, 1 hr (ii) 25°, 48 hr	m.p. 187–188°	44.7	19b
4-NMe$_2$	BzO$_2$H			133
4-NHCH$_2$Ph	H$_2$O$_2$, AcOH	m.p. 247–250°		127
2-CH=CH$_2$	Peracetic acid in EtOAc			21
2-Me-5-CH=CH$_2$	Peracetic acid in EtOAc, 2 hr			21
Poly(2-vinyl)pyridine	30% H$_2$O$_2$, AcOH			22b
2-CH=CHPh	H$_2$O$_2$, AcOH			134
		m.p. 135°		135
		m.p. 161–162°		136
3-CH=CHCO$_2$Pr		m.p. 89–91°	70	106
3-CH=CHCO$_2$Bu		m.p. 67–71°	93	106

Table IV-1 (*Continued*)

Pyridine (R)	Conditions	Properties	Yield (%)	Ref.
β-4-Pyridylacrylic acid	H_2O_2, AcOH, 70°, 18 hr	m.p. 237–240° (decomp.), hemi-acetate, m.p. 237–240° (decomp.)		137
β-4-Pyridylacrylamide	H_2O_2, AcOH, 70°, 18 hr	m.p. 246° (decomp.)		137
2-(5-Nitrofurylvinyl)-pyridine	30% H_2O_2, AcOH, 70–80°, 10 hr	m.p. 249–250°		138
2-(5-Nitrofurylvinyl)-3-picoline	30% H_2O_2, AcOH, 70–80°, 10 hr	m.p. 198–200°		138
6-(5-Nitrofurylvinyl)-3-picoline	30% H_2O_2, AcOH, 70–80°, 10 hr	m.p. 205–207° (decomp.)		138
5-Ethyl-2-(5-nitrofuryl-vinyl)pyridine	30% H_2O_2, AcOH, 70–80°, 10 hr	m.p. 155–157°		138
4-(5-Nitrofurylvinyl)-pyridine	30% H_2O_2, AcOH, 70–80°, 10 hr	m.p. 217–218.5°		138
3-Ethyl-4-(5-nitrofuryl-vinyl)pyridine	30% H_2O_2, AcOH, 70–80°, 10 hr	m.p. 235–236° (decomp.)		138
4-C≡CPh	Perbenzoic acid, 20°, 2 days	m.p. 184.5–185.5°; picrate, m.p. 147–148.5°		139
4-C≡CC$_6$H$_4$Me-*p*	Perbenzoic acid, 20°, 2 days	m.p. 158–159°	52	139
4-C≡CC$_6$H$_4$Cl-*p*	Perbenzoic acid, 20°, 2 days	m.p. 171.5–172.5°	50	139
4-C≡CC$_6$H$_4$NO$_2$-*p*	Perbenzoic acid, 20°, 2 days	m.p. 184.5–185.5°	43	139
2-Ph		m.p. 156°		133
2-(C$_6$H$_4$NO$_2$-*o*)	35% H_2O_2, AcOH, 90°, 9 hr	m.p. 156–157°		140
2-(C$_6$H$_4$NO$_2$-*m*)	30% H_2O_2, AcOH	m.p. 175–177°	78	141
2-(C$_6$H$_4$NO$_2$-*p*)	30% H_2O_2, AcOH	m.p. 212–214°	75	141
3-Ph		m.p. 121°		133
3-(C$_6$H$_4$NO$_2$-*m*)	30% H_2O_2, AcOH	m.p. 174–175°	65	141
4-Ph		m.p. 151°		133
4-(C$_6$H$_4$NO$_2$-*m*)	30% H_2O_2, AcOH	m.p. 208–209°	53	141
4-Ph-2,5-Me$_2$	30% H_2O_2, AcOH, 90°, 10 hr	m.p. 76–79°; HCl, m.p. 181.5–185°; picrate, m.p. 156–158°		103

14

Table IV-1 (*Continued*)

Pyridine (R)	Conditions	Properties	Yield (%)	Ref.
4-(C$_6$H$_4$NHCO$_2$Et-p)	H$_2$O$_2$, AcOH, 70°	m.p. 245–247°		141
2,4,6-Ph$_3$	m-ClC$_6$H$_4$CO$_3$H, CHCl$_3$, dark, 14 days, R.T.	m.p. 186–189°	32	18b
2,6-Ph$_2$-4-C$_6$H$_4$Br-p	m-ClC$_6$H$_4$CO$_3$H, CHCl$_3$, dark, 14 days, R.T.	m.p. 202–204°	10	18b
4-Ph-2,6-(C$_6$H$_4$Br-p)$_2$	m-ClC$_6$H$_4$CO$_3$H, CHCl$_3$, dark, 14 days, R.T.	m.p. 254–256°	5	18b
2,4,6-Ph$_3$-3-Me	m-ClC$_6$H$_4$CO$_3$H, CHCl$_3$, dark, 14 days, R.T.	m.p. 188–189°	57	18b
2,3,4,6-Ph$_4$	m-ClC$_6$H$_4$CO$_3$H, CHCl$_3$, dark, 14 days, R.T.	m.p. 250–251°	42	18b
2,3,5,6-Ph$_4$	m-ClC$_6$H$_4$CO$_3$H, CHCl$_3$, dark, 14 days, R.T.	m.p. 298–302°	53	18b
2,3,4,5,6-Ph$_5$	m-ClC$_6$H$_4$CO$_3$H, CHCl$_3$, dark, 14 days, R.T.	m.p. 245–248°	71	18b
2-SO$_2$C$_6$H$_4$-2,4,6-Cl$_3$	30% H$_2$O$_2$, CF$_3$CO$_2$H	m.p. 209.5–210°	74	15
2-Me-4-SO$_3$H		m.p. 263°		114
3-Me-4-SO$_3$H		m.p. >270°		114
2,6-Me$_2$-4-SO$_3$H		m.p. >270°		114
3-N$_3$	H$_2$O$_2$, AcOH, 75°, 3 hr	m.p. 99–103°		142
4-[H$_2$N(S=)C]-2-Me	H$_2$O$_2$	m.p. 188°		143
4-[H$_2$N(S=)C]-3-Me		m.p. 191–193°		143
4-[H$_2$N(S=)C]-2,6-Me$_2$				143
2-N=NC$_6$H$_4$NO$_2$-p	BzO$_2$H	m.p. 214–215°		144
3-N=NPh	BzO$_2$H	m.p. 85–87°; picrate, m.p. 159–161°		145
4-N=NC$_6$H$_4$NMe$_2$-p	BzO$_2$H			127
2-(4-Pyridyl)oxazole	H$_2$O$_2$, AcOH	m.p. 186°		146
	Perphthalic acid			146b
4-SiMe$_3$	H$_2$O$_2$, AcOH	m.p. 111–113°		146c
4-SiEt$_3$	H$_2$O$_2$, AcOH	m.p. 49–51°		146c
4-GeMe$_3$	H$_2$O$_2$, AcOH	m.p. 116–118°		146c
4-SnMe$_3$	Permaleic acid, CHCl$_3$, R.T.	m.p. 131–133°		146c

15

Side reactions sometimes accompany direct oxidation of pyridines (Table IV-2). Some of these reactions are discussed below.

Oxidation of aminopyridines (**IV-9**) can give nitropyridines (**IV-10**) and

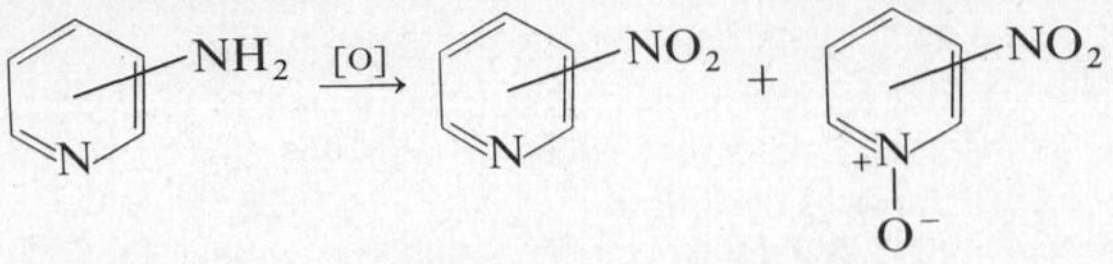

IV-9   IV-10   IV-11

nitropyridine-1-oxides (**IV-11**) (117, 147a). Aminopyridine-1-oxide (**IV-12**) and their derivatives can, however, be obtained by protection of the primary amino function by acylation followed by hydrolysis of the acylamino group after *N*-oxidation (eq. IV-3) (62, 66, 105, 111, 117, 123, 127, 130, 132) (Tables IV-3 and IV-4), or by direct oxidation of 2-aminopyridine hydrochlorides (122) (Table IV-1).

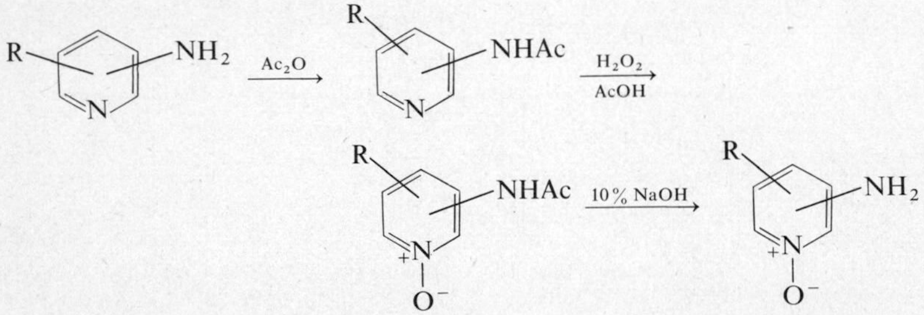

IV-12

Groups such as ethoxycarbonyl may also be used for the protection of an amino function (129, 132). Oxidation of 4-(4-dimethylaminostyryl)pyridine (**IV-13**) with one equivalent of perbenzoic acid resulted in the oxidation of

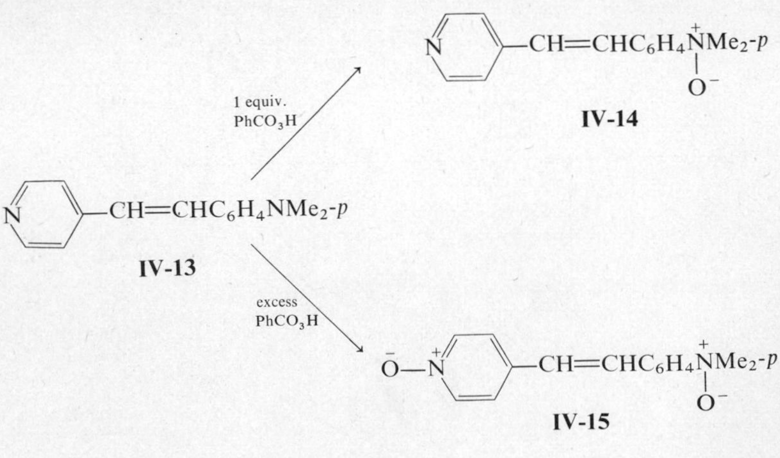

IV-14

IV-13

IV-15

TABLE IV-2. Side Reactions Occurring During Oxidation of Pyridines

Pyridine (R)	Oxidation conditions	Products (yields)	Ref.
3-OCH$_2$CO$_2$Et	H$_2$O$_2$, AcOH	3-Pyridyloxyacetic acid-1-oxide	155
3-OCHCO$_2$Et | CH$_3$	H$_2$O$_2$, AcOH	3-Pyridyloxy-α-propionic acid-1-oxide	155
6-CH$_2$OH-2-Me	30% H$_2$O$_2$, AcOH, 70°	6-Methylpicolinic acid-1-oxide, m.p. 177° (decomp.) (2.7%); 6-hydroxymethyl-2-picoline-1-oxide, m.p. 114° (65%)	161a
2,6-(CH$_2$OH)$_2$	30% H$_2$O$_2$, AcOH, 70°	6-(Hydroxymethyl)picolinic acid-1-oxide, m.p. 195° (trace); dipicolinic acid-1-oxide, m.p. 160° (45.2%); 2,6-bis(hydroxymethyl)pyridine-1-oxide, m.p. 137–138° (5.3%)	161a
2-CH$_2$OH-5-Me	30% H$_2$O$_2$, AcOH, 80–90°	6-Acetoxymethyl-3-picoline-1-oxide	32
2-CH(OAc)$_2$	Perphthalic acid	Pyridine-2-aldehyde-1-oxide, m.p. 74–76°; oxime, m.p. 220–221°	81
2-CH(OAc)$_2$-4-Me	Perphthalic acid	4-Picoline-2-aldehyde-1-oxide, m.p. 127–129°; semicarbazone, m.p. 212–214°	81
6-CH(OAc)$_2$-2-Me	Perphthalic acid	2-Picoline-6-aldehyde-1-oxide, m.p. 82–83°; semicarbazone, m.p. 247–249°, (decomp.)	81
4-CH(OAc)$_2$		Pyridine-4-aldehyde-1-oxide, m.p. 148–150°; semicarbazone, m.p. 82–83°	81
2-Oxocyclohexyl	30% H$_2$O$_2$, AcOH, 70–80°, 3 hr	Picolinic acid-1-oxide	161b
6-Methyl-2-(2-oxocyclohexyl)-pyridine	H$_2$O$_2$, AcOH	6-Methylpicolinic acid-1-oxide, m.p. 174–176°	162
4-Methyl-2-(2-oxocyclohexyl)-pyridine	H$_2$O$_2$, AcOH	4-Methylpicolinic acid-1-oxide, m.p. 164–166°	162
2-CO$_2$Et	30% H$_2$O$_2$, AcOH, 80°, 3–4 hr	Picolinic acid-1-oxide (54–87%)	99
2-CH$_2$CO$_2$Et-4-Me	30% H$_2$O$_2$, AcOH, 100°, 10 hr	4-Methylpicolinic acid-1-oxide	163
2-CH(CH$_3$)CO$_2$Me	m-Chloroperbenzoic acid	Methyl 2-(2-pyridyl)lactate (21%); methyl 2-(2-pyridyl)lactate-1-oxide, m.p. 143–147° (10%); methyl 2-(2-pyridyl)propionate-1-oxide, m.p. 70–75°	78

17

Table IV-2 (*Continued*)

Pyridine (R)	Oxidation conditions	Products (yields)	Ref.
2-CH(Ph)CO$_2$Me	Peracetic acid or *m*-chloroperbenzoic acid	Methyl α-phenyl-2-pyridine-glycolate, m.p. 66–67.5°; methyl α-phenyl-2-pyridineglycolate-1-oxide, m.p. 137°; methyl α-phenyl-2-pyridineacetate-1-oxide (trace)	78
2-CH(Ph)CONH$_2$	*m*-Chloroperbenzoic acid	α-Phenyl-2-pyridineglycolamide, m.p. 131° (57%); α-phenyl-2-pyridineglycolamide-1-oxide, m.p. 205°; α-phenyl-2-pyridine-acetamide-1-oxide, m.p. 152°	78
1,3-Indanedion-2-yl	*m*-Chloroperbenzoic acid	2-Hydroxy-2-(2-pyridyl)-1,3-indanedione, m.p. 212° (54%)	78
2-CN	(i) 50% H$_2$O$_2$, pH 7.5–8.0, 34–40°, 1 hr (ii) 23°, 16 hr	Picolinamide-1-oxide, m.p. 159–160° (70%)	156
2-CN-6-Me	30% H$_2$O$_2$, AcOH, 80–90°, 3 hr	6-Methylpicolinic acid-1-oxide	162
3-CN	50% H$_2$O$_2$, pH 7.5–8.0, 35–40°, 1 hr	Nicotinamide-1-oxide, m.p. 283–284° (decomp.) (44%)	156
2-CH=CH-C$_6$H$_4$NMe$_2$-*p*	Perbenzoic acid	2-(*p*-Dimethylaminostyryl)-pyridine-*N'*-oxide, m.p. 100° (decomp.)	148
2-CH=CH-C$_6$H$_4$NMe$_2$-*p*	Excess perbenzoic acid	2-(*p*-Dimethylaminostyryl)-pyridine-1,*N'*-dioxide, m.p. 148°	148
3-CH=CH-C$_6$H$_4$NMe$_2$-*p*	Perbenzoic acid	3-(*p*-Dimethylaminostyryl)-pyridine-*N'*-oxide	164
4-CH=CH-C$_6$H$_4$NMe$_2$-*p*	Perbenzoic acid	4-(*p*-Dimethylaminostyryl)-pyridine-*N'*-oxide	148
4-CH=CH-C$_6$H$_4$NMe$_2$-*p*	Excess perbenzoic acid	4-(*p*-Dimethylaminostyryl)-pyridine-1,*N'*-dioxide	148
2-NH$_2$	30% H$_2$O$_2$, H$_2$SO$_4$	2-Nitropyridine-1-oxide	147a
	CF$_3$CO$_3$H	2-Nitropyridine-1-oxide, m.p. 89° (20%)	165
2-NH$_2$-5-Br	CF$_3$CO$_3$H	5-Bromo-2-nitropyridine-1-oxide, m.p. 156–157° (26%)	165
3-NH$_2$	30% H$_2$O$_2$, (CF$_3$CO)$_2$O	3-Nitropyridine-1-oxide	117, 147a
4-NH$_2$-3-Me	(i) conc. H$_2$SO$_4$, 10–20° (ii) 30% H$_2$O$_2$, 10–20°, 1 hr	4-Nitro-3-picoline-1-oxide	166
4-NHOH	H$_2$O$_2$, AcOH	4-Azoxypyridine-1,1'-dioxide	167
2-NMe$_2$	AcO$_2$H	2-Dimethylaminopyridine-*N'*-oxide, m.p. 59–60°; picrate, m.p. 132–134°	122

Table IV-2 (*Continued*)

Pyridine (R)	Oxidation conditions	Products (yields)	Ref.
4-($C_6H_4NMe_2$-*p*)	Perbenzoic acid	4-(*p*-Dimethylaminophenyl)-pyridine-1,*N'*-dioxide, m.p. 95°	133
2-N=NPh	AcO_2H	2-Phenylazopyridine-1-oxide, m.p. 109–110°, 2-phenylazoxy-pyridine-1-oxide, m.p. 137–138°	145
	AcO_2H	2-Phenyl-α-azoxypyridine-1-oxide, m.p. 138°	168
	Perbenzoic acid, 0°, 60 hr	2-Phenyl-α-azoxypyridine-1-oxide, m.p. 137–138°; 2-phenyl-azopyridine-1-oxide, m.p. 112°; 2-phenylazopyridine, m.p. 141–142°; 2-phenylazoxypyridine	151
2-N=NC_6H_4Br-*p*	O_3, AcOH	2-(*p*-Bromophenyl)-α-azoxy-pyridine-1-oxide, m.p. 152–153°	151
2-N=NC_6H_4Br-*p*	Perbenzoic acid, $CHCl_3$, 0–5°, 4 days	2-(*p*-Bromophenyl)azopyridine-1-oxide, m.p. 203–205°; 2-(*p*-bromophenyl)-α-azoxypyridine-1-oxide, m.p. 152–153°	151
4-N=NPh	Perbenzoic acid, excess	4-Phenyl-α-azoxypyridine-1-oxide, m.p. 142–143°	151
	O_3, AcOH, 5 hr	4-Phenyl-α-azoxypyridine-1-oxide, m.p. 142–143°	151
		4-Phenylazopyridine-1-oxide, m.p. 150°; 4-phenyl-α-azoxy-pyridine-1-oxide, m.p. 142°	145
4-N=NC_5H_4N	H_2O_2, AcOH	4-Azoxypyridine-1,1'-dioxide	167
2-SOCH_2Ph	Perbenzoic acid	2-Benzylsulfonylpyridine-1-oxide, m.p. 114–115°	169
2-SC_6H_4Cl-*p*	30% H_2O_2, AcOH, 100°, 8 hr	2-(*p*-Chlorobenzenesulfonyl)-pyridine-1-oxide, m.p. 179.5–180° (70%)	15
2-SC_6H_2-2,4,5-Cl_3	H_2O_2, CF_3CO_2H	2-(2,4,5-Trichlorobenzenesulfonyl)-pyridine-1-oxide	15
2-SeH	30% H_2O_2, AcOH	2,2'-Dipyridyl diselenide-1,1'-dioxide	153
  R=CO_2Me	H_2O_2, H_2O	Picolinic acid-1-oxide, m.p. 157–158°	170
5,6-Dimethyl-3-(3-pyridyl)-1,2,4-triazine	H_2O_2, AcOH, 48 hr	Nicotinamide-1-oxide, m.p. 287°	171

TABLE IV-3. Preparation of Aminopyridine-1-oxides

$$\text{IV-9} \quad \xrightarrow{Ac_2O} \quad \text{NHAc} \quad \xrightarrow{[O]} \quad \text{NHAc} \quad \longrightarrow \quad \text{NH}_2$$

IV-9   IV-45   IV-12

Starting amine	Acetamido-1-oxide (yield) (IV-45)	Amine-1-oxide (yield) (IV-12)	Ref.
2-NH$_2$	(85%)	(83%)	111
	m.p. 75° (45%)	m.p. 159° (65%)	147b
2-NH$_2$-5-Br	m.p. 165–168°	m.p. 184°	66
2-NH$_2$-4-CO$_2$H	m.p. 268–271°	m.p. 156–158°	105
2-NH$_2$-5-NO$_2$	m.p. 196–197°	m.p. 245–246°	123
3-NH$_2$	m.p. 209–210.5°; picrate, m.p. 135–137°	m.p. 124–125°	130
	m.p. 216°	m.p. 121–124°	117
	m.p. 208° (51%)	m.p. 122° (24%)	147b
	3-Benzamido, m.p. 209–210°	Hydrochloride	127
	3-Ethylcarbamate, m.p. 197°	Picrate, m.p. 175°	129
3-NH$_2$-4-Me	m.p. 194–196°; picrate, m.p. 157.5–158.5°	Amine	130
3-NH$_2$-2,6-Me$_2$	m.p. 165–167°	Hydrochloride, m.p. 280°	132
3-NH$_2$-2,4,6-Me$_3$		Hydrochloride, m.p. 211°	132
4-NH$_2$	4-Benzamido	Hydrochloride, m.p. 181°	127
	m.p. 150° (77%)	Picrate, m.p. 198° (68%)	147b
4-NH$_2$-2-Cl		Hydrochloride, m.p. 152–153.5°	62
4-C$_6$H$_4$NH$_2$-p	m.p. 245–247°	m.p. 270° (18%)	139

TABLE IV-4. Hydrolysis of Acylaminopyridine-1-oxides to Aminopyridine-1-oxides

Starting 1-oxide	Reaction conditions	Product	Ref.
3-NHCHO-4-Me	20% H$_2$SO$_4$		130
3-NHAc-4-Me	20% H$_2$SO$_4$	m.p. 195–196°	130
4-NHAc-2-OH	MeOH, conc. HCl, Δ, 5 hr	m.p. 230–233° (decomp.) HCl, m.p. 191–193°	62
4-NHAc-2-OH	20% HCl, Δ, 4–5 hr	HCl, m.p. 163°	172
4-NHAc-2-OMe	20% HCl, 4 hr	m.p. 164.5–166°	62

the dimethylamino, but not the ring, nitrogen (**IV-14**); with excess perbenzoic acid both nitrogen atoms are oxidized (**IV-15**) (148). Similarly, 2-dimethyl-aminopyridine undergoes oxidation of the dimethylamino group (122, 149, 150). On the other hand 2-methylaminopyridine gives the 1-oxide (122).

2-And 4-phenylazopyridine (**IV-16**) undergo both 1-oxidation to **IV-17** and oxidation of the azo function to give the azoxy compound (**IV-18**)

IV-16   IV-17   IV-18

(145, 151). 2-(4-Chlorophenylthio)pyridine (**IV-19**) and 30% hydrogen peroxide in glacial acetic acid give the sulfone-1-oxide (**IV-20**) (15). 2,2'-Dipyridyl disulfide and 30% hydrogen peroxide in acetic acid gave 2,2'-

IV-19   IV-20

dipyridylthiosulfinate-N,N'-dioxide (152). When 2-selenopyridine (**IV-21**) is oxidized, diselenide formation accompanies N-oxidation (**IV-22**) (153).

IV-21   IV-22

Oxidation of either ethyl picolinate (**IV-23**) (98) or ethyl (2-pyridyl)acetate (**IV-24**) (134, 154), yields picolinic acid-1-oxide (**IV-25**). Methyl α-phenyl-3-

IV-23   IV-25   IV-24

and 4-pyridylacetate (**IV-26**; R = Me) were oxidized recently with peracetic acid or with m-chloroperbenzoic acid to the corresponding N-oxide (**IV-27**)

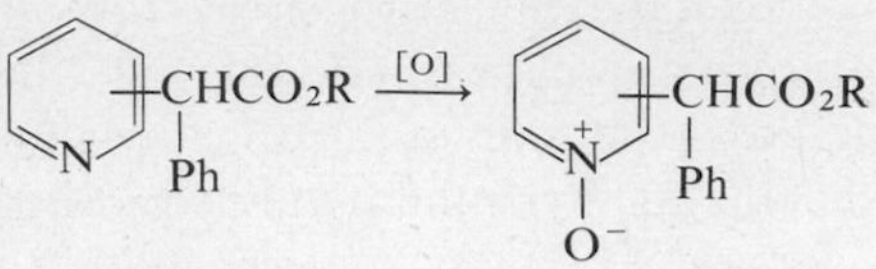

IV-26   IV-27

in high yield (78), but methyl α-phenyl-2-pyridylacetate (**IV-28a**) did not
react normally. Three products were isolated. The main product was the
glycolate (**IV-29a**), the minor products **IV-30a** and **IV-31a**. The net result
was hydroxylation of the α-carbon. Similar results were obtained with
α-phenyl-2-pyridylacetamide (**IV-28b**) (78). In the case of methyl 2-(2-

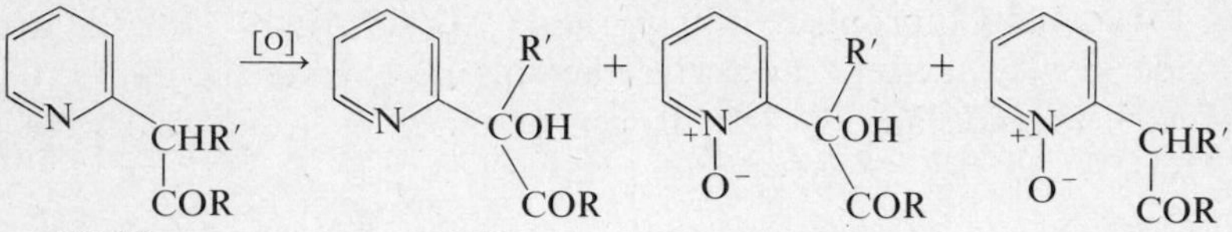

IV-28   IV-29   IV-30   IV-31

a; R = OCH$_3$, R′ = Ph
b; R = NH$_2$, R′ = Ph
c; R = OCH$_3$, R′ = CH$_3$

pyridyl)propionate (**IV-28c**), the 1-oxide **IV-31c** was the major product,
the hydroxy compounds **IV-29c** and **IV-30c** being also isolated (78).
Oxidation of 2-(2-pyridyl)-1,3-indanedione (**IV-32**) (R.I. 1391) gave the
2-hydroxy compound (**IV-33**) (15%) together with at least three other
compounds (78). However, oxidation of 2-benzhydrylpyridine gave the

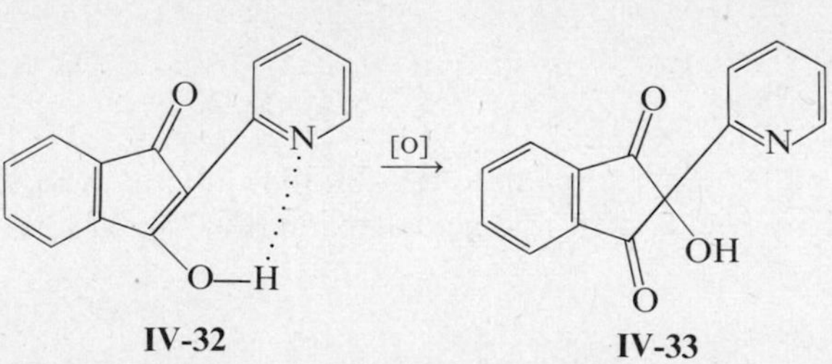

IV-32   IV-33

N-oxide exclusively (78). The much higher rate of α-oxidation of **IV-31** as
compared with attack of the pyridine nitrogen atom shows that rearrangement of initially formed *N*-oxide is not involved. A mechanism was suggested
in which the rate-determining step involves epoxidation of the tautomeric
form **IV-34**, followed by fast opening of the oxirane ring in **IV-35** to give
IV-29 (eq. IV-4) (78). Ethyl 3-pyridyloxyacetate (**IV-36**) is oxidized and

hydrolyzed by hydrogen peroxide in glacial acetic acid to yield the acid
IV-37 (155). 2- And 3-cyanopyridine are oxidized to the corresponding

acetamide with 30% hydrogen peroxide (156). The action of performic or
trifluoroperacetic acid at room temperatures on the *N,N*-disubstituted
2-aminopyridine (**IV-38**) gives the hydroxylamino derivative **IV-39** and not
the expected 1-oxide (16). Hydrogen peroxide in glacial acetic acid oxidation

of nicotine at room temperature gave nicotine-1'-oxide; at 80 to 90°, nicotine-
1,1'-dioxide was obtained (157). Treatment of anabasine with 34% hydrogen
peroxide in glacial acetic acid for 24 hr at 70 to 80°, or with 10% hydrogen
peroxide for 8 days at room temperature gave δ-oximino-δ-(3-pyridyl)valeric
acid (158).

6-Benzyloxy-2-picoline **(IV-40)** and peracetic acid gave 1-benzyloxy-6-methyl-2-pyridone **(IV-41)** (159). *N*-Oxidation of 6-methoxy-2-picoline

IV-40   **IV-41**

(IV-42) with fresh peracetic acid gave **IV-43**; when, however, old peracetic acid (which had been standing for 1 year) was used, a mixture of the 1-oxide **(IV-43)** and of 1-methoxy-6-methyl-2-pyridone **(IV-44)** was obtained (71).

Pyridine-(2-, 3-, or 4-)sulfonic acids are unaffected by hydrogen peroxide in glacial acetic acid (36), but the sodium sulfonates are oxidized under these conditions to give the corresponding *N*-oxides (36).

5-Methylpicolinic acid was obtained from the reaction of (1,2-dimethoxy-carbonyl-6-methyl-3-indolinyl) methoxyacetate with hydrogen peroxide in acetic acid (160).

2. From Nonpyridine Precursors

Only a few methods of preparing pyridine-1-oxides from nonpyridine precursors are known (Table IV-5). A good yield of 4-hydroxylamino-pyridine-1-oxide **(IV-46)** was obtained from a mixture of γ-pyrone **(IV-47)**

TABLE IV-5. Preparation of Pyridine-1-oxides from Nonpyridine Precursors

Starting compound and conditions	Product (yield)	Ref.
Trimethylpyrylium perchlorate, NH_2OH, MeOH, NaOH	Collidine-1-oxide, b.p. 134–135°/9 mm (60%)	175a
Unstable oxime of $CH_3COCH= C(CH_3)CH=C(CH_3)CN$	Collidine-1-oxide, b.p. 118/2 mm, m.p. 45–46° (88%); picrate, m.p. 173°; picrolonate, m.p. 195°; chloroplatinate, m.p. 216° (decomp.); perchlorate, m.p. 176°	177
2-Methyl-4,6-diphenylpyrylium perchlorate, $NH_2OH \cdot HCl$, MeOH, MeONa	4,6-Diphenyl-2-picoline-1-oxide, m.p. 180–182° (85%); picrate, m.p. 125–126.5°	175a
2,4,6-Triphenylpyrylium salt, NH_2OH, AcOH, AcONa	2,4,6-Triphenylpyridine-1-oxide, m.p. 186–189° (51%)	18b
4-(p-Bromophenyl)-2,6-diphenyl-pyrylium salt, NH_2OH, AcOH, AcONa	4-(p-Bromophenyl)-2,6-diphenylpyridine-1-oxide, m.p. 202–204° (35%)	18b
2,6-Di-(p-bromophenyl)-4-phenyl pyrylium salt, NH_2OH, AcOH, AcONa	2,6-Di-(p-bromophenyl)-4-phenylpyridine-1-oxide, m.p. 254–256° (10%)	18b
3-Methyl-2,4,6-triphenylpyrylium salt, NH_2OH, AcOH, AcONa	2,4,6-Triphenyl-3-picoline-1-oxide, m.p. 188–189° (83%)	18b
2,3,4,6-Tetraphenylpyrylium salt, NH_2OH, AcOH, AcONa	2,3,4,6-Tetraphenylpyridine-1-oxide, m.p. 250–251° (98%)	18b
2,3,5,6-Tetraphenylpyrylium salt, NH_2OH, AcOH, AcONa	2,3,5,6-Tetraphenylpyridine-1-oxide, m.p. 298–302° (88%)	18b
2,3,4,5,6-Pentaphenylpyrylium salt, NH_2OH, AcOH, AcONa	2,3,4,5,6-Pentaphenylpyridine-1-oxide, m.p. 245–248° (97%)	18b
γ-Pyrone, NH_2OH, HCl, 20°, 2–3 days	4-Hydroxylaminopyridine-1-oxide	167
Diacetylacetone barium salt, $NH_2OH \cdot HCl$, EtOH, dark, 72 hr	4-Hydroxylamino-2,6-lutidine-1-oxide, m.p. 219–220 (decomp.); HCl, m.p. 190°	173
2,6-Dimethyl-4-pyrone, $NH_2OH \cdot HCl$, EtOH, Δ, 6 hr	4-Hydroxylamino-2,6-lutidine-1-oxide, m.p. 200°	174
2,6-Dimethyl-4-pyrone, $NH_2OH \cdot HCl$, R.T. 5 days	4,4′-Azoxy-2,6-lutidine-1,1′-dioxide	173
Diacetylacetone, 3 molar equiv. NH_2OH, 65°, 5 days	4,4′-Azoxy-2,6-lutidine-1,1′-dioxide (70%)	179b
$Cl(ClC=CCl)_2CHO$, NH_2OH	2,3,4,5-Tetrachloropyridine-1-oxide	176
3-Hydroxy-2-aci-nitropropionitrile, $2N$ H_2SO_4, 24 hr	2,4,6-Trinitropyridine-1-oxide, m.p. 222° (50%)	178
$HOCH_2C(CN)=N(O)OK$, acid	2,4,6-Tricyanopyridine-1-oxide	179a

and hydroxylamine. The monoxime of a 1,5-carbonyl compound was assumed to be the intermediate which underwent condensation cyclization (167). 4-Hydroxylamino-2,6-lutidine-1-oxide (IV-48) was obtained from a mixture of the barium salt of the dienol of diacetylacetone (IV-49) and hydroxylamine hydrochloride in ethanol in the absence of light. It was also obtained from 2,4-dimethyl-γ-pyrone (IV-50) (173, 174).

IV-49   **IV-48**   **IV-50**

4,6-Diphenyl-2-picoline-1-oxide (**IV-51**) and 2,4,6-collidine-1-oxide (**IV-52**) were derived from the reaction of 2-methyl-4,6-diphenylpyrylium perchlorate (**IV-53**) and 2,4,6-trimethylpyrylium perchlorate (**IV-54**), respec-

IV-53   **IV-51**

IV-54   **IV-52**

tively, with hydroxylamine under basic conditions (175a). Treatment of 2,4,6-trimethyl- and 2,6-diethyl-4-methylpyrylium salts with hydroxylamine in aqueous medium gave the respective pyridine-1-oxides in lower yield (175b). The lack of formation of the *N*-oxide from 2,6-diisopropyl-4-methyl- and 2,6-diphenyl-4-methylpyrylium salts was assumed to be due to steric hindrance (175b). Recently, it was found that the reaction between a series of polyarylpyrylium salts and hydroxylamine in acid medium results in the formation of pyridine-1-oxides in good yields (18b). These results showed that higher yields of polyarylpyridine-1-oxides were obtained from the pyrylium salts than from the *N*-oxidation of polyarylpyridines with *m*-chloro-perbenzoic acid (18b).

IV-55   **IV-56**

Reaction of **IV-55** with hydroxylamine gave 2,3,4,5-tetrachloropyridine-1-oxide **(IV-56)** (176). The preparation of 2,4,6-collidine-1-oxide **(IV-52)** by heating an unstable oxime of cyanoketone **IV-57** was reported (177). 2,4,6-

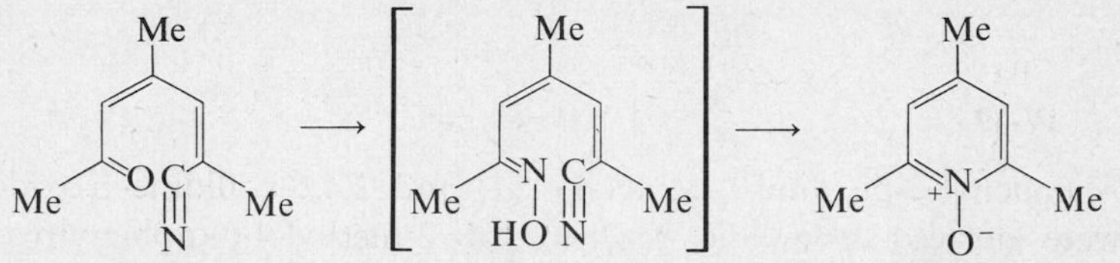

IV-57   IV-52

Tricyanopyridine-1-oxide was prepared (50%) by treating 3-hydroxy-2-*aci*-nitropropionitrile with an excess of 2N sulfuric acid for 24 hr (178). Similarly, 2,4,6-trinitropyridine-1-oxide was obtained from 2,2-dinitro-ethanol in the presence of dilute sulfuric acid (178).

Reaction of **IV-58** with formaldehyde gave an unstable intermediate **(IV-59)** which, on acidification, gave 2,4,6-tricyanopyridine-1-oxide **(IV-60)**

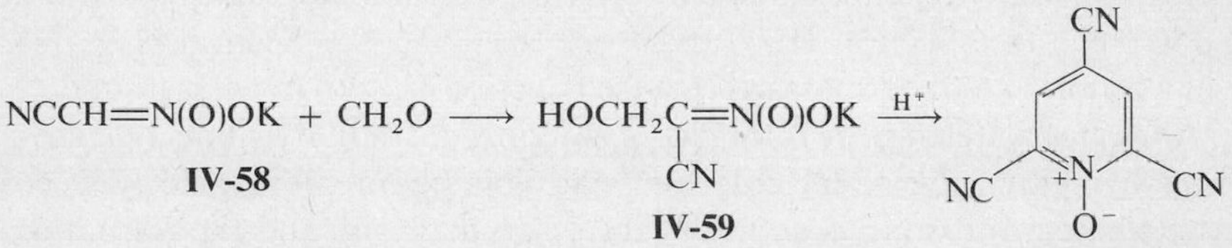

$$NCCH{=}N(O)OK + CH_2O \longrightarrow HOCH_2C{=}N(O)OK \xrightarrow{H^+}$$

IV-58

CN

IV-59

IV-60

(179a). When diacetylacetone was treated with three molar equivalents of hydroxylamine in water at 65° for 5 days, an azoxy compound **(IV-61)** was isolated in 70% yield (179b).

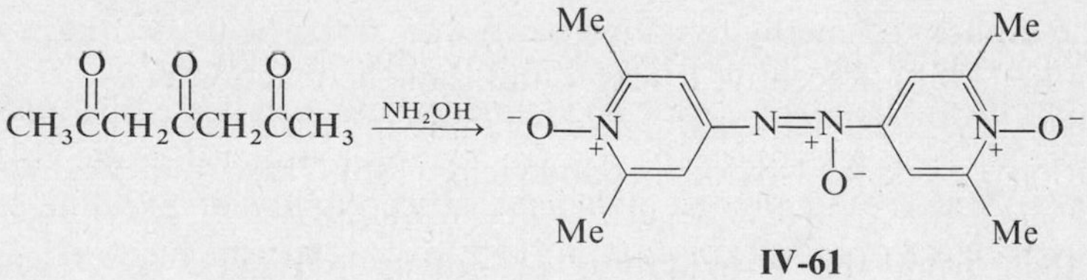

$$CH_3\overset{O}{\overset{\|}{C}}CH_2\overset{O}{\overset{\|}{C}}CH_2\overset{O}{\overset{\|}{C}}CH_3 \xrightarrow{NH_2OH}$$

IV-61

3. From *N*-Arylpyridinium Salts

Treatment of *N*-(2,4-dinitrophenyl)pyridinium chloride **(IV-62)** with hydroxylamine hydrochloride and triethylamine in methanol gave 5-(2,4-dinitroanilino)-2,4-pentadienal oxime **(IV-63)** which was boiled in 4:1

dioxane-water to give pyridine-1-oxide (87–92%) (179c, d). 2- And 3-picoline-1-oxide and 3,5-lutidine-1-oxide were also prepared by this method (179c).

IV-62   IV-63

II. Physical and Spectral Properties of Pyridine-1-oxides

1. Detection of the N-Oxide Function

The N-oxide function in pyridine-1-oxides can be detected by its infrared absorption at 1,200 to 1,300 cm^{-1}, with a maximum at approximately 1,250 cm^{-1} (see Section II.7), and from the U.V. (see Section II.8), NMR (see Section II.6), and mass spectra (see Section II.9) of the compound.

A method of testing for N-oxides was reported (180). Pyridine-1-oxide and dimethylaniline give a blue color on heating in hydrochloric acid. This color is attributed to oxidative action of the N-oxide group and the formation of crystal violet by a series of condensations with formaldehyde, which is formed as a result of this reaction. This color reaction is not specific for an N-oxide function since it is also given by nitrobenzene and m-dinitrobenzene.

2. Dipole Moments

Comparison of the dipole moments of pyridine-1-oxide and pyridine with those of trimethylamine N-oxide and trimethylamine indicate that resonance structure **IV-64** is an important contribution in pyridine-1-oxide (181). On this basis, one would predict increased susceptibility of pyridine-1-oxide nucleus to electrophilic substitution. However, experiments with 4-nitro-pyridine-1-oxide suggested that there must also be contributions from **IV-65** (1). Similarly, nucleophilic substitution of 2- and 4-halopyridine-1-oxides was shown to be easier than that of the corresponding pyridine (36,

IV-64   IV-65

TABLE IV-6. Dipole Moments (and Mesomeric Moments) of 4-Substituted Pyridines, Their N-oxides and BCl_3 Complexes (185, 186)

X	4-X-pyridine	4-X-pyridine-1-oxide	4-X-pyridine-BCl_3	XC_6H_5
NMe_2	4.31 (2.27)	6.76 (2.74)	— —	(1.66)
OMe	2.96 (1.16)	5.08 (1.39)	8.86 (1.35)	(0.96)
Cl	0.78 (0.57)	2.82 (0.59)	6.71 (1.02)	(0.41)
Me	2.61 (0.39)	4.74 (0.50)	8.73 (0.67)	(0.35)
H	2.22 (0.00)	4.24 (0.00)	7.70 (0.00)	(0.00)
CO_2Et	2.53 (0.47)	3.80 (0.93)	— (0.17)	(0.50)
COMe	2.41 (0.21)	3.19 (0.62)	7.74 (—)	(0.56)
NO_2	1.58 (0.55)	ca. 0.7 (0.99)	— (—)	(0.76)
CN	1.65 (0.27)	— —	4.20 (0.10)	(0.45)

182–184). Values for the dipole moments of 4-substituted-pyridines, their N-oxides and their BCl_3 complex are given in Table IV-6 (185, 186).

The mesomeric moments of 4-substituted pyridine-1-oxides, pyridine-boron trichloride complex, and pyridine are compared with that of benzene (Table IV-6) (185, 186). When the substituent is electron-donating the mesomeric moment shifts are positive, and when it is electron-withdrawing the mesomeric moment shifts to negative values. The contribution of resonance structures **IV-64** and **IV-65** depends on the nature of the substituent at the 4-position (187). The dipole moments of 4-substituted pyridine-1-oxides show that the pyridine-1-oxide ring can have either a deficit and a surfeit of electrons at the 4-position depending upon the conditions (188). In 4-p-substituted styryl- and phenylethynylpyridines and their N-oxides, dipole moment measurements show that there is no large interaction in the ground state between the substituent and the heterocyclic ring (189).

The relative contributions of **IV-64** and **IV-65** were examined (187). The dipole moment of pyridine-1-oxide was compared with that of pyridine: boron trihalide complex and this, in turn, was compared with those of the N-oxide and of the boron trihalide complex of trimethylamine in which the contributions of the canonical form comparable to **IV-64** can be disregarded (Table IV-7). From these results the mesomeric moment of the N-oxide could be calculated as follows:

$$\mu_m = \mu \text{ (pyridine-1-oxide)-}\mu \text{ (triethylamine } N\text{-oxide)-}[\mu \text{ (pyridine}$$
$$\text{boron complex)-}\mu \text{ (trimethylamine boron complex)}]$$
$$= [(-4.24) - (-5.02)] - (-1.31)$$
$$= (+0.78) + (1.31)$$
$$= +2.09 \text{ D}$$

From these data it was concluded that, in unsubstituted pyridine-1-oxide,

TABLE IV-7. Comparison of Dipole Moments of Amine Oxide and Boron Trihalide Complex
of an Amine (187)

Adduct for NEt$_3$ or pyridine	NMe$_3$ adduct	Pyridine adduct	Pyridine adduct less NEt$_3$ adduct
Nil	−0.65 ±0.003	−2.22 ±0.02	−1.57
O	−5.02 ±0.02	−4.24 ±0.02	+0.78
BH$_3$	−4.62 ±0.01	−5.86 ±0.01	−1.24
BF$_3$	−5.63 ±0.03	−6.90 ±0.05	−1.27
BCl$_3$	−6.31 ±0.03	−7.70 ±0.02	−1.39
BBr$_3$	−6.57 ±0.01	−7.90 ±0.06	−1.33

the inward drift of electrons originating from the oxygen atom to the ring is
slightly greater than that calculated (181). These results are close to the
value of 2.34 D which was obtained from the difference between the
predicted moment (6.58 D) (181) and the observed moment (4.24 D).

The $\mu_{C-Z} - \mu_{C-H}$ moments for substituents in monosubstituted benzenes,
and 3- and 4-substituents in pyridine, pyridine-1-oxide, and nitrobenzene
indicate the degree of relative interaction between the substituent and the
ring system (190). The dipole moments of 3- and 4-substituted pyridine-1-
oxides observed (190) appear to be more reasonable than previous results
in which care was not taken to protect the compounds from moisture
(Table IV-8). The dipole moments of pyridine-1-oxides were found to be

TABLE IV-8. Observed Dipole Moments of 3- and 4-Substituted
Pyridine-1-oxides (190).

N-oxide	μ	N-oxide	μ
3-Me	4.33	4-Me	4.50
		4-Et	4.54
		4-Pr	4.56 (4.73 D)[a]
		4-isoPr	4.59 (4.74 D)[a]
		4-t-Bu	4.63
3-Cl	3.68	4-Cl	2.83
3-Br	3.69	4-Br	2.90
3-I	3.79		
		4-Ph	4.91

[a] Solutions of the compound prepared in presence of partially dried
air.

very dependent on atmospheric humidity and consistent results could only be obtained when moisture was excluded. The water probably exerts its effect by forming a highly polar hydrate. The value obtained for 4-chloropyridine-1-oxide, which is not particularly hygroscopic, is identical within experimental error with the values previously obtained (185).

Dipole moments for pyridine-1-oxide and of the mono-*N*-oxides of pyrazine, pyrimidine, and pyridazine have been calculated by a simple molecular orbital method (191).

3. X-Ray Diffraction

Recently, crystallographic analysis of several pyridine-1-oxides has been reported.

The N—O distance for pyridine-1-oxide was found to be 1.33 Å (192a).

Three-dimensional X-ray structural analyses has been completed on 2-hydroxymethylpyridine-1-oxide (192b) and the following data obtained: (*a*) 7.079, (*b*) 8.046, (*c*) 10.599 Å, cos B − 0.2254; space group P2, 1c, 4 molecules per cell, and monoclinic crystal symmetry. For 2-hydroxymethylpyridine-1-oxide, the N—O bond distance is 1.332 Å, and the hydroxyl hydrogen atom is bonded to an *N*-oxide oxygen atom of a neighboring, symmetry related molecule with a bond length of 1.99 Å. The pyridine moieties are planar within experimental error (192b).

Figure IV-1 illustrates the results of detailed crystallographic analysis of 4-nitropyridine-1-oxide (192c) and 4,4′-*trans*-azopyridine-1,1′-dioxide (193).

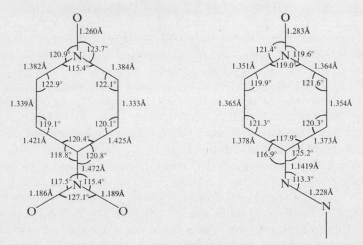

Fig. IV-1. Crystallographic analysis of 4-nitropyridine-1-oxide and 4,4′-*trans*-azopyridine-1,1′-dioxide.

In 4-nitropyridine-1-oxide, the length of the N—O bond of the *N*-oxide group is greater than that in the nitro group. The C-2—C-3 and C-5—C-6 bonds are shorter than the C-3—C-4 and C-4—C-5 bonds, which gives evidence for the important contributions from a *p*-quinonoïd canonical structure in this compound. In 4,4'-azopyridine-1,1'-dioxide the N—O bond in the *N*-oxide function is also longer than that in the nitro group. Comparison of the C-2—C-3 and C-3—C-4 bond lengths in the azo and nitro compound suggests the absence of a quinonoïdal contribution to the structure of the azo compound. X-Ray crystal structure determinations of 4-nitro and 4-chloropyridine-1-oxide (194) were in good agreement with earlier results (192a, 195). Both of these compounds were found to be bipyramidal and rhombic. The crystal structures of various 4-substituted pyridine-1-oxides (R = 4-NO$_2$, 4-Cl, 4-CN, 4,4'-*trans*-azo-bis, 4,4'-*cis*-azo-bis, 4-OH, 4-CH$_2$OH) were studied (195).

2-(2'-Pyridyl-1'-oxido)-1-hydroxypyridine   perchlorate   (**IV-66**)   and 4-bromo-2-methylpicolinic acid-1-oxide (**IV-67**) were investigated by X-ray diffraction (196). In both cases, the space groups were P2, 1m, or P2. The following parameters were determined: (*a*) 12.93, (*b*) 12.57, (*c*) 6.78 Å, β 98° 47′ and *z* = 4 and (*a*) 15.60, (*b*) 17.99, (*c*) 5.89 Å, and *z* = 8,

IV-66   IV-67

respectively. The crystal of the molecular compound between a dibromo-pyridine-1-oxide and mercuric chloride was found to occur as rods 1 to 2 mm long and had the following parameters: (*a*) 9.2 ± 0.1, (*b*) 27.1 ± 0.1, (*c*) 4.08 ± 0.05 Å, β 97 ± 1°; *z* = 4 and a space group P2, 1a (197). The crystal structure of bis-(pyridine-1-oxide)copper (II) nitrate (198) and poly-bis-[μ-(2-picoline-1-oxide)-chlorocopper(II)-di-μ-chloro]diaquo copper (II) (199) have been determined. The crystallographic properties of 4-picoline-1-oxide have been reported (200).

As indicated previously the N—O bond in the 1-hydroxypyridinium ion should show a decrease in double bond character and an increase in bond length. A study of the crystal structure of the hydrochloride and hydrobromide salts of pyridine-1-oxide showed (201) the N—O bond in pyridine-1-oxide, HCl to be 1.3 Å long. Using the heavy-atom method to study the structure of pyridine-1-oxide hydrochloride, the N—O distance was found to be 1.37 Å. When this is compared with trimethylamine-*N*-oxide,

HCl in which the N—O bond length is 1.424 Å, it appears that pyridine-1-oxide hydrochloride still has some contribution from a canonical structure related to **IV-64**. Pyridine-1-oxide HCl and HBr belong to the orthorhombic system with (*a*) 11.06 ± 0.02, (*b*) 7.70 ± 0.02, (*c*) 7.25 ± 0.02 Å and (*a*) 11.40 ± 0.02, (*b*) 7.87 ± 0.02, (*c*) 7.42 ± 0.02 Å, respectively. The configuration of the 1-hydroxypyridinium ion (as its chloride) is found to be almost planar and each chlorine atom is surrounded by six 1-hydroxypyridinium groups as neighbors.

4. Polarography

Various aspects of polarographic reduction of pyridine-1-oxides are discussed by Ochiai (7). In 1943, he indicated that the polarographic reduction of pyridine-1-oxide occurs at a more negative potential than is necessary in simple aliphatic N-oxides due to the greater stability of the N—O bonds in the aromatic N-oxide and the contribution from canonical form **IV-64**.

The polarographic reduction of N-oxides of pyridine and N-methylpiperidide are sufficiently different to allow determination of the location of the N—O group in alkaloids using this method (202).

Ochiai (7) deduced the following facts from the half-wave potentials for the reduction of the N—O bond in nitrogen containing heterocycles: (*a*) aromatic N-oxides are more resistant to reduction than aliphatic ones; (*b*) the reduction potential of the N—O bond in aromatic N-oxides compounds varies with pH and is generally lower in acidic media; (*c*) when a benzene ring is fused to monocyclic N-oxides or a second nitrogen is introduced into the ring, the reduction becomes easier.

It was shown that pyridine-1-oxide is more easily reduced when the conjugation of the N-oxide oxygen with the heterocyclic ring by backdonation is diminished by the presence of an aryl substituent, which itself may conjugate with the heterocyclic ring (203). Since reduction involves a nucleophilic attack by the electrons, the ease of reduction may be understood in items of the increased electrophilic nature of the N-oxide group itself due to stabilization of the energy levels of the lowest vacant molecular orbital in the N-oxide compounds (7). Reduction of the N-oxide group is pH-dependent, and the reduction wave rapidly decreases with increase in pH (204). It has, however, been reported that pyridine and picoline-1-oxide can only be reduced in their protonated form (205). It was suggested (204) that pyridine-1-oxide and its derivatives are reduced as follows:

$$\overset{+}{N}-\bar{O} + H_3O^+ \longrightarrow \overset{+}{N}-OH \xrightarrow{2H^+ + 2e} \left[\overset{+}{N}-H \rightleftharpoons N + H_3O^+ \right]$$

The reduction potential is generally more negative than that required for pyridine-1-oxide when the substituent is an electron donating group, and is more positive when the substituent is an electron-withdrawing group (7).

In some cases, the substituent was reduced before the N—O bond was (206). 4-Nitropyridine-1-oxide undergoes reduction of the nitro group to hydroxylamino followed by reduction of the N-oxide to give 4-hydroxyl-aminopyridine (207). Nonaqueous polarographic reduction of 4-nitro-pyridine-1-oxide shows two waves (Table IV-9) (208). The first wave is attributed to anion radical $(A^{\,\cdot\,-})$ formation from the neutral species (N), and the second wave to dianion (A^{2-}) formation $(N \rightleftharpoons A^{\,\cdot\,-} \rightleftharpoons A^{2-})$. Reduction of azo and azoxy groups precedes that of the N-oxide group (207).

TABLE IV-9. Polarographic Data for 4-Nitropyridine-1-oxide and its Anion Radical (208)

4-Nitropyridine-1-oxide	$-E_{1/2}$ (V)	i_{DC} (μA)	i_{AC} (μmhos)	$-E_{1/2}$ (V)	i_{DC} (μA)	i_{AC} (μmhos)
	First reduction wave			Second reduction wave		
Neutral species	0.80	1.66	294	1.69	2.10	172
	Oxidation wave			Reduction wave		
After electrolysis at -1.10 V	0.77	1.38	288	1.66	1.88	128

In acid medium, the first reduction wave is for the aldehyde group in 2- and 4-formylpyridine-1-oxide, and the second wave is due to the reduction of the 1-oxide group in the presence of a hydrated aldehyde group (209). At all pH values, 3-formylpyridine-1-oxide is reduced normally with the reduction of the aldehyde group preceding that of the N-oxide function. Similar results were obtained on polarographic reduction of acetyl- and benzoylpyridine-1-oxides (210). The reduction of the N-oxide group is a two-electron reduction of the N—O bond (7, 211). The 1-hydroxypyridinium ion is more easily reduced than the free base since in the pyridinium ion, protonation of a lone pair of electrons on oxygen increases the latter's electronegativity and the drift of electrons from oxygen to the nitrogen in the ring is inhibited (7).

5. Dissociation Constants

Pyridine-1-oxides have a lower pK_a than the parent pyridine and a much lower pK_a than aliphatic N-oxides. This supports canonical form **IV-64** for pyridine-1-oxide (7). When electron-donating substituents are present contributions from **IV-68** (e.g., Me, OH, and NH$_2$) increase the pK_a, electron-withdrawing substituents (e.g., CO$_2$H and NO$_2$) (contributor **IV-69**) lower the pK_a.

IV-68   IV-69

While the basicity of many 3- and 4-substituted pyridine-1-oxide could be correlated with the substituent σ-constant (139), for strongly electron-donating substituents (e.g., NH$_2$ and OH) their σ^+ constants had to be used (212). An excellent correlation was also reported between the substituent σ-constant and the pK_a of the N-oxide function in 4-substituted nicotinate-1-oxides (213).

As clearly indicated by nmr spectroscopy (see next section), polar solvents undergo donor-acceptor interaction between the donor N-oxide oxygen of the pyridine-1-oxide and the electron-pair acceptor solvent. It is not surprising, therefore, that whereas σ_{meta} and σ_{para} of the $>\!\overset{+}{N}\!-\!\bar{O}$ group in pyridine-1-oxides are always positive (i.e., they act as an electron-withdrawing group), this is particularly so in polar solvents in which some co-ordination occurs, while in nonpolar solvents (in which the oxygen atom is, presumably, not coordinated with solvent so that back-donation is maximized) the σ-value of the N-oxide function becomes less positive, hence more representative of the "free" N-oxide group (214, 215).

The pK_a's of some substituted pyridine-1-oxides are given in Table IV-10.

TABLE IV-10. pK_a Values of Substituted Pyridine-1-oxides

N-oxide	pK_a	Ref.
Pyridine	1.90	216
2-Me	2.63	216
	1.022	116
3-Me	2.39	216
4-Me	2.86	216
2,6-Me$_2$	1.442	116
3-Cl	1.34	216
2-OH	−0.8	217
1-OMe, 2-oxo	−1.3	217
1-OCH$_2$Ph, 2-oxo	−1.7	217
4-OH	2.45	217
1-OMe, 4-oxo	2.57	217
1-OCH$_2$Ph, 4-oxo	2.58	217

Table IV-10 (*Continued*)

N-oxide	pK_a	Ref.
2-OMe	1.23	217
2-OEt	1.18	217
4-OMe	2.05	217
4-OCH$_2$Ph	1.99	217
4-OMe, 2-Me	9.414	116
4-OEt, 2-Me	2.106	116
2-SH	-1.95	131
4-SH	1.53	131
2-SCH$_2$Ph	-0.23	131
4-SCH$_2$Ph	2.09	131
4-CO$_2$H, 2,6-Me$_2$	-0.015	116
4-CO$_2$Et, 2,6-Me$_2$	-0.126	116
4-CONH$_2$, 2,6-Me$_2$	-0.317	116
4-CN, 2-Me	-0.674	116
4-CN, 2,6-Me$_2$	-0.614	116
4-NO$_2$	1.11	216
4-NO$_2$, 2-Me	-0.968	116
4-NO$_2$, 2,6-Me$_2$	-0.861	116
2-NH$_2$	2.67	217
4-NH$_2$, 2-Me	4.10	116
2-NHMe	2.61	217
2-NMe$_2$	2.27	217
4-NHMe	3.85	217
4-NMe$_2$	3.88	217
2-NHCOMe	-0.42	131
2-NHCOPh	-0.44	131
2-N(Me)COMe	-1.02	131
2-N(Me)COPh	-1.39	131
3-NHCOMe	0.99	131
4-NHCOMe	1.59	131
4-N(Me)COMe	1.36	131
4-N(Me)COPh	1.70	131
2-Ph	0.77	141
2-C$_6$H$_4$NO$_2$-m	0.26	141
2-C$_6$H$_4$NO$_2$-p	0.28	141
2-C$_6$H$_4$NH$_2$-m	3.92, 0.20	141[a]
2-C$_6$H$_4$NH$_2$-p	3.82, 0.25	141[a]
3-Ph	0.74	141
3-C$_6$H$_4$NO$_2$-m	0.47	141
4-Ph	0.83	141
4-C$_6$H$_4$NO$_2$-m	0.58	141
4-C$_6$H$_4$NO$_2$-p	0.58	141
4-C$_6$H$_4$NH$_2$-p	3.64	141

[a] High value for the NH$_2$ group; low value corresponds to $\overset{+}{N}$—$\overset{-}{O}$ group.

6. Nuclear Magnetic Resonance Spectra

A number of studies of the NMR spectra of pyridine-1-oxide have been reported (7, 218–222). Ochiai (7) discusses various aspects of the NMR spectra of pyridine-1-oxides; thus only a few are mentioned here.

The positions of the absorption lines in pyridine-1-oxide depend markedly on the acidity of the solvent, that is, its tendency to coordinate with a lone pair on oxygen in the N-oxide (220). The spectra in carbon tetrachloride solution are the closest to being those of the free bases, while in $18N$ sulfuric acid the species under study is the N-hydroxypyridinium salt. Even in D_2O partial protonation occurs. In the free base, C-3—H absorbs at lower field than C-4—H, which is the reverse of what is observed with pyridine itself, and reflects the back-donation illustrated by **IV-64**. As the medium becomes more "electron-pair demanding," the spectrum gradually shifts to that of the conjugate acid and C-4—H appears at lower field than C-3—H, as expected from **IV-70** (rather than **IV-71**). In the "neutral" pyridine-1-oxide molecules, the order of increasing field strengths at which resonance occurs is C-2—H $\ll$ C-3—H $<$ C-4—H, compared with C-2—H $<$ C-4—H $<$ C-3—H for pyridine itself. Change of solvent from CCl_4 to $18N$ D_2SO_4 causes the signals due to the α-protons to move downfield by ca. 0.7 ppm, and those due to the γ-protons to move in the same direction by

IV-70   IV-71

about 1.3 ppm, with the net effect that the total electron densities at the nuclear carbon atoms in the protonated 1-oxide are in the order C-2 $<$ C-4 $<$ C-3 (220). Proton chemical shifts for various pyridine-1-oxides in CCl_4, D_2O, and $18N$-H_2SO_4 are given in Table IV-11. The chemical shifts for various organometallopyridine-1-oxides are given in Table IV-12 (146c).

7. Infrared Spectra

Pyridine-1-oxides are characterized by two strong absorption bands in the region of 1,200 to 1,300 cm^{-1} and 835 cm^{-1} (233). The position of the absorption at 1,200 to 1,300 cm^{-1} is very sensitive to the presence of a substituent. In deuterated pyridine-1-oxide, the band is shifted from 1,250

TABLE IV-11. Chemical Shifts (τ) for Proton Resonance of Pyridine-1-oxides in CCl_4, D_2O, and $18N$ D_2SO_4 (220)

N-oxide	Solvent	C-2—H	C-6—H	C-3—H	C-5—H	C-4—H	Me–Ar
4-Me	CCl_4	2.13	2.13	3.03	3.03	—	7.70
	D_2O	1.79	1.79	2.55	2.55	—	7.56
	$18N$-D_2SO_4	1.44	1.44	2.18	2.18	—	7.34
4-OEt	CCl_4	2.11	2.11	3.30	3.30	—	5.96, 8.57[a]
	D_2O	1.78	1.78	2.84	2.84	—	5.75, 8.55
	$18N$-D_2SO_4	1.47	1.47	2.64	2.64	—	5.61, 8.55
4-Cl, 3-Me	CCl_4	1.88	2.05	—	2.83	—	7.68
	D_2O	1.72	1.84	—	2.39	—	7.63
	$18N$-D_2SO_4	1.35	1.42	—	1.99	—	7.45
2-Me	CCl_4	—	1.80	←—— 2.78 ± 0.1 ——→			7.62
	D_2O	—	1.58	←—— 2.38 ± 0.1 ——→			7.41
	$18N$-D_2SO_4	—	1.26	(2.04)	(2.04)	1.64	7.14
3-Me	CCl_4	1.94	2.00	—	2.81	2.97	7.73
	D_2O	1.71	1.73	—	2.35	2.35	7.58
	$18N$-D_2SO_4	1.28	1.28	—	1.97	1.65	7.38
2,5-Me$_2$	CCl_4	—	1.99	2.86	—	3.10	7.62[C(2)] 7.76[C(5)]
	D_2O	—	1.78	2.51	—	2.57	7.49[C(2)] 7.64[C(5)]
	$18N$-D_2SO_4	—	1.43	2.18	—	1.83	7.22[C(2)] 7.48[C(5)]
2,6-Me$_2$	CCl_4	—	—	2.83	2.83	3.02	7.59
	D_2O	—	—	2.57	2.57	2.57	7.44
	$18N$-D_2SO_4	—	—	2.26	2.26	1.83	7.20
3,5-Me$_2$	CCl_4	2.26	2.26	—	—	3.27	7.78
	D_2O	2.00	2.00	—	—	2.59	7.70
	$18N$-D_2SO_4	1.62	1.62	—	—	1.96	7.51
3-CH$_2$OH	CCl_4	Insufficiently soluble					
	D_2O	1.56	1.63	—	(2.23)	(2.23)	—
	$18N$-D_2SO_4	1.11	1.18	—	1.87	1.52	—
None	CCl_4	1.90	1.90	(2.72)	(2.72)	(2.92)	—
	D_2O	1.58	1.58	←—— 2.27 ± 0.12 ——→			—
	$18N$-D_2SO_4	1.19	1.19	1.90	1.90	1.51	—
4-NO$_2$, 3-Me	CCl_4	(2.04)	(2.04)	—	(2.04)	—	7.39
	D_2O	1.53	1.62	—	1.75	—	7.39
	$18N$-D_2SO_4	1.06	1.14	—	1.49	—	7.27

[a] 4-OCH_2–CH$_3$ and 4-OCH$_2$CH_3, respectively.

to 1,222 cm^{-1} (224). Hydrogen bonding results in a shift of this band to lower wave numbers. It is the only band strongly altered by hydrogen bonding with methanol and a linear relation was found to exist between various solvents and γ_{N-O} (1,250 cm^{-1}), and the corresponding γ_{S-O} band of Me$_2$SO (224). The band at 835 cm^{-1} decreases in intensity with the introduction of a nuclear methyl group and is absent in pentamethylpyridine-1-oxide.

TABLE IV-12. Chemical Shifts (τ) for Proton Resonance of Organometallopyridine-1-
oxides in $CDCl_3$ (146c)

N-oxide	C-2,6-H	C-3,5-H	Me
4-SiMe$_3$	1.76	2.58	9.70
4-SiEt$_3$	1.76	2.61	9.10
4-GeMe$_3$	1.77	2.62	9.56
4-SnMe$_3$	1.81	2.61	9.65

The infrared spectrum of pyridine-1-oxide in carbon disulfide changes on addition of methanol (225). The band at $1,265 \text{ cm}^{-1}$ gradually disappears with increasing methanol concentration, and a new absorption band at $1,240 \text{ cm}^{-1}$ appears. This is due to a decrease in the double bond character of the N—O group (**IV-64**) with increasing concentration of methanol. The hydroxyl group absorption of methanol is shifted from 3,650 to $3,360 \text{ cm}^{-1}$ in the presence of pyridine-1-oxide. Electron-donating substituents will increase the hydrogen bonding ability of the N-oxide group. In chloroform, the infrared spectrum of pyridine-1-oxide shows a band at $3,067 \text{ cm}^{-1}$ attributed to hydrogen bonded chloroform (**IV-72**) (226).

IV-72

The N—O stretching absorption of pyridine-1-oxides occurs at a higher frequency than that in aliphatic N-oxides, which has been attributed to contributions of **IV-64** in pyridine-1-oxides, which cannot be obtained in the aliphatic series (7). The position and nature of the substituents on pyridine-1-oxide influence the N—O absorption frequency. An electron-withdrawing group shifts the absorption to higher wave numbers: contributions from **IV-69** account for this. Contributions from **IV-68** account for the fact that electron-donating groups shift the absorption to lower wave numbers (215).

In pyridine-1-oxide, the N—O stretching frequency shifts to a lower wave number by 20 to 40 cm^{-1} on addition of methanol to the solution in CS_2, but that in 3-picoline-1-oxide shifts downward by less than 15 cm^{-1}. This is supported by other work (47, 106, 227). In 2- and 4-picoline-1-oxides, N—O is 5 cm^{-1} lower than in pyridine-1-oxides, which is attributed to the electron-donating ability of the methyl group. However, in 3-methyl and 3,5-dimethylpyridine-1-oxide, the N—O frequencies in CS_2 appear to be at 20 and 53 cm^{-1} higher wave numbers, respectively, which is inconsistent

TABLE IV-13. N—O Stretching Frequencies of Substituted Pyridine-1-Oxides

Substituent on N-oxide	Solvent	$\overset{+}{N}$—$\overset{-}{O}$	Ref.
None	CS$_2$	1,265	146c
	CS$_2$	1,265	215
	CCl$_4$	1,278	228
	Nujol	1,244	146a
2-Me	CS$_2$	1,260	215
	CCl$_4$	1,273, 1,230	228
3-Me	CS$_2$	1,285	215
4-Me	CS$_2$	1,260	215
	CCl$_4$	1,265	228
3-Et	CS$_2$	1,276	215
4-Et	CS$_2$	1,260	215
2,6-Me$_2$	CCl$_4$	1,263	228
3,5-Me$_2$	CS$_2$	1,319	215
	CCl$_4$	1,315, 1,294	228
2,4,6-Me$_3$	CCl$_4$	1,255	228
4-C(Me)$_3$	CS$_2$	1,259	146c
	Nujol	1,240, 1,251	146c
3-Cl	CS$_2$	1,293	215
4-Cl	CS$_2$	1,269	215
4-Cl-2,6-Me$_2$	CCl$_4$	1,264	228
3-Br	CS$_2$	1,294	215
4-Br	CS$_2$	1,271	215
4-Br-2,6-Me$_2$	CCl$_4$	1,267	228
4-I-2,6-Me$_2$	CCl$_4$	1,263	228
4-OMe	CS$_2$	1,240	215
4-OCH$_2$Ph	CS$_2$	1,238	215
3-COMe	CS$_2$	1,302	215
4-COMe	CS$_2$	1,258	215
3-CO$_2$Et	CS$_2$	1,302	215
4-CO$_2$Et	CS$_2$	1,267	215
3-CN	CS$_2$	1,307	215
4-CN	CS$_2$	1,301	215
3-NO$_2$	CS$_2$	1,298	215
4-NO$_2$	CS$_2$	1,303	215
4-NO$_2$-2-Me	CDCl$_3$	1,300	228
4-NO$_2$-3-Me	CS$_2$	1,311	215
	CCl$_4$	1,316	228
4-NO$_2$-2,6-Me	CCl$_4$	1,300	228
	CDCl$_3$	1,295	228
4-NH$_2$-2,6-Me$_2$	CCl$_4$	1,197	228
4-SiMe$_3$	CS$_2$	1,275, 1,250	146c
	Nujol	1,260, 1,248	146c
4-SiEt$_3$	CS$_2$	1,275, 1,234	146c
	Nujol	1,256, 1,240	146c
4-GeMe$_3$	CS$_2$	1,270, 1,239	146c
	Nujol	1,259, 1,238	146c
4-SnMe$_3$	CS$_2$	1,268	146c
	Nujol	1,252	146c

with the electron-donating nature of the methyl group (7). The N—O stretching frequencies of substituted pyridine-1-oxides are given in Table IV-13. The Hammett substituent constants were plotted against the N—O stretching vibration frequency of substituted pyridine-1-oxides as measured in dilute carbon disulfide solution, and the results showed a good linear correlation between the values (215). The majority of 3-substituted pyridine-1-oxides deviated from the linearity, and it was assumed that the absorptions assigned to the N—O stretching frequency contains frequencies other than that of the N—O absorption.

The infrared spectra of the hydrohalide salts of pyridine-1-oxides were examined in potassium halide discs (229). Their structure can be represented by $\overset{+}{N}$—OH . . . X⁻, where X is halide ion. The OH stretching vibration showed a high frequency shift in the order of hydrochlorides, hydrobromides, and hydroiodides, while the in-plane and out-of-plane vibrations shifted to lower frequencies, respectively. For a series of N-oxides hydrohalides, OH stretching, in-plane bonding, and out-of-plane bonding vibrations could be assigned with the aid of the spectra of deuterohalides. The hydroiodide, perchlorate, hexachloroantimonate, and hexachlorostannate salts of 2,6-lutidine-1-oxide in dimethylsulfoxide all showed a broad band at 1,830 to 2,050 cm⁻¹ (N⁺O—H vibration) and a strong band at 1,255 cm⁻¹ (N—OH vibration) whereas the free base in dimethylsulfoxide exhibited only the N—O⁻ band (230).

Hydrogen bonding in 1:1 organic acid salts of N-oxides of pyridine derivatives was studied (231). It was suggested that trichloroacetates are involved in a partial transfer of protons from one molecule of acid to two molecules of base to give symmetrical hydrogen bonds IV-73. 2-Picoline-1-oxide forms a 2:1 adduct with HBr which has an infrared spectrum different

$$\left[\langle \rangle \text{N—O} \cdots \text{H} \cdots \text{O—N} \langle \rangle \right]^+ \left[\underset{\text{Cl}_3\text{CCO}}{\overset{\text{O}}{\|}} \cdots \text{H} \cdots \underset{\text{OCCCl}_3}{\overset{\text{O}}{\|}} \right]^-$$

IV-73

from that of the 1:1 adduct. The former is interpreted in terms of the symmetrical hydrogen bond with the proton bridging the two oxygen atoms (232). The infrared spectrum of the 1:1 benzoate adduct of 2,6-lutidine-1-oxide was examined (233). Pyridine-1-oxide hexachloroantimonate showed a weak broad band at 3,553 cm⁻¹ (234). The infrared spectra of 4-substituted-6-methylpicolinic acid-1-oxide in Nujol or KBr indicate that when the 4-substituent is a nitro group, the molecule has structure IV-74. When, however, the 4-substituent is H or OMe, the molecule exists in two tautometic modifications, IV-74 and IV-75. When a 4-amino group is present the structure of the molecule is IV-76 (235).

Pyridine-1-oxides

IV-74   IV-75   IV-76

The infrared spectra of pyridine carboxylic acid-1-oxides have been studied to establish the type of association and hydrogen bonding (98, 161a, 233, 235, 236a, b). Comparison of the infrared spectra of solid (KBr and Nujol) picolinic acid-1-oxide and quinaldinic acid-1-oxide indicated that in the solid state the acid forms a strong intermolecular hydrogen bond, possibly the symmetrical one resulting in the dimer (IV-77) (236a). Recently, the infrared spectra of 5-methylpicolinic acid-1-oxide and deuterated analogs in chloroform have been examined, and this gives more information about the 3,000 to 1,900 cm^{-1} region (236b). The infrared spectra of other

IV-77

substituted pyridine-1-oxides were studied: 2-substituted pyridine-1-oxides (227), 3,4-disubstituted pyridine-1-oxides (37), 4-substituted-2,2'-bipyridyl-1-oxides (237), halopyridine-1-oxides (238, 239), aminopyridine-1-oxides (240), phenylpyridine-1-oxides (241), cyanopyridine-1-oxides (113, 242), pyridine-1-oxide carbonyl compounds (89), pyridine-1-oxide esters (243), and others (244–250).

8. Ultraviolet Absorption Spectra

As mentioned previously, the ground state of pyridine-1-oxide has contributions from polar structures IV-64 and IV-65, which in solutions are stabilized by electrostatic interaction due to the polarity of the solvent and by hydrogen bonding in hydroxylic solvents. The U.V. spectra of pyridine-1-oxides show a strong solvent effect in going from aprotic to hydroxylic solvents (251). In acid solution, where the $\overset{+}{N}$—$\overset{-}{O}$ group is converted into $\overset{+}{N}$—OH, the absorption spectra of these molecules resemble those of the corresponding pyridinium ions (252). Since there is a change in the electronic configuration of the pyridine nucleus on N-oxidation, one cannot compare the parent pyridine and its N-oxide. Red shifts in the absorption and fluorescence spectra are observed in nonpolar solvents, and the vibrational structure of the absorption bands become more pronounced. It was concluded that the electron transfer from the oxygen atom to the aromatic nucleus is promoted in nonpolar solution (252). Protonation of the N—O group results in a blue shift and a hypochromic solvent shift increasing in the order in ether, water, and $1N$ hydrochloric acid solution (in $1N$ HCl solution the conjugate acid is formed). In non-hydrogen-bonding solvents, a blue shift of the $\pi-\pi^*$ band occurs which increases with increasing polarity of the solvent (253). The $\pi-\pi^*$ band of pyridine-1-oxide occurs at 280.5 nm in ether solution. The 330-nm band is considered to be an $n-\pi^*$ transition and is affected by hydroxylic solvents (254). The $n-\pi^*$ transition may be due to excitation of the $2p$ electrons of oxygen to the π^* level of the molecule and has been identified at 330 nm in the vapor spectrum of pyridine-1-oxide (255). These bands are observed in nonpolar solvents but are absent in hydroxylic solvents. In their vapor state, 2- and 3-picoline-1-oxide exhibit $n-\pi^*$ bands (256, 257).

It was found that introduction of a methyl group at C-3 or C-4 results in a shift to a longer wave length, but the opposite effect is observed with a C-2 methyl group (258). A study of the U.V. spectra of pyridine-1-oxides containing basic substituents, for example, NH_2 and NR_2 shows that protonation occurs first at the N-oxide function (7). Due to the strong electron-withdrawing power of the $\overset{+}{N}$—OH group, the second protonation is retarded in 2- and 4-aminopyridine-1-oxides. 3-Aminopyridine-1-oxide is, however, converted to the dication in 62% sulfuric acid. Other studies of the U.V. spectra of pyridine-1-oxides (259–261), picoline-1-oxide:iodine complex (262), alkylpyridine-1-oxides (263, 264), α-hydroxymethylpyridine-1-oxide (82), pyridinecarboxylic acid-1-oxides (265, 266), aminopyridine-1-oxides (122),

TABLE IV-14. Electronic Absorption Spectrum of Pyridine-1-oxide (269b)

Theoretical values				Experimental values		
ΔE(eV)	f	nature	Pol.	ΔE(eV)	f	Pol.
2.98	0.00	$n-\pi^*$	Z			
3.76	0.03	$\pi-\pi^*$	X	3.81	0.02	X
3.94	0.15	$\pi-\pi^*$	Y	4.40	0.21	Y
5.62	0.00	$n-\pi^*$	Z			
5.92	0.25	$\pi-\pi^*$	X	5.72	0.20	
6.05	0.14	$\pi-\pi^*$	Y			
6.56	0.40	$\pi-\pi^*$	Y	6.60	0.25	

phenylpyridine-1-oxides (267, 268), and phenylazopyridine-1-oxides (269a) have also been carried out.

The electronic absorption spectrum of pyridine-1-oxide was calculated using SCF-MO's (see Section II.11) (269b) (Table IV-14). Good agreement between the calculated and observed parameters for four transition bands were obtained. The first transition at $\epsilon = 3.81$ eV is assigned to the lowest $\pi-\pi^*$ excitation, which agrees well with some recent theoretical and experimental conclusions found for pyridine-1-oxide (269c–d).

9. Mass Spectra

In pyridine-1-oxide, large $(M-16)^+$ ions are often of diagnostic value (17, 34, 270–272a–e) (eq. IV-5) (Table IV-15). Loss of oxygen occurs to give

TABLE IV-15. Mass Spectral Data of Some Pyridine-1-oxides

Substituent in N-oxide	Relative abundance (%)				Base peak	Ref.
	M^+	M^+-16	M^+-17	M^+-18		
None	75	29	17	—	39	272b
		87			39	270
2-Me	69	28	100	2	92	272b
		40			92(M—OH)$^+$	270
2-Et		17			106(M—OH)$^+$	270
3-Me	100	25	11	1	109	272b
	100	41	15	—	109	19b
4-Me	100	16	8	4	109	272b
2,6-Me$_2$	94	14	100	3	106	272b
2-CH$_2$OH	7	9	35	17	79	272b
2-CD$_2$OH	18	9	11	42	78	272b
3-CH$_2$OH	60	100	75	34	109	272b
3-F	100	45	10	—	113	19b
3-Br	62	53	—	—	39	19b
2,3,4,5,6-Cl$_4$	—	—	—	—	251(M-16$^+$)	17
3-OMe	100	12	<1	—	125	19b
3-CN	100	27	8	—	120	19b
3-NHSO$_2$CH$_3$	13	51	—	—	93	19b
4-CHO	4	79	27	—	51	272b
2-CO$_2$H	28	0.8	0.2	—	78	272b
3-CO$_2$H	100	72	3	—	139	272b
3-CO$_2$Me	100	10	2	—	153	19b
3-CO$_2$Et	100	14	2	—	167	272b
4-NO$_2$	58	10	—	—	39	272b
		68				272c

the pyridine derivative which undergoes further fragmentation depending on the nature of the substituents present (34, 273). The abundance of $(M-16)^+$

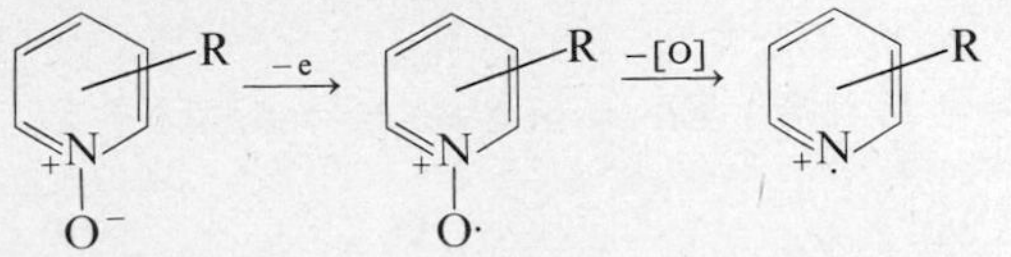

$$(IV-5)$$

ions is drastically decreased by alkyl substitution at the 2-position due to the following reaction leading to loss of OH (eq. IV-6) (270). Thus the base peak

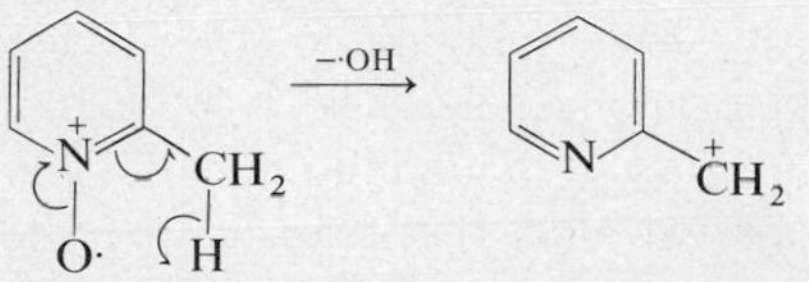

$$(IV-6)$$

in the spectra of the 2-alkylpyridine-1-oxide is the $(M-OH)^+$ ion and it is conceivable that in suitably substituted aromatic N-oxides, the operation of an ortho-effect may result in a further lowering of the abundance of the $(M-16)^+$ ion (270). In addition to the loss of oxygen from 2-alkylpyridine-1-oxides, there is evidence for rearrangement reactions occurring before the oxygen of the N-oxide is lost (34).

In its mass spectrum, pentachloropyridine-1-oxide (**IV-78**) gives a parent ion at m/e 267 and a base peak $(M-16)^+$ ion due to loss of an oxygen atom (17a). At higher temperatures, expulsion of oxygen was not observed but a new base peak at m/e 204 is explained by fragmentation of the N-oxide to give C_4Cl_4N with loss of COCl; this is probably due to rearrangement of the N-oxide to the oxaziridine (**IV-79**) or the oxazepine (**IV-80**).

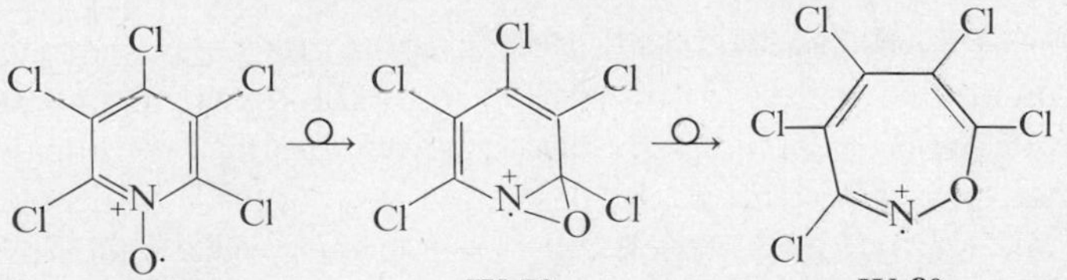

3-Substituted pyridine-1-oxides have been shown to fragment in a generalized manner unless the substituent itself fragments in such a way as to cause other routes to become energetically favorable (19b). The 3-substituted pyridine-1-oxide may undergo fragmentation by loss of an oxygen atom (M-16) or a hydroxyl radical (M-17) or the parent ion may also rearrange to give the oxazepine radical cation or the 2-pyridone radical cation. Loss of hydrogen cyanide from the oxazepine cation intermediate gives a

3-substituted furan ion which undergoes further fragmentation. The 2-pyridone radical cation may lose carbon monoxide to give the pyrrole derivative which can undergo further fragmentation.

10. Electron Spin Resonance

The esr spectrum of the radical anion of 4-nitropyridine-1-oxide and 4-cyanopyridine-1-oxide with and without ^{15}N labeling is reported (274a, b). The spin densities on the ring N and C are larger in 4-nitropyridine-1-oxide than in 4-nitropyridine. Reaction of equimolar amounts of 4-nitrosopyridine-1-oxide with 1-hydroxy-2-phenylindole in ether gave products identified by thin layer chromatography on silica gel as 4,4'-azoxypyridine-1,1'-dioxide, N,N'-dihydroxy-2,2'-diphenyl-3,3'-biindolyl, and another product (IV-81) which exhibited unpaired electrons detected by the esr spectrum (275).

IV-81

The esr spectra of 2- and 4-pyridine-1-oxide carboxaldehyde radical anions produced in base by air oxidation of the corresponding primary alcohol have been obtained (276).

11. Molecular Orbital Theory and Chemical Reactivity

Ochiai (7) has discussed various aspects of the molecular orbital theory of heteroaromatic N-oxides. In this section, only the problems of substitution reactions of pyridine-1-oxides are discussed. Molecular orbital calculations on pyridine and its derivatives are described in Chapter I. Molecular orbital calculations on pyridine-1-oxide have also been carried out (269b, 274a, 277–291).

Based on Hammett's substitution constants, parameters were derived that permitted atom localization energies for electrophilic and nucleophilic substitution in pyridine-1-oxide to be calculated (279). These predicted that electrophilic substitution would occur at C-4 > C-2 > C-3 and nucleophilic substitution would occur at C-4 > C-3 > C-2. Results of π-electron density calculations are given in IV-82a. Using different parameters, the calculated localization energies predicted that the order of reactivity for radical substitution would be $2 > 4 > 3$ and for nucleophilic substitution was $4 > 2 > 3$ (278). The calculated π-electron densities are as in **IV-82b**.

IV-82a   **IV-82b**

Recently CNDO-Cl analysis for an assumed geometry of pyridine-1-oxide was carried out using the following values (269b), and the energy values for

$$d(C\text{—}N) = 1.40 \text{ Å}$$
$$d(N\text{—}O) = 1.28 \text{ Å}$$
$$d(C\text{—}C) = 1.40 \text{ Å}$$
$$d(C\text{—}H) = 1.08 \text{ Å}$$

some characteristic molecular orbitals were obtained (Table IV-16). Computation of the total dipole moment μ showed that the postulated geometry of pyridine-1-oxide was valid. The ground state charge distribution of pyridine-1-oxide calculated on this basis is given in Table IV-17 and from these charge distributions a calculated dipole moment of $+4.78$ D (debyes) is obtained:

$$\mu_{at} = +4.02 \text{ D} \qquad \text{(valence dipole distribution)}$$

$$\mu_{sp} = +0.76 \text{ D} \qquad \text{(atomic dipole distribution)}$$

$$\mu_{tot} = +4.78 \text{ D}$$

Experimentally one finds $\mu_{ex} = +4.78 \pm 0.07 \text{ D}$.

It can be readily seen that neither the π-electron densities as shown in **IV-82a** or **IV-82b** nor the total charges (Table IV-17) predict the observed behavior of pyridine-1-oxide toward either electrophilic or nucleophilic substitution. None of these calculations take into account the nature of the reagent used ("hard" or "soft," strong nucleophile or weak nucleophile)

TABLE IV-16. Energy Levels of Some Characteristic
M.O.'s of Pyridine-1-oxide (269b)

Highest occupied M.O.'s	Lowest vacant M.O.'s
$\psi_{14}(\sigma) - 14.61 \text{ eV}$	$\psi_{19}(\pi^*) - 0.64 \text{ eV}$
$\psi_{15}(\pi) - 13.99 \text{ eV}$	$\psi_{20}(\pi^*) - 0.45 \text{ eV}$
$\psi_{16}(\pi) - 11.67 \text{ eV}$	$\psi_{21}(\pi^*) + 2.01 \text{ eV}$
$\psi_{17}(n) - 11.16 \text{ eV}$	$\psi_{22}(\sigma^*) + 3.78 \text{ eV}$
$\psi_{18}(\pi) - 9.50 \text{ eV}$	$\psi_{23}(\sigma^*) + 4.56 \text{ eV}$

TABLE IV-17. Ground State Charge Distribution in Pyridine-1-oxide (269b)

Atom	A.O. charges				Total charge
	$2s$	$2p_x$	$2p_y$	$2p_z$	
2	1.028	0.988	0.988	1.054	4.058
3	1.039	1.005	1.003	0.957	4.004
4	1.031	0.908	0.963	1.124	4.026
N	1.170	1.133	1.084	1.140	4.526
O	1.770	1.993	1.160	1.644	6.566
H-2, 0.970; H-3, 0.963; H-4, 0.948					

or the nature of the solvent. Thus if a strong nucleophilic reagent such as an organolithium compound is used, the N-oxide oxygen atom will be coordinated with the metal, and it would seem that ground state π-electron density calculations for the pyridinium ion would be more appropriate for predictive purposes. Again, protic solvents coordinate the oxygen atom to a greater or lesser degree (220) and, for example, with a weak nucleophile in such a solvent nucleophilic localization energies varying all the way from those calculated for the free base to those for the O-protonated species may be appropriate in predicting the site of attack. The electron distribution in pyridine-1-oxides and the chemical reactivity of the species are dependent upon the nature of the solvent. For example, in the reaction of an alkoxy-pyridinium ion with cyanide ion to give 2- and 4-cyanopyridine (280, 281), the rate is solvent-dependent. Protic or acidic solvents stabilize the cyanide ion (make it a weaker nucleophilic reagent) so that the transition state resembles the Wheland σ-complex and the 4-isomer is formed preferentially, as expected on the basis of the nucleophilic atom localization energies. In nonpolar solvents, however, the transition state is closer to the ground state and the main product formed is the 2-isomer, consistent with the ground state π-electron density of pyridinium ion but not pyridine-1-oxide (282). Nitration of pyridine-1-oxide to give the 4-nitro-derivative involves nitration of the free N-oxide (283, 284) (see Section IV.1.B). Bromination with excess bromine in 90% sulfuric acid and silver sulfate at 200° may occur at C-2 and C-4, again through the free base (63) (see Section IV.1.C.). On the other hand, sulfonation and bromination in 65% fuming sulfuric acid occurs at C-3 and proceeds *via* the conjugate acid of the N-oxide (see Section IV.1.C and E), as predicted for pyridinium salts (eq. IV-7).

(IV-7)

III. Reactions of the *N*-Oxide Function

1. Electrophilic Addition to Oxygen in Pyridine-1-oxides

A. Salt Formation

Pyridine-1-oxide is a weak base which forms stable salts (e.g. hydro-chlorides, hydrobromides, perchlorates, and picrates) only with strong acids. These salts are used for the purification and characterization of pyridine-1-oxides. Depending on the solvent used, two different types of hydrohalide salts are formed. 2-Picoline-1-oxide in ethanol gave a normal (1:1) hydrohalide salt (**IV-83**), but in benzene, an abnormal (2:1) salt (**IV-84**) is obtained (292). The 1:1 hydrochloride and hydrobromide salts are converted to the 2:1 hydroiodide, on treatment with sodium iodide. An

abnormal salt of 2,6-lutidine-1-oxide and perchloric acid was obtained (292).

The hydrogen iodide, perchlorate, hexachloroantimonate, and tetra-chlorostannate salts of selected pyridine-1-oxides did not show free hydroxyl absorption in the infrared (293). The hydrochloride, hydrobromide, and hydroiodide salts of pyridine-1-oxides were also shown to be abnormal (294). 4-Pyridylhydrazone-1-oxides of a number of ketones gave hydrochloride salts with a base to acid molar ratio of 1:1 and also of 2:1 depending on the amount of acid used. The former exhibited the normal $\overset{+}{\text{N}}$O—H stretching frequency at 2,550 to 2,380 cm^{-1} (**IV-85**) while the latter did not give rise to such a band (**IV-86**) (eq. IV-8) (295) (Table IV-18). 3,4-Lutidine-1-oxide and

TABLE IV-18. Normal and Abnormal Hydrohalide Salts of 4-Pyridylhydrazone-
1-oxides (295)

Ketone	Yield (%)	m.p.	IR(OH) (μ)
Normal salt			
Cyclohexanone	76	213° (decomp.)	3.97
Acetophenone	86	239–240°	4.21
Ethylpyruvate	70	220–222°	4.16
Cyclohexylpyruvamide	89	243–244°	4.12
Propionaldehyde	70	174–175°	4.1
Abnormal salt			
Cyclohexanone	87	207–208°	
Acetophenone	86	209°	
Ethylpyruvate	85	155–157°	

$$(IV\text{-}8)$$

IV-85   IV-86

methanol form a 1:2 complex (in 2,2,4-trimethylpentane solution) as well as
a 1:1 complex, as indicated by the U.V. spectrum of the solution (296).
Gallic acid salts of pyridine-1-oxides and picoline-1-oxides have been
reported (297). 4-Hydroxy-3,5-dinitropyridine-1-oxide forms crystalline
salts with various amines and thus can be used for their characterization
(298).

B. Quaternary Salts

Pyridine-1-oxides add alkyl halides, dialkyl sulfates; and alkyl sulfonates
to form alkoxypyridinium salts, and react with acylhalides or anhydrides to
form acyloxyammonium salts (Scheme IV-1) (Table IV-19). Generally, the

pyridinium salts are formed by heating the two reactants under anhydrous conditions or leaving them at room temperature in the presence of a solvent such as acetonitrile. N-Aryloxypyridinium salts (**IV-87**) can be obtained by

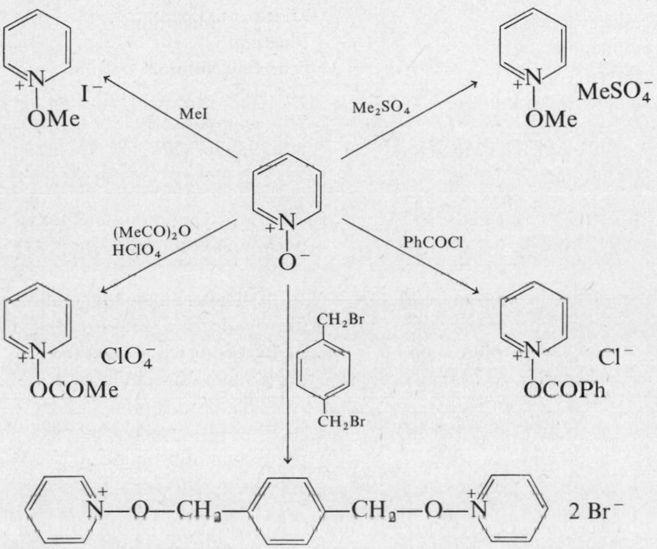

(IV-1)

Scheme IV-1

treatment of the N-oxide in acetonitrile solution with the diazonium tetra-fluoroborate salt (**IV-88**) of an aromatic amine bearing an electron-attracting substituent (eq. IV-9) (299a). Important side-reactions occur when a 4-nitro- or a 4-cyano-group is present in the N-oxide. 1-Hydroxy-2-pyridone

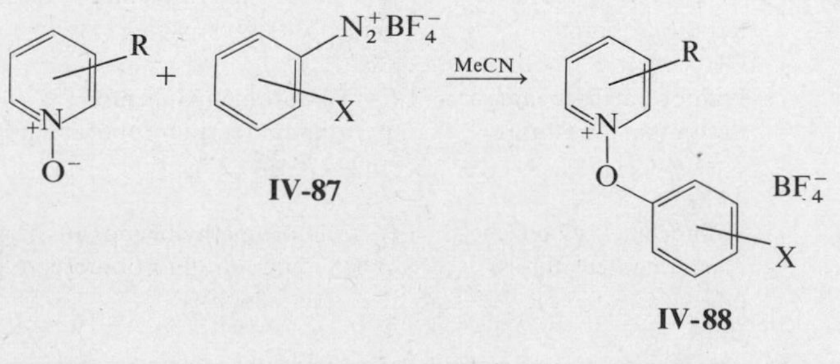

(IV-9)

TABLE IV-19. Preparation of Alkoxy-, Acyloxy-, and Aryloxy-pyridinium Salts

Substituent in pyridine-1-oxide	Reagents and conditions	Products, yields (%)	Ref.
None	MeI	1-Methoxypyridinium iodide, m.p. 90° (decomp.)	306
None	EtBr	1-Ethoxypyridinium bromide, m.p. 168–169°; picrate, m.p. 76–76.5°	307
None	(i) t-BuBr, MeNO$_2$, 2.5 hr (ii) AgClO$_4$, MeNO$_2$, 0°, 4 hr	1-t-Butoxypyridinium perchlorate, m.p. 95–96° (22%)	308
None	Me$_3\overset{+}{N}$CH$_2$CH$_2$Br Br$^-$, MeCN	1-(2-Trimethylammoniumethoxy)-pyridinium dibromide, m.p. 198° (75%)	309
None	Ethylene dibromide MeCN, Δ, 2 hr	N,N-1,2-Dioxyethane bispyridinium dibromide, m.p. 175° (51%)	310
None	p-Xylylene dibromide, MeCN, Δ, 10 mm	1,1'-[Bis-(p-phenylene)methoxy]-pyridinium dibromide, m.p. 188–189° (85%)	310
None	C$_6$H$_4$(CH$_2$CH$_2$Br)$_2$-p, 30 min	1,1'-[Bis-2-(p-phenylene)ethoxy]-pyridinium dibromide, m.p. 104–105° (65.8%)	310
None	PhSO$_2$NSO, anhyd. C$_6$H$_6$	m.p. 162–163° (92%)	305a
None	p-Cyanobenzenediazonium tetrafluoroborate, CH$_3$CN, R.T., 24 hr	1-(p-Cyanophenoxy)pyridinium tetrafluoroborate, m.p. 214–215° (69.5%)	299a
None	o-Nitrobenzenediazonium tetrafluoroborate, CH$_3$CN, R.T., 24 hr	1-(o-Nitrophenoxy)pyridinium tetrafluoroborate, m.p. 161–162° (49.3%)	299a
None	m-Nitrobenzenediazonium tetrafluoroborate, CH$_3$CN, R.T., 24 hr	1-(m-Nitrophenoxy)pyridinium tetrafluoroborate, m.p. 149–150° (24.0%)	299a
None	p-Nitrobenzenediazonium tetrafluoroborate, CH$_3$CN, R.T., 24 hr	1-(p-Nitrophenoxy)pyridinium tetrafluoroborate, m.p. 157.5–159° (83.0%)	299a
None	o-Trifluoromethylbenzene-diazonium tetrafluoroborate, CH$_3$CN, R.T., 24 hr	1-(o-Trifluoromethylphenoxy)-pyridinium tetrafluoroborate, m.p. 169–170° (44.2%)	299a
None	p-Trifluoromethylbenzene-diazonium tetrafluoroborate, CH$_3$CN, R.T., 24 hr	1-(p-Trifluoromethylphenoxy)-pyridinium tetrafluoroborate, m.p. 135–136° (36.5%)	299a

Table IV-19 (*Continued*)

Substituent in pyridine-1-oxide	Reagents and conditions	Products, yields (%)	Ref.
None	(i) *p*-Nitrobenzenesulfenyl chloride, benzene 10° (ii) 60% perchloric acid at 0°	1-(*p*-Nitrobenzenesulfenyloxy)-pyridinium chloride (perchlorate, m.p. 160–170°, decomp.)	305b
None	*p*-Nitrobenzenesulfinyl chloride, benzene, 25°	1-(*p*-Nitrobenzenesulfinyloxy)-pyridinium chloride, m.p. 98–100°	305c
None	Ac$_2$O, HCl, 70% HClO$_4$	1-Acetoxypyridinium perchlorate, m.p. 125–127.5° (79%)	294
2-Me	Me$_2$SO$_4$, 80°	1-Methoxy-2-methylpyridinium methosulfate, m.p. 57–60°	108
	EtBr	1-Ethoxy-2-methylpyridinium bromide, m.p. 104.5–105°; picrate m.p. 78–79°	307
	EtI, Me$_2$CO	1-Ethoxy-2-methylpyridinium iodide, m.p. 105° (decomp.) (84%)	311
	C$_6$H$_5$COCH$_2$Br, 0°	1-Benzoylmethoxy-2-methyl-pyridinium bromide	312
	Me$_3$$\overset{+}{N}CH_2CH_2$BrBr$^-$, MeCN	*N*-(2-Trimethylammoniumethoxy)-2-picolinium dibromide, m.p. 196° (51%)	309
	(i) *p*-Xylylene dibromide, Δ, 10 mm (ii) 24 hr, R.T.	1,1′-[Bis-(*p*-phenylene)methoxy]-2-methylpyridinium dibromide, m.p. 160–161° (77%)	310
	(i) Ac$_2$O, AcOH (ii) HClO$_4$, AcOH	1-Acetoxy-2-methylpyridinium perchlorate, m.p. 148–150° (83%)	313
	Ac$_2$O, HCl, 70% HClO$_4$	1-Acetoxy-2-methylpyridinium perchlorate, m.p. 153–154.5 (80%)	294
	(i) *p*-Nitrobenzenesulfenyl chloride, benzene, 10° (ii) 20% perchloric acid at 0°	1-(*p*-Nitrobenzenesulfenyloxy)-2-methylpyridinium chloride, per-chlorate, m.p. 128–150° (decomp.)	305b
	p-Nitrobenzenesulfinyl chloride, benzene, 25°	1-(*p*-Nitrobenzenesulfinyloxy)-2-methylpyridinium chloride, m.p. 174–177°	305c
2,6-Me$_2$	Me$_2$SO$_4$, 80°	1-Methoxy-2,6-dimethylpyridinium methosulfate, m.p. 95–97°	311
	Me$_3$$\overset{+}{N}CH_2CH_2$BrBr$^-$, MeCN	*N*-(2-Trimethylammoniumethoxy)-2,6-dimethylpyridinium dibromide, m.p. 175–175.5° (95%)	309
	C$_6$H$_4$(CH$_2$CH$_2$Br)$_2$-*p*, 8 hr	1,1′-[Bis-2-(*p*-phenylene)ethoxy]-2,6-dimethylpyridinium dibromide, m.p. 130–131° (43.0%)	310

Table IV-19 (*Continued*)

Substituent in pyridine-1-oxide	Reagents and conditions	Products, yields (%)	Ref.
	p-Xylylene dibromide, 35°, 18 hr	1,1'-[Bis(*p*-phenylenemethoxy]-2,6-dimethylpyridinium dibromide, m.p. 165° (91.5%)	310
	Ac_2O, HCl, 70% $HClO_4$	1-Acetoxy-2,6-dimethylpyridinium perchlorate, m.p. 145–147.5° (96%)	294
2-Me-5-Et	Me_2SO_4, 80°	5-Ethyl-1-methoxy-2-methylpyridinium methosulfate, brown oil	108
2,5-Me_2	$Me_3\overset{+}{N}CH_2CH_2BrBr^-$, MeCN	*N*-(2-Trimethylammoniummethoxy)-2,5-dimethylpyridinium dibromide, m.p. 191° (49%)	309
	p-Xylylene dibromide, Δ, 30 mm	1,1'-[Bis(*p*-phenylene)methoxy]-2,5-dimethylpyridinium dibromide, m.p. 180–180.5° (89.5%)	310
2,4-Me_2	Me_2SO_4, 80°	1-Methoxy-2,4-dimethylpyridinium methosulfate, brown oil	311
	EtBr	1-Ethoxy-2,4-dimethylpyridinium bromide, m.p. 130–131°; picrate m.p. 110.5–111°	307
	$Me_3\overset{+}{N}CH_2CH_2BrBr^-$	*N*-(2-Trimethylammoniummethoxy)-2,4-dimethylpyridinium dibromide, m.p. 195–195.5° (67%)	309
	p-Xylylene dibromide, Δ, 30 mm	1,1'-[Bis-(*p*-phenylene)methoxy]-2,4-dimethylpyridinium dibromide, m.p. 187° (90%)	310
2-Me-3-OH-5-CN	Me_2SO_4, 100–110°, 2 hr	5-Cyano-3-hydroxy-1-methoxy-2-methylpyridinium methosulfate, m.p. 189–190°	69
2-Me-3-OH-5-CN	Et_2SO_4, 100–110°, 2 hr	5-Cyano-1-ethoxy-3-hydroxy-2-methylpyridinium ethosulfate, m.p. 129–130°	69
2-CH_2Ph	(i) Ac_2O, AcOH (ii) $HClO_4$, AcOH	1-Acetoxy-2-benzylpyridinium perchlorate, m.p. 132–135° (94%)	313
2-$CH_2C_6H_4$-NO_2-*p*	(i) Ac_2O, AcOH (ii) $HClO_4$, AcOH	1-Acetoxy-2-(*p*-nitrobenzyl)pyridinium perchlorate, m.p. 161–164° (80%)	313
3-Me	Me_2SO_4, 80°	1-Methoxy-3-methylpyridinium methosulfate, red-brown oil	108
	EtBr	1-Ethoxy-3-methylpyridinium bromide, m.p. 101–102°; picrate, m.p. 111–112°	307
	$Me_3\overset{+}{N}CH_2CH_2Br\ Br^-$, MeCN	*N*-(2-Trimethylammoniummethoxy)-3-methylpyridinium dibromide, m.p. 175° (14%)	309
	Ac_2O, HCl, 70% $HClO_4$	1-Acetoxy-3-methylpyridinium perchlorate, m.p. 143.5–144.5° (85%)	294

Table IV-19 (*Continued*)

Substituent in pyridine-1-oxide	Reagents and conditions	Products, yields (%)	Ref.
	p-Xylylene dibromide (i) Δ, 10 min (ii) R.T., 24 hr	1,1′-[Bis-(p-phenylene)methoxy]-3-methylpyridinium dibromide, m.p. 179–180° (87%)	310
	p-Cyanobenzenediazonium tetrafluoroborate, CH$_3$CN, R.T., 24 hr	1-(p-Cyanophenoxy)-3-methylpyridinium tetrafluoroborate, m.p. 158–159° (75.6%)	299a
	o-Nitrobenzenediazonium tetrafluoroborate, CH$_3$CN, R.T., 24 hr	3-Methyl-1-(o-nitrophenoxy)pyridinium tetrafluoroborate, m.p. 131–132° (41.7%)	299a
3,4-Me$_2$	PhCH$_2$Br	1-Benzyloxy-3,4-dimethylpyridinium bromide·½EtOH, m.p. 125–126° (95% EtOH)	314a
3,5-Me$_2$	Me$_3$N̄CH$_2$CH$_2$Br Br$^-$, MeCN	N-(2-Trimethylammoniumethoxy)-3,5-dimethylpyridinium dibromide, m.p. 196.5–197° (100%)	309
	p-Xylylene dibromide 45 min	1,1′-[Bis(p-phenylene)methoxy]-3,5-dimethylpyridinium dibromide, m.p. 174 (94.8%)	310
4-Me	MeI, Δ, 2 hr	1-Methoxy-4-methylpyridinium iodide, m.p. 90–92°	98
	Me$_2$SO$_4$, 80°	1-Methoxy-4-methylpyridinium methosulfate, m.p. 69–73°	108
	EtBr	1-Ethoxy-4-methylpyridinium bromide, m.p. 103–104°; picrate, m.p. 99°	307
	EtI, Me$_2$CO	1-Ethoxy-4-methylpyridinium iodide, m.p. 114.5° (decomp.) (82%)	311
	Me$_3$N̄CH$_2$CH$_2$Br Br$^-$, CH$_3$CN	N-(2-Trimethylammoniumethoxy)-4-methylpyridinium dibromide, m.p. 190° (50%)	309
	p-Xylylene dibromide (i) Δ, 10 min (ii) R.T., 24 hr	1,1′-[Bis(p-phenylene)methoxy]-4-methylpyridinium dibromide, m.p. 173.5–174.5° (79%)	310
	C$_6$H$_4$(CH$_2$CH$_2$Br)$_2$-p, 35 min	1,1′-[Bis-2-(p-phenylene)ethoxy]-4-methylpyridinium dibromide, m.p. 160–161° (34.5%)	310
	Ac$_2$O, HCl, 70% HClO$_4$	1-Acetoxy-4-methylpyridinium perchlorate, m.p. 84.5–86° (79%)	294
	p-Cyanobenzenediazonium tetrafluoroborate, CH$_3$CN, R.T., 4 hr	1-(p-Cyanophenoxy)-4-methylpyridinium tetrafluoroborate, m.p. 177–179° (36.6%)	299a
	o-Nitrobenzenediazonium tetrafluoroborate, CH$_3$CN, R.T., 4 hr	4-Methyl-1-(o-nitrophenoxy)pyridinium tetrafluoroborate, m.p. 147–148° (33.9%)	299a

Table IV-19 (*Continued*)

Substituent in pyridine-1-oxide	Reagents and conditions	Products, yields (%)	Ref.
	p-Nitrobenzenediazonium tetrafluoroborate, CH_3CN, R.T., 4 hr	4-Methyl-1-(*p*-nitrophenoxy)pyridinium tetrafluoroborate, m.p. 136–137° (22.1%)	299a
4-CH=CHPh	MeI, Δ, dark	1-Methoxy-4-styrylpyridinium iodide, m.p. 118° (decomp.)	306
3-Br	*p*-Cyanobenzenediazonium tetrafluoroborate, CH_3CN, R.T., 24 hr	3-Bromo-1-(*p*-cyanophenoxy)pyridinium tetrafluoroborate, m.p. 175–176° (39.6%)	299a
4-Cl	(i) *p*-Nitrobenzenesulfenyl chloride, $CHCl_3$, R.T. (ii) 69% $HClO_4$	4-Chloro-1-(*p*-nitrobenzenesulfenyloxy)pyridinium perchlorate, m.p. 250–253°	305d
4-OMe	*p*-Nitrobenzenesulfenyl chloride, $CHCl_3$, R.T.	4-Methoxy-1-(*p*-nitrobenzenesulfenyloxy)pyridinium chloride, crude solid	305d
4-OEt	p-$MeC_6H_4SO_2NSO$, dry $CHCl_3$	EtO—⟨ring⟩$\overset{+}{N}$—$O_2S\overset{-}{N}SO_2C_6H_4Me$-$p$- m.p. 135–146°	305a
4-NO₂	Me_2SO_4(1–1.2 equiv.), 65–70°	1-Methoxy-4-nitropyridinium methosulfate; hygroscopic crystals	314b
	p-Nitrobenzenesulfenyl chloride, $CHCl_3$, R.T.	4-Nitro-1-(*p*-nitrobenzenesulfenyloxy)pyridinium chloride, yellow solid	305d
2-Me-4-NO₂	Me_2SO_4(1–1.2 equiv.), 65–70°	1-Methoxy-2-methyl-4-nitropyridinium methosulfate, hygroscopic crystals	314b
2-Me-5-Et-4-NO₂	Me_2SO_4(1–1.2 equiv.), 65–70°	5-Ethyl-1-methoxy-2-methyl-4-nitropyridinium methosulfate, hygroscopic crystals	314b
3-Me-4-NO₂	Me_2SO_4(1–1.2 equiv.), 65–70°	1-Methoxy-3-methyl-4-nitropyridinium methosulfate, hygroscopic crystals	314b
2-NH₂	p-$MeC_6H_4SO_3Me$	2-Amino-1-methoxypyridinium *p*-toluene sulfonate, m.p. 127–129°	217
2-NHAc	p-$MeC_6H_4SO_3Me$, Δ, 12 hr	2-Amino-1-methoxypyridinium *p*-toluene sulfonate, m.p. 123–124°	131
4-NHAc	p-$MeC_6H_4SO_3Me$, Δ, 12 hr	4-Amino-1-methoxypyridinium *p*-toluene sulfonate, m.p. 126–127°	131
4-CN	*p*-Cyanobenzenediazonium tetrafluoroborate CH_3CN, 78°, 2 hr	4-Cyano-1-(*p*-cyanophenoxy)pyridinium tetrafluoroborate, m.p. 166–167° (10.3%)	299a
4-Ph	Me_2SO_4, 100°, 12 hr	1-Methoxy-4-phenylpyridinium methosulfate, m.p. 100.5°	306
	MeI, Δ, dark	1-Methoxy-4-phenylpyridinium iodide, m.p. 99° (decomp.)	306

Table IV-19 (*Continued*)

Substituent in pyridine-1-oxide	Reagents and conditions	Products, yields (%)	Ref.
	p-Cyanobenzenediazonium tetrafluoroborate, CH₃CN, R.T., 24 hr	1-(p-Cyanophenoxy)-4-phenylpyridinium tetrafluoroborate, m.p. 193–194° (75.0%)	299a
	p-Nitrobenzenediazonium tetrafluoroborate, CH₃CN, R.T., 24 hr	1-(p-Nitrophenoxy)-4-phenylpyridinium tetrafluoroborate, m.p. 186–187° (21.2%)	299a

(**IV-89**), 2-ethoxypyridine-1-oxide, and related compounds react with tetra-*O*-acetyl-α-D-glucopyranosyl bromide to yield the corresponding pyridinium glycoside (**IV-90**) (20, 299b). When a suitable 2-hydroxyalkyl-

pyridine-1-oxide (**IV-91**) is heated with a hydrohalic acid, an intramolecular reaction occurs to yield the alkoxypyridinium salt (**IV-92**) (83, 300).

2-(p-Dimethylaminophenylazo)pyridine-1-oxide and 2-(p-hydroxyphenylazo)pyridine-1-oxide were methylated with dimethyl sulfate to give *N*-methoxypyridinium compounds. On the other hand, 2-(p-dimethylaminophenylazo)pyridine-1-oxide reacts with methyl iodide in chloroform to give *N*-methylpyridinium triiodide derivatives (301). Methylation of 2-aminophenylazopyridine-1-oxide and derivatives of the amino group with methyl iodide or dimethyl sulfate was found to give dyes (302). 1-Acyloxy-2(1*H*)-pyridones (**IV-94**) can be prepared by heating 2-ethoxypyridine-1-oxide (**IV-93**) with the appropriate acid chloride (303). Acylation of the thallium

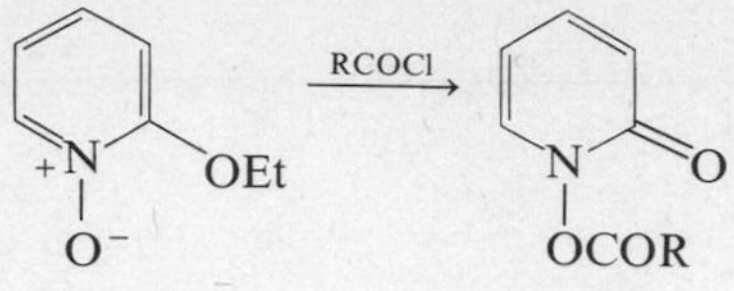

IV-93   **IV-94**

(I) salt of 1-hydroxy-2-(1H)-pyridone **(IV-95)** with an acyl or a sulfonyl chloride yields the "active ester" **(IV-94)** which can be used in peptide synthesis (Table IV-20) (304). Alternatively, treatment of 1-hydroxy-2-(1H)-

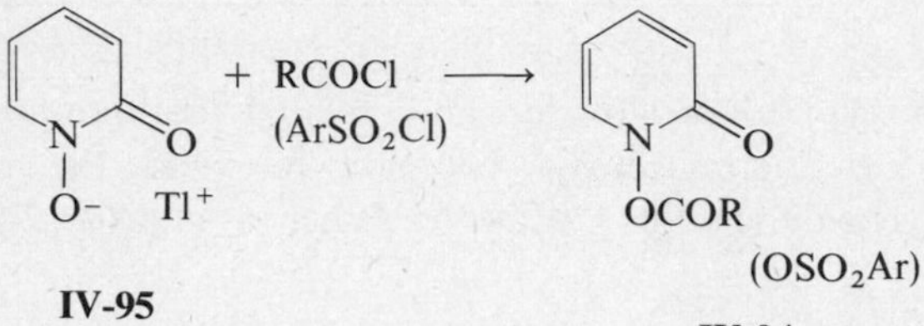

IV-95

IV-94

pyridone with excess thionyl chloride at room temperature yields the N-chlorosulfite **(IV-96)** which, on addition of a thallium (I) carboxylic acid salt in tetrahydrofuran solution gives the 1-acyloxy-2-(1H)-pyridone active esters **(IV-94)** with evolution of sulfur dioxide (304). Yet another approach

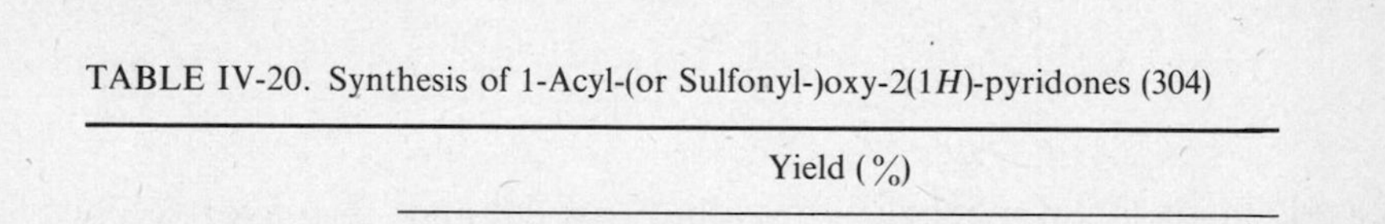

IV-89   **IV-96**   **IV-94**

TABLE IV-20. Synthesis of 1-Acyl-(or Sulfonyl-)oxy-2(1H)-pyridones (304)

	Yield (%)	
1-Acyl group in **(IV-88)**	From thallium(I) salt of 1-hydroxy-2(1H)-pyridone with the acid halide	From 1-hydroxy-2(1H)-pyridone-SOCl$_2$ with thallium(I) carboxylate
CH$_3$CO	95	69
C$_6$H$_5$CO	95	60
p-NO$_2$C$_6$H$_4$CO	98.5	57
C$_6$H$_5$SO$_2$	96	—
p-MeC$_6$H$_4$SO$_2$	95	29

involves the treatment of 2-chloro- or 2-bromopyridine-1-oxide with thallium (I) acetate in tetrahydrofuran containing 20% acetic acid to give **(IV-94, R = CH₃) (304).**

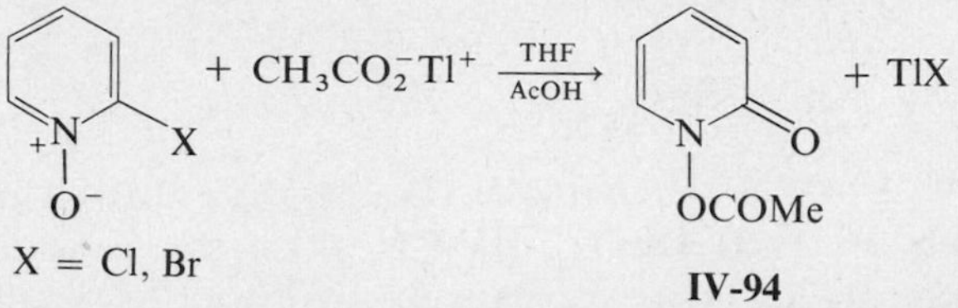

X = Cl, Br

IV-94

Pyridine-1-oxides and *N*-sulfonylarylsulfonamides yield a dipolar 1:1 adduct **(IV-97)** (305a). Pyridine-1-oxides and *p*-nitrobenzenesulfenyl chloride

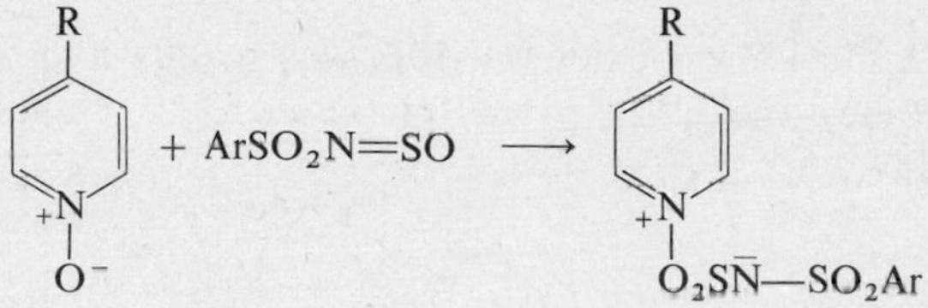

IV-97

or *p*-nitrobenzenesulfinyl chloride react to form the *N*-pyridinium adducts **(IV-98)** and **(IV-99)**, respectively (305b–d).

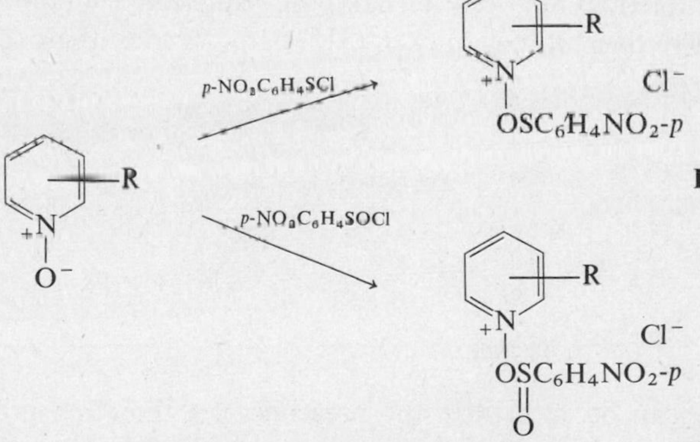

IV-98

IV-99

2. Deoxygenation

A. Catalytic Reduction

Catalytic reduction of pyridine-1-oxide over Raney nickel in alcohol is selective for the reduction of the *N*-oxide group. Under these conditions, the carbon–carbon double bond in 2-styrylpyridine-1-oxide (**IV-100**) and the azo linkage in 4,4′-azopyridine-1,1′-dioxide (**IV-101**) are not reduced, although deoxygenation of the *N*-oxide group occurs (134).

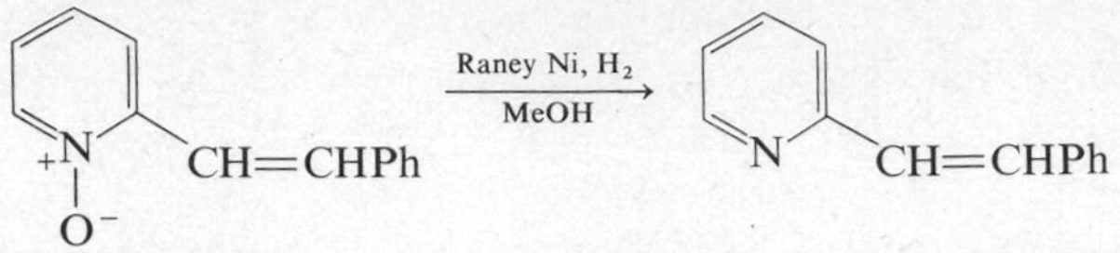

IV-100

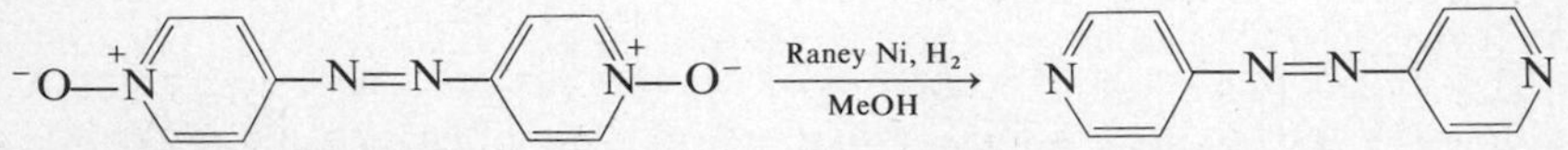

IV-101

4,4′-Azoxy-2,6-lutidine-1-oxide (**IV-102**) was readily hydrogenated in the presence of Raney nickel at room temperature to 4-amino-2,6-lutidine (**IV-103**) (179d).

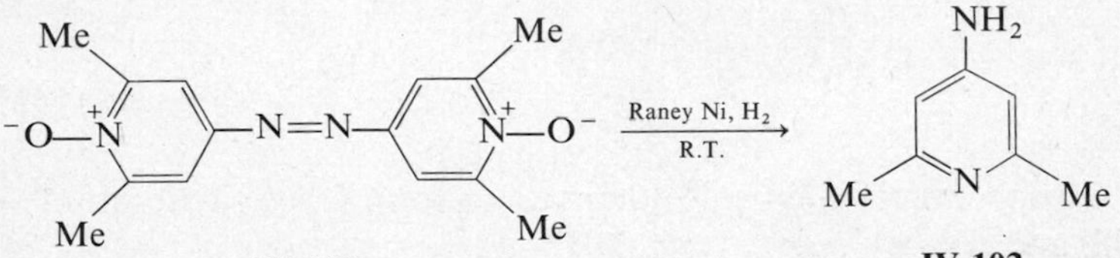

IV-102   **IV-103**

Dehalogenation does not occur on catalytic reduction of 4-chloropyridine-1-oxide (**IV-104**) over Raney nickel (315, 316). Under these conditions,

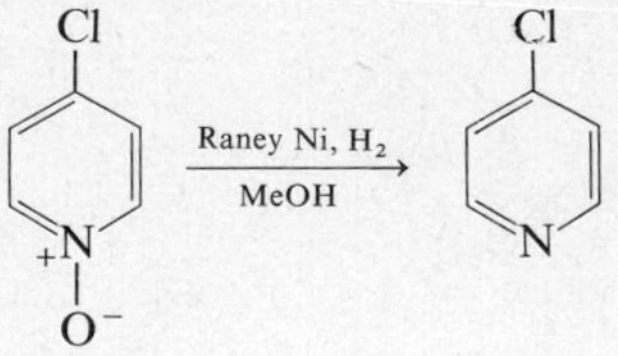

IV-104

deoxygenation can be effected in the presence of a thioether group, as for example, in the reduction of 4-phenylthiopyridine-1-oxide (**IV-105**) or

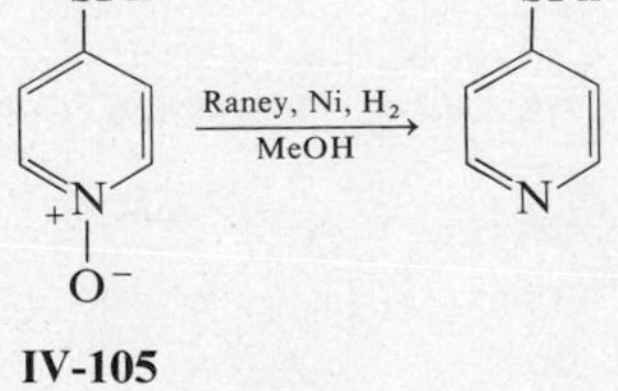

IV-105

4-ethylthiopyridine-1-oxide (317). 2- And 4-(tetra-O-acetyl-β-D-glucopyrano-sylthio)pyridine-1-oxide are deoxygenated using hydrogen over Raney nickel (318). Deoxygenation with hydrogen over Raney nickel is also carried out in glacial acetic acid (319). Catalytic reductions of N-oxides in the presence of a Raney nickel catalyst are summarized in Table IV-21. As compared with the easy deoxygenation of N-oxides in the presence of nickel catalysts, reduction with palladium catalysts in acid solution progresses very gradually. Reduction of 4-nitro-3-picoline-1-oxide (**IV-106**) over a palladium catalyst gives 4-amino-3-picoline-1-oxide (**IV-107**) (323); when the reduction was carried out in the presence of acetic anhydride and glacial acetic acid, deoxygenation occurred

TABLE IV-21. Catalytic Hydrogenation of Pyridine-1-oxides over Raney Nickel

N-oxide	Conditions	Products (yield)	Ref.
3-Me	Raney Ni, H_2, AcOH, Ac_2O	3-Picoline	319
4-Me	Raney Ni, H_2, AcOH, Ac_2O	4-Picoline	319
2-Cl	Raney Ni, H_2, AcOH, Ac_2O	2-Chloropyridine	319
4-Cl	Urushibara Ni, H_2, MeOH	4-Chloropyridine (88%)	316, 319, 320
3-OH	Raney Ni, H_2, AcOH, Ac_2O	3-Hydroxypyridine	319
4-OH	Raney Ni, H_2, MeOH	4-Hydroxypyridine (98%)	316
2-OEt	Raney Ni, H_2, AcOH, Ac_2O	2-Ethoxypyridine	319
2-OCH$_2$Ph	Raney Ni, 1 mole, H_2	2-Benzyloxypyridine	321
2-OCH$_2$Ph	Raney Ni, 2 moles, H_2	2-Hydroxypyridine	321
4-OMe	Urushibara Ni, H_2, MeOH	4-Methoxypyridine (85%)	320
4-OEt	Raney Ni, H_2, AcOH, Ac_2O	4-Ethoxypyridine (86%)	319
4-SEt	Raney Ni, H_2, McOH	4-Ethylthiopyridine (44%)	317
4-SPh	Raney Ni, H_2, MeOH	4-Phenylthiopyridine (82.5%)	317
4-SO$_2$Ph	Raney Ni, H_2, MeOH	4-Benzenesulfonylpyridine (98%)	317
4-CH$_2$Ph	Urushibara Ni, H_2, MeOH, Raney Ni	4-Benzyloxypyridine	320, 321
4-CH$_2$OH	Raney Ni, H_2, AcOH, Ac_2O	4-Hydroxymethylpyridine	319
2-CH$_2$OAc	Raney Ni, H_2	2-Acetoxymethylpyridine	321
2-CH$_2$CH$_2$OMe	Raney Ni, H_2, MeOH, Δ, 1 hr	2-(2′-Methoxyethyl)pyridine	322
4-CO$_2$H	Raney Ni, H_2, AcOH, Ac_2O	4-Isonicotinic acid	319
2-CO$_2$Me	Raney Ni, H_2, AcOH, Ac_2O	Methyl picolinate	319
2-CH=CHPh	Raney Ni, H_2, MeOH	2-Styrylpyridine	134
4-CH=CHPh	Raney Ni, H_2, 1 mole H_2	4-Styrylpyridine	321
4-NO$_2$	Raney Ni, H_2, AcOH	4-Aminopyridine (90%)	316
4-NH$_2$	Raney Ni, H_2, MeOH	4-Aminopyridine (98%)	316
4,4′-N=N—	Raney Ni, 1 mole H_2, MeOH	4,4′-Azopyridine (14%)	134
	Raney Ni, 3 moles H_2, MeOH	4,4′-Hydrazopyridine (93%)	134
4,4′-$\overset{+}{N}$=N— $\overset{|}{O^-}$	Raney Ni, H_2, R.T.	4-Amino-2,6-lutidine	179d

as well as reduction of the nitro group to give 4-acetamido-3-picoline
(IV-108) (323). Using a palladium catalyst, hydrogenolysis and deoxygena-

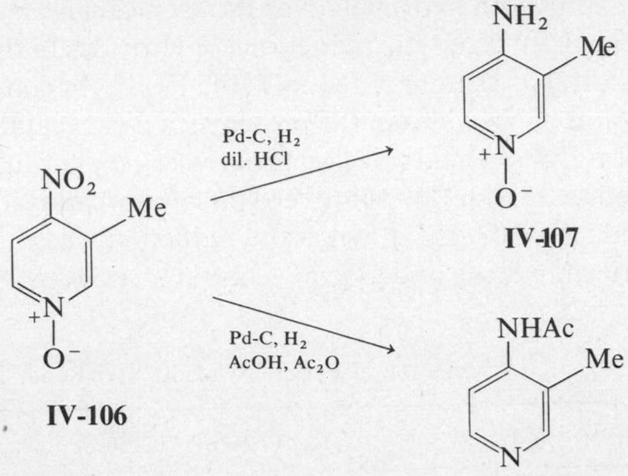

IV-106

IV-107

IV-108

tion of the *N*-oxide group in 4-benzyloxynicotinic acid-1-oxide (IV-109)
occur simultaneously to yield 4-pyridone-3-carboxylic acid (IV-110) (324,
325). With the same catalyst, the carbon–carbon double bond in 4-styryl-

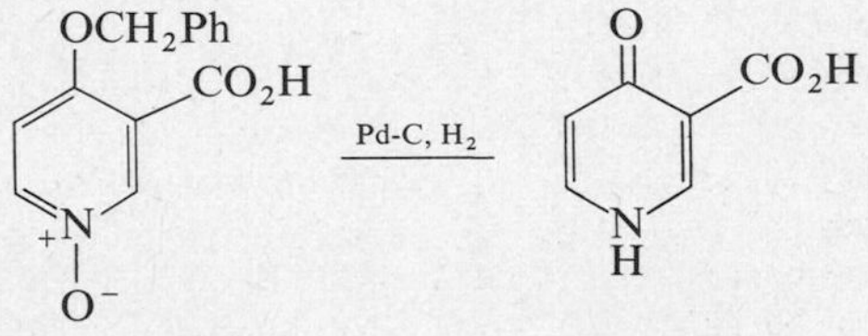

IV-109

IV-110

pyridine-1-oxide (IV-111) and the chloro substituent in 4-chloropyridine-1-
oxide (IV-104) are reduced (325). The use of palladium-charcoal as catalyst

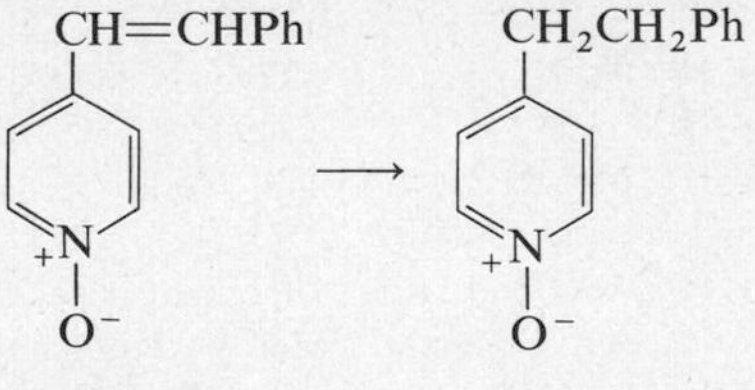

IV-111

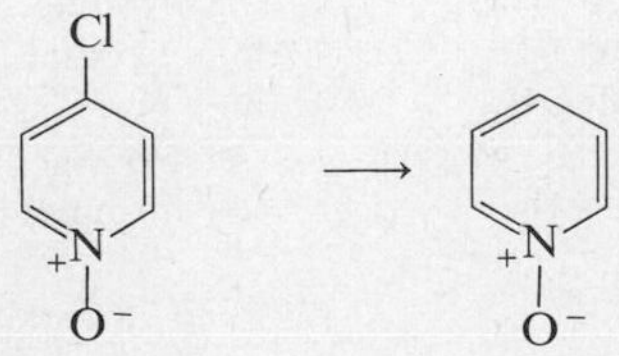

IV-104

TABLE IV-22. Catalytic Reduction of Pyridine-1-oxides over Palladium-Charcoal

Substituent in pyridine-1-oxide	Conditions	Products (yield)	Ref.
None	5% Pd-C, EtOH, R.T.	Pyridine (84%)	325
2-Me	5% Pd-C, EtOH, R.T.	2-Picoline (73%)	325
4-Me	5% Pd-C, EtOH, R.T.	4-Picoline (82%)	325
4-Cl	5% Pd-C, EtOH, R.T.	Pyridine-1-oxide (85%)	325
4-OH	5% Pd-C, EtOH, R.T.	4-Pyridone (85%)	321, 325
4-OH-3-CO_2H	Pd-C, AcOH, 70°	4-Pyridone-3-carboxylic acid	324
2-OCH_2Ph	Pd-C	2-Pyridone	321
4-OMe	5% Pd-C, EtOH, R.T.	4-Methoxypyridine (92%)	325
4-OEt	Pd-C, 0.1N HCl	4-Ethoxypyridine	326
4-OMe-3-$CONH_2$	Pd, silica gel, H_2, 3 hr at 20°/ 1 atm	4-Methoxynicotinamide, m.p. 151° (40–50%)	327
4-OCH_2Ph	5% Pd-C, EtOH	1-Hydroxy-4-pyridone	217
4-OCH_2Ph-3-CO_2H	Pd(OH)$_2$-C, AcOH, 3 atm, H_2, R.T., 22 hr	4-Pyridone-3-carboxylic acid	324
4-OCH_2Ph-3-$CONH_2$	Pd, silica gel, MeOH	4-Pyridone-3-carboxyamide, m.p. 263° (decomp.) (50–80%)	327
2-CH_2Ph	5% Pd-C, EtOH, R.T.	2-Benzylpyridine (84%)	325
2-CH_2OAc	Pd-C	2-Picoline	321
2-CH_2CH_2OMe	5% Pd-C, EtOH, 18–23°, 9 hr	2-(2'-Methoxyethyl)pyridine	322
3-$CH_2CH_2CO_2Et$	5% Pd-C, EtOH, R.T.	3-Carboethoxyethylpyridine (58%)	325
4-$CH_2CH_2CONH_2$	5% Pd-C, EtOH, R.T.	4-Carbamoylethylpyridine (88%)	325
2-CH=CHPh	5% Pd-C, EtOH, R.T.	2-Phenethylpyridine-1-oxide (45%)	325
3-CH=$CHCO_2H$	5% Pd-C, EtOH, R.T.	β-(3-Pyridyl)propionic acid (63%)	325
3-CH=$CHCO_2Et$	5% Pd-C, EtOH, R.T.	Ethyl β-(3-pyridyl)propionate (42%)	325
3-CH=$CHCONH_2$	5% Pd-C, EtOH, R.T.	4-Carbamoylethylpyridine (80%)	325
4-CH=CHPh	5% Pd-C, EtOH, R.T.	4-Phenethylpyridine (89%)	325
4-CH=$CHCO_2Et$	5% Pd-C, EtOH, R.T.	4-Carboethoxyethylpyridine-1-oxide (60–70%)	325
4-CH=$CHCONH_2$	5% Pd-C, EtOH, R.T.	4-Carbamoylethylpyridine (83%)	325
3-COMe	5% Pd-C, EtOH, R.T.	3-Acetylpyridine (70%)	325
4-COMe	5% Pd-C, EtOH, R.T.	4-Acetylpyridine (29%)	325
2-CO_2H-4-COMe	5% Pd-C, H_2O	2-(1-Hydroxyethyl)nicotinic acid lactone m.p. 77.5–79°	328
3-CO_2H-4-NHMe	Pd-C, H_2	4-Methylaminonicotinic acid, m.p. 274–276° (decomp.)	329
3-CO_2H-4-NHPh	Pd-C, MeOH	4-Anilinonicotinic acid	166
3-CO_2H-4-$NHC_6H_4CF_3$-m	Pd-C, 60°	4-(m-Trifluoromethylanilino)- nicotinic acid, m.p. 200–202°; HCl, m.p. 278–284°	330
3-CO_2H-4-NHOH	Pd-C, AcOH, 70°	4-Aminonicotinic acid	324
3-CO_2H-4-NO_2	Pd-C, AcOH, 70°	4-Aminonicotinic acid	324
3-CO_2Et	5% Pd-C, EtOH, R.T.	Ethyl nicotinate (60%)	325

Table IV-22 (*Continued*)

Substituent in pyridine-1-oxide	Conditions	Products (yield)	Ref.
4-CO$_2$Et	5 % Pd-C, EtOH, R.T.	Ethyl isonicotinate	325
2-CONH$_2$	5 % Pd-C, EtOH, 23°, 30 min, 50 lb/in.^{2}	Picolinamide, m.p. 105–106°	156
3-CONH$_2$	5 % Pd-C, EtOH, 23°, 30 min, 50 lb/in.^{2}	Nicotinamide, m.p. 127–128°	156
4-CONHCH$_2$Ph	5 % Pd-C, EtOH, R.T.	4-PhCH$_2$NHCO pyridine (86 %)	325
4-NO$_2$	5 % Pd-C, EtOH, R.T.	4-Aminopyridine (90 %)	325, 326
4-NO$_2$-2-Me	Pd-C, dil. HCl	4-Amino-2-picoline-1-oxide	331
	Pd-C, AcOH, Ac$_2$O	4-Acetamido-2-picoline	331
4-NO$_2$-3-Me	Pd-C, dil. HCl	4-Amino-3-picoline-1-oxide (72 %)	323
	Pd-C, AcOH, Ac$_2$O	4-Acetamido-3-picoline (92 %)	323
2-NH$_2$	5 % Pd-C, EtOH, R.T.	2-Aminopyridine (85 %)	325
4-NH$_2$	5 % Pd-C, EtOH, R.T.	4-Aminopyridine (77 %)	325
4-NHMe-3-Me	Pd-C, AcOH	4-Methylamino-3-picoline	332
2-Ph	5 % Pd-C, EtOH, R.T.	2-Phenylpyridine (87 %)	325
4-Ph	5 % Pd-C, EtOH, R.T.	4-Phenylpyridine (89 %)	325

in the hydrogenation of various pyridine-1-oxides is summarized in Table IV-22. A few examples of deoxygenation of pyridine-1-oxides with platinum-on-carbon (321) and platinum dioxide catalyst (50) show that under these conditions carbon–carbon double bonds in 4-styrylpyridine-1-oxide (**IV-111**) (321) and in 1,2-bis(4-methoxy-3-pyridyl-1-oxide)ethene (**IV-112**) (50) are reduced (Table IV-23).

IV-111

IV-112

Using Raney copper, 4-styrylpyridine-1-oxide underwent deoxygenation to give 4-styrylpyridine (321) without reduction of the olefinic double bond.

TABLE IV-23. Catalytic Reduction of Pyridine-1-oxides over a Platinum-Carbon or Platinum Dioxide Catalyst

N-oxide	Conditions	Products	Ref.
4-CH=CHPh	Pt-C	4-Phenethylpyridine, 4-styrylpyridine (1:1)	321
1,2-Di-(4-OMe-3-pyridyl)ethene-1,1'-dioxide	PtO$_2$, H$_2$, 3 atm 44°, 17 hr	1,2-Di-(4-methoxy-3-pyridyl)ethane- (87%)	50
1,2-Di-(4-Cl-3-pyridyl)ethane 1,1'-dioxide	PtO$_2$, H$_2$, 3 atm 43°, 3 hr	1,2-Di-(3-pyridyl)ethane dihydrochloride, m.p. 304° (decomp.)	50

The products obtained from 1-(1-oxido-2-pyridylmethyl)pyridinium salt **(IV-113)** depend on the reduction conditions and the nature of the counter-ion (333) (Table IV-24).

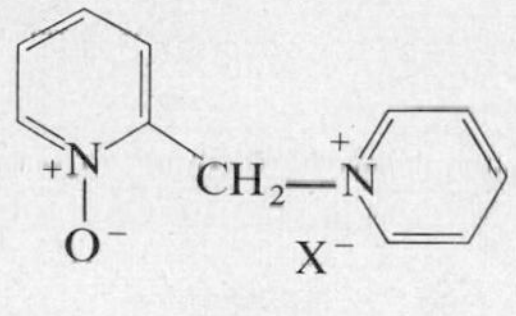

IV-113 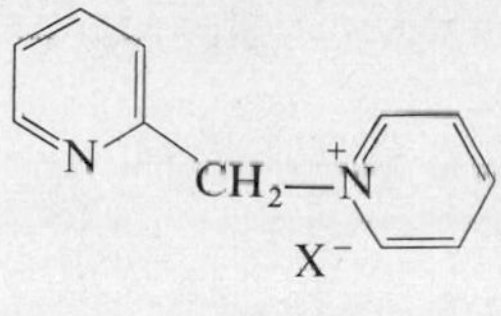   **IV-114**

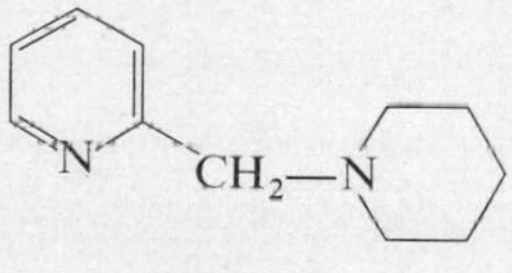

IV-115 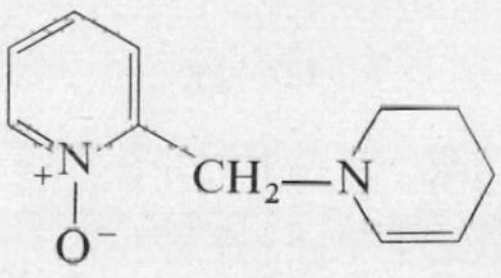   **IV-116**

TABLE IV-24. Reduction of 1-(1-Oxido-2-pyridylmethyl)pyridinium Salts **(IV-103)** under Various Conditions (333)

X$^-$	Conditions	Products (yield)
Cl	Raney Ni, 1 mole H$_2$	**(IV-114)** (43%)
Cl	Raney Ni, 4 moles H$_2$	**(IV-115)** (74–91%), b.p. 100–115°/4 mm; picrate, m.p. 180–182°
OH	Raney Ni, 4 moles H$_2$	**(IV-115)** (71%)
OH	Raney Ni, 4 moles H$_2$, 1% NaOH	**(IV-115)** (91%)
Cl	Pd-C, acid	**(IV-114)** (68%)
Cl	Pd-C, neutral	**(IV-116)** (46%)
I	NaBH$_4$, R.T.	**(IV-116)** (92%)
I	Zn dust, AcOH	2-Picoline (58%) Pyridine (51%)

B. Reduction with Metals or Metal Salts in Acid

Reduction of 5-fluoro-4-nitro-3-picoline-1-oxide (**IV-117**) with iron and glacial acetic acid yielded 4-amino-5-fluoro-3-picoline (**IV-118**) (56). Under these conditions, reduction of a carbon–carbon double bond does not occur

IV-117   IV-118

as shown by the reduction of 4-allyloxypyridine-1-oxide (**IV-119**) (334).

IV-119

Examples of the reduction of pyridine-1-oxides with iron in glacial acetic acid are given in Table IV-25.

Pyridine-1-oxides are deoxygenated by heating them with ferrous oxalate at 300° (140) (Table IV-26).

A number of pyridine-1-oxides have been deoxygenated with zinc and 2N sulfuric acid (345) (Table IV-27).

4-Nitropyridine was obtained in 93% yield by heating its N-oxide with sulfuric acid and nitric acid at 240°, or with nitrosyl sulfate in concentrated sulfuric acid at 170 to 200°. This deoxygenation can be effected by passing nitrogen monoxide gas through the solution of the N-oxide in concentrated sulfuric acid at 200°. Thus 3-bromopyridine and 3-picoline were obtained in good yields in this manner from their N-oxides (346).

TABLE IV-25. Reduction of Pyridine-1-oxides with Iron in Glacial Acetic Acid

N-oxide	Conditions	Products (yield)	Ref.
2,4-I_2-5-OH	Δ, 1 hr	5-Hydroxy-2,4-diiodo-pyridine (51 %)	335
3-F-5-Me-4-NO_2		4-Amino-5-fluoro-3-picoline, m.p. 90° (89 %)	56
3-F-2,6-Me_2-4-NO_2		4-Amino-3-fluoro-2,6-lutidine	336
3-Br-4-NHMe		3-Bromo-4-methylaminopyridine	337
4-Cl	2 hr	4-Chloropyridine	338
4-Cl-2,6-Me_2	70°, 2 hr	4-Chloro-2,6-lutidine (69 %)	339
4-Br-3-Me		4-Bromo-3-picoline	340
4-Br-3-Et		4-Bromo-3-ethylpyridine	340
4-Br-3-isoPr		4-Bromo-3-isopropylpyridine	340
4-Br-2,6-Me_2	70°, 2 hr	4-Bromo-2,6-lutidine (83 %)	339
4-Br-2,5-Me_2	70°	4-Bromo-2,5-lutidine, b.p. 88–90°/14 mm, $n_D^{20°}$ 1.5501	31
4-OCH_2CH=CH_2		4-Allyloxypyridine	334
3-CO_2H-4-SO_3H		4-SO_3H-nicotinic acid, m.p. 257–259°	341
4-NO_2-2,5-Me_2	70°	4-Amino-2,5-lutidine, m.p. 137–139°; picrate, m.p. 173–175°	31
4-NO_2-2-Me-5-Et		4-Amino-5-ethyl-2-picoline (93.7 %)	342
4-NO_2-3-NH_2-2-Me	30 min	3,4-Diamino-2-picoline, m.p. 170° (50 %)	343
4-NO_2-3-NH_2-5-Me		4,5-Diamino-3-picoline (54.5 %)	56
4-NO_2-3-NH_2-2,6-Me_2	30 min	3,4-Diamino-2,6-lutidine	336
4-NHMe-3-Me		4-Methylamino-3-picoline	337
4-NHMe-3-isoPr		4-Methylamino-3-isopropylpyridine	337
4-NHMe-3,5-Me_2		4-Methylamino-3,5-lutidine	337
4-NHMe-2,3,5,6-Me_4		2,3,5,6-Tetramethyl-4-methylaminopyridine	337
2-N=NC_6H_4NHAc-p	Δ, 22 hr	2-(p-Acetamidophenylazo)pyridine	344
2-CH=CHC_6H_4NMe_2-p		2-(p-Dimethylaminostyryl)pyridine	135
4-CH=CHC_6H_4NMe_2-p		4-(p-Dimethylaminostyryl)pyridine (93 %)	148
4-SO_3H-2-Me	Δ, 90 mm	2-Picoline-4-sulfonic acid, m.p. 263° (95 %)	341

TABLE IV-26. Deoxygenation of Pyridine-1-oxides with Ferrous Oxalate (140)

N-oxide	Product (yield)
None	Pyridine (64 %)
2-Me	2-Picoline (62 %)
3-Me	3-Picoline (72 %)
2-NH_2	2-Aminopyridine (45 %)
2-Ph	2-Phenylpyridine (63 %)

TABLE IV-27. Deoxygenation of Pyridine-1-oxides using Zinc and
Sulfuric Acid (345)

N-oxide	Product (yield)
4-OMe-3-Me	4-Methoxy-3-picoline (86.5%)
4-OEt-3-Me	4-Ethoxy-3-picoline (81.6%)
4-OPr-3-Me	4-Propoxy-3-picoline (85.8%)
4-n-BuO-3-Me	4-n-Butoxy-3-picoline (78.6%)
4-O(CH$_2$)$_4$CH$_3$-3-Me	4-n-Pentyloxy-3-picoline (75.4%)
4-O(CH$_2$)$_5$CH$_3$-3-Me	4-n-Hexyloxy-3-picoline (72.2%)
4-O(CH$_2$)$_6$CH$_3$-3-Me	4-n-Heptyloxy-3-picoline (72%)
4-O(CH$_2$)$_7$CH$_3$-3-Me	4-n-Octyloxy-3-picoline (70.1%)
4-isoPrO-3-Me	4-Isopropoxy-3-picoline (82.6%)
4-isoBuO-3-Me	4-Isobutoxy-3-picoline (81.5%)
4-O-isoamyl-3-Me	4-Isoamyl-3-picoline (80.9%)

C. Deoxygenation with Phosphorus Trihalides

Pyridine-1-oxides are readily deoxygenated with phosphorous trihalides,
for example, phosphorous trichloride, to give the parent pyridine (eq. IV-10).
New reactions since the last compilation as summarized in Table IV-28.
When the reaction conditions are controlled, 4-nitro-3-picoline-1-oxide

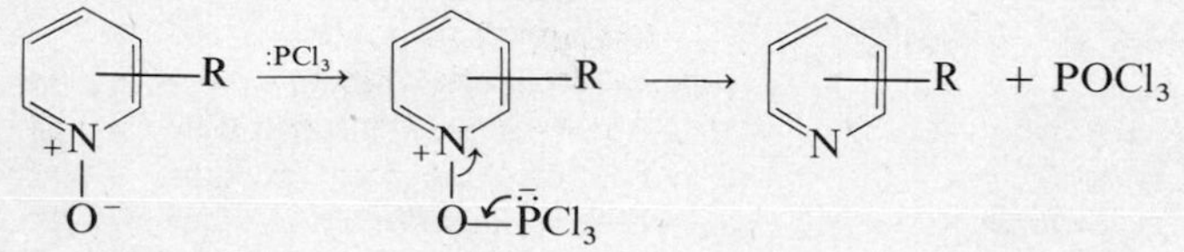

$$ \text{(IV-10)} $$

(**IV-106**) is deoxygenated to yield 4-nitro-3-picoline (**IV-120**) in good yield
(7). If, however, the reaction is prolonged, 4-chloro-3-picoline (**IV-121**) is

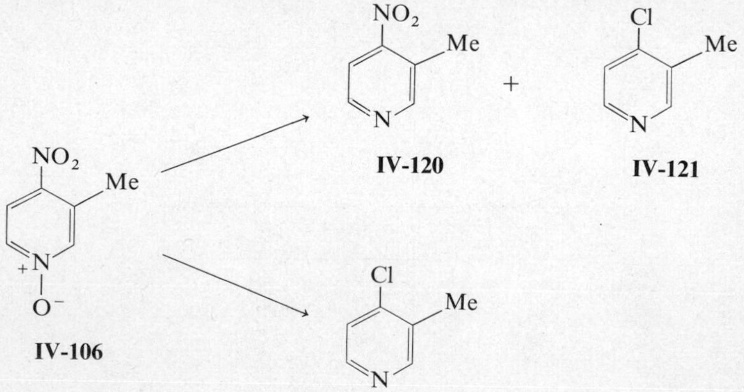

IV-120   IV-121

IV-106

IV-121

TABLE IV-28. Deoxygenation of Pyridine-1-oxide with Phosphorous Trihalide (or Phosphorus Pentachloride without Halogenation)

N-oxide	Conditions	Products (yield)	Ref.
2,6-Me$_2$	PCl$_3$, R.T.	2,6-Lutidine (87%)	349
2-Cl-4-NO$_2$	PCl$_3$, CHCl$_3$	2-Chloro-4-nitropyridine	350
2,3,4,5-Cl$_4$-6-Me	PCl$_3$, CHCl$_3$	3,4,5,6-Tetrachloro-2-picoline	351
2-Br-4-NO$_2$	PBr$_3$	2-Bromo-4-nitropyridine	350
2-I-4-NO$_2$	PCl$_3$	2-Chloro-4-nitropyridine, 2-iodo-4-nitropyridine	350
3-F-4-NO$_2$	PCl$_3$, CHCl$_3$, 1 hr	4-Chloro-3-fluoropyridine	352
3-F-4-NO$_2$	PBr$_3$	4-Bromo-3-fluoropyridine	352
3-Cl-4-OMe	PCl$_3$, CHCl$_3$, 30 min	3-Chloro-4-methoxypyridine (81.4%)	353
3-Cl-4-OEt	PCl$_3$, CHCl$_3$, 30 min	3-Chloro-4-ethoxypyridine, b.p. 118°/20 mm (77.1%)	353
3-Br-4-OMe	PCl$_3$, CHCl$_3$, 30 min	3-Bromo-4-methoxypyridine, b.p. 114–116°/12 mm (81.4%)	353
3-Br-4-OEt	PCl$_3$, CHCl$_3$, 30 min	3-Bromo-4-ethoxypyridine, b.p. 116–118°/12 mm (61.2%)	353
3-I-4-OEt	PCl$_3$, CHCl$_3$, 30 min	4-Ethoxy-3-iodopyridine, b.p. 143°/15 mm (63.8%)	353
4-Cl-3-Me	PCl$_3$, CHCl$_3$, 3°	4-Chloro-3-picoline (51%)	51
4-Cl-2,6-Me$_2$	PCl$_3$	4-Chloro-2,6-lutidine (72%)	339
4-Br-2,6-Me$_2$	PBr$_3$, CHCl$_3$, 70–80°, 3 hr	4-Bromo-2,6-lutidine (70%)	339
4-I	PCl$_3$, CHCl$_3$	4-Iodopyridine (58.5%)	354
4-I-2-Me	PCl$_3$, CHCl$_3$	4-Iodo-2-picoline (92.7%)	354
4-I-2,6-Me$_2$	PCl$_3$, CHCl$_3$	4-Iodo-2,6-lutidine (87%)	354
4-I-2-Me-5-Et	PCl$_3$, CHCl$_3$	5-Ethyl-4-iodo-2-picoline (93.5%)	354
4-I-3-Me	PCl$_3$, CHCl$_3$	4-Iodo-2-picoline (67.4%)	354
3,4,5,6-Cl$_4$-2-Me	PCl$_3$, CHCl$_3$, Δ	3,4,5,6-Tetrachloro-2-picoline, m.p. 91–92°	354b
3,4,5-Cl$_3$-2,6-Me$_2$	PCl$_3$, CHCl$_3$, Δ	3,4,5-Trichloro-2,6-lutidine, m.p. 71–72°	354b
3,4,5-Cl$_3$-2,6-Et$_2$	PCl$_3$, CHCl$_3$, Δ	3,4,5-Trichloro-2,6-diethylpyridine, b.p. 90°/1.5 mm	354b
3,4,5,6-Cl$_4$-2-Ph	PCl$_3$, CHCl$_3$, Δ	3,4,5,6-Tetrachloro-2-phenylpyridine, m.p. 112°	354b
3,4,5-Cl$_3$-2,6-Ph$_2$	PCl$_3$, CHCl$_3$, Δ	3,4,5-Trichloro-2,6-diphenylpyridine, m.p. 166°	354b
3,4,5,6-Cl$_4$-2-OH	PCl$_3$, CHCl$_3$, Δ	3,4,5,6-Trichloro-2-hydroxypyridine, m.p. 224°	354b
2,4-(OMe)$_2$	PCl$_3$, Δ, 1 hr	2,4-Dimethoxypyridine (62.6%)	355, 356
6-OEt-2-CH$_2$-COCO$_2$Et	PCl$_3$, CHCl$_3$	Ethyl 6-ethoxy-2-pyridylpyruvate (26%)	71
4-OMe-2-Me	(i) PCl$_3$, R.T., 1 hr (ii) Δ, 30 min	4-Methoxy-2-picoline	348
4-OEt-2-Me	(i) PCl$_3$, R.T., 1 hr (ii) Δ, 30 min	4-Ethoxy-2-picoline	348
4-OCH$_2$CH=CH$_2$	PCl$_3$, 3 hr	4-Allyloxypyridine (90.5%)	334
4-OCH$_2$Ph	PCl$_3$	1-(4-Pyridyl)-4-pyridone	334

69

Table IV-28 (*Continued*)

N-oxide	Conditions	Products (yields)	Ref.
4-OCH$_2$Ph-2-Me	(i) PCl$_5$, R.T., 1 hr (ii) Δ, 30 min	4-Benzyloxy-2-picoline	348
4-OPh-2-Me	(i) PCl$_5$, R.T., 1 hr (ii) Δ, 30 min	4-Phenoxy-2-picoline	348
1,2-Di-(4-OEt-3-pyridyl)ethane	PCl$_3$	1,2-Di(4-ethoxy-3-pyridyl)ethane	50
4-SPh	(i) PCl$_3$, CHCl$_3$, 0° (ii) 100°, 30 min	4-Phenylthiopyridine	357
2-CH$_2$CH$_2$OMe	PCl$_3$, AcOEt, 80–90°	2-(2'-Methoxyethyl)pyridine	322
2-CH=NOH	PCl$_3$, CHCl$_3$, 0°, 4 hr	2-Cyanopyridine	347
2-CONHCOMe	POCl$_3$, Δ, 30 min	2-(CONHCOMe)pyridine, m.p. 81–82°	358
3-CO$_2$Et-5-NO$_2$	PCl$_3$, CHCl$_3$, R.T.	Ethyl 5-nitronicotinate (70%)	359
3-NO$_2$	PCl$_3$, CHCl$_3$	3-Nitropyridine	117
4-NO$_2$-2,6-Me$_2$	PCl$_3$, CHCl$_3$, 80°, 3 hr	4-Nitro-2,6-lutidine (75%)	120
4-NO$_2$-3-NH$_2$-2,6-Me$_2$	PCl$_3$, CHCl$_3$, 1 hr	3-Amino-4-nitro-2,6-lutidine (82%)	336
4-NO$_2$-2-Me-5-Et	PCl$_3$, CHCl$_3$, 5 hr	4-Chloro-5-ethyl-2-picoline	360
4-NO$_2$, 2,3,5,6-Me$_4$	PCl$_3$, CHCl$_3$	4-Nitro-2,3,5,6-tetramethylpyridine·2H$_2$O	337
4-NO$_2$-2-NMe$_2$	PCl$_3$	2-Dimethylamino-4-nitropyridine	355
4-NO$_2$-2-NHAc	PCl$_3$, CHCl$_3$	2-Acetamido-4-nitropyridine (87.1%)	361
4-NO$_2$-3-Me	PCl$_3$, CHCl$_3$, 10°	4-Nitro-3-picoline (45%)	166
4-NO$_2$-3-Me	(i) PCl$_3$, CHCl$_3$ (ii) 70°, 1 hr	4-Chloro-3-picoline; picrate, m.p. 153–155° (95%)	50
4-NO$_2$-3-Me	PBr$_3$, AcOEt, Δ, 30 min	4-Bromo-3-picoline, b.p. 104–107°/60 mm (36%)	362
4-NO$_2$-3,5-Me$_2$	PCl$_3$, CHCl$_3$	4-Nitro-3,5-lutidine·H$_2$O	337
4-NO$_2$-3-Et	PCl$_3$, CHCl$_3$	3-Ethyl-4-nitropyridine	337
4-NO$_2$-3-isoPr	PCl$_3$, CHCl$_3$	3-Isopropyl-4-nitropyridine	337
1,2-di-(4-NO$_2$-3-pyridyl)ethane	PCl$_3$	1,2-Di-(4-nitro-3-pyridyl)-ethane	50
	PCl$_3$, CHCl$_3$, 60°, 3.5 hr	1,2-Di-(4-nitro-3-pyridyl)ethane; 1,2-Di-(4-chloro-3-pyridyl)ethane	51
3-NH$_2$-4-NO$_2$	PCl$_3$, CHCl$_3$, 30 min	3-Amino-4-nitropyridine, (90%)	352

obtained as a result of nucleophilic displacement of the nitro group by a chloride ion (50, 166). Reaction of 2-pyridine aldoxime-1-oxide (**IV-122**) with phosphorous trichloride yields 2-cyanopyridine (**IV-123**) (347), while

IV-122   IV-123

4-benzyloxypyridine-1-oxide (**IV-124**) gives 1-(4-pyridyl)-4-pyridone (**IV-125**) (334). A series of 4-alkoxy-2-picoline-1-oxides have been deoxygenated with

IV-124   **IV-125**

phosphorus pentachloride without the usual accompanying nuclear chlorination (348).

D. Deoxygenation with Organophosphorus Compounds

Deoxygenation of pyridine-1-oxides with phosphorous halides has already been discussed (see Section III.2.C). Triphenyl phosphite was found to be less effective than phosphorous trichloride (363). Deoxygenation with triethyl

phosphite in the presence of oxygen and a peroxide was assumed to proceed *via* a free radical mechanism (364). Triphenylphosphine was found to be effective for the deoxygenation of pyridine-1-oxide (365). Results of the deoxygenation of pyridine-1-oxides with organophosphorous compounds are given in Table IV-29.

TABLE IV-29. Deoxygenation of Pyridine-1-oxides with Organophosphorus Compounds

Substituent in N-oxide	Conditions	Product (yield)	Ref.
None	Et_3P, 48 hr	Pyridine (70%)	366
None	$(EtO)_3P$, $(EtOCH_2CH_2)_2O$, R.T.	Pyridine	364
None	Ph_3P, 230–233°, 0.6 hr	Pyridine (89.8%)	365
None	Ph_3P, AcOEt	Pyridine (72%)	363
2-Me	Ph_3P, AcOEt, 0.6 hr	2-Picoline (51%)	365
4-Me	Ph_3P, 275–280°, 0.7 hr	4-Picoline (92.8%)	365
4-OH	Ph_3P, AcOEt	4-Pyridone (20%)	363
4-OMe	$(EtO)_3P$, $(EtOCH_2CH_2)_2O$	4-Methoxypyridine	364
4-OMe	Ph_3P, 270–275°, 0.8 hr	4-Methoxypyridine (61.6%)	365
2-CH_2CH_2OMe	Ph_3P, 210°	2-(2′-Methoxyethyl)pyridine	322
4-NO_2	Ph_3P	No 4-nitropyridine	365
4-NH_2	Ph_3P, AcOEt	4-Methoxypyridine	364

E. Deoxygenation with Sulfur and its Derivatives

Mercapto compounds, sulfides, and thiourea are used, as is sulfur, to effect deoxygenations to pyridine and its derivatives, the oxide oxygen generally being either bonded to sulfur or released as water (367a, b).

Oxidation of mercaptans by pyridine-1-oxide possibly occurs by the initial protonation of pyridine-1-oxide to form **IV-126** which then reacts with mercaptide ion to give pyridine and the unstable sulfinic acid (eq. IV-11)

(367a). When 4-picoline-1-oxide was heated with sulfur at 145° for 18 hr, an unusual product, tetra-(4-pyridyl)thiophene (IV-127) was obtained (367a).

The results obtained on deoxygenation of pyridine-1-oxides with sulfur and alkyl sulfides are given in Table IV-30. Pyridine-1-oxides and picoline-1-oxides were deoxygenated in 21 to 78% yield by passing a slow stream of sulfur dioxide into their solution in water or dioxan for a period of 3 hr (368).

TABLE IV-30. Deoxygenation of Pyridine-1-oxides with Sulfur and Alkyl Sulfides

Substituent in *N*-oxide	Conditions	Products (yield)	Ref.
None	S_8, 150°, 10 hr	Pyridine	367a
2-Me	S_8, o-$Cl_2C_6H_4$, 150°, 64 hr	2-Picoline (42%)	367a
2-Me	Bu_2S, 150°, 24 hr	2-Picoline, 2-picoline-1-oxide	367a
2-CH_2CH_2OMe	S_8, 150°, 3 hr	2-(2'-Methoxyethyl)pyridine	322

Under these conditions, 4-nitro-2-picoline-1-oxide and methyl isonicotinate-1-oxide were not deoxygenated, indicating that electron-withdrawing substituents at C-4 hinder the reaction (369) (Table IV-31). Deoxygenation can also be effected with lower oxides of sulfur, such as sulfurous acid, in the presence of an acylating agent (370). It was reported that pyridine-1-oxide

was deoxygenated with sodium hydrosulfite or sodium sulfite (371) (eq. IV-12). Pyridine-1-oxide and its derivatives are reduced with thionyl chloride

TABLE IV-31. Reduction of Substituted Pyridine-1-oxides with Sulfur Dioxide (369)

Substituent in *N*-oxide	Product (yield)
None	Pyridine (60%)
2-Me	2-Picoline (34%)
4-Me	4-Picoline (31%)
2,6-Me_2	2,6-Lutidine (67%)
2,5-Me_2	2,5-Lutidine (65%)
2-Me-5-Et	5-Ethyl-2-picoline (63%)
3-Cl	3-Chloropyridine (21%)
4-Cl-2-Me	4-Chloro-2-picoline (41%)
4-OMe-2-Me	4-Methoxy-2-picoline (62%)
5-CO_2Et-2-Me	Ethyl 6-methylnicotinate (68%)
6-NHAc-2-Me	6-Acetamido-2-picoline (65%)
3-NO_2-2,6-Me_2	3-Nitro-2,6-lutidine (31%)

$$\text{(structure with NO}_2\text{, Me, Me, N}^+\text{, O}^-) \xrightarrow[\text{Na}_2\text{S}_2\text{O}_3]{\substack{\text{NaHSO}_3 \text{ or} \\ \text{Na}_2\text{SO}_3 \text{ or}}} \text{(NH}_2\text{, Me, Me, N)} + \text{(NH}_2\text{, SO}_3\text{H, Me, Me, N)} \qquad \text{(IV-12)}$$

(65%)

$$\text{(structure N}^+\text{O}^-\text{, CH}_2\text{CH}_2\text{OMe)} \xrightarrow[\text{80–90}^\circ,\ 15\ \text{min.}]{\text{Na}_2\text{S}_2\text{O}_3,\ \text{H}_2\text{O}} \text{(structure N, CH}_2\text{CH}_2\text{OMe *)}$$

at 50° in yields of 60 to 85% (293). Deoxygenation of 3-α-hydroxymethyl-pyridine-1-oxide (IV-128) was accompanied by side-chain chlorination to give 3-α-chloromethylpyridine (IV-129) (372). Treatment of 2-picoline-1-oxide

$$\text{(CH}_2\text{OH, N}^+\text{, O}^-)} \xrightarrow[50^\circ]{\text{SOCl}_2} \text{(CH}_2\text{Cl, N)}$$

IV-128   IV-129

with thionyl chloride in benzene solution at 15 to 20° gave the hydrochloride salt IV-130, while at 50° deoxygenation occurred to yield 2-picoline (292).

$$\text{(N}^+\text{O}^-\text{, Me)} \xrightarrow[\substack{\text{SOCl}_2 \\ \text{C}_6\text{H}_6}]{\substack{15\text{–}20^\circ \\ 50^\circ}} \begin{cases} \text{(N}^+\text{, Me, OH, Cl}^-\text{) IV-130} \quad (34\%) \\ \\ \text{(N, Me)} \quad (92\%) \end{cases}$$

Pyridine-1-oxides are deoxygenated by diaryl disulfides (373). With symmetrical disulfides, for example, di-(p-nitrophenyl)disulfide and pyridine-1-oxide, the products are pyridine, di-(p-nitrophenyl)disulfide, and p-nitro-benzenesulfonic acids. Reaction of unsymmetrical disulfides, for example, aryl p-nitrophenyl disulfide and pyridine-1-oxide, gives pyridine, diaryl

* See Ref. 322.

TABLE IV-32. Reaction of Pyridine-1-oxide with Diaryl Disulfides (373)

Disulfide	Conditions	Product (in addition to pyridine)	Second-order rate constant in bromobenzene k_2 [l/(mole)(sec)]
$PhSSC_6H_4NO_2$-p	150–160°, 1 hr	$(p$-$O_2NC_6H_4S)_2$, $(PhS)_2$, $PhSO_3H$	6.27×10^{-2}
$(p$-$NO_2C_6H_4S)_2$	180°, 4 hr	$(p$-$O_2NC_6H_4S)_2$, p-$O_2NC_6H_4SO_3H$	3.75×10^{-2}
p-$MeOC_6H_4S$-$SC_6H_4NO_2$-p	160°, 3 hr	$(p$-$O_2NC_6H_4S)_2$, $(p$-$MeOC_6H_4S)_2$, p-$MeOC_6H_4SO_3H$	3.10×10^{-1}
p-ClC_6H_4SS-$C_6H_4NO_2$-p	160°, 3 hr	$(p$-$O_2NC_6H_4S)_2$, $(p$-$ClC_6H_4S)_2$, $ClC_6H_4SO_3H$	1.18×10^{-1}

disulfide, di-(p-nitrophenyl)disulfide, and aryl sulfonic acid. p-Nitro-benzene sulfonic acid was not obtained, which suggests that the attack by the N-oxide takes place on the sulfur atom attached to the less electron-withdrawing aryl group (eq. IV-13) (Table IV-32).

(IV-13)

F. Other Methods of Deoxygenation

Pyridine-1-oxides undergo partial deoxygenation on vapor phase chromatography on 20% SE 30 on chromosorb W; the relative amounts of the pyridine and its N-oxide depend on the age of the column (374).

Reduction of substituted pyridine-1-oxides and the N-methoxypyridinium salts **(IV-131)** with sodium in ethanol gives predominately the pyridine base, together with substituted piperidine (IV-132) and substituted 3-piperideine (IV-133) (Table IV-33) (375). When 4-substituents were present in the ring, an increased amount of 3-piperideine was observed.

IV-131   IV-132   IV-133

Sodium borohydride reductions of pyridine-1-oxides yield approximately 30% yields of mixtures containing a number of components (375). In the case of 1-alkoxypyridinium salts the products (pyridine, piperidine, and 3-piperideine) are present in almost equal amounts (Table IV-33) (375). Reduction of substituted pyridine-1-oxides and N-alkoxypyridinium salts with aluminum hydride gives a mixture of products in which the substituted 3-piperideine (IV-133) is the major product (Table IV-33) (375).

Pyridine-1-oxide is smoothly deoxygenated by sodium borohydride in the presence of aluminum chloride in diglyme at 25°; at 75° further reduction to dihydropyridine occurs (376). 4-Chloro-2,6-lutidine-1-oxide (IV-134) can be deoxygenated with NABH₄-AlCl₃ in diglyme at room temperature (371, 377).

IV-134

Pyridine-1-oxides are rapidly reduced by lithium aluminum hydride (378). Reduction of pyridine-1-oxides with lithium aluminum hydride has been reported to give small amounts of piperidines along with the corresponding 3-piperideines and pyridines (in the ratio 1:2) (Table IV-33) (375).

Bis-(β-methoxyethoxy)sodium aluminum hydride reduction of pyridine-1-oxides gave mixtures of pyridine and 3-piperideines and of piperidines (Table IV-33) (375). The 1-alkoxypyridinium salts are reduced more readily to give up to 40% yields of piperidines (Table IV-33) (375). Lithium trimethoxyaluminohydride has been reported to deoxygenate 4-picoline-1-oxide (379). Pyridine-1-oxide is reduced electrolytically to pyridine which is then reduced further (Table IV-33) (375). The products of electrolytic reductions of 1-alkoxypyridinium salts are very similar to those obtained

TABLE IV-33. Reduction of Pyridine-1-oxides and 1-Alkoxypyridinium Salts (**IV-131**) with Sodium in Ethanol, Sodium Borohydride, Lithium Aluminum Hydride, Aluminum Hydride, Bis(β-methoxyethoxy)sodium Aluminum Hydride or by Electrolysis (375).

Substituents in N-oxide	Conditions	Products (yield)
None (R = H)	Na, EtOH, Δ	Piperidine (28), 3-piperideine (22), pyridine (48), unidentified compounds (2) (65 %)
	LiAlH$_4$, Δ, 6 hr	Piperidine (1), 3-piperideine (28), pyridine (61), unidentified compounds (10) (46 %)
	AlH$_3$, Δ, 2 hr	Piperidine (19), 3-piperideine (69), pyridine (11), unidentified compounds (1) (80 %)
	NaAlH$_2$(OCH$_2$—CH$_2$OCH$_3$)$_2$	Piperidine (9), 3-piperideine (35), pyridine (51), unidentified compounds (5) (45 %)
	Electroreduction	Piperidine (35), 3-piperideine (45), pyridine (17), unidentified compounds (3) (85 %)
R = H, R' = Me, X = CH$_3$OSO$_3$	Na, EtOH, Δ	Piperidine (8), pyridine (85), unidentified components (7) (70 %)
	NaBH$_4$	Piperidine (27), 3-piperideine (26), pyridine (33), unidentified compounds (14) (56 %)
	NaAlH$_2$(OCH$_2$—CH$_2$OCH$_3$)$_2$	Piperidine (43), 3-piperideine (21), pyridine (26), unidentified components (10) (72 %)
	Electroreduction	Piperidine (41), 3-piperideine (25), pyridine (22), unidentified components (12) (67 %)
R = H, R' = Me, X = OH	NaBH$_4$	Piperidine (36), pyridine (49), unidentified components (15) (65 %)
R = H, R' = n-C$_3$H$_7$, X = I	LiAlH$_4$, Δ, 6 hr	Piperidine (1), 3-piperideine (66), pyridine (15), unidentified components (18) (75 %)
	AlH$_3$, Δ, 2 hr	Piperidine (35), 3-piperideine (54), pyridine (5), unidentified components (6) (95 %)
R = 3–Me	Na, EtOH, Δ	3-Methylpiperidine (20), 5-methyl-3-piperideine (4), 3-methyl-3-piperideine (20), 3-picoline (40), unidentified components (16) (66 %)
	LiAlH$_4$, Δ, 6 hr	3-Methylpiperidine (1), 5-methyl-3-piperideine (9), 3-methyl-3-piperideine (34), 3-picoline (56) (63 %)
	AlH$_3$, Δ, 2 hr	3-Methylpiperidine (21), 5-methyl-3-piperideine (23), 3-methyl-3-piperideine (50), 3-picoline (6) (87 %)
	NaAlH$_2$(OCH$_2$—CH$_2$OCH$_3$)$_2$	3-Methylpiperidine (11), 5-methyl-3-piperidine (8), 3-methyl-3-piperideine (45), 3-picoline (35), unidentified components (1) (45 %)
	Electroreduction	3-Methylpiperidine (26), 5-methyl-3-piperideine (19), 3-methyl-3-piperideine (33), 3-picoline (22) (85 %)
R = 3–Me, R' = Me, X = CH$_3$OSO$_3$	Na, EtOH, Δ	3-Methylpiperidine (6), 5-methyl-3-piperideine (2), 3-methyl-3-piperideine (14), 3-picoline (66), unidentified components (12) (73 %)
	NaBH$_4$	3-Methylpiperidine (25), 5-methyl-3-piperideine (2), 3-methyl-3-piperideine (23), 3-picoline (42), unidentified components (8) (56 %)

Table IV-33 (*Continued*)

Substituents in N-oxide	Conditions	Products (yield)
	NaAlH$_2$(OCH$_2$—CH$_2$OCH$_3$)$_2$	3-Methylpiperidine (40), 5-methyl-3-piperideine (14), 3-methyl-3-piperideine (36), 3-picoline (5), unidentified components (5) (65%)
	Electroreduction	3-Methylpiperidine (35), 5-methyl-3-piperideine (19), 3-methyl-3-piperideine (41), 3-picoline (5) (90%)
R = 3–Me, R' = CH$_3$, X = OH	NaBH$_4$	3-Methylpiperidine (22), 3-picoline (72), unidentified components (6) (55%)
R = 3–Me, R' = n–C$_3$H$_7$, X = I	LiAlH$_4$, Δ, 6 hr	3-Methylpiperidine (12), 5-methyl-3-piperideine (15), 3-methyl-3-piperideine (49), 3-picoline (24) (92%)
	AlH$_3$	3-Methylpiperidine (22), 5-methyl-3-piperideine (20), 3-methyl-3-piperideine (50), 3-picoline (8) (95%)
4-Me	Na, EtOH, Δ	4-Methylpiperidine (19), 4-methyl-3-piperideine (40), 4-picoline (33), unidentified components (8) (63%)
	LiAlH$_4$, Δ, 6 hr	4-Methylpiperidine (3), 4-methyl-3-piperideine (32), 4-picoline (63), unidentified components (2) (70%)
	AlH$_3$	4-Methylpiperidine (15), 4-methyl-3-piperideine (80), 4-picoline (2), unidentified components (3) (95%)
	NaAlH$_2$(OCH$_2$—CH$_2$OCH$_3$)$_2$	4-Methylpiperidine (0.5), 4-methyl-3-piperideine (16), 4-picoline (82), unidentified components (1.5) (48%)
	Electroreduction	4-Methylpiperidine (25), 4-methyl-3-piperideine (51), 4-picoline (24) (86%)
R = 4-Me, R' = Me, X = CH$_3$OSO$_3$	Na, EtOH, Δ	4-Methylpiperidine (16), 4-methyl-3-piperideine (29), 4-picoline (54), unidentified components (1) (70%)
	NaBH$_4$	4-Methylpiperidine (32), 4-methyl-3-piperideine (25), 4-picoline (37), unidentified components (6) (80%)
	NaAlH$_2$(OCH$_2$—CH$_2$OCH$_3$)$_2$	4-Methylpiperidine (40), 4-methyl-3-piperideine (25), 4-picoline (34), unidentified components (1) (75%)
	Electroreduction	4-Methylpiperidine (22), 4-methyl-3-piperideine (41), 4-picoline (32), unidentified components (5) (91%)
R = 4-Me, R' = n–C$_3$H$_7$, X = I	LiAlH$_4$	4-Methylpiperidine (2), 4-methyl-3-piperideine (69), 4-picoline (29) (90%)
	AlH$_3$	4-Methylpiperidine (4), 4-methyl-3-piperideine (70), 4-picoline (26) (97%)
R = 4–Me, R' = CH$_3$, X = OH	NaBH$_4$	4-Methylpiperidine (14), 4-picoline (85), unidentified components (1) (53%)

from pyridine-1-oxide (Table IV-33) (375). Epoxides undergo oxidation with amine oxides, such as pyridine-1-oxide; for example, 1,2-butylene oxide in benzene gave formaldehyde and propionaldehyde (380). In the presence of copper stearate, pyridine-1-oxide and ethyl diazoacetate gave pyridine and ethyl glyoxalate. The carbenoïd [CuCHCO₂Et] is apparently formed which abstracts oxygen from pyridine-1-oxide (381a) (eq. IV-14). Fluorenyl carbene has been used for the deoxygenation of pyridine-1-oxides (381b). A facile

$$
\begin{array}{ccc}
\text{pyridine-1-oxide} + \underset{\text{Cu}}{\overset{\text{CHCO}_2\text{Et}}{\|}} & \xrightarrow{-\text{Cu}} & \underset{\text{O}\overset{\frown}{-}\text{CHCO}_2\text{Et}}{\text{pyridinium}} \longrightarrow \text{pyridine} + \text{OHCCO}_2\text{Et} \quad \text{(IV-14)}
\end{array}
$$

reaction of pyridine-1-oxide with hydrazobenzene yields pyridine, azo-benzene, and water (367a). Oxidation of hydrazobenzene may involve a two-step mechanism, with an initial protonation of the *N*-oxide by hydrazo-benzene (eq. IV-15). Alternatively, a concerted reaction involving a cyclic

$$
\text{pyridine-1-oxide} + \text{Ph}-\overset{..}{\underset{H}{N}}-\overset{..}{\underset{H}{N}}-\text{Ph} \rightleftharpoons \text{pyridinium-OH} + \text{Ph}-\overset{..}{\underset{H}{N}}-\overset{..}{N}-\text{Ph} \longrightarrow
$$

$$
\text{pyridine} + \text{Ph}-\overset{}{\underset{HO}{N}}-\overset{}{\underset{H}{N}}-\text{Ph} \longrightarrow \text{PhN}{=}\text{NPh} + \text{H}_2\text{O} \qquad \text{(IV-15)}
$$

transition state may obtain, similar to the mechanism for the oxidation of hydrazobenzene by molecular oxygen (382). When the 2-picoline-1-oxide-iodine complex is heated with excess pyridine-1-oxide, 2-pyridine aldehyde and 2-picoline are obtained (see Section V.1). Picoline-1-oxides and its derivatives underwent deoxygenation as well as oxidation to give other products on heating in dimethyl sulfoxide (see Section V.1) (383). Four-electron reduction eliminates the two *N*-oxide groups of bis-(2-pyridyl-1-oxide)disulfide, and in the presence of a complex-forming agent, a six-electron reduction of the *N*-oxide and disulfide groups occurs (384). Other methods used in the deoxygenation of pyridine-1-oxides are summarized in Table IV-34).

Photolysis of pyridine-1-oxides results in deoxygenation to give pyridines as well as the other photolysis products which are discussed in Section VII. Irradiation (2,537 Å) of a solution of naphthalene (10 mg) and pyridine-1-oxide (100 mg) in dichloromethane (4 ml) showed the presence of 1,2-naphthalene oxide (0.14 mg) and naphthol (0.13 g). This constitutes the

TABLE IV-34. Miscellaneous Methods of Deoxygenation of Pyridine-1-oxides

Substituent in N-oxide	Conditions	Products (yield)	Ref.
None	PhNHNHPh	Pyridine, azobenzene, water	367a
	PhHgCCl$_3$, Δ, 44 hr	Pyridine (23%)	385
	CCl$_3$CO$_2$Me, NaOMe, C$_6$H$_6$, 10–15°	Pyridine (40%)	385
	I$_2$, 160°	Pyridine (50%)	386
2-Me	80% NH$_2$NH$_2$, Cu, 180°, diethylene glycol	2-Picoline (82%)	387
	I$_2$, 130°	2-Picoline (50%)	386
2-Cl-4-OMe	NH$_4$OH, FeSO$_4$	2-Chloro-4-methoxypyridine (83.4%)	355
2-I-4-NO$_2$	NH$_2$NH$_2$, H$_2$O, EtOH	4-Amino-2-iodopyridine	355
2-CH$_2$OH	SeO$_2$, 90–100°, 8 hr	Pyridine-2-aldehyde, picolinic acid	81
2-CH$_2$OH-6-Me	SeO$_2$, 90–100°, 8 hr	6-Methylpyridine-2-aldehyde, 6-methylpyridine-2-aldehyde-1-oxide	81

first example of chemical epoxidation of an aromatic double bond (388) (eq. IV-16).

$$(IV-16)$$

3. Reactions of O-Alkylated and O-Acylated Pyridinium Salts

The preparation of N-alkoxy- and N-acyloxypyridinium salts has been discussed (see Section IV.1.C). Reaction of 1-alkoxypyridinium salts with nucleophiles can occur by one of four possible pathways, A–D (308, 389) (Scheme IV-2). A fifth mode of reaction is conceivable, which has not yet been reported for such salts, but has been observed with a proposed O-complexed intermediate species, is nuclear proton abstraction from the α-position with a strong base at low temperatures in a nonprotic solvent (path E) (299a). Alkyloxypyridinium salts decompose on warming in basic solution to yield the parent pyridine and a carbonyl compound (path A) (390, 391). This reaction was used for the preparation of aldehydes. For example, benzaldehyde was obtained from pyridine-1-oxide and benzyl

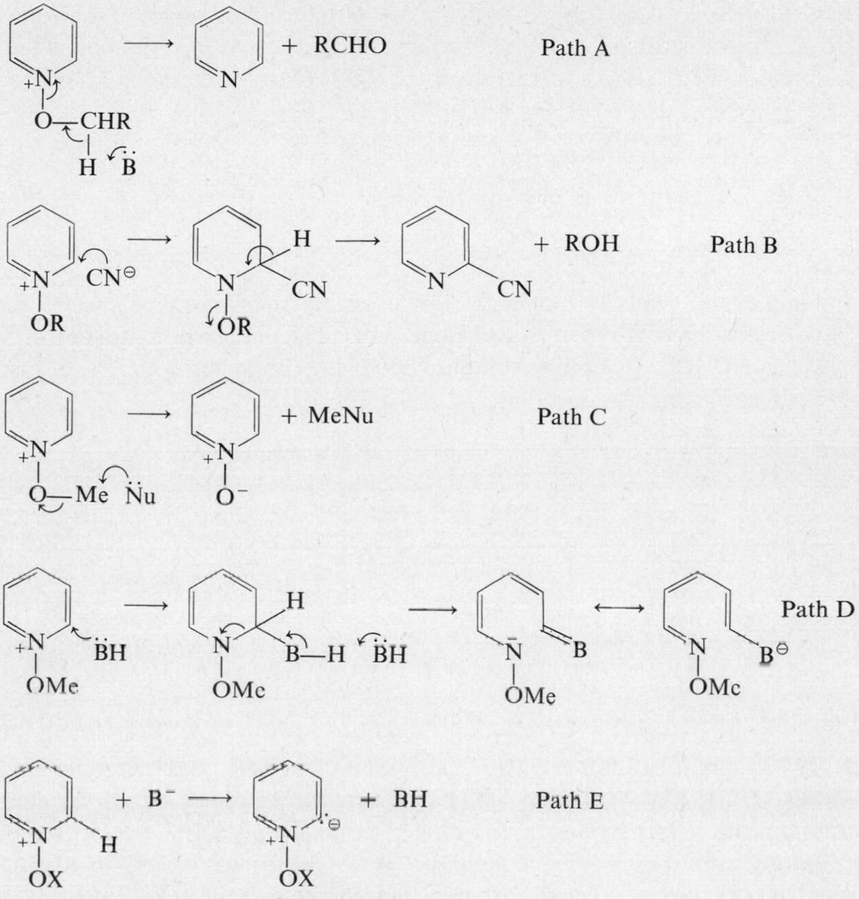

Scheme IV-2

bromide followed by treatment of the pyridinium salt with dilute aqueous sodium hydroxide (392) (eq. IV-17). An intermediate in the synthesis of a

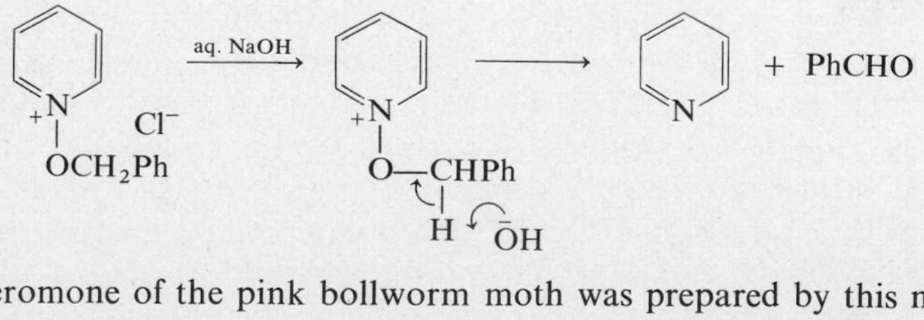

(IV-17)

sex pheromone of the pink bollworm moth was prepared by this method. 5-Bromopentylacetate (or its iodo derivative) was treated with pyridine-1-oxide; in the presence of sodium bicarbonate the *N*-alkoxypyridinium salt

gave 5-acetoxyvaleraldehyde (40 % yield) (393). Decomposition of the adduct
of picoline-1-oxide and α-haloketones is a convenient synthesis of α-keto-
aldehydes (312) (IV-18). Heating an excess of pyridine-1-oxide with α-

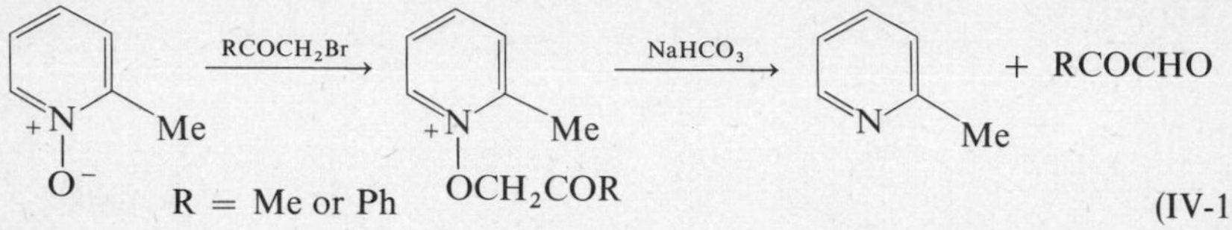

R = Me or Ph

(IV-18)

haloacids also leads to aldehydes; for example, α-bromophenylacetic acid
gave benzaldehyde (49 %), phenylacetic acid (5 %), and carbon dioxide (46 %)
(394) (eq. IV-19). In boiling toluene, however, the yields were 58, 58, and
11 %, respectively.

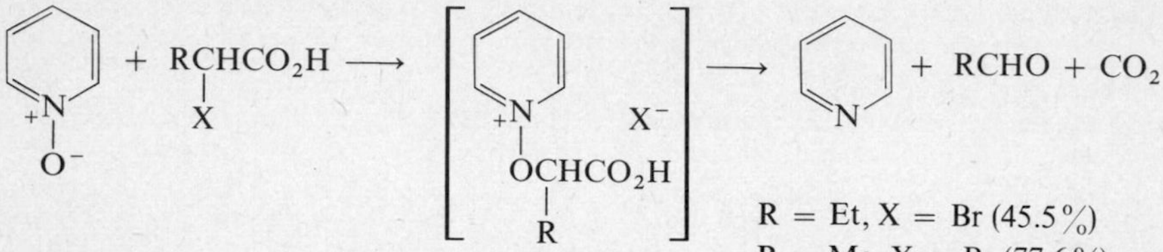

IV-135

R = Et, X = Br (45.5 %)
R = Me, X = Br (77.6 %)
R = Pr, X = Br (67.4 %)
R = H, X = Cl (65 %)

(IV-19)

The proposed reaction scheme involves nucleophilic attack of pyridine-1-
oxide on the α-halo acid to give IV-135; this provides a synthesis of aldehydes
and ketones (394). 4-Hydroxypyridine-1-oxide (IV-136) and 4-hydroxy-3,5-
diiodopyridine-1-oxide react similarly with chloroacetic acid in alkaline
solution to give 1-[4-oxo-1(H)-pyridyl]oxyacetic acid (IV-137) and the
corresponding 3,5-diiodo compound, respectively (395). Treatment of 1-
[4-oxo-1(H)-pyridyl]oxyacetic acid (IV-137) with diazomethane resulted in
decomposition to 4-methoxypyridine (IV-138). Carboxylic acid anhydrides

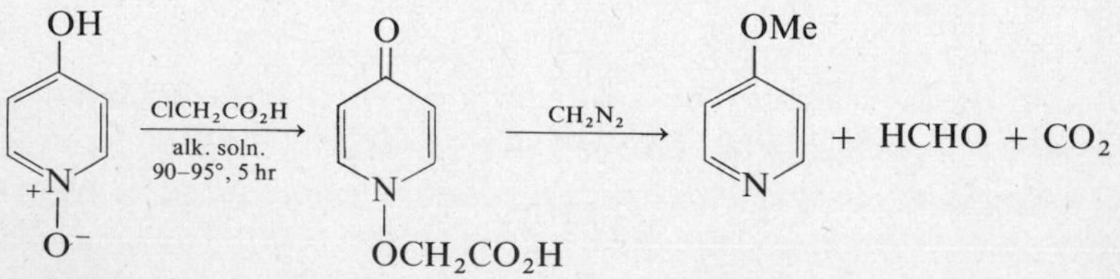

IV-136   IV-137   IV-138

or carboxylic acids of the type $RR'CHCO_2H$, in the presence of acetic
anhydride, undergo oxidative decarboxylation with two equivalents of

TABLE IV-35. Reaction of Pyridine-1-oxide with Carboxylic Acid Anhydrides $(RR^1CHCO)_2O$

R, R^1	Conditions	Product (yield)	Ref.
Me, Me	PhMe	Me_2CO (39%)	399
Me, Me	PhCl, 120°, 6 hr, 1.0 moles/l.	Me_2CO, CO_2, (0.16 moles/l.)	399
Et, H	Xylene	EtCHO (12.4%)	396
CH_2=CH, H	PhCl, 100°, 0.1 hr, 1.1 moles/l.	CH_2=CHCHO, 0.17 moles/l.; CO_2, 0.43 moles/l.	399
Ph, H	C_6H_6	PhCHO (69%)	396
$PhCH_2CO_2H(Ac_2O)$	C_6H_6	PhCHO (68%)	396
Ph, H	C_6H_6, 80°, 16 hr, 0.5 moles/l.	PhCHO, 0.46 moles/l.; CO_2, 0.47 moles/l.	399
Ph, H	PhCl, 120°, 24 hr, 0.25 moles/l.	PhCHO, CO_2, 0.73 moles/l.	399
Ph, H	MeCN, 80°, 39 hr, 0.5 moles/l.	PhCHO, CO_2, 0.56 moles/l.	399
p-MeOC$_6$H$_4$, H	PhCl, 120°, 40 hr, 0.25 moles/l.	p-MeOC$_6$H$_4$CHO, 0.69 moles/l., CO_2, 0.65 moles/l.	399
p-MeOC$_6$H$_4$, H	MeCN, 80°, 23 hr, 0.50 moles/l.	p-MeOC$_6$H$_4$CHO, CO_2, 0.65 moles/l.	399
p-NO$_2$C$_6$H$_4$, H	MeCN, 80°, 23 hr, 0.50 moles/l.	p-NO$_2$C$_6$H$_4$CHO, 0.20 moles/l., CO_2, 0.21 moles/l.	399
Ph, Ph	C_6H_6	Ph_2CO (62%)	396
Ph, Ph	PhCl, 80°, 18 hr, 0.11 moles/l.	Ph_2CO, 0.52 moles/l.; CO_2, 0.77 moles/l.	399
Ph, Ph	PhCl, 80°, 21 hr, 1 mole/l	Ph_2CO, CO_2, 0.93 moles/l.	399

pyridine-1-oxide to give the corresponding aldehyde or ketone (396) (eq. IV-20) (Table IV-35). It was suggested that attack by pyridine-1-oxide

$$2 \, C_5H_5NO + \begin{matrix} R \\ \diagdown \\ CHCO_2H \\ \diagup \\ R' \end{matrix} + Ac_2O \longrightarrow 2\,C_5H_5N + \begin{matrix} R \\ \diagdown \\ C{=}O \\ \diagup \\ R' \end{matrix} + CO_2 + 2\,AcOH$$

(IV-20)

becomes more favorable as the positive charge on the electrophile increases (397). When phenylacetic acid was used, important side-products obtained were phenylglyoxylic acid and diphenylmaleic anhydride, and in the presence of acetic acid, acetylmandelic acid was obtained (398). These results were interpreted in terms of a novel mechanism which involves nucleophilic substitution by pyridine-1-oxide for hydrogen at the α-position of the acid and subsequent decomposition of the intermediate to benzaldehyde and phenylglyoxylic acid (398). Other workers have discussed a mechanism involving ionic fragmentation of N-acyloxypyridinium ions, giving rise to oxidizable carbonium ions (399). Under similar conditions, 2- and 4-picoline-1-oxides yield less carbon dioxide and carbonyl compounds than does pyridine-1-oxide. In view of these results, pyridine-1-oxide and acid anhydrides $(RCO)_2O$, where $R = EtCH{=}CH{-}$ and $MeCH{=}\overset{|}{C}Me$ were heated in a suitable solvent at 140° for 25 to 400 min to give $MeCH{=}CHCHO$ and $H_2C{=}HCCOMe$, respectively. When R was $CH{=}CHPh$ or $CH{=}CHCHMe_2$, no isomeric allylic products were obtained on decarboxylation, thus refuting the possible intermediacy of carbonium ions (400). It was suggested that the first step is a "carboxy inversion" in the intermediate N-acyloxypyridinium ion (400). s-Alkane-(or cycloalkane)carboxylic acid anhydrides undergo oxidative decarboxylation in good yield (401).

The reactions of pyridine-1-oxide with isopropylisonitrile in the presence of chlorine, bromine, or iodine to yield the isocyanate were studied (402). Reaction with bromine was said to give an initial adduct formulated as **IV-139**, which then gave the isocyanate (eq. IV-21). Reaction of pyridine-1-

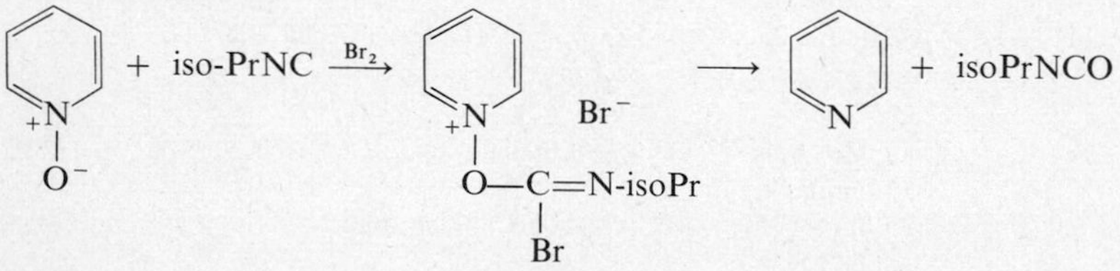

IV-139

oxide with diethyl or dimethyl bromomalonate at 0° for 3 days gave pyridine hydrobromide and tetraethyl 1-hydroxy-2-bromoethane-1,1,2,2-tetracarboxylate **(IV-140)** (or the corresponding ester). When this reaction was carried out in the presence of ethoxide ion in ethanol, the product was tetramethyl 1,2-epoxyethane-1,1,2,2-tetracarboxylate **(IV-141)** (403). Pyridine-1-oxide and styrene oxide react at high temperature to give small

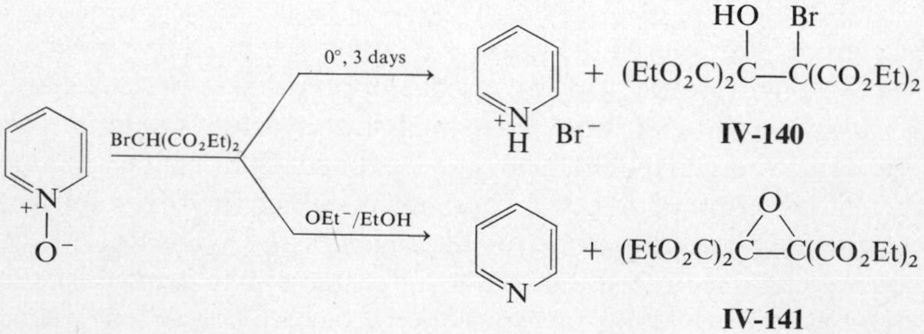

IV-140

IV-141

yields of phenacyl alcohol and benzaldehyde (380, 404), **IV-142** being the proposed intermediate (eq. IV-22). In the presence of strong acid the same

$$\text{(pyridine-}N\text{-oxide)} + \text{PhCH}\overset{\text{O}}{-}\text{CH}_2 \xrightarrow{\Delta} \text{(pyridinium, } \overset{O^-}{N}\text{, OCH}_2\text{CHPh)} \longrightarrow \text{(pyridine)} + \text{PhCCH}_2\text{OH} + \text{PhCHO}$$

$$\text{(IV-22)}$$

IV-142

two reactants gave an alkoxy ammonium intermediate. If an α-methyl group is present, decomposition of the intermediate may proceed *via* an initial abstraction of a methyl hydrogen (ylid formation), followed by intramolecular abstraction of the benzylic hydrogen through a sterically favored transition state **IV-143** (404).

$$\text{A}^- = \text{CF}_3\text{CO}^- \text{ or } \text{ClO}_4^-$$

IV-143

Decomposition of the adduct from 2,6-lutidine-1-oxide, 1,2-epoxycyclohexane, and perchloric acid by aqueous base yields 2,6-lutidine, adipoin, and 1-cyclopentenecarboxaldehyde (405, 406). Similarly, 2,6-lutidine-1-oxide styrene oxide adducts decompose in pyridine or in alkaline solution to 2,6-lutidine, phenacyl alcohol, benzaldehyde, and formaldehyde (406). Base-induced decomposition of the 2,6-lutidine-1-oxide adducts give products in which the amine oxide oxygen is assumed to become the carbonyl oxygen (eq. IV-23 and IV-24) (380, 406).

$$\underset{\underset{O-NR_3}{|}}{R-CH}\overset{O^-}{\underset{|}{-}}CHR' \longrightarrow RCHO + R'CHO + NR_3 \qquad \text{(IV-23)}$$

$$\underset{\underset{O-NR_3}{|}}{R-CH}\overset{O^- \ H}{\underset{|}{-}}C-R \longrightarrow \underset{}{RCH}\overset{OH \ O}{\underset{}{-}}CR + NR_3 \qquad \text{(IV-24)}$$

TABLE IV-36. Deoxygenation of Pyridine-1-oxides with *p*-Nitrobenzenesulfenyl and Sulfinyl Chlorides

Substituent in N-oxide	Conditions	Products	Ref.
None	(i) p-$NO_2C_6H_4SCl$ (ii) 150–160°, 5 hr	Pyridine, $(p$-$NO_2C_6H_4S)_2$, p-$NO_2C_6H_4SOSC_6H_4NO_2$-p, p-$NO_2C_6H_4SO_3H$	305b
2-Me	(i) p-$NO_2C_6H_4SOCl$, C_6H_6, 25° (ii) 180–200°, 2 hr, N_2 atm	Pyridine, $(p$-$NO_2C_6H_4S)_2$, p-$NO_2C_6H_4Cl$, SO_2	305c
	(i) p-$NO_2C_6H_4SCl$ (ii) Δ	2-Picoline, $(p$-$NO_2C_6H_4S)_2$, p-$NO_2C_6H_4SO_2SC_6H_4NO_2$-$p$, p-$NO_2C_6H_4SO_3H$	305b
	(i) p-$NO_2C_6H_4SOCl$ (ii) 140–150°, 2 hr, N_2 atm	2-Picoline, $(p$-$NO_2C_6H_4S)_2$, p-$NO_2C_6H_4SO_3H$	305c
	(i) PhSOCl, 160°, 1 hr (ii) 60°, 1 hr	2-Picoline, $(PhS)_2$, $PhSO_3H$	305c
4-Cl	(i) p-$NO_2C_6H_4SOCl$, $CHCl_3$, R.T.	1-(4-Pyridyl)-4-pyridone,	305d
	(ii) 200°, 10 hr	[structure: Cl—pyridinium N—$O_2SC_6H_4NO_2$-p, $^-SC_6H_4NO_2$-p] $(p$-$NO_2C_6H_4S)_2$, benzenethiol sulfonate, p-$NO_2C_6H_4SO_3H$	375
4-OMe	p-$NO_2C_6H_4SOCl$, $CHCl_3$, Δ, 10 hr	4-methoxypyridine, p-$NO_2C_6H_4SO_3H$, p-nitrophenyl p-nitrobenzene-thiosulfonate	
	p-$NO_2C_6H_4SOCl$, 200°, 5 hr	4-Methoxypyridine, $(p$-$NO_2C_6H_4S)_2$, p-$NO_2C_6H_4SO_3H$	305d
4-NO_2	(i) p-$NO_2C_6H_4SOCl$, $CHCl_3$, R.T. (ii) 80°, —NO_2 (iii) 180–200°, 5 hr	1-(4-pyridyl)-4-chloropyridinium chloride [structure: Cl—pyridinium N—$O_2SC_6H_4NO_2$-p, $^-SC_6H_4NO_2$-p]	305d

The 2,6-lutidine-1-oxide 1,2-epoxycyclohexane adduct may be acetylated to **IV-144** which, with pyridine or aqueous sodium carbonate, yields 2,6-lutidine and 2-acetoxycyclohexanone (**IV-148**). Direct abstraction of the hydrogen atom *alpha* to the amine oxide group in **IV-144** does not occur. Decomposition of **IV-144** in Na_2CO_3/D_2O gave 2,6-lutidine with a high deuterium content in the methyl groups. H−D exchange occurs in adduct **IV-144** *via* the resonance-stabilized ylid **IV-145**. Thus,when the methyl groups in **IV-144** are fully deuterated (**IV-146**), it decomposes in Na_2CO_3/D_2O or in pyridine to 2,6-lutidine containing exactly one proton in its methyl group

(**IV-147**). It was concluded that intramolecular hydrogen transfer occurs *via* a six-membered transition state (**IV-146**) (Scheme IV-3) (405). Similar results are reported for 2,6-lutidine-1-oxide styrene oxide adduct (406).

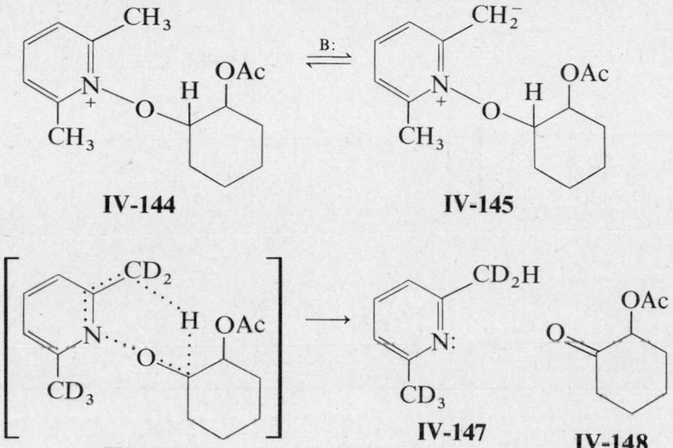

IV-144 **IV-145**

IV-146 **IV-147** **IV-148**

Scheme IV-3

Pyridine-1-oxides and *p*-nitrobenzenesulfenyl chloride or *p*-nitrobenzenesulfinyl chloride react to form the *N*-pyridinium adducts which, on heating, decompose to give the parent pyridine together with disulfide and other products (305b–d) (Table IV-36).

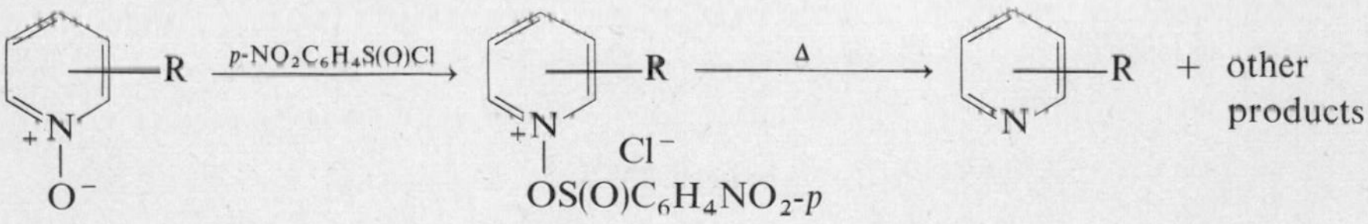

Equivalent amounts of pyridine-1-oxide and *p*-nitrobenzenesulfenyl chloride in benzene gave 1-(*p*-nitrobenzenesulfenyloxy)pyridinium chloride (**IV-98**) which was heated at 150° for 5 hr to yield pyridine, *p*,*p*′-dinitrodiphenyl disulfide, *p*-nitrophenyl *p*-nitrobenzenethiosulfonate, and *p*-nitrobenzenesulfonic acid (Scheme IV-4) (305b). Similar results were obtained with 2-methyl-, 4-chloro-, 4-methoxy-, and 4-nitropyridine-1-oxide (305b, d). Neither a ring substituted rearrangement product nor unreacted pyridine-1-oxide were found among the products. On the basis of the reaction products and their distribution, it has been proposed that the N—O bond of **IV-4** cleaves homolytically (305b).

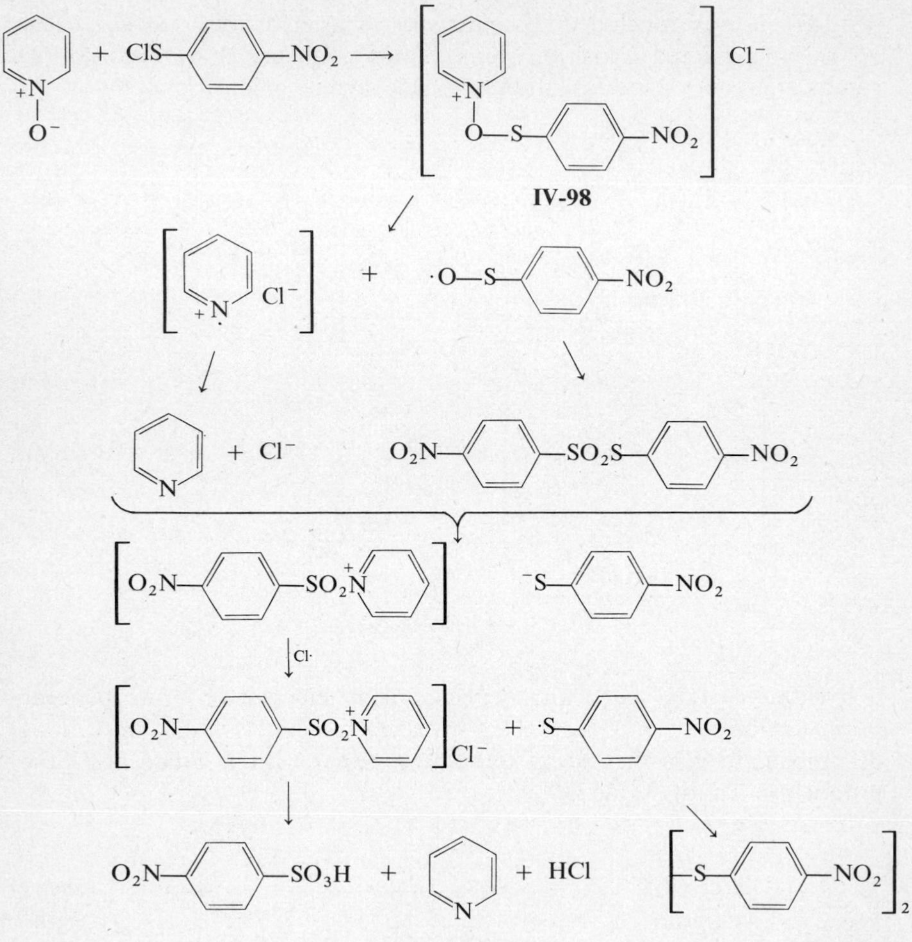

Scheme IV-4

When equivalent amounts of pyridine-1-oxide and *p*-nitrobenzenesulfinyl chloride in benzene were mixed, 1-(*p*-nitrobenzenesulfinyloxy)pyridinium chloride **(IV-99)** was obtained (305c). On heating this salt at 180 to 200° for 2 hr under nitrogen, several products were obtained: pyridine, *p*-nitrochlorobenzene, *p*-nitrobenzenesulfonic acid, and *p,p'*-dinitrodiphenyl-disulfide (Scheme IV-5). As in the previous example, neither a ring-substituted rearrangement product, nor unreacted pyridine-1-oxide were isolated among the other products, and the reaction is initiated by the cleavage of the N—O bond of the salt **(IV-99)**.

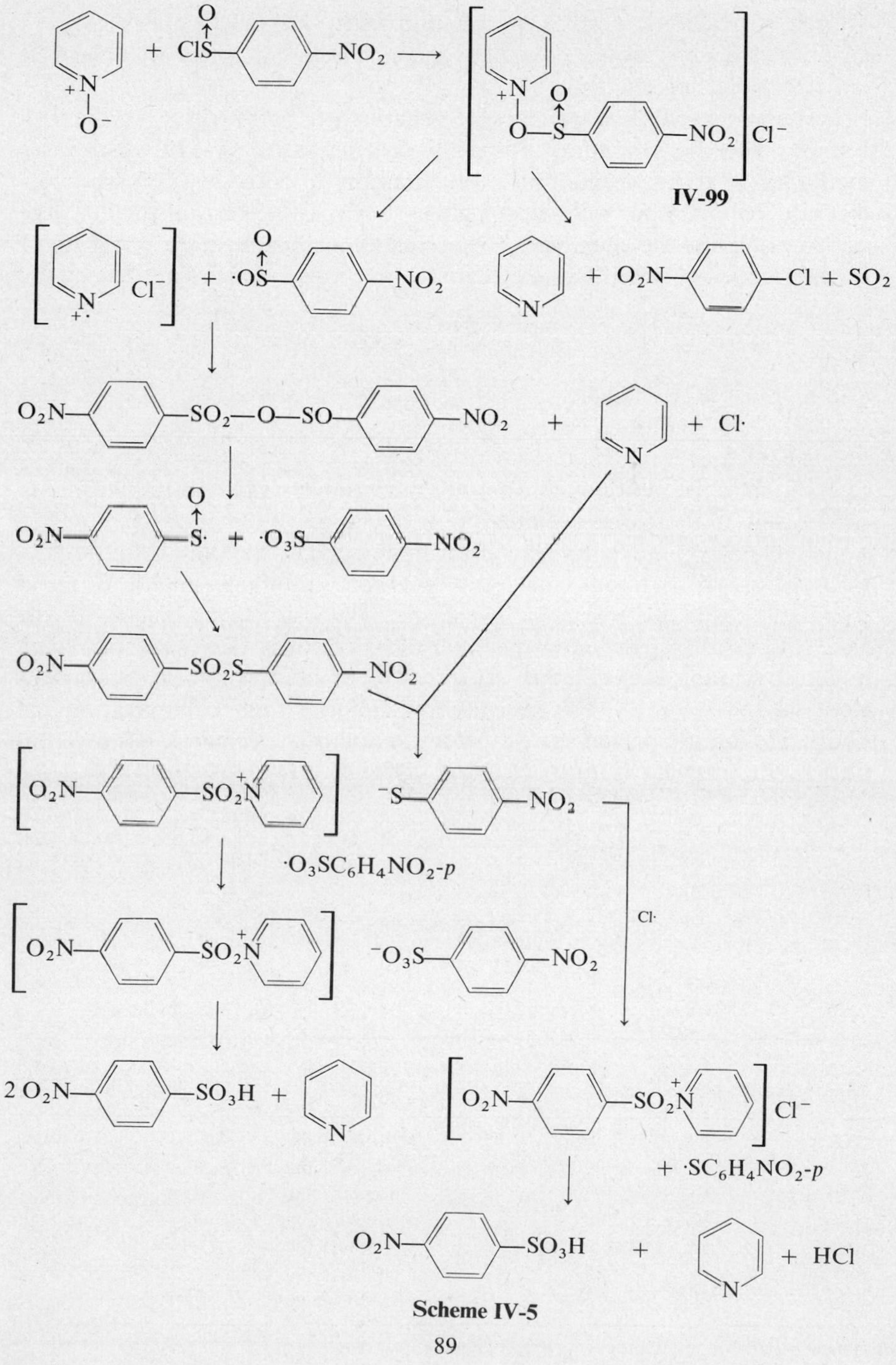

Scheme IV-5

89

Path B reactions of alkoxypyridinium salts with nucleophiles such as Grignard reagents, alkyl mercaptans, cyanide ion, and others are described in detail in Section IV.2.C.

1-Methoxypyridinium perchlorate reacts with borohydride anion irreversibly with the formation of the dihydro derivative **IV-149** which loses methanol to give pyridine (308). All attempts to trap this dihydro-intermediate by reaction with electrophiles such as acetic anhydride were unsuccessful. In the presence of excess sodium borohydride products of further reduction were obtained. Type C reactions are characterized by the

IV-149

interaction of the pyridinium salts with nucleophiles and lead to pyridine-1-oxides and, in most cases, alkylation or acylation of the nucleophile. Thus 1-methoxypyridinium salts react with nucleophiles such as nitrite, iodide, thiosulfate, azide, thiocyanate, and benzene sulphinate anions to yield pyridine-1-oxide and occasionally the methylated nucleophile (308) (Table IV-37). 1-*t*-Butoxypyridinium perchlorate (**IV-150**) is slowly solvolyzed in aqueous solution to pyridine-1-oxide and *t*-butanol *via* an S_N1 decomposition, as shown by NMR spectroscopy (308). If high concentrations of hydroxide ion are present, ring opening by path D competes. No reaction occurs at room temperature in aprotic solvents. If **IV-150** is heated in polar

IV-150

TABLE IV-37. Reaction of 1-Methoxypyridinium Perchlorates with Nucleophiles (308)

Nucleophile	Conditions	Product	N-methyl nucleophile (m.p.)
NO_2^-	DMF, 24°, 24 hr	Pyridine-1-oxide	—
I^-	EtOH-acetone, 25°, 6 hr	Pyridine-1-oxide	—
$S_2O_3^=$	H_2O, 80°, 12 hr	Complex mixture	—
N_3^-	H_2O, 85°, 9 hr	Pyridine-1-oxide	—
NCS^-	EtOH-acetone, 25°, 1.5 hr	Pyridine-1-oxide	—
$PhSO_2^-$	H_2O, 100°, 24 hr	Pyridine-1-oxide	87°
$PhSO_2^-$	DMF, 24°, 120 hr	Pyridine-1-oxide	85–87°

TABLE IV-38. Solvolysis of 1-Aralkoxypyridinium Salts (**IV-151**) (407)

Ar	R	Conditions	Product
Ph	Ph	H_2O, HBr, alc.	Ph_2CHOEt
Ph	H	H_2O, HBr, alc.	$PhCH_2OEt$
Ph	Ph	H_2O, Δ	Ph_2CHOH
Ph	H	H_2O, Δ	$PhCH_2OH$
Ph	Ph	PhOH	$Ph_2CHC_6H_4OH$-*p*
Ph	H	PhOH	$PhCH_2C_6H_4OH$-*p* and *o*-isomer
Ph	Ph	$BrCH(CO_2Me)_2$	$(MeO_2C)_2C(OH)C(Br)(CO_2Me)_2$ (40%), Ph_2CHBr (85%), pyridine HBr (70%)
Ph	H	$BrCH_2(CO_2Me)_2$	$(MeO_2C)_2C(OH)C(Br)(CO_2Me)_2$, $PhCH_2Br$ (85%), pyridine HBr

solvents, the *N*-oxide and *t*-butylene were obtained (308). When 1-aralkoxy-pyridinium salts (**IV-151**) are heated in aqueous ethanolic hydrobromic acid, the C—O bond is cleaved to form pyridine-1-oxide hydrobromide and the corresponding aralkoxy ether (407) (Table IV-38).

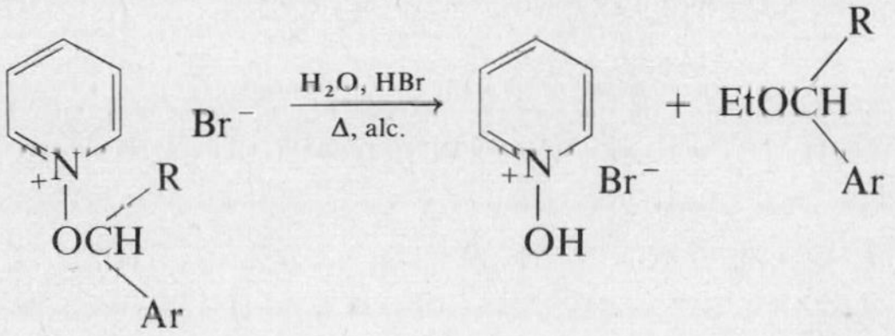

IV-151

The 1,2-epoxycyclohexane and 1,2-epoxybutane 2,6-lutidine-1-oxide adducts are stable in hot water (100°) overnight, but the styrene oxide adduct decomposed to phenyl-1,2-ethanediol and amine oxide (eq. **IV-25**) (406).

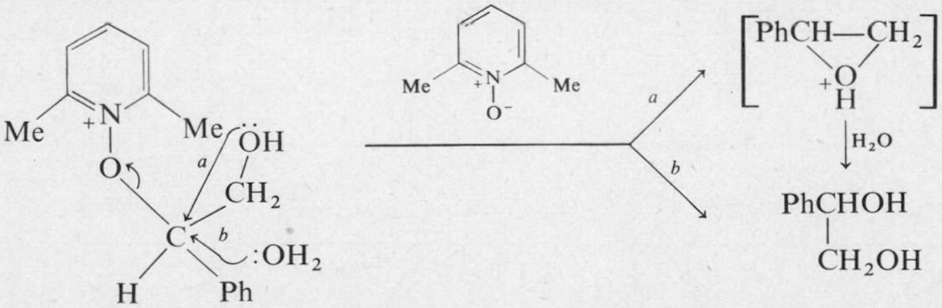

(IV-25)

Thermolysis of 2,6-lutidine-1-oxide styrene oxide perchlorate adduct under vacuum produces 2,6-lutidine-1-oxide perchlorate and phenylacetaldehyde (Scheme IV-6) (406). Thermolysis of 2,6-lutidine-1-oxide 1,2-epoxycyclo-hexane perchlorate adduct under vacuum gave a complex mixture of products

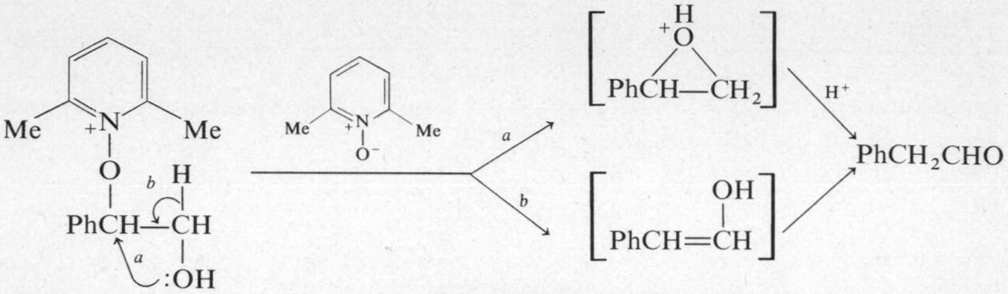

Scheme IV-6

including (2,6-lutidine-1-oxide)$_2$HClO$_4$ and cyclohexanone (406). 1-Methoxypyridinium salts (**IV-152**) are methylating agents (180). For example, N-methylaniline was isolated as its N-p-toluenesulfonyl derivative from the reaction of aniline with the 1-methoxypyridinium salt. A few examples of

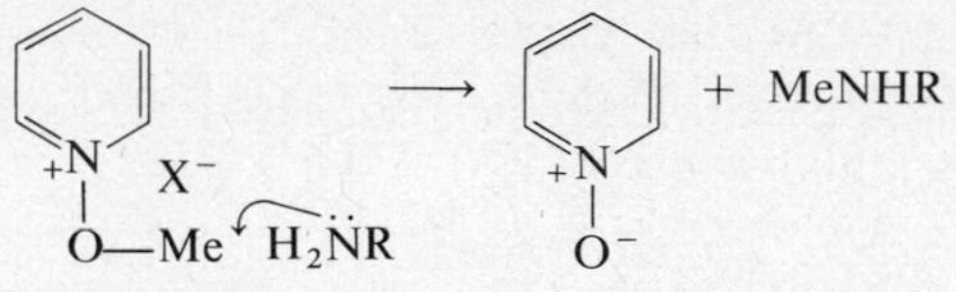

IV-152

the decomposition of cyclic alkoxyammonium salts are shown below (83, 408).

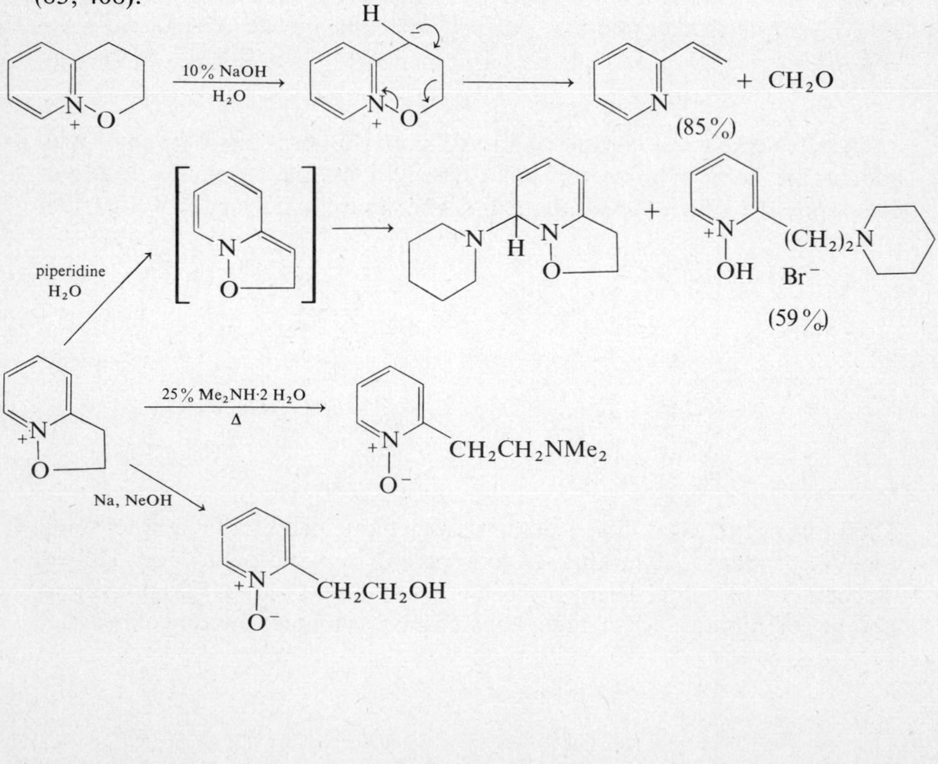

The pyridinium chloride **IV-153** loses carbon dioxide to give 4-picoline-1-oxide and benzyl chloride. When **IV-153** is treated with tetraethyl-ammonium phenylacetate it gives 4-picoline-1-oxide, carbon dioxide, and benzyl phenylacetate (409) (Scheme IV-7).

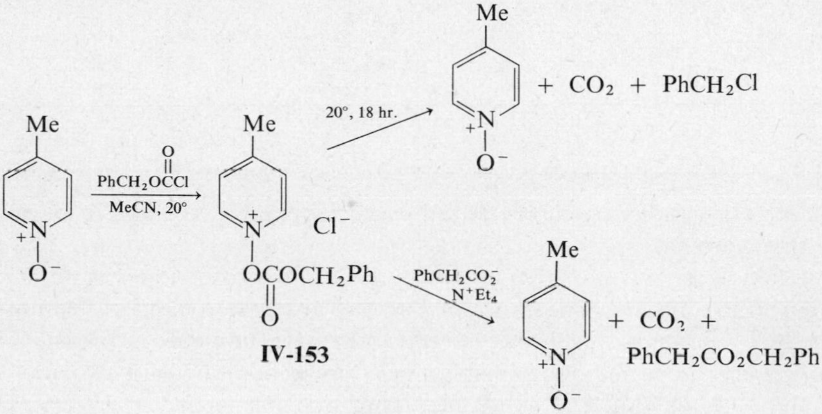

Scheme IV-7

1-Benzoyloxynitro-2-pyridones (**IV-154**) were prepared by the reaction of 2-chloro-3-nitro- and 2-chloro-5-nitropyridine-1-oxide with substituted benzoic acids (13, 410, 411). They react with amines (13, 410) and alcohols (411) to give amides and esters, respectively (Scheme IV-8). Their reactivity towards nucleophilic reagents has been related with the position of the ester

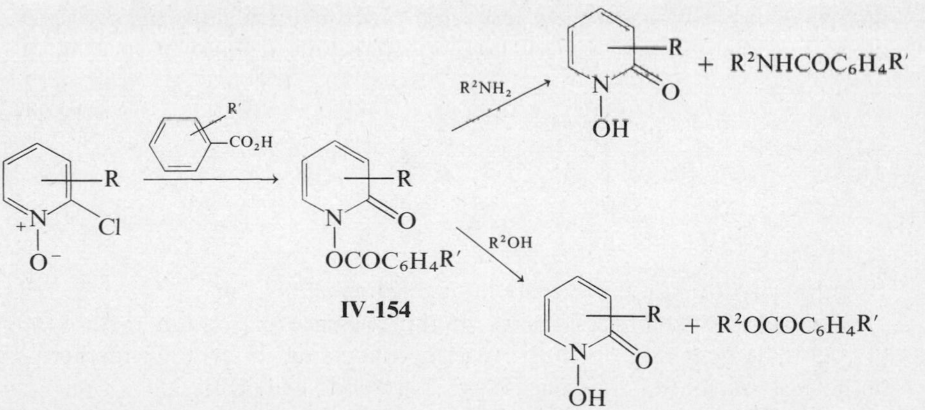

IV-154

carbonyl group in the infrared (411). A number of substituted pyridine-1-oxides and other oxides have been used as catalysts in the reaction of *m*-chloroaniline with benzoyl chloride. The reaction rate constants are recorded in Table IV-39 (412a, b).

TABLE IV-39. Rate Constants for the Reaction of *m*-Chloroaniline with Benzoyl Chloride in the Presence of a Pyridine-1-oxide and Other Oxides (412a)

Pyridine-1-oxide	k_N(mole^{-1} sec^{-1})	Oxide	k_N(mole^{-1} sec^{-1})
H	63 $\pm$ 9	Me$_2$SO	2.0
3-Me	188 $\pm$ 17	Ph$_2$SO	0.29
3-Cl	7.6 $\pm$ 0.8	Bu$_3$PO	9.97 $\pm$ 0.28
4-OMe	1,036	Ph$_3$PO	2.38 $\pm$ 0.14
4-OCH$_2$Ph	1,581 $\pm$ 187	(Me$_2$N)$_3$PO	4.17 $\pm$ 0.23
4-NO$_2$	0.52 $\pm$ 0.02		

Path D reactions are characterized by nucleophilic attack at C-2 followed by ring-opening.

Reaction of 1-methoxypyridinium salts with hydroxide ion was observed to give the transient appearance of a strong absorption band at 340 nm in the U.V. spectrum and hysterisis effects in potentiometric titration. This can be accounted for by reversible nucleophilic addition of hydroxide ion to the pyridine ring and subsequent ring opening (389) (eq. IV-26). The strong absorbance at 343 nm in alkaline solution is attributed to the inter-mediate **IV-156**, the enolate of the *O*-methyl ether of glutaconic aldehyde mono-oxime (389). Acidification of the solution results in loss of the 343 nm absorption and regeneration of **IV-155**. Reaction of **IV-155** with methoxide ion in methanol yields pyridine and gives rise to no U.V. absorption above 275 nm (389). The reaction of 1-methoxypyridinium perchlorate with aqueous hydroxide ion *via* path D to yield the ring-opened compound is independent of alkali strength. U.V. and NMR spectra show the presence of an unstable glutaconic derivative which results from ring opening of

(IV-26)

1-methoxypyridinium perchlorate by piperidine (308).

With substituted nitromethanes in the presence of sodium methoxide **IV-155** undergoes ring opening of the pyridine ring to form a 1-methoxy-imino-6-*aci*-nitrohexa-2,4-diene (**IV-157**) in good yield (413).

Piperidine and dimethylamine react with the isoxazolinopyridinium cation (**IV-158**) to yield 2-ω-aminobutadienylisoxazoles (**IV-159**) (414). The

IV-158		**IV-159**

nature of the reaction between 1-methoxypyridinium salts and nucleophiles depends on several factors (308): (*a*) structure of the nucleophile — ring-opening by path D requires abstraction of an acidic hydrogen by a nucleophilic reagent; (*b*) "hardness" or "softness" of the nucleophile—the ability of the nucleophile to remove the C-2 proton from the dihydro intermediate in path B, a process which will also be affected by the electron-withdrawing or donating properties of the substituted nucleophilic group; (*c*) differentiation between attack at the C-2 or C-4 position is determined by kinetic versus thermodynamic control.

For nuclear attack by paths B or D, the initial step leading to the dihydropyridine intermediate is reversible whenever the entering nucleophile is a good leaving group. The strength of the nucleophile determines the ability of the nucleophile to initiate ring attack, which involves a considerable loss of resonance energy. Thus weak nucleophiles are either unable to initiate the first step in B or the equilibrium lies too far toward reactants.

IV. Substitution at Carbon

1. Electrophilic Substitution

A. Acid-Catalyzed H–D Exchange

2,4,6-Trimethylpyridine-1-oxide and 2,6-lutidine-1-oxide undergo hydrogen-deuterium exchange in acid solution as their conjugate acids. Exchange takes place at C-3 and C-5 (415). In the case of 2,6-lutidine-1-oxide, the free base can undergo exchange at C-4 at sufficiently low acidities. Comparison of the acid-catalyzed H—D exchange rates of 2,4,6-trimethylpyridine and its *N*-oxide indicates that ⟩N—OH is slightly more deactivating than ⟩NH, which is probably due to the electron attracting effect of the hydroxyl group (415, 416). It is reported that after 2 hr in 96% D_2SO_4 at 220° pyridine-1-oxide does not undergo hydrogen-deuterium exchange; on the other hand, after 3 hr at 180° in neutral solution (in D_2O) exchange occurred at C-2 and C-6 (417).

In contrast to 2,6-lutidine- and 2,4,6-collidine-1-oxide, 3,5-lutidine-1-oxide undergoes acid-catalyzed H—D exchange at C-2, C-4, and C-6 as the free base, while 3-hydroxypyridine-1-oxide also reacts at C-2 as the free base (418).

B. Nitration

As is well known (5–7), pyridine-1-oxides undergo nitration in fuming sulfuric acid with potassium nitrate or nitric acid at 100 to 130°, or with fuming nitric acid in sulfuric acid at 90° to give 4-nitropyridine-1-oxides in good yields (Table IV-40). Charge distribution and localization energy calculations predict that electrophilic substitution of pyridine-1-oxide should occur at C-2 (278), but, in fact, attack at C-4 is observed. In the case of pyridine-1-oxide, a small amount of 2-nitropyridine (**IV-160**) was obtained in addition to 4-nitropyridine-1-oxide (**IV-161**) (419). The 4-nitro compound

IV-161   **IV-160**

also results from the direct *in situ* nitration with concentrated nitric and sulfuric acids of pyridine-1-oxide formed by heating pyridine with 30% H_2O_2 in glacial acetic acid (Table IV-41) (eq. IV-27). Alkyl groups at C-2

facilitate nitration at C-4. On the other hand, a 3-alkyl substituent tends to exert a slight steric hindrance. There is no significant steric hindrance by halogen groups on the nitration of halopyridine-1-oxide (7).

TABLE IV-40. Nitration of Pyridine-1-oxides

Substituent in *N*-oxide	Conditions	Products (yield)	Ref.
None	Fuming HNO_3, 80°, 7 hr	4-Nitropyridine-1-oxide, m.p. 157–159°, m.p. 162° (63%)	427 428
None	$BzONO_2$	3-Nitropyridine-1-oxide, m.p. 165–169° (minor), pyridine-1-oxide (major)	118
None	(i) p-$O_2NC_6H_5COCl$, $AgNO_3$, 0°; 1 hr, R.T., 1 hr (ii) sat. HCl in EtOH	Pyridine-1-oxide, HCl, picrate, m.p. 176–178°; 3,5-dinitropyridine-1-oxide, m.p. 183–185°; 3-nitropyridine-1-oxide, m.p. 169–170°	423
None, HCl	HNO_3, H_2SO_4	4-Nitropyridine-1-oxide, m.p. 158–159°	338
None, 2-^{14}C	HNO_3, H_2SO_4, 90°, 4 hr	4-Nitropyridine-2-^{14}C-1-oxide, m.p. 155–156°	28
2,6-d_2	KNO_3, H_2SO_4	4-Nitropyridine-1-oxide-2,6-d_2, m.p. 163°	115
2-Me	Fuming HNO_3, 80°, 7 hr	4-Nitro-2-picoline-1-oxide, m.p. 154–153°; 4-nitro-2-picoline-1-oxide, m.p. 156° (90%)	427 341
2-Et	(i) HNO_3, H_2SO_4, 95°, 3 hr (ii) 95–98°, 2 hr	2-ethyl-4-nitropyridine-1-oxide, m.p. 94–95°	429
	HNO_3, H_2SO_4, 1.5–2 hr	2-Ethyl-4-nitropyridine-1-oxide, m.p. 94–96° (49%)	430
2,6-Me_2	HNO_3 ($d1.41$) H_2SO_4 ($d1.84$)	4-Nitro-2,6-lutidine-1-oxide, m.p. 163° (85%)	120
2,3-Me_2	HNO_3, H_2SO_4, Δ, 3.5 hr	4-Nitro-2,3-lutidine-1-oxide, m.p. 91.5–93°	30
2,5-Me_2	Fuming HNO_3, 80°, 7 hr	4-Nitro-2,5-lutidine-1-oxide, m.p. 151–152°	427
	Fuming HNO_3, conc. H_2SO_4, 120°, 1 hr	4-Nitro-2,5-lutidine-1-oxide, m.p. 151–152°	30
	HNO_3 ($d1.5$) conc. H_2SO_4 0°, 3 hr	4-Nitro-2,5-lutidine-1-oxide, m.p. 148–149°	33
2-Me-5-Et	(i) HNO_3 ($d1.5$) H_2SO_4, 0° (ii) 75°, 1 hr	5-Ethyl-4-nitro-2-picoline-1-oxide, m.p. 84°	110
	HNO_3, H_2SO_4, 130°, 2.5 hr	5-Ethyl-4-nitro-2-picoline-1-oxide, m.p. 78–79° (61.9%)	360
	Fuming HNO_3, 80°, 7 hr	5-Ethyl-4-nitro-2-picoline-1-oxide, m.p. 78–79°	427
2,3,5,6-Me_4	HNO_3, H_2SO_4, 90–100°, 3.5 hr	2,3,5,6-Tetramethyl-4-nitropyridine-1-oxide, m.p. 115–116°; picrate, m.p. 160–161°	337

Table IV-40 (*Continued*)

Substituent in *N*-oxide	Conditions	Products (yield)	Ref.
3-Me	Fuming HNO_3, 80°, 7 hr	4-Nitro-3-picoline-1-oxide, m.p. 137°	427
	Fuming HNO_3, H_2SO_4	4-Nitro-3-picoline-1-oxide, m.p. 130–135°	341
	Fuming HNO_3, H_2SO_4, 90°, 2 hr	4-Nitro-3-picoline-1-oxide, m.p. 136–137°	329
	HNO_3, H_2SO_4, 80–85°	4-Nitro-3-picoline-1-oxide, m.p. 136°	345
3-Et	Fuming HNO_3, H_2SO_4, Δ, 1 hr	3-Ethyl-4-nitropyridine-1-oxide, m.p. 63–64°	79
	HNO_3, H_2SO_4, 90–100°, 3.5 hr	3-Ethyl-4-nitropyridine-1-oxide	337
3-isoPr	HNO_3, H_2SO_4 90–100°, 3.5 hr	3-Isopropyl-4-nitropyridine-1-oxide. m.p. 138–139°	337
3,5-Me$_2$	HNO_3, H_2SO_4 90–100°, 3.5 hr	4-Nitro-3,5-lutidine-1-oxide, m.p. 174–175°; picrate, m.p. 137.5–138.5°	337
2-Cl	Fuming HNO_3, conc. H_2SO_4, 90°, 1 hr	2-Chloro-4-nitropyridine-1-oxide, m.p. 153–153.5°	59
	Fuming HNO_3, H_2SO_4, 128–130°, 3.5 hr	2-Chloro-4-nitropyridine-1-oxide, m.p. 140–143°	431
2-Br	Fuming HNO_3, H_2SO_4	2-Bromo-4-nitropyridine-1-oxide, m.p. 140–141.5°	395
	HNO_3, H_2SO_4, 128–130°, 3.5 hr	2-Bromo-4-nitropyridine-1-oxide, m.p. 143–144°	437
2,6-Br$_2$	Fuming HNO_3, H_2SO_4	2,6-Dibromo-4-nitropyridine-1-oxide (65%)	395
	Fuming HNO_3, H_2SO_4, 90°, 2 hr	2,6-Dibromo-4-nitropyridine-1-oxide, m.p. 219–221°	11
3-F	HNO_3 (*d*1.5), H_2SO_4, 1 hr	3-Fluoro-4-nitropyridine-1-oxide, m.p. 128°	352
3-F-2-Me	Fuming HNO_3, H_2SO_4, 1 hr	3-Fluoro-4-nitro-2-picoline-1-oxide, m.p. 183°	343
3-F-5-Me	HNO_3, conc. H_2SO_4	5-Fluoro-4-nitro-3-picoline-1-oxide, m.p. 119°	56
3-F-2,6-Me$_2$	HNO_3, H_2SO_4	3-Fluoro-4-nitro-2,6-lutidine-1-oxide, m.p. 136°	336
3-Cl	HNO_3, H_2SO_4	3-Chloro-4-nitropyridine-1-oxide, m.p. 115° (84.5%)	428
3-Br	HNO_3, H_2SO_4 90–100°, 3.5 hr	3-Bromo-4-nitropyridine-1-oxide, m.p. 156–157°	337, 432
4-Cl-2-OH	HNO_3, AcOH	4-Chloro-1-hydroxy-5-nitro-2-pyridone, m.p. 197–198°	62
2-OH-4-OMe	HNO_3, AcOH, 10 min	1-Hydroxy-4-methoxy-5-nitro-2-pyridone, m.p. 216–217°	62, 422

Table IV-40 (*Continued*)

Substituent in *N*-oxide	Conditions	Products (yield)	Ref.
2-OH-4-CO$_2$Et	HNO$_3$, AcOH	Ethyl 1-hydroxy-5-nitro-2-pyridone-4-carboxylate, m.p. 185°	62, 422
2-OH-4-CN	HNO$_3$, AcOH, 10 min	4-Cyano-1-hydroxy-5-nitro-2-pyridone, m.p. 209–210°	62, 422
3-OMe-5-Et-2-Me	NaNO$_3$, conc. H$_2$SO$_4$	5-Ethyl-3-methoxy-4-nitro-2-picoline-1-oxide, m.p. 183–184°	72
2-CH$_2$CH$_2$OAc	HNO$_3$ (*d*1.5), AcOH, 2 hr	2-(2-Hydroxyethyl)-4-nitro-pyridine-1-oxide	433
3-CH$_2$CH$_2$CO$_2$H	Fuming HNO$_3$, H$_2$SO$_4$, Δ, 1 hr	β-(4-Nitro-3-pyridyl)propionic acid-1-oxide, m.p. 169–171°	79
2-CH$_2$Ph	(i) HNO$_3$, H$_2$SO$_4$, 0° (ii) 100°, 2 hr	2-(4-Nitrobenzyl)pyridine-1-oxide, m.p. 166.5–167° (55%); picrate, m.p. 134–134.5°	306
4-CH$_2$Ph	(i) HNO$_3$, H$_2$SO$_4$, 0° (ii) 100°, 1.5 hr	4-(4-Nitrobenzyl)pyridine-1-oxide, m.p. 165.5–166°; picrate, m.p. 156–156.5°	306
2-CH=CHPh	(i) HNO$_3$, H$_2$SO$_4$, 0° (ii) 100°, 1.5 hr	*x*, *y*-Dinitro-2-styrylpyridine-1-oxide, m.p. 193–195°	306
2-CO$_2$H	HNO$_3$ (*d*1.5) conc. H$_2$SO$_4$, 135°, 4 hr	4-Nitropicolinic acid-1-oxide	99
	Fuming HNO$_3$, conc. H$_2$SO$_4$, 120°, 1 hr	4-Nitropicolinic acid-1-oxide, m.p. 148° (45.5%)	95
2-CO$_2$H-5-Me	(i) HNO$_3$, conc. H$_2$SO$_4$, 0° (ii) 90°, 3 hr	5-Methyl-4-nitropicolinic acid-1-oxide, m.p. 145°	34
3-CO$_2$Et	(i) *p*-O$_2$NC$_6$H$_4$COCl, AgNO$_3$, 1 hr (ii) 40–50°, 2.5 hr (iii) 55°, 2 hr	Ethyl 5-nitronicotinate-1-oxide, m.p. 154°	359
3-CN	(i) *p*-O$_2$NC$_6$H$_4$COCl, AgNO$_3$, 1 hr (ii) 40–50°, 2.5 hr (iii) 55°, 2 hr	3-Cyano-5-nitropyridine-1-oxide, m.p. 205° (decomp.) (36.5%)	359
2-NMe$_2$	(i) Ac$_2$O, AcOH, 0–5° (ii) HNO$_3$ (*d*1.52) <2° (iii) R.T. 3 hr	2-Dimethylamine-5-nitropyridine-1-oxide, m.p. 149–51°; picrate, m.p. 141–145°	149
2-Ph	(i) HNO$_3$, H$_2$SO$_4$, 0° (ii) 100°, 12 hr	2-(3-Nitrophenyl)pyridine-1-oxide, m.p. 73° (54%); 2-(2-nitrophenyl)pyridine-1-oxide, m.p. 59–60°	306
3-Ph	(i) HNO$_3$, H$_2$SO$_4$, 0° (ii) 100°, 12 hr	3-(4-Nitrophenyl)pyridine-1-oxide, m.p. 148° (38%)	306
4-Ph	(i) HNO$_3$, H$_2$SO$_4$, 0° (ii) 100°, 12 hr	4-(3-Nitrophenyl)pyridine-1-oxide, nitrate hemihydrate, m.p. 215°	306

TABLE IV-41. Oxidation and Nitration without Isolation of the Intermediate N-oxide

Substituent in N-oxide	Conditions	Products (yield)	Ref.
None	(i) 30% H_2O_2, Ac_2O, R.T. 5 hr (ii) 60–65°, 3.0 hr (iii) HNO_3, H_2SO_4	4-Nitropyridine-1-oxide, m.p. 162° (63%)	428
None	(i) 30% H_2O_2, AcOH (ii) 63% HNO_3, conc. H_2SO_4, 90–95°, 20 hr	4-Nitropyridine-1-oxide, m.p. 158–159° (76–79%)	434
2-^{14}C	(i) 40% AcO_3H, <80°, 6 hr (ii) HNO_3, H_2SO_4, 1 hr	4-Nitropyridine-2-^{14}C-1-oxide, m.p. 155–156°	28
2-Me	(i) 30% H_2O_2, AcOH (ii) 63% HNO_3, conc. H_2SO_4, 90–95°, 20 hr	4-Nitro-2-picoline-1-oxide, m.p. 154° (78%)	434
2,6-Me_2	(i) 30% H_2O_2, AcOH (ii) 63% HNO_3, conc. H_2SO_4, 90–95°, 20 hr	4-Nitro-2,6-lutidine-1-oxide, m.p. 163° (80%)	434
2-Me-5-Et	(i) 30% H_2O_2, AcOH (ii) 63% HNO_3, conc. H_2SO_4, 90–95°, 20 hr	5-Ethyl-4-nitro-2-picoline-1-oxide, m.p. 87–88° (90%)	434
2-Et	(i) Perhydrol, glac. AcOH, 75°, 20 hr (ii) HNO_3, H_2SO_4, 95°, 3 hr	2-Ethyl-4-nitropyridine-1-oxide, m.p. 94–95° m.p. 94–96° (49%)	429 430
3-Me	(i) 30% H_2O_2, AcOH, 50°, 48 hr (ii) Fuming HNO_3, conc. H_2SO_4, 90°, 2 hr	4-Nitro-3-picoline-1-oxide, m.p. 135–137° (87%)	329
3-Me	(i) 30% H_2O_2, AcOH (ii) 63% HNO_3, conc. H_2SO_4, 90–95°, 20 hr	4-Nitro-3-picoline-1-oxide, m.p. 135° (76.5%)	434
3-F	(i) H_2O_2, Ac_2O, 60°, 24 hr (ii) HNO_3, (d1.5), H_2SO_4, 1 hr	3-Fluoro-4-nitropyridine-1-oxide, m.p. 128° (58%)	352
3-F-2-Me	(i) 30% H_2O_2, Ac_2O, 65°, 24 hr (ii) Fuming HNO_3, H_2SO_4, 1 hr	3-Fluoro-4-nitro-2-picoline-1-oxide, m.p. 183°	343
3-F-2,6-Me_2	(i) H_2O_2, Ac_2O, 60°, 30 hr (ii) HNO_3, H_2SO_4, waterbath, 1.5 hr	3-Fluoro-4-nitro-2,6-lutidine-1-oxide	336
3-F-5-Me	(i) 30% H_2O_2, Ac_2O, 60°, 24 hr (ii) HNO_3, conc. H_2SO_4	5-Fluoro-4-nitro-3-picoline-1-oxide, m.p. 119°	56
3-Cl	(i) 30% H_2O_2, Ac_2O, R.T. 5 hr (ii) 60–65°, 30 hr (iii) HNO_3, H_2SO_4	3-Chloro-4-nitropyridine-1-oxide, m.p. 115° (84.5%)	428

Table IV-41 (*Continued*)

Substituent in N-oxide	Conditions	Products (yield)	Ref.
2-CH$_2$CH$_2$OH	(i) Ac$_2$O, 75°, 2 hr (ii) 50% H$_2$O$_2$, 75–85°, 6–10 hr (iii) conc. HNO$_3$, Δ, 2 hr	2-(1-Oxido-4-nitro-2-pyridyl)ethanol	433
2-CH$_2$CH$_2$OH-5-Et	(i) Ac$_2$O, 75°, 2 hr (ii) 50% H$_2$O$_2$, 75–85°, 6–10 hr (iii) Conc. HNO$_3$, Δ, 2 hr	2-(1-Oxido-4-nitro-4-ethyl-2-pyridyl)ethanol	433
2-CH$_2$CH$_2$OH-3,6-Me$_2$	(i) Ac$_2$O, 75°, 2 hr (ii) 50% H$_2$O$_2$, 75–85°, 6–10 hr (iii) conc. HNO$_3$, Δ, 2 hr	2-(1-Oxido-3,6-dimethyl-4-nitro-2-pyridyl)ethanol	433
2-CH$_2$CH$_2$CH$_2$OH	(i) Ac$_2$O, 75°, 2 hr (ii) 50% H$_2$O$_2$, 75–85°, 6–10 hr (iii) Conc. HNO$_3$, Δ, 2 hr	3-(1-Oxido-4-nitro-2-pyridyl)propanol	433
3-CH$_2$OH	(i) Ac$_2$O, 75°, 2 hr (ii) 50% H$_2$O$_2$, 75–85°, 6–10 hr (iii) Conc. HNO$_3$, Δ, 2 hr	3-Hydroxymethyl-4-nitropyridine-1-oxide	433

Picolinic acid-1-oxide (**IV-25**) is nitrated with fuming nitric acid in concentrated sulfuric acid to yield 4-nitropicolinic acid-1-oxide (**IV-162**) (95, 99).

The orienting effect of the N-oxide group dominates that of an alkoxy group. On the other hand, a hydroxyl group takes over the orientation from the N-oxide function, and nitration occurs *ortho* and *para* to the hydroxyl group (420). Electrophilic substitution in 3-hydroxypyridine-1-oxide (**IV-163**) first occurs at C-2 and then at C-4 or C-6 (421). Nitration of 4-substituted 1-hydroxy-2-pyridone (**IV-164**) with nitric acid in acetic acid gives the 4-substituted-1-hydroxy-5-nitro-2-pyridone (**IV-165**) (62, 422). Similarly, nitra-

IV-164   IV-165

tion of 2-dimethylaminopyridine-1-oxide (**IV-166**) yields the 5-nitro-derivative (**IV-167**) (149). The phenyl groups in phenylpyridine-1-oxides and 4-benzylpyridine-1-oxide undergo nitration, rather than the pyridine ring.

IV-166   IV-167

When pyridine-1-oxide was nitrated with $BzONO_2$, a very low yield of 3-nitropyridine-1-oxide was obtained (118). Nitration of 3-cyano- and 3-ethoxycarbonylpyridine-1-oxide with p-nitrobenzoyl chloride and silver nitrate yielded the 5-nitro-derivative (359). With pyridine-1-oxide itself under these conditions, 3-nitro- (**IV-168**) and 3,5-dinitropyridine-1-oxide (**IV-169**), as well as unreacted pyridine-1-oxide (as the hydrochloride salt) were obtained (423). The mechanism of the reaction is discussed in Chapter I.

IV-168   IV-169

Whereas nitration of isoquinoline-2-oxide at 25° occurred *via* the conjugate acid, a comparison with the nitration of pyridine-1-oxide at 125° and 1-methoxypyridinium salts indicated that nitration of pyridine-1-oxide proceeded *via* the free base (424). Nitration in sulfuric acid at 80° (and of 2,6-lutidine-1-oxide) to give the 4-nitro-compound does not involve the protonated N-oxides but may possibly proceed by nitration of the small equilibrium concentration of free bases (425). Other mechanisms were discussed. Kinetic evidence also showed that the nitration of pyridine-1-oxide and its derivatives at C-4 occurs on the free base (426). On the other hand, 2,6-dimethyl, 4-methoxy-2,6-dimethoxy-, and 2,4,6-trimethoxypyridine-1-oxides undergo nitration at C-3; thus nitration of the conjugate acid is occurring (426).

C. Halogenation

Bromination of pyridine-1-oxides is as difficult to effect as is that of pyridine itself. When pyridine-1-oxide was brominated with excess bromine in 90% sulfuric acid and silver sulfate at 200°, a 10% yield of 2- and 4-bromopyridine-1-oxide, in the ratio of 1:2, was obtained (415). Bromination in 65% fuming sulfuric acid, however, gave 3-bromopyridine (ca. 60%) and minor amounts of the 2- and 4-bromo- compounds, together with 2,5-dibromo- (ca. 35%) and 3,4- and 2,3-dibromo compounds (ca. 5%) (63). To explain the attack at C-3 it was suggested that in fuming sulfuric acid, a complex (IV-170) is formed between pyridine-1-oxide and sulfur trioxide (6, 7). The 2-, 4-, and 6- positions are thus deactivated toward electrophilic attack, and bromination occurs at C-3. It was found that pyridine-1-oxide

IV-170

in acetic anhydride and sodium acetate reacted with bromine to give 3,5-dibromopyridine-1-oxide (IV-171) in 35% yield (435, 436). Bromination of 3-bromopyridine-1-oxide with bromine in 90% sulfuric acid containing

IV-171

silver sulfate at 160° for 20 hr gave a 5 to 10% yield of products which consisted of 2,5- (61%), 3,4- (16%), and 2,5-dibromo (16%) derivatives. It was suggested that in fuming sulfuric acid, the 3-bromine atom takes over the orientation from the N-oxide and directs ortho and para to itself (63). When 2-bromopyridine-1-oxide was brominated in 90% sulfuric acid, the

2,4-:2,6-isomer ratio was 4.5:1; in fuming sulfuric acid, however, the 2,3-
and 2,5-dibromopyridine-1-oxides were obtained. 3,4-Dibromopyridine-
1-oxide (84%) and 3,4,5-tribromopyridine-1-oxide (10%) were obtained as
a result of bromination of 4-bromopyridine-1-oxide in fuming sulfuric acid
(63). Iodination of hydroxypyridine-1-oxide with iodine or iodine-potassium
iodide results in attack of iodine *ortho* and *para* to the pyridone function
(335, 395) (Scheme IV-9).

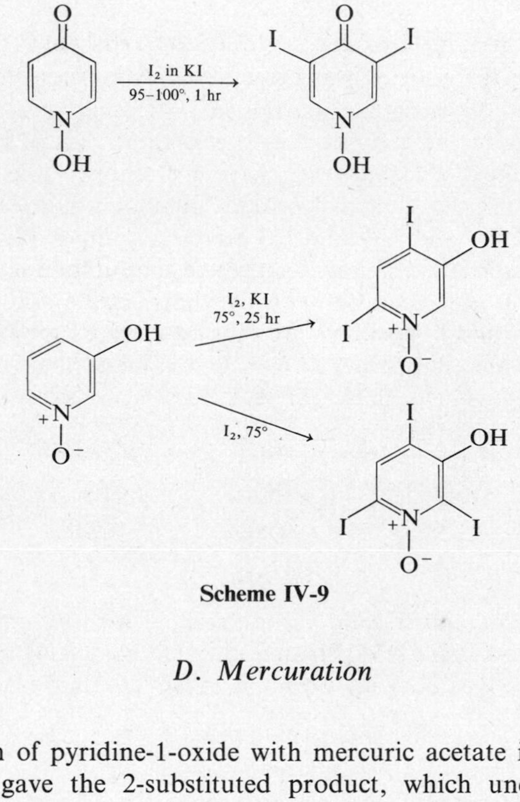

Scheme IV-9

D. Mercuration

Mercuration of pyridine-1-oxide with mercuric acetate in glacial acetic
acid at 130° gave the 2-substituted product, which undergoes further
substitution as shown at top of page 105 (eq. IV-28) (437, 438).

E. Sulfonation

Pyridine-1-oxides undergo sulfonation under vigorous conditions. For
example, pyridine-3-sulfonic acid-1-oxide (**IV-172**) was the major product
obtained on heating pyridine-1-oxide with catalytic amounts of mercuric
sulfate in 20% sulfuric acid at 220 to 240° (36, 439). Substitution at C-3 is

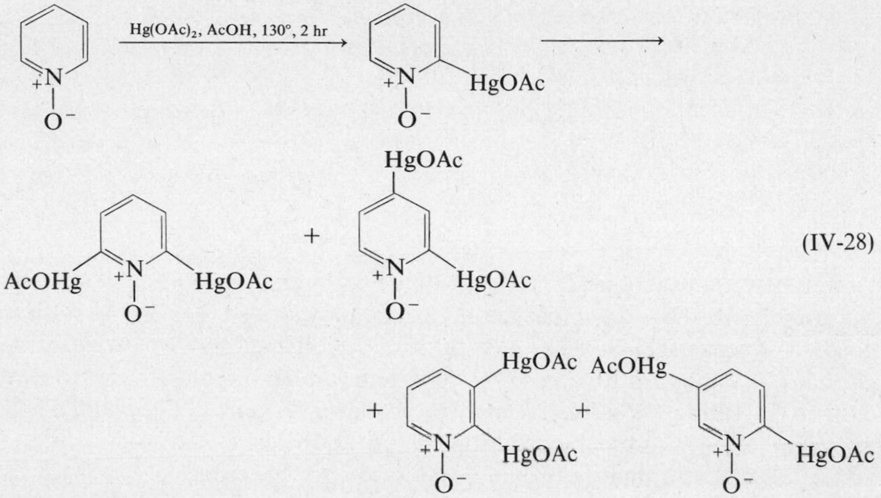

(IV-28)

probably due to the formation of the pyridine-1-oxide-sulfur trioxide complex as discussed previously. Pyridine-2-sulfonic acid-1-oxide (0.5–1 %) and the 4-sulfonic acid (2–2.5 %) were isolated as minor products. Under similar conditions, 2,6-lutidine-1-oxide undergoes substitution at C-3 (36).

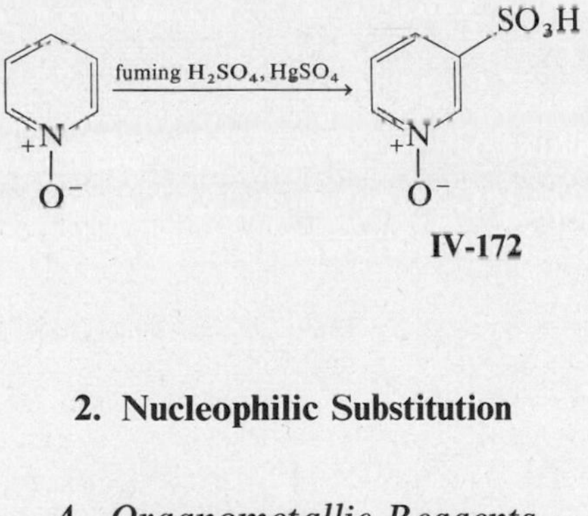

IV-172

2. Nucleophilic Substitution

A. Organometallic Reagents

Grignard reagents react with the —N(O)=CH— group converting it into —N=CR—. With pyridine-1-oxide, low yields of the products were obtained (6, 7). However, both N-methylanabasine-N'-oxide (IV-173) and the N,N'-dioxide reacted with methylmagnesium iodide to give 6-methyl-3-(1-methyl-2-piperidyl)pyridine (IV-174) (440, 441). The observed orientation is probably due to steric hindrance by the 3-substituent (complexed with organometallic reagent) to the approach of the Grignard compound.

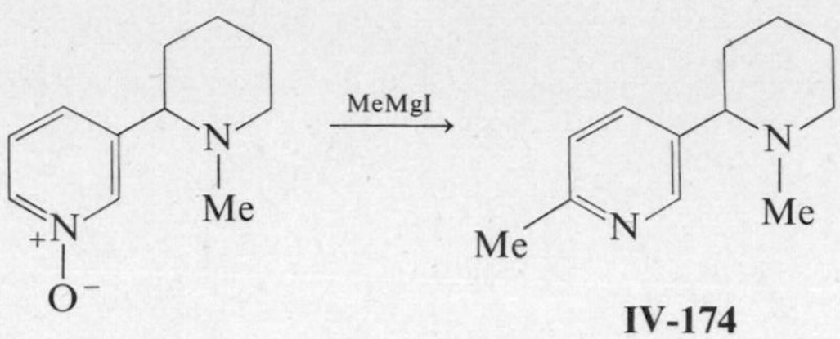

IV-173

IV-174

Recently, the reaction of pyridine-1-oxide with phenylmagnesium bromide was reinvestigated, except that tetrahydrofuran was used (442) as the solvent instead of benzene (443a). The main product of this reaction was 1-hydroxy-2-phenyl-1,2-dihydropyridine (**IV-175**) which, on heating, lost water to give 2-phenylpyridine (**IV-176**). 2-Phenylpyridine (5%) and 2,2′-diphenyl-4,4′-bipyridyl (**IV-177**) were by-products of the addition. 2- And 4-picoline-1-oxide, as well as quinoline-1-oxide, also undergo this reaction. Later it was

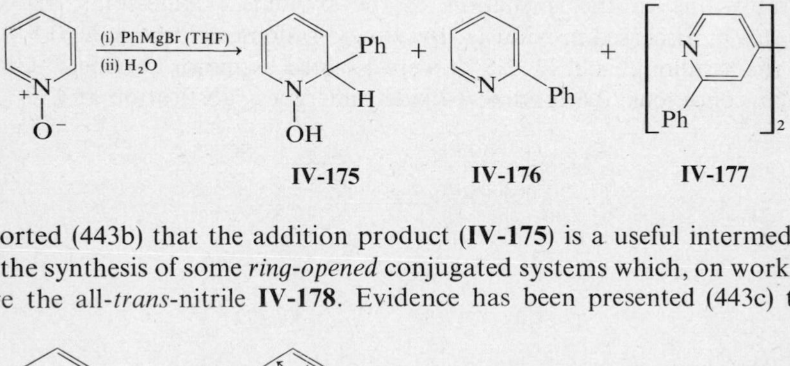

IV-175   IV-176   IV-177

reported (443b) that the addition product (**IV-175**) is a useful intermediate for the synthesis of some *ring-opened* conjugated systems which, on work up, gave the all-*trans*-nitrile **IV-178**. Evidence has been presented (443c) that

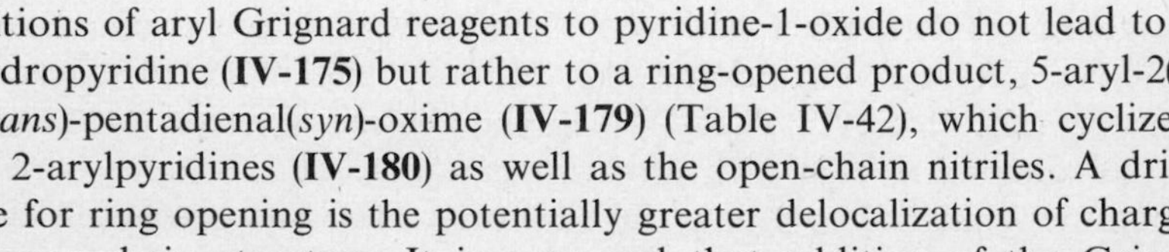

$$Ph(CH=CH)_2C=NO_2CPh$$

$$\downarrow \, -PhCO_2H$$

$$Ph(CH=CH)_2CN$$

IV-175   IV-178

additions of aryl Grignard reagents to pyridine-1-oxide do not lead to 1,2-dihydropyridine (**IV-175**) but rather to a ring-opened product, 5-aryl-2(*cis*), 4-(*trans*)-pentadienal(*syn*)-oxime (**IV-179**) (Table IV-42), which cyclizes to give 2-arylpyridines (**IV-180**) as well as the open-chain nitriles. A driving force for ring opening is the potentially greater delocalization of charge in the open-chain structure. It is supposed that addition of the Grignard reagent perpendicular to the pyridine ring is followed by immediate disrotatory opening of **IV-175a** leading directly to the observed stereochemistry (eq. IV-29). Treatment of **IV-179** with acetic anhydride gives 2-arylpyridine (**IV-180**) and a small amount of open-chain nitrile.

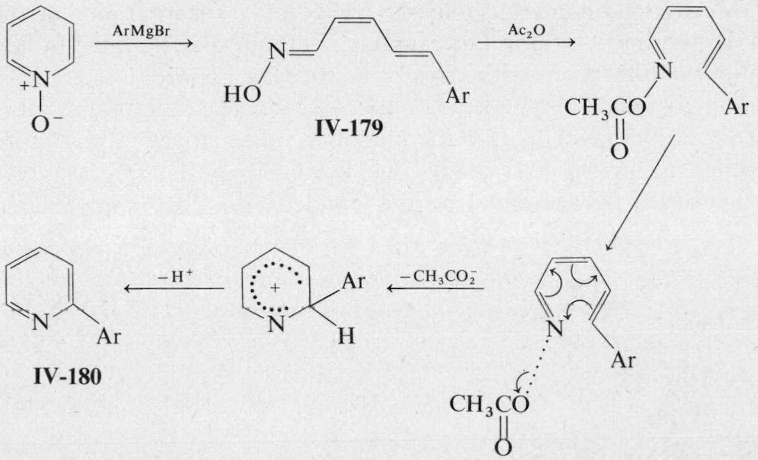

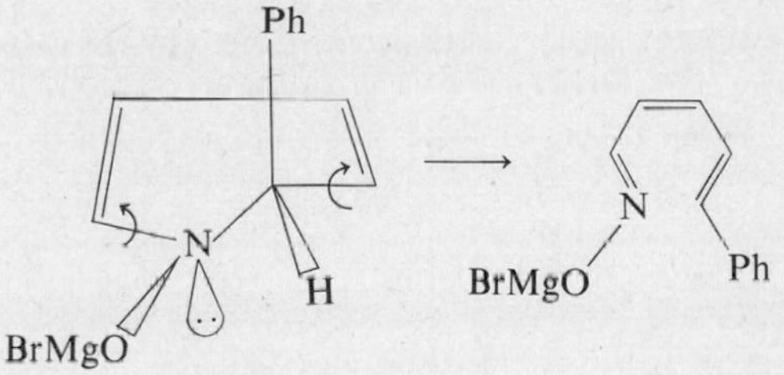

(IV-29)

IV-175a

TABLE IV-42. Reaction of Pyridine-1-oxide with Grignard Reagents to Yield Oximes (443c)

Conditions	Products
PhMgX (1.5 equiv.), THF, R.T., 1 hr	5-Phenyl-2(*cis*),4(*trans*)-pentadienal (*syn*)-oxime (28%)
C_6D_5MgX (1.5 equiv.), THF, R.T., 1 hr	5-Perdeuteriophenyl-2(*cis*),4(*trans*)-pentadienal (*syn*)-oxime (27%)
p-MeC$_6$H$_4$MgX (1.5 equiv.), THF, R.T., 1 hr	5-*p*-Tolyl-2(*cis*),4(*trans*)-pentadienal (*syn*)-oxime (45%)
p-AnisylMgX (1.5 equiv.), THF, R.T., 1 hr	5-*p*-Anisyl-2(*cis*),4(*trans*)-pentadienal (*syn*)-oxime (10%)
2-ThienylMgX (1.5 equiv.), THF, R.T., 1 hr	5-(2-Thienyl)-2(*cis*),4(*trans*)-pentadienal (*syn*)-oxime (48%)

N-Alkoxypyridinium salts undergo nucleophilic substitution more readily than do pyridine-1-oxides. For example, 1-ethoxy-3-substituted pyridinium bromide (**IV-181**) reacts with alkyl Grignard reagents with loss of the ethoxy group to give the 2-alkyl-3-substituted pyridine. The pyridinium salt is also partially decomposed to give the parent pyridine, together with a dialkyl carbinol (eq. IV-30) (307, 444). The key step in the synthesis of (±)-desethyldasycarpidone involves the condensation of indolylmagnesium

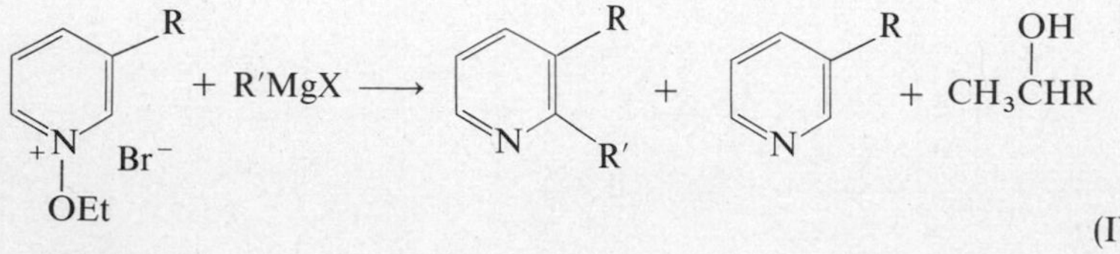

(IV-30)

IV-181

bromide (**IV-182**) with methyl isonicotinate-1-oxide (**IV-183**) in the presence of benzoyl chloride in tetrahydrofuran-methylene dichloride to give intermediate **IV-184**, which yields the pyridylindole compound **IV-185** (12%) as a mixture of stereoisomers (109b). When ether-methylene dichloride was used as solvent, **IV-182** (1%) and a byproduct (**IV-186**) were obtained. Using ether as solvent only 3-benzoylindole (**IV-187**) was formed. The

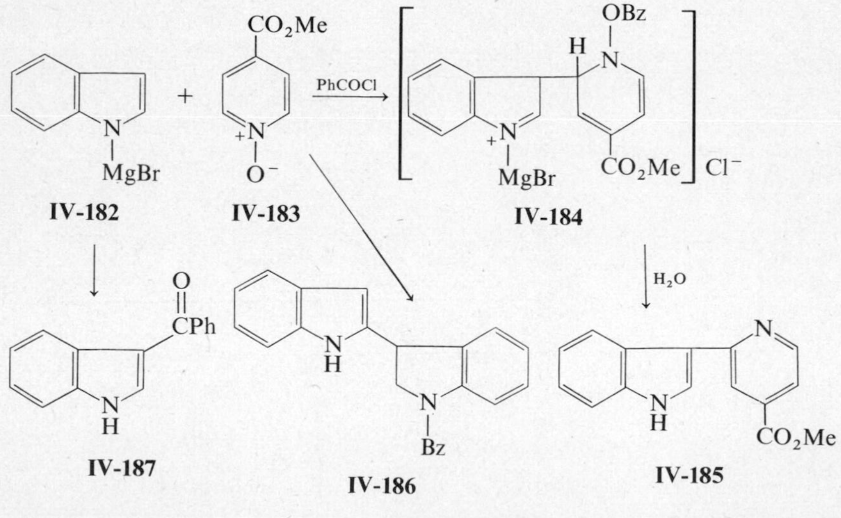

IV-182 **IV-183** **IV-184**

IV-187 **IV-186** **IV-185**

reaction of pyridine-1-oxide with phenyllithium was reported to give 2-phenylpyridine and other unidentified products (445).

It was reported that 2,3,4,5,6-pentachloropyridine-1-oxide (**IV-188**) in ether reacted with one equivalent of methylmagnesium iodide to give 3,4,5,6-tetrachloro-2-picoline-1-oxide (**IV-189**) (40%). In the presence of excess Grignard reagent (two equivalents), 3,4,5,6-tetrachloro-2-picoline-1-oxide (**IV-189a**) (37 %) and 3,4,5-trichloro-2,6-lutidine-1-oxide (**IV-190b**) (29 %) were obtained (Table IV-43) (351, 354b). When both of these reactions were carried out in boiling ether, deoxygenation of the starting material became the main reaction and the methylated 1-oxides were obtained in less than 10 % yield. The reactions of pentachloropyridine-1-oxide (**IV-188**) with phenyl- and ethylmagnesium bromide occurred more readily in THF

TABLE IV-43. Products from the Reaction of Pentachloropyridine-1-oxide with Grignard Reagents (354b)

Conditions	Products (%)
MeMgI (1 equiv.), ether, 25°, 0.5 hr	Pentachloropyridine-1-oxide (63 %); pentachloropyridine (2 %); 3,4,5,6-tetrachloro-2-picoline-1-oxide, m.p. 137° (32 %); 3,4,5-trichloro-2,6-lutidine-1-oxide, m.p. 135° (3 %)
MeMgI (2 equiv.), ether, 25°, 2 hr	Pentachloropyridine-1-oxide (23 %); pentachloropyridine (2 %); 3,4,5,6-tetrachloro-2-picoline-1-oxide (37 %); 3,4,5-trichloro-2,6-lutidine-1-oxide (29 %); polymeric tars (13 %)
EtMgBr (1 equiv.), ether, 25°, 10 hr	Pentachloropyridine-1-oxide (75 %); 3,4,5,6-tetrachloro-2-ethylpyridine-1-oxide, m.p. 109° (4 %); 3,4,5-trichloro-2,6-diethylpyridine-1-oxide, m.p. 108° (8 %); 3,4,5,6-tetrachloro-2-hydroxypyridine-1-oxide (13 %)
EtMgBr (1 equiv.), THF, 60°, 2 hr	Pentachloropyridine-1-oxide (56 %); 3,4,5,6-tetrachloro-2-ethylpyridine-1-oxide (4 %); 3,4,5-trichloro-2,6-diethylpyridine-1-oxide (10 %); 3,4,5,6-tetrachloro-2-hydroxypyridine-1-oxide (30 %)
EtMgBr (2 equiv.), THF, 60°, 2 hr	Pentachloropyridine-1-oxide (30 %); 3,4,5,6-tetrachloro-2-ethylpyridine-1-oxide (7 %); 3,4,5-trichloro-2,6-diethylpyridine-1-oxide (14 %); 3,4,5,6-tetrachloro-2-hydroxypyridine-1-oxide (50 %)
EtMgBr (4 equiv.), THF, 60°, 2 hr	3,4,5-Trichloro-2,6-diethylpyridine-1-oxide (72 %); 3,4,5,6-tetrachloro-2-hydroxypyridine-1-oxide (21 %)
PhMgBr (1 equiv.), THF, 60°, 2 hr	Pentachloropyridine-1-oxide (62 %); 3,4,5,6-tetrachloro-2-phenylpyridine-1-oxide, m.p. 112° (14 %); 3,4,5-trichloro-2,6-diphenylpyridine-1-oxide, m.p. > 200° (4 %); 3,4,5,6-tetrachloro-2-hydroxypyridine-1-oxide, m.p. 180° (decomp.) (20 %)
PhMgBr (2 equiv.), THF, 60°, 2 hr	Pentachloropyridine-1-oxide (38 %); 3,4,5,6-tetrachloro-2-phenylpyridine-1-oxide (30 %); 3,4,5-trichloro-2,6-diphenylpyridine-1-oxide (3 %); 3,4,5,6-tetrachloro-2-hydroxypyridine-1-oxide (30 %)
PhMgBr (4 equiv.), THF, 60°, 3 hr	3,4,5,6-Tetrachloro-2-phenylpyridine-1-oxide (75 %); 3,4,5-trichloro-2,6-diphenylpyridine-1-oxide (8 %); 3,4,5,6-tetrachloro-2-hydroxypyridine-1-oxide (10 %); polymeric tars (5 %)

than in ether (354b). With a large excess of Grignard reagent (four equivalents) the expected 2,6-diphenyl- **(IV-190b)** or diethyl- **(IV-190c)** trichloropyridine-1-oxide was the major product (354b). With one to two equivalents of Grignard reagent, a mixture of starting material, 2-mono- **(IV-189)** and 2,6-diphenyl- **(IV-190b)** or ethyl-pyridine-1-oxides **(IV-190c)** and an unexpected 3,4,5,6-tetrachloro-2-hydroxypyridine-1-oxide **(IV-191)** were obtained (Table IV-43) (354b). 3,4,5,6-Tetrachloro-2-hydroxypyridine-1-oxide **(IV-191)** was

(a; R = Me; b; R = Ph; c; R = Et)

produced only in the reaction with phenyl- and ethylmagnesium bromide and is formed from the intermediate exchange Grignard reagent **(IV-192)**, which suffers atmospheric oxidation followed by hydrolysis. In the case of methylmagnesium iodide, the methyl halide generated was capable of exchange to give the monomethylated-1-oxide **(IV-189a)**; phenyl and ethyl halides are, however, less prone to react with the complex **IV-192**. When the reaction was carried out under a nitrogen atmosphere, only a small amount of the 2-hydroxy compound (< 5%) was obtained.

Attempts to carry out similar reactions with n-BuLi, MeLi, and PhLi under various conditions failed. Only starting material or tars were isolated (351).

2-Ethynylpyridine-1-oxide **(IV-193a)** was obtained in good yield from pyridine-1-oxide and ethynyl sodium in dimethyl sulfoxide at room temperature (446). 2-(Phenylethynyl)pyridine-1-oxide **(IV-193b)** was similarly prepared (477).

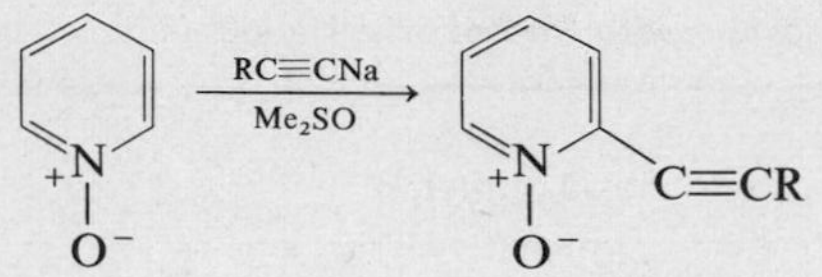

a; R = H (90%)
b; R = Ph (40%)

IV-193

B. *Inorganic Acid Halides*

2- And 4-chloropyridines were obtained on chlorination of pyridine-1-oxide with phosphorus pentachloride, phosphorus oxychloride, or sulfuryl chloride (5, 7). The inorganic acid halide complexes with the *N*-oxide function to give a complex which undergoes nucleophillic attack by the halide ion to give 2- and 4-halopyridines (Scheme IV-10).

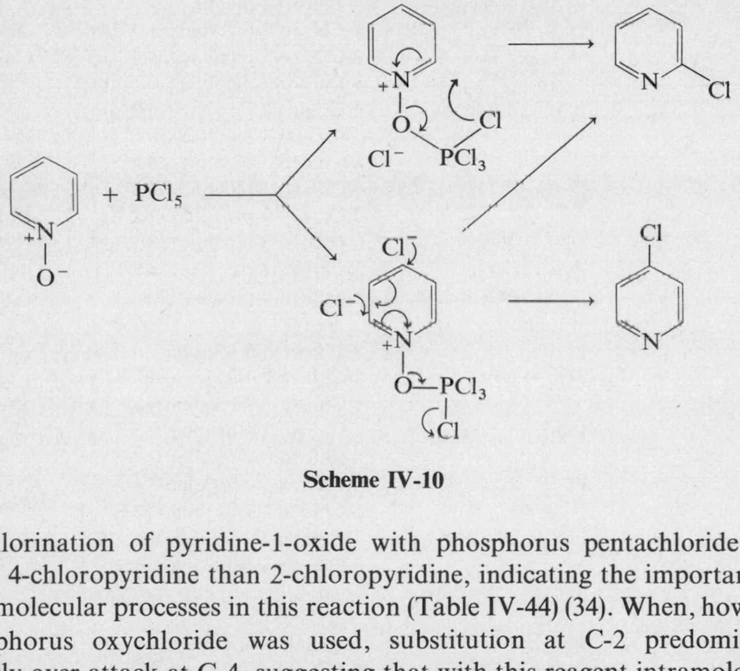

Scheme IV-10

Chlorination of pyridine-1-oxide with phosphorus pentachloride gave more 4-chloropyridine than 2-chloropyridine, indicating the importance of intermolecular processes in this reaction (Table IV-44) (34). When, however, phosphorus oxychloride was used, substitution at C-2 predominated slightly over attack at C-4, suggesting that with this reagent intramolecular attack is important. The results obtained on chlorination of 3-picoline- and 3,4-lutidine-1-oxide are summarized in Table IV-44 (34).

When the 2- or 4-position of pyridine-1-oxide is occupied, the chlorine atom can enter the 4- or 6-positions. For example, chlorination of 2-chloropyridine-1-oxide with phosphorus oxychloride gave 2,6-dichloropyridine (358). A 2-methyl group is occasionally attacked as well. For example,

TABLE IV-44. Reaction of Pyridine-1-oxides with Phosphorus Halides

Substituent in N-oxide	Conditions	Products (yield)		Ref.
None	PCl$_5$, 1.5 hr	4-Chloropyridine (58.5%) 2-Chloropyridine (41.5%)	(13.0%)	34
	PCl$_5$, 3.5 hr	4-Chloropyridine (48.5%) 2-Chloropyridine (51.5%)	(8.8%)	34
	POCl$_3$, 1.5 hr	4-Chloropyridine (30.1%) 2-Chloropyridine (69.9%)	(52.2%)	34
	POCl$_3$, 3.5 hr	4-Chloropyridine (31.8%) 2-Chloropyridine (68.2%)	(62.5%)	34
	PCl$_5$—POCl$_3$ (1:4.5), 1.5 hr	4-Chloropyridine (57.6%) 2-Chloropyridine (42.4%)	(8.2%)	34
	PCl$_5$—POCl$_3$ (1:4.5), 3.5 hr	4-Chloropyridine (53.5%) 2-Chloropyridine (46.5%)	(13.1%)	34
2-Me	POCl$_3$, toluene, 4 hr	4-Chloro-2-picoline		348
3-Me	PCl$_5$, 1.5 hr	4-Chloro-3-picoline (51.6%) 2-Chloro-3-picoline (28.3%) 6-Chloro-3-picoline (20.1%)	(30.01%)	34
	POCl$_3$, 1.5 hr	4-Chloro-3-picoline (44.0%) 2-Chloro-3-picoline (30.6%) 6-Chloro-3-picoline (25.4%)	(96.5%)	34
	PCl$_5$—POCl$_3$ (1:4.5), 1 hr	4-Chloro-3-picoline (44.8%) 2-Chloro-3-picoline (31.2%) 6-Chloro-3-picoline (23.9%)	(51.03%)	34
4-Me	POCl$_3$, 1.5 hr	2-Chloro-4-picoline (34.2%)		34
	PCl$_5$—POCl$_3$ (1:4.5), 1.5 hr	2-Chloro-4-picoline (7.5%)		34
2,6-Me$_2$	POCl$_3$	4-Chloro-2,6-lutidine		61
3,4-Me$_2$	PCl$_5$, 1.5 hr	2-Chloro-3,4-lutidine (63.7%) 6-Chloro-3,4-lutidine (36.3%)	(0.9%)	34
	POCl$_3$, 1.5 hr	2-Chloro-3,4-lutidine (45.3%) 6-Chloro-3,4-lutidine (54.7%)	(69.2%)	34
	PCl$_5$—POCl$_3$ (1:4.5), 1.5 hr	2-Chloro-3,4-lutidine (44.0%) 6-Chloro-3,4-lutidine (56.0%)	(11.3%)	34
	POCl$_3$, NaCl	2-Chloro-3,4-lutidine (46.1%) 6-Chloro-3,4-lutidine (53.9%)	(76.2%)	34
3-t-Bu	SO$_2$Cl$_2$, 110–120°, 2 hr	3-t-Butylpyridine; picrate, m.p. 152–153°		337
2-Cl	POCl$_3$	2,6-Dichloropyridine (91%)		357
4-Cl-3-NO$_2$	POCl$_3$, 10 hr	2,4-Dichloro-3-nitropyridine; 2,4-dichloro-5-nitropyridine		449
3-OAc-2-Me-4,5-(CH$_2$OAc)$_2$	(i) POCl$_3$, 100°, 0.5 hr (ii) 120°, 0.5 hr	3-Acetoxy-6-chloro-4,5-diacetoxy-methyl-2-picoline		73
3-CH$_2$Cl	POCl$_3$	2-Chloro-3-chloromethylpyridine, b.p. 115°/13 mm		372
3-CH$_2$CN	POCl$_3$	2-Chloro-3-cyanomethylpyridine		80a

Table IV-44 (*Continued*)

Substituent in *N*-oxide	Conditions	Products (yield)	Ref.
3-CO$_2$H	PCl$_5$, POCl$_3$	2-Chloronicotinic acid, nicotinic acid	450
4-CO$_2$Et	(i) POCl$_3$, 130–140° (ii) KOH	2-Chloroisonicotinic acid (70%)	451
3-CO$_2$Me-2-Me-4,5-(CH$_2$CO$_2$Me)$_2$	POCl$_3$, 100°	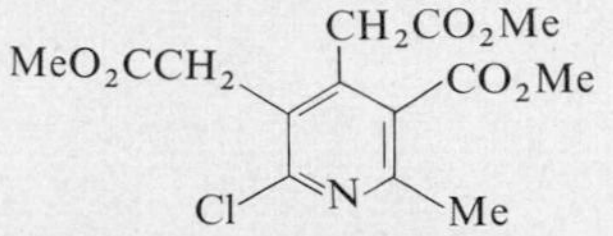	73
3,4-(CO$_2$Me)$_2$-6-Me	POCl$_3$, 120°	Dimethyl 6-chloro-2-picoline-3,4-dicarboxylate (14%); dimethyl 2-chloromethylpyridine-4,5-dicarboxylate (40%)	448
4-CN-3-Et	POCl$_3$, CHCl$_3$, Δ	2-Chloro-4-cyano-3-ethylpyridine; 2-chloro-4-cyano-5-ethylpyridine	452
4-CONH$_2$	PCl$_5$, POCl$_3$, 120–130°, 15 hr	2-Chloro-4-cyanopyridine	143
	PCl$_5$, POCl$_3$, Δ, 1.5 hr	2-Chloro-4-cyanopyridine, m.p. 60–65° (50%)	110
3-NO$_2$	POCl$_3$, PCl$_5$, 150°, 5 hr	2-Chloro-3-nitropyridine (45%)	423
	POCl$_3$, Δ, 1.5 hr	2-Chloro-3-nitropyridine (30%); 2-chloro-5-nitropyridine (8.4%)	117
3,5-(NO$_2$)$_2$	PCl$_5$, POCl$_3$, 150°, 5 hr	2-Chloro-3,5-dinitropyridine	423

dimethyl 2-picoline-4,5-dicarboxylate-1-oxide (**IV-194**) reacts with phosphorus oxychloride to give dimethyl 6-chloro-2-picoline-4,5-dicarboxylate (**IV-195**) and the picolyl chloride derivative (**IV-196**) (448). In the nuclear

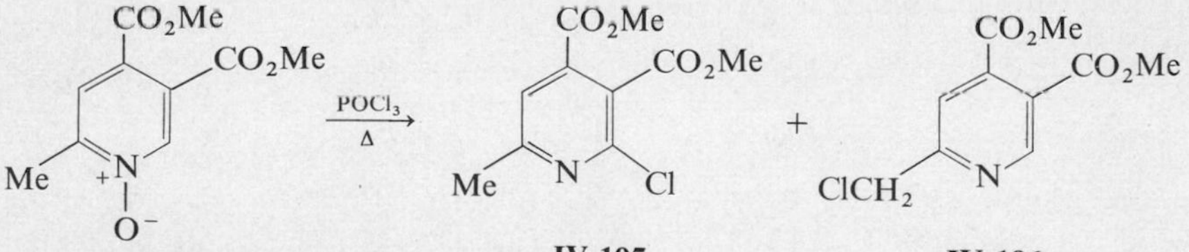

chlorination of pyridyl alcohol-1-oxides, it is usually necessary to protect the alcoholic hydroxyl group by acylation (eq. IV-31) (73). With the exception

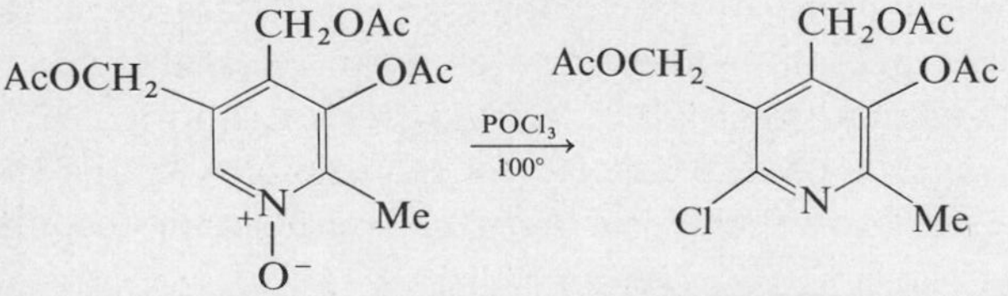

(IV-31)

of 3-bromopyridine-1-oxide, the other 3-substituted pyridine-1-oxides direct the entering chlorine atom to the 2-position. For example, chlorination of 3-nitropyridine-1-oxide with phosphorus oxychloride gave 2-chloro-3-nitropyridine (30%) and 2-chloro-5-nitropyridine (8.4%) (117). Chlorination of 4-isonicotinamide-1-oxide (IV-197) with phosphorus oxychloride and phosphorus pentachloride gives 2-chloro-4-cyanopyridine (IV-198) (144).

IV-197   IV-198

Some oxides, for example 1-hydroxy-4-pyridone, will undergo nucleophilic displacement rather than attack of the nucleus to give 4-chloropyridine-1-oxide. These reactions are discussed in the section on side chain-reactions (see Section V.7).

C. Nucleophilic Attack on 1-Alkoxy and 1-Acyloxypyridinium Salts

The Reissert reaction is unsuccessful in the pyridine-1-oxide series except in the case of 4-chloropyridine-1-oxide (eq. IV-32) (453, 454). However, a

(IV-32)

cyano group was introduced at the 2- and 4-positions of a pyridine ring which was activated by a 1-alkoxy group (Table IV-45). The results show that an electron-withdrawing group at the 2- or 4-position accelerates this reaction and an electron-donating group has the opposite effect. With the exception of a 3-carboxylic ester, a 3-substituent (either electron-donating or attracting) directs the entering nucleophile predominately to C-2. External factors such as temperature, pH, and solvent, as well as an internal factor, such as the size of the O-alkylating group, affect the isomer ratio (6, 7). A two-stage mechanism was proposed to account for the results. The first step is postulated to be rapidly reversible, the dihydro intermediate decomposing with

TABLE IV-45. Substitution of Pyridine-1-oxides and 1-Alkoxypyridinium Salts with Cyanide Ion

Substituent in N-oxide	Conditions	Products (yield)	Ref.
None		2-Cyanopyridine (49%) (48%); 4-cyanopyridine (32%) (24%)	280, 455
2,6-d_2	(i) MeI (ii) KCN, D_2O, dioxane	2-Cyanopyridine-6d; 4-cyanopyridine-2,6-d_2	115
2-Me		6-Cyano-2-picoline (48%) (45%); 4-cyano-2-picoline (10%) (18%)	280, 455
2-Me	(i) Me_2SO_4 (ii) KCN, 0–10°, overnight	6-Cyano-2-picoline, m.p. 71–73°	108
2-Et	(i) Me_2SO_4 (ii) NaCN	4-Cyano-2-ethylpyridine, b.p. 110–130°/30 mm; 2-cyano-6-ethylpyridine, b.p. 130–140°/30 mm	40
2,6-Me_2	Me_2SO_4, KCN, 0–10°, overnight	4-Cyano-2,6-lutidine, m.p. 83–85°, (40%)	108, 455
		4-Cyano-2,6-lutidine (13%); 6-cyanomethyl-2-picoline (33%)	280
2,4-Me_2	Me_2SO_4, KCN, 0–10°, overnight	6-Cyano-2,4-lutidine, m.p. 55–56° (73%)	108, 455
3-Me		2-Cyano-3-picoline (30%); 4-cyano-3-picoline (15%)	280
		2-Cyano-3-picoline (36%); 4-cyano-3-picoline (6%); 6-cyano-3-picoline (6%)	458
	Me_2SO_4, KCN, 0–10°, overnight	2-Cyano-3-picoline, m.p. 87–90°	108
3-$SiMe_3$	Me_2SO_4, KCN, 0–10°, overnight	4-Cyano-3-trimethylsilylpyridine, b.p. 223°; picrate, m.p. 151°; 3-trimethylsilylpicolinamide, m.p. 87°; 5-trimethylsilylpicolinamide, m.p. 128°;	46
4-Me		2-Cyano-4-picoline (30–40%)	280, 455
	Me_2SO_4, KCN, 0–10°, overnight	2-Cyano-4-picoline, m.p. 89–91°	108
	(i) MeI, Δ, 2 hr (ii) KCN, 2 hr	2-Cyano-4-picoline	98
4-Et	(i) MeI, MeOH (ii) KCN, MeOH	2-Cyano-4-ethylpyridine, b.p. 115–120°/10 mm	459
2-Cl	Me_2SO_4, KCN	2-Chloro-6-cyanopyridine, m.p. 86–68° (46.7%)	460
3,5-Br_2	Me_2SO_4, KCN	2-Cyano-3,5-dibromopyridine, (70%)	458
4-Cl	Me_2SO_4, KCN	4-Chloro-2-cyanopyridine, m.p. 85–86° (55.6%); 4-chloropicolinamide, m.p. 160–162°	460

Table IV-45 (*Continued*)

Substituent in N-oxide	Conditions	Products (yield)	Ref.
2-OMe	Me$_2$SO$_4$, KCN	6-Cyano-2-methoxypyridine, m.p. 66–68° (37.4%); 2,6-dicyanopyridine, m.p. 126–127° (14.6%); 6-cyanopicolinamide, m.p. 184–186° (trace)	460
3-OMe	Me$_2$SO$_4$, KCN	2-Cyano-3-methoxypyridine (67.8%)	460
4-OMe	Me$_2$SO$_4$, KCN	2,4-Dicyanopyridine, m.p. 90–91° (40%); 4-cyanopicolinamide, m.p. 256–258° (decomp.)	460
4-OMe-2-(2-C$_5$H$_4$N)	Me$_2$SO$_4$, KCN	2-Cyano-4-methoxy-6-(2-pyridyl)pyridine	457
3-CH$_2$CO$_2$Et	Me$_2$SO$_4$, KCN, 23–25°	Ethyl 2-cyano-3-pyridylacetate (7.4%); ethyl 4-cyano-3-pyridylacetate (36%);	76
2-CO$_2$Me	Me$_2$SO$_4$, KCN	Methyl 6-cyanopicolinate m.p. 111–113.5° (50.3%); methyl 6-carbamoylpicolinate, m.p. 136–138°	460
3-CO$_2$Me	Me$_2$SO$_4$, KCN	Methyl 2-cyanonicotinate (19%); methyl 4-cyanonicotinate (31.6%)	460
3-CO$_2$Et		Ethyl 6-cyanonicotinate (19%); ethyl 4-cyanonicotinate (32%)	281, 458
5-CO$_2$Et-2-Me	Me$_2$SO$_4$, KCN 0–10°, overnight	Ethyl 4-cyano-6-methylnicotinate, m.p. 89–90.5°	108
4-CO$_2$Me	Me$_2$SO$_4$, KCN	Methyl 2-cyanoisonicotinate, m.p. 107–109° (69.2%)	460
2-CN	Me$_2$SO$_4$, KCN	2,6-Dicyanopyridine, m.p. 126–127° (83.7%)	460
3-CN		2,3-Dicyanopyridine (27.8%); 2,6-dicyanopyridine (17.6%)	458, 460
4-CN		2,4-Dicyanopyridine (54%), m.p. 90–91° 70%; m.p. 88–91°	455, 460, 108
4-NO$_2$	Me$_2$SO$_4$, KCN R.T., overnight	2-Cyano-4-nitropyridine, m.p. 72.4° (53.7%); 4-nitro-picolinamide, m.p. 154–158°	460
	(i) Me$_2$SO$_4$, 65–70° (ii) aq. NaCN	2-Cyano-4-nitropyridine, m.p. 73–74° (56.9%)	314b
2-Me-4-NO$_2$	(i) Me$_2$SO$_4$, 65–70° (ii) aq. NaCN	2-Cyano-6-methyl-4-nitropyridine, m.p. 76.5–77.5° (67.1%)	314b
5-Et-2-Me-4-NO$_2$	(i) Me$_2$SO$_4$, 65–70° (ii) aq. NaCN	2-Cyano-3-ethyl-6-methyl-4-nitropyridine, b.p. 94°/0.05 mm (82.4%)	314b
3-Me-4-NO$_2$	(i) Me$_2$SO$_4$, 65–70° (ii) aq. NaCN	2-Cyano-3-methyl-4-nitropyridine, m.p. 64–65° (85.7%)	314b

elimination of alkoxide ion to give the cyanopyridine in the rate-determining and product-determining step (eq. IV-33) (281, 455).

(IV-33)

3-Nitropyridine-1-oxide reacts with silver cyanide in the presence of benzyl chloride to give 2-cyano-3-nitropyridine and 2-cyano-5-nitropyridine (456).

The Reissert-Kaufmann reaction of substituted 1-methoxy-4-nitropyridinium methosulfate with an aqueous solution of sodium cyanide under a nitrogen atmosphere gave the corresponding 2-cyano-4-nitropyridines in satisfactory yields (314b). Under these conditions, 3-methyl-4-nitropyridine-1-oxide gave the corresponding 2-cyano compound but no 6-cyano isomer. When 1,4-dimethoxy-2-(2-pyridyl)pyridinium methosulfate (IV-199) was treated with potassium cyanide, 6-cyano-4-methoxy-2-(2-pyridyl)pyridine (IV-200) was obtained (457).

IV-199   IV-200

The reaction of pyridine-1-oxide and cyclohexanone piperidine (or morpholine) enamine in the presence of benzoyl chloride gave a good yield of 2-(2′-pyridyl)cyclohexanone (IV-201) (461). Other substituted pyridine-1-oxides also undergo this reaction (eq. IV-34) (162, 462) (Table IV-46).

Pyridine-1-oxides undergo substitution accompanied by deoxygenation with alkyl mercaptans in the presence of acylating agents, to form alkyl pyridyl sulfides.

1-Ethoxypyridinium ethosulfate (IV-202) was treated with sodium n-propylmercaptide in a mixture (10:1 w/w) of 1-propanethiol and ethanol,

TABLE IV-46. Preparation of 2-(2'-Pyridyl)cycloalkanones

Substituent in N-oxide	Conditions	Product (yield)	Ref.
None	PhCOCl, cyclopentanone morpholine enamine	2-(2-Pyridyl)cyclopentanone, orange oil, b.p. 54–56°/0.04 mm (40%)	463
	PhCOCl, cyclohexanone morpholine enamine	2-(2-Pyridyl)cyclohexanone	461
	PhCOCl, cycloheptanone morpholine enamine	2-(2-Pyridyl)cycloheptanone, yellow oil, b.p. 84–88°/10–15 mm (90%)	463
2-Me	PhCOCl, cyclohexanone morpholine enamine	2-(6-Methyl-2-pyridyl)cyclohexanone b.p. 115–120°/0.2 mm (82%)	162
4-Me	PhCOCl, cyclohexanone morpholine enamine	2-(4-Methyl-2-pyridyl)cyclohexanone, b.p. 115–120°/0.12 mm	162
2-Cl	PhCOCl, cyclohexanone morpholine enamine	2-(6-Chloro-2-pyridyl)cyclohexanone, m.p. 166–167.5°	462
4-Cl	PhCOCl, cyclohexanone morpholine enamine	2-(4-Chloro-2-pyridyl)cyclohexanone, b.p. 125–127°/0.17 mm; picrate, m.p. 172–174°	462

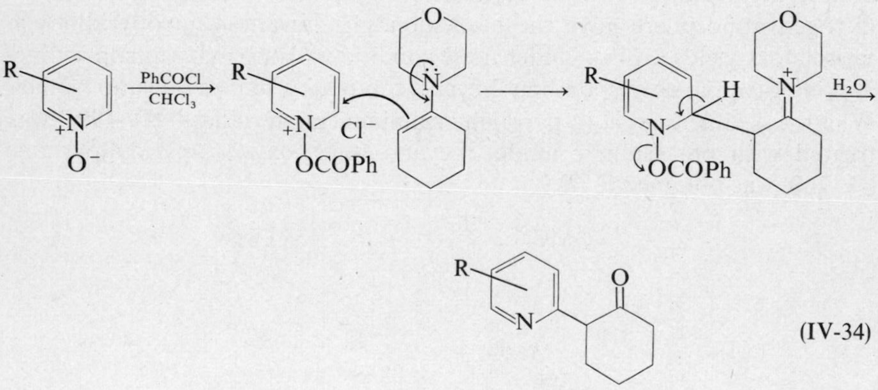

(IV-34)

IV-201

to give pyridine (70%) and a 6:1 mixture of 3- (IV-203) and 4-pyridyl propyl sulfides (30%) (IV-204) (464). Reaction of the O-alkylated salts of 2- or

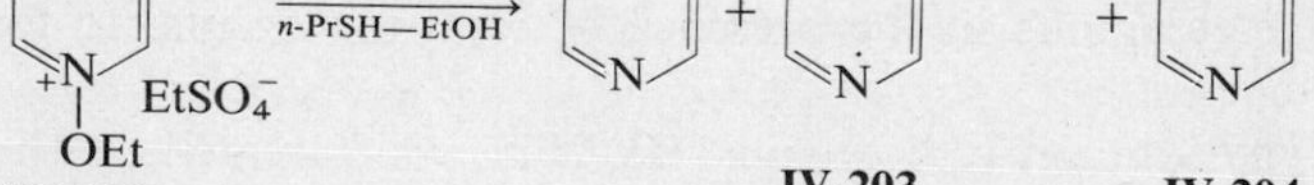

IV-202   IV-203   IV-204

4-picoline-1-oxides (**IV-205**) with sodium thiophenoxide gave the picoline (**IV-206**), the picoline-1-oxide (**IV-207**), and the phenylthiomethylpyridine (**IV-208**) (465). Reaction of a 1-methoxy-4-picolinium salt with sodium

IV-205   **IV-206**   **IV-207**   **IV-208**

n-propyl mercaptide in *n*-propyl mercaptan gave a variety of products (eq. IV-35) (465). The reaction of butyl mercaptan with various types of

quaternary salts of pyridine-1-oxide was examined (eq. IV-36) (Table IV-47) (466). Recently, substitution of pyridine-1-oxide with methyl, *n*-propyl,

and *t*-butyl mercaptan in acetic anhydride at 95° for 3 hr was reported to yield 2- and 3-pyridyl sulfides (Table IV-48) (467). Other *N*-oxides of 2-, 3-, and 4-picolines, 4-*t*-butylpyridine, and 2,6-, 3,4-, and 3,5-lutidine were treated with *t*-butyl mercaptan in acetic anhydride to determine if a bulky C-4 substituent in the pyridine ring would prevent the entry of *t*-butyl mercaptide at C-3. From these results, a 3-methyl group does not appear to exert any steric effect upon nucleophilic attack at C-2 [C-2 (ca 45%), C-6 (ca 20%)]. In the case of 2-picoline-1-oxide, no substitution occurred at C-3: for *β*-substitution to occur a free adjacent *α*-position is necessary (467). The

TABLE IV-47. Reaction of Various 1-Alkoxy- and 1-Acyloxy-Pyridinium Salts with Butyl Mercaptan (eq. IV-35) (466)

RX	Reagent	Overall yield of sulfide (%)	Ratios of isomeric n-butylthiopyridines		
			2-	3-	4-
Et$_2$SO$_4$	NaSBu–BuSH	15	16	60	24
TsOEt	NaSBu–BuSH	11	11	74	15
Ac$_2$O	BuSH	67	61	39	—
AcCl	NaSBu–BuSH	10	89	9	2
PhCOCl	BuSH	19	81	18	1
PhCOCl	NaSBu–BuSH	16	81	15	4
PhSO$_2$Cl	BuSH	32	50	50	—

mechanism proposed for these reactions is shown below (Scheme IV-11) (467). The presence of a 1-acetyl-1,2,3,6-tetrahydropyridine in which a

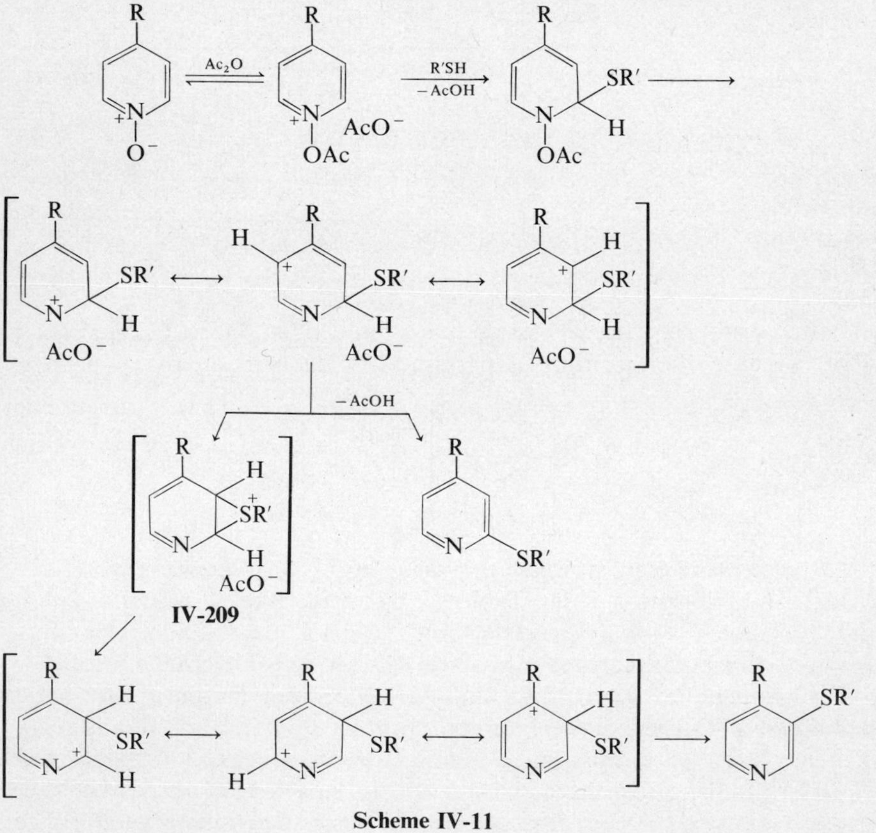

Scheme IV-11

TABLE IV-48. Reaction of Pyridine-1-oxides with Alkyl Mercaptans in the Presence of Acetic Anhydride (467)

			Isomer distribution of pyridyl sulfides			
R	R'	Yield (%)	2-	3-	5-	6-
H	Me	38	52	48	—	—
H	n-Pr	46	76	24	—	—
H	n-Bu	67	61	39	—	—
H	t-Bu	62	70	30	—	—
2-Me	t-Bu	32	—	—	16	84
3-Me	t-Bu	66	45	—	36	19
4-Me	n-Pr	31	50	50	—	—
4-Me	t-Bu	41	71	29	—	—
4-t-Bu	t-Bu	48	88	17	—	—
4-Ph	t-Bu	18	44	56	—	—
2,6-Me₂	t-Bu	0	—	—	—	—
3,4-Me₂	t-Bu	35	48	—	29	—
3,5-Me₂	t-Bu	66	100	—	—	—

sulfide group was attached to C-3 and the α-positions were substituted by either sulfide or acetoxy groups was established (468), and the episulfonium intermediate **IV-209** was proposed to account for the formation of these products. A number of 1,2,3,6-tetrahydropyridines (**IV-210**) were subsequently isolated from the reaction of pyridine-1-oxide with alkyl mercaptans in acetic anhydride and the stereochemistry of these compounds was discussed (469). Still more recently, a series of these reactions was carried out

R = H, Me, n-Pr, isoPr, t-Bu

IV-210

in the presence of triethylamine, which led to the formation of a higher percentage of the 2-isomer (Table IV-49) (470). This was accounted for by suggesting that acetate **IV-211** dissociates into ion-pairs **IV-212** and then **IV-213**. The former can lose the 2-proton to yield the 2-butylthio-derivative (**IV-214**), the latter can rearrange to the 3-isomer (**IV-215**). Since triethylamine is a stronger base than acetate ion, proton abstraction from C-2

TABLE IV-49. Reaction of Pyridine-1-oxide with *t*-Butyl Mercaptan in Acetic Anhydride (470)

Substituent	Yield (%)	With triethylamine; isomer distribution		Yield (%)	Without triethylamine; isomer distribution	
		2-	3-		2-	3-
None	41	90	10	62	70	30
4-Me	33	82	18	41	71	29
4-Et	32	87	13	—	—	—
4-*n*-Pr	45	70	30	—	—	—
4-isoPr	39	80	20	—	—	—
4-*t*-Bu	48	96	4	48	83	17

would be enhanced, thus leading to increased yields of **IV-214** (470). In the presence of triethylamine, these reactions produced new tetrahydropyridines.

For example, when 4-picoline-1-oxide was treated with *t*-butyl mercaptan and acetic anhydride in the presence of triethylamine, a tetrahydropyridine was obtained which, on pyrolysis, yielded 3-*t*-butylmercapto-4-picoline. On the basis of its infrared, NMR, and mass spectra, this tetrahydropyridine was assigned structure **IV-216**. Similarly, 4-ethylpyridine-1-oxide, *t*-butyl mercaptan, and acetic anhydride gave, in the presence of triethylamine, the expected sulfides, as well as the tetrahydropyridine analogous to **IV-216**. In the case of 4-*n*-propyl- and 4-isopropylpyridine-1-oxides, the expected sulfides were obtained, but this time tetrahydropyridines (**IV-210**) were also isolated (468, 469). Formation of **IV-216** is visualized to occur *via* the

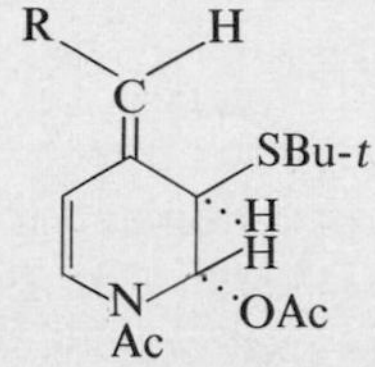

IV-216

episulfonium salt **IV-213** which, with acetate ion, yields **IV-217**. Quarterniza-tion with acetic anhydride to give **IV-218** followed by abstraction of the acidic proton of the active methylene group at the 4-position by triethylamine could give **IV-216**. Since 4-*n*-propyl and 4-isopropylpyridine-1-oxide gave **IV-210** and not **IV-216** it was proposed that triethylamine may either be incapable of (or very slow in) abstracting the acidic proton of 4-*n*-propyl- or 4-isopropylpyridine-1-oxide, or have difficulty in approaching the acidic proton to form the anion. Nucleophilic attack by *t*-butyl mercaptan at C-6 is now faster than proton-abstraction and **IV-210** is formed. When pyridine-1-oxide itself was treated with *t*-butyl mercaptan in acetic anhydride and

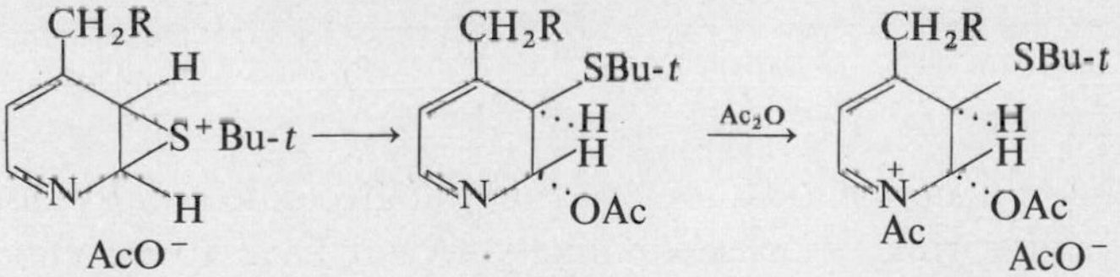

IV-213   **IV-217**   **IV-218**

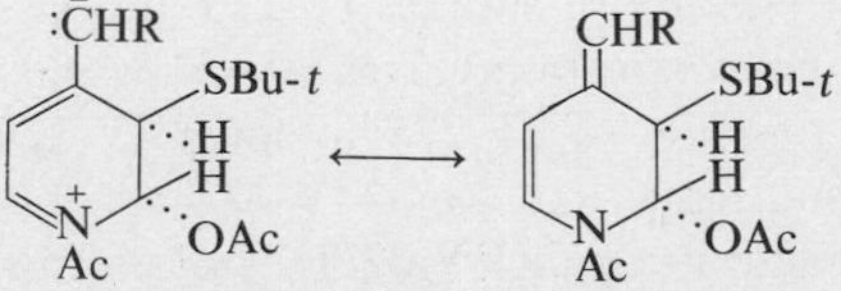

IV-216

triethylamine, still a different tetrahydropyridine was obtained and, on the basis of its infrared, NMR, and mass spectra, structure **IV-219** was assigned to it (470). Attack by acetate upon the episulfonium ion **IV-213** (R = H) to

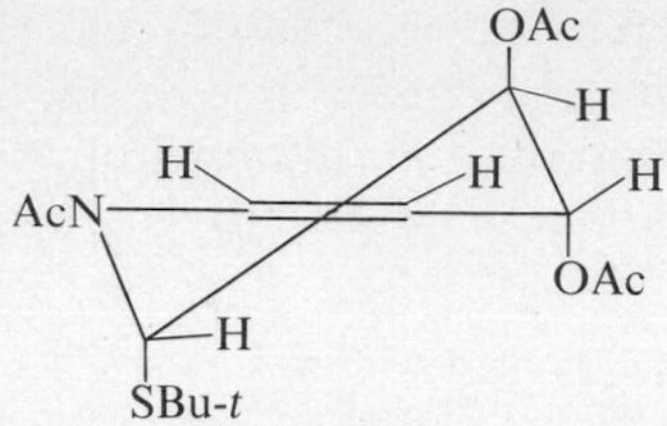

IV-219

give **IV-220** followed by quaternization and nucleophilic addition of acetate at the 4-position would yield **IV-219** (470). Explanations not involving the intervention of episulfonium ions are also possible, since nucleophilic

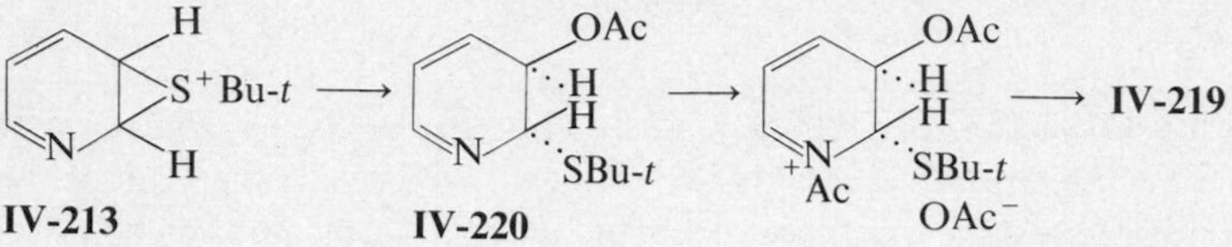

attack at the β-position of a 1,2-dihydropyridine are also known (see Chapter I). 3-Picoline-1-oxide and *t*-butyl mercaptan in acetic anhydride gave **IV-221**, with or without triethylamine (470).

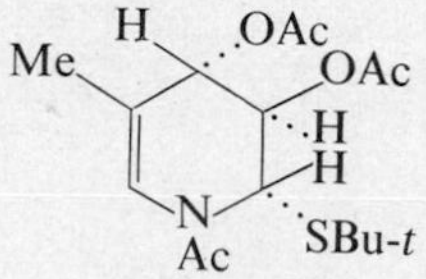

IV-221

   To promote nucleophic attack by ethyl thioglycolate and by thiophenols on the pyridine ring, its electrophilicity would have to be enhanced by changing the acylating agent from acetic anhydride to a sulfonyl halide. Thus deoxidative substitution of substituted pyridine-1-oxides (**IV-222**) took place to form a mixture of sulfides **IV-223** and **IV-224** (Table IV-50) (471).
   The proposed mechanism involves attack by the thiophenol on the highly electrophilic α-position of **IV-225** to form **IV-226** followed by the facile departure of sulfonate ion from **IV-227** to form the highly energetic nitrenium-sulfonate ion pair (**IV-227**). Though this would appear to be an energetically unlikely process, evidence has been given in support of the tight solvent cage **IV-227**. On going from benzene to chloroform, the isomer ratio for a particular reaction remains constant. A bulky group (*t*-butyl) at the γ-position offers little steric hindrance to nucleophilic attack at the β-position. Changing from benzenesulfonyl chloride to methanesulfonyl

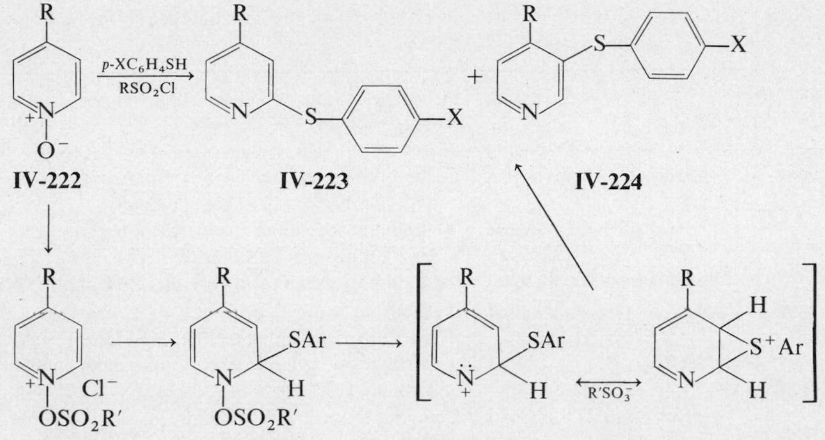

IV-222   **IV-223**   **IV-224**

IV-225   **IV-226**   **IV-227**

chloride results in a higher proportion of **IV-224**. All of the reactions are postulated to proceed *via* 1-sulfonyloxy-2-arylthio-1,2-dihydropyridine intermediates (**IV-226**) with the exception of that with 2,6-lutidine-1-oxide, when either a 1,2- or 1,4-dihydropyridine can be involved.

When pyridine-1-oxide and *N*-phenylbenzimidoyl chloride were heated in the presence of benzenethiol, 3-phenylthiopyridine and other products were obtained (472). This reaction is discussed in Section IV.5.

N-Methoxypyridinium salts react with alkali metal derivatives of diethyl-phosphonate to form pyridine-2-phosphonate (**IV-228**) (Table IV-51) (473). It has been proposed that the exclusive 2-substitution suggests that a dipolar complex (**IV-229**) precedes formation of the dihydropyridine intermediate (**IV-230**) (473). 3-Alkyl substituted alkoxypyridinium salts (**IV-231**) react with

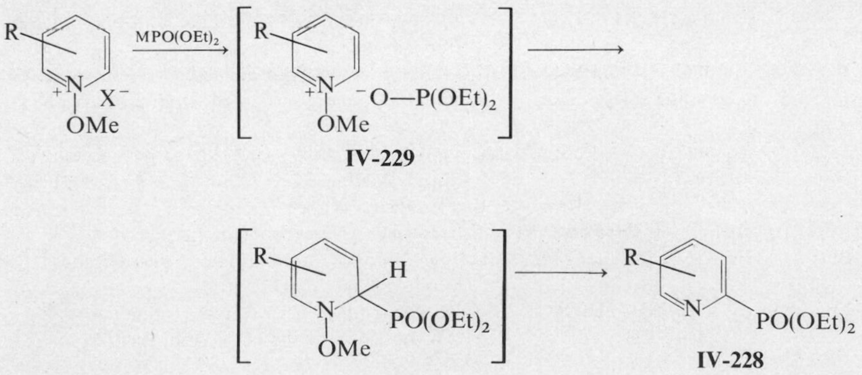

IV-229

IV-230   **IV-228**

TABLE IV-50. Deoxidative Substitution of Pyridine-1-oxides by Thiophenols in the Presence of Sulfonyl Halides (471)

Substituent in N-oxide	Conditions	Products (yield)
None	PhSH, MeSO$_2$Cl, benzene	2-Phenylthiopyridine: 3-phenylthiopyridine (40:60; 27.2%); diphenyl disulfide (60.2%)
	PhSH, PhSO$_2$Cl, benzene	2-Phenylthiopyridine: 3-phenylthiopyridine (41:59; 30.2%); diphenyl disulfide (52.8%)
	PhSH, PhSO$_2$Cl, CHCl$_3$	2-Phenylthiopyridine: 3-phenylthiopyridine (39:61; 45.7%)
	p-ClC$_6$H$_4$SH, PhSO$_2$Cl, CHCl$_3$	2-p-Chlorophenylthiopyridine: 3-p-chlorophenylthiopyridine (37:63; 50.0%); di-p-chlorophenyl disulfide (35.6%)
	p-t-BuC$_6$H$_4$SH, PhSO$_2$Cl, benzene	2-p-t-Butylphenylthiopyridine: 3-p-t-butylphenylthiopyridine (32:68; 33.0%); di-p-t-butylphenyl disulfide (30.9%)
	p-t-BuC$_6$H$_4$SH, PhSO$_2$Cl, CHCl$_3$	2-p-t-Butylphenylthiopyridine: 3-p-t-butylphenylthiopyridine (38:62, 36.1%)
	EtO$_2$CCH$_2$SH, PhSO$_2$Cl	Ethyl (2-pyridinethio)acetate (14.2%)
4-Me	PhSH, MeSO$_2$Cl, benzene	2-Phenylthio-4-picoline: 3-phenylthio-4-picoline (56:44; 41.0%); diphenyl disulfide (37.4%)
	PhSH, MeSO$_2$Cl, CHCl$_3$	2-Phenylthio-4-picoline: 3-phenylthio-4-picoline (54:46; 62.3%); diphenyl disulfide (19.8%)
	PhSH, PhSO$_2$Cl, benzene	2-Phenylthio-4-picoline: 3-phenylthio-4-picoline (67:33; 51.1%), diphenyl disulfide (51.1%)
	PhSH, PhSO$_2$Cl, CHCl$_3$	2-Phenylthio-4-picoline: 3-phenylthio-4-picoline (66:34; 72.3%), diphenyl disulfide (21.8%)
	p-ClC$_6$H$_4$SH, MeSO$_2$Cl, benzene	2-p-Chlorophenylthio-4-picoline: 3-p-chlorophenylthio-4-picoline (46:54; 49.2%); di-p-chlorophenyl disulfide (27.4%)
	p-ClC$_6$H$_4$SH, PhSO$_2$Cl, CHCl$_3$	2-p-Chlorophenylthio-4-picoline: 3-p-chlorophenylthio-4-picoline (44:56; 73.2%); di-p-chlorophenyl disulfide (19.3%)
	p-t-BuC$_6$H$_4$SH, MeSO$_2$Cl, CHCl$_3$	2-p-t-Butylphenylthio-4-picoline: 3-p-t-butylphenylthio-4-picoline (27:73; 55.5%)
	p-t-BuC$_6$H$_4$SH, PhSO$_2$Cl, benzene	2-p-t-Butylphenylthio-4-picoline: 3-p-t-butylphenylthio-4-picoline (38:62; 40.4%); di-p-t-butylphenyl disulfide (43.0%)
	p-t-BuC$_6$H$_4$SH, PhSO$_2$Cl, CHCl$_3$	2-p-t-Butylphenylthio-4-picoline: 3-p-t-butylphenylthio-4-picoline (34:66; 53.6%); di-p-t-butylphenyl disulfide (27.4%)
	EtO$_2$CCH$_2$SH, PhSO$_2$Cl	Ethyl (4-methyl-2-pyridinethio)acetate
t-Bu	PhSH, PhSO$_2$Cl, CHCl$_3$	2-Phenylthio-4-t-butylpyridine: 3-phenylthio-4-t-butylpyridine (57:43; 25.0%)
	p-ClC$_6$H$_4$SH, PhSO$_2$Cl, CHCl$_3$	2-p-Chlorophenylthio-4-t-butylpyridine: 3-p-chlorophenylthio-4-t-butylpyridine (49:51; 26.6%)

Table IV-50 (*Continued*)

Substituent in N-oxide	Conditions	Products (yield)
	p-t-BuC$_6$H$_4$SH, PhSO$_2$Cl, benzene	2-p-t-Butylphenylthio-4-t-butylpyridine: 3-p-t-butylphenylthio-4-t-butylpyridine (37:63; 34.8%); di-p-t-butylphenyl disulfide (60.2%)
	p-t-BuC$_6$H$_4$SH, PhSO$_2$Cl, CHCl$_3$	2-p-t-Butylphenylthio-4-t-butylpyridine: 3-p-t-butylphenylthio-4-t-butylpyridine (40:60, 28.9%); di-p-t-butylphenyl disulfide (58.7%)
2,6-Me$_2$	PhSH, PhSO$_2$Cl	3-Phenylthio-2,6-lutidine: 4-phenylthio-2,6-lutidine (82:18; 36%), diphenyl disulfide (56.8%)
	p-t-BuC$_6$H$_4$SH, PhSO$_2$Cl ·	3-p-t-Butylphenylthio-2,6-lutidine: 4-p-t-butyl-phenylthio-2,6-lutidine (83:17; 11.7%); di-p-t-butylphenyl disulfide (82%)
3,5-Me$_2$	p-t-BuC$_6$H$_4$SH, PhSO$_2$Cl	2-p-t-Butylphenylthio-3,5-lutidine (24.4%); di-p-t-butylphenyl disulfide (61.5%)
2,4,6-Me$_3$	PhSH, PhSO$_2$Cl	3-Phenylthio-2,4,6-collidine (9.5%)

alkali metal salts of diethylphosphonate to yield a mixture of diethyl 3-alkylpyridine-2-phosphonate (**IV-232**) and diethyl 3-alkylpyridine-6-phos-phonate (**IV-233**) (Table IV-51) (473). The observed isomer ratio is typical of nucleophilic aromatic substitutions in 3-methylpyridine derivatives (6). Reaction of diethyl sodiophosphonate on N-methoxy-2,6-dimethylpyri-dinium methosulfate gave 2,6-lutidine, 2,6-lutidine-1-oxide and diethyl

TABLE IV-51. Preparation of Dialkyl Pyridine-2-phosphonates (473)

Substituent in N-oxide	Conditions	Products (yield)
None	(i) Me$_2$SO$_4$ (ii) LiPO(OEt)$_2$, $-15°$, 1 hr	Diethyl pyridine-2-phosphonate, b.p. 105–122°/0.8 mm (67%)
2-Me	(i) Me$_2$SO$_4$ (ii) NaPO(OEt)$_2$, 10–20°, 70 min	Diethyl 2-picoline-6-phosphonate, b.p. 125–127°/0.1 mm (30%)
3-Me	(i) Me$_2$SO$_4$ (ii) NaPO(OEt)$_2$, 10–20°, 70 min	Diethyl 3-picoline-2-phosphonate, diethyl 3-picoline-6-phosphonate (ratio 6:1; 48%)
4-Me	(i) Me$_2$SO$_4$ (ii) NaPO(OEt)$_2$, 10–20°, 70 min	Diethyl 4-picoline-2-phosphonate, b.p. 109–112°/10.05 mm
2,6-Me$_2$	(i) Me$_2$SO$_4$ (ii) LiPO(OEt)$_2$, -5 to 0°, 1.25 hr (iii) 70°, 2 hr	2,6-Lutidine (47%), 2,6-lutidine-1-oxide (6%), diethyl 2,6-lutidine-4-phosphonate (24%)
3,5-Me$_2$	(i) Me$_2$SO$_4$ (ii) LiPO(OEt)$_2$, 10–20°, 70 min	Diethyl 3,5-lutidine-2-phosphonate (54%)
3,4-Me$_2$	(i) Me$_2$SO$_4$ (ii) LiPO(OEt)$_2$ $-15°$, 1 hr	Diethyl 3,4-lutidine-2-phosphonate, diethyl 3,4-lutidine-6-phosphonate (ratio 3:1; 47.5%)

IV-231   **IV-232**   **IV-233**

2,6-lutidine-4-phosphonate, the only example of 4-substitution, which is not
unexpected since the 2-positions are blocked in the substrate.

D. Reaction with Tosyl Chloride

When pyridine-1-oxide was heated with tosyl chloride at 200 to 205°
3-tosyloxypyridine was formed (474). This reaction was also subsequently
found to give 2,3'-dipyridyl ether, N-(2-pyridyl)-2-pyridone, N-(2-pyridyl)-
3-chloro-2-pyridone, and N-(2-pyridyl)-5-chloro-2-pyridone (475). Reinvestigation of the reaction using labeled $MeC_6H_4S^{18}O_2Cl$ suggested that
an intimate ion-pair mechanism was operating since the ^{18}O concentration
in the 3-hydroxypyridine formed was normal and excess ^{18}O was retained
in the p-$MeC_6H_4SO_3H$, both of which were obtained on hydrolysis of the
3-tosyloxypyridine (**IV-234**) (476). A 1,5-sigmatropic shift in the 1,2-dihydro
derivative could also explain the observations.

IV-234

When 2-picoline-1-oxide was heated with tosyl chloride, 2-chloromethyl-
pyridine was obtained in good yield (35, 477a, b).

R = 6-Me (72%)
R = 5-Et (70.6%)

Picoline acid-1-oxides, for example pyridine-2,4-dicarboxylic acid-1-oxide
and pyridine-2,5-dicarboxylic acid-1-oxide, underwent decarboxylation
to give 2-pyridones on heating with tosyl chloride (478).

Pyridine-1-oxide reacts with tosyl chloride in pyridine solution to give a mixture of 2- and 4-pyridylpyridinium chloride (479). 2- and 4-chloro-pyridine-1-oxide and 1-hydroxy-2- and 4-pyridone undergo related reactions

(480). The active methylene group in 2- and 4-picoline-1-oxide also gets involved in this reaction so that both ring and side-chain N-pyridinium salts are formed. In chloroform solution, only the ω-substituted compound is formed. The pyridinium salts react with amines to give the corresponding aminopyridines. The 4-alkylpyridinium salt **IV-235** reacts with aniline to give the expected product which immediately rearranges to 4-p-amino-benzylpyridine (**IV-236**) (481, 482) (eq. IV-37).

(IV-37)

E. Acid Anhydrides

The mechanism of the reaction of pyridine-1-oxide with acid anhydrides has been discussed previously (5–7, 483a), but only results obtained since 1958 are summarized here. It will also be more convenient to treat *both* nuclear and side-chain substitution under this heading (see also Chapter XII).

Reaction of 3-picoline-1-oxide with acetic anhydride was said to give only 3-methyl-2-pyridone (483b). Similarly, 3-fluoro-, 3-chloro-, and 3-bromopyridine-1-oxide (**IV-237**) gave exclusively the 3-halo-2-pyridone

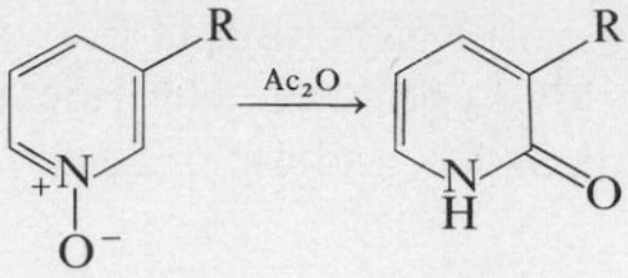

IV-237   **IV-238**

(**IV-238**) (55). 3-Nitro-2-pyridone was the only product of the reaction of 3-nitropyridine-1-oxide and acetic anhydride (117), but nicotinic acid-1-oxide and acetic anhydride gave 2-pyridone-3-carboxylic acid (18%), 2-pyridone-5-carboxylic acid (3%), and 2-acetylnicotinic acid-1-oxide (30%) (328). The formation of the 2-acetyl compound was attributed to intramolecular rearrangement of a mixed anhydride, as shown below (484) (eq. IV-38). Reinvestigation of the reaction with 3-picoline-1-oxide showed

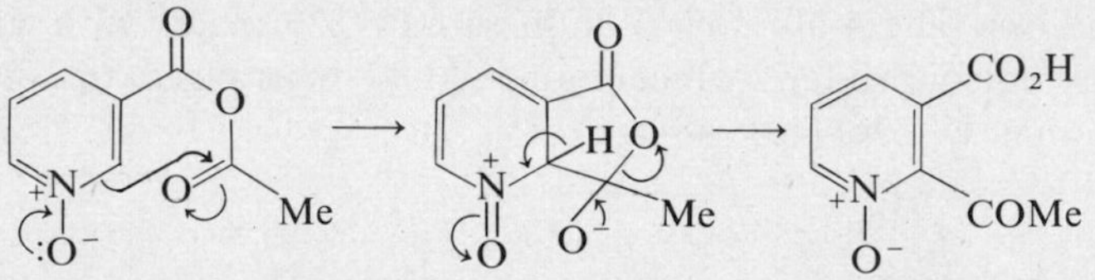

(IV-38)

that both 3- and 5-methyl-2-pyridone were obtained (35 to 40% each), as well as 1-(5-methyl-2-pyridyl)-3-methyl-2-pyridone (4%) (484) and 2-amino-3-picoline (485). 3-Hydroxypyridine-1-oxide was found to give only 3-hydroxy-2-pyridone (328). Methyl picolinate-1-oxide gives the 6-pyridone (34%) (96), and methyl isonicotinate-1-oxide forms the 2-pyridone (56%) (96, 486). Picolinic acid-1-oxide reacts with acetic anhydride in acetonitrile to yield 2-pyridone and pyridine-1-oxide accompanied by the quantitative evolution of carbon dioxide (328). In view of the side-chain acetoxylation observed in the reaction of 2-picoline-1-oxide (487) (see below), the reaction of 6-methylpicolinic acid-1-oxide was looked at. It was first reported that the reaction of 6-methylpicolinic acid-1-oxide gave 2-acetoxymethylpyridine (49). More recently, this structure was revised to 6-acetoxy-2-picoline (488). This was confirmed by hydrolysis of the product to 6-methyl-2-pyridone. It was concluded that decarboxylation, N—O bond fission, and acetoxylation at the 2-position must take place in concert. 6-Acetoxy-2-picoline can react further with acetic acid during the reaction or during distillation to give 6-methyl-2-pyridone (eq. IV-39) (488).

2- And 4-alkylpyridine-1-oxides react differently than the non-alkylated derivatives. 2-Picoline-1-oxide and acetic anhydride did give some 6-methyl-2-pyridone, but the major product was 2-pyridylmethyl acetate (483b, 489). It has also been reported that the reaction of 2-picoline-1-oxide with acetic anhydride gives three isomers: 3-acetoxy-2-picoline, 5-acetoxy-2-picoline,

$$(IV-39)$$

and 2-acetoxymethylpyridine in the ratio of 15:18:67, respectively (487). Thus side-chain oxidation occurs. Reaction of 4-picoline-1-oxide with acetic anhydride yields 4-picolyl acetate and 3-acetoxy-4-picoline (490, 491), together with a minor product, 1,2-di-(4-pyridyl)ethylene (492). A possible mechanism for the formation of **IV-239** from **IV-240** involves the intermediacy of the *N*-acetoxypyridinium salt (**IV-241**) followed by proton-abstraction from the side-chain by acetate and finally nucleophilic addition of acetate to the activated olefinic double bond in **IV-242** (313) (eq. IV-40). Structure **IV-241** was isolated as the perchlorate salt from the reaction of

$$(IV-40)$$

IV-240 with acetic anhydride in perchloric acid, and as the picrate from its reaction with picryl acetate. Treatment of the latter salt with triethylamine gave **IV-239** and triethylammonium picrate so that **IV-241** is an acceptable intermediate in these reactions (313).

Treatment of 1-acetoxy-2-benzylpyridinium perchlorate and 1-acetoxy-2-*p*-nitrobenzylpyridinium perchlorate with triethylamine in acetonitrile did not give the anhydro bases but instead gave the acetates. Treatment of 1-acetoxy-2-(α,α-dideuterobenzyl)pyridinium perchlorate (**IV-244**) with sodium acetate in acetic acid and acetonitrile lead to 50 % reaction: no loss of deuterium was observed in the rearranged ester **IV-245** and the *N*-oxide (**IV-243**) formed from **IV-244** by hydrolysis (313) (eq IV-41). The mechanism

of the reaction requires a slow rate-controlling conversion of **IV-241** to **IV-242** with a rapid rearrangement of **IV-242** to **IV-239**. This conclusion is consistent with the absence of accumulation of anhydro base (**IV-242**), the absence of deuterium exchange during the reaction, and the effect of base strength on the reaction.

Reaction of 2-picoline-1-oxide and phenylacetic anhydride in boiling benzene gave 2-pyridylmethyl phenylacetate (32.2 %), carbon dioxide (15 %), 2-β-phenylethylpyridine (15 %), and a mixture probably consisting of 3-benzyl- and 5-benzyl-2-picoline (3.3 %). 4-Picoline-1-oxide and phenylacetic anhydride yielded 4-pyridylmethyl phenylacetate and 3-phenylacetoxy-4-picoline (2.5 %), carbon dioxide (28.4 %), 4-β-phenylethylpyridine (6.5 %), 2-benzyl-4-picoline (1.9 %), and 3-benzyl-4-picoline (0.9 %) (493–495).

Evidence that the acetate free radical is the source of the carbon dioxide, methane, and methyl acetate was presented (490, 496). A 4-picolyl radical is apparently formed in the reaction of 4-picoline-1-oxide with acetic anhydride, which could account for the 4-ethylpyridine (0.6 %) formed. That it was not a key intermediate in the reaction, however, was shown by the fact that *m*-dinitrobenzene, a free radical scavenger, had little effect on the amount of 4-picolyl acetate formed, and that treatment of 4-picoline-1-oxide with butyric anhydride in the presence of sodium acetate gave 4-picolyl butyrate (and not acetate). On this basis, Traynelis and Martello (490) excluded a radical chain mechanism and proposed an intramolecular rearrangement of the intermediate anhydro base, with the possibility of a second allylic rearrangement to account for both 3-acetoxy-4-picoline and 4-picolyl acetate (eq. IV-42). Attempts to trap the intermediate in the reaction of 4-picoline-1-oxide with acetic anhydride in anisole gave a mixture of three picolylanisoles (20 %), in addition to the usual ester products (497). The *meta/para* ratio (0.25) of 4-picolylanisoles produced was consistent with a cationic attack of a ring bearing an *ortho-para* directing substituent, and would exclude an alkyl radical intermediate. When benzonitrile was used as solvent, the picolyl residue attacked the cyanide nitrogen exclusively to give **IV-246**, probably *via* **IV-247**. When a mixture (1:1) of anisole and benzonitrile was used as the solvent, essentially all of the substitution occurred in the anisole nucleus. These results indicate that the reaction of 4-picoline-1-oxide and acetic anhydride generates the picolyl cation in what is apparently a side reaction.

IV-247   IV-246

No reaction was observed between 2-picoline-1-oxide and either phenyl acetate or 2-chlorophenyl acetate (498). In the cases of 4-nitro-, 2,4-dinitro-, and 2,4,6-trinitrophenyl acetates, 2-pyridylmethyl acetate (5 to 43 %) and the respective phenols were obtained, but 2-(aryloxymethyl)pyridine was not.

The reaction of 2-picoline-1-oxides with acetic anhydride in chloroform solution proceeded more smoothly if an electron-donating group (rather than an electron-withdrawing group) was present at C-4 (116). The reaction rate constants agreed with a *pseudo* first-order reaction. The first step of the reaction is the addition of the acetyl group to the oxygen of the N-oxide, which is dependent on acetic anhydride concentration, followed by the rate-determining step, the cleavage of the nitrogen-oxygen bond (116, 499). An intermolecular mechanism involving the free ions (**IV-248**) or an inter-molecular or concerted intramolecular mechanism of the ion pairs (**IV-249**) would explain the observed kinetic results (499). Making use of ^{18}O-labeled

IV-249   IV-248

acetic anhydride, Oae and his co-workers have probed the mechanism of
the reaction of 2-picoline-1-oxide with acetic anhydride. They concluded
that a free-radical pathway (and not an intramolecular rearrangement or
an intermolecular ionic mechanism) was followed. Since the nature of the
products was independent of the presence or absence of solvent, a solvent
caged radical was proposed. Only half the label was found in the 2-hydroxy-
methylpyridine formed following hydrolysis of the product.

The large kinetic isotope effect found for the reaction of 2- and 4-picoline-
1-oxide and 2-benzylpyridine-1-oxide with acetic anhydride-^{18}O suggested
that proton removal is the rate-determining step (501). The uneven ^{18}O
distribution between the alcohol and carbonyl-O-atoms of the ester formed
appears to result from the conformational preference of the anhydro base
intermediate (501).

Intramolecular scrambling of the oxygen atoms of the acetoxyl group in
such reactions was confirmed (502). On the other hand, investigation of a
variety of 2-alkylpyridine-1-oxides gave results which were interpreted in
terms of an ion pair, rather than of a radical-pair, intermediate. 2-Cyclo-
pentylmethylpyridine-1-oxide and 2-neopentylpyridine-1-oxide gave products which were said to arise via a carbonium ion rearrangement (502, 503).
This agreed with the earlier observation that cleavage of the N—O bond
in the rearrangement of the anhydro base from the reaction of 2-picoline-1-
oxide with acetic anhydride was a heterolytic process since the yield of
carbon dioxide did not significantly increase on going from acetic to phenylacetic or to trichloroacetic anhydride (eq. IV-43) (504a).

(IV-43)

Recently, it has been reported that the reaction of 2-picoline-1-oxide with
acetic anhydride in benzene in the NMR probe at 70 to 130° gave intense
emission lines at 3.44 and 0.77 ppm due to the MeO-protons of MeOAc and

to the ethane also formed (504b). Since no emission or absorption due to 2-acetoxymethylpyridine was observed, it suggested that this product was not formed by a radical mechanism but by a [1,3]- or a [3,3]-sigmatropic shift (504b).

Examination of the ^{18}O content of 4-acetoxymethylpyridine and 3-acetoxy-4-picoline gave results consistent with an intermolecular mechanism involving nucleophilic attack by acetate ion (eq. IV-44) (501).

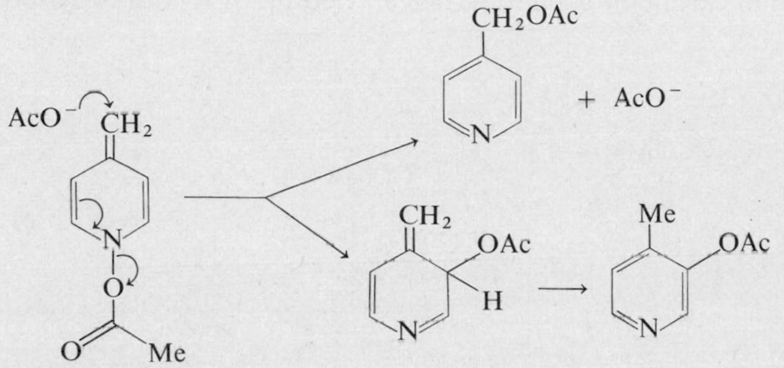

(IV-44)

The observation of chemically induced nuclear spin polarization emission in the reaction of 4-picoline-1-oxide with acetic anhydride gave direct evidence for the formation of radical pairs (504c). This evidence is in favor of the intermediacy of some free radicals in the formation of 4-acetoxymethyl-pyridine, 4-ethylpyridine, and methane. It was concluded that, together with the already established ^{18}O scrambling, intramolecularity of rearrangement in aromatic solvents and rate-determining deprotonation mechanism, the present findings are compatible with a dual mechanism in which the anhydro base cleaves to give both radical and ion pairs (504c).

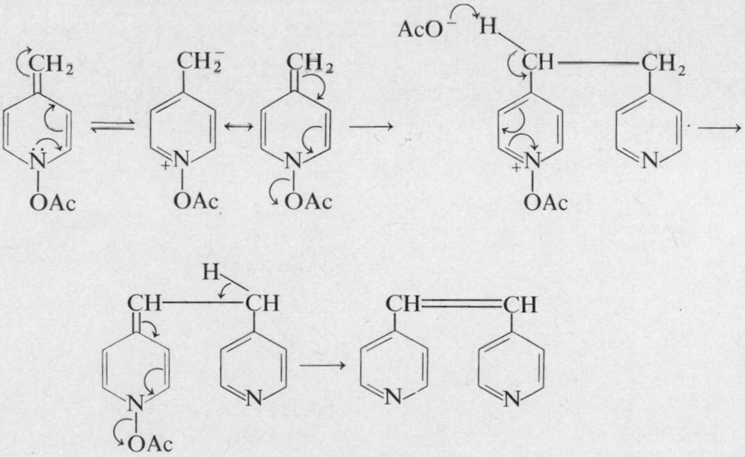

Scheme IV-12

1,2-**Bis**-(6-methyl-2-pyridyl)ethane was formed as a by-product in the reaction of 2,6-lutidine-1-oxide with acetic anhydride (68). Small amounts of the corresponding dipyridylethylene derivative were also formed from 2- and 4-picoline-1-oxide and could arise as outlined in Scheme IV-12 (505).

No appreciable reaction occurred between pyridine-1-oxide and trichloroacetic anhydride in boiling chloroform. When, however, acetonitrile was used as the solvent, an immediate exothermic reaction occurred at 0° and evolution of carbon dioxide took place. A complex mixture of products was

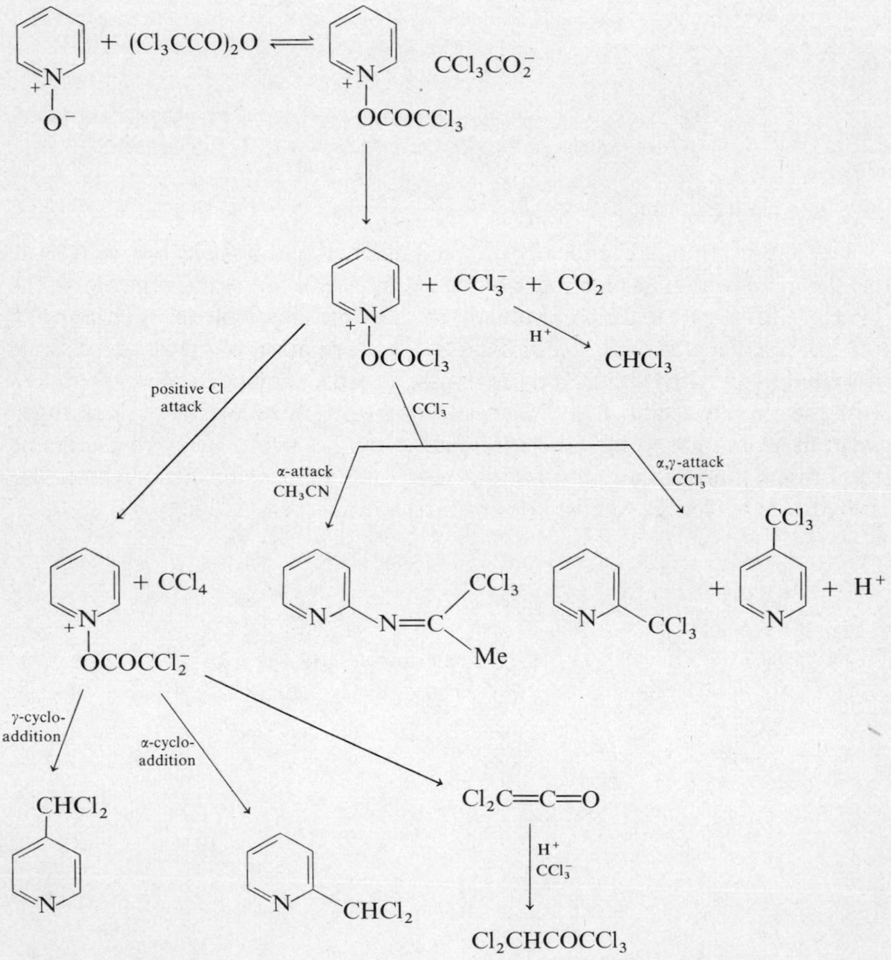

Scheme IV-13

obtained: CCl_4 (40%), $CHCl_3$ (20%), pentachloroacetone (20%), 2-trichloro-methylpyridine (10 to 20%), 4-trichloromethylpyridine (1 to 4%), 2-dichloro-methylpyridine (trace), 4-dichloromethylpyridine (5%), and 2-aminopyridine (4%), which has been formulated (506) as in Scheme IV-13. Homolytic cleavage of the N—O bond of the trichloroacetylated N-oxide does not occur since very little carbon dioxide is obtained when the N-oxide is treated with trichloroacetyl chloride. A possible explanation is that the driving force for the reaction is the exothermic loss of carbon dioxide from free trichloroacetate ion which is not complexed with its counterion. The increased polarity due to the presence of acetonitrile probably allows formation of free trichloroacetate. Radical and carbene mechanisms are also possible.

The reaction of $2\text{-}(X \cdot C_6H_4CH_2)C_5H_4NO$ (X = $p\text{-}NO_2$, $p\text{-}Cl$, H, $m\text{-}Me$, $p\text{-}Me$) with a large excess of acetic anhydride in dioxan at 30° was studied (507). With 2-benzylpyridine-1-oxide itself and its $\alpha,\alpha\text{-}d_2$ derivative, a large kinetic isotope effect was observed, confirming that the rate-determining step in the reaction is the deprotonation. In agreement with this is the rate enhancement by electron-withdrawing substituents in the benzyl group. If another anionic group is present in the reaction, substitution by the anionic

$$\text{(IV-45)}$$

R = H (26%)
R = Me (17%)

group may be observed (eq. IV-45) (508). The decomposition of N-acyloxy-pyridinium salts to give aldehydes and ketones was discussed in Section III.3.

The products obtained from the reactions of pyridine-1-oxides with acid anhydrides are given in Table IV-52.

Ketene has been used in the presence of a catalytic quantity of concentrated sulfuric acid instead of acetic anhydride (eq. IV-46) (Table IV-53) (517, 518).

IV-250   IV-251   IV-252

TABLE IV-52. Reaction of Pyridine-1-oxides with Acid Anhydrides

Substituent in N-oxide	Conditions	Products (yield)	Ref.
None	Ac_2O	2-Acetoxypyridine	499
	(i) $Ac^{18}O$, Δ, 8 hr (ii) $H_2^{18}O$	2-Pyridone, m.p. 104–105°, (no ^{18}O)	26
	Ac_2O	2-Pyridone, m.p. 106–107°, pyridine hydrochloride, 2-aminopyridine, 2,4-dipyridyl; picrate, m.p. 210–213°; 1-(2-pyridyl)-2-pyridone, m.p. 50–52°; picrate, m.p. 116–118°	509
	Ac_2O	2-Pyridone, 2-acetoxypyridine (60%), 3-acetoxypyridine, 2-acetylpyridine, 2-aminopyridine, 1-(2-pyridyl)-2-pyridone	510
2-Me	Ac_2O	2-Pyridylmethyl acetate	496
	Ac_2O	2-Pyridylmethyl acetate (60%), 3-acetoxy-2-picoline, 5-acetoxy-2-picoline (ratio 67:15:18)	487
	$(MeC^{18}O)_2O$, 140°, (0.78 atom $\%^{18}O$)	2-Pyridylmethyl acetate (0.50 atom $\%^{18}O$)	500
	Ac_2O	2-Pyridylmethanol, 6-methyl-2-pyridone	511
	$(PhCH_2CO)_2O$	2-Pyridylmethyl phenylacetate, 2-phenethylpyridine (1.5:1)	495
2,6-Me$_2$	Ac_2O, 100°, 10 hr	6-Acetoxymethyl-3-picoline, b.p. 110–114°/15 mm; picrate, m.p. 110–112°	41
2,6-Me$_2$	(i) Ac_2O, Δ (ii) 10% HCl, Δ	6-Hydroxymethyl-2-picoline, 3-hydroxy-2,6-lutidine, 1,2-di-(6-methyl-2-pyridyl) ethane	68
2,6-Me$_2$-3-OH	(i) Ac_2O, 100°, 3 hr (ii) HCl	3-Hydroxy-6-hydroxymethyl-2-picoline, m.p. 153.5°; picrate, m.p. 190–191°, 5-hydroxy-6-hydroxymethyl-2-picoline	68
2,6-Me$_2$-4-CH(OH)Ph	(i) Ac_2O, Δ (ii) 10% HCl	4-(α-Hydroxybenzyl)-6-hydroxymethyl-2-picoline (20%), 4-benzoyl-2,6-lutidine (65%)	512
2,6-Me$_2$-4-CN	(i) Ac_2O, Δ (ii) 10% HCl	4-Cyano-6-hydroxymethyl-2-picoline (74.5%)	116, 513
2,6-Me$_2$-4-NH$_2$	(i) Ac_2O, 120–130° (ii) KOH, EtOH, Δ	4-Acetamido-2,6-lutidine (50%)	116, 513
2,5-Me$_2$	Ac_2O, 70–80°, 2 hr	6-Acetoxy-3-picoline, b.p. 133–136°/14 mm, n_D^{20} 1.4952; picrate, m.p. 122–125°	32
2,4-Me$_2$	Ac_2O, Δ, 15 min	2-Acetoxy-4-picoline, b.p. 95–101°/5 mm	459
2-Me-3-n-Pr	Ac_2O	2-Acetoxymethyl-3-n-propylpyridine, b.p. 131–132°/13 mm (66.9%)	41
2-Me-3-n-Bu	Ac_2O	2-Acetoxymethyl-3-n-butylpyridine, b.p. 135–136°/9 mm (60.8%)	41
2-Me-3-n-Amyl	Ac_2O	2-Acetoxymethyl-3-n-amylpyridine, b.p. 150–152°/11 mm (78.2%)	41
2-Me-3-n-hexyl	Ac_2O	2-Acetoxymethyl-3-n-hexylpyridine, (74.1%)	41

Table IV-52 (*Continued*)

Substituent in *N*-oxide	Conditions	Products (yield)	Ref.
2-Me-6-CH$_2$OH	(i) Ac$_2$O, Δ (ii) 10% HCl	2,6-Di-(hydroxymethyl)pyridine, 6-methylpyridine-2-aldehyde, 5-hydroxy-6-hydroxymethyl-2-picoline	68
2-Me-6-CH$_2$OH	Ac$_2$O, Δ	2,6-Di-(hydroxymethyl)pyridine (79.5%), 5-acetoxy-6-acetoxymethyl-2-picoline (18.5%), 3-acetoxy-2-acetoxymethyl-2-picoline (20%), 1,2-di-(6-acetoxymethyl-2-pyridyl)ethylene	505
2-Me-5-CH$_2$OH	Ac$_2$O	2,5-Di-(acetoxymethyl)pyridine	86
2-Me-3-CH$_2$OMe	Ac$_2$O	2-Acetoxymethyl-3-methoxymethylpyridine	87
2-Me-3-OAc-4,5-(CH$_2$OAc)$_2$	Ac$_2$O, Δ	3-Acetoxy-2,4,5-triacetoxymethylpyridine	73
2,4,5-(CH$_2$OH)$_3$-3-OH	Ac$_2$O, 110–120°, 6 hr	3-Acetoxy-2,4,5-triacetoxymethylpyridine	73
2-Me-6-Cl	Ac$_2$O, Δ, 7 hr	6-Methyl-2-pyridone, 6-chloro-2-hydroxymethylpyridine, 3- or 5-hydroxy-6-chloro-2-picoline	358
2-Me-4-NH$_2$	Ac$_2$O, 120–130°	2-Acetoxymethyl-4-acetamidopyridine (58.7%)	513
2-Me-3-CH$_2$CO$_2$Et	Ac$_2$O, Δ	Ethyl 2-acetoxymethyl-3-pyridylacetate	77
2-Me-6-CO$_2$H	Ac$_2$O	1-(6-Methyl-2-pyridyl)-6-methyl-2-pyridone, m.p. 202°	512
2-Me-3-CO$_2$Et	Ac$_2$O	Ethyl 2-acetoxymethylnicotinate (60.5%); ethyl 5-acetoxymethylnicotinate (trace)	77
	(i) Ac$_2$O (ii) 25% HCl, 110–120°, 3 hr	6-Hydroxy-2-methylnicotinic acid, m.p. >300°	107
2,6-Me$_2$-3-CO$_2$Et	Ac$_2$O, 90°	Ethyl 2-acetoxy-6-methylnicotinate, m.p. 52–54°	107
2-Me-4-CO$_2$Et	Ac$_2$O, 125°, 3 hr	Ethyl 2-acetoxymethylisonicotinate, b.p. 134–136°/2.5 mm (57.4%)	110
2-Me-3-CONHCH$_2$Ph	Ac$_2$O, Δ	2-Acetoxymethyl-*N*-benzylnicotinamide	77
2,6-Et$_2$	Ac$_2$O, 100°, 10 hr	2-Ethyl-6-(α-acetoxyethyl)pyridine, b.p. 123–125°/5 mm (75%)	41
2-Et-4-CO$_2$Me	Ac$_2$O, 125°, 3 hr	Methyl 2-(α-acetoxyethyl)isonicotinate, b.p. 142–147°/6 mm (61%)	110
2,6-Pr$_2$	Ac$_2$O, 100°	2-Propyl-6-(α-acetoxypropyl)pyridine, b.p. 122–124°/10 mm (92.9%)	41
2-Pr-4-CO$_2$Me	Ac$_2$O, 125°, 3 hr	Methyl 2-(α-acetoxypropyl)isonicotinate, b.p. 155–162°/9 mm (66%)	110
2,6-Bu$_2$	Ac$_2$O, 100°	2-Butyl-6-(α-acetoxybutyl)pyridine, b.p. 146–149°/10 mm (90%)	41
2,6-*n*-Amyl	Ac$_2$O, 100°	2-*n*-Amyl-6-(α-acetoxy-*n*-amyl)pyridine, b.p. 170–172°/4 mm (88.4%)	41

Table IV-52 (*Continued*)

Substituent in N-oxide	Conditions	Products (yield)	Ref.
2-CH$_2$CH$_2$CH$_2$OH	Ac$_2$O, Δ, 10 hr	2-(1,3-Diacetoxypropyl)pyridine	84
2-CH=CHPh	(i) Ac$_2$O, Δ (ii) 10% HCl	1,2-Di-(2-pyridyl)glycol (39%), 3-hydroxy-2-styrylpyridine (8%), 5-hydroxy-2-styrylpyridine (5%), 2-styrylpyridine (6.5%)	136
2-CH=CHMe	(i) Ac$_2$O, Δ, 3 hr, hydroquinone (ii) NaOH	2-(CH=CHMe)pyridine-1-oxide, 2-[1-(1,2-dihydroxy)propyl]pyridine, 3- (or 5-) hydroxy-2-(CH=CHMe)pyridine	136
2-Cl	Ac$_2$O, Δ	1-Hydroxy-2-pyridone (71%)	110
2,3,4,5-Cl$_4$	Ac$_2$O, 95%, AcOH, Δ, 1 hr	3,4,5-Trichloro-2-pyridone	176
2-OEt	Ac$_2$O, Δ	1-Hydroxy-2-pyridone (61%)	358
2-OPh	Ac$_2$O, Δ	1-Hydroxy-2-pyridone (84%)	358
2-CO$_2$H	Ac$_2$O, 50°	Pyridine-1-oxide, 2-pyridone, 1-(2-pyridyl)-2-pyridone	96, 358, 510
	Ac$_2$O, MeCN, 75–80°, 1 hr	Pyridine-1-oxide; picrate, m.p. 175–178°; 2-acetoxypyridine, styphnate; m.p. 179–180°	96
	Ac$_2$O, MeCN, 160°	Pyridine-1-oxide (60%), 2-acetoxypyridine (18%)	96
2-CO$_2$H-5- CO$_2$Me	Ac$_2$O, 45°	Methyl nicotinate	514
2-CO$_2$Me	Ac$_2$O, Δ	6-Methoxycarbonyl-2-pyridone (34%)	96
3-Me	Ac$_2$O	3-Methyl-2-pyridone (35–40%), 5-methyl-2- pyridone (35–40%), 1-(5-methyl-2-pyridyl)- 3-methyl-2-pyridone (4%)	484
	Ac$_2$O, 140–150°, 4 hr	3-Picoline; picrate, m.p. 145–147°; 1-(5- methyl-2-pyridyl)-3-methyl-2-pyridone, m.p. 107–108.5°, 2-amino-3-picoline, 5-methyl-2-pyridone, m.p. 185–186°; picrate, m.p. 146–147.5°; 3-methyl-2- pyridone	485
3,5-Me$_2$	Ac$_2$O, 160°, 2.5 hr	3,5-Lutidine-1-oxide, 3,5-dimethyl-2-pyridone, m.p. 118.5–119.5°, 1-(3,5-dimethyl-2- pyridyl)-3,5-dimethyl-2-pyridone, m.p. 104–105°	485
5-Me-2-CH$_2$OAc	Ac$_2$O	6-Diacetoxymethyl-3-picoline	32
3-F	Ac$_2$O, Δ	2-Acetoxy-3-fluoropyridine (65%)	55
3-Cl	Ac$_2$O, Δ, 4 hr	2-Acetoxy-3-chloropyridine (61%)	55
3-Br	Ac$_2$O, Δ	2-Acetoxy-3-bromopyridine (50%)	55
3-NO$_2$	Ac$_2$O, Δ, 24 hr	3-Nitro-2-pyridone	514
3-SiMe$_3$	Ac$_2$O	3-Trimethylsilyl-2-pyridone, 5-trimethylsilyl- 2-pyridone	46

Table IV-52 (*Continued*)

Substituent in N-oxide	Conditions	Products (yield)	Ref.
3-CO$_2$H	Ac$_2$O, Δ, 6 hr	2-Acetoxynicotinic acid-1-oxide	500
	Ac$_2$O, Δ	2-Acetylnicotinic acid-1-oxide (30%), 2-pyridone-5-carboxylic acid (3%), 2-pyridone-3-carboxylic acid (18%)	484
	(EtCO)$_2$O, Δ	2-Pyridone-3-carboxylic acid (18%), 2-pyridone-5-carboxylic acid (2%), 5,7-dioxo-6-methylcyclopenteno[*b*]pyridine-1-oxide (8%)	484
3-CO$_2$Me	Ac$_2$O, Δ	3-Methoxycarbonyl-2-pyridone (28%), 5-methoxycarbonyl-2-pyridone (16%)	96
3-CO$_2$Et	Ac$_2$O, 12 hr	2-Ethoxycarbonyl-2-pyridone, m.p. 143–144°	96
4-Me	Ac$_2$O	4-Acetoxymethylpyridine, b.p. 80–100°/6 mm	515
	Ac$_2$O, AcOH, 40 min, 140–150°, dry N$_2$ atm	4-Pyridylmethyl acetate (88%) ⎱ (65%) 3-Acetoxy-4-picoline (12%) ⎰ 4-Picoline (2.9%), 2,4-lutidine (0.2%), 4-ethylpyridine (0.6%)	490
	Ac$_2$O, PhNO$_2$	4-Hydroxymethylpyridine (89%) ⎱ (30%) 3-Acetoxy-4-picoline (11%) ⎰	490
	Ac$_2$O, styrene	4-Hydroxymethylpyridine (89%) ⎱ 42% 3-Acetoxy-4-picoline (11%) ⎰ 30% Polystyrene, 10% unreacted styrene	490
	Ac$_2$O, Δ	4-Pyridylmethyl acetate (64.2%), 3-acetoxy-4-picoline (35.8%), 1,2-bis(4-pyridyl)ethylene (ca. 13%)	505
	(MeC^{18}O)$_2$^{18}O, Δ	3-Acetoxy-4-picoline (33%), (0.71 atm % ^{18}O), 4-pyridylmethyl acetate (67%), (0.71 atm % ^{18}O)	500
	Ac$_2$^{18}O	3-Acetoxy-4-picoline (30%), 4-pyridylmethyl acetate (70%)	516
	(PhCH$_2$CO)$_2$O	4-Pyridylmethyl phenylacetate: 4-phenethylpyridine (0.06:1)	495
4-Me-2-CH$_2$OAc	(i) Ac$_2$O (ii) HCl	4-Picoline-2-carboxaldehyde	459
4-Et	(i) Ac$_2$O, Δ (ii) 10% HCl	4-(α-Hydroxyethyl)pyridine	486
4-CO$_2$H	Ac$_2$O, Δ	Isonicotinic acid (22%), 2-pyridone-4-carboxylic acid (7%)	328, 484
4-CO$_2$Me	Ac$_2$O, Δ	4-Methoxycarbonyl-2-pyridone (56%)	96
4-CO$_2$Et	Ac$_2$O, 6 hr	4-Ethoxycarbonyl-2-pyridone, m.p. 211–212°	96
4-Me-2-Cl	Ac$_2$O	2-Chloro-4-hydroxymethylpyridine, 3- or 5-hydroxy-2-chloro-4-picoline, 1-hydroxy-4-methyl-2-pyridone	358

141

TABLE IV-53. Reaction of Pyridine-1-oxides with Ketene (517, 518)

N-oxide	Product (yield %)			Recovered N-oxide
	IV-250	IV-251	IV-252	
2-Me	50–60	Trace	—	—
2,6-Me$_2$	33	17	4	—
3-Me	—	—	3	10
4-Me	5	0.3	4	—
4-NO$_2$	—	—	—	91.3
4-NO$_2$-2,6-Me$_2$	8	2^a	3	41

^{a} 4-Acetoxy-6-acetoxymethyl-2-picoline.

A number of mechanisms have been considered for the pyridine-1-oxide-mediated oxidative decarboxylation of acyl residues containing α-H atoms, for example, the reaction of diphenylacetic anhydride with pyridine-1-oxide (519a). The favored mechanism involves an "α-lactone-like" (IV-253) intermediate.

The reaction of pyridine-1-oxide and diphenylketene under aprotic conditions might be expected to proceed by initial formation of IV-253 and have subsequent steps in common with the anhydride reaction. The reaction of diphenylketene was studied in two ways: (a) by addition of solutions of purified ketene to pyridine-1-oxide solutions, and (b) by the in situ formation of the ketene from the acid chloride through the action of triethylamine without removal of the triethylammonium chloride. At room temperature, the reaction gave high yields of carbon dioxide and low yields of ketone (benzophenone) (Table IV-54). When the reaction was carried out by the slow addition of diphenylacetic anhydride to the hot N-oxide, the results were similar to those obtained with the ketene. When the ketene was formed

TABLE IV-54. Reaction of Pyridine-1-oxides with Ketenes (519a)

N-oxide	Ketene	Ratio of N-oxide to ketene	Conditions	Products (moles/mole ketene)
Pyridine (0.5M)	Diphenyl (purified)	3	Benzene, Δ	CO_2 (0.70); benzophenone, (0.46); polymer
Pyridine (0.8M)	Diphenyl (purified)	4	Benzene, R.T.	CO_2 (0.46); benzophenone
Pyridine (0.7M)	Diphenyl (purified)	2	Acetonitrile, Δ	CO_2 (0.63); benzophenone (0.46); polymer
Pyridine (ref. 519b)	Dimethyl		Benzene, 2 days	Pyridine, 4-isopropylpyridine, acetone, bicyclic compound (**IV-255** or **IV-256**)
4-Picoline (1.4M)	Diphenyl (purified)	4	Acetonitrile, Δ	CO_2 (0.84); benzophenone (0.48); polymer
4-Picoline (1.4M)	Diphenyl (purified)	2	Acetonitrile, R.T.	CO_2 (0.60); benzophenone (0.50); polymer
4-Picoline (1.2M)	Diphenyl (in situ)	1	Acetonitrile, R.T.	CO_2 (0.15); benzophenone (0.25); picolinium salt (0.35)
Pyridine (1.9M)	Diphenyl (in situ)	2	Acetonitrile, R.T.	CO_2 (0.82); benzophenone (0.36); picolinium salt (0.19)
4-Picoline (0.6M) pyridine	Diphenyl (in situ)	3	Acetonitrile, R.T.	CO_2 (0.23); benzophenone (0.34); picolinium salt (0.05); pyridinium salt (0.20)
4-Picoline (0.4M)	Fluorenyl (purified)	8	Acetonitrile R.T.	CO_2 (0.61); fluorenone (0.31); bis-biphenylene succinic anhydride
4-Picoline (0.7M)	Fluorenyl (purified)	6	Acetonitrile, R.T.	CO_2 (0.67); fluorenone (0.45); bis-biphenylene succinic anhydride

in situ, a low yield of isolable material was obtained (Table IV-55) and in addition 1-benzhydryl picolinium and pyridinium salts were isolated. A common ylid (**IV-254**) was suggested. In the presence of pyridine, both picolinium and pyridinium salts were isolated, and the relative yields depended on the relative amounts of pyridine to 4-picoline-1-oxide which were initially present. Since pyridine does not react with benzhydryl chloride

TABLE IV-55. Reaction of 4-Picoline-1-oxide with Acid Chlorides in Acetonitrile (519a)

Amine	Acid chloride	Ratio of acid chloride to N-oxide	Products (moles/mole)
Triethylamine	Diphenylacetyl	1.05	CO_2 (0.52); benzophenone (0.33); picolinium salt (0.065); diphenylacetic acid (0.11–0.20 equiv.)
Triethylamine	Diphenylacetyl	2.07	CO_2 (0.52); benzophenone (0.39); picolinium salt (0.068); diphenylacetic acid (0.11–0.20 equiv.)
Triethylamine	Diphenylacetyl	4.19	CO_2 (0.69); benzophenone (0.66); picolinium salt (0.030); diphenylacetic acid (0.11–0.20 equiv.)
Triethylamine	Diphenylacetyl	9.00	CO_2 (0.69); benzophenone (0.65)
Pyridine	Diphenylacetyl	0.97	CO_2 (0.56); benzophenone (0.21); picolinium salt (0.065); N-benzhydrylpyridinium chloride (0.26)
Pyridine	Diphenylacetyl	1.88	CO_2 (0.63); benzophenone (0.50); picolinium salt (0.093); N-benzhydrylpyridinium chloride (0.11)
Pyridine	Diphenylacetyl	3.87	CO_2 (0.60); benzophenone (0.62); picolinium salt (0.062); N-benzhydrylpyridinium chloride (0.06)
Pyridine	Diphenylacetyl	10.68	CO_2 (0.47); benzophenone, (0.61)
None	Diphenylacetyl	3.0	CO_2 (0.78); benzophenone (0.57)
Triethylamine	9-Fluorenyl	10.0	Fluorenone (0.72)
none	9-Fluorenyl	8.0	CO_2 (0.64); fluorenone (0.46)
Triethylamine	2,3-Diphenyl-cyclopropenyl	2.29	CO_2 (0.51); ketone (nil), 2,3-diphenyl-propenal, bis-(2,3-diphenyl)cyclopropenyl ether

rapidly enough to account for the formation of the pyridinium salt, inter-mediates such as **IV-254** are postulated. The reaction of biphenylene ketene

IV-254

with pyridine-1-oxide gave the ylid directly and a reduction in polymer formation was observed. A new product was isolated and identified as bis-biphenylene succinic anhydride (519a).

The data in Table IV-54 and IV-55 indicate that an intermediate is formed in all of these reactions which is capable of reacting with both free amine and amine-1-oxide. This accounts for the less than quantitative yields of ketones formed.

Carbon dioxide was slowly evolved from a cold solution of diphenylacetyl chloride and pyridine-1-oxide in acetonitrile to give an N-acetoxonium chloride salt ($v_{C=O}$ 1830 cm^{-1}) (519a). Treatment of these solutions with triethylamine resulted in evolution of carbon dioxide and products similar to those obtained from the reaction of the *in situ* formed diphenyl ketene with pyridine-1-oxide were isolated. In the presence of pyridine, the reaction was slower but similar products were isolated (Table IV-55). From these results, it was observed that the yield of ketone increases with increasing N-oxide concentration and the yield of the pyridinium salts decreases. When fluorene-9-carbonyl chloride was added to the pyridine-1-oxide, the reaction proceeded rapidly without the addition of a catalyst to give products similar to those observed with the corresponding ketene. In the case of the reaction of 2,3-diphenylpropene-1-carbonyl chloride under these conditions, the rate of gas evolution did not increase when triethylamine was added. No ketone was obtained; instead, the ether and aldehyde which result from the solvolysis of the corresponding cyclopropenoyl chloride were isolated (519a) (eq. IV-47). Reaction with ^{18}O-enriched diphenyl ketene gave carbon dioxide in which one oxygen atom came from the ketene and one from the

(IV-47)

N-oxide; the ketone arose solely by oxygen transfer from the N-oxide to the α-carbon of the ketene (Table IV-56).

A dipolar addition-elimination process could actually be involved in the formation of the ketone and ylid (eqs. IV-48 and IV-49) (519a).

This reaction appears to be different from the anhydride-N-oxide reaction in that very little ylid or products derived from it appear to be formed when the anhydride is present in appreciable concentration (519a).

TABLE IV-56. Oxygen-18 distribution[a] in Reactions with 4-Picoline-1-oxide (519a)

Compound		CO_2	Benzophenone
Diphenylacetyl chloride	0.788[a]	1.07	0.208
Diphenylketene	0.788[a]	1.06	0.209

[a] Atom % ^{18}O per molecule.

$$\longrightarrow \qquad \longrightarrow \quad + CO_2 + RCOR \qquad (\text{IV-48})$$

$$\longrightarrow \qquad \longrightarrow \qquad\qquad\qquad (\text{IV-49})$$

The reaction of pyridine-1-oxide with dimethylketene in benzene solution gave a mixture of products: pyridine, 4-isopropylpyridine, acetone, and a compound with empirical formula $C_{13}H_{17}NO_4$ to which was assigned structure **IV-255** or **IV-256** (519b). Scheme IV-14 was proposed to account for the formation of **IV-255** or **IV-256**. 4-Isopropylpyridine (**IV-259**) could arise *via* **IV-257** and **IV-258** either from the hydrolysis, 1,4-elimination and decarboxylation of **IV-255** or **IV-256** or by spontaneous decomposition of a possible 1:1 adduct (**IV-260**) (Scheme IV-14). Reaction of dimethylketene with 2-picoline-1-oxide, 4-picoline-1-oxide, and 2,6-lutidine-1-oxide gave the alkylpyridine as the only characterizable product; numerous other products were formed in low yield (519b).

3. Homolytic Substitution

Homolytic substitution of pyridine-1-oxide has not been studied systematically. Homolytic arylation of pyridine-1-oxide with benzenediazonium tetrafluoroborate in homogeneous solution was reported (520). After deoxygenation of the *N*-oxides with thiourea, the observed ratio of the isomeric 2-, 3-, and 4-phenylpyridines was 66.2:2.5:31.3. In contrast, values of 76.4:8.3:15.3 were given for the phenylation of pyridine-1-oxide using phenyldiazoaminobenzene at 131° (521a). Recently, phenylation of pyridine-1-oxide using diazoaminobenzene at 131° was performed to obtain the total rate ratio (Table IV-57) (521b). Using electrolytic reduction of benzenediazonium tetrafluoroborate in acetonitrile to phenylate pyridine-1-oxide at 0°, relatively high yields of phenylated pyridine-1-oxide (ca. 35%) have been obtained. Competitive reactions at 0° gave total and partial rates close

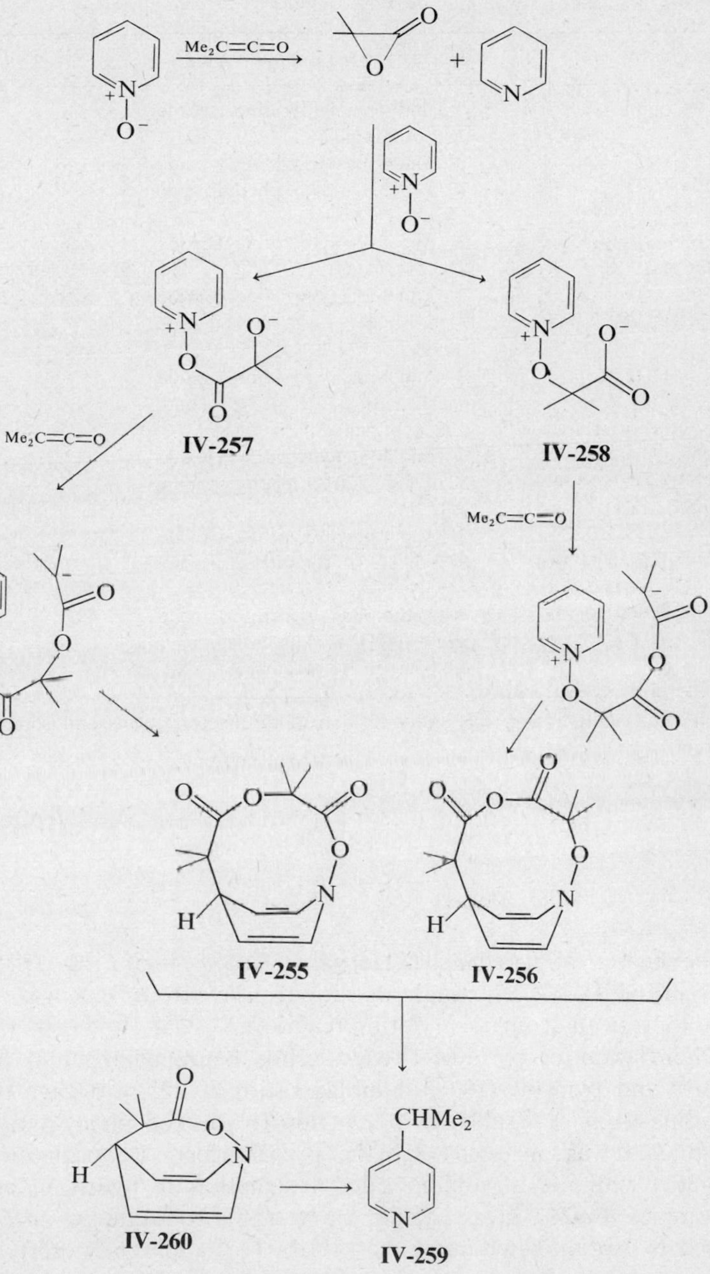
Me2C=C=O
Me2C=C=O
Me2C=C=O
IV-257
IV-258
IV-255
IV-256
IV-260
IV-259
CHMe2

Scheme IV-14

TABLE IV-57. Radical Phenylation of Pyridine-1-oxides[a]

Reaction conditions	Products (isomer ratio and yield)	$\overset{+}{\underset{C-H}{N-O}}K^b$	$F_r{}^c$	Ref.
$PhNHN_2Ph + Py^+\text{-}O^-$, 131°	2-, 3-, And 4-phenylpyridine-1-oxide (76.2:7.8:16) (27%)			521a
$PhNHN_2Ph + Py^+\text{-}O^-$, 181°	2-, 3-, And 4-phenylpyridine-1-oxide (79:7.7:13.3) (28%); phenylpyridines (6%)			521a
$PhNHN_2Ph + Py^+\text{-}O^-$ ($+ PhOCH_3$), 131°	2-, 3-, And 4-phenylpyridine-1-oxide (82:4:14) (45%)	20.6	2-, 2.5 3-, 7.3	521b
$PhNHN_2Ph + Py^+\text{-}O^-$ ($+ PhOCH_3$), 131°	2-, 3-, And 4-phenylpyridine-1-oxide (82:4:14) (45%)	16.2	2-, 39.85 3-, 1.95 4-, 13.6	521b
$PhN_2BF_4 + C_5H_5N +$ $Py^+\text{-}O^-$ (1:1) $+$ CH_3CN, R.T. $-60°$	2-, 3-, And 4-phenylpyridine-1-oxide (66.2:2.5:31.3) (0.9%); 2-, 3-, and 4-phenylpyridine (65.5:11.8:22.7) (6%)			520
$PhN_2BF_4 + C_5H_5N +$ $Py^+\text{-}O^-$ (0.02:0.04:1.0) on Hg pool in CH_3CN, 20°	2-, 3-, And 4-phenylpyridine-1-oxide (87: <1:12) (20%); phenylpyridines (<1%)			521b
$PhN_2BF_4 + C_5H_5N +$ $Py^+\text{-}O^-$ (0.02:0.04:1.0) on Hg pool in CH_3CN	2-, 3-, And 4-phenylpyridine-1-oxide (87: <1:12) (20%); phenylpyridines (<1%)			521b
Elect. redn. of PhN_2BF_4 in $Py^+\text{-}O^-$, CH_3CN ($+ PhH$), 0°	2-, 3-, And 4-phenylpyridine-1-oxide (89: <1:10) (35%) (from isomer ratio and competitive reactions)	52	2-, 139 3-, 1.5 4-, 31.2	521b

[a] Experimental isomer ratios, total rate ratios, partial rate factors, yields, and isomer ratios.

[b] $\overset{+}{\underset{C-H}{N-O}}K$ = total rate ratio of pyridine-1-oxide relative to benzene.

[c] Calculated from isomer ratios and $\overset{+}{\underset{C-H}{N-O}}K$.

to the theoretically predicted (521c) values ($\overset{+}{\underset{C-H}{N-O}}K$, 48.5; F_r: 2-, 134; 3-, 1.2; 4-, 19.7) (Table IV-57). The total rate ratio (relative to benzene was remarkably high (52) for an aromatic arylation reaction.

Phenylation of pyridine-1-oxide using benzenediazonium tetrafluoroborate and pyridine (100:2:4 molar ratio) at 20° also gave the phenylpyridine-1-oxides (Table IV-57). Only traces of phenylpyridines were obtained. It was suggested (521b) that pyridine forms a coordination complex (IV-261) with the diazonium salt. Abstraction of hydrogen atoms from σ-complex IV-262 gives pyridinium tetrafluoroborate as originally proposed by Abramovitch and Saha (521d). Hg° apparently exerts a catalytic effect by facilitating the one-electron transfer from the nitrogen lone pair. Hence the rate of decomposition is increased when mercury is present.

$$ArN_2^+BF_4^- + n Py \longrightarrow [ArN_2(Py)n]^+BF_4^- \longrightarrow$$

IV-261

$$Py^{\cdot\,+} + ArN_2^{\cdot}$$

$$\downarrow$$

$$Ar^{\cdot} + N_2$$

$$Ar{-}Ar' \longleftarrow \quad \Big\downarrow Ar'H$$

$$Ar'H \equiv Py^+{-}O^-$$

$$[Ar{-}Ar'H]^{\cdot}$$

IV-262

4. Nuclear Proton Abstraction

Base-catalyzed hydrogen-deuterium exchange of 3-bromopyridine-1-oxide in NaOD—D_2O showed that the rate of exchange for 3-bromo-pyridine-1-oxide to be in the order of 2- > 6- > 4- ≫ 5- (eq. IV-50) (522). The *pseudo* first-order rate constants for 3-bromopyridine-1-oxide were $k_{H-2}^{5^\circ} = 1.7 \times 10^{-4}$ sec^{-1}; $k_{H-6}^{50^\circ} = 3.9 \times 10^{-5}$ sec^{-1}; $k_{H-4}^{50^\circ} = 4.6 \times 10^{-6}$ sec^{-1} (522). A similar study (523) with 3-chloropyridine-1-oxide showed the

(IV-50)

same order of reactivity, as contrasted with the rates of exchange of pyridine in strong base (4 > 3 > 2) (524). The rates of exchange for deuterated pyridine-1-oxide, 3-chloropyridine-1-oxide, and 3,5-dichloropyridine-1-oxide in MeONa—MeOH were reported (525). At 138°, the relative rates for pyridine-1-oxide are 2.6-:3.5-:4 = 1,500:10:1.0. At 50°, the relative rates of exchange for 3-chloropyridine-1-oxide are 2-:6-:4- = 1,840:12.2:0.37 and for 3,5-dichloropyridine-1-oxide, 2,6-:4- = 11,800:1,370, all corrected statistically. The kinetic data for nuclear H—D exchange in 1-methylpyridinium chloride (**IV-263**) and pyridine-1-oxide in $D_2PO_4^- = DPO_4^{-2}$ buffer indicated that the rates of deprotonation of these molecules are qualitatively similar to that for the decarboxylation of 1-methylpyridinium carboxylate (**IV-264**) (526). These results indicate that a carbanion intermediate is involved in the base-catalyzed H—D exchange in these

IV-263   IV-264

compounds and in the decarboxylation of 1-methylpyridinium-2-carboxylic acid salt. By applying the Hammick reaction (527), the carbanions of 1-methylpyridinium salts (528) and pyridine-1-oxide (96) were trapped. The decarboxylation of pyridine dicarboxylic acid-1-oxides was reported, but the intermediate carbanions were not trapped (101, 102, 529a).

Recently, the rates of H—D exchange at the equivalent 2,6-position of N-substituted pyridinium ions in D_2O at 75.0° were obtained by use of NMR (529b). Under these conditions, no other annular protons underwent detectable exchange, and in all cases, exchange is first order in deuteroxide ion. The substituent on nitrogen has a pronounced influence on the reactivity, as indicated by the following *relative* second-order rate constants at 75° for groups O^-, Me, and OMe: 1.0, 1.4 × 10^2, and 1.3 × 10^6, respectively.

The H—D exchange of the side chain methyl protons in substituted picoline-1-oxides under basic conditions has been reported (530–534). Recently, it was found that nuclear α-proton exchange accompanied methyl proton exchange in the presence of sodium carbonate or triethylamine and sodium deuteroxide at 100° (535). The relative amount of ring and side-chain exchange varied with reaction time and the nature of base (Table IV-58).

TABLE IV-58. Base-Catalyzed H—D Exchange in Alkylpyridine-1-oxides in Deuterium Oxide (535)

N-oxide	Conditions	Exchange (%)	
		Nuclear α-protons	Side-chain protons
2-Me	Na_2CO_3, 0.5 hr	12	66
	Na_2CO_3, 5 hr	45	97
	NaOD, 0.5 hr	86	86
2-Me	Et_3N, 72 hr	40	20
	Na_2CO_3, 0.5 hr	80	50
		Exchange	Equivalent
2-CH$_2$Ph	Et_3N, 21 hr	3	100
	Na_2CO_3, 0.5 hr	3	78
	Na_2CO_3, 5 hr	30	100
4-CH$_2$Ph	Et_3N, 20 hr	17	95
	Na_2CO_3, 0.5 hr	18	94
	Na_2CO_3, 5 hr	100	100

More side-chain than ring proton exchange occurred with 2-picoline-1-oxide. In contrast, with 4-picoline-1-oxide the amount of ring H—D exchange exceeded that of side-chain exchange. Under the conditions above, 2- and 4-benzylpyridine-1-oxides also undergo base-catalyzed exchange at both the benzylic hydrogens and the 2-protons of the pyridine nucleus (535).

Pyridyl-1-oxide carbanions have been generated and trapped in non-protic solvents (536). For example, 4-chloro-3-picoline-1-oxide (**IV-265**; X = Cl, R = Me) and *n*-butyl lithium at −65° gave the carbanion (**IV-266**; X = Cl, R = Me) which, with cyclohexanone, gave the tertiary alcohol (**IV-267**; X = Cl, R = Me) in which the 4-chloro group and the *N*-oxide function

IV-265   **IV-266**

IV-267   **IV-268**   (IV-51)

are retained (eq. IV-51). Some 4-chloro-2,6-di-(1-hydroxycyclohexyl)-3-picoline-1-oxide (**IV-268**; X = Cl, R = Me) was also formed. The 2,6-disubstituted product probably results from an attack at C-2 occurring after attack at C-6. A possible explanation for this could be steric hindrance by the methyl group to the approach of the base in the plane of the ring (34). Formation of the 2,6-disubstituted product (**IV-268**) can be envisaged as the result of the formation of a dilithio intermediate (**IV-269**) or the formation of the 6-monosubstituted intermediate (**IV-270**) in which the RO—Li can abstract a proton from C-2 of another molecule of **IV-270** to give the lithium derivative **IV-271**, which can then react with more cyclohexanone to give the 2,6-disubstituted compound (**IV-268**) (34). With pyridine-1-oxide and 4-substituted pyridine-1-oxides, where there is no steric hindrance to the approach of the reagent, the mono- (**IV-267**; R = H) and di- (**IV-268**; R = H) tertiary alcohols were obtained (Table IV-59). The lithium derivatives of substituted pyridine-1-oxides were also treated with other ketones to give tertiary alcohols and with aldehydes to give secondary alcohols

IV-269   IV-268

IV-270   IV-271

(Table IV-59) (34, 536a, b). It was found that pyridyl-1-oxide carbanions could be generated from a cold ($-65°$) suspension of the pyridine-1-oxide in ether containing two equivalents of *n*-butyllithium and then warming to room temperature over a period of 20 min before reaction with the ketone. Generally higher yields of the alcohols were obtained (Table IV-59) (536b). A variety of bases, other than *n*-butyllithium were investigated for their ability to produce the pyridyl-1-oxide carbanion (Table IV-59) (536b). Both side-chain and proton abstraction occur with 2-picoline-1-oxide and butyllithium.

The reaction of 3,4-lutidine-1-oxide with lithium bis-trimethylsilylamide and cyclohexanone in ether gave 2-(1-hydroxycyclohexyl-4,5-dimethyl-pyridine-1-oxide (**IV-267**; R = X = Me) and a new product, 4-(1-hydroxy-cyclohexylmethyl)-3-picoline-1-oxide (**IV-272**) (Table IV-59) (536b).

IV-272

TABLE IV-59. Reaction of 2-Lithiopyridine-1-oxides with Aldehydes and Ketones (34, 536a, b)

Substituent in N-oxide	Conditions	Products (yield)
None	(i) n-BuLi, THF, −65° (ii) acetaldehyde	2-(1-Hydroxyethyl)pyridine-1-oxide, m.p. 97–98° (36.3%); 2,6-di-(1-hydroxyethyl)pyridine-1-oxide, m.p. 70–72° (30.1%)
	(i) n-BuLi, ether, −65° (ii) cyclohexanone	2-(1-Hydroxycyclohexyl)pyridine-1-oxide, m.p. 93–94° (7.4%)
	(i) n-BuLi, THF: ether, (2:1 v/v), −65° (ii) Cyclohexanone	2-(1-Hydroxycyclohexyl)pyridine-1-oxide, m.p. 89–91° (4.6%); 2,6-di-(1-hydroxycyclohexyl)pyridine-1-oxide, m.p. 158° (14.8%)
	(i) n-BuLi, ether, R.T. (ii) Cyclohexanone	2-(1-Hydroxycyclohexyl)pyridine-1-oxide, m.p. 89–91° (12.5%); 2,6-di-(1-hydroxycyclohexyl)pyridine-1-oxide, m.p. 159–161° (35.5%)
	(i) Na bis-trimethyl-silylamide, benzene, Δ (ii) Cyclohexanone	2-(1-Hydroxycyclohexyl)pyridine-1-oxide, m.p. 89–91° (0.85%)
	(i) n-BuLi, THF, −65° (ii) Acetone	2,6-Di-(1-methyl-1-hydroxyethyl)pyridine-1-oxide, m.p. 118° (17.8%)
2-Me	(i) n-BuLi, THF, −65° (ii) Cyclohexanone	1-(2-Methylpyridyl)cyclohexanol-1-oxide, m.p. 113–114° (4.3%); α,6-di-(1-hydroxycyclohexyl)-2-methylpyridine-1-oxide, m.p. 111° (19.6%)
3-Me	(i) n-BuLi, THF: ether (1:4), −65° (ii) Cyclohexanone	6-(1-Hydroxycyclohexyl)-3-picoline-1-oxide, m.p. 123–124° (25.1%); 2,6-di-(1-hydroxycyclohexyl)-3-picoline-1-oxide, m.p. 138–139° (4.6%)
4-Me	(i) n-BuLi, THF, −65° (ii) Cyclohexanone	2-(1-Hydroxycyclohexyl)-4-picoline-1-oxide, m.p. 115° (21.1%); 2,6-Di-(1-hydroxycyclohexyl)-4-picoline-1-oxide, m.p. 198–199° (27.3%)
	(i) n-BuLi, ether, R.T. (ii) Cyclohexanone	2-(1-Hydroxycyclohexyl)-4-picoline-1-oxide, m.p. 113–115° (19.8%); 2,6-di-(1-hydroxycyclohexyl)-4-picoline-1-oxide, m.p. 202–204° (24.9%)
3,4-Me$_2$	(i) n-BuLi, ether, −65° (ii) n-Butyraldehyde	(4,5-Dimethyl-2-pyridyl)-n-propylcarbinol-1-oxide, m.p. 114° (14.7%)
	(i) n-BuLi, THF, −65° (ii) Cyclohexanone	6-(1-Hydroxycyclohexyl)-3,4-lutidine-1-oxide, m.p. 148–149° (38.7%); 2,6-di-(1-hydroxycyclohexyl)-3,4-lutidine-1-oxide, m.p. 189–190° (12.1%)
	(i) n-BuLi, TMEDA, THF, −65° (ii) Cyclohexanone	6-(1-Hydroxycyclohexyl)-3,4-lutidine-1-oxide (28.5%); 2,6-di-(1-hydroxycyclohexyl)-3,4-lutidine-1-oxide (6.7%)
	(i) MeLi, THF, −65° (ii) Cyclohexanone	6-(1-Hydroxycyclohexyl)-3,4-lutidine-1-oxide (47.3%); 2,6-di-(1-hydroxycyclohexyl)-3,4-lutidine-1-oxide (8.9%)
	(i) n-BuLi, ether, −65° (ii) Cyclohexanone	6-(1-Hydroxycyclohexyl)-3,4-lutidine-1-oxide (56.3%); 2,6-di-(1-hydroxycyclohexyl)-3,4-lutidine-1-oxide (15.6%)

Table IV-59 (*Continued*)

Substituent in N-oxide	Conditions	Products (yield)
	(i) Na bis-trimethyl-silylamide, benzene, Δ (ii) Cyclohexanone	6-(1-Hydroxycyclohexyl)-3,4-lutidine-1-oxide (1.7%)
	(i) Na bis-trimethyl-silylamide, THF, R.T. (ii) Cyclohexanone	6-(1-Hydroxycyclohexyl)-3,4-lutidine-1-oxide (1.4%)
	(i) Na bis-trimethyl-silylamide, THF, Δ (ii) Cyclohexanone	6-(1-Hydroxycyclohexyl)-3,4-lutidine-1-oxide (1.4%)
	(i) Li bis-trimethyl-silylamide, ether, R.T. (ii) Cyclohexanone	6-(1-Hydroxycyclohexyl)-3,4-lutidine-1-oxide (4.0%); 4-(1-hydroxycyclohexylmethyl)-3-picoline-1-oxide, m.p. 217–219° (2.6%)
	(i) Li bis-trimethyl-silylamide, ether, Δ (ii) Cyclohexanone	6-(1-Hydroxycyclohexyl)-3,4-lutidine-1-oxide (4.5%); 4-(1-hydroxycyclohexylmethyl)-3-picoline-1-oxide (3.3%)
	(i) n-BuLi, THF, −65° (ii) Acetone	Dimethyl (4,5-dimethyl-2-pyridyl)carbinol-1-oxide, m.p. 129° (40.3%)
	(i) n-BuLi, THF, −65° (ii) Acetophenone	1-(4,5-Dimethyl-2-pyridyl)-1-phenylethanol-1-oxide, m.p. 141° (39.4%)
4-Cl	(i) n-BuLi, ether, −100° (ii) Cyclohexanone	2,6-Di-(1-hydroxycyclohexyl)-4-chloropyridine-1-oxide, m.p. 195° (10.9%)
	(i) n-BuLi, ether, −65° (ii) Cyclohexanone	2-(1-Hydroxycyclohexyl)-4-chloropyridine-1-oxide, m.p. 113–114° (35.6%); 2,6-di-(1-hydroxycyclohexyl)-4-chloropyridine-1-oxide (20.7%)
	(i) n-BuLi, THF, −65° (ii) Cyclohexanone	2,6-Di-(1-hydroxycyclohexyl)-4-chloropyridine-1-oxide (10.5%)
	(i) n-BuLi, ether, −15° (ii) Cyclohexanone	2,6-Di-(1-hydroxycyclohexyl)-4-chloropyridine-1-oxide (13.2%)
	(i) n-BuLi, ether, 0° (ii) Cyclohexanone	2,6-Di-(1-hydroxycyclohexyl)-4-chloropyridine-1-oxide (7.3%)
4-Cl-3-Me	(i) n-BuLi, ether, −65° (ii) PhCHO	4-Chloro-6-(α-hydroxybenzyl)-3-picoline-1-oxide, m.p. 134–135° (9.4%)
	(i) n-BuLi, ether, −65° (ii) Cyclohexanone	4-Chloro-6-(1-hydroxycyclohexyl)-3-picoline-1-oxide, m.p. 164–165° (43.8%); 4-chloro-2,6-di-(1-hydroxycyclohexyl)-3-picoline-1-oxide, m.p. 168–169° (4.7%)
4-OEt	(i) n-BuLi, ether, −65° (ii) Cyclohexanone	4-Ethoxy-2-(1-hydroxycyclohexyl)pyridine-1-oxide, m.p. 128° (3.7%); 4-ethoxy-2,6-di-(1-hydroxycyclohexyl)pyridine-1-oxide, m.p. 166° (12.5%)
	(i) n-BuLi, THF, −65° (ii) Cyclohexanone	4-Ethoxy-2-(1-hydroxycyclohexyl)pyridine-1-oxide, (19.9%); 4-ethoxy-2,6-di-(1-hydroxycyclohexyl)pyridine-1-oxide (20.7%)
	(i) NaH, THF, Δ (ii) Cyclohexanone	4-Ethoxy-2-(1-hydroxycyclohexyl)pyridine-1-oxide (7.5%)

Reaction of 2-lithiopyridine-1-oxide with epoxides gave only vinyl polymers and no β-alcohols (314a).

Treatment of 2-lithiopyridine-1-oxides with an excess of an ester gives the desired ketone. If a 3-substituent is present, both possible isomeric ketones are obtained (eq. IV-52) (Table IV-60) (34).

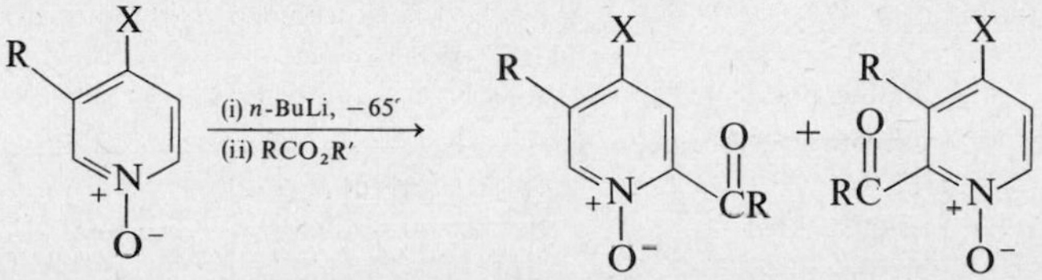

(IV-52)

The 6-lithio derivative of 3,4-lutidine-1-oxide was treated with N,N-dimethylacetamide to give 6'-acetyl-3',4,4',5-tetramethyl-2,2'-dipyridyl-1-oxide (IV-273) (34). The formation of IV-274 can be explained by the formation of intermediate IV-274, which then undergoes nucleophilic attack at C-2 by more 6-lithio-3,4-lutidine-1-oxide to give IV-275. This, on hydrolysis, gives the dipyridyl derivative (IV-273). 4-Chloro-6-lithio-3-picoline-1-oxide

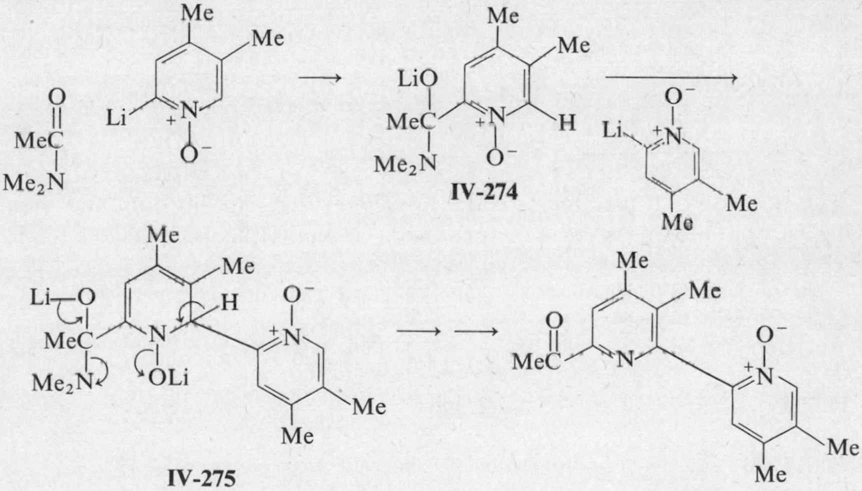

IV-274

IV-275

IV-273

and benzonitrile gave 6'-benzoyl-4,4'-dichloro-3',5-dimethyl-2,2'-dipyridyl-1-oxide (IV-276) (34). The substituted pyridine-1-oxide anions react with

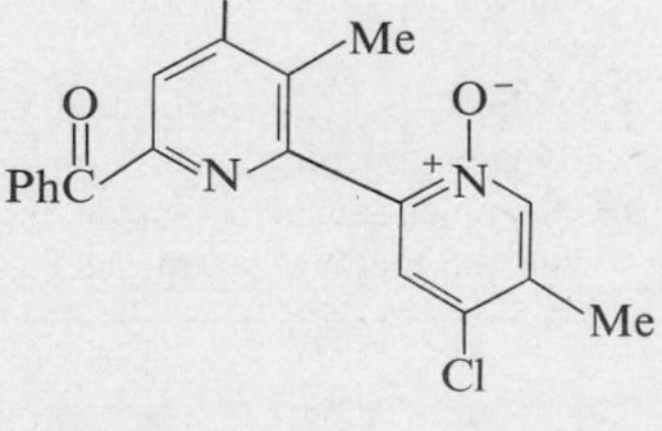

IV-276

TABLE IV-60. Ketones from Substituted 2-Lithiopyridine-1-oxides (34)

Substituent in N-oxide	Conditions	Products (yield)
3,4-Me$_2$	Ethyl acetate, THF	6-Acetyl-3,4-lutidine-1-oxide, m.p. 61–62° (19.6%); 2-acetyl-3,4-lutidine-1-oxide, m.p. 109–110° (6.3%)
	N-Acetylmorpholine, THF	2-Acetyl-3,4-lutidine-1-oxide (2.8%)
	N,N-Dimethylacetamide, THF	6'-Acetyl-3',4,4',5-tetramethyl-2,2'-dipyridyl-1-oxide, m.p. 217° (12.9%)
4-Cl-3-Me	Benzonitrile, ether	6'-Benzoyl-4,4'-dichloro-3',5-dimethyl-2,2'-dipyridyl-1-oxide, m.p. 235° (11.5%)
4-OEt	Ethyl butyrate	2,6-Di-n-butyroyl-4-ethoxypyridine-1-oxide, m.p. 64–65° (16.5%)

carbon dioxide to give substituted picolinic acid-1-oxides (**IV-277**) (Table IV-61) (34, 536a). Under these conditions, 4-picoline-1-oxide gave only the dicarboxylic acid (**IV-278**).

$$\text{IV-277} \qquad \text{IV-278}$$

The reaction of 2-picoline-1-oxide (**IV-279**) with n-butyllithium and then with carbon dioxide gave two products: 6-methylpicolinic acid-1-oxide

$$\text{IV-279} \qquad \text{IV-280} \qquad \text{IV-281}$$

TABLE IV-61. Reaction of 2-Lithiopyridine-1-oxide with Carbon Dioxide (34, 536a)

Substituent in N-oxide	Conditions	Products (yield)
2-Me	CO$_2$, THF	6-Methylpicolinic acid-1-oxide, m.p. 181–182° (13.5%); 2-methylpyridine-1-oxide-α,6-dicarboxylic acid, m.p. 177° (10.1%)
4-Me	CO$_2$, THF	4-Picoline-2,6-dicarboxylic acid-1-oxide, m.p. 160° (48.0%)
3,4-Me$_2$	CO$_2$, THF	4,5-Dimethylpicolinic acid-1-oxide, m.p. 180–181° (17.9%)
4-Cl	CO$_2$, THF	4-Chloropicolinic acid-1-oxide, m.p. 136° (49.0%)
4-Cl-3-Me	CO$_2$, THF	4-Chloro-5-methylpicolinic acid-1-oxide, m.p. 160° (23.8%)

(**IV-280**) and 2-picoline-α,6-dicarboxylic acid-1-oxide (**IV-281**) (34). Treatment of 2-lithiopyridine-1-oxides with a Schiff base, such as *N*-benzylideneaniline gave the corresponding substituted secondary amines (eq. IV-53) (Table IV-62) (314a).

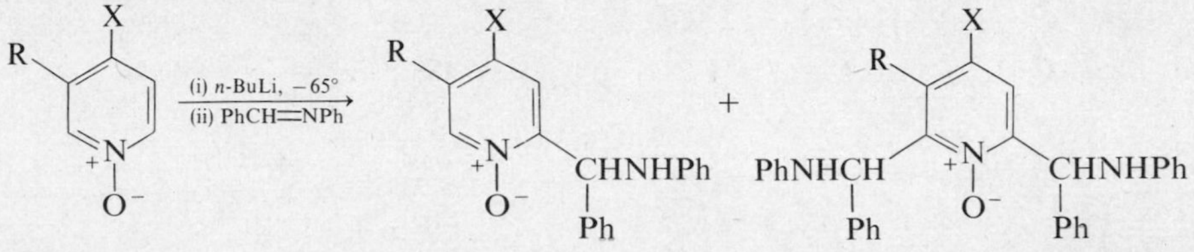

(IV-53)

TABLE IV-62. Reaction of 2-Lithiopyridine-1-oxides with Benzylideneaniline (314a)

Substituent in *N*-oxide	Conditions	Products (yield)
None	THF, −65°	2,6-Bis-(α-*N*-phenylaminobenzyl)pyridine-1-oxide, m.p. 105° (40.9%)
3,4-Me$_2$	THF, −65°	2,6-Bis-(α-*N*-phenylaminobenzyl)-3,4-lutidine-1-oxide, m.p. 125° (61.8%); 2-(α-*N*-phenylaminobenzyl)-4,5-dimethylpyridine-1-oxide, m.p. 212–213° (12.1%)
	ether, R.T.	2,6-Bis-(α-*N*-phenylaminobenzyl)-3,4-lutidine-1-oxide (31%); 2-(α-*N*-Phenylaminobenzyl)-4,5-dimethylpyridine-1-oxide (17.5%)

Addition of sulfur to lithiopyridine-1-oxide and its simple derivatives is a convenient route to 1-hydroxy-2-pyridinethiones and related compounds (**IV-282**) (Table IV-63) (537a). In the case of 3,4-lutidine-1-oxide (**IV-283**),

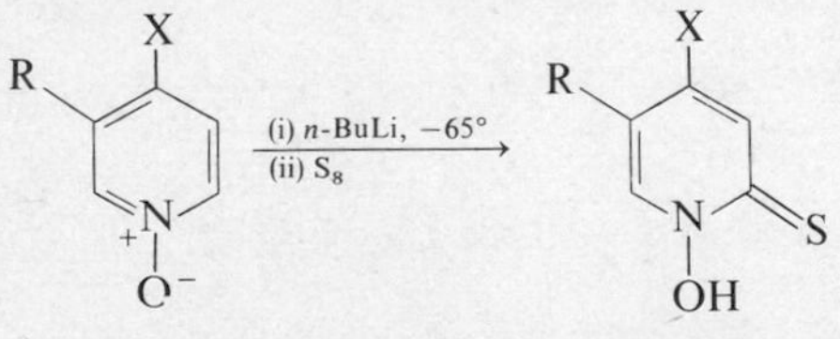

IV-282

both possible thiohydroxamic acids (**IV-284**) and (**IV-285**) were obtained, together with a dimeric product (**IV-286**), which, on reduction, gave the 1-hydroxythiolthione **IV-287** (537a). Similarly, the two possible hydroxamic acids **IV-288** and **IV-289** were obtained in poor yield from the reaction of

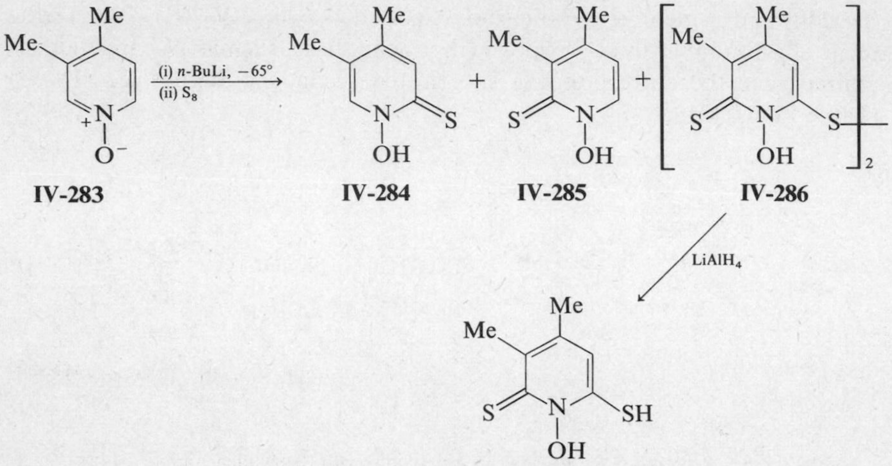

IV-283   IV-284   IV-285   IV-286

LiAlH₄

IV-287

substituted lithiopyridine-1-oxides with oxygen (Table IV-64) (537a). Pyridine-1-oxide itself did not yield a hydroxamic acid (537a).

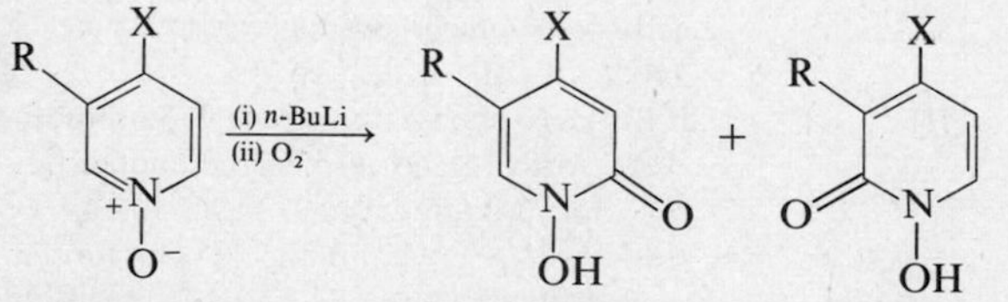

IV-288   IV-289

The reaction of the lithiopyridine-1-oxides with bromine gave a variety of dihalogenated products (537a). In the case of pyridine-1-oxide, 2,6-dibromopyridine-1-oxide (**IV-290**), 6,6′-dibromo-2,2′-dipyridyl-1,1′-dioxide (**IV-291**), and 6,6′-dibromo-2,2′-dipyridyl-1-oxide (**IV-292**) were formed

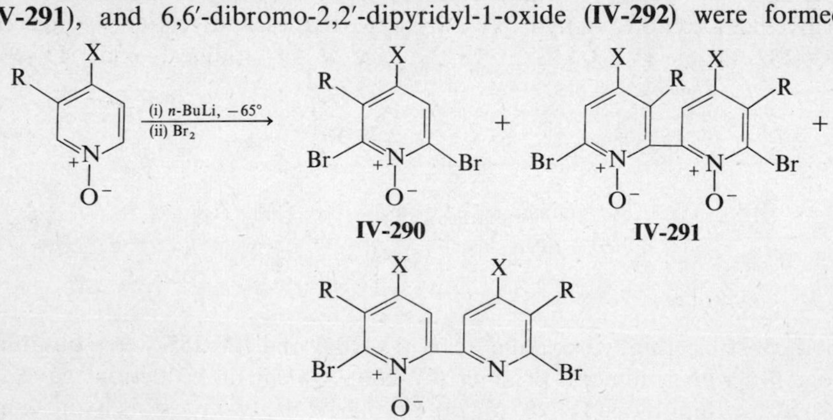

IV-290   IV-291

IV-292

TABLE IV-63. Reaction of 2-Lithiopyridine-1-oxides with Sulfur and Related Compounds (314a, 537a)

Substituent in N-oxide	Conditions	Products (yield)
None	n-BuLi, THF	1-Hydroxy-2-pyridinethione, m.p. 68° (7.9%)
	n-BuLi, ether, R.T.	1-Hydroxy-2-pyridinethione (10.1%)
	LiH, dimethoxyethane, 80°, 18 hr	1-Hydroxy-2-pyridinethione (12%)
	LiH, dimethoxyethane: 2-(2-methoxyethoxy)-ethanol (50:4 v/v), 80°, 18 hr	1-Hydroxy-2-pyridinethione (19.45%)
	LiH, dimethoxyethane: 2-methoxyethanol (50:4 v/v), 80°, 18 hr	1-Hydroxy-2-pyridinethione (21.5%)
4-Me	n-BuLi, THF	1-Hydroxy-4-methyl-2-pyridinethione, m.p. 59° (38.7%)
	LiH, dimethoxyethane: 2-methoxyethanol (50:4 v/v)	1-Hydroxy-4-methyl-2-pyridinethione (5.2%)
4-Cl-3-Me	n-BuLi, THF	4-Chloro-1-hydroxy-5-methyl-2-pyridinethione, m.p. 99–101° (11.45%)
3,4-Me$_2$	n-BuLi, THF	2,2'-(1,1'-Dihydroxy-4,4',5,5'-tetramethyldipyridyl-6,6'-dithione)disulfide, m.p. 186–187° (37.4%); 1-hydroxy-3,4-dimethyl-2-pyridinethione, m.p. 128–129° (12.5%); 1-hydroxy-3,4-dimethyl-6-pyridinethione, m.p. 121–122° (24.1%)
	n-BuLi, ether, R.T.	2,2'-(1,1'-Dihydroxy-4,4',5,5'-tetramethyldipyridyl-6,6'-dithione)disulfide, m.p. 186–187° (21.1%); 1-hydroxy-3,4-dimethyl-2-pyridinethione, m.p. 126–127° (7.9%); 1-hydroxy-3,4-dimethyl-6-pyridinethione, m.p. 120–121° (7.3%)
	LiH, dimethoxyethane, Δ, 20 hr	1-Hydroxy-3,4-dimethyl-2-pyridinethione (0.9%)
	LiH, dimethoxyethane: 2-methoxyethanol (50:4 v/v), 80°, 4 hr	1-Hydroxy-3,4-dimethyl-2-pyridinethione (5.2%)
	n-BuLi, sulfur monochloride, THF	1-Hydroxy-3,4-dimethyl-2-pyridinethione (4.6%); 1-hydroxy-3,4-dimethyl-6-pyridinethione (7.5%)
	n-BuLi, ethylene sulfide, ether	1-Hydroxy-3,4-dimethyl-2-pyridinethione (7.7%); 1-hydroxy-3,4-dimethyl-2-pyridinethione (11%)

TABLE IV-64. Reaction of 2-Lithiopyridine-1-oxide with Oxygen (537a)

Substituent in N-oxide	Products (yield)
None	Tars
4-Me	1-Hydroxy-4-methyl-2-pyridone, m.p. 131–132° (12.7%)
3,4-Me$_2$	1-Hydroxy-3,4-dimethyl-2-pyridone, m.p. 169–170° (10%); 1-hydroxy-3,4-dimethyl-6-pyridone, m.p. 195° (14.9%)

160 Pyridine-1-oxides

TABLE IV-65. Reaction of 2-Lithiopyridine-1-oxides with Bromine (537a)

Substituents in N-oxide	Conditions	Products (yield)
None	(i) THF-ether (1:1 v/v), (ii) Phenol	2,6-Dibromopyridine-1-oxide, m.p. 187–188° (decomp.) (4.0%); 6,6'-dibromo-2,2'-dipyridyl-1,1'-dioxide, m.p. 232–234° (decomp.) (4.1%)
None	THF-ether (1:1 v/v)	6,6'-Dibromo-2,2'-dipyridyl-1-oxide, m.p. 209–211° (decomp.) (6.2%); 2,6-dibromopyridine-1-oxide, m.p. 187–188° (decomp.) (3.1%); 6,6'-dibromo-2,2'-dipyridyl-1,1'-dioxide, m.p. 230° (decomp.) (1.3%)
None	Ether-HMPA (40:1 v/v)	2,6-Dibromopyridine-1-oxide, m.p. 186° (decomp.) (0.8%); 6,6'-dibromo-2,2'-dipyridyl-1,1'-dioxide, m.p. 229–232° (decomp.) (1.15%)
4-Me	THF	6,6'-Dibromo-4,4'-dimethyl-2,2'-dipyridyl-1-oxide, m.p. 166–167° (18.1%); 2,6-dibromo-4-methylpyridine-1-oxide, m.p. 154–155° (4.6%); 6,6'-dibromo-4,4'-dimethyl-2,2'-dipyridyl-1,1'-dioxide, m.p. 219–222° (decomp.) (12.7%)
3,4-Me$_2$	THF, 3 hr	6,6'-Dibromo-3',4,4',5-tetramethyl-2,2'-dipyridyl-1,1'-dioxide, m.p. 200–202° (3.2%); 2,6-dibromo-3,4-lutidine-1-oxide, m.p. 144° (12.8%)
	(i) THF, 15 min (ii) Phenol	2,6-Dibromo-3,4-lutidine-1-oxide (23.3%); 6,6'-dibromo-3',4,4',5-tetramethyl-2,2'-dipyridyl-1,1'-dioxide (1.6%)
	(i) THF, 15 min excess bromine (ii) Phenol	2,6-Dibromo-3,4-lutidine-1-oxide (8.9%); 6,6'-dibromo-3',4,4',5-tetramethyl-2,2'-dipyridyl-1,1'-dioxide (2.4%)

(Table IV-65). Addition of chlorine gas to lithiopyridine-1-oxide gave 2,6-dichloropyridine-1-oxide in poor yield but no bimolecular products (Table IV-66) (537a).

Mercuration of 2-lithiopyridine-1-oxide with mercuric chloride give mixtures of the mercuric chloride and dipyridyl mercury derivatives which could not be resolved but were analyzed by mass spectrometry. Bromination

TABLE IV-66. Reaction of 2-Lithiopyridine-1-oxide with Chlorine (537a)

Substituents in N-oxide	Conditions	Products (yield)
None	THF-ether (1:1 v/v)	2,6-Dichloropyridine-1-oxide, m.p. 139–140° (4.5%)
3,4-Me$_2$	(i) THF (ii) Phenol	2,6-Dichloro-3,4-lutidine-1-oxide, m.p. 165–166° (8.8%)

yielded the corresponding 2-bromo- and 2,6-dibromopyridine-1-oxides (Scheme IV-15) (Table IV-67) (537b).

TABLE IV-67. Reaction of 2-Lithiopyridine-1-oxides with Mercuric Chloride Followed by Bromination (537b)

Substituent in N-oxide	Conditions	Products (yield)
None	(i) HgCl$_2$ (1 equiv.) (ii) aq. NaBr, Br$_2$, 50°	2-Bromopyridine-1-oxide, m.p. 133–135° (13%); 2,6-dibromopyridine-1-oxide, m.p. 188–190° (12%)
	(i) HgCl$_2$ (2 equiv.) (ii) aq. NaBr, Br$_2$, 50°	2,6-Dibromopyridine-1-oxide (17%)
4-Me	(i) HgCl$_2$ (1 equiv.) (ii) aq. NaBr, Br$_2$, 50°	2-Bromo-4-picoline-1-oxide, m.p. 145–147° (14%); 2,6-dibromo-4-picoline-1-oxide, m.p. 155–156° (29%)
	(i) HgCl$_2$ (2 equiv.) (ii) aq. NaBr, Br$_2$, 50°	2,6-Dibromo-4-picoline-1-oxide (32%)
3,4-Me$_2$	(i) HgCl$_2$ (2 equiv.) (ii) aq. NaBr, Br$_2$, 50°	2,6-Dibromo-3,4-lutidine-1-oxide, m.p. 141–142° (36%)

Scheme IV-15

5. Intramolecular Reactions

2-Alkenyloxypyridine-1-oxides **(IV-293)** rearrange slowly at room temperature to the 1-alkenyloxy-2-pyridones **(IV-294)** (Table IV-68) (538) (see also

IV-293   IV-294

Chapter XII). 2-Crotonyloxypyridine-1-oxide gave 1-crotonyloxy-2-pyri-
done, 2-α-methylallyloxypyridine-1-oxide gave 1-α-methylallyloxy-2-pyri-
done, and 2-(crotonyloxy-1-*d*)pyridine-1-oxide gave only 1-(crotonyloxy-
1-*d*)-2-pyridone, indicating that the carbon attached to oxygen was the same
in both starting material and product. This shows that the reaction proceeds
without a 1,3-allylic bond shift, and a cyclic seven-membered transition state
is not involved. Both 2-α-methylallyloxypyridine-1-oxide and 2-α-methyl-
allyloxy-5-methylpyridine-1-oxide exhibit a large rate enhancement in
comparison to the other 2-alkenyloxypyridine-1-oxides. This increased
reactivity is consistent with an ion pair process.

It was observed that only 30% of the product in the rearrangement of
2-α-methylallyloxy-5-methylpyridine-1-oxide (IV-295) was the expected

TABLE IV-68. Thermal Rearrangement of 2-Alkenyloxypyridine-1-oxide (538)

N-oxide (IV-293)	Conditions	Products (IV-294) (yield)
R = H, R′ = CH₂CH=CH₂	Diglyme, 83–84°	R = H, R′ = CH₂CH=CH₂ (100%)
R = H, R′ = CH₂CH=CHCH₃	Diglyme, 83–84°	R = H, R′ = CH₂CH=CHCH₃ (98%); R = H, R′ = CH(CH₃)CH=CH₂ (2%)
R = H, R′ = CH(CH₃)CH=CH₂	Neat, R.T.	R = H, R′ = CH(CH₃)CH=CH₂ (80%); R = H, R′ = CH₂CH=CHCH₃ (10%)
R = H, R′ = CH(D)CH=CHCH₃	Diglyme, 83–84°	R = H, R′ = CH(D)CH=CHCH₃ (100%)
R = H, R′ = CH₂—◁	Diglyme, 126–127°	R = H, R′ = CH₂—◁ (96%); R = H, R′ = —◇ (4%)
R = H, R′ = CH(CH₃)CH=CHCH₃	Diglyme, 83–84°	R = H, R′ = CH(CH₃)CH=CHCH₃ (100%)
R = Me, R′ = CH₂CH=CH₂	Diglyme, 83–84°	R = Me, R′ = CH₂CH=CH₂ (100%)
R = Me, R′ = CH₂CH=CHCH₃	Diglyme, 83–84°	R = Me, R′ = CH₂CH=CHCH₃ (100%)
R = Me, R′ = CH(CH₃)CH=CH₂	Neat, R.T.	R = Me, R′ = CH(CH₃)CH=CH₂ (92%)
R = Me, R′ = CH(CH₃)CH=CH₂	Diglyme, 83–84°	R = Me, R′ = CH(CH₃)CH=CH₂ (30%)

1-α-methylallyloxy-5-methyl-2-pyridone **(IV-296)** and the corresponding crotyloxyl-derivative **(IV-297)**; 60% were *cis*- and *trans*-3-crotyl-1-hydroxy-5-methyl-2-pyridones **(IV-298)** arising from rearrangement to the ring carbon (538). Table IV-69 illustrates the strong temperature dependence of these reactions. The rearrangement to the ring carbon to yield **IV-298** is more energy-demanding but a less ordered process, and it can compete favorably with other rearrangements at high temperatures.

IV-295   **IV-298**   +

IV-296   **IV-297**

Rearrangements to carbon had not been observed in any other reactions at 83°. Heating 2-allyloxypyridine-1-oxide in diglyme at 125° gave 50% 3-allyl-1-hydroxy-2-pyridone and 43% 1-allyloxy-2-pyridone; at 137° the amounts were 55 and 39%, respectively. Experimental results show that the 3-alkenyl-1-hydroxy-2-pyridones do not arise from 1-alkenyloxy-2-pyridones. It was concluded that they are formed directly from the 1-oxide.

N-Acetoacetyl, *N*-phenylacetyl, and *N*-cyanoacetyl derivatives of 2- and 4-pyridinesulfonamide-1-oxide **(IV-299a, b)** undergo intramolecular nucleophilic alkylation in the presence of alkali to give the compounds reported

TABLE IV-69. Thermal Rearrangement of 6-α-Methylallyloxy-3-picoline (538)

		Products (yield)		
Temp. (°C)	$10^4 k$ (sec^{-1})	**IV-296**	**IV-297**	**IV-298**
83	1.2	8%	59%	32%
108	6.8	6%	71%	22%
137	25.0	5%	75%	20%

in Table IV-70 (539–542). The rearrangement of the 2-sulfonamido derivative is accompanied by participation of the 1-oxide oxygen atom to form a 4-substituted pyrido[1,2-*b*]isoxazol-5-one (R.I. 7,951). This reaction does not occur with pyridines not bearing an *N*-oxide function (543). The rate of rearrangement was measured by the determination of the sulfurous acid formed. The following mechanism (Scheme IV-16) accounts for the products of these reactions.

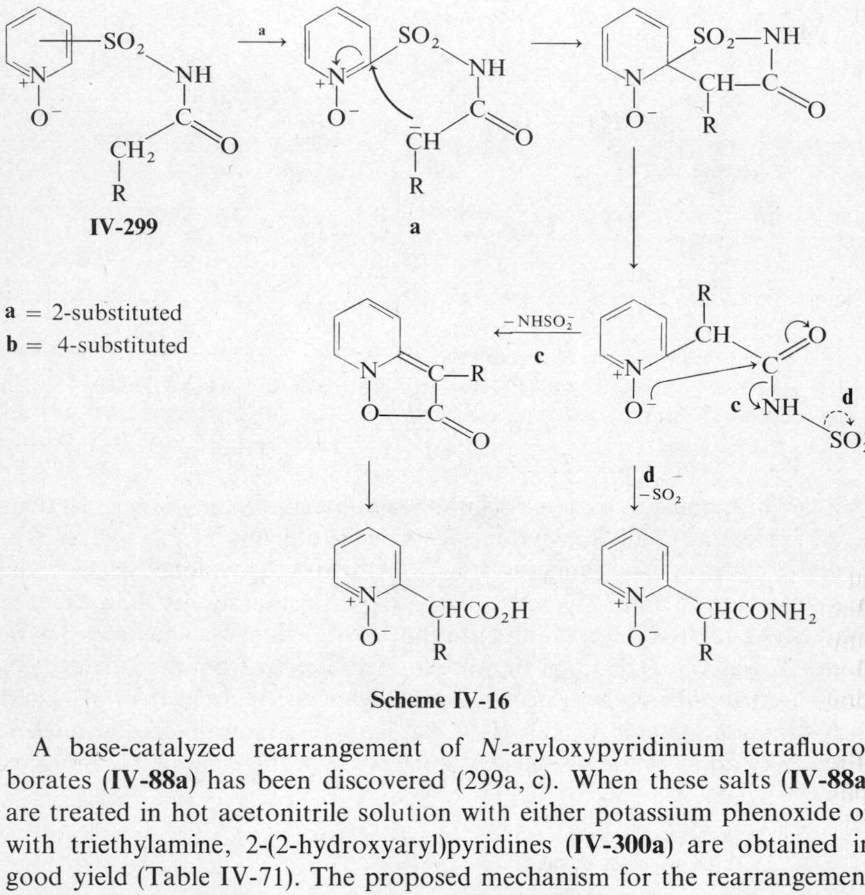

a = 2-substituted
b = 4-substituted

Scheme IV-16

A base-catalyzed rearrangement of *N*-aryloxypyridinium tetrafluoroborates (**IV-88a**) has been discovered (299a, c). When these salts (**IV-88a**) are treated in hot acetonitrile solution with either potassium phenoxide or with triethylamine, 2-(2-hydroxyaryl)pyridines (**IV-300a**) are obtained in good yield (Table IV-71). The proposed mechanism for the rearrangement involves base-catalyzed proton abstraction from the 2-position of the pyridinium ring followed by nucleophilic attack at the *ortho* position of the benzene ring (eq. IV-54).

4-Pyridylhydrazone-1-oxides undergo indolization in the presence of zinc chloride at 195 to 200° with retention of the *N*-oxide function (295, 544) (Table IV-72). Under these conditions, *N*-cyclohexylpyruvamide 4-pyridyl-hydrazone-1-oxide did not cyclize, but gave pyruvamide 4-pyridylhydrazone-1-oxide (295). Heating cyclohexanone 4-pyridylhydrazone-1-oxide (**IV-301**)

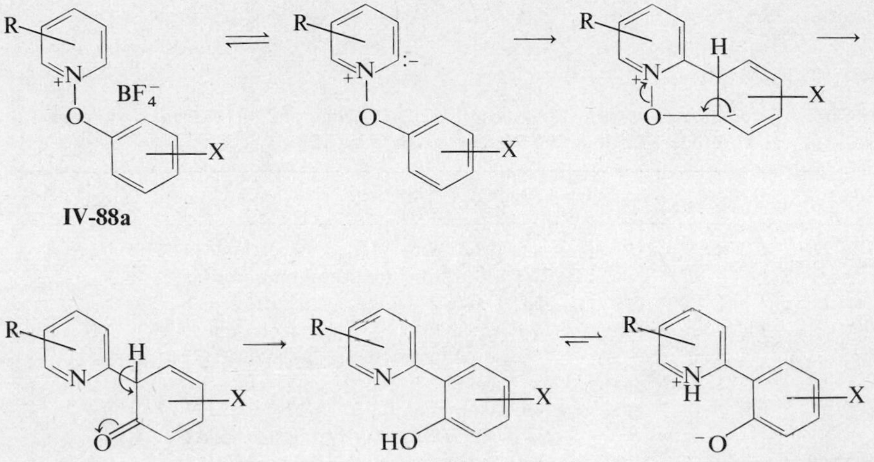

IV-88a

IV-300a   (IV-54)

in diethylene glycol yielded 6,7,8,9-tetrahydro-γ-carboline (**IV-302**) (545) (RI 2,890), in contrast to the retention of the N-oxide in the previously mentioned reaction (295). When acetophenone 4-pyridylhydrazone-1-oxide was treated similarly, only deoxygenation of the N-oxide occurred and no Fisher cyclization (545).

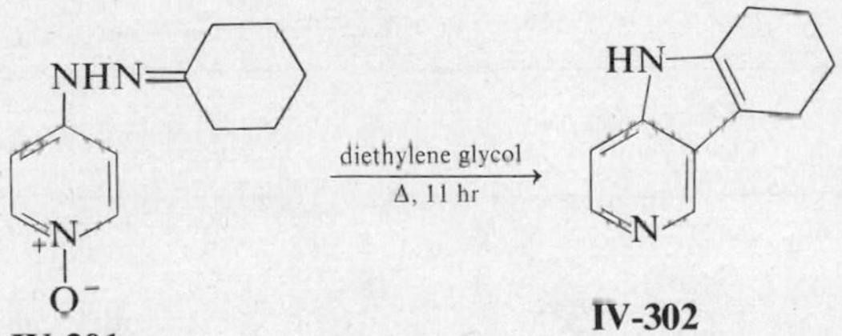

IV-301   **IV-302**

2-(2-Azidophenyl)pyridine-1-oxide (**IV-303**) undergoes intramolecular cyclization in hot decalin to give δ-carboline (**IV-304**) (RII 2,888) and δ-carboline-1-oxide (major) (**IV-305**) (140).

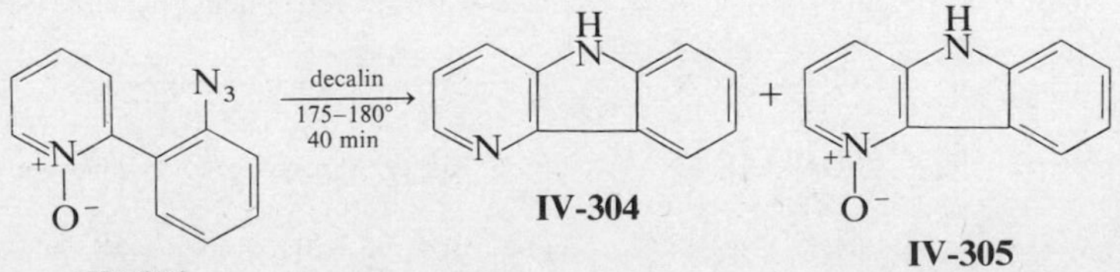

IV-303   **IV-304**   **IV-305**

Heating pyridine-1-oxide with 2-bromopyridine or 2-bromoquinoline yields 1-(2-pyridyl)-2-pyridone (**IV-306**) and its monobromo derivative, and 1-(2-pyridyl)carbostyril, respectively (546). A number of pyridine-1-oxides

TABLE IV-70. Rearrangement of N-Substituted 2-(**IV-299a**) and 4-(**IV-299b**) pyridinesulfon-amide-1-oxide in the Presence of 10% NaOH (539–542)

N-oxide	R	Temp. (°C)	Products (yield)
IV-299a	COMe	20–36	α-(1-Oxido-2-pyridyl)acetoacetamide (42.8%); 4-acetyl-pyrido[1,2-b]isoxazol-5-one (56.8%)
		90–95	2-(1-Oxido-2-pyridyl)acetoacetic acid (92.2%)
	Ph	20–36	4-Phenylpyrido[1,2-b]isoxazol-5-one (6.7%); α-(1-oxido-2-pyridyl)phenylacetic acid (77.0%)
		90–95	α-(1-Oxido-2-pyridyl)phenylacetic acid (83.5%)
	CN	20–36	4-Cyanopyrido[1,2-b]isoxazol-5-one (70.7%)
		90–95	α-(1-Oxido-2-pyridyl)cyanoacetic acid (82.0%)
IV-299b	COMe	20–36	α-(1-Oxido-4-pyridyl)acetoacetamide (78.8%); α-(1-oxido-4-pyridyl)acetoacetic acid (8.9%)
		90–95	α-(1-Oxido-4-pyridyl)acetoacetic acid (82.0%)
	Ph	20–36	α-(1-Oxido-4-pyridyl)phenylacetic acid (84.8%)
		90–95	α-(1-Oxido-4-pyridyl)phenylacetic acid (82.0%)

TABLE IV-71. Preparation of 2-(2-Hydroxyaryl)pyridines (299a, c)

N-Aryloxypyri-dinium salt (**IV-88**)	Conditions	Products (yield)
R = H, X = 4-CN	PhOK, CH$_3$CN, Δ, 1 hr	2-(5-Cyano-2-hydroxyphenyl)pyridine, m.p. 198–199° (69%)
R = H, X = 2-NO$_2$	PhOK, CH$_3$CN, Δ, 1 hr	2-(2-Hydroxy-3-nitrophenyl)pyridine, m.p. 167–168° (66%)
R = H, X = 3-NO$_2$	PhOK, CH$_3$CN, Δ, 1 hr	2-(2-Hydroxy-X-nitrophenyl)pyridine, m.p. 187–188° (33%)
R = H, X = 4-NO$_2$	PhOK, CH$_3$CN, Δ, 1 hr	2-(2-Hydroxy-5-nitrophenyl)pyridine, m.p. 216–217° (52%)
	Et$_3$N, CH$_3$CN, Δ, 1 hr	2-(2-Hydroxy-5-nitrophenyl)pyridine (43%)
	KOH, CH$_3$CN, Δ, 1 hr	2-(2-Hydroxy-5-nitrophenyl)pyridine (37%)
R = H, X = 2-CF$_3$	PhOK, CH$_3$CN, Δ, 1 hr	2-(2-Hydroxy-3-trifluoromethylphenyl)pyridine, m.p. 87–88° (49%)
R = H, X = 4-CF$_3$	PhOK, CH$_3$CN, Δ, 1 hr	2-(2-Hydroxy-5-trifluoromethylphenyl)pyridine, m.p. 93–94° (40%)
R = 4-Ph, X = CN	PhOK, CH$_3$CN, Δ, 1 hr	2-(5-Cyano-2-hydroxyphenyl)-4-phenylpyridine, m.p. 165–166° (53%)

TABLE IV-72. Indolization of 4-Pyridylhydrazone-1-oxides

Hydrazone	Conditions	Products (yield)	Ref.
Cyclohexanone 4-pyridylhydrazone-1-oxide, HCl	(i) $ZnCl_2$, NaCl, 180° (ii) PtO_2, H_2	6,7,8,9-Tetrahydro-γ-carboline, m.p. 269–271°	295
2,3-Dioxopiperidine 3-(4-pyridylhydrazone-1-oxide), HCl	$ZnCl_2$, NaCl 195–200°	1-Hydroxy-3,4-dihydro-6-aza-β-carboline-6-oxide, HCl, m.p. 264–265° (RII. 2,885)	544
2,3-Dioxopiperidine 3-(2-methyl-4-pyridyl-hydrazone-1-oxide)	$ZnCl_2$, 195–200°	1-Hydroxy-3,4-dihydro-5-methyl-6-aza-β-carboline-6-oxide, HCl, m.p. >300°, (28%)	295
2,3-Dioxopiperidine 3-(5-ethyl-2-methyl-4-pyridylhydrazone-1-oxide)	$ZnCl_2$, 195–200°	1-Hydroxy-3,4-dihydro-5-methyl-8-ethyl-6-aza-β-carboline-6-oxide, HCl, m.p. >300°	295

have been shown to undergo this reaction (Table IV-73) (547–549). In the case of 3-picoline-1-oxide, only the 5′-methyl derivative is formed. Steric hindrance by the methyl group in 3-picoline-1-oxide is believed to prevent cyclization at the 2-position. The likely mode of formation of the products obtained in this reaction is outlined in Scheme IV-17. The unidentified

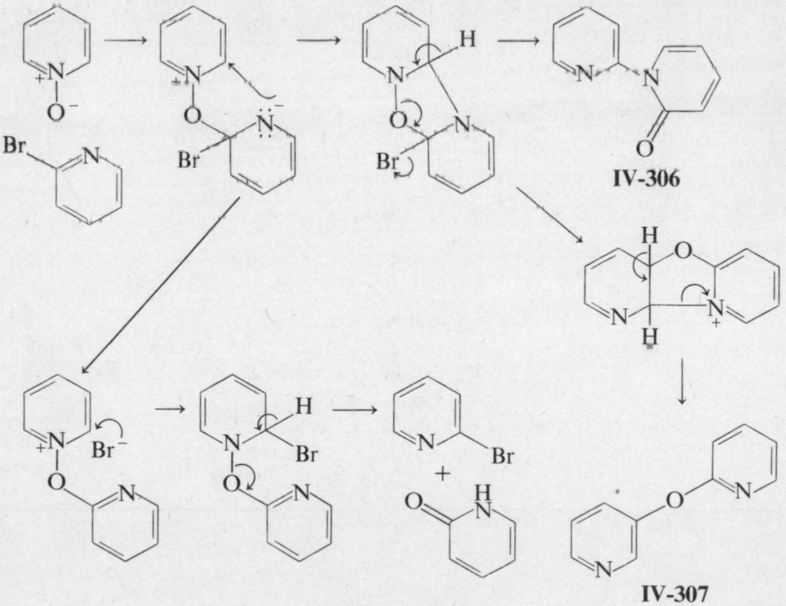

Scheme IV-17

TABLE IV-73. Reaction of Pyridine-1-oxides with Bromopyridines

Substituent in N-oxide	Conditions	Products (yield)	Ref.
None	2-BrC$_5$H$_4$N, N$_2$, 100–250°, 2.5 hr	Pyridine (7.6%), 3-bromo-1-(2-pyridyl)-2-pyridone (21%); 1-(2-pyridyl)-2-pyridone (24%)	547
	2-BrC$_5$H$_4$N, PhMe, 30% HBr in AcOH, 105°, 6 hr	Pyridine (trace), 1-(2-pyridyl)-2-pyridone (53%)	547
	2-BrC$_5$H$_4$N, 100°, 4 hr	Pyridine (27%), 3,5-dibromo-1-(2-pyridyl)-2-pyridone (19%); 3-bromo-1-(2-pyridyl)-2-pyridone (12%); 1-(2-pyridyl)-2-pyridone (3.5%)	547
	2-BrC$_5$H$_4$N, 3 hr, dioxane	2-Pyridone (3.1%); 1-(2-pyridyl)-2-pyridone (60.2%); 2,3'-dipyridyl ether (2.6%)	340
	1-Methyl-2-pyridone, 200–210°, 4 hr	Pyridine (48%), 3,5-dibromo-1-methyl-2-pyridone, (trace)	547
	2-Bromoquinoline	1-(2-Pyridyl)carbostyril (61.7%); carbostyril (18.9%); 3-pyridyl-2-quinolyl ether (0.9%)	340
	4-BrC$_5$H$_4$N, 3 hr, 110–112°, dioxane, sealed tube	Pyridine (30%), 1-(4-pyridyl)-4-pyridone, 4-pyridone	548
	4-Br-3-Me-C$_5$H$_3$N, 110–120°, 3 hr, dioxane	3-Methyl-1-(3-methyl-2-pyridyl)-4-pyridone (51%); 3-methyl-4-pyridone (30.5%)	340
2-Me	2-BrC$_5$H$_4$N, 100°, 9 hr, N$_2$	2-Picoline (40%); 1-(6-methyl-2-pyridyl)-2-pyridone (9.5%); 2-pyridone (26%)	547
	2-BrC$_5$H$_4$N, 30% HBr—AcOH, PhMe, 100°, 10 hr	No reaction	547
	2-BrC$_5$H$_4$N, 30% HBr—AcOH, PhMe, Δ, 6.5 hr	1-(6-Methyl-2-pyridyl)-2-pyridone (16.1%); 2-picoline-1-oxide (29%); 2-pyridone (1.1%)	340
2,6-Me$_2$	2-BrC$_5$H$_4$N, 30% HBr—AcOH, 100°, 10 hr	Mixture darkens	547
2-OMe	2-BrC$_5$H$_4$N	1-Methoxy-2-pyridone (43.8%); 2-methoxy-pyridine-1-oxide (25.5%)	340
2-NHAc	2-BrC$_5$H$_4$N	2,2'-Dipyridylamine (3%); N-[6-(2-pyridylamino)-2-pyridyl]-2-pyridone (10.6%); 2,2'-dipyridyl-amine-1-oxide (2.8%)	340
3-Me	2-BrC$_5$H$_4$N, 100°, 2.5 hr	3-Picoline (3.2%); 3-bromo-1-(5-methyl-2-pyridyl)-2-pyridone (20.4%); 1-(5-methyl-2-pyridyl-2-pyridone (17.8%)	547

Table IV-73 (*Continued*)

Substituent in *N*-oxide	Conditions	Products (yield)	Ref.
	2-BrC$_5$H$_4$N, HBr—AcOH, 40°, 4 hr	3-Picoline (4.6%); 3-bromo-1-(5-methyl-2-pyridyl)-2-pyridone (4.5%); 1-(5-methyl-2-pyridyl)-2-pyridone (51%)	547
	2-BrC$_5$H$_4$N, 30% HBr—AcOH, PhMe, Δ, 6.5 hr	1-(5-Methyl-2-pyridyl)-2-pyridone (42.4%); 3-bromo-1-(5-methyl-2-pyridyl)-2-pyridone (13%); 1-(3-methyl-2-pyridyl)-2-pyridone (11.3%); 2-pyridone (1.3%)	340
	4-BrC$_5$H$_4$N, 110°, 3 hr, dioxane	3-Picoline-1-oxide (50%); 3-picoline (0.4%); 1-(4-pyridyl)-4-pyridone, 4-pyridone	548
3-NHAc	2-BrC$_5$H$_4$N, 30% HBr—AcOH, Δ, 6 hr	Dipyrido[1,2-*a*; 3′,2′-*d*]-imidazole	550
	2-BrC$_5$H$_4$N	1-(5-Acetamido-2-pyridyl)-2-pyridone (21.2%); 1-(3-acetamido-2-pyridyl)-2-pyridone (8.6%); dipyrido[1,2-*a*; 2′,3′-*d*]imidazole (6.2%); 1-(5-amino-2-pyridyl)-2-pyridone (9.4%); 1-(3-amino-2-pyridyl)-2-pyridone (8.6%)	340
3-OMe	2-BrC$_5$H$_4$N	1-(5-Methoxy-2-pyridyl)-2-pyridone (13.3%); 1-(3-methoxy-2-pyridyl)-2-pyridone, unidentified C$_{10}$H$_{10}$N$_2$O$_2$, m.p. 232–233° (27.5%)	340
3-CN	2-BrC$_5$H$_4$N	3-Cyanopyridine (1.4%); 3-bromo-1-(5-cyano-2-pyridyl-2-pyridone (1.1%); 1-(5-cyano-2-pyridyl)-2-pyridone (31.3%), 2-pyridone (0.2%)	340
3-NO$_2$	2-BrC$_5$H$_4$N	3-Nitropyridine (22.4%); 3-bromo-1-(5-nitro-2-pyridyl)-2-pyridone (13.5%); 1-(5-nitro-2-pyridyl)-2-pyridone (27%); 1-(3-nitro-2-pyridyl)-2-pyridone (3%); 2-pyridone (0.9%)	340
4-Me	2-BrC$_5$H$_4$N, Δ, N$_2$	4-Picoline (17.4%); 1-(4-methyl-2-pyridyl)-2-pyridone, 2-pyridone	547
	2-BrC$_5$H$_4$N, 30% HBr—AcOH, 100°, N$_2$	4-Picoline (1%); 1-(4-methyl-2-pyridyl)-2-pyridone (30%); 2-pyridone	547
	2-BrC$_5$H$_4$N, 30% HBr—AcOH, PhMe, Δ, 6.5 hr	1-(4-Methyl-2-pyridyl)-2-pyridone (31%); 2-pyridone (0.8%)	340
4-OMe	2-BrC$_5$H$_4$N	1-(4-Methoxy-2-pyridyl)-2-pyridone (13.5%); 2-pyridone (2.7%)	340
4-NHAc	2-BrC$_5$H$_4$N	1-(4-Acetamido-2-pyridyl)-2-pyridone (39.6%); 1-(4-amino-2-pyridyl)-2-pyridone (8.1%); 2-pyridone	340

product $C_{11}H_{10}N_2O_2$ obtained from 3-methoxypyridine-1-oxide and 2-bromopyridine is probably the 5-methoxy derivative of **IV-307**.

On the other hand, 1-(4-pyridyl)-4-pyridone results from the reaction of pyridine-1-oxide and 4-bromopyridine, indicating that in this case the *N*-oxide bearing ring is not a partner in this reaction [1-(4-pyridyl)-4-pyridone is the hydrolysis product of the dimer of 4-bromopyridine] but probably functions as a base. The same product is obtained from the reaction of 3-picoline-1-oxide with 4-bromopyridine (340, 548, 549). When 3-picoline-1-oxide and 2-acetoxypyridine were heated at 100° for 6 days, a mixture of 1-(5′-methyl-2′-pyridyl)-2-pyridone (**IV-308**) (10%), 1-(3′-methyl-2′-pyridyl)-2-pyridone (**IV-309**) (4%), and 2-pyridone (**IV-310**) were obtained (328).

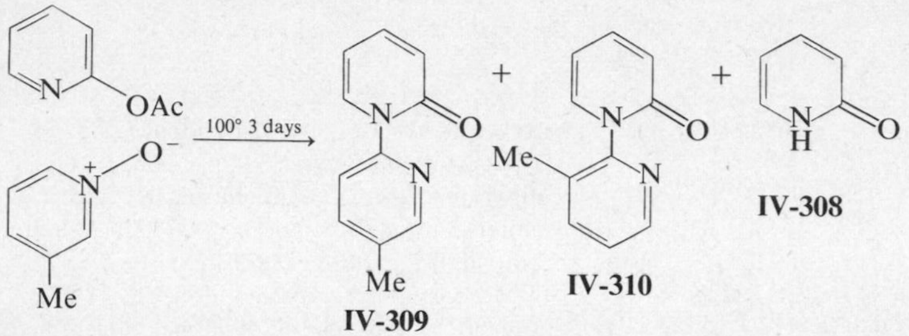

The reaction of 2-bromopyridine with pyridine-1-oxides is very similar to the reaction of pyridine-1-oxide with tosyl chloride (551a) (see Section

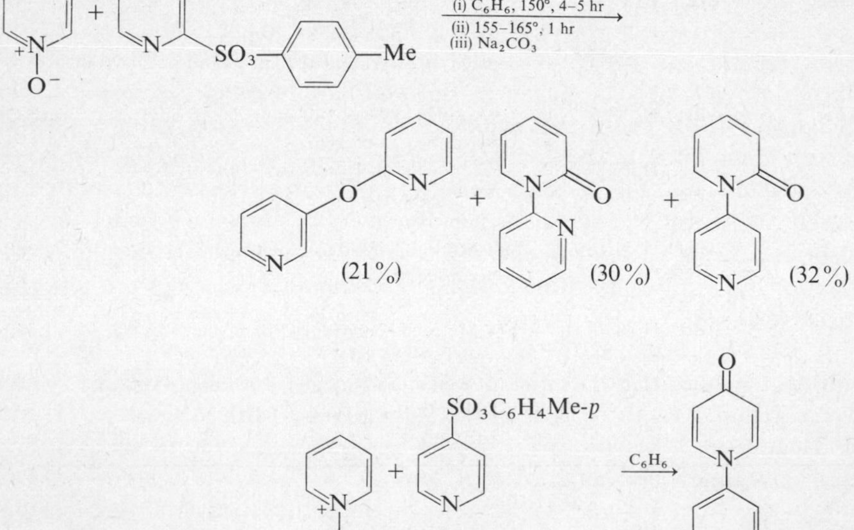

Scheme IV-18

IV.2.C) or with 2-pyridyl *p*-toluenesulfonate (551b) (Scheme IV-18). Treatment of pyridine-1-oxides with perfluoropropene yields 2-(1,2,2,2-tetrafluoroethyl)pyridine (**IV-311**) (552), presumably *via* **IV-312** (eq. IV-55). A new method of alkyl- and aryl-amidation of pyridine-1-oxides has been

reported (553a). For example, pyridine-1-oxide is heated with an imidoyl chloride, *N*-phenylbenzimidoyl chloride (or the corresponding nitrilium salt), in a polar non-protic solvent to give *N*-benzoyl-2-anilinopyridine (**IV-313**; X = Y = H, R = R' = Ph) which, on hydrolysis, yields 2-anilinopyridine (**IV-314**; X = Y = H', R' = Ph) (Table IV-74). Other approaches involve the formation of the nitrilium salt $[RC\equiv N^+R]Z^-$ (553a) prior to addition of the *N*-oxide or the decomposition of an aromatic diazonium salt in acetonitrile in the presence of pyridine-1-oxide (554a) (Scheme IV-19). From the reaction of pyridine-1-oxide with *N*-phenylbenzimidoyl chloride, benzanilide was formed as a by-product and could be isolated *before* the addition of water. 3-Chloropyridine was isolated by gas chromatography and can be accounted for by pathway **b** (or **b'**) which is also an alternative route to **IV-313** and **IV-314** (553a). The results of the reactions of various pyridine-1-oxides with imidoyl chlorides are reported in Table IV-74.

It was observed that the reaction of pyridine-1-oxide with *N*-phenyl-4-substituted benzimidoyl chlorides (**IV-315**) gave various products which were depended on the nature of the substituent (554b). When X = H, Me, or OMe (+I, +M or +M > −I), 2-anilinopyridine (50 to 60%) and 3-chloropyridine were obtained, but when X = Cl or NO$_2$ (−I > +M or −I, −M), 2-chloro- and 3-chloropyridine were isolated and no 2-anilinopyridine was formed (Table IV-74) (Scheme IV-20).

The effect of a 3-substituent in the pyridine-1-oxide ring upon the orientation of the direct acylamination has been studied (Table IV-75) (eq. IV-56)

TABLE IV-74. Reaction of Pyridine-1-oxides with Imidoyl Chlorides

Substituent in *N*-oxide	Conditions	Products (yield)	Ref.
None	PhC(Cl)=NPh, ethylene chloride, Δ, 10 hr	Pyridine-1-oxide hydrochloride (47%); 3-chloropyridine (28.1%); 2-anilinopyridine (3.1%); 3-benzoyloxypyridine (1.2%); benzanilide (36.8%); 2-(*N*-benzoylanilino)pyridine (57.2%)	553a, b
	PhC≡$\overset{+}{N}$Ph SbCl$_6^-$, ethylene chloride, R.T., 10 hr	Benzanilide; 2-(*N*-benzoylanilino)pyridine (34%)	553b
	PhC(Cl)=NC$_6$H$_4$-*p*-Me, ethylene chloride, Δ, 10 hr	Pyridine-1-oxide hydrochloride (68%); *N*-benzoyl-*p*-toluidide (42%); 2-(*N*-benzoyl-*p*-toluidino)pyridine (57%)	553b
	(i) PhC(Cl)=NC$_6$H$_4$-*p*-Me, ethylene chloride, Δ, 11 hr (ii) 5% NaOH, Δ, 4 hr	*N*-Benzoyl-*p*-toluidide (11%); 2-*N*-*p*-toluidino-pyridine (80%)	553b
	PhC(Cl)=NC$_6$H$_4$-*p*-OMe, ethylene chloride, Δ, 7 hr	Pyridine-1-oxide hydrochloride (54%); *N*-benzoyl-*p*-anisidide (trace), 2-(*N*-benzoyl-*p*-anisidino)pyridine (44%)	553b
	(i) PhC(Cl)=NC$_6$H$_4$-*p*-OMe, ethylene chloride, Δ, 7 hr (ii) 2*N* HCl, Δ, 9 hr	Pyridine-1-oxide hydrochloride (54%); 2-*N*-*p*-anisidinopyridine (64%)	553b
	PhC(Cl)=NC$_6$H$_4$-*p*-Cl, chlorobenzene, Δ, 9.5 hr	Pyridine-1-oxide hydrochloride (69%); 2-(*N*-benzoyl-*p*-chloroanilino)pyridine (29%)	553b
	(i) PhC(Cl)=NC$_6$H$_4$-*p*-Cl, chlorobenzene, Δ, 10 hr (ii) 2*N* HCl, Δ, 9 hr	2-*N*-*p*-chloroanilinopyridine (53%)	553b
	PhC(Cl)=NC$_6$H$_4$-*p*-NO$_2$, chlorobenzene, Δ, 32 hr	Pyridine-1-oxide hydrochloride (61%); *N*-benzoyl-*p*-nitroanilide (19%); 2-(*N*-benzoyl-*p*-nitroanilino)pyridine (33%)	553b
	(i) PhC(Cl)=NC$_6$H$_4$-*p*-NO$_2$, chlorobenzene, Δ, 9.5 hr (ii) 5% NaOH, Δ, 3 hr	2-*N*-*p*-Nitroanilinopyridine (71%)	553b
	(i) PhC≡$\overset{+}{N}$C$_6$H$_4$-*p*-NO$_2$-SbCl$_6^-$, ethylene chloride, Δ, 17 hr (ii) 2*N* HCl, Δ, 2 hr	2-*N*-*p*-Nitroanilinopyridine (89%)	553a, b
	PhC(Cl)=NCH$_2$Ph, ethylene chloride, Δ, 5 hr	Pyridine-1-oxide hydrochloride (75%); 2-(*N*-benzoylbenzylamino)pyridine (54%)	553a, b
	(i) PhC(Cl)=NCH$_2$Ph, ethylene chloride, Δ, 5 hr (ii) 2*N* HCl, Δ	Pyridine-1-oxide hydrochloride (75%); *N*-benzoylbenzylamide (38%); 2-*N*-benzyl-aminopyridine (50%)	553b

Table IV-74 (*Continued*)

Substituent in N-oxide	Conditions	Products (yield)	Ref.
	PhC(Cl)=N—⟨cyclohexyl⟩ ethylene chloride, Δ, 10 hr	Pyridine-1-oxide hydrochloride (84%); N-benzoylcyclohexylamide (17%); 2-(N-benzoylcyclohexylamino)pyridine (64%)	553b
	Caprolactam pseudochloride, CHCl₃, Δ, 15 hr	Pyridine-1-oxide hydrochloride (61%); N-2-pyridylcaprolactam (11%)	553b
	Saccharin pseudochloride, chlorobenzene, Δ, 18 hr	N-2-Pyridyl-O-carbethoxybenzenesulfonamide (10.4%); pyridine-1-oxide (39%)	553a, b
	(i) Saccharin pseudochloride, chlorobenzene, Δ, 50 hr (ii) Crystallized from ethanol	N-2-Pyridylsaccharin (38.4%)	553b
	(i) Saccharin pseudochloride, chlorobenzene, Δ, 50 hr (ii) Chromatographed on silica gel	N-2-Pyridyl-O-carbethoxybenzenesulfonamide (21%)	553b
	PhC(Cl)=NPh, PhCONHC₆H₄-*p*-Me (1:1 molar ratio) ethylene chloride, Δ, 8 hr	Pyridine-1-oxide hydrochloride (39%); 2-(N-benzoylanilino)pyridine (23.6%); 2-(N-benzoyl-*p*-toluidino)pyridine (24.7%)	553b
	p-MeC₆H₄C(Cl)=NPh, (CH₂Cl)₂, Δ	2-Anilinopyridine (58.8%); 3-chloropyridine (1.7%)	554b
	p-MeOC₆H₄C(Cl)=NPh, (CH₂Cl)₂, Δ	2-Anilinopyridine (52.3%); 3-chloropyridine (14.7%)	554b
	p-ClC₆H₄C(Cl)=NPh, (CH₂Cl)₂, Δ	3-Chloropyridine (11.3%); 2-chloropyridine (13.4%)	554b
	p-NO₂C₆H₄C(Cl)=NPh, (CH₂Cl)₂, Δ	3-Chloropyridine (15%); 2-chloropyridine (21.8%)	554b
2-Me	PhC(Cl)=NPh, ethylene chloride, Δ, 6 hr	Benzanilide, 2-(N-benzoylanilino)-6-methylpyridine (42%)	553b
2,6-Me₂	PhC(Cl)=NPh, CHCl₃, Δ, 48 hr	O-(2,6-Dimethyl-3-pyridyl)-N-phenylbenzimidate (35.4%); 3-chloro-2,6-lutidine (13.3%); 2-chloromethyl-6-methylpyridine (8.7%); benzanilide (32.2%)	19b, 555a
	PhC(Cl)=NPh, CHCl₃	O-(2,6-Dimethyl-3-pyridyl)-N-phenylbenzimidate (59%)	555b
	PhC(Cl)=NPh, (CH₂Cl)₂, Δ, 48 hr	O-(2,6-Dimethyl-3-pyridyl)-N-phenylbenzimidate (37.3%); 3-chloro-2,6-lutidine (20.8%); 2-chloromethyl-6-methylpyridine (12.3%); benzanilide (44.9%)	19b, 555a
4-Me	PhC(Cl)=NPh, ethylene chloride, Δ, 3 hr	Benzanilide (36%); 2-(N-benzoylanilino)-4-picoline (45%); 4-picoline-1-oxide (18%)	553b

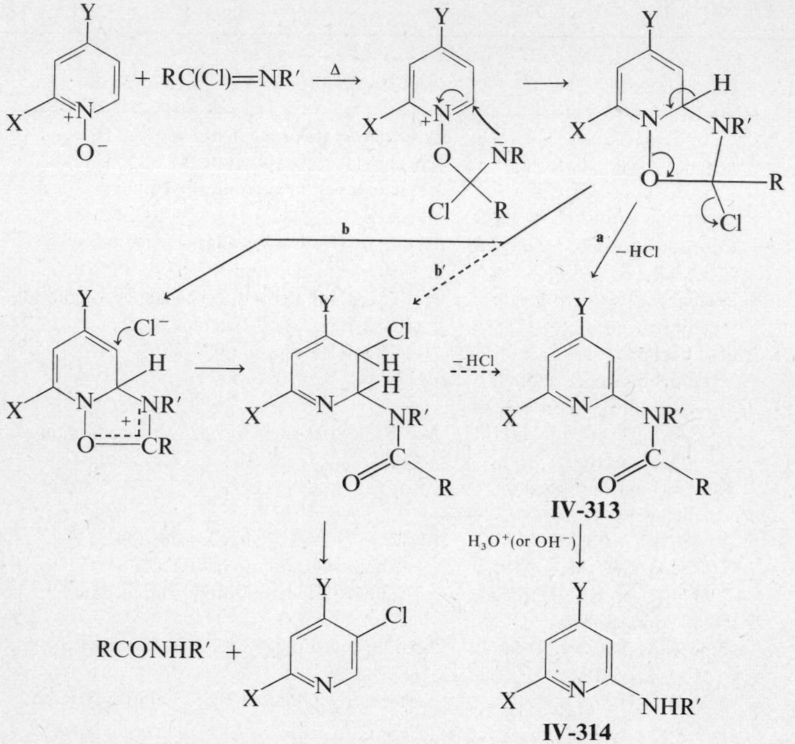

Scheme IV-19

(19b). The expected 2-(N-benzoylanilino)-3-mesylaminopyridine was not obtained when 3-mesylaminopyridine-1-oxide and N-phenylbenzimidoyl

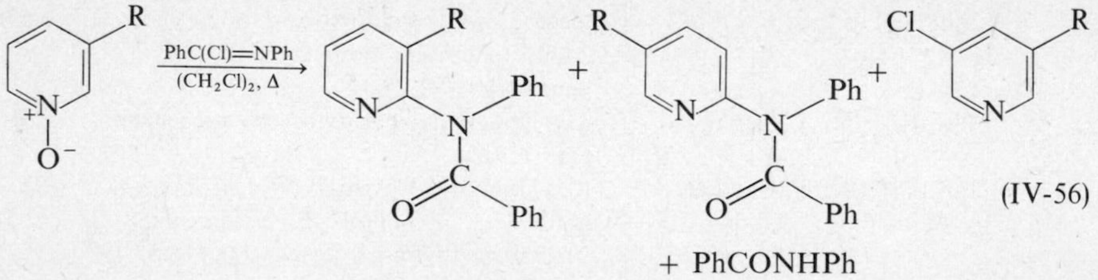

$$ (IV\text{-}56) $$

chloride were heated in ethylene chloride. Instead, 1,2-diphenyl-7-aza-benzimidazole R.I. 1,193 (IV-316) was obtained (19b), together with the expected 2,5-isomer. Another interesting feature is the absence of any 5-chloro-3-mesylaminopyridine among the products but the formation of a small amount of 2-chloro-5-mesylaminopyridine (IV-317).

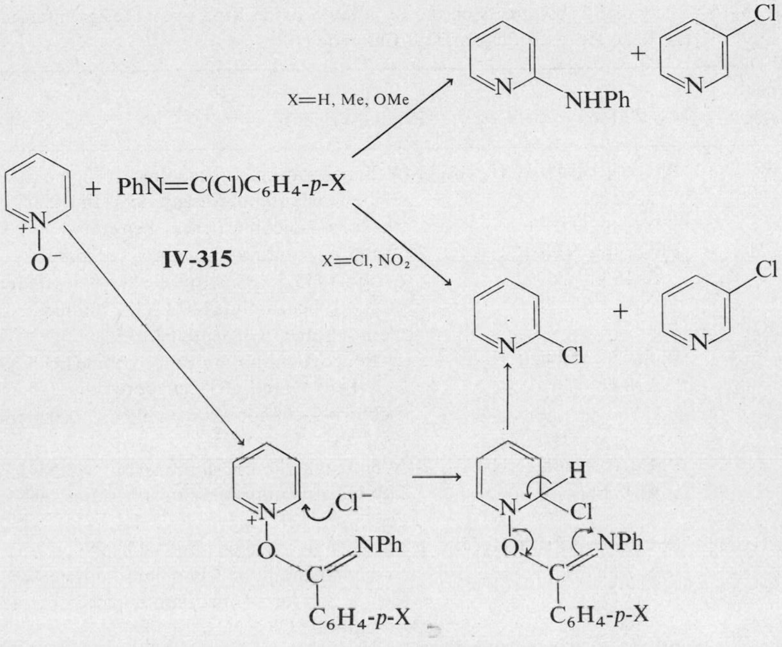

Scheme IV-20

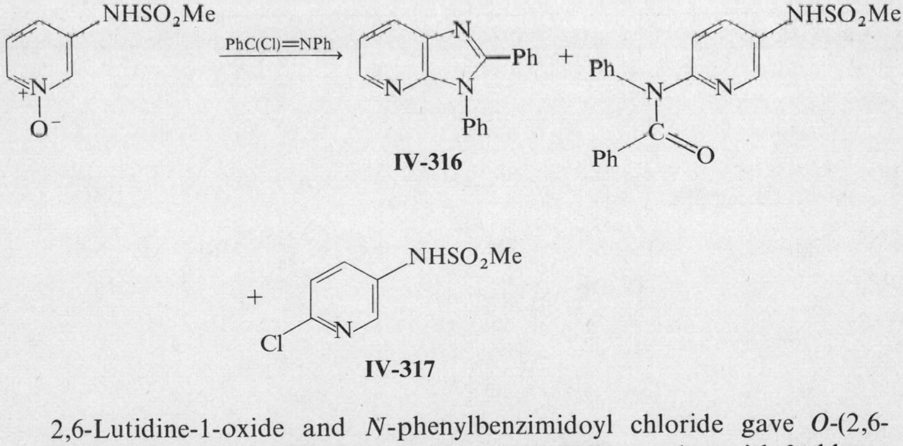

IV-316

IV-317

2,6-Lutidine-1-oxide and *N*-phenylbenzimidoyl chloride gave *O*-(2,6-dimethyl-3-pyridyl)-*N*-phenylbenzimidate (**IV-318**), together with 3-chloro-2,6-lutidine and 2-chloromethyl-6-methylpyridine (Table IV-74) (eq. IV-57) (19b). A proposed mechanism which would lead to the formation of **IV-318** involves a 1,5-sigmatropic shift.

TABLE IV-75. Effect of 3-Substituents on the Pyridine-1-oxide Ring upon the Orientation in the Reaction with Benzimidoyl Chloride (19b)

Substituent in N-oxide	Conditions	Products (yield)
3-Me	PhC(Cl)=NPh, (CH$_2$Cl)$_2$, Δ, 12 hr	2-(N-Benzoylanilino)-3-picoline (2.3%); 2-(N-benzoylanilino)-5-methylpyridine (82.3%); 5-chloro-3-picoline (trace), benzanilide (12.3%)
3-CO$_2$CH$_3$	PhC(Cl)=NPh, (CH$_2$Cl)$_2$, Δ, 16 hr	2-(N-Benzoylanilino)-5-methoxycarbonyl-pyridine (53.5%); methyl 5-chloronicotinate (36.5%); benzanilide (40.1%); 2-anilino-5-methoxycarbonylpyridine (4.1%)
3-CN	PhC(Cl)=NPh, (CH$_2$Cl)$_2$, Δ, 8 hr	2-(N-Benzoylanilino)-3-cyanopyridine (33.4%); 2-(N-benzoylanilino)-5-cyanopyridine (19.8%); 5-chloro-3-cyanopyridine (30.6%); benzanilide (36.7%)
3-F	PhC(Cl)=NPh, (CH$_2$Cl)$_2$, Δ, 48 hr	2-(N-Benzoylanilino)-3-fluoropyridine (25.9%); 2-(N-benzoylanilino)-5-fluoropyridine (55.9%); benzanilide (10%)
3-NHSO$_2$CH$_3$	PhC(Cl)=NPh, (CH$_2$Cl)$_2$, Δ, 7 hr	1,2-Diphenyl-7-azabenzimidazole (29%); 2-(N-benzoylanilino)-5-mesylaminopyridine (26%); 2-chloro-5-mesylaminopyridine (trace), benzanilide (28%)
3-OMe	PhC(Cl)=NPh, (CH$_2$Cl)$_2$, Δ, 14 hr	2-(N-Benzoylanilino)-3-methoxypyridine (76.2%); 2-(N-benzoylanilino)-5-methoxy-pyridine (18.9%); benzanilide (4.2%)
3-Br[a]	PhC(Cl)=NPh, (CH$_2$Cl)$_2$, Δ, 12 hr	2-(N-Benzoylanilino)-3-bromopyridine (30%)

[a] Ref. 374.

(IV-57)

The reaction of pyridine-1-oxide with alkyl mercaptans in the presence of acetic anhydrides and with aryl mercaptans in the presence of alkyl or aryl sulfonyl halides has been described previously (see Section IV.2.C). Reaction of an aryl mercaptan, for example thiophenol, with pyridine-1-oxide and benzimidoyl chloride gives 3-phenylthiopyridine in addition to 2-anilinopyridine and 3-chloropyridine (eq. IV-58) (Table IV-76) (472). In the presence of alkyl mercaptans only small amounts of 3-alkylthiopyridines were obtained.

TABLE IV-76. Reaction of Alkyl and Aryl Mercaptans with Pyridine-1-oxide Imide Chloride Complex (472)

Conditions	Products (yield)
(i) PhC(Cl)=NPh, ethylene chloride, R.T., 1 hr (ii) PhSH, Δ, 8 hr	3-Phenylthiopyridine (35.6%); 3-chloropyridine (1.3%); 2-anilinopyridine (49.2%)
(i) PhC(Cl)=NPh, R.T., ethylene chloride, 1 hr (ii) p-MeOC₆H₄SH, Δ, 24 hr	3-(p-Methoxyphenylthio)pyridine (30%); 3-chloropyridine (traces), 2-anilinopyridine (43%)
(i) PhC(Cl)=NPh, R.T., ethylene chloride, 1 hr (ii) t-BuSH, Δ	3-t-Butylthiopyridine (4.1%); 3-chloropyridine (1.5%); 2-anilinopyridine (46.0%)
(i) PhC(Cl)=NPh, R.T., ethylene chloride, 1 hr (ii) n-decylmercaptan, Δ	3-n-Decylthiopyridine (0.9%); 3-chloropyridine (1.7%); 2-anilinopyridine (47.3%)

(IV-58)

Pyridine-1-oxide undergoes a 1,3-dipolar addition with phenylisocyanate to form an adduct **IV-319** (not isolated)* which decomposes to give 2-anilinopyridine in good yield (556) (eq. IV-59). In contrast, it should be mentioned that pyridine-1-oxide is an effective catalyst for the reaction of phenyl

* The adducts from 3-picoline-1-oxide have recently been isolated (T. Hisano, S. Yoshikawa, and K. Muraoka, *Org. Prep. Proc. Intern.*, **5**, 95 (1973)).

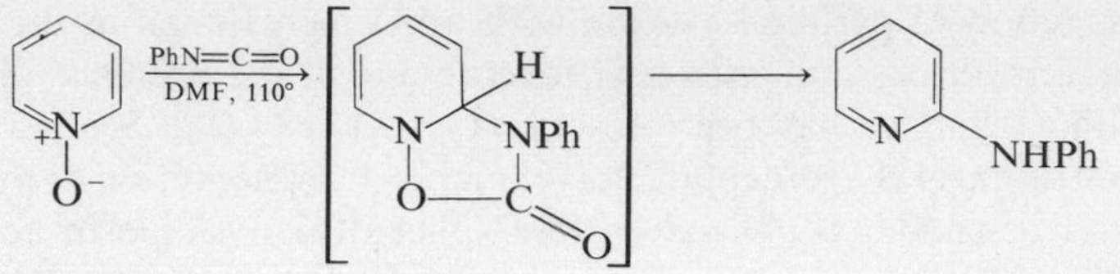

(IV-59)

IV-319

isocyanate with butanol in toluene at 39° and the combination of pyridine-1-oxide with alkylene oxide causes the trimerization of phenylisocyanate (557). Pyridine-1-oxide was also shown to be an effective catalyst for the conversion of isocyanates to carbodiimides. This may be attributed to the simple polarization of the $-N=C=O$ by dipole-dipole interaction or *via* an intermediate such as **IV-320**, which reacts as such with another molecule of isocyanate with the simultaneous elimination of carbon dioxide and regeneration of the catalyst (558).

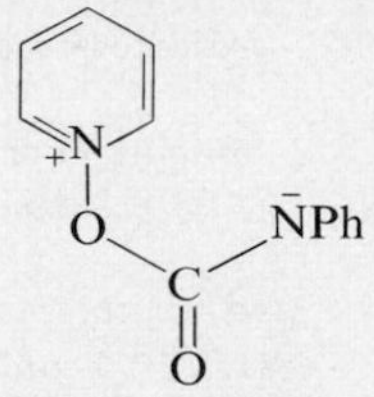

IV-320

N-(p-Dimethylaminophenyl)-α-(1-oxido-2-pyridyl)nitrone **(IV-321)** and thionyl chloride react to give a variety of products (eq. IV-60) (559). Nuclear chlorination is thought to occur probably through intermediate **IV-322**.

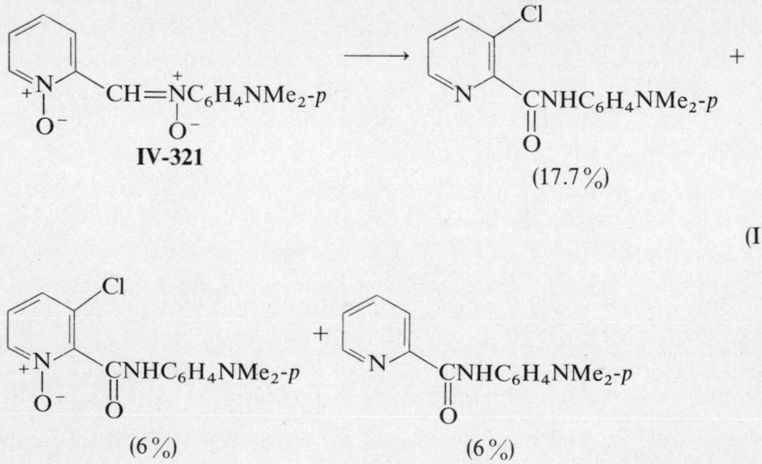

IV-321

(17.7%)

(IV-60)

(6%)   (6%)

IV-322

6. Pyridyne-1-oxides (Dehydropyridine-1-oxides)

Various aspects of the chemistry of pyridyne-1-oxides have already been reviewed (560, 561).

2-Chloropyridine-1-oxide reacts with potassium amide in liquid ammonia to give very low yields of 2- and 3-aminopyridine-1-oxide (562, 563). This provides evidence for the intervention of a 2,3-pyridyne-1-oxide intermediate, although the 2-amino compound probably arises mainly by a direct S_N2Ar attack (eq. IV-61). Both 3-chloro- and 3-bromopyridine-1-oxide were converted almost exclusively into 3-aminopyridine-1-oxide, but

$$\text{(IV-61)}$$

3-fluoropyridine-1-oxide remains unchanged when treated with a solution of potassium amide in liquid ammonia (31). This evidence excludes the AE mechanism for the animation of 3-chloro- and 3-bromopyridine-1-oxide as in aromatic compounds undergoing this reaction fluorine atoms are more easily substituted than other halogen atoms. Since no 4-aminopyridine-1-oxide was obtained from the aminations of 3-chloro- and 3-bromopyridine-1-oxide, these aminations did not proceed via 3,4-pyridyne-1-oxide (31), as predicted on the basis of the base-catalyzed H—D exchange results (522). Aminations of 3-chloro- and 3-bromopyridine-1-oxides via 2,3-pyridyne-1-oxide were confirmed by the fact that together with 3-aminopyridine-1-oxide, some 2-aminopyridine-1-oxide was formed. Under these conditions, 4-chloropyridine-1-oxide gave only 4-aminopyridine-1-oxide indicating that this reaction probably does not proceed via a dehydro intermediate, otherwise the formation of some 3-amino derivative might have been expected (562). In the case of 2-bromo- and 3-bromo-4-ethoxypyridine-1-oxide, 3-amino-4-ethoxypyridine-1-oxide was formed (562). This reaction probably involves 4-ethoxy-2,3-pyridyne-1-oxide and the addition of the amide ion to this intermediate is directed by the inductive effects of the N-oxide function (eq. IV-62).

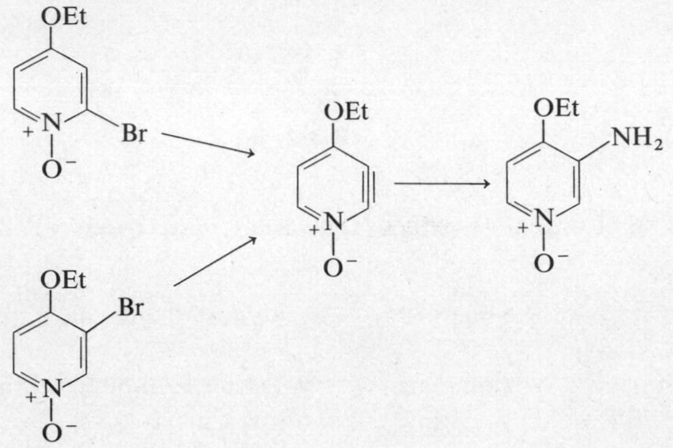

(IV-62)

4-Halopyridine-1-oxides yield 4-aminopyridine-1-oxides either by direct S_N2 substitution or, less likely, *via* a 3,4-pyridyne-1-oxide. If a 2-methyl substituent is present in the 3,4-pyridyne-1-oxide, addition of the nucleophile occurs at both the 3- and 4-position (eq. IV-63) (61). 2- And 4-chloropyridine-

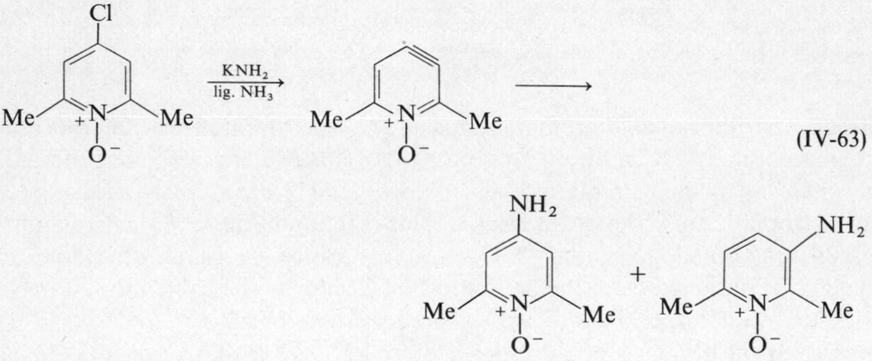

(IV-63)

1-oxide were treated with piperidine at 100° to yield 2- and 4-(N-piperidino)-pyridine-1-oxides, respectively (564). Under these conditions, 3-chloro-pyridine-1-oxide gave 80% of unreacted starting material and 20% of products composed of 3-piperidinopyridine-1-oxide (82%), the corresponding 4-isomer (4%), and unidentified products. Thus 3,4-pyridyne-1-oxide appears to be an intermediate in this reaction (564).

The results of various reactions thought to involve a pyridyne-1-oxide intermediate are given in Table IV-77.

TABLE IV-77. Reactions Which May Involve Dehydropyridine-1-oxides

N-oxides	Conditions	Products (yield)	Ref.
2-Br	(i) KNH$_2$, liq. NH$_3$, 4 min (ii) Reduction	2-Aminopyridine (3%); 3-aminopyridine (6%)	31
2-Br-4-Me	(i) KNH$_2$, liq. NH$_3$, 30 min (ii) Reduction	2-Amino-4-picoline (trace); 3-amino-4-picoline (10–15%)	31
2-Br-4-OEt	(i) KNH$_2$, liq. NH$_3$, 10 min (ii) Fe, AcOH	2-Amino-4-ethoxypyridine (trace); 3-amino-4-ethoxypyridine (10–15%); 4-ethoxypyridine (trace); 3-bromo-4-ethoxypyridine [3-amino-4-ethoxypyridine-1-oxide (estimated yield 13–14%)]	562
2-Cl	(i) KNH$_2$, liq. NH$_3$ (ii) Fe, AcOH	2- And 3-Aminopyridine (ratio 5:1), 2-chloropyridine [2-aminopyridine-1-oxide (est. 5%), 3-aminopyridine-1-oxide (est. 1%)]	562
3-Br	(i) KNH$_2$, liq. NH$_3$, 4 min (ii) Reduction	2-Aminopyridine (trace), 3-aminopyridine (75–80%)	31
3-Br-4-Me	(i) KNH$_2$, liq. NH$_3$, 30 min (ii) Reduction	2-Amino-4-picoline (trace), 3-amino-4-picoline (30–35%)	31
3-Br-2,5-Me$_2$	(i) KNH$_2$, liq. NH$_3$, 1 hr	3-Amino-2,5-lutidine (10–15%); 4-amino-2,5-lutidine (55–60%)	31
3-Br-4-OEt	(i) KNH$_2$, liq. NH$_3$, 10 min (ii) Fe, AcOH	3-Amino-4-ethoxypyridine, 4-ethoxypyridine [3-amino-4-ethoxypyridine-1-oxide (est. 60%)]	562
3-Br-5-OEt	(i) KNH$_2$, liq. NH$_3$, 10 min (ii) Reduction	3-Amino-5-ethoxypyridine (60–65%); 4-amino-5-ethoxypyridine (5%)	31
3-Cl	(i) KNH$_2$, liq. NH$_3$ (ii) Fe, AcOH	3-Aminopyridine, 3-chloropyridine (trace) [3-aminopyridine-1-oxide (est. 80%)]	562
	Piperidine, benzene, 100°	3-(N-Piperidino)pyridine-1-oxide, 4-(N-piperidino)pyridine-1-oxide, (ratio 19:1) (86%)	564
3-Cl-2,6-Me$_2$	KNH$_2$, liq. NH$_3$	3-Amino-2,6-lutidine-1-oxide, 4-amino-2,6-lutidine-1-oxide (ratio 0.26:1) (63%)	61
3-F	KNH$_2$, liq. NH$_3$	No reaction	31
4-Br	(i) KNH$_2$, liq. NH$_3$, 4 min (ii) Reduction	4-Aminopyridine (60–65%)	31
4-Br-2,5-Me$_2$	(i) KNH$_2$, liq. NH$_3$, 1 hr (ii) Reduction	3-Amino-2,5-lutidine (trace), 4-amino-2,5-lutidine (60–65%)	31
4-Br-5-OEt	(i) KNH$_2$, liq. NH$_3$, 10 min (ii) Reduction	3-Amino-5-ethoxypyridine (trace), 4-amino-5-ethoxypyridine (60–65%)	31
4-Cl	(i) KNH$_2$, liq. NH$_3$ (ii) Fe, AcOH	4-Aminopyridine, 3-aminopyridine (trace) [4-aminopyridine-1-oxide (est. 60–70%)]	562, 563
4-Cl-2,6-Me$_2$	KNH$_2$, liq. NH$_3$	3-Amino-2,6-lutidine-1-oxide, 4-amino-2,6-lutidine-1-oxide (ratio 0.19:1) (40%)	61

V. Side-chain and Substituent Reactions

1. Methyl Groups

Condensation of 2- (**IV-279**) and 4-picoline-1-oxide with ethyl oxalate in the presence of potassium ethoxide yields the ethyl 2-(**IV-323**) and 4-pyridyl-pyruvate-1-oxides (565). This reaction has been reported with substituted pyridine-1-oxide (71, 566). (See also Chapter XI, Section I.2.E.b). Using the conditions reported by Adams and Miyano (565) for the preparation of

R = Me, OMe, OEt

IV-279   **IV-323**

ethyl 2-pyridylpyruvate-1-oxide (**IV-323**), a compound having the same melting point was obtained (34), but whose NMR spectrum was not in accord with the proposed structure. Structure **IV-324** was favoured for this compound which appeared to be oxidized to **IV-325**. Recent repetition of this work failed to give products corresponding to **IV-324** and **IV-325** but gave instead the desired **IV-323**, oxidation of which with 10% NaOH and 30% H_2O_2 gave picolinic acid-1-oxide.

IV-324   **IV-325**

When 4-nitro-3-picoline-1-oxide (**IV-106**) was treated with ethyl oxalate in the presence of potassium ethoxide, it gave 4,4'-dinitro-3,3'-dipicolyl-1,1'-dioxide (**IV-326**) (12a), which was also prepared by the oxidative dimerization of 4-nitro-3-picoline-1-oxide (**IV-106**) in the presence of butyl nitrite followed by treatment with sodium methoxide. Since this reaction also occurs in the absence of butyl nitrite, it has been suggested that either the nitro or the N-oxide group can function as an oxidizing agent (50, 567). When this reaction was carried out in the presence of oxygen in 30% methanolic potassium hydroxide solution, deoxygenation accompanied by nucleophilic substitution of the nitro group by methoxyl occurred to give 1,2-bis(4-methoxy-3-pyridyl)ethylene (**IV-327**) (50) (eq. IV-64). Under the same conditions, 4-ethoxy-3-picoline-1-oxide does not undergo dimerization, so that

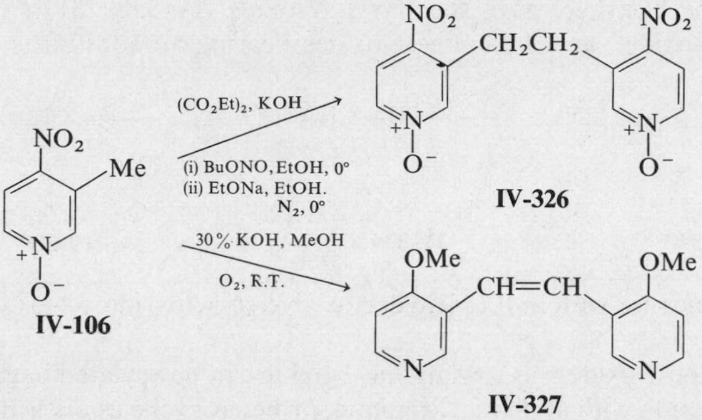

(IV-64)

activation of the methyl by the nitro group is necessary for this reaction to proceed. Table IV-78 summarizes these results.

2- And 4-picoline-1-oxides do not condense with p-dimethylamino-benzaldehyde in the presence of zinc chloride or concentrated hydrochloric acid, but do so in the presence of piperidinium acetate in toluene (568). Similarly, 2-picoline-1-oxide underwent condensation with aldehydes in the presence of pyridine and potassium hydroxide (135, 137, 569) (eq. IV-65).

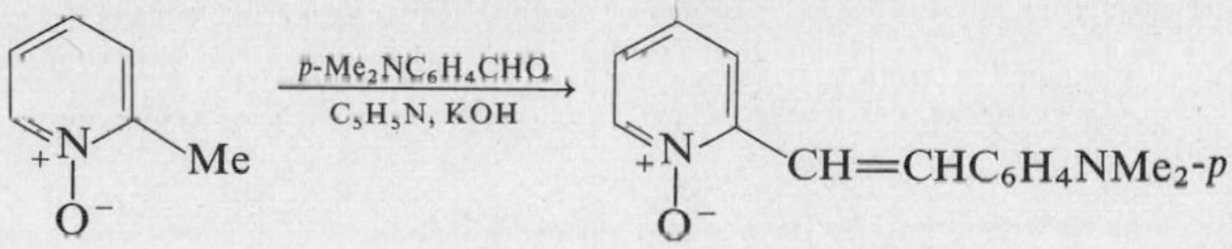

(IV-65)

The methyl groups in 1-ethoxy-2- or 4-methylpyridinium salts (IV-328) readily condenses with benzaldehyde or anisaldehyde to give IV-329 which,

TABLE IV-78. Reaction of 4-Nitro-3-picoline-1-oxides with Base (50)

Conditions	Products (yield)
KOH, MeOH	1,2-Bis(1,1'-dioxido-4-methoxy-3-pyridyl)-ethylene, m.p. 242° (decomp.); picrate, m.p. 237–238° (decomp.)
Na, EtOH, 5°	1,2-Bis(1,1'-dioxido-4-ethoxy-3-pyridyl)ethylene, m.p. 231–232° (decomp.) (42%); picrate, m.p. 211–212°
KOH, MeOH, R.T., 10 hr, air	1,2-Bis(4-methoxy-3-pyridyl)ethylene
Na, EtOH, R.T., 24 hr	1,2-Bis(4-ethoxy-3-pyridyl)ethylene

on alkaline cleavage, gave the styryl *N*-oxide (**IV-330**) (311). 2- And 4-picoline-1-oxides, and 3-picoline-1-oxides bearing an additional activating

IV-328   IV-329   IV-330

group, condense with aryl aldehydes to give styrylpyridine-1-oxides (Table IV-79).

2-Picoline-1-oxide and 2,6-lutidine-1-oxide can be acylated in good yields by the reaction with aliphatic, aromatic, or heterocyclic esters using sodium hydride in liquid ammonia (575).

4-Picoline-1-oxide was condensed with 1,4-dibromobutane in the presence of sodamide in liquid ammonia to give 1,6-di-(4-pyridyl)hexane-1,1'-dioxide in low yield (48). Heating 2-picoline-1-oxide and 4-chloro-2-methylquinoline together in benzene in the presence of potassium *t*-butoxide gave 2-methyl-4-(2-pyridylmethyl-1-oxido)quinoline (**IV-331**) (576).

IV-331

The benzylic-type chlorine atom of 2-chloromethylpyridine-1-oxide (**IV-332**) can be readily replaced by other groups by reaction of **IV-332** with nucleophiles to give **IV-333** (Table IV-80) (80b). 2-Picoline-1-oxide undergoes

IV-332   IV-333

chloromethylation and deoxygenation with trichloroacetyl chloride to yield 2-chloromethylpyridine (**IV-334**) (577). Addition of trichloroacetyl chloride to a boiling solution of 2-picoline-1-oxide resulted in the rapid evolution of carbon dioxide (20%) and after 1 hr the NMR spectrum showed the presence of the trichloroacetate ester of 2-pyridylmethanol (**IV-335**) (40%). After

TABLE IV-79. Condensation of Picoline-1-oxides with Aryl Aldehydes

Substituent in N-oxide	Conditions	Products (yield)	Ref.
2-Me	PhCHO, MeOK, MeOH, Δ, 3 hr	2-Styrylpyridine-1-oxide, m.p. 160° (22%)	137
	PhCHO, MeOK	2-Styrylpyridine-1-oxide, m.p. 162–163°	568
	PhCHO, MeONa, MeOH, Δ, 4 hr	2-Styrylpyridine-1-oxide, m.p. 164° (28%)	570
	p-MeOC$_6$H$_4$CHO	2-(p-Methoxystyryl)pyridine-1-oxide, m.p. 157–158° (46%)	570
	p-ClC$_6$H$_4$CHO	2-(p-Chlorostyryl)pyridine-1-oxide, m.p. 147° (41%)	570
	2,4-Cl$_2$C$_6$H$_3$CHO	2-(2,4-Dichlorostyryl)pyridine-1-oxide, m.p. 140° (15%)	570
	3,4-Cl$_2$C$_6$H$_3$CHO	2-(3,4-Dichlorostyryl)pyridine-1-oxide, m.p. 160° (15%)	570
	p-Me$_2$NC$_6$H$_4$CHO, KOH, C$_5$H$_5$N, Δ	2-(p-Dimethylaminostyryl)pyridine-1-oxide, m.p. 200–201° (57%)	568
	p-Me$_2$NC$_6$H$_4$CHO	2-(p-Dimethylaminostyryl)pyridine-1-oxide, m.p. 200° (50%)	570
2,3-Me$_2$	5-NO$_2$-2-CHO-C$_4$H$_2$O, Ac$_2$O, R.T., 40 hr, pH 8	2-[2-(5-Nitrofuryl)vinyl]-3-picoline-1-oxide, m.p. 198–200°	571
2,6-Me$_2$	5-NO$_2$-2-CHO-C$_4$H$_2$O, Ac$_2$O, R.T., 40 hr, pH 8	2,6-[2-(5-Nitrofuryl)vinyl]-3-picoline-1-oxide, m.p. 207–208°	571
2,4-Me$_2$	5-NO$_2$-2-CHO-C$_4$H$_2$O, Ac$_2$O, 40 hr, pH 8	2-[2-(5-Nitrofuryl)vinyl]-4-picoline-1-oxide, m.p. 213° (decomp.)	571
2-Me-4-CH$_2$OH	5-NO$_2$-2-CHO-C$_4$H$_2$O, Ac$_2$O, 110–130°	4-Hydroxymethyl-2-[2-(5-nitrofuryl)vinyl]-pyridine-1-oxide, m.p. 217–218°	572
2-Me-4-CH$_2$OAc	5-NO$_2$-2-CHO-C$_4$H$_2$O, Ac$_2$O, 110–130°	4-Acetoxymethyl-2-[2-(5-nitrofuryl)vinyl]-pyridine-1-oxide, m.p. 188–189°	572
2,5-Me$_2$	5-NO$_2$-2-CHO-C$_4$H$_2$O, Ac$_2$O, 70–80°, 4 hr	6-[2-(5-Nitrofuryl)vinyl]-3-picoline-1-oxide, m.p. 205–207° (decomp.)	573
2-Me-5-Et	5-NO$_2$-2-CHO-C$_4$H$_2$O, Ac$_2$O, 70–80°, 4 hr	5-Ethyl-2-[2-(5-nitrofuryl)vinyl]pyridine-1-oxide, m.p. 155–157°	573
2-Me-5-CH$_2$OH	5-NO$_2$-2-CHO-C$_4$H$_2$O, Ac$_2$O, 110–130°	5-Hydroxymethyl-2-[2-(5-nitrofuryl)vinyl]-pyridine-1-oxide, m.p. 210–211°	572
2-Me-5-CH$_2$OAc	5-NO$_2$-2-CHO-C$_4$H$_2$O, Ac$_2$O, 110–130°	5-Acetoxymethyl-2-[2-(5-nitrofuryl)vinyl]-pyridine-1-oxide, m.p. 178–179°	572
2-Me-6-CH$_2$OH	5-NO$_2$-2-CHO-C$_4$H$_2$O, Ac$_2$O, 110–130°	6-Hydroxymethyl-2-[2-(5-nitrofuryl)vinyl]-pyridine-1-oxide, m.p. 208–209°	572
	5-NO$_2$-2-CHO-C$_4$H$_2$O, Ac$_2$O, 110–130°	6-Acetoxymethyl-2-[2-(5-nitrofuryl)vinyl]-pyridine-1-oxide, m.p. 195–197°	572
3-Me-4-NO$_2$	PhCHO, piperidine, 3 hr	4-Nitro-3-styryl pyridine-1-oxide, m.p. 179–180°	117

185

Table IV-79 (*Continued*)

Substituent in *N*-oxide	Conditions	Products (yield)	Ref.
	o-HOC$_6$H$_4$CHO, Δ, piperidine	3-(*o*-Hydroxystyryl)-4-nitropyridine-1-oxide, m.p. 254° (decomp.) (47%)	117
	p-HOC$_6$H$_4$CHO, piperidine	3-(*p*-Hydroxystyryl)-4-nitropyridine-1-oxide, m.p. >360°	574
	3,4-(HO)$_2$C$_6$H$_3$CHO	3-(3,4-Dihydroxystyryl)-4-nitropyridine-1-oxide, m.p. >360° (11%)	117
	3-MeO-4-HO-C$_6$H$_3$CHO, Δ, 4 hr	3-(4-Hydroxy-3-methoxystyryl)-4-nitropyridine-1-oxide, m.p. 237° (decomp.) (27%)	117
	p-MeOC$_6$H$_4$CHO, Δ, 16 hr	3-(*p*-Methoxystyryl)-4-nitropyridine-1-oxide, m.p. 196° (68%)	117
	3,4-CH$_2$O$_2$C$_6$H$_3$CHO Δ, 11 hr	3-(3,4-Methylenedioxystyryl)-4-nitropyridine-1-oxide, m.p. 238° (decomp.) (67%)	117
	p-ClC$_6$H$_4$CHO	3-(*p*-Chlorostyryl)-4-nitropyridine-1-oxide, m.p. 216–217° (23%)	117
	p-Me$_2$NC$_6$H$_4$CHO, MeOH, piperidine	3-(*p*-Dimethylaminostyryl)-4-nitropyridine-1-oxide, m.p. 208°	164, 574
	1-C$_{10}$H$_7$CHO, Δ, 3 hr	3-(1-Naphthylvinyl)-4-nitropyridine-1-oxide, m.p. 254–255° (decomp.) (47%)	117
3-CH$_2$CO$_2$H	*p*-Me$_2$NC$_6$H$_4$CHO, 160°, 3 hr	3-(4-Dimethylaminostyryl)pyridine-1-oxide, m.p. 176°	164
4-Me	PhCHO, MeONa, MeOH, Δ, 4 hr	4-Styrylpyridine-1-oxide, m.p. 173° (17%)	570
	p-MeC$_6$H$_4$CHO, KOH, MeOH	4-(*p*-Methylstyryl)pyridine-1-oxide, m.p. 185–187° (35%)	139
	p-MeOC$_6$H$_4$CHO, KOH, MeOH	4-(*p*-Methoxystyryl)pyridine-1-oxide, m.p. 159–160°	139
	p-ClC$_6$H$_4$CHO, KOH, MeOH	4-(*p*-Chlorostyryl)pyridine-1-oxide, m.p. 164–166° (40%)	139
	p-ClC$_6$H$_4$CHO, MeONa, MeOH, Δ, 4 hr	4-(*p*-Chlorostyryl)pyridine-1-oxide, m.p. 178°	570
	2,4-Cl$_2$C$_6$H$_3$CHO, MeONa, MeOH	4-(2,4-Dichlorostyryl)pyridine-1-oxide, m.p. 150° (28%)	570
	3,4-Cl$_2$C$_6$H$_3$CHO, MeONa, MeOH	4-(3,4-Dichlorostyryl)pyridine-1-oxide, m.p. 171° (29%)	570
	p-Me$_2$NC$_6$H$_4$CHO, KOH, MeOH	4-(*p*-Dimethylaminostyryl)pyridine-1-oxide, m.p. 257–258° (11%)	139
	p-Me$_2$NC$_6$H$_4$CHO, KOH, pyridine	4-(*p*-Dimethylaminostyryl)pyridine-1-oxide, m.p. 256–257° (78%)	148
	p-Me$_2$NC$_6$H$_4$CHO, MeONa, MeOH	4-(*p*-Dimethylaminostyryl)pyridine-1-oxide, m.p. 240° (29%)	570

TABLE IV-80. Reaction of 2-Chloromethylpyridine-1-oxide with Nucleophiles (80b)

Conditions	Products (yield)
KCN, MeOH, H_2O, 20–25°, 48 hr	2-Cyanomethylpyridine-1-oxide, m.p. 130–132° (64%)
$NaCH(CO_2Et)_2$, C_6H_6, 20–25°, 2 hr	2-[2′,2′-Di(ethoxycarbonyl)]ethylpyridine-1-oxide, b.p. 140°/0.4 mm (34%)
Et_2NH, acetone, Δ, 2 hr	2-N,N-Diethylaminomethylpyridine-1-oxide, b.p. 100°/0.2 mm (60%), n_D^{20} 1.5470
Et_3N, acetone, 20–25°, 4 days	2-N,N,N-Triethylammoniummethylpyridine-1-oxide, hydrate, m.p. 111–113° (70%)
EtONa, EtOH	2-Ethoxymethylpyridine-1-oxide, b.p. 140°/12 mm (62%), n_D^{20} 1.5400
EtSNa, EtOH, 20–25°, 12 hr	2-Thioethoxymethylpyridine-1-oxide, b.p. 110°/0.4 mm (76%), n_D^{20} 1.6094
MeCOSNa, MeOH, 20–25°, 12 hr	2-Thioacetylmethylpyridine-1-oxide, m.p. 51–52°, b.p. 120°/0.4 mm (64%)
Na_2S, H_2O, 20–25°, 12 hr	Di-(1-oxido-2-pyridylmethyl)sulfide, m.p. 183–185° (48%)
NaOH, H_2O, Δ, 2 hr	Di-(1-oxido-2-pyridylmethyl)ether, hydrate, m.p. 128–129° (70%)
Thiourea, EtOH, Δ, 30 min	S-(1-Oxido-2-pyridylmethyl)isothiouronium chloride, m.p. 198–199° (90%)
$NaP(O)(OEt)_3$, C_6H_6, Δ, 2 hr	Diethyl 2-pyridylmethylphosphonate-1-oxide, oil (40%)

boiling the mixture for 12 hr the yield of carbon dioxide was 83%, and IV-335 was almost completely consumed with the formation of 2-chloromethylpyridine (IV-334). The ester IV-335 gives IV-334 on treatment with hydrochloric acid and boiling the solution in chloroform or acetonitrile (577). These results indicate that trichloroacetyl chloride can act as a chlorinating

agent in the same way as thionyl chloride does with amines. It was concluded
earlier that the reaction of picoline-1-oxides with acid chlorides (577) and
acid anhydrides (see Section III.3.) does not involve a radical-pair
intermediate.

1-(1-Oxido-2-pyridylmethyl)pyridinium iodide (**IV-116**) is obtained from
the reaction of 2-picoline-1-oxide in the presence of iodine and excess
pyridine (74, 333). This product was also obtained when 2-bromomethyl-
pyridine-1-oxide (**IV-336**) was heated with pyridine and sodium iodide.
These pyridinium salts (**IV-116**) undergo condensation with N,N-dimethyl-p-
nitrosoaniline in the presence of a basic catalyst to give the nitrone (**IV-321**);
when this reaction is carried out in the presence of sodium cyanide, the
cyanoanil **IV-337** is formed (578) (eq. IV-66). The methyl substituent in

IV-338 condenses with p-dimethylaminonitrosobenzene to yield the anil
(**IV-339**) (73). When the pyridinium salt (**IV-116**, X = I) is heated in the

presence of hydrogen iodide or iodine in xylene, pyrido[1′,2′-3,4]imidazo-
[1,2-a]pyridinium iodide (**IV-340**) (RRI 12,463) is obtained. However, when
the bromide (**IV-116**; X = Br) is used no reaction occurs (578).

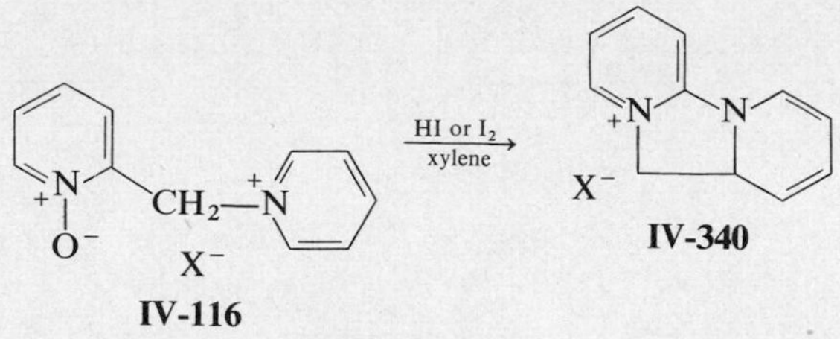
HI or I_2
xylene
X^-
IV-116
IV-340

Reaction of 2-picoline-1-oxide with acyl nitrite or with acetyl chloride and silver nitrite at room temperature gives 2-cyanopyridine (579). With amyl nitrite in liquid ammonia-potassium amide, it gives pyridine-2-carboxaldehyde oxime-1-oxide and picolinamide-1-oxide (580). In the case of 4-picoline-1-oxide, 4-cyanopyridine (2%) is formed in addition to the oxime and amide (580) (Scheme IV-21). 2- And 4-pyridylaldoxime-1-oxides were obtained using similar conditions except that sodium hydride was used

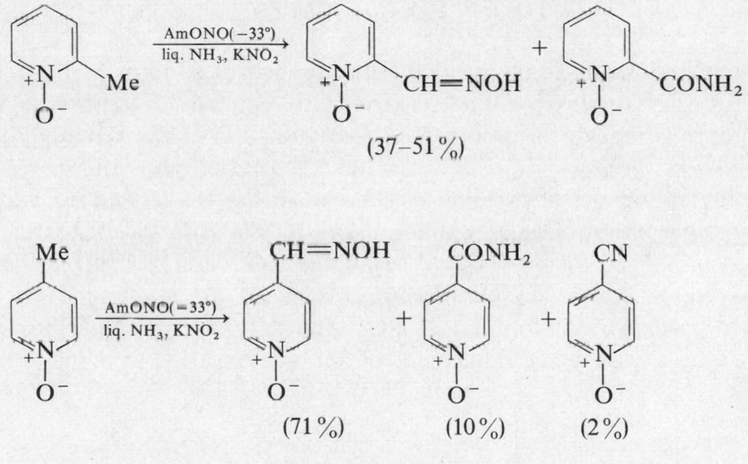
AmONO(−33°)
liq. NH_3, KNO_2
Me
CH=NOH
$CONH_2$
(37–51%)
CN
(71%)
(10%)
(2%)

Scheme IV-21

instead of KNH_2 (581). Treatment of pyridine-2-carboxaldehyde-1-oxide with hydroxylamine hydrochloride in sodium hydroxide at 0° gave the expected aldoxime (**IV-341**) which, on treatment with acetic acid, gave 2-cyanopyridine-1-oxide (347). Pyridine-2-aldoxime is obtained classically from 2-picoline-1-oxide and acetic anhydride as shown in eq. IV-67 (483). 2-Hydroxymethylpyridine, prepared as described above, was converted to the *N*-oxide (**IV-342**) which, on treatment with phenylhydrazine in

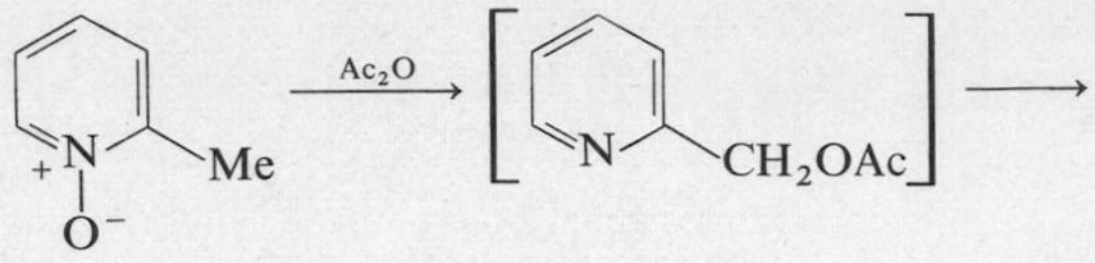

(IV-67)

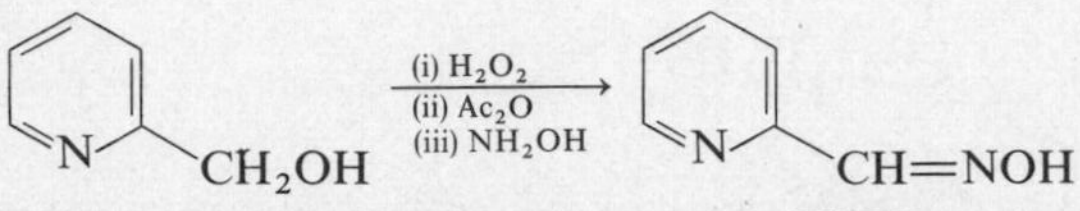

IV-341

dilute sodium hydroxide, gave pyridine-2-carboxaldehyde phenylhydrazone
(IV-343) (79 %) (582). Phenylhydrazine is not an oxidant in this reaction and

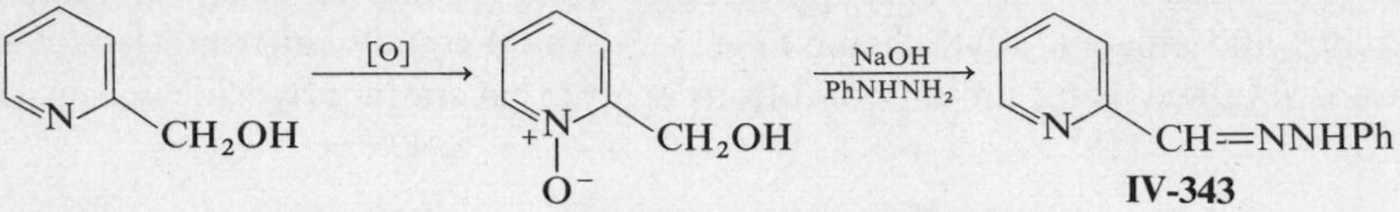

IV-342

no phenylhydrazone is obtained in the absence of sodium hydroxide, or
when 2-hydroxymethylpyridine itself was treated under these conditions
without prior *N*-oxide formation (582). Reaction of **IV-342**, phenylhydrazine,
and alkali in the presence of a 20-fold excess of 2-ethylpyridine gave only
the phenylhydrazone of pyridine-2-carboxaldehyde (61 %) and no evidence
of a possible intermolecular oxidation product. Oxidation of the alcohol
must occur at the expense of the *N*-oxide ın a base-catalyzed intramolecular
process, possibly *via* an elimination involving an enolamine **(IV-344)**
followed by deprotonation and N—O bond fission (582). No ketone was
obtained by acylative rearrangement of 2-(α-acetoxyethyl)pyridine-1-oxide;

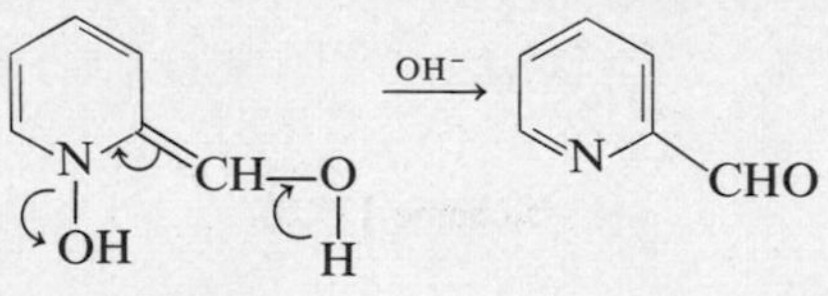

IV-344

a ketone was obtained, however, by base-catalyzed *N*-oxide fission of the
corresponding alcohol (582).

When equal weights of 2-picoline-1-oxide and iodine are heated at 95 to
100° a gum is formed which decomposes at 140 to 150° to yield pyridine-2-
aldehyde and 2-picoline. Alkaline hydrolysis of the gum gave similar results
(583) (eq. IV-68). Excess 2-picoline-1-oxide or pyridine-1-oxide act as a

$$\text{[pyridine-1-oxide-Me]} + I_2 \xrightarrow[\text{(ii) } 140-150°]{\text{(i) } 95-100°} \text{[pyridine-CHO]} + \text{[pyridine-Me]} \qquad \text{(IV-68)}$$

(16%)   (37%)

source of nucleophilic oxygen. Similar results were obtained with 5-ethyl-2-picoline-1-oxide-iodine complex using pyridine-1-oxide as the nucleophilic oxygen source. The proposed intermediate, 2-picolyl-2-picolinium (or pyridinium) iodide is formed, which eliminates pyridine on heating or on basic hydrolysis (386). 4-Picoline-1-oxide and its simple derivative gave the parent pyridine and a variety of other products on heating in DMSO—carbonyl, olefinic, and dimeric compounds—the nature of which depended on the starting pyridine (383) (Table IV-81). The formation of olefins suggests the intervention of the 2-(4-pyridyl)-2-propyl cation, for the reaction with 4-isopropylpyridine-1-oxide. This intermediate would also account for the ketonic products formed. A radical species could arise in the reaction of 4-alkylpyridine-1-oxides with DMSO by the homolytic cleavage of the N—O bond in an oxypyridinium ion formed by nucleophilic attack of N-oxide oxygen upon the electrophilic S atom of DMSO. Deprotonation of the 4-pyridylmethyl radical cation would give the pyridylmethyl radical. It was concluded that both ionic and radical species are generated in this reaction (383) (Scheme IV-22). Ortoteva-King reactions on 2- and 4-picoline-1-oxides, respectively, were carried out in the absence of pyridine to give 2- and 4-pyridine-carboxaldehydes, isolated as their sodium bisulfide adducts (584).

TABLE IV-81. Reaction of 4-Alkylpyridine-1-oxides with DMSO (383)

4-Alkyl group	Conditions	Products (yield)
4-Me	Me_2SO (1:5 mole ratio)[a] Δ, 4 hr, N_2	4-Picoline (35%); pyridine-4-carboxaldehyde (<1%); 4-picoline-1-oxide (35%)
4-CH$_2$Ph	Me_2SO (1:10 mole ratio) Δ, 1 hr, N_2	4-Benzylpyridine (29%); 4-benzoylpyridine (41%); 1,2-diphenyl-1,2-di-(4-pyridyl)-ethanol (9%); diastereomeric products, m.p. 219.5–222.1° and m.p. 270–272°
4-CH$_2$C$_6$H$_4$NO$_2$-p	Me_2SO, 1 hr	4-(p-Nitrobenzoyl)pyridine (51%); 1,2-di-(p-nitrophenyl)-1,2-di-(4-pyridyl)ethane (6%)
4-CHMe$_2$	Me_2SO (1:8 mole ratio)	4-Isopropylpyridine (16%); 4-isopropenyl-pyridine (14%); 2,3-dimethyl-2,3-di-(4-pyridyl)butane (9%); 4-isopropyl-pyridine-1-oxide (8%)

[a] Mole ratio refers to N-oxide to DMSO.

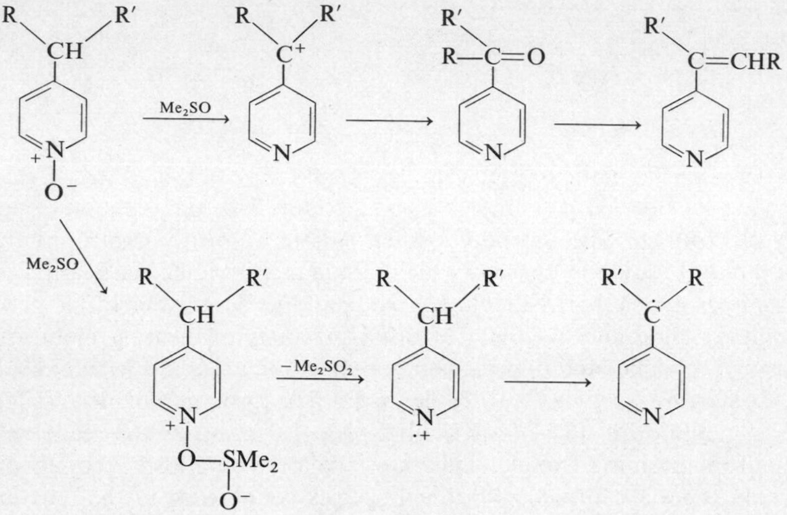

Scheme IV-22

2-, 3-, And 4-pyridinecarboxaldehyde-1-oxides (**IV-346**) were prepared by the oxidation of the appropriate dipyridylglycol dioxide (**IV-346**) with lead tetraacetate (53). Picoline-1-oxides may be oxidized with selenium

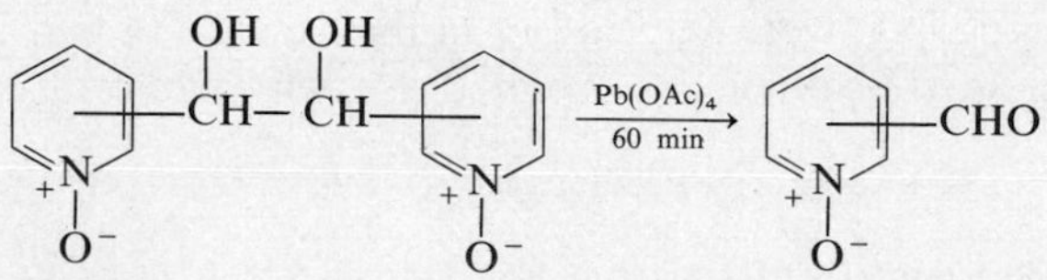

dioxide in boiling dioxane or pyridine. 3- And 5-methyl groups are resistant to these conditions, however. A 2-methyl group is oxidized to the aldehyde stage while a 4-methyl group is oxidized to the corresponding carboxylic acid (585) (eq. IV-69). 4-Substituted-6-methylpicolinic acid-1-oxides (**IV-280**)

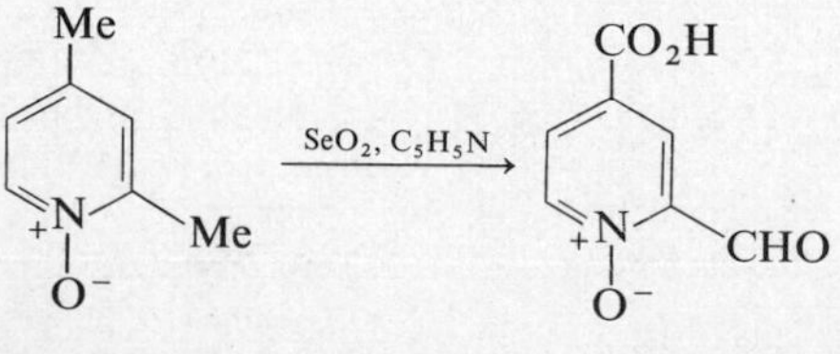

$$(IV-69)$$

are obtained by selective oxidation of the appropriate 2,6-lutidine-1-oxides with aqueous potassium permanganate (586). Others report that 6-methyl-picolinic acid-1-oxide (41.8%) and dipicolinic acid-1-oxide (8.1%) have been isolated from this reaction (161a). Oxidation of 6-methylpicolinic acid-1-oxide with aqueous potassium permanganate gave dipicolinic acid-1-oxide (30.5%) (161a). The methyl group in 3-picoline-1-oxide-4-sulfonic acid

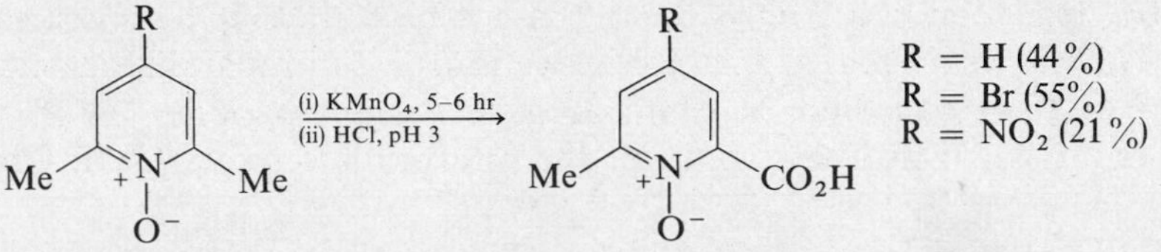

$$\text{(i) KMnO}_4, \text{5–6 hr} \quad \text{(ii) HCl, pH 3}$$

R = H (44%)
R = Br (55%)
R = NO$_2$ (21%)

IV-280

(**IV-347**) is oxidized with potassium dichromate in concentrated sulfuric acid at 20 to 30° to give the nicotinic acid (**IV-348**) (341). 4-Nitro-3-picoline-1-oxide is similarly oxidized to 4-nitronicotinic acid-1-oxide (166), and 4-

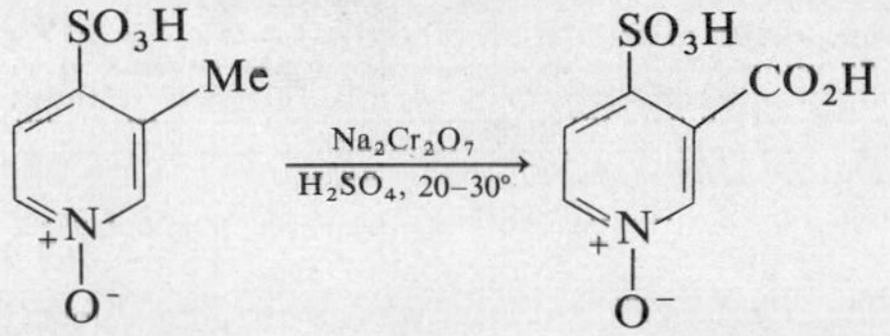

$$\frac{\text{Na}_2\text{Cr}_2\text{O}_7}{\text{H}_2\text{SO}_4, \text{20–30}°}$$

IV-347   IV-348

methylpicolinic acid-1-oxide is oxidized by potassium permanganate to pyridine-2,4-dicarboxylic acid-1-oxide (98).

2-(2-Substituted-ethyl)pyridine-1-oxides were obtained from the reaction of 2-vinylpyridine-1-oxide with organo-sodium compounds or a base (Scheme IV-23) (408).

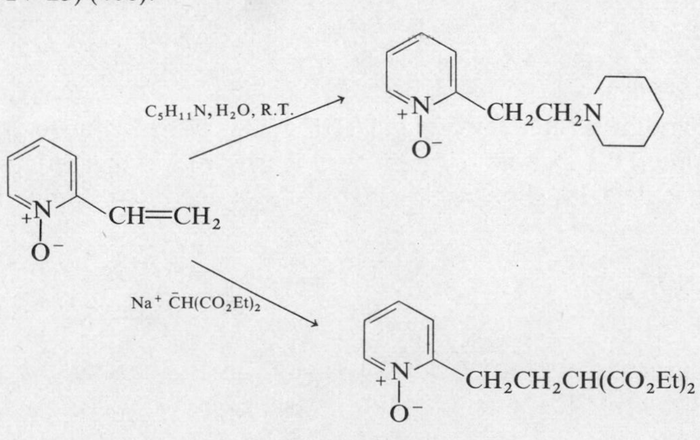

Scheme IV-23

2. Reduction of Nitro Groups

The reduction of a 2-nitro group has not been thoroughly investigated. In 1957, it was reported that the reduction of 2-nitropyridine-1-oxide in ethanol over palladium-on-charcoal gave 2-aminopyridine (**IV-9**). The reduction could not be stopped at the intermediate stages (587). On the

other hand, 4-nitropyridine-1-oxides (**IV-349**) are reduced with palladium-on-charcoal to yield the 4-aminopyridine-1-oxide (**IV-350**) (Table IV-82) (587). Reduction over platinum oxide or Raney nickel in methanol gave

4-aminopyridine (Table IV-82) (167, 315, 588). When 4-nitro-2-picolinic acid-1-oxide (**IV-162**) was reduced over Raney nickel in acetic acid 4,4'-azopicolinic acid-1,1'-dioxide (**IV-351**) was obtained (95). Similarly, when

TABLE IV-82. Reduction of Nitropyridine-1-oxides

N-oxide	Conditions	Products (yield)	Ref.
3-NO$_2$	PtO$_2$, 16 hr, 40 lb/in., R.T.	3-Aminopyridine; picrate, m.p. 201–202°	117
3-NO$_2$-2,4-Me$_2$	FeSO$_4$, NH$_4$OH	3-Amino-2,4-lutidine-1-oxide, m.p. 187–188°	121
3-NO$_2$-2,6-Me$_2$	50% Pd–Norit	3-Amino-2,6-lutidine-1-oxide	61
	FeSO$_4$, NH$_4$OH	3-Amino-2,6-lutidine-1-oxide, m.p. 160–162°	121
4-NO$_2$	Pd–C, EtOH	4-Aminopyridine-1-oxide; picrate, m.p. 191–192°	431
	Pd–C, EtOH normal pressure	4-Aminopyridine-1-oxide; HCl, m.p. 181–183° (decomp.)	544
	2% Pd–SrCO$_3$, H$_2$, EtOH, 60°	4-Aminopyridine-1-oxide, m.p. 235°; picrate, m.p. 203–204°; picrolonate, m.p. 241–244° (decomp.)	431
	PtO$_2$, H$_2$ (in H$_2$O)	4-Aminopyridine, m.p. 160–161°	167
	H$_2$, Raney-Ni (in MeOH)	4-Aminopyridine	315
	H$_2$, Urushibara, Ni, MeOH, Ac$_2$O	4,4'-Hydrazopyridine, m.p. 210–216°; 4,4'-azopyridine-1,1'-dioxide, m.p. 242°; 4,4'-azopyridine, m.p. 107°	320
	Zn, AcOH, 2 days	4,4'-Azoxypyridine-1,1'-dioxide, m.p. 235–236°	167
	80% NH$_2$NH$_2$·H$_2$O, Cu	4,4'-Azoxypyridine-1,1'-dioxide, m.p. 234–235° (decomp.)	387
	(i) 80% NH$_2$NH$_2$·H$_2$O, Cu, 3 hr (ii) Picric acid	4-Aminopyridine-1-oxide; picrate, m.p. 199–200° (decomp.), 4,4'-azoxypyridine-1,1'-dioxide	387
	80% NH$_2$NH$_2$·H$_2$O, O(CH$_2$CH$_2$OH)$_2$, 180–200°, 2 hr	4-Aminopyridine, m.p. 154–156°	387
	PhNHNH$_2$ alc. solution	4-Hydroxylaminopyridine-1-oxide, m.p. 237° (decomp.)	589, 591
	Hg light, EtOH; or PrOH, O$_2$	4-Hydroxypyridine-1-oxide, m.p. 97–100°	592
	Hg immersion lamp, abs. EtOH	4-Hydroxylaminopyridine-1-oxide, m.p. 219° (decomp.)	590
4-NO$_2$-2-Me	Alc. 10% Pd–C, H$_2$, 45 lb/in.^{2}	4-Amino-2-picoline-1-oxide, m.p. 181–183° (90%)	593
	Pd–C, R.T., 3 moles H$_2$	4-Amino-2-picoline-1-oxide; HCl, m.p. 191° (77%)	295
	NH$_2$NH$_2$, H$_2$O, 2 hr, EtOH	4,4'-Azo-2-picoline-1,1'-dioxide, m.p. 224°	387
	NH$_2$NH$_2$·H$_2$O, Cu	4,4'-Azo-2-picoline-1,1'-dioxide, 4-amino-2-picoline-1-oxide, m.p. 180–182°	387
	NH$_2$NH$_2$, Cu, O(CH$_2$CH$_2$OH)$_2$	4-Amino-2-picoline-1-oxide; picrate, m.p. 193°	387
	Hg lamp, abs. EtOH	4-Hydroxylamino-2-picoline-1-oxide, m.p. 226° (decomp.)	590

195

Table IV-82 (*Continued*)

N-oxide	Conditions	Products (yield)	Ref.
4-NO$_2$-2-Et	Raney Ni, MeOH, H$_2$, autoclave	4-Amino-2-ethylpyridine (80%)	429
	PtO$_2$ in AcOH, H$_2$; or Zn, HCl	4-Amino-2-ethylpyridine, m.p. 42–43°; picrate, m.p. 202–203°	79
	Fe filings AcOH, H$_2$SO$_4$	4-Amino-2-ethylpyridine, b.p. 128–130°/ 4–5 mm (89%)	430
4-NO$_2$-3-Me	Pd–C, AcOH	4-Amino-3-picoline	332
	Raney Ni, EtOH	4-Amino-3-picoline-1-oxide, 4,4′-azo-3-picoline	332
	(i) Raney Ni	4-Amino-3-picoline, m.p. 108–109°, 4,4′-azo-	332
	(ii) p-MeC$_6$H$_4$SO$_2$Cl, anhydr. EtOH, Δ, 24 hr	3-picoline	
	(iii) H$_2$SO$_4$, 3 hr		
	p-MeC$_6$H$_4$SO$_2$Cl, EtOH, 10 min	4,4′-Azoxy-3-picoline-1,1′-dioxide	332
	Pd–C, atm. press., R.T., 3 moles H$_2$	4-Amino-3-picoline-1-oxide; HCl, m.p. 219–220° (80%)	295
	Hg lamp, abs., EtOH	4-Hydroxylamino-3-picoline-1-oxide	590
4-NO$_2$-2,6-Me$_2$	Hydrog., 50% Pd–Norit	4-Amino-2,6-lutidine-1-oxide, m.p. 264–266° (decomp.); picrate, m.p. 216° (decomp.), 4-amino-3-picoline	61
	5% Pd–C, EtOH	4-Amino-2,6-lutidine-1-oxide, m.p. 265° (decomp.)	174
	Zn, aq. soln.	4,4′-Azoxy-2,6-lutidine-1,1′-dioxide	174
	Zn dust, H$_2$O, AcOH, 35°, 2 hr	4,4′-Azo-2,6-lutidine-1,1′-dioxide, m.p. 244° (decomp.)	120
	SnCl$_2$—HCl, R.T.	4,4′-Hydrazo-2,6-lutidine-1,1′-dioxide, HCl (80%)	120
	H$_2$S	4-Amino-2,6-lutidine-1-oxide, HCl	120
	∠5% NH$_4$OH, H$_2$S, Δ	4-Amino-2,6-lutidine-1-oxide, m.p. 265°	120
	H$_2$S, NH$_4$OH, 60°, 1 hr	4,4′-Hydrazo-2,6-lutidine-1,1′-dioxide; HCl, m.p. 266–267° (75%), 4,4′-hydrazo-2,6-lutidine-1,1′-dioxide, m.p. 244°	120
4-NO$_2$-3,5-Me$_2$	H$_2$, Raney Ni, MeOH	4-Amino-3,5-lutidine, 2H$_2$O	337
4-NO$_2$-2,6-Me$_2$	Sn, HCl, 100°	4-Amino-2,6-lutidine-1-oxide, HCl	120
4-NO$_2$-2-Me-5-Et	Pd–C, atm. press., H$_2$, 3 moles, R.T.	4-Amino-5-ethyl-2-picoline-1-oxide; HCl, m.p. 181–183°; picrate, m.p. 181–182°	295
4-NO$_2$-2,3,5,6-Me$_4$	Raney Ni, MeOH	4-Amino-2,3,5,6-tetramethylpyridine, $\frac{1}{2}$H$_2$O, m.p. 196–197°; picrate, m.p. 225–226°	337
4-NO$_2$-3-CH=CHPh	Fe, AcOH, 90–95°	4-Amino-3-styrylpyridine, m.p. 179°	164
4-NO$_2$-2,6-Br$_2$	Fe, AcOH, 100°, 1 hr	4-Amino-2,6-dibromopyridine	437
4-NO$_2$-3-NMe$_2$	Fe, AcOH	4-Amino-3-dimethylaminopyridine, m.p. 78–80°; picrate, m.p. 185°	353
4-NO$_2$-3-NHPh	Fe, AcOH	4-Amino-3-anilinopyridine, m.p. 132° (87.4%)	353

Table IV-82 (*Continued*)

N-oxide	Conditions	Products (yield)	Ref.
4-NO$_2$-3-NHC$_6$H$_{11}$	Fe, AcOH	4-Amino-3-cyclohexylaminopyridine, m.p. 164° (71.8%); picrate, m.p. 199°	353
4-NO$_2$-3-CH$_2$CH$_2$CO$_2$H	(i) Fe, AcOH, Δ, 0.5 hr (ii) NH$_4$OH	1,2,3,4-Tetrahydro-2-oxo-1,6-naphthyridine, m.p. 208°	79
4-NO$_2$-2-CO$_2$H	Raney Ni, AcOH, H$_2$, atm. press.	4,4'-Azopicolinic acid-1,1'-dioxide, m.p. 232° (55.2%)	95
	Na$_2$SO$_3$, Δ, 45 min	4-Aminopicolinic acid-1-oxide, m.p. 217 Δ (59.1%), sodium 2-carboxypyridine-4-sulfonate-1-oxide, H$_2$O, m.p. 317° (11.6%)	95
4-NO$_2$-6-Me-2-CO$_2$H	NaHSO$_3$, H$_2$O, Δ, 9 hr	4-Amino-6-methylpicolinic acid-1-oxide, m.p. 299°	586
2-(2-NO$_2$C$_6$H$_4$)	5% Pd–C, AcOH, H$_2$, 1 atm, 1 hr	2-(2-Aminophenyl)pyridine-1-oxide, m.p. 181.5–182.5° (97%)	140
2-(2-NO$_2$C$_6$H$_4$)	5% Pd–Al$_2$O$_3$, EtOH	2-(2-Aminophenyl)pyridine-1-oxide, m.p. 195–200° (20%)	141
2-(3-NO$_2$C$_6$H$_4$)	5% Pd–Al$_2$O$_3$, EtOH	2-(3-Aminophenyl)pyridine-1-oxide, m.p. 172–173° (25%)	141

4-nitropyridine-1-oxide was reduced over Uruşhibara nickel in methanol and acetic anhydride, 4,4'-hydrazopyridine, 4,4'-azopyridine-1,1'-dioxide, and 4,4'-azopyridine were obtained (320) (eq. IV-70).

3-Nitro-2,6-lutidine-1-oxide (**IV-352**) was reduced with ferrous sulfate and ammonia to give the 3-amine (**IV-353**) (121). Reduction of **IV-354** with iron

in acetic acid yields 4-amino-2,6-dibromopyridine (**IV-355**) (437), while zinc in acetic acid reduced 4-nitropyridine-1-oxide (**IV-149**) to 4,4′-azoxypyridine-1,1′-dioxide (**IV-356**) (167).

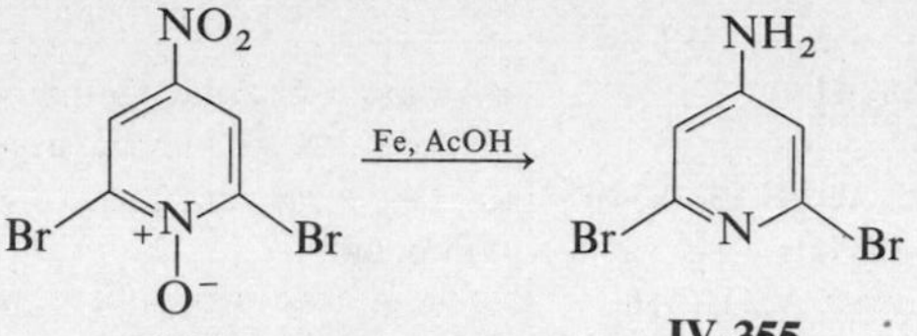

IV-354   **IV-355**

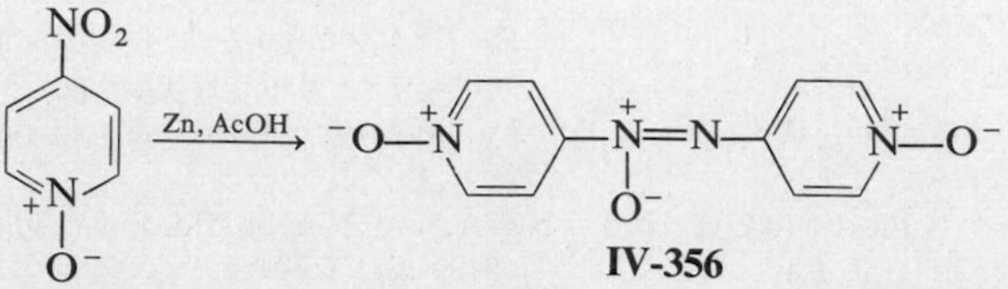

IV-149   **IV-356**

Hydrazine hydrate in ethanol has been used to reduce 4-nitro-2-picoline-1-oxide (**IV-357**) to 4,4′-azo-2-picoline-1,1′-dioxide (**IV-358**). When this reduction is carried out in the presence of copper, 4-amino-2-picoline-1-oxide (**IV-359**) is obtained in addition to the azo-compound (387). 4-Nitro-

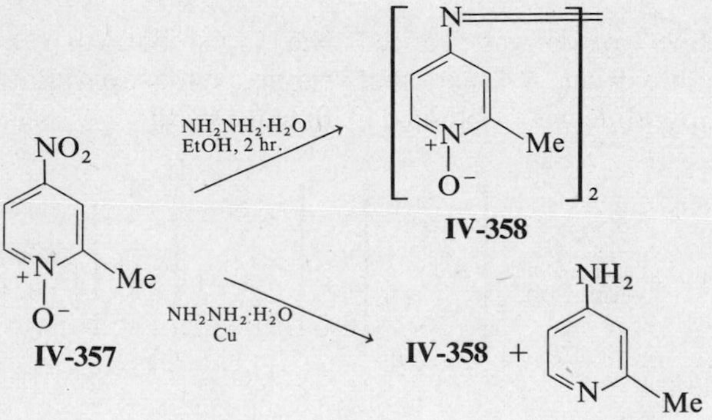

IV-358

IV-357   **IV-358** +

IV-359

pyridine-1-oxide (**IV-149**) can be reduced to 4-hydroxylaminopyridine-1-oxide (**IV-146**) with phenylhydrazine in alcohol (589). A similar reduction could be effected by photolysis of 4-nitropyridine-1-oxide in alcohol (590).

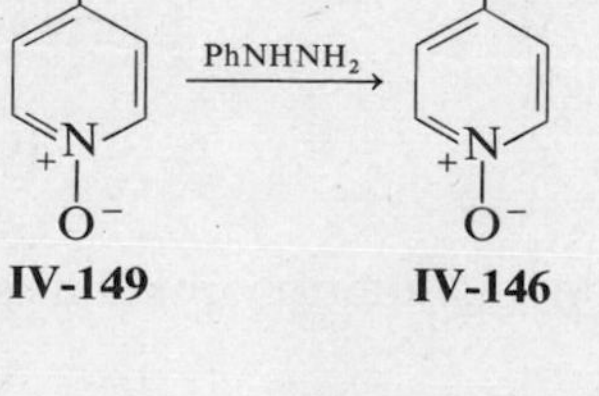

IV-149   **IV-146**

The products obtained from the reduction of nitropyridine-1-oxides under various conditions are summarized in Table IV-82.

3. Displacement of Nitro Groups by Nucleophilic Reagents

A. Halogens

The nitro group in 2-nitropyridine-1-oxide undergoes substitution on treatment with cold acetyl chloride to give the 2-chloro compound; its mobility in this reaction is greater than that in 4-nitropyridine-1-oxide (587). Substituted 4-nitropyridine-1-oxides (**IV-349**) also react with acetyl chloride

R = H, 3-Me, 5-Me, or 6-Me

to yield the corresponding 4-chloro derivatives (**IV-360**) (Table IV-83). The acetyl nitrite produced during this process may react with an active methyl

group in the 2-position to effect its nitrosation, further dehydration of which can lead to the formation of a nitrile as a by-product (eq. IV-71) (594, 595).

Acetyl bromide reacts similarly with 4-nitronicotinamide-1-oxide (**IV-361**) to give **IV-362** (339, 340, 596) (Table IV-83). The nitro group in 4-nitro-3-picoline-1-oxide (**IV-106**) is displaced by chloride when dry hydrogen chloride

TABLE IV-83. Reaction of 4-Nitropyridine-1-oxides with Acetyl Halides and Hydrohalic Acids

N-oxide	Conditions	Products (yield)	Ref.
4-NO$_2$	AcCl, Δ, 3 hr	4-Chloropyridine-1-oxide, m.p. 167–168° (84%)	338, 597
	AcCl	4-Chloropyridine-1-oxide, m.p. 164–166°	133
	48% HBr, 17 hr	4-Bromopyridine-1-oxide, HBr, m.p. 143–144°; picrate, m.p. 142–143°	431
2-Me-4-NO$_2$	AcCl (with cooling)	4-Chloro-2-picoline-1-oxide (68%); HCl, m.p. 133–134°, 4-chloro-2-pyridine aldoxime-1-oxide, m.p. 185° (decomp.) (12%)	594
	AcCl (on steam-bath)	4-Chloro-2-picoline-1-oxide (4.2%), 4-chloro-2-cyanopyridine, m.p. 130.5–131° (57%); 4-chloro-2-picoline, m.p. 183–183.5° (63%)	594
	AcCl in CHCl$_3$ at R.T.	4-Chloro-2-picoline-1-oxide (17%); 4-nitro-2-picoline-1-oxide (42%)	594
3-Me-4-NO$_2$	(i) AcCl, 0°, 45 min (ii) 40°, 30 min (iii) 50°, 30 min	4-Chloro-3-picoline-1-oxide, m.p. 120–122°	329
	AcCl	4-Chloro-3-picoline-1-oxide	34
	HCl gas	4-Chloro-3-picoline-1-oxide (69.5%)	345
	AcBr, 45 min, 65–70°	4-Bromo-3-picoline-1-oxide, m.p. 109° (59%)	340
	HBr, Δ, 24 hr	4-Bromo-3-picoline-1-oxide, m.p. 112–113° (75%)	362
3-Et-4-NO$_2$	AcCl	4-Chloro-3-ethylpyridine-1-oxide, m.p. 86°; picrate, m.p. 137–138°	337
3-isoPr-4-NO$_2$	AcCl	4-Chloro-3-isopropylpyridine-1-oxide, m.p. 87–88°; picrate, m.p. 130–131°	337
	AcBr, 45 min, 65–70°	4-Bromo-3-isopropylpyridine-1-oxide, m.p. 110–112.5° (75.5%)	340
2,5-Me$_2$-4-NO$_2$	AcBr, 0°	4-Bromo-2,5-lutidine-1-oxide, m.p. 92–94°	31
2-Me-5-Et-4-NO$_2$	AcCl	4-Chloro-5-ethyl-2-picoline-1-oxide	348
2,6-Me$_2$-4-NO$_2$	(i) AcBr, 20–30° (ii) 50–60°, 2 hr	4-Bromo-2,6-lutidine-1-oxide, m.p. 80.5–81.5°; HCl, m.p. 212–213°	339
2,6-Me$_2$-4-NO$_2$	AcCl, 50–60°, 5 hr	4-Chloro-2,6-lutidine-1-oxide, m.p. 95.5–97°; HCl, m.p. 220–221°	339
3,5-Me$_2$-4-NO$_2$	AcCl	4-Chloro-3,5-lutidine-1-oxide, m.p. 201–202°; picrate, m.p. 142–143°	337
2,3,5,6-Me$_4$-4-NO$_2$	AcCl	4-Chloro-2,3,5,6-tetramethylpyridine-1-oxide, m.p. 153–154°; picrate, m.p. 154–155°	337
2-Cl-4-NO$_2$	AcCl, H$_2$NSO$_3$H, 50°	2,4-Dichloropyridine-1-oxide, m.p. 52–55° (77%)	62
3-Br-4-NO$_2$	AcCl	3-Bromo-4-chloropyridine-1-oxide, m.p. 153.5–154.5°; picrate, m.p. 120–121°	337
2-OH-4-NO$_2$	AcCl	4-Chloro-1-hydroxy-2-pyridone	62
3-CH=CHPh-4-NO$_2$	AcCl, Δ, 3 hr	4-Chloro-3-styrylpyridine-1-oxide	598
2-CO$_2$Me-4-NO$_2$	AcCl, MeOH	4-Chloro-2-picolinic acid-1-oxide, m.p. 144° (57.6%); Me ester, m.p. 102°	95
3-CO$_2$H-4-NO$_2$	AcCl, 70–74°, 4 hr	4-Chloropicolinic acid-1-oxide, m.p. 169–170°	324

is passed through a boiling solution in methanol or ethanol (99, 345), and by bromide on heating with concentrated hydrobromic acid (362).

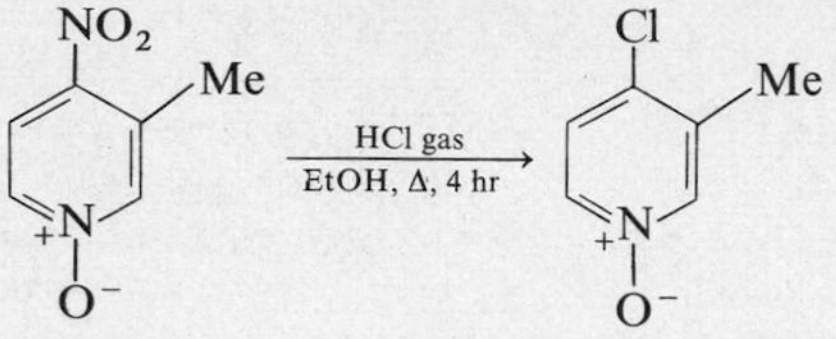

NO$_2$ ⟶ Me HCl gas, EtOH, Δ, 4 hr ⟶ Cl ⟶ Me

IV-106

B. Substitution by Oxygen or Sulfur-containing Groups

The nitro group in 4-nitropyridine-1-oxides may be substituted by alkoxide, phenoxide, and thiophenoxide ions (7) (eq. IV-72) (Table IV-84).

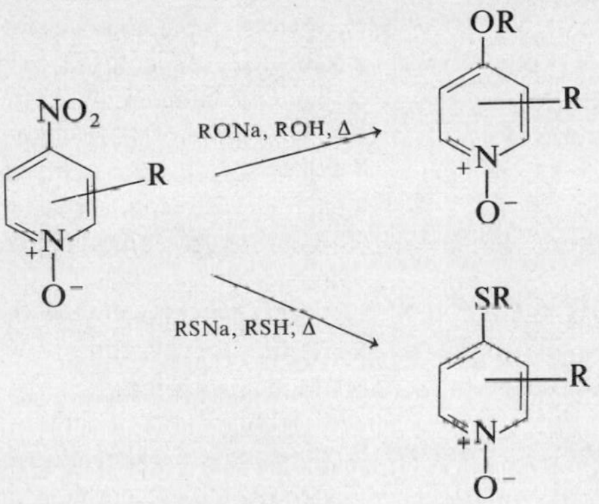

RONa, ROH, Δ ⟶ OR, R

RSNa, RSH, Δ ⟶ SR, R

(IV-72)

The 4-nitro group in the pyridine-1-oxide ring is more susceptible to substitution by alkoxide ions than is a 2-chloro substituent as shown in the following example given in eq. IV-73 (355).

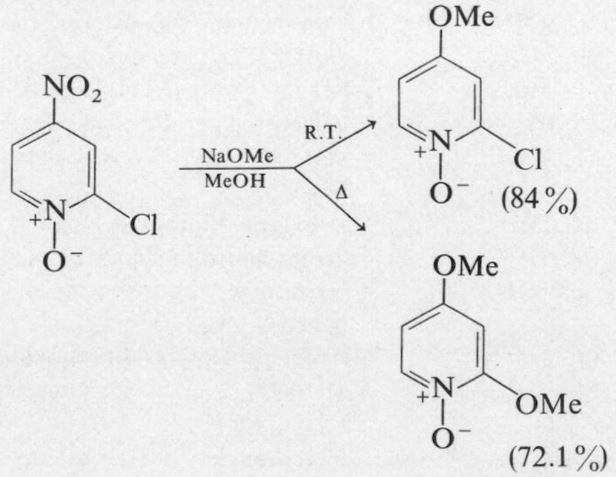

NaOMe, MeOH

R.T. ⟶ OMe, Cl (84%)

Δ ⟶ OMe, OMe (72.1%)

(IV-73)

TABLE IV-84. Reaction of 4-Nitropyridine-1-oxides with Alkoxide, Phenoxide, and Thiophenoxide Ions

N-oxide	Conditions	Products (yield)	Ref.
4-NO$_2$	MeONa, MeOH	4-Methoxypyridine-1-oxide·H$_2$O, m.p. 80–82°; picrate, m.p. 142–144°	364
	EtOH, K$_2$CO$_3$, Δ, 8 hr	4-Ethoxypyridine-1-oxide, m.p. 126° (82.5%)	434
	isoPrOH, Na, 15 hr	4-Isopropoxypyridine-1-oxide, b.p. 135–137°/ 0.3 mm (66.6%)	434
	CH$_2$=CHCH$_2$OH, Na, R.T., 15 hr	4-Allyloxypyridine-1-oxide	334
4-NO$_2$	PhSH, Na, EtOH	4-Phenylthiopyridine-1-oxide, m.p. 137°	357
2-Me-4-NO$_2$	MeONa, MeOH	4-Methoxy-2-picoline-1-oxide	348
	MeOH, Na	4-Methoxy-2-picoline-1-oxide, m.p. 86° (87%)	434
	EtONa, EtOH	4-Ethoxy-2-picoline-1-oxide	348
	EtOH, Na, 15 hr	4-Ethoxy-2-picoline-1-oxide, m.p. 78° (84%)	434
	PrOH, Na, 15 hr	4-Propoxy-2-picoline-1-oxide, m.p. 52° (81%)	434
	Amyl alcohol, Na, 15 hr	4-Amyloxy-2-picoline-1-oxide, b.p. 146°/0.1 mm	434
	PhCH$_2$ONa, PhCH$_2$OH	4-Benzyloxy-2-picoline-1-oxide	348
	PhONa, PhOH	4-Phenoxy-2-picoline-1-oxide	348
5-Et-2-Me-4-NO$_2$	MeONa, MeOH	5-Ethyl-4-methoxy-2-picoline-1-oxide	348
	EtONa, EtOH	4-Ethoxy-5-ethyl-2-picoline-1-oxide	348
3-Me-4-NO$_2$	MeOH, Na, 15 hr	4-Methoxy-3-picoline-1-oxide, m.p. 75–77° (53%)	434
3-F-4-NO$_2$	EtONa, EtOH, R.T., 10 min	3,4-Diethoxypyridine-1-oxide, m.p. 105°; picrate, m.p. 171°	352
2-Cl-4-NO$_2$	MeOH, Na, R.T., 24 hr	2-Chloro-4-methoxypyridine-1-oxide (84.1%)	355
	MeOH, Na, Δ, 5 hr	2,4-Dimethoxypyridine-1-oxide	355
3-Cl-4-NO$_2$	MeONa, MeOH	3-Chloro-4-methoxypyridine-1-oxide, m.p. 99–100° (87%); picrate, m.p. 152–153°	353
	EtONa, EtOH	3-Chloro-4-ethoxypyridine-1-oxide, m.p. 145–146° (90.5%); picrate, m.p. 121°	353
3-Br-4-NO$_2$	MeONa, MeOH	3-Bromo-4-methoxypyridine-1-oxide, m.p. 92–93° (83.3%); picrate, m.p. 142–143°	353
	EtONa, EtOH	3-Bromo-4-ethoxypyridine-1-oxide, m.p. 150° (85.4%); picrate, m.p. 125–126°	353
3-I-4-NO$_2$	MeONa, MeOH	3-Iodo-4-methoxypyridine-1-oxide, m.p. 191–192° (90.1%); picrate, m.p. 156°	353
3-OMe-4-NO$_2$	MeONa, MeOH	3,4-Dimethoxypyridine-1-oxide (93.2%)	352
3-OEt-4-NO$_2$	EtONa, EtOH	3,4-Diethoxypyridine-1-oxide (94.5%)	352
2-CO$_2$H-4-NO$_2$	NaOH, MeOH	4-Methoxypicolinic acid-1-oxide, m.p. 154° (92.5%)	95
	NaOH, EtOH	4-Ethoxypicolinic acid-1-oxide, m.p. 144°	95
	NaOH, PrOH	4-Propoxypicolinic acid-1-oxide, m.p. 107°	95
	NaOH, isoPrOH	4-Isopropoxypicolinic acid-1-oxide, m.p. 147° (43.6%)	95
	30% H$_2$O$_2$, NaOH	4-Hydroxypicolinic acid-1-oxide (56.8%)	95
2-CO$_2$H-6-Me-4-NO$_2$	NaOH, MeOH, Δ, 1 hr	4-Methoxy-6-methylpicolinic acid-1-oxide, m.p. 172° (73.6%)	586

Dimethyl sulfoxide was found to increase the rate of replacement of the nitro group in 4-nitropyridine-1-oxide by hydroxyl ion (599a). 4-Nitropicolinic acid-1-oxide reacts with 30 % H_2O_2 in sodium hydroxide solution to give 4-hydroxypicolinic acid-1-oxide (56.8 %) (95).

C. Substitution by Amines

4-Nitropicolinic acid-1-oxide (**IV-363**) reacts with aniline to give 4-anilinonicotinic acid-1-oxide (**IV-364**) (Table IV-85) (166, 329). In contrast to the reaction of 2-chloro-4-nitropyridine-1-oxide with methoxide ion,

IV-363   IV-364

2-bromo-4-nitropyridine-1-oxide (**IV-365**) and diethylamine in hot ethanol give 2-diethylamino-4-nitropyridine-1-oxide (**IV-366**) (355). Similarly, methyl 4-nitropicolinate-1-oxide (**IV-367**) reacts with an amine only at the ester

IV-365   IV-366

function to give an amide (**IV-368**), and no displacement of the nitro group occurs (99).

TABLE IV-85. Reaction of 4-Nitronicotinic acid-1-oxide with Amines

Conditions	Products (yield)	Ref.
$PhNH_2$	4-Anilinonicotinic acid-1-oxide, m.p. 252–254°, m.p. 242–244°	166, 329
m-$F_3CC_6H_4NH_2$	4-(m-Trifluoroanilino)nicotinic acid-1-oxide, m.p. 277–280°	330

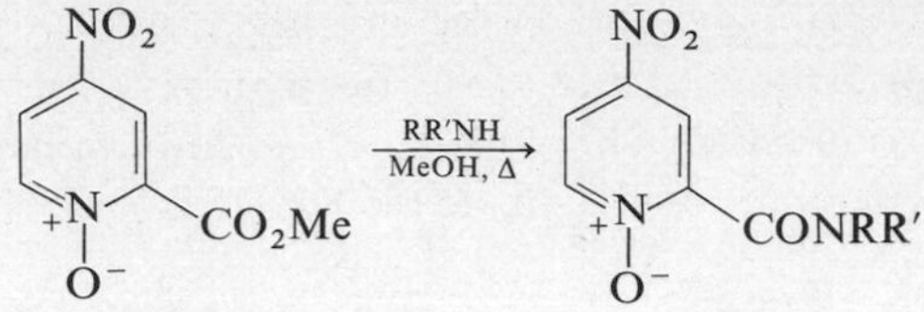

IV-367   **IV-368**

D. *Substitution by Trialkoxyphosphites*

Reaction of 2-nitropyridine-1-oxide with triethyl phosphite gave diethyl pyridine-2-phosphonate rather than deoxygenation of the nitro group (599b). Deoxygenation of the *N*-oxide function did occur, however (eq. IV-74).

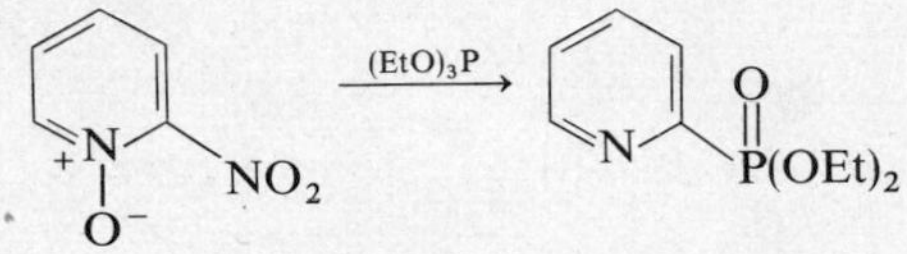

(IV-74)

4. Displacement of Halogens by Nucleophilic Reagents

A. *By Oxygen Nucleophiles*

Halogen groups at C-2 and C-4 undergo substitution by alkoxide or phenoxide ions (50) (eq. IV-75) (Table IV-86). 2-Halo-4-nitropyridine-1-oxides react with two equivalents of sodium methoxide in methanol with

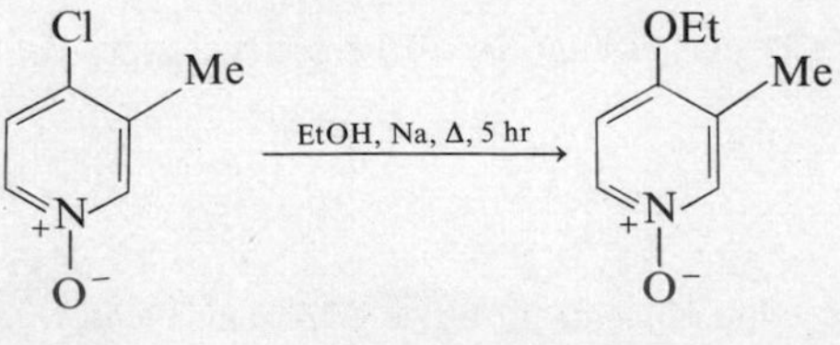

(IV-75)

substitution of both functional groups to give 2,4-dimethoxypyridine-1-oxide; when only one equivalent of sodium methoxide is used, however, only the nitro group is displaced (355, 600) (see also Section V.3.B). With 3-fluoro-4-nitropyridine-1-oxide (**IV-369**) only the 3-fluoro substituent undergoes nucleophilic substitution by alkoxide ion to give **IV-370** (352) (Table IV-86). 2-Aminoethanol and 4-chloro-3-picoline-1-oxide give the

TABLE IV-86. Reaction of Halopyridine-1-oxides with Alkoxide and Phenoxide Ions

N-oxide	Conditions	Products (yield)	Ref.
2-Cl	MeONa	2-Methoxypyridine-1-oxide, m.p. 78–79°	217
	PhONa, 4 hr, 100°, sealed tube	2-Phenoxypyridine-1-oxide, m.p. 49–51° (64%); methiodide, m.p. 177–178°	603
	CH_2=CH—CH_2ONa,	2-Allyloxypyridine-1-oxide (77%)	604
	Crotyl alcohol, Na, 41–45°, 1 hr	2-Crotyloxypyridine-1-oxide, m.p. 82.0–84.5° (96.9%)	538
	PhCH$_2$ONa	1-Benzyloxy-2-pyridone	604
	Crotyl alcohol-1-d, NaH, THF	2-(Crotyloxy-1-d)pyridine-1-oxide, m.p. 82–85°	538
	CH_3CH=CHCH(Me)OH, NaH, THF	2-α,γ-Dimethylallyloxypyridine-1-oxide, m.p. 49.5° (21.5%)	538
	▷—CH$_2$OH, NaH, NaH, THF	2-Cyclopropylcarbinyloxypyridine-1-oxide, m.p. 78–81° (30%)	538
2-Cl-5-Me	3-Buten-2-ol, NaOH, THF, 40°, 2 hr	6-α-Methylallyloxy-3-picoline-1-oxide (50%)	538
2-Cl-5-Me	Crotyl alcohol, Na	6-Crotyloxy-3-picoline-1-oxide, m.p. 89–91° (34%)	538
2-Cl-5-Me	Allyl alcohol, Na	6-Allyloxy-3-picoline-1-oxide, liquid (31%)	538
2,4-Cl$_2$	(i) MeONa, MeOH (ii) 10% HCl—MeOH	2-Chloro-4-methoxypyridine-1-oxide, HCl, m.p. 108–110° (58%)	62
2-Br-6-Me	MeONa	6-Methoxy-2-picoline-1-oxide, m.p. 62° (66.5%)	71
2-Br-3-Me	MeOK, MeOH, Δ	2-Methoxy-3-picoline-1-oxide, hygroscopic, picrate, m.p. 167–168°	602
2-Br-5-Me	MeOK, MeOH, Δ	6-Methoxy-3-picoline-1-oxide, hygroscopic, picrate, m.p. 156–157°	602
	NaOEt, 1 hr	6-Ethoxy-3-picoline (73.5%)	71
2-X-4-NO$_2$	MeONa (1 equiv.)	4-Methoxy-2-X-pyridine-1-oxide	600
	MeONa (2 equiv.)	2,4-Dimethoxypyridine-1-oxide	600
3-F-4-NO$_2$	MeOH, MeONa, 2 hr, R.T.	3-Methoxy-4-nitropyridine-1-oxide, m.p. 134° (100%)	352
	MeOH, MeONa, Δ, 1 hr	3,4-Dimethoxypyridine-1-oxide, m.p. 103–104° (100%); picrate, m.p. 167°	352, 605
	EtOH, EtONa, R.T., 2 hr	3-Ethoxy-4-nitropyridine-1-oxide, m.p. 134° (100%)	352, 605
	PhOH, K$_2$CO$_3$	4-Nitro-3-phenoxypyridine-1-oxide, m.p. 111° (40.8%)	605, 606
	p-O$_2$NC$_6$H$_4$OH, MeONa	4-Nitro-3-(p-nitrophenoxy)pyridine-1-oxide, m.p. 196–198° (72.2%)	352, 605

Table IV-86 (*Continued*)

N-oxide	Conditions	Products (yield)	Ref.
3-F-2-Me-4-NO$_2$	MeOH, MeONa	3-Methoxy-4-nitro-2-picoline-1-oxide, m.p. 70° (54.3%)	343, 605
	EtOH, EtONa	3-Ethoxy-4-nitro-2-picoline-1-oxide, m.p. 76° (50.5%)	343, 605
	H$_2$NCH$_2$CH$_2$OH	3-(β-Aminoethoxy)-4-nitro-2-picoline-1-oxide, m.p. 110° (80%)	343
3-F-2,6-Me$_2$-4-NO$_2$	MeONa, Δ, 1 hr	3-Methoxy-4-nitro-2,6-lutidine-1-oxide, m.p. 137° (84.5%)	336, 605
	EtONa, Δ, 1 hr	3-Ethoxy-4-nitro-2,6-lutidine-1-oxide, m.p. 103° (94.2%)	336, 605
3-F-5-Me-4-NO$_2$	MeONa	5-Methoxy-4-nitro-3-picoline-1-oxide, m.p. 146° (65.2%)	605, 606
	EtONa	5-Ethoxy-4-nitro-3-picoline-1-oxide, m.p. 130° (60.6%)	605, 606
3-Cl	MeONa	3-Methoxypyridine-1-oxide, m.p. 100–101°; picrate, m.p. 155–156°	106
	EtONa	3-Ethoxypyridine-1-oxide, m.p. 85–86°, b.p. 175°/0.05 mm; picrate, m.p. 125–125.5°	106
4-Cl	EtOH, EtONa,	4-Ethoxypyridine-1-oxide, m.p. 71–73°; picrate, m.p. 110–111°	217
	PhCH$_2$ONa, Δ, 5 min	4-Phenoxypyridine-1-oxide, m.p. 76–78° (58%)	217
4-Cl-3-Me	EtOH/Na, Δ, 5 hr	4-Ethoxy-3-picoline-1-oxide, m.p. 138–139°	50
	PrOH/Na, Δ, 2 hr	4-Propoxy-3-picoline-1-oxide, m.p. 77° (80%)	345
	BuOH/Na, Δ, 2 hr	4-Butoxy-3-picoline-1-oxide, m.p. 68° (79.3%)	345
	isoPrOH/Na, Δ, 2 hr	4-Isopropoxy-3-picoline-1-oxide, m.p. 73° (83%)	345
	n-AmylOH/Na	4-n-Amyloxy-3-picoline-1-oxide, m.p. 58° (71%)	345
	IsoamylOH/Na, Δ, 2 hr	4-Isoamyloxy-3-picoline-1-oxide, m.p. 61° (66.2%)	345
	CH$_3$(CH$_2$)$_5$OH/Na, Δ, 2 hr	4-Hexyloxy-3-picoline-1-oxide, m.p. 49° (64%)	345
	isoBuOH/Na, Δ, 2 hr	4-Isobutoxy-3-picoline-1-oxide, m.p. 79–80° (74%)	345
	CH$_3$(CH$_2$)$_6$OH/Na	4-Heptyloxy-3-picoline-1-oxide, m.p. 56° (57.9%)	345
	CH$_3$(CH$_2$)$_7$OH/Na	4-Octyloxy-3-picoline-1-oxide, m.p. 69° (50.3%)	345
	H$_2$NCH$_2$CH$_2$OH/Na, 2 hr	4-(β-Aminoethoxy)-3-picoline-1-oxide, HCl, m.p. 159° (70.4%)	345

IV-369   **IV-370**

ether **IV-371** (345). Various 2-alkenyloxypyridine-1-oxides were prepared (538). For example, 2-chloropyridine-1-oxide, crotyl alcohol, and sodium

IV-371

were heated together to yield 2-crotyloxypyridine-1-oxide (eq. IV-76) (Table IV-86).

$$(IV-76)$$

As expected, bromopyridine-1-oxides are more reactive than the parent pyridine bases (183). 2-Bromopyridine-1-oxide reacts 760 times faster than does 2-bromopyridine with methoxide ion in methanol 110° and $\Delta E_{act} = 7$ kcal/mole. 2-Bromopyridine-1-oxide reacts faster with NaOEt in EtOH than it does with KOMe in MeOH (601). In the N-oxide, the order of halogen mobility is 2-chloro-1-oxide > 2-bromo-1-oxide, which is the reverse of that observed with the corresponding halopyridines. A 3- or 5-methyl substituent decreases the reactivity and increases E_{act}, by 2.0–2.4 kcal/mole, and the $o:p$ ratio is close to 1 ($10^4 k_2^{110°}$; 2-Br-3-Me, 144.5; 2-Br-5-Me, 128.8). It appears that steric hindrance by both the methyl group and the N-oxide function to attack in 2-bromo-3-picoline-1-oxide cancels out any ion-dipole attractive interaction between the methyl group and the attacking methoxide nucleophile (602). 2- And 4-halopyridine-1-oxides react with sodium or potassium hydroxide to yield the respective 1-hydroxypyridones (Scheme IV-24) (324, 607) (Table IV-87). In the case of a 3-fluoro-4-nitro-pyridine-1-oxide, the fluorine atom again undergoes substitution by hydroxyl while the nitro group is unaffected (336) (eq. IV-77) (Table IV-87).

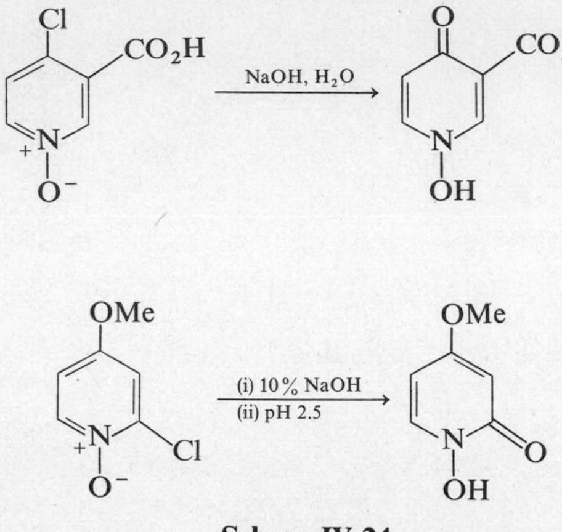

Scheme IV-24

TABLE IV-87. Reaction of Halopyridine-1-oxides with Hydroxide Ion

N-oxide	Conditions	Products (yield)	Ref.
2-Cl-6-Me	NaOH, Δ, 3 hr	1-Hydroxy-6-methyl-2-pyridone, m.p. 143–144°	358
2-Br-6-Me	5% NaOH, 0.5 hr	1-Hydroxy-6-methyl-2-pyridone, m.p. 144–145° (78%)	71
2-Cl-4-Me	20% NaOH, Δ, 4 hr	1-Hydroxy-4-methyl-2-pyridone, m.p. 133–135°	358
2-Cl-4-OMe	10% NaOH, Δ, 3 hr	1-Hydroxy-4-methoxy-2-pyridone, m.p. 175.5–176° (40%)	62
2-Cl-4-OMe, HCl	(i) 10% NaOH (ii) pH 2.5	1-Hydroxy-4-methoxy-2-pyridone, m.p. 175.5–176°	607
2-Cl-4-OEt, HCl	(i) 10% NaOH (ii) pH 2.5	1-Hydroxy-4-ethoxy-2-pyridone m.p. 136–137°	607
	10% NaOH, Δ, 3 hr	1-Hydroxy-4-ethoxy-2-pyridone, m.p. 136–137° (36%)	62
3-F-4-NO$_2$	NaOH	3-Hydroxy-4-nitropyridine-1-oxide, m.p. 173° (91.1%)	605
3-F-2-Me-4-NO$_2$	NaOH	3-Hydroxy-4-nitro-2-picoline-1-oxide, m.p. 182° (decomp.) (50.6%)	343, 605
3-F-2,6-Me$_2$-4-NO$_2$	30% H$_2$O$_2$, KOH, H$_2$O	3-Hydroxy-4-nitro-2,6-lutidine-1-oxide, m.p. 99°	336, 605
3-F-5-Me-4-NO$_2$	30% H$_2$O$_2$	5-Hydroxy-4-nitro-3-picoline-1-oxide, m.p. 224° (81%)	56, 605
4-Cl-3-Me	KOH, Δ, 16 hr	1-Hydroxy-3-methyl-4-pyridone, m.p. 215–216°	608
4-Cl-3-CO$_2$H	NaOH, H$_2$O	4-Hydroxynicotinic acid-1-oxide, m.p. 260° (56%)	324

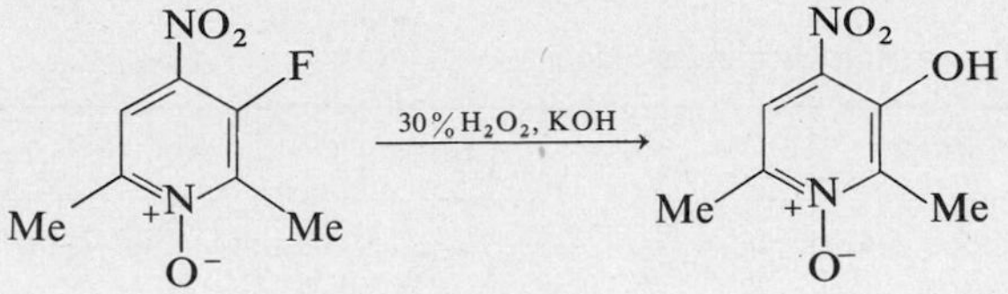

$$ \text{(IV-77)} $$

B. By Sulphur and Selenium Nucleophiles

A halogen in a pyridine-1-oxide undergoes substitution by a thiophenoxide or a thioalkoxide ion (605, 606) (eq. IV-78) (Table IV-88). Similarly, 4-chloronicotinic acid-1-oxide undergoes nucleophilic substitution with

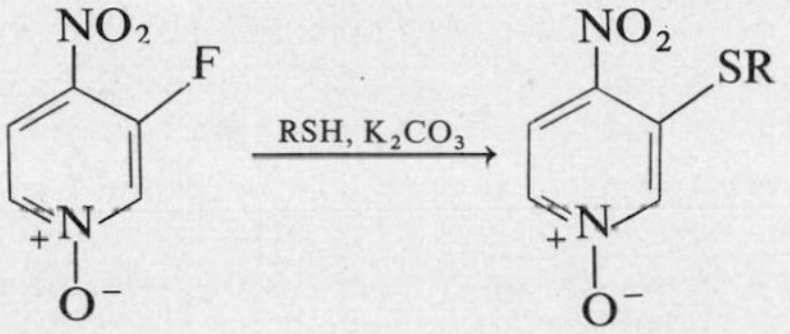

$$ \text{(IV-78)} $$

KHS/MeOH to yield nicotinic acid-1-oxide-4-thiol (324) (Table IV-88) (eq. IV-79). 2-Selenopyridine-1-oxide (**IV-372**) was prepared by saturating a

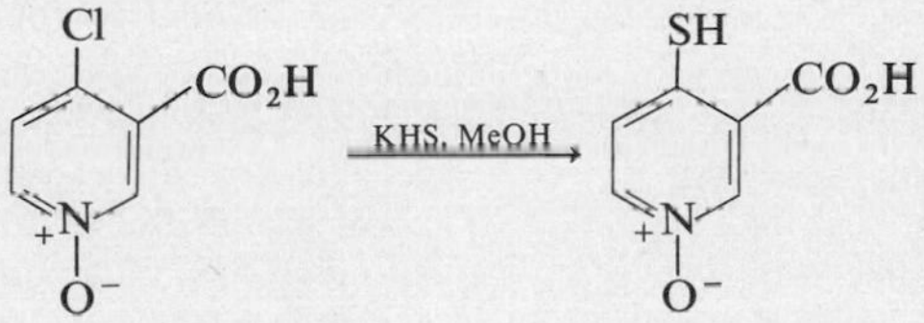

$$ \text{(IV-79)} $$

sodium ethoxide-ethanol solution of 2-bromopyridine-1-oxide with hydrogen selenide (153). Pyridinesulfonic acid-1-oxides (**IV-172**) were isolated

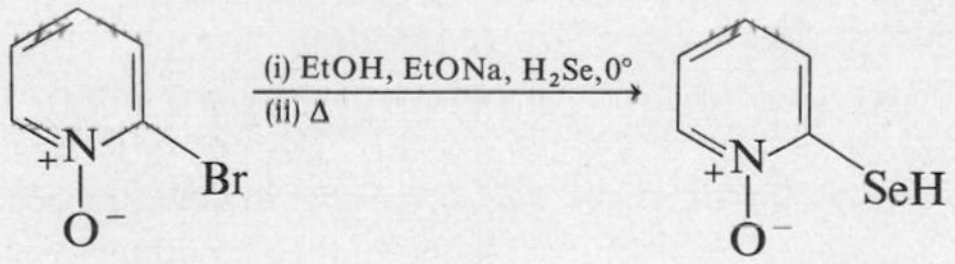

IV-372

from the reaction of halogen substituted pyridine-1-oxides with sodium sulfite (36). 2-Halopyridine-1-oxides react with thiourea to give the *S*-

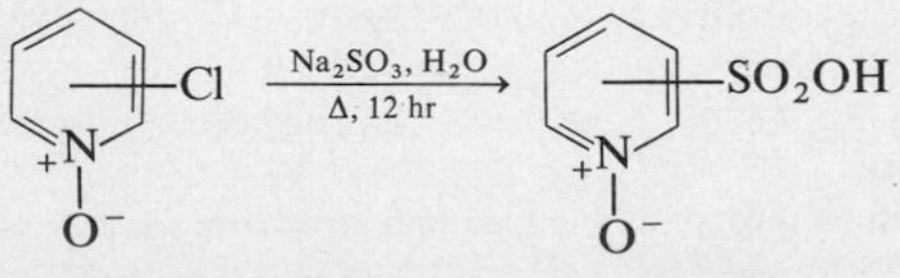

IV-172

(2-pyridyl)isothiouronium chloride-1-oxides (**IV-373**) which, on treatment with sodium carbonate, yield 1-hydroxy-2-pyridinethiones (**IV-374**) (131,

TABLE IV-88. Reaction of Halopyridine-1-oxides with Thiols

N-oxide	Conditions	Products (yield)	Ref.
2-F-3-Et	1-Dodecylimidazolidine-2-thione	2-(1-Dodecyl-2-imidazolinylthio)-3-ethyl-pyridine-1-oxide, HF	611
2-Cl	Na$_2$S, NaHS in H$_2$O, 95°, 30 min	1-Hydroxy-2-pyridinethione	612
	EtSNa, 100°, 3 hr	2-Ethylthiopyridine-1-oxide, m.p. 104–105°; picrate, m.p. 212–213°	603
	C$_6$H$_5$CH$_2$SNa, EtONa, Δ, 1 hr	2-Benzylthiopyridine-1-oxide, m.p. 168–170° (89%)	131
	PhSNa, 100°, 4 hr	2-Phenylthiopyridine-1-oxide, m.p. 111–112°; picrate, m.p. 139–140°	603
	CS(NH$_2$)$_2$, EtOH	S-(2-Pyridyl)isothiuronium chloride-1-oxide, m.p. 156–157°	131, 603
2-Br	H$_2$Se, Na, EtOC$_2$H$_4$OH	2-Selenopyridine-1-oxide, m.p. 72.5–73.0° (59%)	153
	3-NaSCH$_2$CH$_2$C$_5$H$_4$N	2-(3-Pyridyl-β-ethylthio)pyridine-1-oxide	613
	4-NaSCH$_2$CH$_2$C$_5$H$_4$N	2-(4-Pyridyl-β-ethylthio)pyridine-1-oxide	613
	4,6-Me$_2$-2-NaS-CH$_2$CH$_2$C$_5$H$_2$N	2-(4,6-Dimethyl-2-pyridyl-β-ethylthio)-pyridine-1-oxide	613
	2-Me-5-NaSCH$_2$CH$_2$C$_5$H$_3$N	2-(2-Methylpyridyl)-β-ethylthiopyridine-1-oxide	613
	1-(β-Ethylhexyl)hexahydro-diazepine-2-thione	2-{2-[1-(β-Ethylhexyl)]-tetrahydrodiazepenylthio}pyridine-1-oxide, HBr	611
	Imidazoline-2-thione, Δ	2-(2-Imidazolinylthio)pyridine-1-oxide	611
	Imidazoline-2-thione	2-(2-Imidazolinylthio)pyridine-1-oxide, HBr, m.p. 153.0–156.5° (decomp.)	611
	1-β-Ethylhexylimidazolidine-2-thione	2-{2-[1-(2-Ethylhexyl)]imidazolinylthio}-pyridine-1-oxide, HBr	611
2-Br·HCl	KSCN, KOH, EtOH, 3 hr	2-Thiocyanatopyridine-1-oxide, m.p. 158–160°	431
2-Br·HBr	Hexahydropyrimidine-2-thione	2-(2-Tetrahydropyrimidylthio)pyridine-1-oxide, HBr, m.p. 160–161°	611
	1-(2-Phenylethyl)imidazo-lidine-2-thione	2-{1-[4-(2-Phenylethyl)]-imidazolinyl-mercapto}pyridine-1-oxide, HBr	611
	Hexahydrodiazepine-2-thione	2-(2-Tetrahydrodiazepenylthio)pyridine-1-oxide, HBr, m.p. 140–141°	611
2-Br-5-Me	5-Et-2-HSCH$_2$CH$_2$C$_5$H$_3$N	2-(5-Ethyl-2-pyridyl-β-ethylthio)-5-methyl-pyridine-1-oxide	613
2-Br-4,6-Me$_2$	(i) CS(NH$_2$)$_2$, EtOH, Δ, 16 hr (ii) NaOH	1-Hydroxy-4,6-dimethyl-2-pyridinethione	64
2-Br-3-Et-6-Me	(i) CS(NH$_2$)$_2$, EtOH, Δ, 16 hr (ii) NaOH	1-Hydroxy-2-ethyl-6-methyl-2-pyridinethione	64
2-Br-3-I(CH$_2$)$_3$O·HCl	1-Ethylimidazolidine-2-thione	2-(1-Ethyl-2-imidazolinylthio)-5-iodopro-poxypyridine-1-oxide, HCl	611
2-Br-5-Cl	2,4,5-Cl$_3$C$_6$H$_2$SH, dioxane, Δ	2-(2,4,5-Trichlorophenylthio)-5-chloro-pyridine-1-oxide, m.p. 148.5–150° (71%)	354a, 614

Table IV-88 (*Continued*)

N-oxide	Conditions	Products (yield)	Ref.
2-Br-5-Cl	4-ClC$_6$H$_4$SH, dioxane, Δ	2-(4-Chlorophenylthio)-5-chloropyridine-1-oxide, m.p. 156–157.5° (66.5%)	354a, 614
2,3,5-Br$_3$·HBr	1-Isopropylhexahydropyrimidine-2-thione	2-(1-Isopropyl-2-tetrahydropyrimidylthio)-3,5-dibromopyridine-1-oxide, HBr	611
2-I	Hexahydropyrimidine-2-thione	2-(2-Tetrahydropyrimidylthio)pyridine-1-oxide, HBr, m.p. 160–161° (decomp.)	611
2-I-3-Pr	5-Butylhexahydropyrimidine-2-thione	2-(5-Butyl-2-tetrahydropyrimidylthio)-3-propylpyridine-1-oxide	611
2-I-3-F, HCl	Hexahydropyrimidine-2-thione	2-(2-Tetrahydropyrimidylthio)-5-fluoro-pyridine-1-oxide, HCl	611
3-F-4-NO$_2$	PhSH, K$_2$CO$_3$	4-Nitro-3-phenylthiopyridine-1-oxide, m.p. 148°	606
	NH$_4$SCN	4-Nitro-3-thiocyanatopyridine-1-oxide, m.p. 134° (55%)	606
	Na$_2$SO$_3$	Sodium 4-nitropyridine-3-sulfonate-1-oxide (69%)	606
	CS(NH$_2$)$_2$, Me$_2$CO, NaHCO$_3$	Bis-(4-nitro-3-pyridyl-1-oxide)sulfide, m.p. 222°	606
	MeCSNH$_2$	Bis-(4-nitro-3-pyridyl-1-oxide)sulfide (89.2%)	606
3-Cl	Na$_2$SO$_3$, H$_2$O, 143°, 10 hr	Pyridine-3-sulfonic acid-1-oxide, m.p. 238–243° (63%)	36
3-Br·HCl	CS(NH$_2$)$_2$, KOH, EtOH	S-(1-Oxido-3-pyridyl)isothiouronium bromide, m.p. 140 142°	431
3-Br	CS(NH$_2$)$_2$, EtOH, Δ, 5 hr	S-(1-Oxido-3-pyridyl)isothiouronium bromide, m.p. 145–147°	65
3,5-Br$_2$	CS(NH$_2$)$_2$, EtOH, 5 hr, Δ	S-(5-Bromo-1-oxido-3-pyridyl)isothiouronium bromide, m.p. 162–163°	65
4-Cl	Na$_2$S, 9H$_2$O, Δ, 5 hr	4-Thiopyridine-1-oxide, m.p. 142–143° (89%)	609
4-Cl (2 equiv.)	Na$_2$S, 9H$_2$O, Δ, 5 hr	Bis-(4-pyridyl-1-oxide)disulfide, m.p. 228–230°; picrate, m.p. 181–183°	609
4-Cl	PhSH, NaOH	4-Phenylthiopyridine-1-oxide, m.p. 137.5–138° (96%)	317
	Na$_2$SO$_4$, H$_2$O, Δ, 4 hr	Pyridine-4-sulfonic acid-1-oxide, m.p. >270°, m.p. 307–309° (80%)	114, 615
	PhCH$_2$SNa, Δ, 45 min	4-Benzylthiopyridine-1-oxide, m.p. 147–149° (46%)	47
	2-Br-5-MeC$_6$H$_3$SH, NaOH, 100°	4-(2-Bromo-5-methylphenylthio)pyridine-1-oxide, m.p. 120°	616
	3-Br-4-MeC$_6$H$_3$SH, NaOH, 100°	4-(3-Bromo-4-methylphenylthio)pyridine-1-oxide, m.p. 168°	616
	3-Cl-4-MeC$_6$H$_3$SH, NaOH, 100°	4-(3-Chloro-4-methylphenylthio)pyridine-1-oxide, m.p. 151°	616

211

Table IV-88 (*Continued*)

N-oxide	Conditions	Products (yield)	Ref.
4-Cl-3-Br	KSCN, Δ, 3 hr	3-Bromo-4-thiocyanatopyridine-1-oxide, m.p. 159–160° (77%)	609
4-Cl-3-NO₂	KSCN, EtOH, Δ, 3 hr	3-Nitro-4-thiocyanatopyridine-1-oxide, m.p. 169–170° (47.5%)	609
4-Cl-3-CH=CHPh	(i) Thiourea, EtOH (ii) NaOH	3-Styrylpyridine-1-oxide-4-thiol, m.p. 145–146°	598
4-Cl-3-CO₂H	KHS, MeOH, Δ, 21 hr	Nicotinic acid-1-oxide-4-thiol, m.p. 204° (47%)	324
4-Cl-2-CO₂H	H₂S, satd., alc. KOH, 150°, 1 hr, (CH₂OH)₂	Picolinic acid-1-oxide-4-thiol, m.p. 155°; bis-(4-picolinic acid-1-oxide)disulfide, m.p. 178° (15.2%)	95
4-Br, HBr	alc. HSCN	No reaction	431

603). (For a more convenient preparation of these compounds see Section IV.4). The reaction of halopyridine-1-oxides with thiocyanate is illustrated

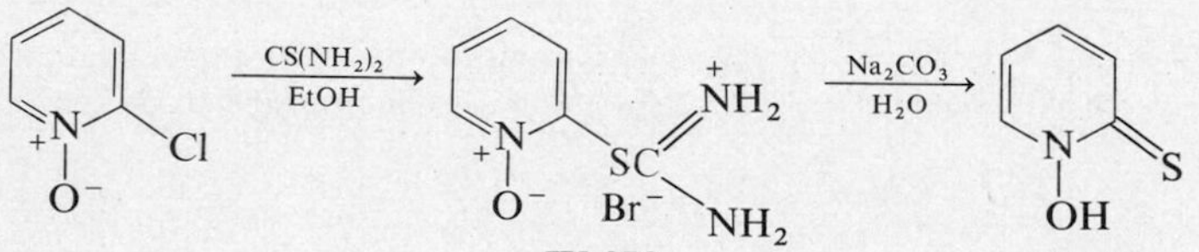

by that between 3-bromo-4-chloropyridine-1-oxide and potassium thiocyanate to yield 3-bromo-4-thiocyanatopyridine-1-oxide (eq. IV-80) (609). 2- And 4-(2,3,4,6-tetra-*O*-acetyl-β-D-glucopyranosylthio)pyridine-1-oxide are

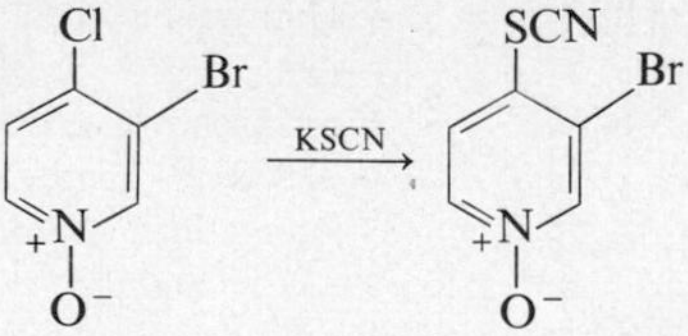

(IV-80)

prepared from 2- and 4-halopyridine-1-oxides, respectively, and 2,3,4,6-tetra-*O*-acetyl-1-thio-β-D-glucopyranose in base (318, 610).

C. By Amines

The halogen atom in 2- and 4-halopyridine-1-oxides (eq. IV-375) is displaced by amines to give **IV-376** (337) (Table IV-89). Again, it is the

IV-375 MeNH₂, 140°, 18 hr → **IV-376**

3-fluoro substituent in 3-fluoro-4-nitropyridine-1-oxide (**IV-377**) which is substituted (see Section V.4.A) to give the corresponding amine (**IV-378**) (605). Substituted 3-fluoro and 3-bromo-4-nitropyridine-1-oxides react with

IV-377 RNH₂ → **IV-378**

aminoacids to give crystalline derivatives (Table IV-89). A route to 3-alkylamino-4-aminopyridines is illustrated by the preparation of 4-amino-3-methylaminopyridine from 3-bromo-4-nitropyridine-1-oxide (617). Pentachloropyridine-1-oxide (**IV-188**) reacts with primary amines (molar ratio 1:2) in dioxane, ethanol, or benzene at room temperature to give the 2-amino derivative (**IV-379**) exclusively (618). With excess amine present and under reflux in dioxane, only the 2,6-disubstituted derivative (**IV-380**) was obtained (618) (Table IV-89) (see also Chapter VI).

IV-188 RNH₂, R.T. → **IV-379** R = Me, Et

RNH₂, Δ → **IV-380**

TABLE IV-89. Reaction of Halopyridine-1-oxides with Amines

N-oxide	Conditions	Products (yield)	Ref.
2-Cl	Me$_2$NH	2-Dimethylaminopyridine-1-oxide	149
	Piperidine, PhNO$_2$, sealed tube, 130–150°, 5 hr	2-Piperidinopyridine-1-oxide, m.p. 125.7° (75 %); picrate, m.p. 132–134°	603
	Morpholine	2-Morpholinopyridine-1-oxide, m.p. 148–151°; picrate, m.p. 148–151°	603
2-Cl-4-NO$_2$	(i) NH$_3$, EtOH, 140°, 4 hr (ii) Ac$_2$O	2-Acetamido-4-nitropyridine-1-oxide, m.p. 206° (decomp.) (38.9 %)	361
2,3,4,5,6-Cl$_5$	MeNH$_2$, R.T., 16 hr	2-Methylamino-3,4,5,6-tetrachloro-pyridine-1-oxide, m.p. 152° (75 %)	618
	EtNH$_2$, R.T., 16 hr	2-Ethylamino-3,4,5,6-tetrachloropyridine-1-oxide, m.p. 115° (85 %)	618
	Excess MeNH$_2$, Δ	2,6-Di-(methylamino)-4,5,6-trichloro-pyridine-1-oxide, m.p. 154° (100 %)	618
	Excess n-BuNH$_2$, Δ	2,6-Di-(n-butylamino)-4,5,6-trichloro-pyridine-1-oxide, m.p. 47.5° (65 %)	618
2-Br-4-NO$_2$	dl-Alanine	dl-3-N-Alanyl-4-nitropyridine-1-oxide, m.p. 216° (decomp.) (52.4 %)	619
	β-Alanine	3-N-β-Alanyl-4-nitropyridine-1-oxide, m.p. 228° (decomp.) (52.4 %)	619
	dl-Valine	dl-3-N-Valyl-4-nitropyridine-1-oxide, m.p. 214° (decomp.) (64.4 %)	619
	dl-Norvaline	dl-3-N-Norvalyl-4-nitropyridine-1-oxide, m.p. 164° (29.2 %)	619
	l-Isoleucine	l-3-N-Isoleucyl-4-nitropyridine-1-oxide, m.p. 217° (decomp.) (88.8 %)	619
	dl-Isoleucine	dl-3-N-Isoleucyl-4-nitropyridine-1-oxide, m.p. 179° (74.0 %)	619
	dl-Norleucine	dl-3-N-Norleucyl-4-nitropyridine-1-oxide, m.p. 164° (74.0 %)	619
	dl-Proline	dl-3-N-Prolyl-4-nitropyridine-1-oxide, m.p. 173° (decomp.) (63.0 %)	619
	dl-Serine	dl-3-N-Seryl-4-nitropyridine-1-oxide, m.p. 210° (decomp.) (33.5 %)	619
	dl-Methionine	dl-3-N-Methionyl-4-nitropyridine-1-oxide, m.p. 159° (55.5 %)	619
	dl-Aspartic acid	dl-3-N-Aspartyl-4-nitropyridine-1-oxide, m.p. 214° (decomp.) (44.2 %)	619
	dl-Glutamic acid	dl-3-N-Glutamyl-4-nitropyridine-1-oxide, m.p. 203° (decomp.) (48.9 %)	619
	dl-Phenylalanine	dl-3-N-Phenylalanylpyridine-1-oxide, m.p. 202° (decomp.) (78.8 %)	619
2-X-4-NO$_2$	Me$_2$NH	2-Dimethylamino-4-nitropyridine-1-oxide	600
	Et$_2$NH	2-Diethylamino-4-nitropyridine-1-oxide	600

Table IV-89 (*Continued*)

N-oxide	Conditions	Products (yield)	Ref.
3-F	Piperidine, Δ, 4 hr	3-Piperidinopyridine-1-oxide, m.p. 85–87°; picrate, m.p. 193°	54
	NH_2NH_2, H_2O	3-Hydrazopyridine-1-oxide, m.p. 148°	54
3-F-4-NO$_2$	NH_3, EtOH	3-Amino-4-nitropyridine-1-oxide, m.p. 235° (100%)	352, 605
	NH_2NH_2, H_2O, MeOH	3-Hydrazo-4-nitropyridine-1-oxide, m.p. 192° (98.5%)	605, 606
	$EtNH_2$	3-Ethylamino-4-nitropyridine-1-oxide, m.p. 154° (98.4%)	605
	Et_2NH, MeOH, 50°	3-Diethylamino-4-nitropyridine-1-oxide, m.p. 74° (95.9%)	605, 606
	$HOCH_2CH_2NH_2$	3-β-Hydroxyethylamino-4-nitropyridine-1-oxide, m.p. 165° (82.6%)	605, 606
	$PhNH_2$	3-Anilino-4-nitropyridine-1-oxide, m.p. 198° (89%)	605
	o-MeC$_6$H$_4$NH$_2$	4-Nitro-3-o-toluidinopyridine-1-oxide, m.p. 162° (90.3%)	605, 606
	m-MeC$_6$H$_4$NH$_2$	4-Nitro-3-m-toluidinopyridine-1-oxide, m.p. 160° (99.3%)	605, 606
	p-MeC$_6$H$_4$NH$_2$	4-Nitro-3-p-toluidinopyridine-1-oxide, m.p. 204° (96.7%)	605, 606
	Glycine	3-N-Glycyl-4-nitropyridine-1-oxide, m.p. 216° (93.9%)	620
	dl-Serine	*dl*-3-N-Seryl-4-nitropyridine-1-oxide, m.p. 212° (decomp.) (52.5%)	620
	dl-Cysteine	*dl*-3-N-Cysteinyl-4-nitropyridine-1-oxide, m.p. 173° (decomp.) (61.7%)	620
	dl-α-Alanine	*dl*-3-α-Alanyl-4-nitropyridine-1-oxide, m.p. 216° (decomp.) (87.1%)	620
	β-Alanine	3-β-N-Alanyl-4-nitropyridine-1-oxide, m.p. 228° (decomp.) (78.4%)	620
	dl-Aspartic acid	*dl*-3-N-Aspartyl-4-nitropyridine-1-oxide, m.p. 214° (decomp.) (90.4%)	620
	dl-Threonine	*dl*-3-N-Threonyl-4-nitropyridine-1-oxide, m.p. 203° (decomp.) (54%)	620
	dl-Proline	*dl*-3-N-Prolyl-4-nitropyridine-1-oxide, m.p. 175° (decomp.) (80.8%)	620
	dl-Glutamic acid	*dl*-3-N-Glutamyl-4-nitropyridine-1-oxide, m.p. 205° (decomp.) (77.4%)	620
	dl-Valine	*dl*-3-N-Valyl-4-nitropyridine-1-oxide, m.p. 219° (decomp.) (76.7%)	620
	dl-Norvaline	*dl*-3-N-Norvalyl-4-nitropyridine-1-oxide, m.p. 168° (69%)	620
	dl-Methionine	*dl*-3-N-Methionyl-4-nitropyridine-1-oxide, m.p. 162° (94.4%)	620

Table IV-89 (*Continued*)

N-oxide	Conditions	Products (yield)	Ref.
	dl-Histidine	*dl*-3-*N*-Histidyl-4-nitropyridine-1-oxide, m.p. 199° (decomp.) (84.6%)	620
	l-Leucine	*l*-3-*N*-Leucyl-4-nitropyridine-1-oxide, m.p. 182° (89.2%)	620
	dl-Isoleucine	*dl*-3-*N*-Isoleucyl-4-nitropyridine-1-oxide, m.p. 179° (89.2%)	620
	dl-Norleucine	*dl*-3-*N*-Norleucyl-4-nitropyridine-1-oxide, m.p. 164° (57.8%)	620
	dl-Lysine	*dl*-3-*N*-Lysyl-4-nitropyridine-1-oxide, m.p. 185° (decomp.) (91.6%)	620
	dl-Tyrosine	*dl*-3-*N*-Tyrosyl-4-nitropyridine-1-oxide, m.p. 145° (decomp.) (94.1%)	620
	dl-Phenylalanine	*dl*-3-*N*-Phenylalanyl-4-nitropyridine-1-oxide, m.p. 202° (decomp.) (99%)	620
	dl-Tryptophan	*dl*-3-*N*-Tryptophyl-4-nitropyridine-1-oxide, m.p. 216° (decomp.) (83.5%)	620
	dl-Cystine	*dl*-3-*N*-Cystyl-4-nitropyridine-1-oxide, m.p. 172° (decomp.) (46.6%)	620
3-F-4-NO$_2$-2-Me	NH$_3$, MeOH, autoclaved, 100°, 3 hr	3-Amino-4-nitro-2-picoline-1-oxide, m.p. 218° (decomp.) (81.4%)	343, 605
	PhNH$_2$	3-Anilino-4-nitro-2-picoline-1-oxide, m.p. 169° (84.2%)	343, 605
	HOCH$_2$CH$_2$NH$_2$	3-β-Hydroxyethylamino-4-nitro-2-picoline-1-oxide, m.p. 110° (80.7%)	605
	Glycine	3-*N*-Glycyl-4-nitro-2-picoline-1-oxide, m.p. 199° (decomp.) (88.1%)	343, 605
	H$_2$NCH(CO$_2$H)-(CH$_2$)$_2$CONH$_2$	(structure: 4-nitro-2-methylpyridine-1-oxide with 3-NHCH(CO$_2$H)CH$_2$CH$_2$CONH$_2$), m.p. 265° (70.4%)	343, 605
	Glutamic acid	3-*N*-Glutamyl-4-nitro-2-picoline-1-oxide, m.p. 119° (66.9%)	343, 605
	Methionine	3-*N*-Methionyl-4-nitro-2-picoline-1-oxide, m.p. 148° (66.4%)	343, 605
	Proline	4-Nitro-3-*N*-prolylpicoline-1-oxide, m.p. 148° (59.9%)	343
3-F-4-NO$_2$-5-Me	NH$_3$	5-Amino-4-nitro-3-picoline-1-oxide, m.p. 244° (decomp.) (71.2%)	605
	NH$_2$NH$_2$, H$_2$O, MeOH	5-Hydrazo-4-nitro-3-picoline-1-oxide, m.p. 192° (84%)	56, 605
	PhNH$_2$	5-Anilino-4-nitro-3-picoline-1-oxide, m.p. 180° (42.1%)	56, 605

Table IV-89 (*Continued*)

N-oxide	Conditions	Products (yield)	Ref.
	EtNH$_2$	5-Ethylamino-4-nitro-3-picoline-1-oxide, m.p. 108° (87.3%)	56, 605
	HOCH$_2$CH$_2$NH$_2$	5-β-Hydroxyethylamino-4-nitro-3-picoline-1-oxide, m.p. 175° (81.3%)	56, 605
	Glycine	5-N-Glycyl-4-nitro-3-picoline-1-oxide, m.p. 226° (52.8%)	56, 605
	Glutamic acid	5-N-Glutamyl-4-nitro-3-picoline-1-oxide, m.p. 130° (55.5%)	56, 605
	Arginine	5-N-Arginyl-4-nitro-3-picoline-1-oxide, m.p. 251° (61.3%)	56, 605
	Methionine	5-N-Methionyl-4-nitro-3-picoline-1-oxide, m.p. 157° (93%)	56, 605
3-F-4-NO$_2$-2,6-Me$_2$	Alc. NH$_3$, 0°, autoclave, 120–130°, 4 hr	3-Amino-4-nitro-2,6-lutidine-1-oxide, m.p. 189–190° (71.2%)	336, 605
	PhNH$_2$, 120–130°, 2 hr	3-Anilino-4-nitro-2,6-lutidine-1-oxide, m.p. 164° (90%)	336, 605
3-Br-4-Cl	3% MeNH$_2$, 140°, 18 hr	4-Methylamino-3-bromopyridine-1-oxide (hygroscopic); picrate, m.p. 189–191°	337
3-Br-4-NO$_2$	(i) NH$_3$, autoclave, 120°, 90 min (ii) Δ, 3 hr	3-Amino-4-nitropyridine-1-oxide, m.p. 237° (decomp.) (28.8%)	353
	NH$_2$NH$_2$, H$_2$O, 50°	3-Hydrazo-4-nitropyridine-1-oxide, m.p. 192° (decomp.)	353
	Me$_2$NH, 3 hr	3-Dimethylamino-4-nitropyridine-1-oxide, m.p. 146° (83.3%)	353
	Et$_2$NH, 1 hr	3-Diethylamino-4-nitropyridine-1-oxide, m.p. 73–74° (57.1%)	353
	PhNH$_2$, 4 hr	3-Anilino-4-nitropyridine-1-oxide, m.p. 197–198° (47.4%)	353
	PhCH$_2$NH$_2$, 1 hr	3-Benzylamino-4-nitropyridine-1-oxide, m.p. 189° (81.3%)	353
	C$_6$H$_{11}$NH$_2$	3-Cyclohexylamino-4-nitropyridine-1-oxide, m.p. 197–199° (75.6%)	353
	HOCH$_2$CH$_2$NH$_2$	3-β-Hydroxyethylamino-4-nitropyridine-1-oxide, m.p. 164° (73.4%)	353
4-Cl	25% MeNH$_2$, 140°, 20 hr	4-Methylaminopyridine-1-oxide, m.p. 192–194°; picrate, m.p. 193–194°; picrolonate, m.p. 211–212°	217
	Morpholine	4-N-Morpholinopyridine-1-oxide, m.p. 91–92°	597
4-Cl-3-Me	3% MeNH$_2$, 140°, 18 hr	4-Methylamino-3-picoline-1-oxide, m.p. 106–107°; picrate, m.p. 184–185°	337
4-Cl-3-Et	3% MeNH$_2$, 140°, 18 hr	4-Methylamino-3-ethylpyridine-1-oxide, m.p. 117–118°, b.p. 120–122°/0.5 mm; picrate, m.p. 182–183°	337

217

Table IV-89 (*Continued*)

N-oxide	Conditions	Products (yield)	Ref.
4-Cl-3-isoBu	3% MeNH$_2$,140°, 18 hr	4-Methylamino-3-isobutylpyridine-1-oxide; picrate, m.p. 164–165°	337
4-Cl-3,5-Me$_2$	Aq. NH$_3$, 140°, 18 hr	4-Amino-3,5-lutidine-1-oxide·2H$_2$O, m.p. 227–229°; picrate, m.p. 221–223°	337
	3% MeNH$_2$, 140°, 18 hr	4-Methylamino-3,5-lutidine-1-oxide, m.p. 94.5–95.5°; picrate, m.p. 172–173°	337
4-Cl-2,3,5,6-Me$_4$	3% MeNH$_2$, 140°, 18 hr	4-Methylamino-2,3,5,6-tetramethyl-pyridine-1-oxide (hygroscopic); picrate, m.p. 140–141°	337
4-Cl-3-CO$_2$H	NH$_4$OH, 100°, 18 hr	4-Aminonicotinic acid-1-oxide, m.p. 270.5° (49%)	324
	NH$_2$NH$_2$, H$_2$O, EtOH, 6 hr	4-Hydrazinonicotinic acid-1-oxide, m.p. 230–231° (91%)	621
4-Cl-3-CO$_2$H	PhNHNH$_2$, EtOH, Δ, 4.5 hr	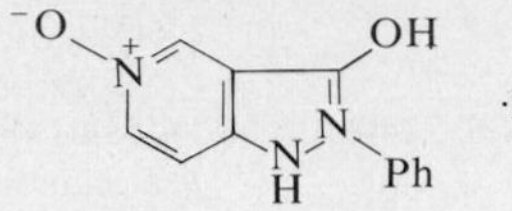 ·HCl m.p. 230–232° (91%)	621
	MeNHNH$_2$, EtOH, Δ, 6 hr	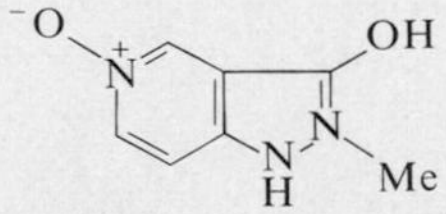 ·HCl (85.5%)	621

5. Reactions of Thiol and Sulfide Groups*

Oxidation of substituted pyridine-1-oxide-2-thiols (**IV-381**) (in equilibrium with the 1-hydroxy-2-pyridinethiones) with 30% H$_2$O$_2$—H$_2$O gave the corresponding disulfides (**IV-382**) (622) (see also Section I). Lithium aluminum hydride reduction of 2,2′-(1,1′-dihydroxy-4,4′,5,5′-tetramethyldipyridyl-6,6′-dithione)disulfide (**IV-382**; R = 4-Me, R′ = 5-Me) in tetrahydrofuran

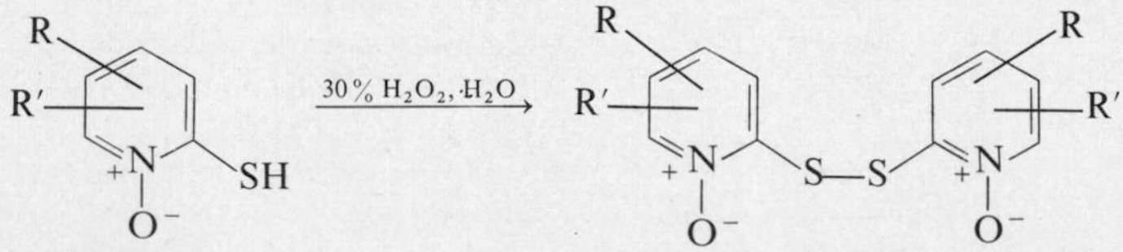

IV-381   VI-382

R = R′ = H, m.p. 200°
R = 4-Me, R′ = H, m.p. 195–196°
R = 4-Me, R′ = 6-Me, m.p. 228–229°

* See also Chapter XV.

gave 1-hydroxy-2-sulfhydro-3,4-dimethyl-6-pyridinethione **(IV-381; R =
4-Me, R′ = 5-Me)** (314). 4-Phenylthiopyridine-1-oxide was further oxidized
with hydrogen peroxide in acetic acid to yield 4-benzenesulfonylpyridine-
1-oxide **(IV-383)** (317). Similarly, (4-nitro-1-oxido-3-pyridyl)sulfide **(IV-384)**

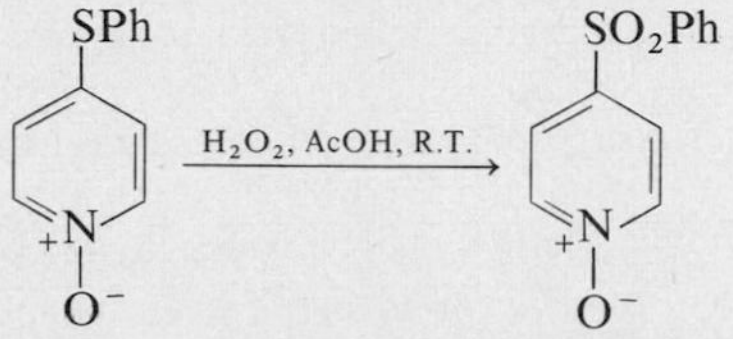

IV-383

was oxidized with hydrogen peroxide in trifluoroacetic acid to give the
sulfone **(IV-385)** (606).

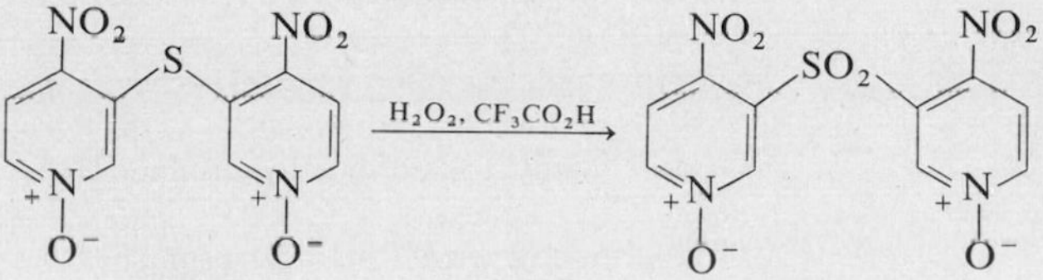

IV-384   **IV-385**

1-Hydroxy-2-pyridinethiones **(IV-381)** react with aryl or alkyl halides to
form the corresponding sulfides **(IV-386)** (Table IV-90). The sodium salt of

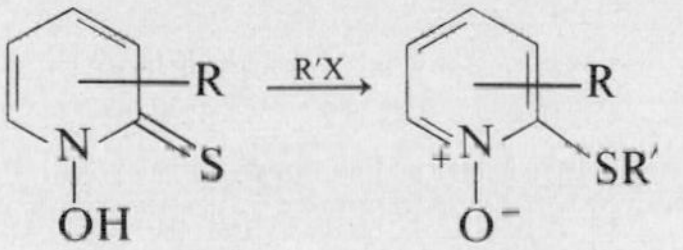

IV-381   **IV-386**

IV-381 reacts with paraformaldehyde in ethylene dichloride to yield the S-
hydroxymethylated product **(IV-387)** (623). 1-Hydroxy-2-pyridinethione

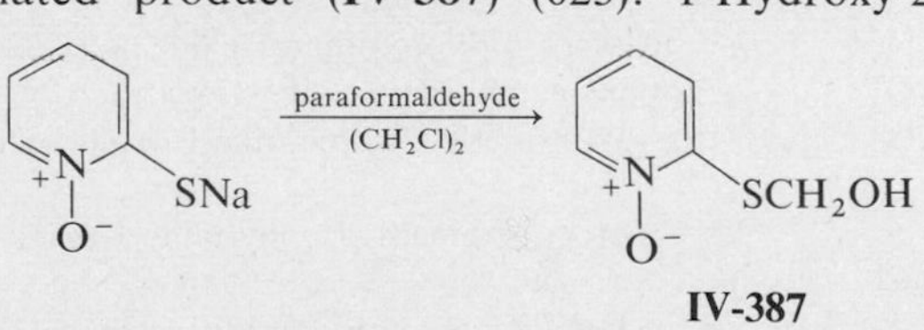

IV-387

(IV-374) and phenylisocyanate react in the presence of triethylamine to give

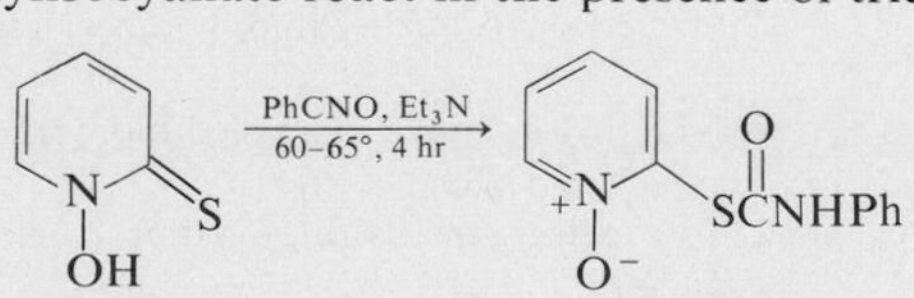

IV-374   **IV-388**

TABLE IV-90. Reaction of Pyridine-1-oxide-2-thiol or with Alkyl Halides

Conditions	Products (yield)	Ref.
ClCH$_2$OMe, 61–62°, 1.5–3.0 hr	2-Methoxymethylthiopyridine-1-oxide, m.p. 129.5–135.0°	625
ClCH$_2$OEt	2-Ethoxymethylthiopyridine-1-oxide, m.p. 69.5–70° (69%)	625
ClCH$_2$O-isoPr	2-Isopropoxymethylthiopyridine-1-oxide, m.p. 101.5–105.5° (80%)	625
ClCH$_2$O(CH$_2$)$_5$CH(CH$_3$)$_2$	2-(6-Methyl-n-heptyloxy)methylthiopyridine-1-oxide, m.p. 75–76° (80%)	625
ClCH$_2$OC$_{12}$H$_{25}$	2-n-Dodecyloxymethylthiopyridine-1-oxide, m.p. 78.0–79.5° (87.5%)	625
ClCH$_2$OC$_{18}$H$_{37}$	2-Octadecyloxymethylthiopyridine-1-oxide, m.p. 90–93° (63%)	625
ClCH$_2$SMe	2-(α-Methylthio)methylthiopyridine-1-oxide, m.p. 105–107° (63%)	625
ClCH$_2$SPr	2-(α-Propylthio)methylthiopyridine-1-oxide, m.p. 98.5–100.5° (69%)	625
Cl$_3$CSH, 20–24°	1-Oxido-2-pyridyl trichloromethyl disulfide·HCl, m.p. 99°	626
Cl$_3$CSH	1-Oxido-2-pyridyl trichloromethyl disulfide·HCl, m.p. 85–119°	627
2,4-Cl$_2$C$_6$H$_3$OCH$_2$COCl	1-Oxido-2-pyridyl-2',4'-dichlorophenoxythioacetate, m.p. 90–91°	628
4-ClC$_6$H$_4$OCH$_2$COCl	1-Oxido-2-pyridyl-4'-chlorophenoxythioacetate, m.p. 124–125°	628
α-C$_{10}$H$_7$CH$_2$COCl	1-Oxido-2-pyridyl-α'-naphthylthioacetate, oil	628
2,4,5-Cl$_3$C$_6$H$_2$OCH$_2$COCl	1-Oxido-2-pyridyl-2',4',5'-trichlorophenoxythioacetate, m.p. 121–126°	628
2-(CH$_2$Cl)-4-NO$_2$-furan	2-(4-Nitro-2-furylmethylthio)pyridine-1-oxide, m.p. 175° (decomp.)	629
2-Cl-4-NO$_2$-furan	2-(4-Nitro-2-furylthio)pyridine-1-oxide	630
Sodium salt, ClCH$_2$OMe, dioxane 55°, 1 hr	2-Methoxymethylthiopyridine-1-oxide (87.8%)	623
Sodium salt, ClCH$_2$OEt	2-Ethoxymethylthiopyridine-1-oxide, m.p. 62.6° (97%)	623
Sodium salt, ClCH$_2$OPr	2-Propoxymethylthiopyridine-1-oxide (86.2%)	623
Sodium salt, ClCH$_2$O-isoPr	2-Isopropoxymethylthiopyridine-1-oxide, oil (90%)	623
Sodium salt, ClCH$_2$OCH$_2$CH$_2$Cl	2-β-Chloroethoxymethylthiopyridine-1-oxide (69.3%)	623
Sodium salt, ClCH$_2$OC$_{12}$H$_{25}$	2-n-Dodecyloxymethylthiopyridine-1-oxide	623
Sodium salt, ClCH$_2$OC$_{18}$H$_{37}$	2-n-Octadecyloxymethylthiopyridine-1-oxide, m.p. 90–93° (63%)	623
Sodium salt, ClCH$_2$O-(CH$_2$)$_5$CH(CH$_3$)$_2$	2-(6-Methyl-n-heptyloxy)methylthiopyridine-1-oxide, m.p. 75–76° (80%)	623
Sodium salt, ClCH$_2$SPr	2-(α-Propylthio)methylthiopyridine-1-oxide, m.p. 97.5–99° (68%)	623

Table IV-90 (*Continued*)

Conditions	Products (yield)	Ref.
Sodium salt (i) ClCH₂CH=CH₂, 31–34°, 4 hr (ii) 39°, 5 hr	2-Allylthiopyridine-1-oxide, m.p. 73–74°, m.p. 57–63°	627, 631
Sodium salt, methallyl chloride	2-(Methallylthio)pyridine-1-oxide, clear oil (81.1 %)	627
Sodium salt, 2,3-dichloro-propene	2-(2-Chloroallylthio)pyridine-1-oxide, clear oil (70.5 %) m.p. 114–115°	627, 631
Sodium salt, 2-bromo-3-chloropropene	2-(2-Bromoallylthio)pyridine-1-oxide, m.p. 122–123°	631
Sodium salt, 1,1,3-trichloro-propene	2-(3,3-Dichloroallylthio)pyridine-1-oxide	631
Sodium salt, 1,3-dichloro-propene	2-(3-Chloroallylthio)pyridine-1-oxide	631
Sodium salt, PhCH₂Cl	2-Benzylthiopyridine-1-oxide, m.p. 170–172°	632
Sodium salt, 3-MeC₆H₄-CH₂Cl,50–60°, 5 hr	2-(3-Methylbenzylthio)pyridine-1-oxide, m.p. 108–110°	632
Sodium salt, 2,6-Cl₂C₆H₃-CH₂Cl, 50–60°, 5 hr	2-(2,6-Dichlorobenzylthio)pyridine-1-oxide, m.p. 213–216°	632
Sodium salt, 2,3,6-Cl₃-C₆H₂CH₂Cl, 50–60°, 5 hr	2-(2,3,6-Trichlorobenzylthio)pyridine-1-oxide, m.p. 237–238°	632

S-(1-oxido-2-pyridyl)-N-phenylthiocarbamate (**IV-388**) (624). 1-Hydroxy-2- and 4-pyridinethione and 1,2,3,4,6-penta-O-acetyl-α-D-glucopyranosyl bromide in the presence of base gave 2- and 4-(2,3,4,6-tetra-O-acetyl-β-D-glucopyranosylthio)pyridine-1-oxide (318, 610).

6. Amino and Hydroxylamino Compounds

The U.V. spectrum of 2-aminopyridine-1-oxide (**IV-12**) shows the molecule to exist mainly in the amino and not in its imino (**IV-389**) tautomeric form (633a).

IV-12   IV-389

The stability of the diazonium salt of 2-aminopyridine-1-oxide as compared to 2-aminopyridine may be due to contributions of structures **IV-390** and **IV-391** and possibly also **IV-392** (7). When 2-aminopyridine-1-oxide is diazotized in the presence of tetrafluoroboric acid the cyclic diazotate salt **(IV-393)** is said to be formed which coupled with alkaline β-naphthol to give

the corresponding azo-compound (633b). Similarly, the diazonium tetrafluoroborate of 4-aminopyridine-1-oxide was formed and was coupled with β-naphthol (633b). Diazotization of 2-, 3-, and 4-aminopyridine-1-oxides **(IV-12)** followed by coupling with phenols and with N,N-dialkylaminobenzenes gives the corresponding azo-compounds **(IV-394)** (Tables IV-91

TABLE IV-91. Diazotization of Aminopyridine-1-oxides and Their Coupling with Phenols

N-oxide	Phenol	Product (yield)	Ref.
2-NH$_2$	β-HOC$_{10}$H$_7$, 0°	2-(Azo-2'-hydroxy-1'-naphthyl)pyridine-1-oxide, m.p. 211°	633b
4-NH$_2$	β-HOC$_{10}$H$_7$, 0°	4-(Azo-2'-hydroxy-1'-naphthyl)pyridine-1-oxide, m.p. 231°	633b
4-NH$_2$-2-CO$_2$H	PhOH	4-Azo-4'-hydroxyphenylpicolinic acid-1-oxide, m.p. 278° (decomp.) (42.2%)	95
4-NH$_2$-2-CO$_2$H	m-MeC$_6$H$_4$OH	4-Azo-4'-hydroxy-3'-methylphenylpicolinic acid-1-oxide, m.p. 239°	95
4-NH$_2$-2-CO$_2$H	β-HOC$_{10}$H$_7$	4-Azo-2'-hydroxy-1'-naphthylpicolinic acid-1-oxide, m.p. 230°	95
4-NH$_2$-2-CO$_2$H	α-HOC$_{10}$H$_7$	4-Azo-1'-hydroxy-4'-naphthylpicolinic acid-1-oxide	95

TABLE IV-92. Diazotization of Aminopyridine-1-oxides and Coupling with N,N-Dialkyl-aminobenzenes

N-oxide	ArNR'R''	Properties of IV-394	Ref.
2-NH$_2$	Ar=Ph, R' = R'' = Me		302
	Ar=Ph, R' = H, R'' = CH$_2$SO$_3$Na		634
2-NH$_2$-5-NO$_2$	Ar=Ph, R' = R'' = Et	m.p. 154–156°; blue violet	123
	Ar=3-MePh, R' = R'' = Et	m.p. 152–154°; blue	123
	Ar=3-MeOPh, R' = R'' = Et	m.p. 147–149°; navy blue	123
	Ar=Ph, R' = R'' = CH$_2$CH$_2$OH	m.p. 210°; blue violet	123
	Ar=3-MePh, R' = R'' = CH$_2$CH$_2$OH	m.p. 160–163° (decomp.); blue violet	123
	Ar=Ph, R' = CH$_2$CH$_2$OH, R'' = CH$_2$CH$_2$CN	m.p. 124–126° (decomp.); red violet	123
3-NH$_2$	Ar=Ph, R' = R'' = Me	m.p. 186°	127
4-NH$_2$	Ar=Ph, R' = R'' = Me	m.p. 212°	127
4-NO$_2$-2-^{14}C	(i) Pd–C, H$_2$	m.p. 218–219°	28
	(ii) diazotize and couple with Ar=Ph, R' = R'' = Me		

and IV-92), respectively. The diazonium salt of 2-aminopyridine-1-oxide (**IV-12**) did not undergo a Gomberg-Hey reaction with benzene (6). 4-Halopyridine-1-oxides and pyridine-4-sulfonic acid-1-oxides are formed when the 4-diazonium salt is treated with cuprous halide, potassium iodide or SO$_2$-CuCl$_2$·2H$_2$O, respectively (339, 341, 354a) (Table IV-93). The diazonium salt from aminopyridine-1-oxides couples with azide ion to give azidopyridine-1-oxide (**IV-395**) (Table IV-94) (147b). Diazotization of 4-

IV-395

amino-5-ethyl-2-picoline-1-oxide (**IV-396**) followed by coupling with 3-carbethoxy-2-piperidone yields 2,3-dioxopiperidino-3-[4-(5-ethyl-1-oxido-2-picolyl)hydrazone] (**IV-397**) (295). 4-Amino-2-picoline-1-oxide reacts similarly. Using Pschorr's phenanthrene synthesis, an azaphenanthrene

IV-396   IV-397

TABLE IV-93. Preparation of 4-Halopyridine-1-oxides and Pyridine-4-sulfonic acid-1-oxides

N-oxide	Conditions	Product (yield)	Ref.
4-NH$_2$	(i) Diazotize (ii) KI	4-Iodopyridine-1-oxide, m.p. 171° (41.5%)	339, 354a
4-NH$_2$-2-Me	(i) Diazotize (ii) KI	4-Iodo-2-picoline-1-oxide, m.p. 101–102° (40.5%)	354a
4-NH$_2$-2-Me	(i) Diazotize (ii) CuCl$_2$·2H$_2$O sat. SO$_2$	2-Picoline-4-sulfonic acid-1-oxide, m.p. 257–258° (60%)	341
4-NH$_2$-2,6-Me$_2$	(i) Diazotize (ii) CuCl	4-Chloro-2,6-lutidine-1-oxide (33%)	339
	(i) Diazotize (ii) CuBr	4-Bromo-2,6-lutidine-1-oxide (49%)	339
	(i) Diazotize (ii) KI	4-Iodo-2,6-lutidine-1-oxide, m.p. 104–105° (46.3%)	354a
4-NH$_2$-2-Me- 5-Et	(i) Diazotize (ii) KI	5-Ethyl-4-iodo-2-picoline-1-oxide, m.p. 143–144° (47.2%)	354a
4-NH$_2$-3-Me	(i) Diazotize (ii) KI	4-Iodo-3-picoline-1-oxide, m.p. 170–171° (74.9%)	354a
	(i) Diazotize (ii) CuCl$_2$·2H$_2$O sat. SO$_2$	3-Picoline-4-sulfonic acid-1-oxide, m.p. 270°	341

(IV-398) (RRI 3,526) was obtained from the appropriate 4-aminopyridine-1-oxide (IV-399) (569).

IV-399   IV-398

TABLE IV-94. Preparation of Azidopyridine-1-oxide (147b)

N-oxide	Conditions	Product (yield)
2-NH$_2$	(i) NaNO$_2$, HCl, 0° (ii) NaN$_3$, 0–5° (iii) R.T., 1 hr	2-Azidopyridine-1-oxide, m.p. 83° (decomp.) (54%)
2-NH$_2$-3-Me	(i) NaNO$_2$, HCl, 0° (ii) NaN$_3$, 0–5° (iii) R.T., 1 hr	2-Azido-3-picoline-1-oxide, m.p. 88° (decomp.) (48%)
3-NH$_2$	(i) NaNO$_2$, HCl, 0° (ii) NaN$_3$, 0–5° (iii) R.T., 1 hr	3-Azidopyridine-1-oxide, m.p. 98° (decomp.) (33%)
4-NH$_2$	(i) NaNO$_2$, HCl, 0° (ii) NaN$_3$, 0–5° (iii) R.T., 1 hr	4-Azidopyridine-1-oxide, m.p. 143° (decomp.) (37%)

TABLE IV-95. Acylation of Aminopyridine-1-oxides

Substituent in N-oxide	Conditions	Product	Ref.
2-NH$_2$	Ac$_2$O	2-Acetamidopyridine-1-oxide, m.p. 139–140°	131
	Diketene, Δ, 15 min	2-Acetoacetamidopyridine-1-oxide, m.p. 126–128° (80%)	129
	Diketene, CHCl$_3$, Δ, 30 min	2-[(6-Acetoacetamido-2-pyridyl)-methyl]-6-methyl-4H-pyran-4-one, m.p. 190° (decomp.)	129
2-NHMe	Ac$_2$O	2-N-Methylacetamidopyridine-1-oxide, m.p. 97–98°	131
	PhCOCl, C$_6$H$_6$, NEt$_3$	2-N-Methylbenzamidopyridine-1-oxide, m.p. 152–153° (78%)	131
3-NH$_2$	Diketene, CHCl$_3$, R.T., 2 hr	3-Acetoacetamidopyridine-1-oxide, m.p. 162° (93%)	129
4-NH$_2$	Ac$_2$O, ethyl methyl ketone, Δ, 45 min	4-Acetamidopyridine-1-oxide, m.p. 260–261°	131
4-NHMe	Ac$_2$O, AcOH, Δ, 2 hr	4-(N-Methylacetamido)pyridine-1-oxide, m.p. 145–147° (38%)	131
	PhCOCl, ethyl methyl ketone, R.T., 20 hr	4-(N-Methylbenzamido)pyridine-1-oxide, m.p. 116–118° (55%)	131

4-Aminopyridine-1-oxides (**IV-400**; R = R' = H) undergo the Skraup reaction to give low yields of 1,6-naphthyridine-1-oxide (**IV-401**) (RII 1,681). In the case of 4-amino-2-picoline-1-oxide (**IV-400**; R = H, R' = Me), cyclization occurred at the 3-position (635, 636). As expected, amino-pyridine-1-oxides undergo acylations and alkylations (66, 131) (Table IV-95),

R = R' = H (6%)
R = R' = Me (9%)
R = H, R" = Me (5%)

IV-400   IV-401

as well as reaction with sulfonyl chlorides to yield the sulfonamides (121) (Table IV-96). Diketene was also used to acylate aminopyridine-1-oxides (129).

4-Hydroxylaminopyridine-1-oxides (**IV-405**) undergo catalytic reduction to give 4-aminopyridine-1-oxide (167) as well as 4-aminopyridine (324). In the presence of 10% KOH, 4-hydroxylaminopyridine-1-oxide yields 4,4'-azopyridine-1,1'-dioxide (**IV-406**) (167, 173). Oxidation of 4-hydroxylamino-pyridine-1-oxide with 30% hydrogen peroxide in glacial acetic acid gave 4,4'-azoxypyridine-1,1'-dioxide (**IV-407**) (167, 173). Heating 4-hydroxyl-aminopyridine-1-oxide in water gave **IV-406** and **IV-407** (167). Oxidation of

TABLE IV-96. Sulfonamidopyridine-1-oxides (121)

IV-402	Conditions	IV-403 (yield)	IV-404 (yield)
3-NH$_2$-2,6-Me$_2$	p-AcNHC$_6$H$_4$SO$_2$Cl	m.p. 269–271° (57.7%)	m.p. 246–247°
3-NH$_2$-2,6-Me$_2$	p-O$_2$NC$_6$H$_4$SO$_2$Cl	m.p. 269–270° (19.6%)	—
3-NH$_2$-2,4-Me$_2$	p-AcNHC$_6$H$_4$SO$_2$Cl	m.p. 188–196° (71%)	—
5-NH$_2$-2,4-Me$_2$	p-AcNHC$_6$H$_4$SO$_2$Cl	m.p. 265–266° (82.3%)	m.p. 218–219° (quant.)

IV-405 with potassium permanganate in 20% sulfuric acid gave 4-nitroso-pyridine-1-oxide (**IV-408**) (167, 173). Heating 4-hydroxylaminopyridine-1-oxide and 4-nitrosopyridine-1-oxide in water for 3 to 4 min gave 4,4′-azoxypyridine-1,1′-dioxide (**IV-407**) (167).

7. Hydroxyl and Alkoxyl Groups

The behavior of hydroxy and alkoxypyridine-1-oxides are discussed comprehensively in Chapter XII, and only a few reactions are considered here.

Chlorination of 1-hydroxy-4-pyridones with phosphorus oxychloride occurs under mild conditions. For example, 1-hydroxy-3-nitro-4-pyridone gave 4-chloro-3-nitropyridine-1-oxide (39%) (449). In dimethylformamide, deoxygenation also occurred to give 4-chloro-3-nitropyridine (637). 1-Acetoxy-3-nitro-4-pyridone (IV-409) underwent chlorination and deacetylation to give 4-chloro-3-nitropyridine-1-oxide (IV-410) in 63% yield (449).

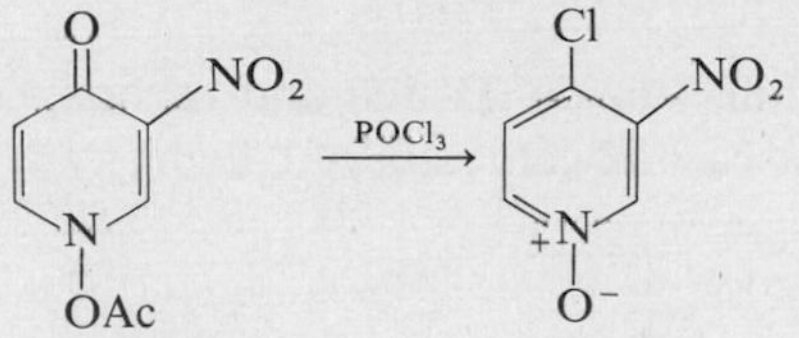

IV-409   IV-410

1-Hydroxy-2-pyridones (IV-411) can be obtained from 4-substituted-2-methoxypyridine-1-oxides (IV-412) by treatment with acetyl chloride, followed by heating in 2 to 4% hydrochloric acid solution for 10 hr (638). 4-Acetamido-2-methoxypyridine-1-oxide may be heated with acetyl chloride and then with ethanol to give 4-acetamido-1-hydroxy-2-pyridone (639).

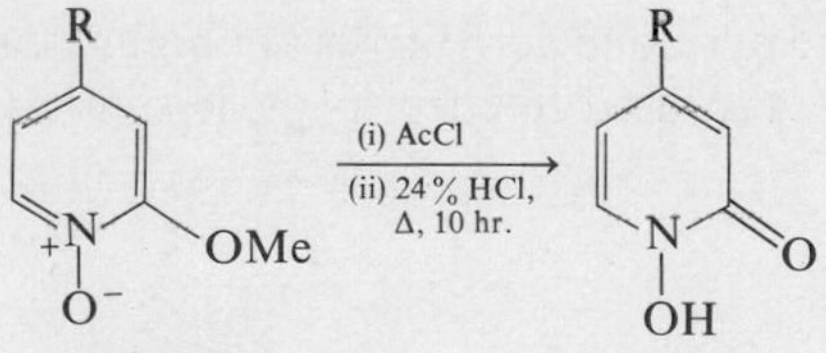

IV-412   IV-411

R = Cl, m.p. 145.6°
R = NO$_2$, m.p. 169–171°
R = CO$_2$H, m.p. >250°
R = NHCOMe, m.p. 260–261°

Hydrolysis of 2-benzyloxypyridine-1-oxide (IV-413) with hydrochloric acid gives 1-hydroxy-2-pyridone (IV-89) as the main product and 1-benzyloxy-2-pyridone (IV-414) as a by-product (640a, b). 2-Alkoxypyridine-1-oxides

IV-413 → **IV-89** + **IV-414**

(**IV-415**) undergo rearrangement to the corresponding 1-alkoxy-2-pyridones (**IV-416**) in good yield at 100° (604). As previously described, 2-alkenyloxy-pyridine-1-oxides rearrange to 1-alkenyloxy-2-pyridones and other products

$$R = CH_2CH\!=\!CH_2 \ (83\%)$$
$$R = CH_2Ph \ (92\%)$$
$$R = Me \ (89\%)$$

IV-415   **IV-416**

(538). 2-Ethoxypyridine-1-oxide (**IV-93**) and carboxylic acid chlorides give ethyl chloride and the carboxylate ester of 1-hydroxy-2(1*H*)-pyridone (**IV-94**)

IV-93   **IV-94**

(411, 640c). These pyridone derivatives are highly susceptible to nucleophilic attack and have been used to effect peptide synthesis without racemization (see Section III.3). Both the amino and *N*-oxide groups in 2-aminopyridine-1-oxide **(IV-12)** are acylated by phosgene to yield pyridino[1′,2′:2,3]-1,2,4-oxadiazoline-5-one **(IV-417)** (RI 7,936) (640d). Alternatively, **IV-417** (R = Me) was prepared by cyclization of 6-(ethylcarboxamido)-3-picoline-1-oxide **(IV-418)**. Treatment of **IV-417** with ammonia gave **IV-419** (125).

It has been shown that sodium iodide in acetic acid is an effective agent for the dealkylation of some alkoxypyrimidines (641). In the case of 4-methoxypyridine-1-oxide, dealkylation occurred only to a small extent. When ethyl 6-hydroxynicotinate-1-oxide **(IV-420)** and $(EtO)_2P(S)Cl$ were heated in the presence of potassium carbonate, ethyl 6-diethoxyphosphino-thionyloxypyridine-1-oxide **(IV-421)** was obtained (642).

8. Miscellaneous Side-chain Reactions

N-(*p*-Dimethylaminophenyl)-α-(1-oxido-2-pyridyl)nitrone **(IV-321)** reacts with sulfur dioxide to give the amide **(IV-422)** and with phosphorous trichloride to yield **IV-423** (559). When the nitrone **(IV-321)** and phosphorus oxychloride were heated together both **IV-423** (14 %) and **IV-422** (5 %) were isolated, while treatment with various acylating agents—for example, acetic anhydride, acetyl chloride, tosyl chloride, or benzoyl chloride—gave **IV-423**

(559). With hydroxylamine hydrochloride and sodium acetate in acetic acid it gave the aldoxime **(IV-424)** and with phenylhydrazine the phenylhydrazone **(IV-425)** (347). Isonicotinamide-1-oxide and nicotinamide-1-oxide give 4-and 3-cyanopyridine-1-oxide, respectively, on treatment with phosphorus

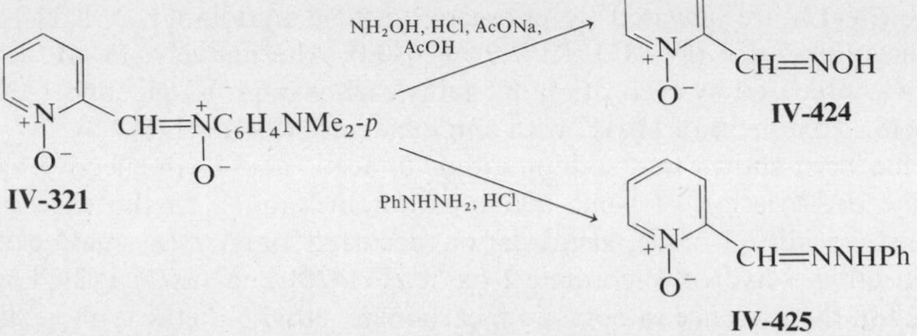

IV-321

IV-424

IV-425

oxychloride (643). Under these conditions, picolinamide-1-oxide (**IV-426**) underwent concommitent deoxygenation to yield 2-cyanopyridine (**IV-123**) (65). In the presence of phosphorus pentachloride, ring chlorination also occurs (see Section IV.2.B).

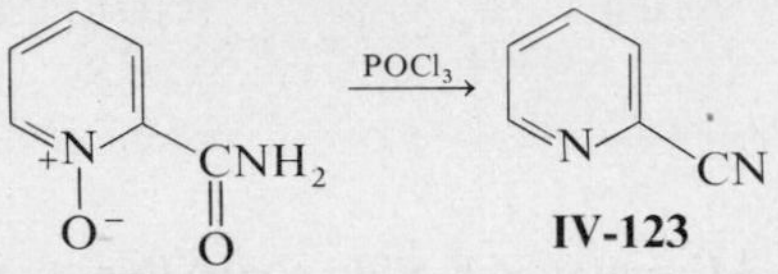

IV-426   **IV-123**

4-(2-Bromophenylthio)-3-aminopyridine-1-oxide (**IV-427**; R = H), a useful intermediate for the synthesis of 2-azaphenothiazine, was prepared by the reaction of 4-(2-bromophenylthio)nicotinamide-1-oxide (**IV-428**; R = H) with sodium hypochlorite (644).

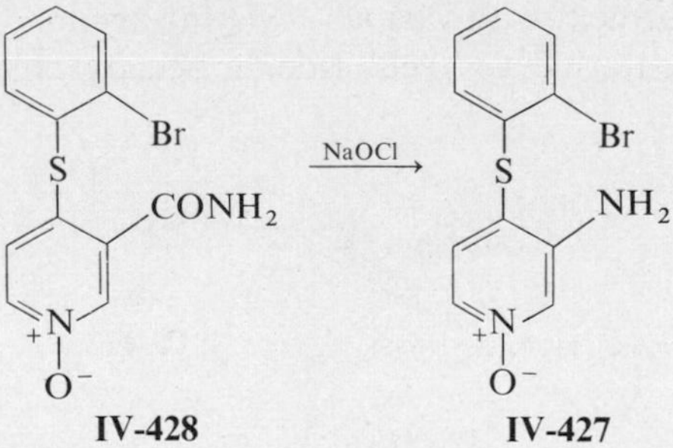

IV-428   **IV-427**

Pyridine-1-oxide-4-sulfonylhydrazide reacts with aldehydes and ketones to give the hydrazones (645) (Table IV-97).

When pyridine-1-oxide-4-sulfonyl chloride is treated with isonicotinic acid hydrazide the corresponding hydrazide is obtained (645). With amines, the expected sulfonamides are formed (Table IV-98). The sodium salt of pyridine-4-sulfonic acid-1-oxide gives 4-cyanopyridine and its N-oxide on heating with potassium cyanide at 300 to 330° (114).

Pyridine-1-oxide carboxylic esters are listed in Table IV-99 and amides in Table IV-100.

Aldehyde or ketone	Hydrazone, m.p. (°C)
EtCOEt	183
EtCOPr	175
$CH_2\!\!=\!\!CHCH_2CH_2COCH_3$	177
Cyclohexanone	165
PhCOMe	199
p-MeC$_6$H$_4$COMe	212
PhCH$=$CHCHO	193
PhCHO	191
p-MeOC$_6$H$_4$CHO	195
p-Me$_2$NC$_6$H$_4$CHO	202
3,4-(MeO)$_2$C$_6$H$_3$CHO	206
Piperonal	—
Vanillin	220
CH$_3$COCH$_2$CO$_2$Et	163
CH$_3$COCH$_2$COCH$_3$	165

TABLE IV-98. Reaction of Pyridine-1-oxide-4-sulfonyl chloride with Amines

R	R'	Amine Sulfonamide, m.p. (°C)	Ref.
H	Et	200–205	615
Et	Et	175	615
H	Me$_2$CH	210–212	615
H	Bu	175	615
H	PhCH$_2$	195, 192	615, 645
H	PhMeCH	125	615
R,R' = -CH$_2$CH$_2$OCH$_2$CH$_2$		143, 190	615, 645
R,R' = -(CH$_2$)$_5$-		164	645
R,R' = -(CH$_2$)$_6$-		124	615
H	Ph	164	645
Et	Ph	144	645
H	o-MeC$_6$H$_4$	178	645
H	p-MeC$_6$H$_4$	181	645
H	o-ClC$_6$H$_4$	176	645
H	m-ClC$_6$H$_4$	228	645
H	p-ClC$_6$H$_4$	203	645
H	o-MeOC$_6$H$_4$	125	615
H	p-EtOC$_6$H$_4$	145	615
H	3,4-HO(EtO$_2$C)C$_6$H$_3$	177	615
H	3,4-HO(H$_2$NSO)$_2$C$_6$H$_3$	228	615
H	m-EtO$_2$CC$_6$H$_4$	185	645
H	p-EtO$_2$CC$_6$H$_4$	164, 191	615, 645

TABLE IV-99. Esterification of Pyridinecarboxylic Acid-1-oxides

N-oxide	Conditions	Products (yield)	Ref.
2-CO$_2$H	MeOH, HCl	Methyl picolinate-1-oxide, b.p. 175–180°/0.1 mm, m.p. 73–74°	89, 99
	MeOH, SOCl$_2$	Methyl picolinate-1-oxide, b.p. 75–79°/0.01–0.05 mm; picrate m.p. 165°	95
	EtOH, HCl	Ethyl picolinate-1-oxide, b.p. 170–180°/0.4 mm	89
2-CO$_2$H-4-OMe	CH$_2$N$_2$, MeOH	Methyl 4-methoxypicolinate-1-oxide	95
2-CO$_2$H-4-NO$_2$	MeOH, HCl	Methyl 4-chloropicolinate-1-oxide, m.p. 104° (65%)	99
	CH$_2$N$_2$	Methyl 4-nitropicolinate-1-oxide, m.p. 135°	99
3-CO$_2$H-4-OCOPh	MeOH, HCl	Methyl 4-benzoyloxynicotinate-1-oxide, m.p. 154–155°	327
3-CO$_2$H-4-(o-BrC$_6$H$_4$S)	(i) HCl, MeOH, 0–10°, 1 hr (ii) Δ, 2 hr	Methyl 4-(o-bromophenylthio)nicotinate-1-oxide, m.p. 126–128° (77.5%); picrate, m.p. 144.5–146°	646
3-CO$_2$H-4-(2-Br-4-Cl-C$_6$H$_3$S)	(i) HCl, MeOH, 0–10°, 1 hr (ii) Δ, 2 hr	Methyl 4-(2-bromo-4-chlorophenylthio)-nicotinate-1-oxide, m.p. 62–63°	646
3-CO$_2$H-4-(2-Cl-5-OMe-C$_6$H$_3$S)	(i) HCl, MeOH, 0–10°, 1 hr (ii) Δ, 2 hr	Methyl 4-(2-chloro-5-methoxyphenylthio)-nicotinate-1-oxide, m.p. 158°	646
3-CO$_2$H-4-NO$_2$	(i) NH$_4$OH, AgNO$_3$ (ii) MeOH, MeI, Δ, 16 hr	Methyl 4-nitronicotinate-1-oxide, m.p. 158–159° (decomp.)	166
4-CO$_2$H	MeOH, H$_2$SO$_4$	Methyl isonicotinate-1-oxide, m.p. 121°; picrate, m.p. 107°	486
4-CO$_2$H-2,6-Me	EtOH, H$_2$SO$_4$	Ethyl 2,6-dimethylnicotinate-1-oxide, m.p. 93°	116
4-CO$_2$H-2-OH	EtOH, SOCl$_2$	Ethyl 2-hydroxyisonicotinate-1-oxide, m.p. 147.5–148.5°	647
4-CO$_2$H-2-OMe	CH$_2$N$_2$	Methyl 2-methoxyisonicotinate-1-oxide, m.p. 129–130° (55%)	62

TABLE IV-100. Preparation of Pyridinecarboxamide-1-oxides

N-oxide	Conditions	Products (yield)	Ref.
(a) From acids			
2-CO$_2$H-4-OMe	(i) SOCl$_2$ (ii) NH$_3$	4-Methoxypicolinamide-1-oxide, m.p. 179° (74%)	95
2-CO$_2$H-4-OEt	(i) SOCl$_2$ (ii) NH$_3$	4-Ethoxypicolinamide-1-oxide, m.p. 169°	95
2-CO$_2$H-4-OPr	(i) SOCl$_2$ (ii) NH$_3$	4-Propoxypicolinamide-1-oxide, m.p. 126°	95
2-CO$_2$H-4-Cl	Conc. NH$_4$OH	4-Chloropicolinamide-1-oxide, m.p. 146–148° (81%)	99

Table IV-100 (*Continued*)

N-oxide	Conditions	Products (yield)	Ref.
(b) From acid chlorides			
2-COCl-4-OMe	NH_2OH, abs. alc.	4-Methoxy-2-picolylhydroxamic acid-1-oxide, m.p. 211°	95
	$NH_2NH_2 \cdot H_2O$, C_5H_5N	4-Methoxy-2-picolinoylhydrazide-1-oxide, m.p. 207–208°	95
	$NH_2(CH_2)_2NEt_2$, MeOH, K_2CO_3, −10°	4-Methoxy-N-(diethylaminoethyl)-picolinamide, m.p. 81° (51.3%)	95
2-COCl-4-OEt	NH_2OH, abs. alc.	4-Ethoxy-2-picolylhydroxamic acid-1-oxide, m.p. 205°	95
	$NH_2NH_2 \cdot H_2O$, C_5H_5N	4-Ethoxy-2-picolinoylhydrazide-1-oxide, m.p. 184°	95
	$NH_2(CH_2)_2NEt_2$, MeOH, K_2CO_3, −10°	4-Ethoxy-N-(diethylaminoethyl)-picolinamide, m.p. 55°	95
2-COCl-4-OPr	NH_2OH, abs. alc.	4-Propoxy-2-picolylhydroxamic acid-1-oxide, m.p. 179°	95
	$NH_2NH_2 \cdot H_2O$, C_5H_5N	4-Propoxy-2-picolinoylhydrazide-1-oxide, m.p. 103°	95
	$NH_2(CH_2)_2NEt_2$, MeOH, K_2CO_3, −10°	4-Propoxy-N-(diethylaminoethyl)-picolinamide, m.p. 53°	95
2-COCl-4-Cl	NH_2OH, abs. alc.	4-Chloro-2-picolylhydroxamic acid-1-oxide, m.p. 206°	95
	$NH_2(CH_2)_2NEt_2$, MeOH, K_2CO_3, −10°	4-Chloro-N-(diethylaminoethyl)-picolinamide, m.p. 53°	95
2-COCl-4-NO₂	NH_2OH, abs. alc.	4-Nitro-2-picolylhydroxamic acid-1-oxide, m.p. 163–164°	95
	$NH_2(CH_2)_2NEt_2$, MeOH, K_2CO_3, −10°	4-Nitro-N-(diethylaminoethyl)-picolinamide, m.p. 68°	95
(c) From esters			
2-CO₂Me-4-NO₂	NH_4OH	4-Nitropicolinamide-1-oxide, m.p. 246° (78%)	95, 99
	(i) AcCl (ii) NH_3	4-Chloropicolinamide-1-oxide, m.p. 156–158° (69.5%)	95
	$NH_2NH_2 \cdot H_2O$, Δ, 1 hr	4-Nitro-2-picolinoylhydrazide-1-oxide, m.p. 184–185° (90%)	99
	Et_2NH, MeOH, 3–5 hr	4-Nitro-N,N-diethylpicolinamide-1-oxide·H_2O, m.p. 126° (decomp.) (65%)	99
	Pr_2NH	4-Nitro-N,N-dipropylpicolinamide-1-oxide·H_2O, m.p. 119–120° (55%)	99
	Bu_2NH	4-Nitro-N,N-dibutylpicolinamide-1-oxide·H_2O, m.p. 128–129° (decomp.) (63%)	99
	Piperidine	N-(4-Nitro-1-oxido-2-picolinoyl)-piperidide, m.p. 139° (decomp.) (55%)	99
	$NH_2CH_2CO_2Et$	Ethyl N-(4-nitro-1-oxido-2-picolinoyl)-glycinate, m.p. 126° (91%)	99
	Alanine	N-(4-Nitro-1-oxido-2-picolinoyl)alanine, m.p. 128–129° (88%)	99

Table IV-100 (*Continued*)

N-oxide	Conditions	Products (yield)	Ref.
	α-Aminobutyric acid	N-(4-Nitro-1-oxido-2-picolinoyl)-α-aminobutyric acid, m.p. 109–110° (90%)	99
	Norvaline	N-(4-Nitro-1-oxido-2-picolinoyl)norvaline, m.p. 124° (89%)	99
	Norleucine	N-(4-Nitro-1-oxido-2-picolinoyl)-norleucine, m.p. 83° (96–97%)	99
3-CO$_2$Et	Aq. or alc. NH$_3$	Nicotinamide-1-oxide, m.p. 282° (90%)	648
	NH$_2$NH$_2$·H$_2$O	Nicotinoylhydrazide-1-oxide, m.p. 221.5°	171, 648
3-CO$_2$Me-4-O$_2$CPh	Liq. NH$_3$, 8 hr	4-Benzoyloxynicotinamide-1-oxide, m.p. 147° (94%)	327
4-CO$_2$Et-2-OH	NH$_4$OH, 3 hr, R.T.	2-Hydroxyisonicotinamide-1-oxide, m.p. 270–272°	62, 649

VI. Photolysis

Recently, the electronic spectrum of pyridine-1-oxide was computed on the basis of the 30 lowest monoexcited singlet states starting with the SCF—MO's (Table IV-14) (269b). The calculated and observed parameters for the four transition bands observed are in good to excellent agreement.

To explain the observed solvent effects on the absorption spectrum of pyridine-1-oxide, the overall dipole moments for the first excited singlet states have been computed (Table IV-101). It has been shown that a decrease in the total dipole moment of the excited state induces a blue shift (650). In the two π–π^* states of lower energy (S_2^* and S_3^*) the dipole moment is reduced, a prediction which agrees with the blue shift observed (651, 652). Only the S_0–S_5 transition, which is observed at 217 nm (5.72 eV) is slightly red-shifted with an increase in solvent polarity. The n–n^* transition leading to singlet states S_1 and S_4 results in a drastic reversal of the direction of dipole moment ($\Delta\mu = -8.5$ for the S_0–S_1 transition) which may account for the fact that this transition does not show up in solution. The nature and

TABLE IV-101. Dipole Moments of Some Singlet States of Pyridine-1-oxide (269b)

Electronic states	S_0	S_1^*	S_2^*	S_3^*	S_4^*	S_5^*	S_6^*	S_7^*
Total μ (debyes)	+4.78	−3.73	+0.41	+1.34	−6.96	+5.78	+0.72	+1.44

TABLE IV-102. Charge Transfer in Singlet State (296b)

States	π Transfer	σ Transfer	Total transfer
S_0	+0.216	−0.308	−0.092
S_1	−0.170	+0.504	+0.334
S_2	+0.528	−0.307	+0.220
S_3	+0.269	−0.303	−0.034
S_4	+0.216	+0.692	+0.808
S_5	−0.090	−0.304	−0.394
S_6	+0.366	−0.261	+0.105
S_7	+0.370	−0.294	+0.077

importance of the intramolecular charge transfer in ground and excited states is given in Table IV-102. Among the $\pi-\pi^*$ singlet states, S_2 and S_5 states correspond to high transfer hands.

To compute the energies and the properties of the excited triplet state, the interaction of the thirty lowest monoexcited triplets was considered (Table IV-103).

The photolysis of pyridine-1-oxides has been recently reviewed (653). Results of the photolysis of substituted pyridine-1-oxides are given in Table IV-104.

Irradiation of pyridine-1-oxide and 3-picoline-1-oxide in the gas phase results chiefly in deoxygenation (257). In the gas phase, photolysis of pyridine-1-oxide by irradiation at the 3,261 or 2,537 Å resonance lines corresponds to $n \rightarrow \pi^*$ and to $\pi \rightarrow \pi^*$ transitions, respectively. On irradiation at 3,261 Å, the quantum yield of pyridine increases with temperature but is independent of temperature when 2,537 Å light is used (256). In the case of 2-picoline-1-oxide (**IV-279**), irradiation at 3,261 Å (corresponding to its $n \rightarrow \pi^*$ transition) gave 2-pyridylmethanol, while irradiation at 2,537 Å ($\pi \rightarrow \pi$ transition) led to fission of the $\overset{+}{N}$—$\overset{-}{O}$ bond to give 2-picoline (257).

TABLE IV-103. Triplet State Properties of Pyridine-1-oxide (269b)

States	ΔE (eV)	Nature	Pol.	Dipole moment (D)
T_1	1.67	$\pi-\pi^*$	Y	+1.51
T_2	2.97	$\pi-\pi^*$	X	−0.75
T_3	2.98	$n-\pi^*$	Z	−4.14
T_4	3.90	$\pi-\pi^*$	Y	−1.70
T_5	4.59	$\pi-\pi^*$	Y	+7.01

TABLE IV-104. Photolysis of Pyridine-1-oxides

Substituent in N-oxide	Conditions	Products	Ref.
None	Inert solvent	2-Formylpyrrole (10%)	654
	MeOH or EtOH	Pyridine (6.8%); 2-formylpyrrole (1.3%); N-formylpyrrole (trace) dimethyl or diethyl acetal of N-formylpyrrole	659
	Ether, 40 hr	2-Formylpyrrole (2%); pyridine (30%)	656
2-Me	Inert solvent	2-Formyl-5-methylpyrrole (8%); 6-methyl-2-pyridone	654
	Ether	2-Picoline; 2-formyl-5-methylpyrrole; 3-hydroxy-2-picoline; 5-hydroxy-2-picoline	655
	Ether, 24 hr	2-Picoline; 2-formyl-5-methylpyrrole (5%) 3-Hydroxy-2-picoline 5-Hydroxy-2-picoline } (5%) 6-Methyl-2-pyridone (1%)	656
4-Me		2-Formyl-3-methylpyrrole	657
		2-Formyl-3-methylpyrrole (3%); 4-picoline (35%)	656
2,4-Me$_2$		2-Formyl-3,5-dimethylpyrrole	657
	Ether, 26 hr	2-Formyl-3,5-dimethylpyrrole (21%); 2,4-lutidine (3.3%); 3-hydroxy-2,4-lutidine (3.3%); 5-hydroxy-2,4-lutidine (6%); 4,6-dimethyl-2-pyridone (8%)	656
2,6-Me$_2$		2-Acetyl-5-methylpyrrole	657
	Benzene, Pyrex, 36 hr	Phenol; 2-acetyl-5-methylpyrrole; 2,6-lutidine; 3-hydroxy-2,6-lutidine (7.7%); 2,6-lutidine-1-oxide (62%)	656
	Ether, quartz, 40 hr	2-Acetyl-5-methylpyrrole (9%); 3-acetyl-2-methylpyrrole (6.7%); 3-formyl-2,5-dimethylpyrrole (4%); 3-hydroxy-2,6-lutidine (16%)	656
2,3,6-Me$_3$	Ether, 6 hr	2-Acetyl-4,5-dimethylpyrrole (10%); 2,3,6-collidine (4%); 3-acetyl-2,5-dimethylpyrrole (10%); 3-hydroxy-2,5,6-collidine (16%)	656
2,4,6-Me$_3$	Ether	2-Acetyl-3,5-dimethylpyrrole (15%); 2,4,6-collidine (5%); 3-hydroxy-2,4,6-collidine (21%)	656
3,5-Me$_2$	Pyridine, benzene (1:1 v/v)	Phenol (5%)	656
3,5-Me	Ether, 60 hr	3,5-Lutidine (5%); 3-acetyl-4-methylpyrrole (20%)	656
2-(1-Cyclohexenyl)-4,5-Me$_2$	2,537 Å, isoPrOH, 24 hr	2-(1-Cyclohexenyl)-3-hydroxy-4,5-dimethylpyridine (26%); 6-(1-cyclohexenyl)-3,4-dimethyl-2-pyridone ($\sim$1.5%)	666

Table IV-104 (*Continued*)

Substituent in *N*-oxide	Conditions	Products	Ref.
	3,000 Å, isoPrOH, 8 hr	2-(1-Cyclohexenyl)-3-hydroxy-4,5-dimethylpyridine (40%); 6-(1-cyclohexenyl)-3,4-dimethyl-2-pyridone (~ 1–2%)	666
	2,537 Å, MeCN 40 hr	2-(1-Cyclohexenyl)-3-hydroxy-4,5-dimethylpyridine (45.5%); 6-(1-cyclohexenyl)-3,4-dimethyl-2-pyridone (traces)	666
	3,000 Å, MeCN, 8 hr	2-(1-Cyclohexenyl)-3-hydroxy-4,5-dimethylpyridine (50.5%); 6-(1-cyclohexenyl)-3,4-dimethyl-2-pyridone (traces)	666
	3,000 Å, cyclo-hexane, 6 hr	2-(1-Cyclohexenyl)-3-hydroxy-4,5-dimethylpyridine (41%); 6-(1-cyclohexenyl)-3,4-dimethyl-2-pyridone (traces)	666
2-Ph	Acetone	2-Phenylpyridine; 2-formyl-5-phenylpyrrole (25%)	655
2,4,6-Ph$_3$	Acetone	2,4,6-Triphenylpyridine (12%); 2-benzoyl-3,5-diphenylpyrrole (56%); 3-hydroxy-2,4,6-triphenylpyridine (10%); 2,4,6-triphenyl-1,3-oxazepine (8%);	661
	McOH	2,4,6-Triphenylpyridine; 2,4,6-triphenyl-1,3-oxazepine 3-hydroxy-2,4,6-triphenylpyridine (38%); 2-benzoyl-3,5-diphenylpyrrole (37%)	661
2-CN	Benzene, 48 hr	Phenol (31.5%); 2-cyanopyridine (46%)	656
2-CN-6-Me	*m*-Xylene, 50 hr	2,6-Dimethylphenol; 2,4-dimethylphenol; 6-cyano-2-picoline	656
4-CN	CH$_2$Cl$_2$	4-Cyanopyridine (12%); 6-cyano-1,3-oxazepine (30%); 2-formyl-3-cyanopyrrole (40%)	665
4-CN-2,6-Me$_2$	CH$_2$Cl$_2$	4-Cyano-2,6-lutidine (36%); 6-cyano-2,4-dimethyl-1,3-oxazepine (16%); 2-acetyl-3-cyano-5-methylpyrrole (29%)	665
4-NO$_2$	EtOH, N$_2$ atm	4-Hydroxylaminopyridine-1-oxide	590, 662
	EtOH, O$_2$ atm	1-Hydroxyl-4-pyridone HNO$_3$ (60–65%)	592, 662
4-NO$_2$-2-Me	EtOH, N$_2$ atm	4-Hydroxylamino-2-picoline-1-oxide	590
4-NO$_2$-2,6-Me$_2$	EtOH, N$_2$ atm	4-Hydroxylamino-2,6-lutidine-1-oxide	590
4-NO$_2$-3-Me	EtOH, N$_2$ atm	4-Hydroxylamino-3-picoline-1-oxide	590
4-NO$_2$-3,5-Me$_2$	EtOH, N$_2$, O$_2$ atm	3,5-Dimethyl-4-pyridone (30%)	667
4-NO	EtOH, O$_2$ atm	4-Hydroxylaminopyridine-1-oxide	592
4-NHOH	EtOH, O$_2$ atm	4-Nitropyridine-1-oxide	592
4-N$_3$	Acetone, N$_2$ atm	4,4'-Azopyridine-1-oxide (37%)	664

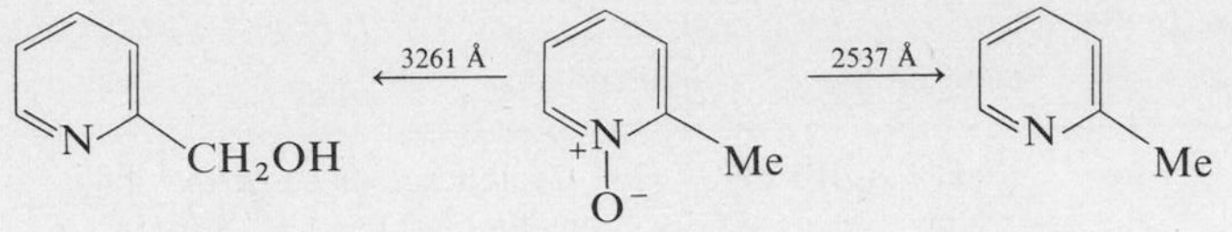

IV-279

Irradiation of a solution of 2-picoline-1-oxide in an inert solvent at 15° with a 125-W high-pressure mercury lamp led to the formation of 2-formyl-5-methylpyrrole (**IV-429**) and 6-methyl-2-pyridone (654). Under the same conditions, pyridine-1-oxide gave a 10% yield of 2-formylpyrrole (654). A recent study of the irradiation of 2-picoline-1-oxide in ether showed that the

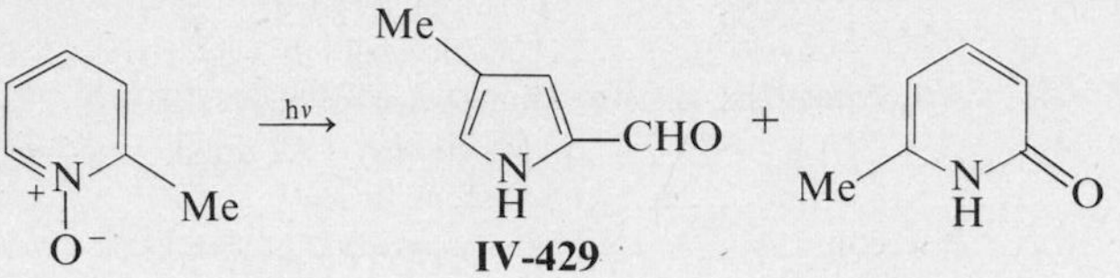

IV-429

expected formylpyrrole derivative was formed, together with 3- or 5-hydroxy-2-picoline (655–657). Similarly, irradiation of 2,6-lutidine-1-oxide in ether yielded 3-hydroxy-2,6-lutidine (655–656).

When pyridine-1-oxide is irradiated in benzene solution, pyridine is obtained and oxygen is transferred (perhaps as atomic oxygen) to the benzene ring at ordinary temperature to give phenol in a moderate yield (15%) (654). 2-Cyanopyridine-1-oxide in benzene is photochemically deoxygenated to yield 2-cyanopyridine and phenol (eq. IV-81) (Table IV-105)

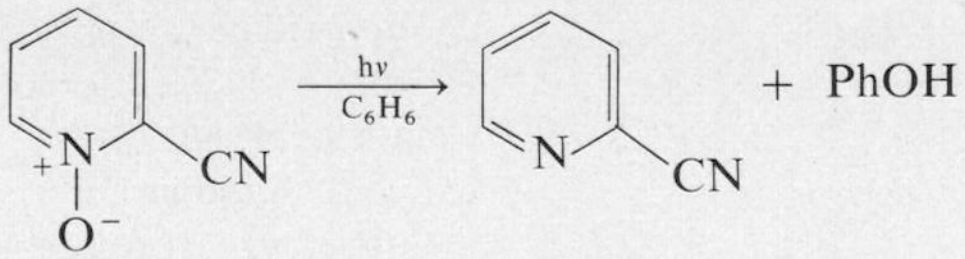

(656, 658). Irradiation of 6-cyano-2-picoline-1-oxide in *m*-xylene yields 2,6-xylenol, 2,4-xylenol, and 6-cyano-2-picoline (657). The yield of 2-cyano-pyridine decreases in the presence of oxygen. A triplet-state intermediate during the photolytic cleavage was suggested. Quenching the triplet state of pyridine-1-oxide should lead to an increased yield of 2-formylpyrrole. In the presence of a transition metal complex (e.g., copper nitrate complex), irradiation of pyridine-1-oxide gave an increased yield of 2-formylpyrrole (Table IV-106). When pyridine-1-oxide complexes of Zn^{2+}, Fe^{3+}, Ni^{2+}, and Co^{2+} salts were irradiated under the same experimental conditions, no

TABLE IV-105. Photolysis of 2-Cyanopyridine-1-oxide in the Presence or Absence of Sensitizers (658)

| Lamp | Conditions | | Yield of 2-cyano-pyridine (%) |
	Sensitizer (conc.)	Filter	
Phillips HPK125	None	Pyrex	47.3
Phillips HPK125	Xanthone (E_T 74 kcal/mol) ($4 \times 10^{-2}M$)	Pyrex	57.6
Phillips HPK125	Xanthone ($8 \times 10^{-2}M$)	Pyrex	63.0
Phillips HPK125	Xanthone ($1.6 \times 10^{-1}M$)	Pyrex	71.0
Hanaw NK 6/20	Tetrabutylammonium phenyltetrazolide (E_T 79 kcal/mol) ($3 \times 10^{-2}M$)	Quartz	71.7
Phillips HPK125	Oxygen bubbled through the solution	Pyrex	33.0

increase in 2-formylpyrrole formation could be detected. The effect of Cu^{II} salts cannot therefore be rationalized on grounds of a simple triplet quenching process by transition metal ions.

TABLE IV-106. Irradiation of Pyridine-1-oxide in H_2O in the Presence of Copper Salts (Vycor Filter) (658)

Reactant	Conditions[a] (hr)	Yield of 2-formylpyrrole (%)
Pyridine-1-oxide ($2 \times 10^{-2}M$)	100	2
$[PyO]_2Cu[NO_3]_2$ $10^{-2}M$	100	32
Pyridine-1-oxide ($2 \times 10^{-2}M$) + $Cu[ClO_4]_2$ ($2 \times 10^{-1}M$)	6	40

[a] Phillips HPK125 lamp was used.

As previously mentioned (see Section III.2.F), irradiation of pyridine-1-oxide and naphthalene in dichloromethane gave 1,2-naphthalene oxide and naphthol (388).

Photolysis of pyridine-1-oxide in methanol or ethanol gave a variety of products as shown in Scheme IV-25 (659). When a Pyrex filter was used,

		yield or (moles)		
Pyrex (6 hr)	0.68	0.27	0.13	trace
Quartz (10 min)	0.69	0.36	0.50	trace

Scheme IV-25

50% of the pyridine-1-oxide was recovered, but when a quartz filter was used only 10% of the pyridine-1-oxide was recovered. The first step in the photochemical reaction of pyridine-1-oxides is the formation of a nonisolable, unstable oxaziridine (IV-430) which undergoes rearrangement (Scheme IV-26) (653, 659, 660).

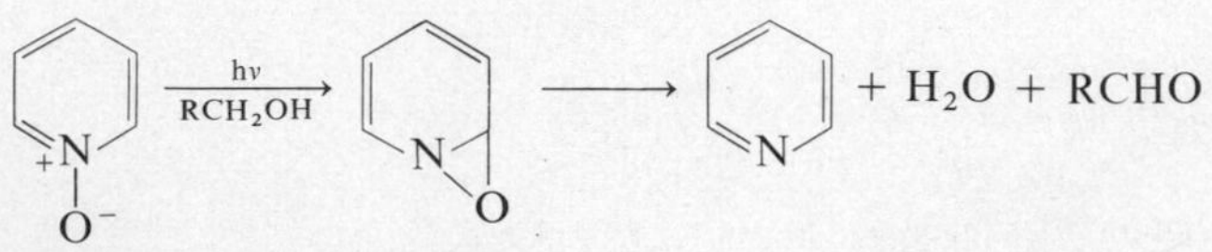

IV-430

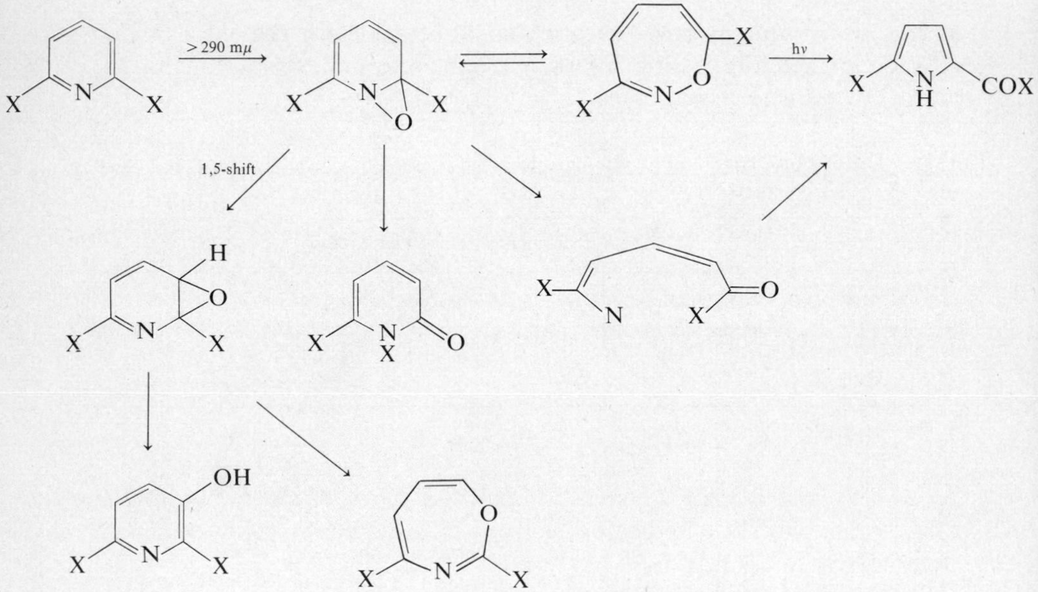

Scheme IV-26

Photolysis of 2,4,6-triphenylpyridine-1-oxide (IV-431) in acetone at >3,000 Å gave a 21% yield of a mixture of 2,4,6-triphenylpyridine (IV-432) (60%) and of the 1,3-oxazepine (RI 339) (IV-433) (40%), as well as 3-hydroxy-2,4,6-triphenylpyridine (IV-434) (10%) (661). When this reaction was carried out in methanol, a different distribution of the same products was obtained (IV-432) and (IV-433) (8%); (IV-434) (38%); and (IV-435) (37%). Irradiation in ethanol in the presence of benzophenone gave only IV-432.

Irradiation of 4-nitropyridine-1-oxide (IV-149) in ethanol or propanol in

IV-431 → **IV-432** + **IV-433**

IV-434 + **IV-435**

the presence of oxygen gave a 60 to 65 % yield of the nitrate of 1-hydroxy-4-pyridone (**IV-436**) (592). The primary photochemical process in 4-nitropyridine-1-oxide was investigated by means of both steady light irradiation and flash photolysis (662). Depending on the conditions 1-hydroxy-4-pyridone nitrate (**IV-436**) (process I) or 4-hydroxylaminopyridine-1-oxide (**IV-46**) (process II) were obtained. The reactions proceed through a common intermediate (**IV-388**). Process I involves a bimolecular interaction between the intermediate and the unexcited 4-nitropyridine-1-oxide, while process II involved a hydrogen-abstracting reaction from a solvent molecule. In the

absence of oxygen, 4-hydroxylaminopyridine-1-oxide is photostable, but in the presence of oxygen 4-nitropyridine-1-oxide is obtained (592). Irradiation of 4-nitrosopyridine-1-oxide with or without oxygen present gave 4-hydroxyl-aminopyridine-1-oxide (592).

Irradiation accelerates the reaction of 4-nitropyridine-1-oxide with piperidine (displacement of the nitro group by piperidine) (663).

Photolysis of 4-azidopyridine-1-oxide in acetone gives 4,4'-azopyridine-1-oxide (664).

2,6-Dicyanopyridine-1-oxide was irradiated at >290 mμ in dichloromethane to yield 2,6-dicyanopyridine (20%), 5-cyano-2-pyrrole carbonyl cyanide (20%), and an oxazepine (35%), tentatively assigned the structure 2,4-dicyano-1,3-oxazepine (665).

2-p-Dimethylaminostyryl-3-nitropyridine-1-oxide **(IV-437)** exposed to sunlight for 14 days in benzene gave pyrrolopyridine-1-oxide **(IV-438)** (RII 1,268). The 4-styryl derivative behaved similarly (574).

IV-437   IV-438

Photolysis of 2-(1-cyclohexenyl)-4,5-dimethylpyridine-1-oxide **(IV-439)** in isopropanol, acetonitrile, or cyclohexane gave 2-(1-cyclohexenyl)-3-hydroxy-4,5-dimethylpyridine **(IV-440)** as the main product together with small amounts of 6-(1-cyclohexenyl)-3,4-dimethyl-2-pyridone **(IV-441)** (666).

IV-439   IV-440   IV-441

References

1. E. Ochiai, *J. Org. Chem.*, **18**, 534 (1953).
2. H. J. den Hertog, M. van Ammers, and S. Schakking, *Rec. Trav. Chim. Pays-Bas*, **74**, 1171 (1955); *Chem. Abstr.*, **50**, 5655i (1956).
3. M. Colonna, *Gazz. Chim. Ital.*, **86**, 705 (1956).
4. (a) A. R. Katritzky, *Quart. Rev.*, **10**, 395 (1956); (b) A. R. Katritzky and J. M. Lagowski, "Chemistry of Heterocyclic N-Oxides," Academic, New York, 1971.
5. E. N. Shaw, in "Pyridine and Its Derivatives," Vol. 2, E. Klingsberg, Ed., Interscience, New York, 1961, p. 97.
6. R. A. Abramovitch and J. G. Saha, in "Advances in Heterocyclic Chemistry," Vol. 6, A. R. Katritzky, Ed., Academic, New York, 1966, p. 229.

7. E. Ochiai, "Aromatic Amine Oxides," Elsevier, Amsterdam, 1967.
8. J. Meisenheimer, *Chem. Ber.*, **59**, 1848 (1926).
9. B. Bobranski, L. Kochanska, and A. Kowalewska, *Chem. Ber.*, **71**, 2385 (1938).
10. E. Ochiai and Z. Sai, *J. Pharm. Soc. Jap.*, **65A**, 63 (1945).
11. R. F. Evans, M. van Ammers, and H. J. den Hertog, *Rec. Trav. Chim. Pays-Bas*, **78**, 408 (1959).
12. E. Smith, U.S. Patent 3,357,984, 1967; *Chem. Abstr.*, **68**,105008 (1968).
13. V. Torlorella, F. Macioi, and G. Poma, *Farmaco Ed. Sci.* **23**, 236 (1968); *Chem. Abstr.*, **69**, 59058 (1968).
14. P. A. van Zwieten, J. A. van Velthuijsen, and H. O. Huisman, *Rec. Trav. Chim. Pays-Bas*, **80**, 1066 (1961).
15. P. A. van Zwieten, M. Gerstenfeld, and H. O. Huisman, *Rec. Trav. Chim. Pays-Bas*, **81**, 604 (1962).
16. S. M. Roberts and H. Suschitzky, *J. Chem. Soc.*, *C*, 1537 (1968).
17. (a) J. M. Roberts and H. Suschitzky, *Chem. Comm.*, 893 (1967); (b) G. E. Chivers and H. Suschitzky, *Chem. Comm.*, 28 (1971).
18. (a) O. E. Edwards and D. C. Gillespie, *Tetrahedron Lett.*, 4867 (1966); (b) C. Pedersen, N. Harrit, and O. Buchardt, *Acta Chem. Scand.*, **24**, 2435 (1970).
19. (a) J. R. Kirchner, U.S. Patent 3,203,957, 1965; *Chem. Abstr.*, **63**, 14821e (1965); (b) R. B. Rogers, Ph.D. Dissertation, University of Alabama, 1971.
20. G. Wagner and H. Pischel, *Arch. Pharm.* (Weinheim), **295**, 897 (1962); *Chem. Abstr.*, **59**, 1744c (1963).
21. J. T. Dunn and D. L. Heywood, U.S. Patent 3,062,824, 1962; *Chem. Abstr.*, **58**, 9030g (1963).
22. (a) J. T. Dunn and D. T. Manning, U.S. Patent 3,159,611, 1964; *Chem. Abstr.*, **62**, 7892ab (1965); (b) M. Lieflaender and K. Stalder, *Z. Physiol. Chem.*, **346**, 69 (1966).
23. B. Plesnicar and G. A. Russel, *Angew. Chem. Internat. Ed.*, **9**, 797 (1970).
24. L. Thunus and J. Delarge, *J. Pharm. Belg.*, **21**, 485 (1966); *Chem. Abstr.*, **66**, 37736h (1967).
25. Ruetgerswerke and Teerverwertung, French Patent 1,376,232, 1964; *Chem. Abstr.*, **62**, 9115d (1965).
26. S. Oae and S. Kozuka, *Tetrahedron*, **21**, 1971 (1965).
27. R. C. Whitman, U.S. Patent 3,047,579, 1962; *Chem. Abstr.*, **58**, 7916d (1963).
28. R. L. Lagomarsino and E. V. Brown, *Trans. Ky. Acad. Sci.*, **27**, 39 (1966); *Chem. Abstr.*, **68**, 87116t (1968).
29. Yu. I. Chumakov, *Metody Poluchemya Khim. Reaktivou Preparatov, Gos. Kom. Sov. Min. SSSR Po Khim.*, **7**, 58 (1963); *Chem. Abstr.*, **61**, 10652 (1964).
30. R. F. Evans and W. Kynaston, *J. Chem. Soc.*, 5556 (1961).
31. R. T. Martens and H. J. den Hertog, *Rec. Trav. Chim. Pays-Bas*, **86**, 655 (1967).
32. N. S. Prostakov, N. M. Mikhailova, S. Ya. Ivanova, and L. A. Shakhparonova, *Khim. Geterotsikl. Soedin, Akad. Nauk. Latu.*, **55R**, 738 (1965); *Chem. Abstr.*, **64**, 11168e (1966).
33. N. S. Prostakov, L. A. Gaivononskaya, N. N. Mikheeva, and N. P. Filippov, *Zh. Obshch. Khim.*, **33**, 2928 (1963); *Chem. Abstr.*, **59**, 1688d (1963).
34. E. M. Smith, Ph.D. Dissertation, University of Saskatchewan, 1969.
35. E. Matsumura, T. Hirooka, and K. Imagawa, *J. Chem. Soc. Jap.*, **82**, 616 (1961); *Chem. Abstr.*, **57**, 12466f (1962).
36. R. F. Evans and H. C. Brown, *J. Org. Chem.*, **27**, 1329 (1962).
37. R. A. Jones and R. P. Rao, *Aust. J. Chem.*, **18**, 583 (1965).
38. J. Bolle and J. Tomaszew, German Patent 1,194,583; *Chem. Abstr.*, **63**, 6648c (1965).
39. Richardson-Merrel Inc., British Patent 975,972, 1964; *Chem. Abstr.*, **62**, 11786h (1965).
40. J. M. Nascimento and M. H. Venda, *Rev. Port. Farm.*, **12**, 321 (1962); *Chem. Abstr.*, **59**, 558e (1963).

41. T. Nakashima, *J. Pharm. Soc. Jap.*, **78**, 666 (1958); *Chem. Abstr.*, **52**, 18399h (1958).
42. T. R. Govindachari and S. Rajappa, *J. Chem. Soc.*, 1306 (1958).
43. D. E. Ames and T. F. Gey, British Patent 838,746, 1960; *Chem. Abstr.*, **55**, 1661 (1961).
44. I. L. Kotlyarevskii, L. N. Korolenok, L. G. Stadnikova, and T. G. Shiskmakova, *Izv. Akad. Nauk. SSR, Ser. Khim.*, 1224 (1966); *Chem. Abstr.*, **65**, 16932h (1966).
45. P. F. Holt and E. T. Nasrallah, *J. Chem. Soc.*, C, 823 (1969).
46. Y. Sakata, K. Adachi, Y. Akahori, and E. Hayashi, *J. Pharm. Soc. Jap.*, **87**, 1374 (1967); *Chem. Abstr.*, **68**, 59649a (1968).
47. A. R. Katritzky and J. N. Gardener, *J. Chem. Soc.*, 2192 (1958).
48. D. E. Ames and J. L. Archibald, *J. Chem. Soc.*, 1475 (1962).
49. W. Baker, K. M. Buggle, J. F. W. McOmie, and D. A. M. Watkins, *J. Chem. Soc.*, 3594 (1958).
50. E. C. Taylor and J. S. Discoll, *J. Org. Chem.*, **26**, 3796 (1961).
51. J. Michalski, K. Wojaezynski, and H. Zajac, *Bull. Acad. Polon. Sci., Ser. Sci. Chim.*, **8**, 285 (1960); *Chem. Abstr.*, **56**, 7263f (1962).
52. G. Buchmann and H. Franz, *Z. Chem.*, **6**, 107 (1966).
53. W. Mathes and W. Sauermilch, *Ann. Chem.*, **618**, 152 (1958).
54. M. Bellas and H. Suschitzky, *J. Chem. Soc.*, 4007 (1963).
55. M. P. Cava and B. Weinstein, *J. Org. Chem.*, **23**, 1616 (1958).
56. T. Talik and Z. Talik, *Rocz. Chem.*, **40**, 1457 (1966).
57. J. R. Kirchner, French Patent 1,393,620, 1965; *Chem. Abstr.*, **63**, 9923h (1965).
58. D. A. Shermer, U.S. Patent 2,951,844, 1960; *Chem. Abstr.*, **55**, 4539 (1961).
59. G. C. Finge and L. D. Starr, *J. Amer. Chem. Soc.*, **81**, 2674 (1959).
60. V. Tortorella, *Chem. Comm.*, 308 (1966).
61. T. Kato and T. Niitsuma, *Chem. Pharm. Bull.* (Tokyo), **13**, 963 (1965); *Chem. Abstr.*, **63**, 14804e (1965).
62. S. Mizukami, E. Hirai, and M. Morimoto, *Shionogi Kenkysho Nempo*, **16**, 29 (1966); *Chem. Abstr.*, **66**, 10827g (1967).
63. M. van Ammers, H. J. den Hertog, and B. Haase, *Tetrahedron*, **18**, 227 (1961).
64. J. Bernstein and K. A. Losee, U.S. Patent 2,809,971, 1957; *Chem. Abstr.*, **52**, 2932g (1958).
65. T. S. Gardner, E. Wenis, and J. Lee, *J. Org. Chem.*, **22**, 984 (1957).
66. D. V. Maier, U.S. Patent 3,163,655, 1964; *Chem. Abstr.*, **62**, 10419c (1965).
67. S. J. Norton and P. T. Sullivan, *J. Heterocycl. Chem.*, **7**, 699 (1970).
68. T. Kato, T. Kitagawa, T. Shibata, and K. Nakai, *J. Pharm. Soc. Jap.*, **82**, 1642 (1962); *Chem. Abstr.*, **59**, 559b (1963).
69. T. Yoshikawa, F. Ishikawa, and T. Naito, *Chem. Pharm. Bull.* (Tokyo), **13**, 878 (1965); *Chem. Abstr.*, **63**, 11490a (1965).
70. H. S. Habib and C. W. Rees, *J. Chem. Soc.*, 3371 (1960).
71. R. Adams and W. Reifschneider, *J. Amer. Chem. Soc.*, **81**, 2537 (1959).
72. T. Teshigawara, T. Okamoto, and H. Tani, Japanese Patent, 21,089, 1963; *Chem. Abstr.*, **60**, 2906h (1964).
73. G. R. Bedford, A. R. Katritzky, and H. M. Wuest, *J. Chem. Soc.*, 4600 (1963).
74. M. Hamana, B. Umezawa, Y. Goto, and K. Noda, *Chem. Pharm. Bull.* (Tokyo), **8**, 692 (1960); *Chem. Abstr.*, **55**, 18723e (1961).
75. Dainippon Pharmaceutical Co. Ltd., British Patent 1,053,730, 1967; *Chem. Abstr.*, **66**, 115605f (1967).
76. R. Tan and A. Taurins, *Tetrahedron Lett.*, 2737 (1965).
77. Y. Sato, T. Iwashige, and T. Miyadera, *Chem. Pharm. Bull.* (Tokyo), **8**, 427 (1960); *Chem. Abstr.*, **55**, 12402a (1961).
78. L. Skattebøl and B. Boulette, *J. Org. Chem.*, **34**, 4150 (1969).
79. B. M. Ferrier and N. Campbell, *Proc. Roy. Soc. Edinburgh, Sect.*, **A65**, 231 (1959–1960).

80. (a) S. Okuda and M. M. Robinson, *J. Amer. Chem. Soc.*, **81**, 740 (1959); (b) P. J. Majewski, R. Bodalski, and J. Michalshi, *Synthesis*, 140 (1971).
81. S. Furukawa, *J. Pharm. Soc. Jap.*, **78**, 957 (1958); *Chem. Abstr.*, **53**, 3219g (1959).
82. N. Hata, *Bull. Chem. Soc. Jap.*, **31**, 224 (1958).
83. V. Boekelheide and W. Feely, *J. Amer. Chem. Soc.*, **80**, 2217 (1958).
84. V. Boekelheide and W. Feely, *J. Org. Chem.*, **22**, 589 (1957).
85. E. Profft and H. W. Linke, *Chem. Ber.*, **93**, 2591 (1960).
86. M. Yoshida and H. Kumagae, *J. Chem. Soc. Jap.*, **81**, 345 (1960); *Chem. Abstr.*, **55**, 6477g (1961).
87. Y. Sato, *Chem. Pharm. Bull.* (Tokyo), **6**, 222 (1958); *Chem. Abstr.*, **52**, 20164i (1958).
88. G. Queguiner, M. Alas, and P. Pastour, *C.R. Acad. Sci.*, *Paris*, *Ser. C*, **265**, 824 (1967).
89. A. R. Katritzky, A. M. Monroe, J. A. T. Beard, D. P. Dearnaley, and N. J. Earl, *J. Chem. Soc.*, 2182 (1958).
90. A. Risaliti and L. Lolli, *Farmaco, Ed. Sci.*, **12**, 705 (1957); *Chem. Abstr.*, **52**, 11038f (1958).
91. K. B. Augustinsson, H. Hasselquist, and L. Larsson, *Acta. Chem. Scand.*, **14**, 1253 (1960).
92. T. Kato, Y. Goto, and T. Chiba, *J. Pharm. Soc. Jap.*, **86**, 1022 (1966); *Chem. Abstr.*, **66**, 85680g (1967).
93. L. Pentimalli and P. Bruni, *Ann. Chim.* (Rome), **54**, 180 (1964).
94. W. Steinke, German (East) Patent 23,754, 1962; *Chem. Abstr.*, **59**, 8713b (1963).
95. E. Profft and W. Steinke, *J. Prakt. Chem.*, **13**, 58 (1961).
96. V. Boekelheide and W. L. Lehn, *J. Org. Chem.*, **26**, 428 (1961).
97. R. M. Acheson and G. A. Taylor, *J. Chem. Soc.*, 1691 (1960).
98. M. Szafran and Z. Sarbak, *Rocz. Chem.*, **43**, 309 (1969).
99. H. W. Krause and W. Langenbeck, *Chem. Ber.*, **92**, 155 (1959).
100. J. T. Dunn and D. L. Heywood, U.S. Patent 3,048,624, 1962; *Chem. Abstr.*, **59**, 6369f (1963).
101. D. L. Heywood and J. T. Dunn, *J. Org. Chem.*, **24**, 1569 (1959).
102. J. T. Dunn and D. L. Heywood, U.S. Patent 2,953,572, 1960; *Chem. Abstr.*, **55**, 4542i (1961).
103. N. S. Prostakov and L. A. Gaivoronskaya, *Zh. Obshch. Khim.*, **32**, 76 (1962); *Chem. Abstr.*, **57**, 12426i (1962).
104. H. Euler and H. Hasselquist, *Ark. Kemi*, **13**, 225 (1959).
105. A. M. Municio and A. Ribera, *Anales Real. Soc. Espa. Fis. Quim.* (Madrid), Ser. B59, 179 (1963); *Chem. Abstr.*, **59**, 9971a (1963).
106. A. R. Katritzky, J. A. T. Beard, and N. A. Coates, *J. Chem. Soc.*, 3680 (1959).
107. H. J. Rimek, *Ann. Chem.*, **670**, 69 (1963).
108. W. E. Feely, U.S. Patent 2,991,285, 1958; *Chem. Abstr.*, **56**, 7282a (1962).
109. (a) W. Treibs and J. Beger, *Ann. Chem.*, **652**, 192 (1962); (b) T. Kametani and T. Suzuki, *J. Chem. Soc.*, *C*, 1053 (1971).
110. D. Libermann, N. Rist, F. Grumback, S. Cals, M. Moyeux, and A. Rouaix, *Bull. Soc. Chim. Fr.*, 687 (1958).
111. J. Delarge and L. Thunus, *Farmaco, Ed. Sci.*, **21**, 846 (1968); *Chem. Abstr.*, **66**, 75889e (1967).
112. H. von Euler and H. Hasselquist, *Ark. Kemi*, **48**, 559 (1958).
113. A. R. Katritzky, *Rec. Trav. Chim. Pays-Bas*, **78**, 995 (1959).
114. Y. Suzuki, *Yakugaku Zasshi*, **81**, 917 (1961); *Chem. Abstr.*, **55**, 27305i (1961).
115. T. Kubota, K. Nishikida, H. Miyazaki, K. Iwatani, and Y. Oishi, *J. Amer. Chem. Soc.*, **90**, 5080 (1968).
116. S. Furukawa, *Yakugaku, Zasshi*, **59**, 492 (1959); *Chem. Abstr.*, **53**, 18029b (1959).
117. E. C. Taylor and J. S. Driscoll, *J. Org. Chem.*, **25**, 1716 (1960).

118. E. Ochiai and C. Kaneko, *Pharm. Bull.* (Tokyo), **5**, 56 (1957); *Chem. Abstr.*, **52**, 1165h (1958).

119. L. Achremovicz, T. Batkowski, and Z. Skrowaczewska, *Rocz. Chem.*, **38**, 1317 (1964).

120. J. Suszko and M. Szafran, *Rocz. Chem.*, **39**, 709 (1965).

121. W. Sliwa and E. Plazek, *Acta Polon. Pharm.*, **23**, 221 (1966); *Chem. Abstr.*, **66**, 10832f (1967).

122. L. Pentimalli, *Gazz. Chim. Ital.*, **94**, 458 (1964).

123. J. W. Dehn, Jr., and A. J. Salina, U.S. Patent 3,249,597, 1966; *Chem. Abstr.*, **65**, 9061b, c (1966).

124. W. Faessinge and E. V. Brown, *Trans. Ky. Acad. Sci.*, **24**, 106 (1963); *Chem. Abstr.*, **60**, 14465g (1964).

125. A. R. Katritzky, *J. Chem. Soc.*, 4385 (1957).

126. (a) S. Kajihara, *J. Chem. Soc. Jap.*, **86**, 839 (1965); *Chem. Abstr.*, **65**, 16935c (1966); (b) D. M. Bailey, R. E. Johnson, J. D. Connolly, and R. A. Ferrai, *J. Med. Chem.*, **14**, 439 (1971).

127. L. Pentimalli, *Gazz. Chim. Ital.*, **89**, 1843 (1959).

128. D. M. Bailey, U.S. Patent 3,450,707, 1969; *Chem. Abstr.*, **71**, 70506j (1969).

129. T. Kato, H. Yamanaka, T. Niitsuma, K. Wagatsuma, and M. Oizumi, *Chem. Pharm. Bull.* (Tokyo), **12**, 910 (1964); *Chem. Abstr.*, **63**, 2949d (1963).

130. W. Herz and D. R. K. Murty, *J. Org. Chem.*, **25**, 2242 (1960).

131. R. A. Jones and A. R. Katritzky, *J. Chem. Soc.*, 2937 (1960).

132. T. Batkowski and E. Plazek, *Rocz. Chem.*, **31**, 273 (1963).

133. L. Pentimalli, *Gazz. Chim. Ital.*, **94**, 902 (1964).

134. E. Hayashi, H. Yamanaka, C. Iijima, and S. Matsushita, *Chem. Pharm. Bull.* (Tokyo), **8**, 649 (1960); *Chem. Abstr.*, **55**, 15480c (1961).

135. L. Pentimalli, *Tetrahedron*, **14**, 151 (1961).

136. S. Furukawa, *J. Pharm. Soc. Jap.*, **59**, 487 (1959); *Chem. Abstr.*, **53**, 18028f (1959).

137. A. R. Katritzky and A. M. Monroe, *J. Chem. Soc.*, 150 (1958).

138. H. Takamatsu, S. Minami, A. Fujita, J. Aritomi, K. Fujimoto, and M. Shimizu, Japanese Patent 26,407, 1964; *Chem. Abstr.*, **62**, 10421c (1965).

139. A. R. Katritzky, D. J. Short, and A. J. Boulton, *J. Chem. Soc.*, 1516 (1960).

140. R. A. Abramovitch and K. A. H. Adams, *Can. J. Chem.*, **39**, 2516 (1961).

141. A. R. Katritzky and P. Simmons, *J. Chem. Soc.*, 1511 (1960).

142. T. Itai and S. Kamiya, *Chem. Pharm. Bull.* (Tokyo), **11**, 1059 (1963); *Chem. Abstr.*, **59**, 12802f (1963).

143. Y. Suzuki, *J. Pharm. Soc. Jap.*, **81**, 1204 (1961); *Chem. Abstr.*, **56**, 3445c (1962).

144. L. Pentimalli, *Gazz. Chim. Ital.*, **93**, 404 (1963).

145. L. Pentimalli, *Ann. Chim.* (Rome), **55**, 435 (1965).

146. (a) M. Brufani, G. Giacomello, and M. L. Stein, *Gazz. Chim. Ital.*, **91**, 767 (1961); (b) G. Quequiner, C. Fugier, and P. Pastour, *Bull. Soc. Chim. Fr.*, 3616 (1970); (c) M. Weiner, *J. Organometal. Chem.*, **23**, C20 (1970).

147. (a) C. D. Johnson, A. R. Katritzky, B. J. Ridgewell, N. Shakiv, and A. M. White, *Tetrahedron*, **21**, 1055 (1965); (b) R. A. Abramovitch and B. W. Cue, *J. Org. Chem.*, **38**, 173 (1973).

148. L. Pentimalli, *Gazz. Chim. Ital.*, **91**, 991 (1961).

149. J. S. Wieczorek and E. Plazek, *Rec. Trav. Chim. Pays-Bas*, **83**, 249 (1964).

150. J. S. Wieczorek and E. Plazek, *Rec. Trav. Chim. Pays-Bas*, **84**, 785 (1965).

151. M. Colonna, A. Risaliti, and L. Pentimalli, *Gazz. Chim. Ital.*, **86**, 1067 (1956).

152. T. Kametani, K. Fukumoto, and O. Umezawa, *Jap. J. Pharm. Chem.*, **31**, 132 (1959); *Chem. Abstr.*, **54**, 11019b (1960).

153. H. G. Mautner, S. H. Chu, and C. M. Lee, *J. Org. Chem.*, **27**, 3671 (1962).

154. E. C. Taylor and A. J. Crovetti, *J. Org. Chem.*, **19**, 1633 (1954).

155. M. P. Mertes, R. F. Borne, and L. E. Hare, *J. Heterocycl. Chem.*, **5**, 281 (1968).

156. G. B. Payne and P. H. Williams, *J. Org. Chem.*, **26**, 668 (1961).

157. E. C. Taylor and N. E. Boyer, *J. Org. Chem.*, **24**, 275 (1959).

158. Ya. L. Gol'dfarb, F. D. Alashev, and V. K. Zvorykina, *Izv. Akad. Nauk. SSSR, Otd. Khim. Nauk.*, 2209 (1962); *Chem. Abstr.*, **58**, 14014e (1963).

159. R. B. Greenwald and C. L. Zirkle, *J. Org. Chem.*, **33**, 2118 (1968).

160. A. Crabtree, A. W. Johnson, and T. C. Tebby, *J. Chem. Soc.*, 3497 (1961).

161. (a) M. Szafran and B. Brzezinzki, *Rocz. Chem.*, **43**, 459 (1969); *Chem. Abstr.*, **71**, 81102a (1969); (b) M. Hamana and H. Noda, *Chem. Pharm. Bull.* (Tokyo), **13**, 912 (1965); *Chem. Abstr.*, **63**, 13204b (1965).

162. M. Hamana and H. Noda, *Chem. Pharm. Bull.* (Tokyo), **14**, 762 (1965); *Chem. Abstr.*, **65**, 16938a (1966).

163. J. Izdebski, *Rocz. Chem.*, **39**, 1625 (1965).

164. L. Pentimalli, *Ann. Chim.* (Rome), **53**, 1123 (1963).

165. M. D. Coburn, *J. Heterocycl. Chem.*, **7**, 455 (1970).

166. W. Herz and D. R. K. Murty, *J. Org. Chem.*, **26**, 122 (1961).

167. F. Paris, P. Bovina, and A. Quilico, *Gazz. Chim. Ital.*, **90**, 903 (1960).

168. L. Pentimalli, *Gazz. Chim. Ital.*, **89**, 1843 (1969).

169. W. Walter, J. Voss, and J. Curts, *Ann. Chem.*, **695**, 77 (1966).

170. R. M. Acheson and G. A. Taylor, *J. Chem. Soc.*, 4600 (1960).

171. H. Hasselquist, *Ark. Kemi*, **15**, 387 (1960).

172. S. Minakami and E. Hirai, Japanese Patent 6,723,194, 1967; *Chem. Abstr.*, **69**, 35967 × (1968).

173. F. Parisi, P. Bovina, and A. Quilico, *Gazz. Chim. Ital.*, **92**, 1138 (1962).

174. P. Yates, M. J. Jorgenson, and S. K. Roy, *Can. J. Chem.*, **40**, 2146 (1962).

175. (a) E. Schmitz, *Chem. Ber.*, **91**, 1488 (1958); (b) A. T. Balaban and C. D. Nenitzeseu, *Ann. Chem.*, **625**, 74 (1959).

176. A. Roedig, R. Kohlhaupt, and G. Maerkl, *Chem. Ber.*, **99**, 698 (1966).

177. A. T. Balaban and C. D. Nenitzescu, *J. Chem. Soc.*, 3566 (1961).

178. K. D. Gundermann and H. U. Alles, *Angew. Chem.*, **78**, 906 (1966).

179. (a) K. D. Gundermann and H. U. Alles, *Chem. Ber.*, **102**, 3014 (1969); (b) L. G. Wideman, *Chem. Comm.*, 1309 (1970); (c) Y. Tamura, N. Tsujimoto, and M. Mano, *Chem. Pharm. Bull.* (Tokyo), **19**, 130 (1971); *Chem. Abstr.*, **74**, 125362d (1971); (d) Y. Tamura and N. Tsukimoto, *Chem. Ind.* (London), 926 (1970).

180. N. A. Coates and A. R. Katritzky, *J. Org. Chem.*, **24**, 1836 (1959).

181. E. P. Linton, *J. Amer. Chem. Soc.*, **62**, 1945 (1940).

182. T. Okamoto, H. Hayatsu, and Y. Baba, *Chem. Pharm. Bull.* (Tokyo), **8**, 892 (1960); *Chem. Abstr.*, **55**, 23528h (1961).

183. M. Liveris and J. Miller, *J. Chem. Soc.*, 3486 (1963).

184. G. Coppens, F. Declerck, C. Gillet, and J. Nasielski, *Bull. Soc. Chim. Belges*, **72**, 572 (1963).

185. A. R. Katritzky, E. W. Randal, and L. E. Sutton, *J. Chem. Soc.*, 1769 (1957).

186. C. M. Bax, A. R. Katritzky, and L. E. Sutton, *J. Chem. Soc.*, 1254 (1958).

187. C. M. Bax, A. R. Katritzky, and L. E. Sutton, *J. Chem. Soc.*, 1258 (1958).

188. A. R. Katritzky, A. M. Monroe, and J. A. T. Beard, *J. Chem. Soc.*, 3721 (1958).

189. A. J. Boulton, G. M. Glover, M. H. Hutchinson, A. R. Katritzky, D. J. Short, and L. E. Sutton, *J. Chem. Soc.*, B, 822 (1966).

190. A. N. Sharpe and S. J. Walker, *J. Chem. Soc.*, 4522 (1961).

191. T. Kubota and H. Watanabe, *Bull. Chem. Soc. Jap.*, **36**, 1093 (1963).

192. (a) J. C. Morrow, D. Ulka, and B. P. Huddle, American Crystallographic Association,

Winter Meeting Abstracts, March, 1970; *Chem. Abstr.*, **74**, 147553w (1971); (b) R. Desiderato and J. C. Terry, *Tetrahedron Lett.*, 3204 (1970); (c) E. L. Eichhorn, *Acta Crystallog.*, **9**, 787 (1956).

193. E. J. Eichhorn, *Acta Crystallog.*, **12**, 746 (1959).
194. L. Favretto, *Univ. Studi Trieste Fac. Sci. Ist. Mineral*, **24**, 6 (1959); *Chem. Abstr.*, **64**, 11631e (1960).
195. E. L. Eichhorn and K. Hoogsteen, *Acta Crystallog.*, **10**, 382 (1957).
196. Z. S. Kostrukevich, *Zh. Strukt. Khim.*, **5**, 323 (1964); *Chem. Abstr.*, **61**, 2562f (1964).
197. F. Genet, J. C. Leguen, and G. Tsoucaris, *C.R. Acad. Sci., Paris, Ser. C*, **262**, 989 (1966).
198. S. Scavnicar and B. Matkovic, *Acta Crystallog.*, *Sect. B*, **B25** (1969).
199. R. S. Sager and W. H. Watson, *Inorg. Chem.*, **7**, 2035 (1968).
200. H. R. Rose, *Acta. Crystallog.*, **14**, 895 (1961).
201. Y. Namba, T. Oda, H. Ito, and T. Watanabe, *Bull. Chem. Soc. Jap.*, **33**, 1618 (1960).
202. S. G. Mairanovskii, N. V. Barashkova, F. D. Alashev, and V. K. Zvorykina, *Izv. Akad. Nauk. SSSR, Otdel. Khim. Nauk.*, 938 (1960); *Chem. Abstr.*, **54**, 22103a (1960).
203. A. Foffani and E. Fornasari, *Atti. Accad. Nazl. Lincei, Rend., Classe Sci. Fis., Mat. Nat.*, **23**, 62 (1957); *Chem. Abstr.*, **53**, 1960i (1959).
204. T. Kubota and H. Miyazaki, *Bull. Chem. Soc. Jap.*, **35**, 1549 (1962).
205. G. Horn, *Acta Chim. Acad. Sci. Hung.*, **27**, 123 (1961); *Chem. Abstr.*, **55**, 23928e (1960).
206. T. R. Emerson and C. W. Rees, *J. Chem. Soc.*, 1923 (1962).
207. T. Kubota and H. Miyazaki, *Bull. Chem. Soc. Jap.*, **39**, 2057 (1966).
208. H. Miyazaki and T. Kubota, *Bull. Chem. Soc. Jap.*, **44**, 279 (1971); *Chem. Abstr.*, **74**, 82478b (1971).
209. E. Laviron and R. Gavasso, *Talanta*, **16**, 293 (1969).
210. E. Laviron, R. Gavasso, and M. Pay, *Talanta*, **17**, 747 (1970).
211. G. Horn, *Monatsber. Deut. Akad. Wiss. Berlin*, **3**, 386 (1961); *Chem. Abstr.*, **57**, 3196e (1962).
212. H. H. Jaffe, *J. Org. Chem.*, **23**, 1790 (1958).
213. J. S. Driscoll. W. Pfleiderer, and E. C. Taylor. *J. Org. Chem.*, **26**, 5230 (1961).
214. H. Shindo, *Pharm. Bull. Jap.*, **4**, 460 (1956); *Chem. Abstr.*, **51**, 12661b (1957).
215. H. Shindo, *Chem Pharm. Bull.* (Tokyo), **6**, 117 (1958); *Chem. Abstr.*. **52**, 19457b (1958).
216. I. I. Grandberg, G. K. Faizova, and A. N. Kost, *Chem. Heterocycl. Cpds.*, **2**, 421 (1968).
217. J. N. Gardner and A. R. Katritzky, *J. Chem. Soc.*, 4375 (1957).
218. R. A. Y. Jones, A. R. Katritzky, and J. M. Lagowski, *Chem. Ind.* (London), 870 (1960).
219. A. R. Katritzky and J. M. Lagowski. *J. Chem. Soc.*. 43 (1961).
220. R. A. Abramovitch and J. B. Davis, *J. Chem. Soc.*, *B*, 1137 (1966).
221. S. Gastellano and R. Kostelnik. *J. Amer. Chem. Soc.*, **90**, 141 (1968).
222. T. Okamoto, Y. Kawazoe, H. Hotta, and M. Itoh, "Abstr. Papers, Symposium on N-oxide Chemistry at the Institute of Physical and Chemical Research," Tokyo, December 1962.
223. K. Nakanishi, "Infrared Absorption Spectroscopy," Holden-day, San Francisco, 1966, p. 51.
224. A. Mangini, S. Ghersetti, and A. Lusa, *Corsi Semin. Chim.*, 67 (1968); *Chem. Abstr.*, **71**, 117972a (1069).
225. H. Shindo, *Chem. Pharm. Bull.* (Tokyo), **7**, 791 (1959); E. Ochiai, "Aromatic Amine Oxides." Elsevier, Amsterdam, 1967, p. 115, Ref. 90.
226. R. H. Wiley and S. C. Slaymaker, *J. Amer. Chem. Soc.*, **79**, 2233 (1957).
227. A. R. Katritzky and A. R. Hands, *J. Chem. Soc.*, 2195 (1958).
228. J. Suszko and M. Szafran *Bull. Acad. Polon. Sci.. Ser. Sci. Chim.*, **10**, 233 (1962); *Chem. Abstr.*, **58**, 7519d (1963).

229. Y. Matsui and T. Kubota, *J. Chem. Soc. Jap.*, **83**, 985 (1962); *Chem. Abstr.*, **58**, 10872g (1963).
230. M. Szafran, *Bull. Acad. Polon. Sci.*, *Ser. Sci. Chim.*, **11**, 169 (1963); *Chem. Abstr.*, **59**, 13479e (1963).
231. M. Szafran, *Bull. Acad. Polon. Sci.*, *Ser. Sci. Chim.*, **12**, 383 (1964); *Chem. Abstr.*, **62**, 3904d (1965).
232. D. Hadzi, *J. Chem. Soc.*, 5128 (1962).
233. M. Szafran, *Bull. Acad. Polon. Sci.*, *Ser. Sci. Chim.*, **12**, 387 (1964); *Chem. Abstr.* **62**, 3904e (1965).
234. C. L. Bell, J. Shoffner, and L. Bauer, *Chem. Ind.* (London), 1435 (1963).
235. M. Szafran, *Bull. Acad. Polon. Sci.*, *Ser. Sci. Chim.*, **10**, 479 (1962); *Chem. Abstr.*, **59**, 1585c (1963).
236. (a) M. Szafran, *Bull. Acad. Polon. Sci.*, *Ser. Sci. Chim.*, **13**, 245 (1965); *Chem. Abstr.*, **63**, 14669e (1965); (b) M. Szafran and B. Brzezinski, *Bull. Acad. Polon. Sci.*, *Ser. Sci. Chim.*, **18**, 247 (1970).
237. R. A. Jones, B. D. Roney, W. H. F. Sasse, and K. O. Wade, *J. Chem. Soc.*, *B*, 106 (1967).
238. A. R. Katritzky and J. M. Lagowski, *J. Chem. Soc.*, 2421 (1960).
239. M. Bellas and H. Suschitzky, *J. Chem. Soc.*, 2096 (1965).
240. A. R. Katritzky and R. A. Jones, *J. Chem. Soc.*, 3674 (1959).
241. A. R. Katritzky and J. M. Lagowski, *J. Chem. Soc.*, 4155 (1958).
242. H. Tani and K. Fukushima, *J. Pharm. Soc. Jap.*, **81**, 27 (1961); *Chem. Abstr.*, **55**, 12040a (1961).
243. A. R. Katritzky, J. M. Lagowski and J. A. T. Beard, *Spectrochim. Acta*, **16**, 954, 964 (1960).
244. M. Ito and N. Hata, *Bull. Chem. Soc. Jap.*, **28**, 353 (1955).
245. Y. Kakruchi and T. Shimanouchi, *Proc. Intern. Symp. Mol. Struct. Spectry.* (Tokyo), C241 (1962); *Chem. Abstr.*, **61**, 3818b (1964).
246. P. Mirone, *Atti Accad. Nazl. Lincei. Rend., Classe Sci. Fis., Mat. Nat.*, **35**, 530 (1963).
247. P. Mirone and B. Fortunato, *Atti Accad. Nazl. Lincei. Rend., Classe Sci. Fis., Mat. Nat.*, **34**, 168 (1963); *Chem. Abstr.*, **60**, 14028e (1964).
248. K. Ramaiah and V. R. Srinivasan, *Proc. Indian Acad. Sci.*, **55A**, 221 (1962); *Chem. Abstr.*, **57**, 6761f (1962).
249. K. Ramaiah and V. R. Srinivasan, *Current Sci.* (India), **27**, 340 (1958); *Chem. Abstr.*, **53**, 7763e (1959).
250. V. Ramaiah and V. R. Srinivasan, *Proc. Indian Acad. Sci.*, **50A**, 213 (1959); *Chem. Abstr.*, **54**, 5247i (1960).
251. S. Ghersetti, G. Maccagnani, A. Mangini, and F. Montanari, *J. Heterocycl. Chem.*, **6**, 859 (1969).
252. T. Kubota, *J. Chem. Soc. Jap.*, **79**, 930 (1958); *Chem. Abstr.*, **53**, 15761a (1959).
253. T. Kubota and M. Yamakawa, *Bull. Chem. Soc. Jap.*, **35**, 555 (1962).
254. A. Mangini and F. Montanari, *Proc. 4th Int. Meeting Mol. Spectry.*, Bologna, 1959, **1**, 458 (1962); *Chem. Abstr.*, **59**, 5946f (1963).
255. M. Ito and N. Hata, *Bull. Chem. Soc. Jap.*, **28**, 260 (1955).
256. N. Hata and I. Tanaka, *J. Chem. Phys.*, **36**, 2072 (1962).
257. N. Hata, *Bull. Chem. Soc. Jap.*, **34**, 1440 (1961).
258. N. Ikekawa and Y. Sato, *Pharm. Bull.* (Japan), **2**, 400 (1954); *Chem. Abstr.*, **50**, 11819g (1956).
259. A. Mangini and F. Montanari, *Bull. Sci. Fac. Chim. Ind. Bologna*, **17**, 19 (1959); *Chem. Abstr.*, **53**, 21153a (1959).
260. R. Saison and H. H. Jaffé, *C.R. Acad. Sci.*, Paris, *Ser. C*, **257**, 677 (1963).

261. A. R. Katritzky, A. J. Boulton, and D. J. Short, *J. Chem. Soc.*, 2954 (1960).
262. T. Kubota, *J. Chem. Soc. Jap.*, **78**, 196 (1957); *Chem. Abstr.*, **53**, 2777e (1959).
263. W. M. Schubert and J. Robins, *J. Amer. Chem. Soc.*, **80**, 559 (1958).
264. D. D. M. Casoni, A. Mangini, and F. Montanari, *Gazz. Chim. Ital.*, **88**, 1035 (1958).
265. N. Hata, *Bull. Chem. Soc. Jap.*, **31**, 255 (1958).
266. A. Foffani and U. Mazzucato, *Ricerca Sci.*, **26**, 2409 (1956); *Chem. Abstr.*, **53**, 823f (1959).
267. M. Colonna and A. Risaliti, *Ann. Chim.* (Rome), **48**, 1395 (1958).
268. A. R. Katritzky and P. Simmons, *J. Chem. Soc.*, 4901 (1960).
269. (a) L. Pentimalli, *Gazz. Chim. Ital.*, **90**, 1203 (1960); (b) C. Leibovici and J. Streith, *Tetrahedron Lett.*, 387 (1971); (c) K. Seibold, G. Wagniere, and H. Labhart, *Helv. Chim. Acta*, **52**, 789 (1969); (d) E. M. Evleth, *Theor. Chim. Acta*, **11**, 145 (1968).
270. R. Grigg and B. G. Odell, *J. Chem. Soc.*, *B*, 218 (1966).
271. H. Budzikiewicz, C. Djerassi, and D. H. Williams, "Mass Spectrometry of Organic Compounds," Holden-Day, San Francisco, 1967, p. 328.
272. (a) J. H. Bowie, R. G. Cooks, N. C. Jamieson, and G. E. Lewis, *Aust. J. Chem.*, **26**, 2545 (1967); (b) N. Bild and M. Hess, *Helv. Chim. Acta*, **50**, 1885 (1967); (c) T. A. Bryce and J. R. Maxwell, *Chem. Comm.*, 206 (1965); (d) W. W. Paudler and S. A. Humphrey, *Org. Mass Spectrom.*, **4**, 513 (1970); (e) D. A. Lightner, R. Nicoletti, G. B. Quistand, and E. Irwin, *Org. Mass Spectrom.*, **4**, 517 (1970).
273. G. Spiteller, *Adv. Heterocycl. Chem.*, **7**, 317 (1967).
274. (a) M. Itoh, T. Okamoto, and S. Nagakura, *Bull. Chem. Soc. Jap.*, **36**, 1665 (1963); (b) T. Kubota, Y. Oishi, K. Nishikida, H. Miyazaki, *Bull. Chem. Soc. Jap.*, **43**, 1622 (1970); *Chem. Abstr.*, **73**, 61082g (1970).
275. M. Colonna and P. Bruni, *Atti Accad. Nazl. Lincei. Rend., Classe Sci. Fis., Mat. Nat.*, **40**, 872 (1966); *Chem. Abstr.*, **66**, 55345 (1967).
276. E. G. Janzen and J. W. Happ, *J. Phys. Chem.*, **73**, 2335 (1969).
277. J. Devanneaux and J.-F. Labarre, *J. Chim. Phys.*, **66**, 1174 (1969).
278. R. A. Barnes, *J. Amer. Chem. Soc.*, **81**, 1935 (1959).
279. H. H. Jaffe, *J. Amer. Chem. Soc.*, **76**, 3527 (1954).
280. T. Okamoto and H. Tani, *Chem. Pharm. Bull.* (Tokyo), **7**, 925 (1959); E. Ochiai, "Aromatic Amine Oxides," Elsevier, Amsterdam, 1967, p. 157, Ref. 169.
281. H. Tani, *Chem. Pharm. Bull.* (Tokyo), **7**, 930 (1959); *Chem. Abstr.*, **55**, 6477i (1961).
282. M. J. S. Dewar and P. M. Maitlis, *J. Chem. Soc.*, 2521 (1957).
283. E. Ochiai and T. Okamoto, *J. Pharm. Soc. Jap.*, **70**, 384 (1950); *Chem. Abstr.*, **45**, 2476 (1951).
284. E. Ochiai and K. Satake, *J. Pharm. Soc. Jap.*, **71**, 1078 (1951); *Chem. Abstr.*, **46**, 5045f (1952).
285. R. D. Brown, *Quart. Rev.*, **6**, 63 (1952).
286. S. Basu and L. Saha, *Naturwissenschaften*, **44**, 633 (1957).
287. T. Kubota, *J. Pharm. Soc. Jap.*, **79**, 388 (1959); *Chem. Abstr.*, **53**, 13765e (1959).
288. T. Kubota, *J. Chem. Soc. Jap.*, **80**, 578 (1959); *Chem. Abstr.*, **53**, 13765d (1959).
289. G. Tscoucaris, *J. Chim. Phys.*, **58**, 613 (1961).
290. M. N. Adamov and I. F. Tupitsyn, *Lietuvos TSR Aukstosios Mokyklos*, **3**, 277 (1963); *Chem. Abstr.*, **60**, 15165d (1964).
291. S. Kabinata and S. Nagakura, *Theor. Chim. Acta*, **14**, 415 (1969).
292. J. F. Vossa, *J. Org. Chem.*, **27**, 3856 (1962).
293. M. Szafran, *Bull. Acad. Polon. Sci., Ser. Sci. Chim.*, **11**, 111 (1963); *Chem. Abstr.*, **59**, 11216 (1963).
294. C. W. Murth and R. S. Darlak, *J. Org. Chem.*, **30**, 1909 (1965).
295. G. Tacconi and S. Pietra, *Ann. Chim.* (Rome), **55**, 810 (1965).

296. A. Trombetti, *J. Chem. Soc., B*, 1578 (1968).
297. K. Yamanaka, M. Maeda, M. Kitamura, R. Koizumi, and S. Okuma, Japanese Patent 1847, 1957; *Chem. Abstr.*, **52**, 5483g (1958).
298. E. Hayashi, *J. Pharm. Soc. Jap.*, **70**, 142 (1950); *Chem. Abstr.*, **44**, 5880i (1950).
299. (a) R. A. Abramovitch, S. Kato, and G. M. Singer, *J. Amer. Chem. Soc.*, **93**, 3074 (1971); (b) G. Wagner and H. Pischel, *Z. Chem.*, **2**, 308 (1962); (c) R. A. Abramovitch and S. Kato, unpublished results.
300. W. G. M. Jones and M. A. Stevens, British Patent 974,168, 1964; *Chem. Abstr.*, **62**, 4015b (1965).
301. T. A. Liss, *J. Org. Chem.*, **32**, 1146 (1967).
302. F. Brody and W. J. Sydor, U.S. Patent 3,118,871, 1964; *Chem. Abstr.*, **60**, 1602g (1964).
303. L. A. Paquette, *J. Amer. Chem. Soc.*, **87**, 5186 (1960).
304. E. C. Taylor, F. Kienzle, and A. McKillop, *J. Org. Chem.*, **35**, 1672 (1970).
305. (a) R. Albrecht and G. Kresze, *Chem. Ber.*, **98**, 1205 (1965); (b) S. Oae and K. Ikura, *Bull. Chem. Soc. Jap.*, **38**, 58 (1965); (c) *Bull. Chem. Soc. Jap.*, **39**, 1306 (1966); (d) *Bull. Chem. Soc. Jap.*, **40**, 1420 (1967).
306. A. R. Hands and A. R. Katritzky, *J. Chem. Soc.*, 1754 (1958).
307. O. Cervinka, *Collect. Czech. Chem. Comm.*, **27**, 567 (1962).
308. A. R. Katritzky and E. Lunt, *Tetrahedron*, **25**, 4291 (1969).
309. K. B. Augustinsson and H. Hasselquist, *Acta Chem. Scand.*, **15**, 817 (1961).
310. K. B. Augustinsson and H. Hasselquist, *Acta Chem. Scand.*, **17**, 953 (1963).
311. C. V. D. M. Brink and P. J. de Jager, *Tydskr. Natuurwetenskappe*, **3**, 74 (1963); *Chem. Abstr.*, **60**, 5447f (1964).
312. T. Kato, Y. Goto, and Y. Yamomoto, *J. Pharm. Soc. Jap.*, **84**, 287 (1964).
313. V. J. Traynelis and P. L. Pacini, *J. Amer. Chem. Soc.*, **86**, 4917 (1964).
314. (a) E. E. Knaus, Ph.D. Dissertation, University of Saskatchewan, 1970; (b) E. Matsumura, M. Ariga, and T. Ohfuji, *Bull. Chem. Soc. Jap.*, **43**, 3210 (1970).
315. E. Hayashi, H. Yamanaka, and K. Shimizu, *Chem. Pharm. Bull.* (Tokyo), **6**, 323 (1958); *Chem. Abstr.*, **53**, 375i (1959).
316. E. Hayashi, H. Yamanaka, and K. Shimizu, *Chem. Pharm. Bull.* (Tokyo), **7**, 141 (1959); *Chem. Abstr.*, **54**, 22665b (1960).
317. E. Hayashi, H. Yamanaka, and C. Iijima, *J. Pharm. Soc. Jap.*, **80**, 1145 (1960); *Chem. Abstr.*, **55**, 546d (1961).
318. G. Wagner, H. Pischel, and R. Schmidt, *Z. Chem.*, **2**, 86 (1962).
319. D. Jerchel and W. Melloh, *Ann. Chem.*, **613**, 144 (1958).
320. E. Hayashi, H. Yamanaka, and C. Iijima, *J. Pharm. Soc. Jap.*, **80**, 839 (1960); *Chem. Abstr.*, **54**, 24730a (1960).
321. S. Mitsui, T. Sakai, and H. Saito, *J. Chem. Soc. Jap.*, **84**, 409 (1965); *Chem. Abstr.*, **63**, 8149a (1965).
322. N. Greenhalgh and M. A. Stevens, British Patent 946,880, 1964; *Chem. Abstr.*, **60**, 10655g (1964).
323. T. Itai and H. Ogura, *J. Pharm. Soc. Jap.*, **75**, 292 (1955); *Chem. Abstr.*, **50**, 1808g (1956).
324. E. C. Taylor and J. S. Driscoll, *J. Amer. Chem. Soc.*, **82**, 3141 (1960).
325. A. R. Katritzky and A. M. Monro, *J. Chem. Soc.*, 1263 (1958).
326. E. Ochiai and M. Katada, *J. Pharm. Soc. Jap.*, **63**, 265 (1943); *Chem. Abstr.*, **45**, 5152e (1951).
327. H. Biener and T. Wieland, *Chem. Ber.*, **95**, 277 (1962).
328. B. M. Bain and J. E. Saxton, *J. Chem. Soc.*, 5216 (1961).
329. G. M. Badger and R. P. Rao, *Aust. J. Chem.*, **17**, 1399 (1964).
330. American Cyanamid Co., Neth. Appl. 6,604,123, 1966; *Chem. Abstr.*, **55**, 104902z (1967).

331. E. Ochiai and I. Suzuki, *J. Pharm. Soc. Jap.*, **67**, 158 (1947); *Chem. Abstr.*, **45**, 9541d (1951).
332. R. P. Rao, *Aust. J. Chem.*, **17**, 1434 (1964).
333. H. Hamana, B. Umezawa, and K. Noda, *Chem. Pharm. Bull.* (Tokyo), **11**, 694 (1963); *Chem. Abstr.*, **59**, 10023g (1963).
334. T. B. S. Thyagarajan, K. K. Balasubramanian, and R. B. Rao, *Chem. Ber.*, **99**, 368 (1966).
335. K. Lewicka and E. Plazek, *Rocz. Chem.*, **40**, 1875 (1966).
336. Z. Talik, T. Talik, and A. Puszynski, *Rocz. Chem.*, **39**, 601 (1965).
337. J. M. Essery and K. Schofield, *J. Chem. Soc.*, 4953 (1960).
338. N. N. Vereshchagina and I. Ya. Postovskii, *Trudy Ural. Politekh. Inst. Im. S. M. Kirova*, **94**, 24 (1960); *Chem. Abstr.*, **56**, 8681d (1962).
339. J. Suszko and M. Szafran, *Rocz. Chem.*, **39**, 1045 (1965).
340. S. Kajihara, *J. Chem. Soc. Jap.*, **86**, 1060 (1965); *Chem. Abstr.*, **65**, 16936e (1966).
341. J. Delarge. *Farmaco.. Ed. Sci.*, **22**, 99 (1967); *Chem. Abstr.*, **66**, 94891d (1967).
342. T. Nakashima and M. Endo, *J. Chem. Soc. Jap.*, **81**, 816 (1960); *Chem. Abstr.*, **56**, 5923d (1962).
343. Z. Talik and B. Brekiesz, *Rocz. Chem.*, **41**, 279 (1967).
344. S. M. Gerber, U.S. Patent 3,386,991, 1968; *Chem. Abstr.*, **69**, 96486f (1968).
345. E. Profft and G. Schulz, *Arch. Pharm.* (Weinheim), **294**, 292 (1961); *Chem. Abstr.*, **55**, 18725a (1961).
346. F. Kronhke and H. Schafer, *Chem. Ber.*, **95**, 1098 (1962).
347. M. Hamana, B. Umezawa, and Y. Goto, *J. Pharm. Soc. Jap.*, **80**, 1519 (1960); *Chem. Abstr.*, **55**, 8405b (1961).
348. M. Endo and T. Nakashima, *J. Pharm. Soc. Jap.*, **80**, 875 (1960); *Chem. Abstr.*, **54**, 24705e (1960).
349. D. Jerchel and H. Hippchen, German Patent 1,019,652, 1957; *Chem. Abstr.*, **54**, 1557i (1960).
350. Z. Talik and T. Talik, *Rocz. Chem.*, **36**, 417 (1962).
351. F. Binns and H. Suschitzky, *Chem. Comm.*, 750 (1970).
352. T. Talik and Z. Talik, *Rocz. Chem.*, **38**, 777 (1964).
353. T. Talik, *Rocz. Chem.*, **36**, 1465 (1962).
354. (a) Y. Suzuki, *J. Pharm. Soc. Jap.*, **81**, 1206 (1961); *Chem. Abstr.*, **56**, 34446e (1962); (b) H. Suschitzky and F. Binns, *J. Chem. Soc.*, **C**, 1223 (1971).
355. Z. Talik, *Rocz. Chem.*, **35**, 475 (1961).
356. E. Hayashi and Y. Hotta, *J. Pharm. Soc. Jap.*, **80**, 834 (1960); *Chem. Abstr.*, **54**, 24597e (1960).
357. T. Takahashi, I. Yamashita, and J. Iwai, *J. Pharm. Soc. Jap.*, **78**, 943 (1958); *Chem. Abstr.*, **52**, 20144c (1958).
358. M. Hamana and M. Yamazaki, *J. Pharm. Soc. Jap.*, **81**, 574 (1961); *Chem. Abstr.*, **55**, 24743a (1961).
359. M. Nakadate, Y. Takano, T. Hirayama, S. Sakaizawa, T. Hirano, K. Okamoto, K. Hirao, T. Kawanura, and M. Kimura, *Chem. Pharm. Bull.* (Tokyo), **13**, 113 (1965); *Chem. Abstr.*, **62**, 14620c (1965).
360. C. Hansch and W. Carpenter, *J. Org. Chem.*, **22**, 936 (1957).
361. T. Talik and Z. Talik, *Nitro Compds.*, *Proc. Intern. Symp. Warsaw*, **81**, 1963 (1964); *Chem. Abstr.*, **64**, 2046e (1966).
362. R. A. Abramovitch and M. Saha, *Can. J. Chem.*, **44**, 1765 (1966).
363. M. Hamana, *J. Pharm. Soc. Jap.*, **75**, 139 (1955); *Chem. Abstr.*, **50**, 1817a (1956).
364. T. R. Emerson and C. W. Rees, *J. Chem. Soc.*, 1917 (1962).

365. E. Howard, Jr. and W. F. Olszewsdi, *J. Amer. Chem. Soc.*, **81**, 1483 (1959).
366. T. R. Emerson and C. W. Rees, *Proc. Chem. Soc.*, 418 (1960).
367. (a) D. I. Relyea, P. O. Tawney, and A. R. Williams, *J. Org. Chem.*, **27**, 477 (1962); (b) T. Hisano and H. Koga, *J. Pharm. Soc. Jap.*, **90**, 552 (1970); *Chem. Abstr.*, **73**, 45267r (1970).
368. F. A. Daniher and B. E. Hackley, Jr., *J. Org. Chem.*, **31**, 4267 (1966).
369. F. A. Daniher and B. E. Hackley, Jr., U.S. Patent 3,467,659, 1969; *Chem. Abstr.*, **71**, 124246 (1969).
370. M. Hamana and K. Funakoshi, *J. Pharm. Soc. Jap.*, **80**, 1027 (1960); *Chem. Abstr.*, **55**, 542a (1961).
371. R. E. Evans and H. C. Brown, *J. Org. Chem.*, **27**, 1665 (1962).
372. F. Gadient, E. Tucker, A. Lindenman, and M. Taeschler, *Helv. Chim. Acta*, **45**, 1860 (1962).
373. K. Ikura and S. Oae, *Tetrahedron Lett.*, 3791 (1968).
374. R. A. Abramovitch and E. M. Smith, unpublished results.
375. M. Jankovsky and M. Ferles, *Collect. Czech. Chem. Comm.*, **35**, 2802 (1970).
376. H. C. Brown and B. C. S. Rao, *J. Amer. Chem. Soc.*, **78**, 2582 (1956).
377. H. C. Brown, U.S. Patent 2,856,428, 1958; *Chem. Abstr.*, **53**, 11190h (1959).
378. H. C. Brown and N. M. Yoon, *J. Amer. Chem. Soc.*, **88**, 1464 (1966).
379. H. C. Brown and P. W. Weissman, *J. Amer. Chem. Soc.*, **87**, 5614 (1965).
380. R. Oda, Y. Hayashi, and T. Yoshida, *J. Chem. Soc. Jap.*, **87**, 975 (1966); *Chem. Abstr.*, **65**, 16923a (1966).
381. (a) I. A. D'yakonov, T. V. Mandelshtam, and O. M. Radul, *Zh. Org. Khim.*, **4**, 723 (1968); *Chem. Abstr.*, **69**, 2841y (1968); (b) E. E. Schweizer, G. J. O'Neil, and J. N. Wemple, *J. Org. Chem.*, **29**, 1744 (1964).
382. W. J. Zubyk, Ph.D. Dissertation, University of Delaware, 1957.
383. V. J. Traynelis and K. Yamauchi, *Tetrahedron Lett.*, 3619 (1969).
384. R. T. Brooks and P. D. Sternglanz, *Anal. Chem.*, **31**, 561 (1959).
385. E. E. Schweizer and G. J. O'Neil, *J. Org. Chem.*, **28**, 2460 (1963).
386. A. Markovac, C. L. Stevens, A. B. Ash, and B. E. Hackley, Jr., *J. Org. Chem.*, **35**, 841 (1970).
387. S. Kubota and T. Akita, *J. Pharm. Soc. Jap.*, **78**, 248 (1958).
388. D. M. Jerina, D. R. Derek, and J. W. Daly, *Tetrahedron Lett.*, 457 (1970).
389. R. Eisenthal and A. R. Katritzky, *Tetrahedron*, **21**, 2205 (1965).
390. M. Henze, *Chem. Ber.*, **70**, 1270 (1937).
391. E. Ochiai, T. Waito, and M. Katada, *J. Pharm. Soc. Jap.*, **64**, 210 (1944); *Chem. Abstr.*, **45**, 5154i (1951).
392. V. Boekelheide, W. E. Freely, and W. L. Lehn, *J. Org. Chem.*, **22**, 1135 (1957).
393. J. C. Stowell, *J. Org. Chem.*, **35**, 244 (1970).
394. T. Cohen and I. H. Song, *J. Org. Chem.*, **31**, 3058 (1966).
395. H. N. Bojarska-Dahlig and H. J. den Hertog, *Rec. Trav. Chim. Pays-Bas*, **77**, 331 (1958).
396. T. Cohen, I. H. Song, and J. H. Fager, *Tetrahedron Lett.*, 237 (1965).
397. T. Cohen, G. L. Deets, and J. A. Jenkins, *J. Org. Chem.*, **34**, 2550 (1969).
398. T. Cohen, I. H. Song, J. H. Fager, and G. L. Deets, *J. Amer. Chem. Soc.*, **89**, 4968 (1967).
399. C. Rüchardt, S. Eichler, and O. Kraetz, *Tetrahedron Lett.*, 233 (1965).
400. C. Rüchardt and O. Kraetz, *Tetrahedron Lett.*, 5915 (1966).
401. C. Rüchardt, O. Kraetz, and S. Eichler, *Chem. Ber.*, **102**, 3922 (1969).
402. H. W. Johnson, Jr. and H. Krutzsch, *J. Org. Chem.*, **32**, 1939 (1967).
403. R. M. Titkova and V. A. Mikhalev, *Zh. Organ. Khim.*, **1**, 1121 (1965); *Chem. Abstr.*, **63**, 11483h (1965).

404. W. N. Marmer and D. Swern, *Tetrahedron Lett.*, 531 (1969).
405. W. N. Marmer and D. Swern, "Abstracts of 159th ACS National Meeting," Houston, March 23–27, 1970, p. 45.
406. W. N. Marmer and D. Swern, *J. Amer. Chem. Soc.*, **93**, 2719 (1971).
407. R. M. Titkova and V. A. Mikhalev, *Zh. Obshch. Khim.*, **34**, 4126 (1964); *Chem. Abstr.*, **62**, 9098e (1965).
408. V. Boekelheide and R. Scharrer, *J. Org. Chem.*, **26**, 3802 (1961).
409. T. Koenig, *Tetrahedron Lett.*, 2751 (1967).
410. V. Tortorella, *Corsi Semin. Chim.*, 117 (1968); *Chem. Abstr.*, **72**, 90843r (1970).
411. L. A. Paquette, *J. Amer. Chem. Soc.*, **87**, 1407 (1965).
412. (a) L. M. Litvinenko and G. D. Titskii, *Dokl. Akad. Nauk. SSR*, **177**, 127 (1967); *Chem. Abstr.*, **68**, 7737g (1968); (b) L. M. Litvinenko, G. D. Titskii, and I. V. Shpan'ko, *Zh. Org. Khim.*, **7**, 107 (1971); *Chem. Abstr.*, **74**, 99221k (1971).
413. T. Okamoto, as quoted by E. Ochiai in "Aromatic Amine Oxides," Elsevier, Amsterdam, 1967; p. 307, ref. 199.
414. R. Eisenthal, A. R. Katritzky, and E. Lunt, *Tetrahedron*, **23**, 2775 (1967).
415. H. C. van der Plas, H. J. den Hertog, M. van Ammers, and B. Haase, *Tetrahedron Lett.*, 32 (1961).
416. A. R. Katritzky, B. J. Ridgewell, and A. M. White, *Chem. Ind.* (London), 1576 (1964).
417. Y. Kawazoe, M. Ohnishi, and Y. Yoshioka, *Chem. Pharm. Bull.* (Tokyo), **12**, 1384 (1964); *Chem. Abstr.*, **62**, 5274g (1965).
418. G. P. Bean, P. J. Brignell, C. D. Johnson, A. R. Katritzky, B. J. Ridgewell, H. O. Tarhan, and A. M. White, *J. Chem. Soc.*, *B*, 1222 (1967).
419. E. Ochiai, E. Hayashi, and M. Katada, *J. Pharm. Soc. Jap.*, **67**, 79 (1947); *Chem. Abstr.*, **45**, 9538a (1951).
420. K. Lewicka and E. Plazek, *Rec. Trav. Chim. Pays-Bas*, **78**, 644 (1959).
421. K. Undheim, V. Nordal, and L. Borka, *Acta Chem. Scand.*, **23**, 2075 (1969).
422. S. Minakami and E. Hirai, Japanese Patent 7,006,268, 1970; *Chem. Abstr.*, **72**, 132534y (1970).
423. E. Ochiai and C. Kaneko, *Chem. Pharm. Bull.* (Tokyo), **8**, 28 (1960); *Chem. Abstr.*, **55**, 5491g (1961).
424. R. B. Moodie, K. Schofield, and M. J. Williamson, *Chem. Ind.* (London), 1577 (1964).
425. J. T. Gleghorn, R. B. Moodie, E. A. Quereshi, and K. Schofield, *J. Chem. Soc.*, *B*, 316 (1968).
426. C. D. Johnson, A. R. Katritzky, N. Shakir, and M. Viney, *J. Chem. Soc.*, *B*, 1213 (1967).
427. E. Matsumura, Japanese Patent 6,828,455, 1968; *Chem. Abstr.*, **70**, 77799 (1969).
428. T. Talik and Z. Talik, *Rocz. Chem.*, **36**, 539 (1962).
429. J. Dabrowski, W. Jasiobedzki, T. Jaworski, M. Medon, and J. Terpinski, Polish Patent 45,711, 1962; *Chem. Abstr.*, **58**, 9033b (1963).
430. N. F. Kucherova, R. M. Khomutov, E. I. Budovskii, V. P. Evadokov, and N. K. Kochetkov, *Zh. Obshch. Khim.*, **29**, 915 (1959); *Chem. Abstr.*, **54**, 1515d (1960).
431. F. Leonard and A. Wajnqurt, *J. Org. Chem.*, **21**, 1077 (1956).
432. S. Okuda and M. M. Robison, *J. Org. Chem.*, **24**, 1008 (1959).
433. F. E. Cislak, U.S. Patent 2,868,797, 1959; *Chem. Abstr.*, **53**, 10256e (1959).
434. E. Profft, W. Krueger, P. Kuhn and W. Lietz, German (East) Patent 69,126, 1969; *Chem. Abstr.*, **72**, 90309w (1970).
435. M. Hamana and M. Yamazaki, *Chem. Pharm. Bull.* (Tokyo), **9**, 414 (1961); *Chem. Abstr.*, **55**, 24749c (1961).
436. M. Yamazaki, Y. Chono, K. Noda, and M. Motoyoshi, *J. Pharm. Soc. Jap.*, **85**, 62 (1962); *Chem. Abstr.*, **62**, 10409e (1965).

437. M. van Ammers and H. J. den Hertog, *Rec. Trav. Chim. Pays-Bas*, **77**, 340 (1958).
438. M. van Ammers and H. J. den Hertog, *Rec. Trav. Chim. Pays-Bas*, **81**, 124 (1962).
439. M. van Ammers and H. J. den Hertog, *Rec. Trav. Chim. Pays-Bas*, **78**, 586 (1959).
440. O. S. Otroshchenko, A. S. Sodykov, M. V. Utebacu, and A. I. Isametova, *Zh. Obshch. Khim.*, **33**, 1038 (1963); *Chem. Abstr.*, **59**, 10142 (1963).
441. P. E. Holt and E. I. Nasrallah, *J. Chem. Soc.*, **B**, 400 (1968).
442. T. Kato and H. Yamanaka, *J. Org. Chem.*, **30**, 910 (1965).
443. (a) E. Ochiai and K. Arima, *J. Pharm. Soc. Jap.*, **69**, 51 (1949); (b) T. Kato, H. Yamanaka, T. Acdachi, and H. Hiranuma, *J. Org. Chem.*, **32**, 3788 (1967); (c) T. J. van Bergen and R. M. Kellogg, *J. Org. Chem.*, **36**, 1705 (1971).
444. O. Cervinka, *Chem. Ind.* (London). 1482 (1960).
445. R. A. Abramovitch, C. S. Giam, and G. A. Poulton, *J. Chem. Soc.*, **C**, 128 (1970).
446. J. H. Blumenthal, U.S. Patent 2,874,162, 1959; *Chem. Abstr.*, **53**, 12311b (1959).
447. Ch. Kaneko and T. Miyasaka, *Shika Zairyo, Kenkyusho Hokoku*, **2**, 269 (1963); Ochiai, "Aromatic Amine Oxides," Elsevier, Amsterdam. 1967, p. 253, Ref. 25.
448. H. Tani, *J. Pharm. Soc. Jap.*, **81**, 182 (1961); *Chem. Abstr.*, **55**, 14450b (1961).
449. Y. Mizuno, T. Itoh, K. Saito, *Chem. Pharm. Bull.* (Tokyo), **12**, 866 (1964); *Chem. Abstr.*, **61**, 14661a (1964).
450. H. V. Euler, H. Hasselquist, and O. Heidenberger, *Chem. Ber.*, **92**, 2266 (1959).
451. K. Palat, M. Celadnik, L. Novacek, and M. Polstev, *Cesk. Farm.*, **6**, 369 (1957); *Chem. Abstr.*, **52**, 10071 (1958).
452. G. Büchi, R. E. Manning, and F. A. Hochstein, *J. Amer. Chem. Soc.*, **84**, 3394 (1962).
453. E. Ochiai and I. Nakayama, *J. Pharm. Bull. Jap.*, **65B**, 582 (1945); *Chem. Abstr.*, **45**, 8529e (1951).
454. I. Nakayama, *J. Pharm. Soc. Jap.*, **70**, 355 (1950); *Chem. Abstr.*, **45**, 2945a (1951).
455. W. E. Feely and F. M. Beavers, *J. Amer. Chem. Soc.*, **81**, 4004 (1059).
456. C. Kaneko, *Chem. Pharm. Bull.* (Tokyo), **8**, 286 (1960); *Chem. Abstr.*, **55**, 5493d (1961).
457. S. Ranganathan, B. B. Singh, and P. V. Divekar, *Can. J. Chem.*, **47**, 165 (1969).
458. H. Tani, *J. Pharm. Soc. Jap.*, **81**, 141 (1961); *Chem. Abstr.*, **55**, 1449 (1961).
459. N. Nishimoto and T. Nakashima, *J. Pharm. Soc. Jap.*, **81**, 88 (1961); *Chem. Abstr.*, **55**, 134201 (1961).
460. H. Tani, *J. Pharm. Soc. Jap.*, **80**, 1418 (1960); *Chem. Abstr.*, **55**, 6477i (1961).
461. M. Hamana and H. Noda, *Chem. Pharm. Bull.* (Tokyo), **11**, 1331 (1963); *Chem. Abstr.*, **60**, 2887g (1964).
462. M. Hamana and H. Noda, *Chem. Pharm. Bull.* (Tokyo), **15**, 474 (1967); *Chem. Abstr.*, **68**, 29558c (1968).
463. T. A. Crabb and E. R. Jones, *Tetrahedron*. **26**, 1217 (1970).
464. L. Bauer and L. A. Gardella, *J. Org. Chem.*, **28**, 1320 (1963).
465. L. Bauer and L. A. Gardella, *J. Org. Chem.*, **28**, 1323 (1963).
466. L. Bauer and T. E. Dickerhofe, *J. Org. Chem.*, **29**, 2183 (1964).
467. F. M. Hershenson and L. Bauer. *J. Org. Chem.*, **34**, 655 (1969).
468. F. M. Hershenson and L. Bauer. *J. Org. Chem.*, **34**, 660 (1969).
469. F. M. Hershenson and L. Bauer, *J. Org. Chem.*, **34**, 665 (1969).
470. B. A. Mikrut, F. M. Hershenson, K. F. King, L. Bauer, and R. S. Egan, *J. Org. Chem.*, **36**, 3749 (1971).
471. K. F. King and L. Bauer, *J. Org. Chem.*, **36**, 1641 (1971).
472. R. A. Abramovitch, E. M. Smith, and P. Tomasik, unpublished results.
473. D. Redmore, *J. Org. Chem.*, **35**, 4114 (1970).
474. M. Murakami and E. Matsumura, *J. Chem. Soc. Jap.*, **70**, 393 (1949); *Chem. Abstr.*, **45**, 4698e (1951).

475. H. J. den Hertog, D. J. Buurman, and P. A. deVilliers, *Rec. Trav. Chim. Pays-Bas*, **80**, 325 (1961).

476. S. Oae, T. Kitao, and Y. Kitaoka, *Tetrahedron*, **19**, 827 (1963).

477. (a) E. Matsumura, T. Hirooka, and K. Imagawa, *J. Chem. Soc. Jap.*, **82**, 104 (1961); E. Ochiai, "Aromatic Amine Oxides," Elsevier, Amsterdam, 1967, p. 337, Ref. 122b; (b) E. Matsumura, T. Nashima, and F. Ishibashi, *Bull. Chem. Soc. Jap.*, **43**, 3540 (1970).

478. E. Matsumura, *Mom. Osaka Univ. Liberal Arts Educ.*, **10**, 191 (1962); Ochiai, "Aromatic Amine Oxides," Elsevier, Amsterdam, 1967, p. 337, Ref. 124.

479. M. Hamana and K. Funakoshi, *J. Pharm. Soc. Jap.*, **82**, 512 (1962); *Chem. Abstr.*, **58**, 4512f (1963).

480. M. Hamana and K. Funakoshi, *J. Pharm. Soc. Jap.*, **84**, 23 (1964); *Chem. Abstr.*, **61**, 3068b (1964).

481. M. Hamana and K. Funakoshi, *J. Pharm. Soc. Jap.*, **82**, 518 (1962); *Chem. Abstr.*, **58**, 4512h (1963).

482. M. Hamana and K. Funakoshi, *J. Pharm. Soc. Jap.*, **82**, 523 (1962); Ochiai, "Aromatic Amine Oxides," Elsevier, Amsterdam, 1967, p. 286, Ref. 143.

483. (a) V. J. Traynelis, in *Mechanisms of Molecular Migrations*, Vol. 2, B. S. Thyagarajan, Ed., Interscience, New York, 1969, p. 1; (b) V. Boekelheide and W. J. Linn, *J. Amer. Chem. Soc.*, **76**, 1286 (1954).

484. B. M. Bain and J. E. Saxton, *Chem. Ind.* (London), 402 (1960).

485. G. F. van Rooyen, C. V. D. M. Brink, and P. A. d Villiers, *Tydskr. Naturwetenskappe*, **4**, 182 (1964); *Chem. Abstr.*, **63**, 13202a (1965).

486. T. Takahashi and K. Kariyone, *Chem. Pharm. Bull.* (Tokyo), **8**, 1106 (1960); *Chem. Abstr.*, **55**, 24744b (1961).

487. P. W. Ford and J. M. Swan, *Aust. J. Chem.*, **18**, 867 (1968).

488. Y. Murakami and J. Sunamoto, *Bull. Chem. Soc. Jap.*, **42**, 3350 (1969).

489. O. H. Bullit and J. T. Maynard, *J. Amer. Chem. Soc.*, **76**, 1370 (1954).

490. V. J. Traynelis and R. F. Martello, *J. Amer. Chem. Soc.*, **82**, 2744 (1960).

491. J. A. Berson and T. Cohen, *J. Amer. Chem. Soc.*, **77**, 1281 (1951).

492. R. Graf, *J. Prakt. Chem.*, **133**, 36 (1932).

493. T. Cohen and J. H. Fager, "Abstracts of the XIXth International Congress of Pure and Applied Chemistry," London, 1963, A1-147.

494. T. Cohen, private communication to R. A. Abramovitch and J. G. Saha, *Adv. Heterocycl. Chem.*, **6**, 331 (1966).

495. T. Cohen and J. H. Fager, *J. Amer. Chem. Soc.*, **87**, 5701 (1965).

496. V. J. Traynelis and R. F. Martello, *J. Amer. Chem. Soc.*, **80**, 6590 (1958).

497. T. Cohen and G. L. Deets, *J. Amer. Chem. Soc.*, **89**, 3939 (1967).

498. V. J. Traynelis, A. I. Gallagher, and R. F. Martello, *J. Org. Chem.*, **26**, 4365 (1961).

499. J. H. Markgraf, H. B. Brown, Jr., S. C. Mohr, and R. G. Peterson, *J. Amer. Chem. Soc.*, **85**, 958 (1963).

500. S. Oae, T. Kitao, and Y. Kitaoka, *J. Amer. Chem. Soc.*, **84**, 3359 (1962).

501. S. Oae, S. Tamagaki, T. Negoro, and S. Kosuka, *Tetrahedron*, **26**, 4051 (1970).

502. R. Bodalski and A. R. Katritzky, *J. Chem. Soc.*, B, 831 (1968).

503. R. Bodalski and A. R. Katritzky, *Tetrahedron Lett.*, 257 (1968).

504. (a) T. W. Koenig, *J. Amer. Chem. Soc.*, **88**, 4045 (1966); (b) H. Iwamura, M. Iwamura, T. Nishida, and I. Miura, *Tetrahedron Lett.*, 3117 (1970); (c) H. Iwamura, M. Iwamura, T. Nishida, and S. Sato, *J. Amer. Chem. Soc.*, **92**, 7474 (1970).

505. Ch. Kaneko, S. Yamanda, and I. Yokoe, *Shika Zairyo Kenkyusho*, **2**, 475 (1963); Ochiai, "Aromatic Amine Oxides," Elsevier, Amsterdam, 1967, p. 295, Ref. 184.

506. T. W. Koenig and J. Wieczorek, *J. Org. Chem.*, **35**, 508 (1970).

507. S. Oae, S. Tamagaki, T. Negoro, K. Ogino, and S. Kozuka, *Tetrahedron Lett.*, 917 (1968).
508. M. Hamana and M. Yamazak, *Chem. Pharm. Bull.* (Tokyo), **11**, 415 (1963); *Chem. Abstr.*, **59**, 9974h (1963).
509. D. M. Pretorius and P. A. de Villiers, *J. S. African Chem. Inst.*, **18**, 48 (1965); *Chem. Abstr.*, **63**, 14807f (1965).
510. A. Klaebe and A. Lattes, *J. Chromatogr.*, **27**, 502 (1967).
511. B. K. Varma and A. B. Lal, *J. Indian Chem. Soc.*, **43**, 613 (1966).
512. W. Sauermilch, *Arch. Pharm.* (Weinheim), **293**, 452 (1960); *Chem. Abstr.*, **54**, 24724a (1960).
513. S. Furukawa, *J. Pharm. Soc. Jap.*, **79**, 77 (1959); Ochiai, "Aromatic Amine Oxides," Elsevier, Amsterdam, 1967, p. 297, Ref. 177.
514. K. R. Huffman, F. C. Schaefer, and G. A. Peters, *J. Org. Chem.*, **27**, 554 (1962).
515. M. Hamana, B. Umezawa, and S. Nakashima, *Chem. Pharm. Bull.* (Tokyo), **10**, 961 (1962); *Chem. Abstr.*, **59**, 3886 (1963).
516. S. Oae, T. Kitao, and Y. Kitaoka, *Ann. Rept. Radiation Center Osaka Prefect.*, **2**, 133 (1961); *Chem. Abstr.*, **58**, 2340d (1963).
517. T. Kato, F. Hamaguchi, and T. Oiwa, *J. Pharm. Soc. Jap.*, **78**, 422 (1958); *Chem. Abstr.*, **52**, 14603e (1958).
518. T. Kato, Y. Goto, and Y. Yamamoto, *J. Pharm. Soc. Jap.*, **82**, 1649 (1962); *Chem. Abstr.*, **59**, 2765d (1963).
519. (a) T. Koenig and T. Barklow, *Tetrahedron*, **25**, 4875 (1969); (b) R. N. Pratt, D. P. Stokes, and G. A. Taylor, *J. Chem. Soc.*, *C*, 1472 (1971).
520. R. A. Abramovitch and O. A. Koleoso, *J. Chem. Soc.*, *B*, 1292 (1968).
521. (a) L. K. Dyall and K. H. Pausacker, *J. Chem. Soc.*, 18 (1961); (b) R. M. Elofson, F. F. Gadallah, and K. F. Schulz, *J. Org. Chem.*, **36**, 1526 (1971); (c) S. C. Dickerman, N. Milstein, and J. F. W. McOmie, *J. Amer. Chem. Soc.*, **87**, 5522 (1965); (d) R. A. Abramovitch and J. G. Saha, *Tetrahedron*, **21**, 3297 (1965).
522. R. A. Abramovitch, G. M. Singer, and A. R. Vinutha, *Chem. Comm.*, 55 (1967).
523. J. A. Zoltewicz and G. M. Kauffman, *Tetrahedron Lett.*, 337 (1967).
524. J. A. Zoltewicz, G. M. Kauffman, and C. L. Smith, *J. Amer. Chem. Soc.*, **90**, 5939 (1968).
525. J. A. Zoltewicz and G. M. Kauffman, *J. Org. Chem.*, **34**, 1405 (1969).
526. J. A. Zoltewicz, G. M. Kauffman, and C. L. Smith, *J. Amer. Chem. Soc.*, **89**, 3358 (1967).
527. P. Dyson and D. L. Hammick, *J. Chem. Soc.*, 1724 (1937).
528. R. K. Howe and K. W. Ratts, *Tetrahedron Lett.*, 4743 (1967).
529. (a) K. R. Huffman, F. C. Schaefer, and G. A. Peters, *J. Org. Chem.*, **27**, 551 (1962); (b) J. A. Zoltewicz and L. S. Helmick, *J. Amer. Chem. Soc.*, **92**, 7547 (1970).
530. N. N. Zatsepine, I. F. Tupitsyn, and L. S. Efros, *Zh. Obshch. Khim.*, **33**, 2705 (1963); *Chem. Abstr.*, **60**, 5292d (1964).
531. Y. Kawazoe, M. Ohnishi, and Y. Yoshioka, *Chem. Pharm. Bull.* (Tokyo), **15**, 1225 (1967); *Chem. Abstr.*, **68**, 28919 (1968).
532. N. N. Zatsepine, I. F. Tupitsyn, and L. S. Efros, *J. Gen. Chem. USSR*, **33**, 3636 (1963).
533. N. N. Zatsepine, I. F. Tupitsyn, and L. S. Efros, *J. Gen. Chem. USSR*, **34**, 4124 (1964).
534. N. N. Zatsepine, I. F. Tupitsyn, and L. S. Efros, *Dokl. Akad. Nauk. SSSR*, **154**, 148 (1964); *Chem. Abstr.*, **60**, 9113g (1964).
535. A. I. Gallagher, B. A. Lalinsky, and C. M. Cuper, *J. Org. Chem.*, **35**, 1175 (1970).
536. (a) R. A. Abramovitch, M. Saha, E. M. Smith, and R. T. Coutts, *J. Amer. Chem. Soc.*, **89**, 1537 (1967); (b) R. A. Abramovitch, M. Saha, E. M. Smith, and E. E. Knaus, *J. Org. Chem.*, **37**, (1972).
537. (a) R. A. Abramovitch and E. E. Knaus, *J. Heterocycl. Chem.*, **6**, 989 (1969); (b) R. A. Abramovitch, J. Campbell, E. E. Knaus, and S. Silhankova, *J. Heterocycl. Chem.*, **9**, 1367 (1972).
538. J. E. Litster and H. Tieckelmann, *J. Amer. Chem. Soc.*, **90**, 4361 (1968).

539. T. Naito and R. Dohmori, *Chem. Pharm. Bull.* (Tokyo), **3**, 38 (1955); *Chem. Abstr.*, **50**, 1647h (1956).

540. T. Naito, R. Dohmori, and T. Kotake, *Chem. Pharm. Bull.* (Tokyo), **12**, 588 (1964); *Chem. Abstr.*, **61**, 4305f (1964).

541. R. Dohmori, *Chem. Pharm. Bull.* (Tokyo), **12**, 588 (1964); *Chem. Abstr.*, **61**, 4306a (1964).

542. T. Naito and T. Kotake, Japanese Patent 428, 1959; *Chem. Abstr.*, **54**, 6763f (1960).

543. R. Dohmori, *Chem. Pharm. Bull.* (Tokyo), **12**, 601 (1964); *Chem. Abstr.*, **61**, 4178g (1964).

544. S. Pietra and G. Tacconi, *Farmaco, Ed. Sci.*, **11**, 741 (1964); *Chem. Abstr.*, **62**, 1637e (1965).

545. A. H. Kelly and J. Parrick, *J. Chem. Soc.*, *C*, 303 (1970).

546. K. Takeda, K. Hamamoto, and H. Tone, *J. Pharm. Soc. Jap.*, **72**, 1427 (1952); *Chem. Abstr.*, **47**, 8071 (1953).

547. F. Ramirez and P. W. von Ostwalden, *J. Amer. Chem. Soc.*, **81**, 156 (1959).

548. S. Kajihara, *J. Chem. Soc. Jap.*, **85**, 672 (1964); *Chem. Abstr.*, **62**, 14624e (1965).

549. S. Kajihara, *J. Chem. Soc. Jap.*, **86**, 93 (1965); *Chem. Abstr.*, **65**, 16936e (1966).

550. K. Hamamoto and S. Kajiwara, Japanese Patent 26,730, 1964; *Chem. Abstr.*, **62**, 14638e (1965).

551. (a) P. A. de Villiers and H. J. den Hertog, *Rev. Trav. Chim. Pays-Bas*, **75**, 1303 (1956); (b) P. A. de Villiers and H. J. den Hertog, *Rev. Trav. Chim. Pays-Bas*, **76**, 647 (1957).

552. E. A. Mailey and L. R. Ocone, *J. Org. Chem.*, **33**, 3343 (1968).

553. (a) R. A. Abramovitch and G. M. Singer, *J. Amer. Chem. Soc.*, **91**, 5672 (1969); U.S. Patent 3,624,096, 1971; (b) G. M. Singer, Ph.D. Dissertation, University of Saskatchewan, 1970.

554. (a) F. F. Gadallah, M.Sc. Thesis, University of Saskatchewan, 1967; (b) R. A. Abramovitch and P. Tomasik, unpublished results.

555. (a) R. A. Abramovitch and R. B. Rogers, *Tetrahedron Lett.*, 1951 (1971); (b) W. E. Parham and K. B. Sloan, *Tetrahedron Lett.*, 1947 (1971).

556. R. Huisgen, *Angew. Chem.*, **75**, 628, 742 (1963).

557. J. Burkus, *J. Org. Chem.*, **27**, 474 (1962).

558. J. J. Monagle, *J. Org. Chem.*, **27**, 3851 (1962).

559. B. Umezawa, *Chem. Pharm. Bull.* (Tokyo), **8**, 693 (1960); *Chem. Abstr.*, **55**, 18724g (1961).

560. R. W. Hoffman, "Dehydrobenzene and Cycloalkynes," Academic, New York, 1967.

561. H. J. den Hertog and H. C. van der Plas, *Adv. Heterocycl. Chem.*, **4**, 121 (1965).

562. R. J. Martens and H. J. den Hertog, *Rec. Trav. Chim. Pays-Bas*, **83**, 621 (1964).

563. T. Kato, T. Niitsuma, and N. Kusaka, *J. Pharm. Soc. Jap.*, **84**, 432 (1964); *Chem. Abstr.*, **61**, 4171d (1964).

564. T. Kauffman and R. Wirthwein, *Angew. Chem. Internat. Edit.*, **3**, 806 (1964).

565. R. Adams and S. Miyano, *J. Amer. Chem. Soc.*, **76**, 3168 (1954).

566. V. Boekelheide and R. J. Windgassen, Jr., *J. Amer. Chem. Soc.*, **81**, 1456 (1959).

567. E. C. Taylor, A. J. Crovetti, and N. E. Boyer, *J. Amer. Chem. Soc.*, **79**, 3549 (1957).

568. E. D. Parker and A. Furst, *J. Org. Chem.*, **23**, 201 (1958).

569. W. Herz and D. R. K. Murty, *J. Org. Chem.*, **26**, 418 (1961).

570. K. Ramaiah and V. R. Srinivasan, *Indian J. Chem.*, **1**, 351 (1963).

571. S. Minami, Japanese Patent 10,987, 1967; *Chem. Abstr.*, **68**, 21838y (1968).

572. A. Fujita, M. Nakata, S. Minami, and H. Takamatsu, *J. Pharm. Soc. Jap.*, **86**, 1014 (1966); *Chem. Abstr.*, **66**, 75885p (1967).

573. S. Minami, A. Fujita, and J. Matsumoto, Japanese Patent 28,591, 1965; *Chem. Abstr.*, **64**, 9690h (1966).

574. M. Hooper, D. A. Patterson, and D. G. Wibberley, *J. Pharm. Pharmacol.*, **17**, 734 (1965).

575. D. R. Osborne and R. Levine, *J. Heterocycl. Chem.*, **1**, 138 (1964).

576. J. P. Schaefer, K. S. Kulkarni, R. Costin, J. Higgins, and L. M. Honig, *J. Heterocycl. Chem.*, **7**, 607 (1970).

577. T. Koenig and J. S. Wiezorek, *J. Org. Chem.*, **33**, 1530 (1968).
578. F. Kronke and K. F. Gross, *Chem. Ber.*, **92**, 22 (1959).
579. F. E. Cislack, U.S. Patent 2,989,534, 1961; *Chem. Abstr.*, **56**, 3466e (1962).
580. T. Kato and Y. Goto, *Chem. Pharm. Bull.* (Tokyo), **11**, 461 (1963); *Chem. Abstr.*, **59**, 7472h (1963).
581. T. Teshigawava, Japanese Patent 22,844, 1964; *Chem. Abstr.*, **62**, 10418f (1965).
582. W. S. Chilton and A. K. Butler, *J. Org. Chem.*, **32**, 1270 (1967).
583. L. Mao-Chin and C. Sae-Lee, *Acta Chim. Sinica*, **31**, 30 (1965); cf., Ref. 386.
584. M. Liu and T. Chu, *Hua Hsueh Hsueh Pao*, **31**, 437 (1965); *Chem. Abstr.*, **64**, 8128h (1966).
585. D. Jerchel, J. Heider, and H. Wagner, *Ann. Chem.*, **613**, 153 (1958).
586. J. Suszko and M. Szafran, *Rocz. Chem.*, **38**, 1793 (1964).
587. E. V. Brown, *J. Amer. Chem. Soc.*, **79**, 3565 (1957).
588. T. Tshil, *J. Pharm. Soc. Jap.*, **72**, 1315 (1952); *Chem. Abstr.*, **47**, 12386e (1953).
589. E. Ochiai and H. Mitarashi, *Chem. Pharm. Bull.* (Tokyo), **11**, 1084 (1963); *Chem. Abstr.*, **59**, 12755a (1963).
590. C. Kaneko, S. Yamada, I. Yokoe, N. Hata, and Y. Ubukata, *Tetrahedron Lett.*, 4729 (1966).
591. E. Ochiai and H. Mitarashi, *Itsuu Kenkyusho Nempo*, **13**, 19 (1963); *Chem. Abstr.*, **60**, 5446g (1964).
592. C. Kaneko, Sachiko, and I. Yokoe, *Chem. Pharm. Bull.* (Tokyo), **15**, 356 (1967); *Chem. Abstr.*, **67**, 64212m (1967).
593. G. N. Walker, M. A. Moore, and B. N. Weaver, *J. Org. Chem.*, **26**, 2740 (1961).
594. T. Kato and H. Hayashi, *J. Pharm. Soc. Jap.*, **83**, 352 (1963); *Chem. Abstr.*, **59**, 7473c (1963).
595. M. Hamana, S. Saeki, Y. Hatano, and M. Nagakura, *J. Pharm. Soc. Jap.*, **83**, 348 (1963); *Chem. Abstr.*, **59**, 9976c (1963).
596. T. Wieland and H. Biener, *Chem. Ber.*, **96**, 266 (1963).
597. M. Freifelder and G. R. Stone, *J. Org. Chem.*, **26**, 3805 (1961).
598. E. C. Taylor and A. J. Crovetti, *J. Org. Chem.*, **25**, 850 (1960).
599. (a) J. Murto and L. Kaariainen, *Suomen Kemistilehti*, B, **39**, 40 (1966); *Chem. Abstr.*, **64**, 19365g (1966); (b) J. I. Cadogan, D. J. Sears, and D. M. Smith, *J. Chem. Soc.*, C, 1314 (1969).
600. Z. Talik, *Bull. Acad. Polon. Sci.*, *Ser. Sci. Chem.*, **9**, 561 (1961); *Chem. Abstr.*, **60**, 2884d (1964).
601. R. M. Johnson, *J. Chem. Soc.*, B, 1063 (1966).
602. R. A. Abramovitch, F. Helmer, and M. Liveris, *J. Chem. Soc.*, B, 492 (1968).
603. M. Hamana and M. Yamazaki, *J. Pharm. Soc. Jap.*, **81**, 612 (1961); *Chem. Abstr.*, **55**, 24742g (1961).
604. F. J. Dinan and H. Tieckelman, *J. Org. Chem.*, **29**, 1650 (1964).
605. T. Talik and Z. Talik, *Bull. Acad. Pol. Sci.*, *Ser. Sci. Chim.*, **16**, 1 (1968); *Chem. Abstr.*, **69**, 59059d (1968).
606. T. Talik and Z. Talik, *Rocz. Chem.*, **40**, 1675 (1966).
607. S. Minakami and E. Hirai, Japanese Patent 6,723,195, 1967; *Chem. Abstr.*, **69**, 35965v (1968).
608. E. C. Taylor and J. S. Driscoll, *J. Org. Chem.*, **26**, 3001 (1961).
609. Y. Suzuki, *J. Pharm. Soc. Jap.*, **81**, 1151 (1961); *Chem. Abstr.*, **56**, 3447a (1962).
610. G. Wagner and H. Pischel, *Arch. Pharm.* (Weinheim), **296**, 576 (1963); *Chem. Abstr.*, **60**, 648d (1964).
611. L. H. Conover, A. R. English, and C. E. Larrabee, U.S. Patent 2,921,073, 1960; *Chem. Abstr.*, **54**, 8860h (1960).

612. R. E. McClure and D. A. Shermer, U.S. Patent 3,159,640, 1964; *Chem. Abstr.*, **62,** 7732e (1965).
613. F. E. Cislak, U.S. Patent 2,826,585, 1958; *Chem. Abstr.*, **52,** 11957f (1958).
614. N. V. Phillips, Belgian Patent 618,679, 1962; *Chem. Abstr.*, **59,** 9999b (1963).
615. J. Angulo and A. M. Municio, *Anales Real. Soc. Espan. Fis. Quim. (Madrid)*, **55B,** 527 (1959); *Chem. Abstr.*, **54,** 3292d (1960).
616. K. N. Dixit, S. D. Jolad, and S. Rajogopal, *Montash. Chem.*, **94,** 414 (1963).
617. J. W. Clark-Lewis and R. P. Singh, *J. Chem. Soc.*, 2379 (1962).
618. S. M. Roberts and H. Suschitzky, *J. Chem. Soc.*, C, 2844 (1968).
619. T. Talik, *Rocz. Chem.*, **37,** 495 (1963).
620. T. Talik and Z. Talik, *Rocz. Chem.*, **38,** 785 (1964).
621. G. M. Badger and R. P. Rao, *Aust. J. Chem.*, **18,** 379 (1965).
622. J. Bernstein and K. A. Losee, German Patent 1,224,744, 1966; *Chem. Abstr.*, **66,** 10853z (1967).
623. J. Rockett, U.S. Patent 2,932,647, 1960; *Chem. Abstr.*, **54,** 18557f (1960).
624. B. B. Brown, U.S. Patent 2,940,978, 1960; *Chem. Abstr.*, **54,** 21140c (1960).
625. R. E. McClure and A. Ross, *J. Org. Chem.*, **27,** 304 (1962).
626. Olin Mathieson Chem. Corp., British Patent 897,900 (1962); *Chem. Abstr.*, **57,** 15077b (1962).
627. J. Rockett and B. B. Brown, U.S. Patent 2,922,790, 1960; *Chem. Abstr.*, **54,** 8857g (1960).
628. J. Rockett, U.S. Patent 2,922,792, 1960; *Chem. Abstr.*, **54,** 8857bd (1960).
629. L. H. Conover, U.S. Patent 3,027,379, 1962; *Chem. Abstr.*, **56,** 15619h (1962).
630. C. E. Maxwell, III and P. N. Gordon, U.S. Patent 3,056,798, 1962; *Chem. Abstr.*, **58,** 5645d (1963).
631. H. L. Raulings and J. J. D'Amico, U.S. Patent 3,107,994, 1963; *Chem. Abstr.*, **60,** 6829a (1964).
632. J. J. D'Amico, U.S. Patent 3,155,671, 1964; *Chem. Abstr.*, **62,** 4015a (1965).
633. (a) A. R. Katritzky, *J. Chem. Soc.*, 191 (1957); (b) H. G. O. Becker, H. Boettcher, and H. Haufe, *J. Prakt. Chem.*, **312,** 433 (1970).
634. C. E. Lewis, A. P. Paul, S. M. Tsang, and J. J. Leavitt, U.S. Patent 3,051,697, 1962; *Chem. Abstr.*, **59,** 5291b (1963).
635. T. Kato, F. Hamaguchi, and T. Oiwa, *Pharm. Bull.* (Japan), **4,** 178 (1956); *Chem. Abstr.*, **51,** 7367e (1957).
636. N. Ikekawa, *Chem. Pharm. Bull.* (Tokyo), **6,** 263 (1958); *Chem. Abstr.*, **53,** 371d (1959).
637. A. Signor, E. Scoffone, L. Biondi, and S. Bezzi, *Gazz. Chim. Ital.*, **93,** 65 (1963).
638. S. Minakami and E. Hirai, Japanese Patent 6,925,582, 1969; *Chem. Abstr.*, **72,** 12596y, 1970.
639. S. Minakami and E. Hirai, Japanese Patent 6,917,386, 1969; *Chem. Abstr.*, **71,** 101724k (1969).
640. (a) E. Shaw, *J. Amer. Chem. Soc.*, **71,** 67 (1949); (b) U. Schollkopf and I. Hoppe, *Tetrahedron Lett.*, 4527 (1970); (c) L. A. Paquette, *J. Amer. Chem. Soc.*, **87,** 5186 (1965); (d) K Hoegerle, *Helv. Chim. Acta*, **41,** 548 (1958).
641. T. L. V. Ulbricht, *J. Chem. Soc.*, 3345 (1961).
642. C. Harukawa and K. Konishi, Japanese Patent 7,976, 1962; *Chem. Abstr.*, **59,** 5141g (1963).
643. S. Minakami and E. Hirai, Japanese Patent 7,006,267, 1970; *Chem. Abstr.*, **72,** 132537b (1970).
644. S. Umio, T. Kishimoto, Japanese Patent 6,804,266, 1968; *Chem. Abstr.*, **70,** 19942a (1969).
645. J. Angulo and A. M. Municio, *Anales Real Soc. Espan Fiz Quim. (Madrid)*, **56B,** 395 (1960); *Chem. Abstr.*, **55,** 7409e (1961).

646. S. Umio, T. Kishimoto, and I. Ueda, Japanese Patent 7,003,779, 1970; *Chem. Abstr.*, **72**, 121373u (1970).

647. S. Minakami and E. Hirai, Japanese Patent 6,723,351, 1967; *Chem. Abstr.*, **69**, 35956t (1968).

648. M. Eckstein, M. Gorczyca, A. Kocwa, and A. Zejc, *Dissertations Pharm.*, **9**, 197 (1957); *Chem. Abstr.*, **52**, 6337i (1958).

649. S. Minakami and E. Hirai, Japanese Patent 6,723,196, 1967; *Chem. Abstr.*, **69**, 35968y (1968).

650. C. Leibovici, Thesis, Bordeaux, 1967.

651. M. F. Schultz-Tissot, Thesis, Mulhouse, 1970.

652. E. M. Kosower, *J. Amer. Chem. Soc.*, **80**, 325 (1958).

653. G. G. Spence, E. C. Taylor, and O. Buchardt, *Chem. Rev.*, **70**, 231 (1970).

654. J. Streith and C. Sigwalt, *Tetrahedron Lett.*, 1347 (1966).

655. J. Streith, B. Danner, and C. Sigwalt, *Chem. Comm.*, 979 (1967).

656. J. Streith and C. Sigwalt, *Bull. Soc. Chim. Fr.*, 1157 (1970).

657. J. Streith, H. K. Darrah, and M. Weil, *Tetrahedron Lett.*, 5555 (1966).

658. F. Bellamy, L. G. R. Barragan, and J. Streith, *Chem. Comm.*, 456 (1971).

659. A. Alkaitis and M. Calvin, *Chem. Comm.*, 292 (1968).

660. C. Kaneko, S. Yamada, and M. Ishikawa, "2nd Symposium of Heterocyclic Cpds., Abstracts," 1969, pp. 201–6.

661. P. L. Kumler and O. Buchardt, *Chem. Comm.*, 1321 (1968).

662. H. Hata, E. Okutsu, I. Tanaka, *Bull. Chem. Soc. Jap.*, **41**, 1769 (1968); *Chem. Abstr.*, **70**, 3032h (1969).

663. R. M. Johnson, *J. Chem. Soc.*, *B*, 15 (1967).

664. S. Kamiya, *Chem. Pharm. Bull.* (Tokyo), **10**, 471 (1962); *Chem. Abstr.*, **58**, 4545b (1963).

665. M. Ishikawa, C. Kaneko, I. Yokoe, and S. Yamada, *Tetrahedron*, **25**, 295 (1969).

666. R. A. Abramovitch and A. Silhankova, unpublished results.

667. D. Kaneko, I. Yokoe, and S. Yamada, *Tetrahedron Lett.*, 775 (1967).

CHAPTER V

Alkylpyridines and Arylpyridines

RONALD G. MICETICH

Raylo Chemicals Limited, Edmonton, Alberta, Canada

The format of this chapter has been changed from that in the 1961 edition to include new trends and developments in the chemistry of the alkyl- and arylpyridines. Although still consisting of six main divisions, the content of some of these divisions has been substantially altered. The di- and poly-pyridylalkanes are included with the alkyl- and aralkylpyridines, while the bi- and polypyridines are a part of the arylpyridines.

Since the publication of the 1961 edition of this series, *"Pyrroles and Pyridines"* by K. Schofield (1) (1967) and reviews on pyridine (2) (1968), the synthesis of alkylpyridines (3) (1961), the synthesis and utilization of alkyl-pyridines, bipyridyls, and vinylpyridines (4) (1969), and the manufacture, chemistry, and uses of 5-ethyl-2-methylpyridine (MEP) and 2-methyl-5-vinylpyridine (5) (1968) have appeared. Reviews on other aspects of pyridine chemistry have been published and are referred to in the appropriate sections.

I. Alkyl- and Aralkylpyridines

1. Synthesis from Compounds Not Containing a Pyridine Ring

The synthesis of alkylpyridines utilizing the Hantzsch reaction and liquid and vapor phase amination of acetylenes, aldehydes, ketones, and alcohols is covered extensively in Chapter II (see also Ref. 3). In this section recent developments of particular interest to alkylpyridines are considered.

A five-stage process for the preparation of 3- and 3,5-alkyl substituted pyridines starting with diethyl acetonedicarboxylate was reported (6). Using this approach 3,5-lutidine (**V-7**; $R = R' = CH_3$) was obtained in 43 % overall yield.

The reaction of various 2-ethoxyadamantyl-substituted-3,4-dihydro-2*H*-pyrans with water and ammonia at 360° over a Pt–Al_2O_3 catalyst gave 2-, 3-, and 4-(1-adamantyl)pyridines (**V-8**) and 3-(1-adamantylmethyl)pyridine (**V-9**) *via* intermediate diketones, such as 1-(penta-1,5-dion-1-yl)adamantane (7).

The Diels-Alder reaction of nitriles with dienes is reported to form pyridine derivatives in low yields (see Chapter II). Recent work has shown that the use of tetraphenylcyclopentadienone as the dienophile produces high yields of arylpyridines (8) (see Section IV.1.D). Jaworski and Kwiatkowski found that α-pyrones (**V-10**) reacted with nitriles to produce the pyridine-3-carboxylic esters, which could be hydrolyzed and decarboxylated in high yields (9).

The synthesis of pyridines based on the reaction of oxazoles (**V-11**) with dienophiles (**V-12**) provides a convenient route to biologically important compounds such as Vitamin B_6 and its analogues. The scope and mechanism of this reaction has been recently reviewed (10).

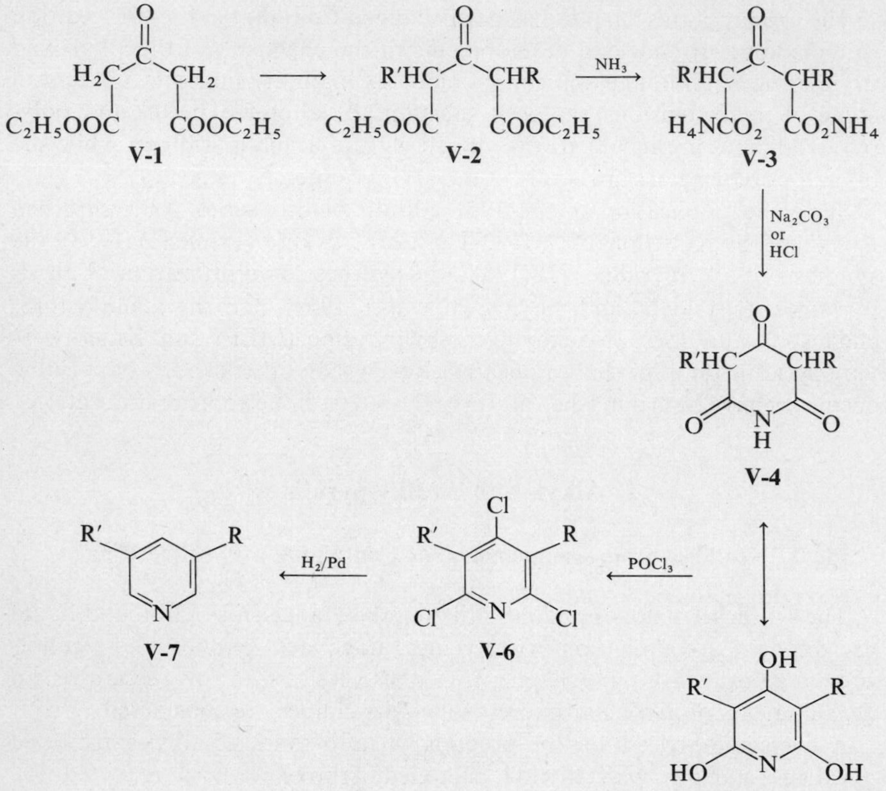

NH₃
Na₂CO₃
or
HCl
POCl₃
H₂/Pd

V-1 V-2 V-3 V-4 V-5 V-6 V-7

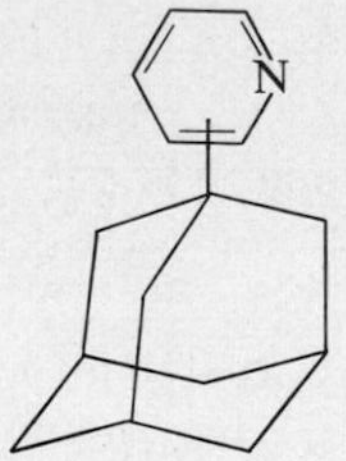

V-8

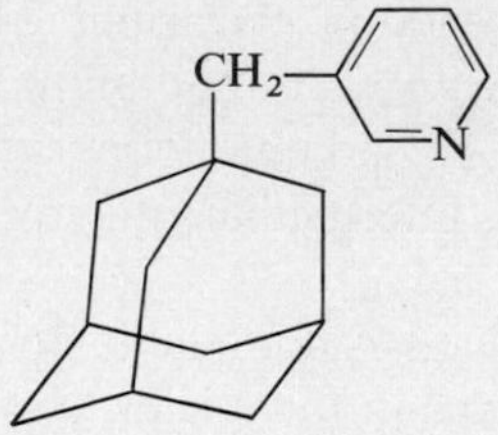

V-9

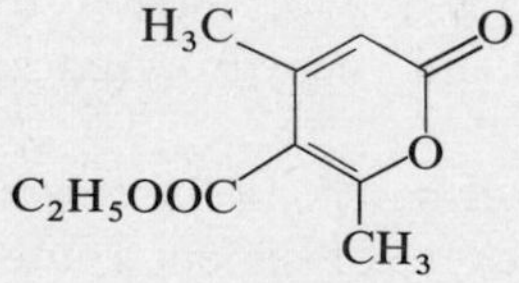

V-10

V-11   V-12   V-13

The bicyclic adduct **V-13** undergoes a facile acid-catalyzed aromatization producing pyridines. Although 3-hydroxypyridines (**V-14**) are the usual products of this reaction, it is possible for elimination of water to occur, forming **V-15**.

V-15 $\xleftarrow{-H_2O}$ **V-13** $\longrightarrow$ V-14

This latter process, the Kondrat'eva Reaction, occurs when R^3 is a group that is not easily eliminated as an anion, such as an alkyl group. The use of acrylic acid as the dienophile produces isonicotinic acid analogues (**V-15**) in which R^1 and R^3 are alkyl groups, R^2 is an aryl or alkyl group, R^4 is COOH, and R^5 is H (11). This method is potentially useful for the preparation of various specifically substituted aryl- and alkylpyridines, since the carboxylic acids obtained may be decarboxylated or converted to alkyl groups by standard methods.

Thiazoles may replace oxazoles as the diene component in these reactions. In this case aromatization is effected by heating with concentrated hydrochloric acid, when hydrogen sulfide is evolved (12).

The hydroformylation of oximes in the presence of a cobalt catalyst produces pyridines in low yields. Thus, heating propionaldoxime and $Co_2(CO)_8$ in benzene at 140° under 120 atm produced 8% 2-ethyl-3,5-dimethylpyridine. Butyraldoxime gave 24% of 3,5-diethyl-2-propylpyridine, and 28% of 2,4,6-trimethylpyridine was obtained from acetoxime (13).

Mohan found that heating certain primary aliphatic nitro compounds (**V-16**) with carbon monoxide under pressure in ethanol or benzene as solvent with Pd or Rh as catalyst in the presence of $FeCl_3$ gave pyridines **V-17** (14). All components were necessary for this reaction.

$$RCH_2CH_2NO_2 + CO \longrightarrow$$

V-16

V-17

Treating 4-(2-oxocyclopentyl)-2-butanone (**V-18**, $n = 1$) with hydroxyl-amine is a convenient route to 6-methyl-2,3-cyclopentenopyridine (**V-19**, $n = 1$). Other cycloalkenopyridines (**V-19**, $n = 2$ and 3) have been prepared by the same route (15).

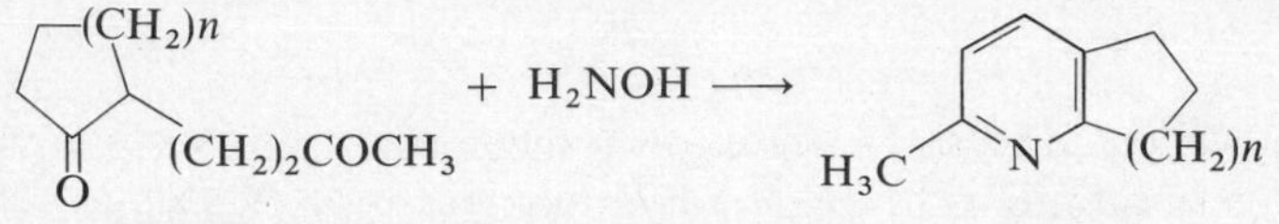

$$\text{V-18} \quad + \text{ H}_2\text{NOH} \longrightarrow \quad \text{V-19}$$

V-18   **V-19**

β-Aminocrotonic esters reacted with α-formylcycloheptanone to give 5-carbethoxy-6-methyl-2,3-cycloheptenopyridine, which was hydrolyzed and decarboxylated in the presence of soda lime to 6-methyl-2,3-cyclohepteno-pyridine (**V-20**) (15).

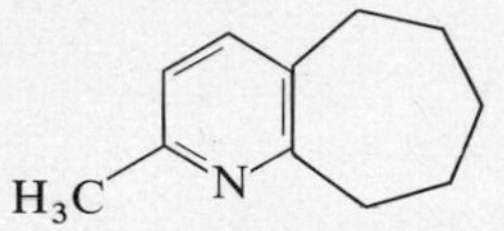

V-20

Lochte and Pittman have described a convenient synthesis of 2,3-cyclo-pentenopyridines (**V-24**) from enamines (**V-21**) (16, 17).

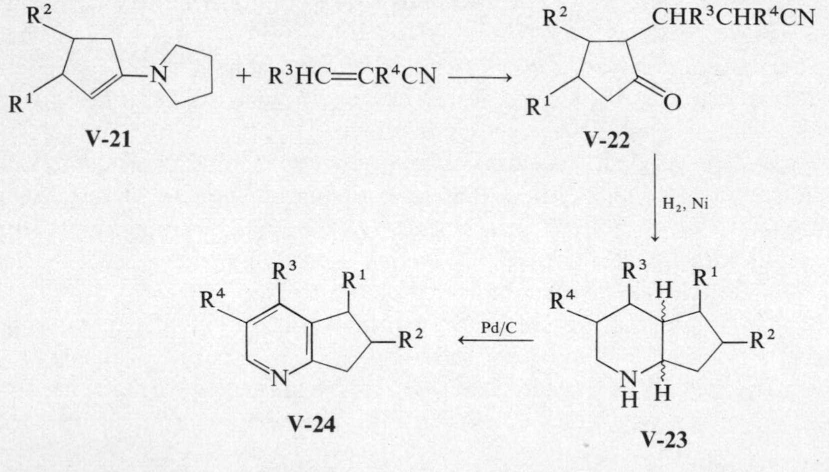

V-21 $+ \text{ R}^3\text{HC}{=}\text{CR}^4\text{CN} \longrightarrow$ **V-22**

$\downarrow$ H$_2$, Ni

V-24 $\xleftarrow{\text{Pd/C}}$ **V-23**

The condensation of 3-aminoacrolein (**V-25**, $R^3 = H$) with compounds containing a —COCH$_2$— group (**V-26**) provides a convenient one step synthesis of 2,3-disubstituted pyridines including 2,3-cycloalkenopyridines (**V-27**). The use of substituted 3-aminoacroleins (**V-25**; $R^3 = CH_3$) produces 2,3,5-trisubstituted pyridines (**V-27**; $R^3 = CH_3$) (18–20).

The isoxazole ring may be considered as a masked β-amino-α,β-unsaturated keto-function, since although stable to various chemical reaction conditions it can be cleaved (at the —N—O— bond) by reduction. Such cleavage of a 4-(3-oxoalkyl)isoxazole **(V-28)** results in the formation of a pyridine **(V-31)** (21, 22).

Adachi found that 2-methyl-3,4-trisubstituted isoxazolium salts **(V-32)** reacted with cyclic enamines **(V-33)** to give 1-methyl-2,3-cycloalkeno-pyridinium salts **(V-34)**, along with benzoylacetamides **(V-35)**. Three different mechanisms have been proposed to account for these products, namely the formation of a keteneimine intermediate, nucleophilic attack at C-5 of the isoxazole ring followed by ring opening and formation of the pyridine ring, or a Diels-Alder type reaction (23).

Cyclic enamines also react with 1,2,4-triazines (**V-36**) by a Diels-Alder cycloaddition to produce pyridines (**V-39**) (24). Yields of **V-39** ($n = 1$) of

over 90% were obtained. The same reaction was observed by using cyclopentene with *p*-benzoquinone in place of the enamine **V-33**. In some instances, the presence of **V-38** was shown by NMR spectroscopic analysis.

The ring expansion of pyrroles (**V-40**) with halocarbenes to form pyridines gave low yields of products and hence appeared to be of academic interest (2) (see Chapter II). These reactions usually lead to two products, a ring expanded pyridine (**V-42**) and a 2*H*-pyrrole (**V-45**). Jones and Rees have presented strong evidence that these two components are produced by different pathways, **V-42** being formed from the cycloaddition of dichlorocarbene to the *neutral* pyrrole molecule followed by aromatization; while electrophilic attack of the carbene on the pyrrole *anion* produces **V-45**. In keeping with this proposal, almost quantitative yields of **V-42** result when the dichlorocarbene is generated by heating sodium trichloroacetate in a *neutral* aprotic solvent (25).

The 2-dichloromethyl-2*H*-pyrroles (**V-45**) underwent base-catalyzed rearrangements in ethanol to form 2-ethoxymethyl-5-methylpyridine (**V-46**; R = H), 2-ethoxymethyl-3,4,5-trimethylpyridine (**V-46**; R = CH$_3$), 4-ethoxymethyl-2,3,5-trimethylpyridine (**V-47**) and *trans*-1,2-di-(5-methyl-2-pyridyl)ethylene (**V-48**). The mechanism of these reactions was discussed (26, 27).

A recent application of this ring expansion reaction of pyrroles is the synthesis of the interesting series of metacyclophanes (**V-49**; n = 6, 4, 2, and 1) from the appropriate indoles. These compounds exist as *dl* pairs since the chlorine prevents the methylene bridge from flipping over (28) (see Section I.6).

A versatile method of synthesizing specifically substituted pyridines is that described by Chumakov and Shetstynk. Pyrans obtained by Diels-Alder reactions of vinyl ethers with α,β-unsaturated aldehydes and ketones are converted to dialkylpyridines of specific orientation *via* diketones. Prepared by this route were 5-*n*-butyl-2-methylpyridine, 3-*n*-butyl-4-methylpyridine, and 5-*n*-hexyl-2-methylpyridine (29).

The action of ammonia on pyrylium salts is another convenient route to pyridines, including the cycloalkenopyridines. The pyrylium salts are readily obtained by treating an olefin or alcohol with an acid anhydride and perchloric acid (30, 31).

2. Preparation from Natural Sources and from Compounds Containing the Pyridine Ring

A. Pyrolysis and Degradation of Natural Products

Many pyridine bases have been isolated from various natural products such as sapropel tar (32–35), nicotine (36), shale oil (37), soybean cake (38), and by the pyrolysis of bone oil (39). New pyridines have also been identified in coal tar bases (40–43).

Table V-1 lists the pyridines thus identified along with the sources. Many of these pyridines were characterized by their physical constants and/or derivatives. These data, when new, are listed in the appropriate table.

TABLE V-1. Alkylpyridines Identified from Natural Products

Alkylpyridine	Source	Ref.
2-Methyl	Soybean cake	38
	Manchurian shale oil	37
	Sapropel tar	32, 33
	Coal tar	41
3-Methyl	Nicotine	36
	Bone oil	39
	Manchurian shale oil	37
	Sapropel tar	32, 33
	Coal tar	41
4-Methyl	Bone oil	39
	Manchurian shale oil	37
	Sapropel tar	32, 33
	Coal tar	41
2-Ethyl	Manchurian shale oil	37
	Sapropel tar	33
3-Ethyl	Nicotine	36
	Sapropel tar	33, 34
	Coal tar	41
4-Ethyl	Sapropel tar	33, 34
	Coal tar	41
3-Isopropyl	Sapropel tar	33, 35
3-Vinyl	Nicotine	36
3-(Buta-1,3-dien-1-yl)	Nicotine	36

Table V-1 (*Continued*)

Alkylpyridine	Source	Ref.
2,3-Dimethyl	Manchurian shale oil	37
	Sapropel tar	32–34
	Coal tar	41
2,4-Dimethyl	Soybean cake	38
	Manchurian shale oil	37
	Sapropel tar	32–34
	Coal tar	41
2,5-Dimethyl	Manchurian shale oil	37
	Sapropel tar	32–34
	Coal tar	41
2,6-Dimethyl	Bone oil	39
	Manchurian shale oil	37
	Sapropel tar	32, 33
	Coal tar	41
3,4-Dimethyl	Manchurian shale oil	37
	Sapropel tar	32, 33
	Coal tar	41
3,5-Dimethyl	Manchurian shale oil	37
	Sapropel tar	32–34
	Coal tar	41
4-Ethyl-2-methyl	Manchurian shale oil	37
	Sapropel tar	32–34
	Coal tar	41
5-Ethyl-2-methyl	Sapropel tar	32–34
6-Ethyl-2-methyl	Manchurian shale oil	37
	Sapropel tar	32, 33
	Coal tar	41
2-Ethyl-4-methyl	Sapropel tar	33, 35
	Coal tar	41
3-Ethyl-4-methyl	Sapropel tar	33, 35
2,3,4-Trimethyl	Sapropel tar	33
2,3,5-Trimethyl	Manchurian shale oil	37
	Sapropel tar	32–34
	Coal tar	41
2,3,6-Trimethyl	Manchurian shale oil	37
	Sapropel tar	32–34
	Coal tar	41
2,4,5-Trimethyl	Manchurian shale oil	37
	Coal tar	41
2,4,6-Trimethyl	Manchurian shale oil	37
	Sapropel tar	32–34
	Coal tar	41
3,4,5-Trimethyl	Coal tar	42
2,6-Dimethyl-4-ethyl	Coal tar	41
2,3,4,5-Tetramethyl	Manchurian shale oil	37
2,3,4,6-Tetramethyl	Manchurian shale oil	37
2,3,5,6-Tetramethyl	Sapropel tar	32, 33
3,4-Cyclopenteno	Coal tar	42
5-Methyl-2,3-cyclopenteno	Coal tar	43
6-Methyl-2,3-cyclopenteno	Coal tar	42

B. Reduction Methods

a. HALOGEN COMPOUNDS. A convenient route to 3,5-dialkylpyridines involves treating an α,α'-dialkylglutarimide with PCl_5 to form the 3,5-dialkyl-2,6-dichloropyridine, which is then reduced by hydrogen in the presence of palladium (44) (see also Section I).

Chloromethylpyridines, produced by the action of thionyl chloride on hydroxymethylpyridines, can also be reduced to methylpyridines using palladium on carbon in the presence of potassium acetate. This general method was used to prepare 3,4-dimethylpyridine in high yields (45). Since pyridine carboxylic esters are reduced by lithium aluminium hydride to the hydroxymethylpyridine, this approach can be employed to convert esters to the methyl derivatives.

Wilson and Harris found that hydrogenation of 3-(β-chloroethyl)-4,6-dichloro-2-methylpyridine (V-50), with palladium on barium sulfate in methanol gave 3-(β-chloroethyl)-2-methylpyridine (V-51), a reaction which they found surprising since they expected the β-chloroethyl group to be attacked (46).

A similar reaction had previously been described by Stevens and co-workers who showed that the nuclear chlorines were preferentially removed from 3-(β-chloromethyl)-2,6-dichloro-4-methylpyridine (V-52), the product being 3-(β-chloroethyl)-4-methylpyridine (V-53) (see Chapter V, 1st ed., p. 159).

The labile nature of the aromatic chlorine to hydrogenation is also illustrated by the fact that 3-cyano-2,6-dichloro-4-methylpyridine (V-54) is reduced to 3-cyano-4-methylpyridine (V-55) (87%) using palladium chloride with sodium acetate in methanol (47), and that ethyl 4,6-dimethylnicotinate was obtained by the hydrogenation of 2-chloro-3-cyano-4,6-dimethyl-pyridine followed by hydrolysis and esterification (48).

CH₃ ... CN ... PdCl₂ / H₂ ... CH₃ ... CN

Cl ... N ... Cl ... N

V-54 ... V-55

b. HYDROXY COMPOUNDS. In general the methods used for converting hydroxy compounds to the alkylpyridines involves the preparation of the halogeno derivatives followed by reduction. However, Holfling and Reckling converted 2-(β-hydroxyethyl)pyridine to 2-ethylpyridine (73%) by hydrogenation over an alumina-nickel oxide catalyst at 320 to 340° (49).

The catalytic reduction of various pyridoxine derivatives was studied by Naito and Veno (50). Lukeš and Pergal prepared 2,6-diethyl-, 2,6-dipropyl-, 2,6-diisopropyl-, and 2,6-di-(2-pentyl)pyridine by reducing the corresponding carbinol with HI and P. The carbinols were prepared by treating 2,6-diacylpyridines with Grignard reagents (51).

Pyridinecarbinols can be reduced electrolytically (see Section I.2.B.c and g).

c. CARBOXYLIC ACIDS. Electrolytic reduction of pyridine carboxylic acids in sulfuric acid at lead or mercury cathodes gave mixtures of picolines and tetra- and hexahydropyridines (52–54) (see also Chapter V, 1st ed. p. 161). Under the same conditions the pyridylcarbinols gave similar products (54). Lund found that 2- and 4-pyridinecarboxamides gave, on electrolysis under controlled conditions of pH and temperature, about 50% yields of the respective aldehydes, which were trapped as the hydrate (55).

d. CARBONYL COMPOUNDS. The Wolff-Kishner reduction has been employed for the preparation of alkylpyridines (51, 56, 57), 1,4-dipyridyl-butanes (58), phenethylpyridines (59), and [n](2,5)pyridinophanes (60) (V-57). The [n](2,5)-pyridinophanones (V-56) required for the preparation of V-57 were obtained by the acid-catalyzed cyclization of bis-(β-aminovinyl)-diketones (60).

(CH₂)n ... C=O ... Wolff-Kishner ... (CH₂)n ... CH₂

N ... N

V-56 ... V-57

Ethyl 5-acetylnicotinate required the Lock modification (61) of the Wolff-Kishner method for reduction to ethyl 5-ethylnicotinate (62). The Lock modification was also effective for the reduction of various 4-acyl-pyridines to the 4-alkylpyridines (63).

e. OLEFINIC COMPOUNDS. The hydrogenation of alkenylpyridines employing Raney nickel was used to prepare 2-, 3- and 4-n-amylpyridines and 2-, 3-, and 4-(5-nonyl)pyridines (64), and a series of 2-alkylpyridines (65). The latter compounds were reduced further by sodium in ethanol to the 2-alkylpiperidines (65).

f. PYRIDINE-1-OXIDES. N-Oxidation modifies the properties of pyridines so that controlled reactions at specific sites are possible with the pyridine-1-oxides (see Chapter IV for more complete discussion). As Abramovitch and co-workers have shown, it is possible to C-2-alkylate certain pyridine-1-oxides via the lithio compounds, a reaction not possible with the pyridines (66). Since the pyridine-1-oxides can be easily deoxygenated (see Chapter IV), this approach is potentially useful for the preparation of specifically substituted pyridines.

Recent methods for the deoxygenation of pyridine-1-oxides have involved the use of sulfur dioxide (67), iron pentacarbonyl (68), benzaldehyde (69), electrolytic reduction (70), and dimethyl sulfoxide (71, 72). Although heating pyridine-1-oxide at 195° for 20 hr with dimethyl sulfoxide containing sulfuric acid produces essentially quantitative yields of pyridine (71), the reaction with alkylpyridine-1-oxides was more complex. 4-Benzylpyridine-1-oxide gave 4-benzylpyridine (29%), 4-benzoylpyridine (41%), and 1,2-diphenyl-1,2-di-(4-pyridyl)ethane (9%). The action of dimethylsulfoxide on 4-methyl-pyridine-1-oxide, 4-p-nitrobenzylpyridine-1-oxide and 4-isopropylpyridine-1-oxide was also studied. It was concluded that both ionic and radical species were produced, the former giving carbonyl and olefinic products while the latter gave dimeric compounds (72). The action of pyridine-1-oxide on trichloroacetic anhydride in acetonitrile gave a mixture containing 2- and 4-trichloromethylpyridine and 2- and 4-dichloromethylpyridine. Although radical or carbene mechanisms are not excluded, it was believed the rapid loss of carbon dioxide from the free trichloroacetate ion was the driving force in this reaction (73).

The reduction of pyridine-1-oxide, 4-methylpyridine-1-oxide, and 3-methylpyridine-1-oxide with sodium in ethanol, sodium borohydride, lithium aluminium hydride, sodium aluminium hydride, and by electrolysis, gave mixtures of the pyridine, piperidine, and 3-piperideine (74).

g. ELECTROLYTIC REDUCTION. The electrolytic reduction of pyridine carbinols and aldehydes gave the corresponding alkyl derivatives along with nuclear reduced products. The results are listed in Table V-2 (55).

TABLE V-2. Electrolytic Reduction of Oxygenated Pyridine Side-Chains (55)

Pyridine compound	Product	Yield (%)
Pyridine-2-aldehyde	2-Picoline	53
Pyridine-3-aldehyde	3-Picoline	4
Pyridine-4-aldehyde	—	0
2-Acetylpyridine	2-Ethylpyridine	3
3-Acetylpyridine	—	0
4-Acetylpyridine	4-Ethylpyridine	30
Pyridine-2-carbinol	2-Picoline	32
Pyridine-3-carbinol	3-Picoline	3.5
Pyridine-4-carbinol	4-Picoline	8.0
Pyridine-3-methylcarbinol	3-Ethylpyridine	3.5
Pyridine-3-dimethylcarbinol	—	0
Pyridine-4-dimethylcarbinol	4-Isopropylpyridine	37

1,4-Bis(pyridyl)butanes were obtained by the electrolytic reductive coupling of 2- and 4-vinylpyridines in solutions containing tetraethyl-ammonium p-toluenesulfonate and dimethylformamide. 2-Vinylpyridine gave 68.8% and 4-vinylpyridine gave 82% of the respective products. This method could also be applied to the higher alkenylpyridines (75, 76).

The electrolysis of a mixture of 3-acetylpyridine and acetophenone gave 2-phenyl-3-(3-pyridyl)-2,3-butanediol (V-58) which was converted to 3-methyl-2-(3-pyridyl)indene (V-59) (RRI 1391) on heating with concentrated hydrochloric acid (77).

V-58   V-59

C. Alkylation Methods

a. CATALYTIC ALKYLATIONS. The vapor phase reaction of pyridine with alcohols at 300 to 400° over catalysts such as activated alumina, alumino-silicate, or H_3BO_3-H_3PO_4 gave a mixture of the 3-alkyl and 3,5-di-alkylpyridines (78–81). Methanol and pyridine in the presence of a nickel–nickel oxide catalyst produced 2-methylpyridine and 2,4-, 2,5-, and 2,6-dimethylpyridines (82).

Methylation of 2-methylpyridine in the vapor phase with methanol or formaldehyde over an alumina catalyst resulted in attack at the C-2 methyl group with formation of 2-ethylpyridine (83, 84). A small amount of 2-vinylpyridine was also produced when formaldehyde was used. 4-Methylpyridine gave 4-ethylpyridine and 4-vinylpyridine with formaldehyde (84).

The catalytic alkylation of alkylpyridines with olefins in the presence of sodium was studied by Pines and co-workers. 3-Ethylpyridine with ethylene gave 3-*sec*-butylpyridine (**V-60**) as the primary product, which reacted further to form **V-61** (R = H and CH$_3$). These reactions were run with catalytic amounts of sodium in 3-ethylpyridine at 40 atm pressure of ethylene and at about 145°C and the products were analyzed by gas liquid chromatography. The effect of reaction time on product distribution was determined (85). This same reaction was applied to the 2- and 4-alkylpyridines and the

V-60

V-61

relative rates of ethylation of various alkylpyridines obtained. A carbanion chain reaction mechanism accounts for these alkylations (86, 87). Alkylations using fluoroölefins and potassium or cesium fluoride at 190° are efficient ways of preparing fluoroalkylpyridines. Thus chlorotrifluoroethylene and pentafluoropyridine gave 1-chloro-1-tetrafluoropyridyltetrafluoroethane (60%), while hexafluoropropene gave perfluoro-4-isopropylpyridine (90%) (88).

b. WIBAUT-ARENS ALKYLATION. The Wibaut-Arens procedure was used to prepare 3,4-diethylpyridine (55%) from 3-ethylpyridine, using iron powder (89). A study of various catalysts showed that iron was the most effective in preparing 4-ethylpyridine (82%) (90). The isolation of 4-acetyl-3,5-dimethylpyridine (**V-63**) (70%), the intermediate in the Wibaut-Arens reaction, in the reaction of 3,5-dimethylpyridine (**V-62**) with acetic anhydride and zinc, was explained as being due to the steric effect of the flanking methyl groups (91).

V-62   V-63

c. USING ORGANOMETALLIC DERIVATIVES. Interest in this area has continued. 2-Thienyllithium proved to be an excellent reagent for the lateral lithiation of 2- and 4-methylpyridines, yields of 98 % of 1,1-diphenyl-2-(2-pyridyl)ethanol (**V-64**) and 94 % of 1,1-diphenyl-2-(4-pyridyl)ethanol, respectively, being realized by the subsequent reaction with benzophenone (92, 93).

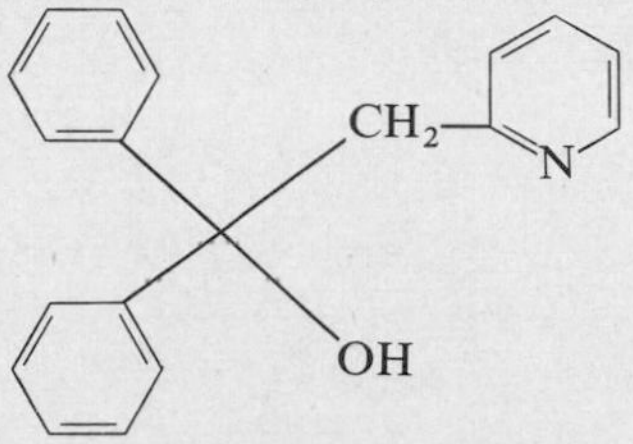

V-64

Jones, Russell, and Skidmore have investigated the action of phenyl-lithium on 2-methyl-, 2,4-dimethyl-, 2,6-dimethyl-, and 2,4,6-trimethyl-pyridine. All these compounds underwent lateral lithiation as shown by an estimate of the benzene recovered after hydrolysis of the product and a quantitative determination of the lithiated product by a counting technique on the nonaqueous product after hydrolysis with tritiated water. Only 2,4-dimethylpyridine showed evidence of concommitant azomethine addition since 5 % of 2,4-dimethyl-6-phenylpyridine was isolated from the reaction (94). These lithio compounds **V-65** with oxygen gave the 1,2-di-(2-pyridyl)-ethanes (**V-66**) and pyridyl-2-carbinols (**V-67**) (95). Treating 2-n-hexyl-pyridine with p-methoxyphenyllithium gave 6,7-di-(2-pyridyl)dodecane (**V-68**) along with 1-(2-pyridyl)hexanol and 2-p-anisyl-6-n-hexylpyridine (96).

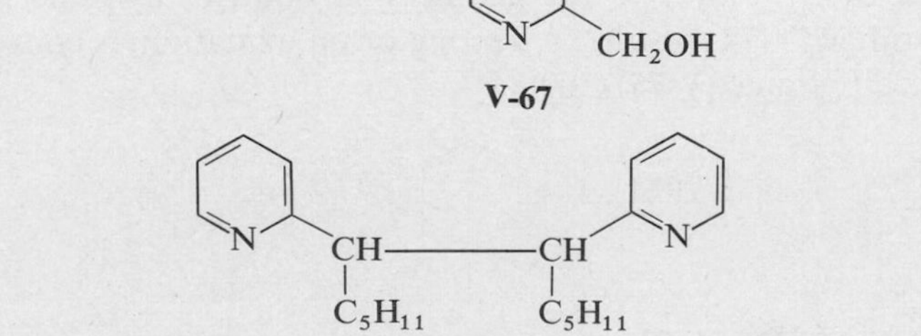

V-65   **V-66**

V-67

V-68

Pyridines with long alkyl side chains in the C-2 position were made by treating **V-65** with alkyl halides. Octyl bromides, for example, gave 2-nonylpyridine (38%) and 9-(2-pyridyl)heptadecane (2%) (97).

2-(Indol-3-ylmethyl)-4-methylpyridine **(V-69)**, (RRI 1286), useful as an analgesic, antitussive, and antiinflammatory agent was prepared by treating 2-lithiomethyl-4-methylpyridine with $BrCH_2CH(OC_2H_5)_2$, and the resulting product with phenylhydrazine (98).

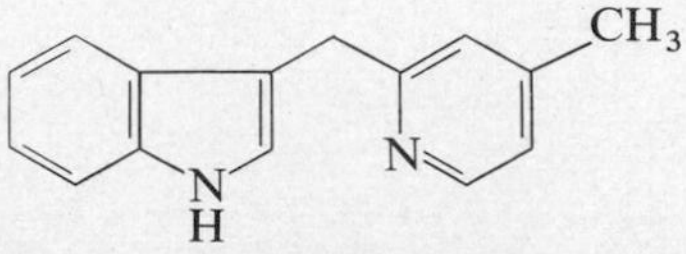

V-69

The preparation of 1-*p*-methoxyphenyl-2-(2-pyridyl)cyclohexane **(V-70)** and 1-(2-pyridyl)-2,3-diethyl-6-methoxyindane **(V-71)** illustrates another useful synthetic approach. Pyridyllithium was treated with the appropriate ketones and the resulting carbinols dehydrated. The olefins thus obtained

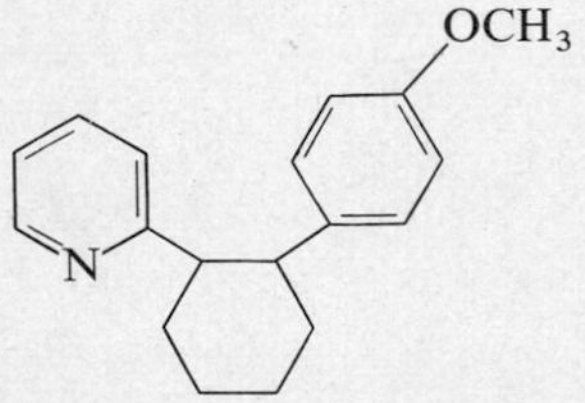

V-70

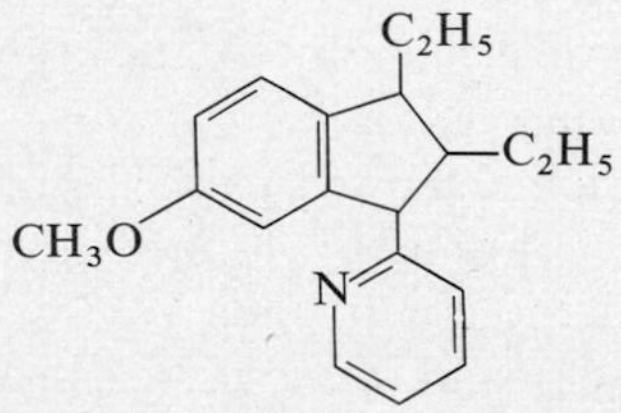

V-71

were hydrogenated. These compounds were tested for their hypocholestermic action (99).

Pyridine and 3-substituted pyridines undergo nucleophilic alkylation or arylation when treated with organometallic compounds, particularly the organolithio derivatives. The effect of substituents in the pyridine ring on this substitution and the mechanism of this reaction were reviewed recently (100). These reactions proceed by an addition-elimination mechanism. The lithium alkyl or aryl **(V-73)** adds to the azomethine bond of pyridine to form a dihydropyridine **(V-74)** which, on heating or on oxidation, is converted to a 2-substituted pyridine **(V-75)** (100).

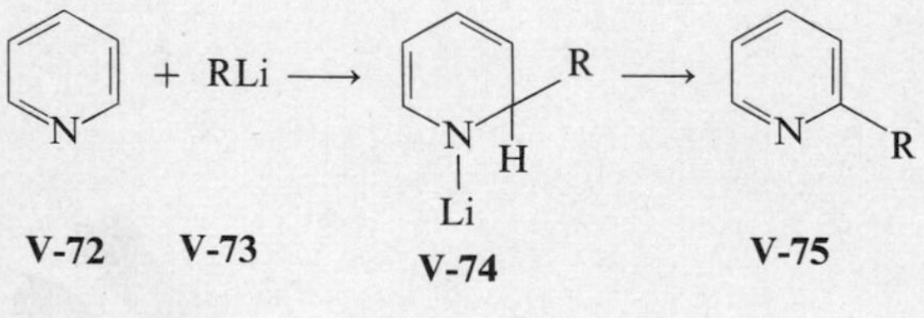

V-72   **V-73**   **V-74**   **V-75**

Abramovitch and co-workers obtained direct evidence for the existence of the intermediate σ-complex **(V-74)** by the isolation of 1,2,5,6-tetrahydro-3-methyl-2-o-tolylpyridine from the reaction of 3-methylpyridine and o-tolyllithium (101, 102). Fraenkel and Cooper presented NMR spectroscopic evidence for the structure of the butyllithium-pyridine adducts **(V-74**; $R = n\text{-}C_4H_9$) (103) and Giam and Stout succeeded in isolating the crystalline σ-complexes (104). Alkylation or arylation of these complexes gave 2,5-disubstituted pyridines **(V-76)** such as 2,5-diphenylpyridine ($R = R' = Ph$), 5-benzyl-2-phenylpyridine ($R = Ph$, $R' = CH_2Ph$), 5-methyl-2-phenylpyridine ($R = Ph$, $R' = CH_3$) and 2-n-butyl-5-methylpyridine ($R = n\text{-}C_4H_9$, $R' = CH_3$) (105).

$$\textbf{V-74} + \text{R'X} \longrightarrow$$

V-76

The alkylation of lithium tetrakis-(N-dihydropyridyl)aluminate, the product from the reaction of lithium aluminium hydride and excess pyridine, provided a convenient, high yield, essentially one step process for preparing 3-alkylpyridines **(V-76**; $R = H$). Giam and Abbott obtained 3-methylpyridine (86%), 3-ethylpyridine (89%), and 3-benzylpyridine (63%) by this method (106). Pyridine and n-butyllithium with tetramethylethylenediamine (TMEDA) followed by the addition of benzophenone gave 2-n-butylpyridine, 5-(2-n-butylpyridyl)diphenylmethanol, 2-(N-methyl-N-dimethylaminoethyl)-aminomethylpyridine, and benzhydrol (107). These results confirm prior observation of 5-substitution in pyridines (105, 852), and the interesting reduction of carbonyl compounds by a σ-complex first reported and studied by Abramovitch and co-workers (853, 854).

The reaction of 3-substituted pyridines with organolithium compounds has been studied fairly extensively. These reactions usually lead to a mixture of the 2,3- and 2,5-isomers. It was shown that the reaction of phenyl-, o-tolyl-, o-ethylphenyl-, and p-methoxyphenyllithium with 3-methylpyridine had little effect on the ratio of the 2,3- and 2,5-isomers (about 95:5) produced (102, 108). Methyllithium with 3-methylpyridine gave an 84.4:15.6 ratio while a 20.5:79.5 ratio resulted from the use of isopropyllithium. This result was ascribed in part to the polymeric nature of isopropyllithium (109). The weaker nucleophile benzyllithium with 3-methylpyridine gave 4-benzyl-3-methylpyridine (41%) and with pyridine gave 4-benzylpyridine (56%) (109).

The action of phenyllithium on 3-methyl- or -ethyl- or -isopropylpyridines resulted in the preferential formation of the 2,3-isomer, while bulky 3-substituents such as t-butyl or phenyl directed substitution to the 6-position The exclusive formation of 2,5-diphenylpyridine in the latter case was attributed to steric inhibition of coplanarity in the transition state preventing the formation of the 2,3-isomer (112). The phenylation of nicotine (V-77) ($R^1 = R^2 = R^3 = H$) resulted in a 1:1 ratio of the 2-phenyl- (V-77; $R^1 = Ph, R^2 = R^3 = H$) and 6-phenyl (V-77; $R^1 = R^2 = H, R^3 = Ph$) isomers (110, 112). Haglid found that when nicotine was methylated with

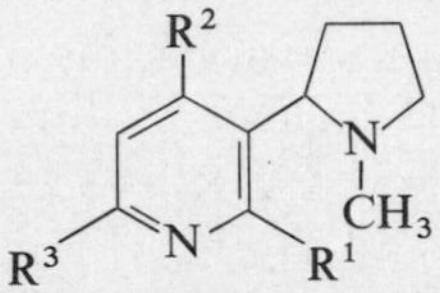

V-77

methyllithium the major product was the 6-methyl isomer (V-77; $R^1 = R^2 = H, R^3 = CH_3$) with some of the 4-methyl isomer (V-77; $R^1 = R^3 = H, R^2 = CH_3$); none of the 2-isomer was detected in this instance. The formation of the 4-isomer in this case was explained by the complexing of the methyllithium with the pyrrolidine nitrogen of nicotine facilitating attack at C-4 (113). Methyllithium reacted with N-(3-methylpyridyl)pyrrolidine (V-78) to form the 2-, 4-, and 6-methyl derivatives, the 4-isomer being present in low yields. The ratio of the isomers was dependent on the solvent (ether-toluene or THF) used (114).

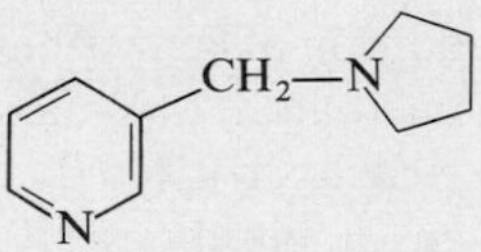

V-78

Sodium derivatives, made by treating alkylpyridines with sodamide in liquid ammonia, were converted to aralkylpyridines by heating with styrenes (115), or aralkyl halides (116) or heteroaralkyl halides (117). 1-(2-Thienyl)-2-(2-pyridyl)ethanes (V-79), useful as anthelmintics, were prepared by this

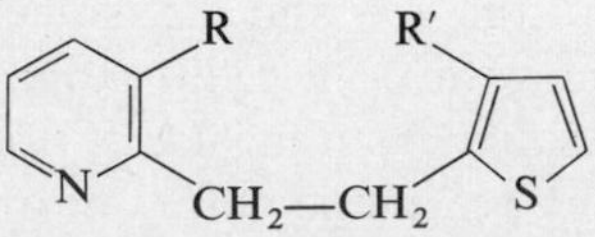

V-79

method (115). Dipyridylalkanes were obtained by using α,ω-dihalogeno-alkanes (118). Cesium compounds, made from the pyridylmercury derivatives, gave the 1-(pyridyl)-2-phenylethane with benzyl chloride (119).

Pyridine reacts with 1-chlorobutane in the presence of magnesium powder to form 4-*n*-butylpyridine (57%) with only traces of the 2-isomer; *n*-butyl-magnesium iodide however produced 2-*n*-butylpyridine (18%), free from the 4-isomer (120). 3-*Sec*-butyl-4,5-dimethylpyridine was prepared by treating 3-cyano-4,5-dimethylpyridine sequentially with methylmagnesium iodide and ethylmagnesium iodide and reducing the resulting *t*-carbinol with HI and red P (121). 4-Alkylpyridines resulted on heating 4-phenoxy-pyridines with alkylmagnesium halides at 150 to 200°C in a nitrogen stream (122). 1-(2-Pyridyl)ethylmagnesium bromide, obtained by the reaction of 2-ethylpyridine and alkylmagnesium bromide, with 2-(2-dimethylamino-ethyl)indan-1-one gave the alcohol which, on heating at 100° for 2 hr in aqueous acid, gave 2-(2-dimethylaminoethyl)-3-[1-(2-pyridyl)ethyl]indene (**V-80**) (123). Quaternary salts of pyridine-1-oxides (**V-81**) with Grignard

V-80

reagents form pyridines (**V-82**), the alkyl group of the Grignard reagent entering the C-2 position of the pyridine ring (124).

V-81   **V-82**

An interesting but rather complicated synthesis of 2,4-disubstituted pyridines (**V-87**) is that of Fraenkel and co-workers who found that Grignard reagents reacted with 4-substituted pyridines (**V-83**) in the presence of chloroformates to produce 2,4-disubstituted-1-alkoxycarbonyl-1,2-dihydro-pyridines (**V-85**). Treatment with butyllithium gave the 1-lithio-1,2-dihydro-pyridine (**V-86**) (confirmed by NMR spectroscopy) which gave the 2,4-disubstituted pyridine (**V-87**), on heating. 2,4-Di-*t*-butylpyridine was made by this method (125). The 2,4-disubstituted pyridines (**V-87**) can, of course, be obtained directly by treating **V-83** with an alkyllithium.

R^2 [pyridine] V-83 + ClCOOR' $\longrightarrow$ R^2 [pyridinium] Cl^- N$^+$ COOR' V-84 $\xrightarrow{R^3MgCl}$ R^2 R^3 N H COOR' V-85

R^{3}Li $\longrightarrow$

BuLi $\downarrow$

R^2 N R^3 V-87 $\xleftarrow{\Delta}$ R^2 R^3 N H Li V-86

d. FREE RADICAL REACTIONS. A comprehensive review covering the homolytic alkylation and arylation of pyridines has appeared (100); hence this discussion is limited to the salient features with reference to the most recent publications in this area.

The relative reactivities of the various positions of the pyridine ring toward homolytic attack have been calculated using atom localization energies (126) and the concept of free valences (127). The calculated reactivity values of Brown (126) agree quite well with the partial rate factors obtained for the homolytic phenylation of pyridine, the order being C-2 > C-4 > C-3 (see Section IV.1.B). Similar calculations show that the C-2 position of the pyridinium *ion* should be more reactive than that in pyridine (128). The overall yield and the isomer distribution of the product from the free radical substitution of pyridine and substituted pyridines depends on the nature of the substituent, the free radical and how it is generated, and the nature of the medium of the reaction. It should be borne in mind that the isomer ratio figures quoted in the earlier work are open to question, since the analysis was performed by fractional distillation or recrystallization of derivatives and not by gas chromatography or IR and NMR spectroscopy, as is now possible.

The earlier work on the alkylation of pyridines showed the presence of the 2- and 4-isomers (129, 130), but subsequent publications reported the presence of the 2-, 3-, and 4-isomers when *t*-butyl peroxide (131), acetyl peroxide, and lead tetra-acetate (132) were reacted with pyridine. Bass and Nababsing found that the methylation of pyridine in an acidic medium

TABLE V-3. Homolytic Alkylation of Pyridines

Radical source	Medium	Overall yield of picolines (%)	Isomer Ratio (%)			Ref.
			2-	3-	4-	
Ac_2O_2	Pyridine/N_2	—	62.7	20.3	17.0	132
	Pyridine/O_2-free	—	62.7	19.5	17.8	132
$Pb(AcO)_4$	Pyridine/N_2	—	62.1	20.5	17.4	132
	Pyridine/N_2	5.0	62.7	21.7	15.6	133
	AcOH/N_2	12.9	76.4	2.9	20.7	133
t-Bu_2O_2	Pyridine	—	58.0	23.0	19.0	131
	Pyridine/N_2	13.5	62.0	22.9	15.1	133
	AcOH/H_2	28.3	77.9	2.7	19.4	133
	AcOH–HCl/N_2	11.8	93.2	0	6.8	133
$(PhCH_2)_2Hg$	Pyridine	—	0	0	0	134
	AcOH	35	73.7	—	26.3	134

resulted in enhancement of the activity of the C-2 and C-4 positions (133) in keeping with predictions (128). They also found that pyridine is *not* benzylated in nonacidic solutions, but gives 2- and 4-benzylpyridine when heated with dibenzylmercury in acetic acid as solvent (134). The reaction of 3-*n*-butylpyridine with lead tetraacetate has been studied using gas chromatography combined with infrared spectroscopy for the analysis of the products. The following products were identified: 3-*n*-butyl-2-methyl-pyridine (60%), 3-*n*-butyl-4-methylpyridine (ca. 20%), 3-*n*-butyl-6-methyl-pyridine (ca. 20%), 3-*n*-butyl-5-methylpyridine (ca. 2%) (135). Table V-3 summarizes the results of these reactions.

In many cases the products contained higher boiling residues in which 2,2′-, 3,3′-, and 4,4′-bipyridyls were detected. These are probably formed as a result of proton abstraction from the pyridine by the attacking free radical with subsequent dimerization. These dimers were not detected when an acidic medium was employed (133). A similar dimerization was observed on treating substituted pyridines (**V-89**) with *t*-butoxy radicals, when **V-90** were produced (136) (see also Ref. 132).

V-89   **V-90**

e. PHOTOCHEMICAL ALKYLATION. Travecedo and Stenberg report that the photolysis of 0.1 molar solutions of pyridine in methanolic HCl solutions in a nitrogen atmosphere gave 2- and 4-methylpyridines, 1-(2-pyridyl)-2-(4-pyridyl)ethane, and 1,2-di-(4-pyridyl)ethane. In the absence of HCl no alkylation was observed (137). Photosubstitution of pyridine (1 %) in cyclohexane produces 2- and 4-cyclohexylpyridines. 2- And 4-methylpyridine also react to form both 4- and 6-cyclohexyl-2-methylpyridine. This indicates that the two picolines react *via* a common azaprismane intermediate (V-91). In this case no acid is required (138).

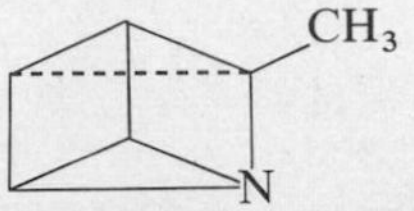

V-91

A near quantitative yield of dimeric salt was obtained by the photolysis of 2-styrylpyridinium methiodide. Sublimation *in vacuo* gave two isomeric dimeric bases (V-92) (139).

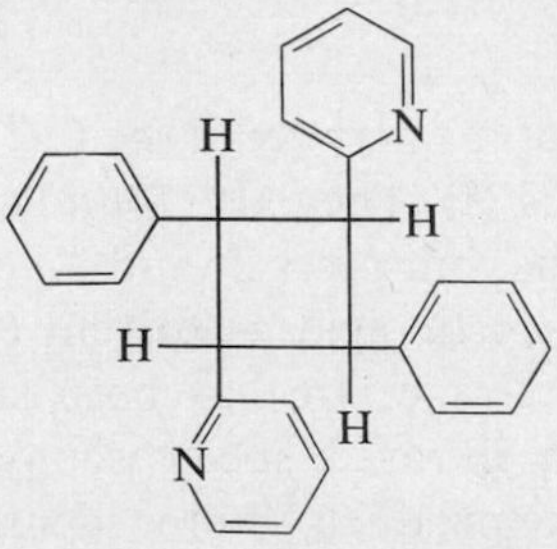

V-92

D. Ladenburg Rearrangement

Kuthan and co-workers studied the thermolysis of the 1-methohalides and the hydrohalides of pyridine and the three picolines at 340 to 360°; they found that alkylation and dealkylation of the ring occurred. The effectiveness of the halide in promoting the alkylation reaction was in the order $I^- > Br^- > Cl^-$. The decomposition of the picoline salts produced small amounts of pyridine, and 2-methylpyridine was detected among the products from the 4- and 3-methylpyridine salts. 3-Methylpyridine was not detected among the products of decomposition of 1-methylpyridinium halides. The mechanism of this reaction was discussed in terms of the Hückel LCAO–MO method (140).

Chumakov and Novikova examined the products of thermolysis (290 to 300°C) of an equimolar mixture of pyridine and methyl iodide by gas chromatography. Their results are summarized in Table V-4.

TABLE V-4. Thermolysis of 1-Methyl-
pyridinium Iodide (141)

Product	Yield (%)
2-Methyl	30.1
3-Methyl	5.8
4-Methyl	34.6
2-Ethyl	3.6
4-Ethyl	1.4
2,4-Dimethyl	12.8
2,5-Dimethyl	1.9
2,6-Dimethyl	9.1

The presence of 3-methylpyridine (see Ref. 140) and the ethylpyridines is indicative of a free radical mechanism for this reaction (141).

Pyridines have been benzylated by heating with benzyl chloride in the presence of copper or a copper salt (141–143), the 2- and 4-benzylpyridines thus formed being separated by recrystallization of their salts (141).

E. Decarboxylation Reactions

2-Methyl-6-phenylpyridine (61%) and 2-methyl-6-(p-chlorophenyl)pyridine (21%) were prepared by heating the respective nicotinic acids with soda lime and distilling the product under reduced pressure (144).

An interesting application of the decarboxylation reaction is the preparation of labeled pyridines. Pyridines containing a C-2–COOD group are decarboxylated in high yields with high specificity, since the positions α to the nitrogen atom are activated to H–D exchange. The C-3 and C-4 acids require higher temperatures and the decarboxylation results in scrambling of the label. NMR data can be used to calculate the mass distribution at various positions in the ring in these samples (145).

3. Properties and Separation of Isomers

A. Properties—General

New data on the physical properties of the alkylpyridines are summarized in Table V-5, of the cycloalkylpyridines in Table V-6, of the cycloalkeno-pyridines in Table V-7, and of the aralkylpyridines in Table V-8.

TABLE V-5. Alkylpyridines

Compound	B.p.	d_4^{20}	n_D^{20}	Picrate (m.p.)	Ref.
2-Ethylpyridine	150°	0.9319			146
3-Ethylpyridine	166°	0.9404			146
4-Ethylpyridine	166.5°	0.9410			146
2-n-Propylpyridine	170°	0.9158	1.4914	75°	87, 146
3-n-Propylpyridine	66°/8 mm	0.9254	1.4950	100.5°	87, 146
4-n-Propylpyridine	69.5°/10 mm	0.9255	1.4977	131°	63, 87, 146
4-Isopropylpyridine	182.5°	0.9258			146
2-n-Butylpyridine	77.5°/15 mm	0.9071			146
3-n-Butylpyridine	82°/7 mm	0.9150			146
4-n-Butylpyridine	84°/8 mm	0.9155	1.4945	111°	63, 146
2-Isobutylpyridine	74°/23 mm	0.8973	1.4862	97°	87
4-Isobutylpyridine	63.66°/8.5 mm		1.4902	120°	87
4-s-Butylpyridine	197°	0.9197			146
4-t-Butylpyridine	197°	0.9219			146
2-n-Pentylpyridine	104.5°/17 mm	0.9009	1.4834	71.8–72.3°	64, 146
3-n-Pentylpyridine	100°/9 mm	0.9077	1.4926	79.5–80°	64, 146
4-n-Pentylpyridine	95°/6 mm	0.9089	1.4925	107.8–108°	63, 64, 146
4-Isopentylpyridine	112.5°/20 mm	0.9122			146
2-(Diethylmethyl)pyridine	197°	0.9017	1.4866	72°	87
4-(Diethylmethyl)pyridine	80–82°/12 mm		1.4918	125°	87
2-n-Hexylpyridine	111°/15 mm	0.8952			146
3-n-Hexylpyridine	113°/7 mm	0.9012			146
4-n-Hexylpyridine	110°/5 mm	0.9014			146
2-n-Heptylpyridine	122.5°/13 mm	0.8906			146
3-n-Heptylpyridine	139.5°/7 mm	0.8954			146
4-n-Heptylpyridine	119°/4 mm	0.8965			146
2-(Di-Isopropylmethyl)pyridine	211–217°	0.8998	1.4870	108°	87
4-(Di-n-propylmethyl)pyridine	110°/8 mm	0.8994			146

Compound	b.p., m.p.	d	n_D	m.p. / derivative	References
2-(Triethylmethyl)pyridine	88–91°/16 mm		1.4970	101–102°	87
2-n-Nonylpyridine	105°/1.5 mm		$1.4866^{21.5}$		97
2-(Di-n-butylmethyl)pyridine	81–83°/0.5 mm	0.8944	1.4912	93.5–94.5°	64
3-(Di-n-butylmethyl)pyridine	83–87°/0.5 mm	0.9257	1.5065		64
4-(Di-n-butylmethyl)pyridine	95–97°/2 mm	0.8911	1.4880	120.2–120.8°	64
9-(2-Pyridyl)heptadecane	161°/1.5 mm		1.4980		97
1-(2-Pyridyl)-2-methyloctane	115°/2 mm		1.4910		97
2,3-Dimethylpyridine	161°, m.p. −15.22°	0.94641			146, 147
2,4-Dimethylpyridine	158°, m.p. −63.96°	0.93102			146, 147
2,5-Dimethylpyridine	157°, m.p. −15.54°	0.92910			146, 147
2,6-Dimethylpyridine	144.5°, m.p. −6.10°	0.92257			146, 147
3,4-Dimethylpyridine	m.p. −11.04°	0.95773			147
3,5-Dimethylpyridine	172°, m.p. −6.50°	0.94234			146, 147
2-Ethyl-3-methylpyridine	69.5°/16 mm		1.5204	139–140° Chloroplatinate 201–202°	846
2-Ethyl-4-methylpyridine	173–174°			122–123°	395
4-Ethyl-3-methylpyridine	201°	0.9477			146
6-Ethyl-2-methylpyridine	158–161°		1.4940	126–127°	847
2,6-Diethylpyridine	171°			116–117°	52
3,4-Diethylpyridine	208–209°		$1.5025^{29.5}$	136–138°	89
3,5-Diethylpyridine	84–84.5°/12 mm	0.9227	1.4991	159.5–160.5°	44, 62
3-Ethyl-2-propylpyridine	93–94.1°/12 mm		1.4975^{25}	120–121.5° 197 (decomp.)	846
2,6-Dipropylpyridine	206–207°			83.5–84°	52
3,5-Dipropylpyridine	100–101°/12 mm		1.4933	93–93.5°	44
2,6-Di-isopropylpyridine	90°/25 mm	0.8788	1.4768^{25}	114°	52
5-n-Butyl-2-methylpyridine	105–106°/19 mm	0.9068	1.4911	137–138°	29
6-t-Butyl-2-methylpyridine					847
3-n-Butyl-4-methylpyridine	119–120°/23 mm	0.9199	1.4986	124.5–125°	29
3,5-Dibutylpyridine	88–90°/2 mm		1.4908	53–54°	44
5-n-Hexyl-2-methylpyridine	140–142°/22 mm	0.8870	1.4868	123–124°	29
2,6-Di-(2'-pentyl)pyridine	111°/12 mm			166–167°	52

Table V-5 (*Continued*)

Compound	B.p.	d_4^{20}	n_D^{20}	Picrate (m.p.)	Ref.
2,3,4-Trimethylpyridine	192°		1.5150	177°	848
2,3,6-Trimethylpyridine	168–170°		1.5030		847
2,4,6-Trimethylpyridine	60–61°/10 mm			157°	13, 30
3,4,5-Trimethylpyridine	211.4°, m.p. 36.8°	0.9471^{20}	1.5132^{37}	178° Picrolonate 232° (decomp.)	42, 848
3,5-Dimethyl-2-ethylpyridine	188°			154–155°	13, 14
3,5-Dimethyl-4-ethylpyridine	218–219°			156.5–157°	91
4,5-Dimethyl-3-ethylpyridine	217°, m.p. −15°		1.5136	133° Methiodide 132° (decomp.)	121, 848
3,4-Dimethyl-5-isopropylpyridine	232–233°		1.5098	143–144°	849
4,5-Dimethyl-3-s-butylpyridine	135°/22 mm		1.5078	131–132°	121
2,6-Diethyl-3-methylpyridine				122°	30
3,5-Diethyl-4-methylpyridine	114–116°/25 mm		1.5119	105°	848
3,5-Diethyl-2-n-propylpyridine	237–239°			121–122°	13, 14
2,5-Dimethyl-5-n-amylpyridine	83–85°/2 mm		1.4959	106–107° Ethobromide 170–171°	850
2,5-Dimethyl-5-n-hexylpyridine	117–119°/2.5 mm	0.9007	1.4960	90–92° Ethobromide 168–170°	850
2,5-Dimethyl-5-n-heptylpyridine	135–137°/2 mm	0.8930	1.4930	89.5–90° Ethobromide 164.5–166°	850
2,5-Dimethyl-5-n-nonylpyridine	151–152°/3 mm			80–82° Ethobromide 160–162°	850
2,5-Dimethyl-5-n-decylpyridine	153–156°/2 mm	0.8973	1.5010	95–97° Ethobromide 162–163.5°	850
2,3,4,6-Tetramethylpyridine	195–203°, m.p. 79–80°			107°	30
2,3,5,6-Tetramethylpyridine				173–174°	847
2,4,5,6-Tetramethylpyridine				177.5°	30
2,6-Diethyl-3,4-dimethylpyridine				137°	30

TABLE V-6. Cycloalkylpyridines

Compound	M.p.	B.p.	n_D^{20}	Picrate (m.p.)	Ref.
2-(1-Adamantyl)pyridine	42–43°			136–137°	7
3-(1-Adamantyl)pyridine		119–120°/0.2 mm	1.5583	209–209.5°	7
4-(Adamantylmethyl)pyridine	101°			222–223°	7
3-(Adamantylmethyl)pyridine	69.5–70°			163–164°	7
1,3-Diphenyl-2,4-	189–190°[a]				
di(2-pyridyl)cyclobutane	117–118°				139

[a] The two m.p.'s refer to two isomers whose exact configuration was not determined.

Vogel and co-workers have purified pyridine and various substituted pyridines and determined the following physical properties of these compounds: refractive indexes at 20° (C, D, F, and G lines), densities, molarity refractivities, and molarity refraction coefficients. Ultraviolet and infrared spectra were also obtained. The compounds included pyridine; 2-, 3-, and 4-alkyl (methyl to n-heptyl) pyridines, 2,3-, 2,5-, 2,6-, and 3,5-dimethylpyridines (146). Coulson and co-workers prepared pure 2,3-, 2,4-, 2,5-, 3,4-, and 3,5-dimethylpyridines and obtained the freezing points, cryoscopic constants, vapor pressure-temperature curves, boiling points, latent heat of vaporization, densities, coefficients of expansion, refractive indices, and infrared absorption spectra of these compounds (147). Pyridine and 2-methylpyridine were purified and the triple points, boiling points, densities, viscosities, surface tensions, and refractive indices determined. In addition, the refractivity intercept, specific dispersion, molarity refraction, molarity volume, parachor, ultraviolet, infrared, and mass spectra were also obtained (148). The effect of the methyl group at various positions on the pyridine ring on the polarizability of the base was studied by Ploquin. Substitution at the 2, 4, or 6 positions increased the polarizability while substitution at the 3 or 5 positions had only a slight effect (149).

The dissociation constants of several 2-substituted pyridines were measured using the half neutralization method at 25°C. Table V-9 summarizes these data (150) (see also Chapter V, 1st ed., p. 176 and Ref. 151).

Bhattacharya has calculated the hyperconjugative effect of the methyl groups in 4-methyl-, 2,6-dimethyl-, and 2,4,6-trimethylpyridines using molecular orbital methods (152).

The isotopic H–D exchange in 2-methylpyridine using CH_3OD in the presence of triethylamine at 130 to 180°C involved only the C-2–CH_3 protons. The reaction was found to be first order, the rate constant being 1.7×10^6 sec^{-1} (153). Group VIII transition metals catalyzed the H–D exchange of the methyl- and dimethylpyridines with D_2O. Platinum was

TABLE V-7. Cycloalkenopyridines(Pyridinophanes)

Compound	B.p.	n_D^{20}	d^{20}	Picrate (m.p.)	Salt (m.p.)	Ref.
[3][2.3]Pyridinophane (2,3-Cyclopentenopyridine)	200–200.5°	1.5445	1.0362	181–182°		16, 17 18, 180, 683
[3][3.4]Pyridinophane	211.8°	1.5439	1.045	144°	Picrolonate 229°	42
2-Methyl[3][2.3]pyridinophane	208°	1.5239^{23}	1.0007	135–136°	Styphnate 156–157°	16
3-Methyl[3][2.3]pyridinophane	207–208°	$1.5360^{24.5}$	1.0270	147–149°	Styphnate 172–173.5°	16
5-Methyl[3][2.3]pyridinophane	226°			169–170°		17
6-Methyl[3][2.3]pyridinophane	222° m.p. 39–41°			204–206°	Picrolonate 199–200.5° (decomp.)	17, 18
7-Methyl[3][2.3]pyridinophane	211.9°, m.p. 31.9°	$1.5297^{37.5}$	0.9910^{33}	155°	Styphnate 138–139°	15, 17, 42
6,8-Dimethyl[3][3,4]pyridinophane				121–122°		31
5,6,8-Trimethyl[3][3,4]-pyridinophane	103–104°/8 mm	1.5360^{25}	0.910	146–147°		31
[4][2.3]Pyridinophane	100–100.5°/15 mm	1.5435^{20}	1.0281^{20}	158.5–159.5°		18, 180, 683
7-Methyl[4][2.3]pyridinophane	46–47°/0.05 mm			182–183°		18
8-Methyl[4][2.3]pyridinophane	109–112°/14 mm			152–153°		15
[5][2.3]Pyridinophane	111.5–112°/16 mm	1.5405	1.0142	138–139°		18, 180, 683
8-Methyl[5][2.3]pyridinophane	55–56°/0.05 mm			162–163°		18
9-Methyl[5][2.3]pyridinophane	116–118°/17 mm			159–160°		15
[6][2.3]Pyridinophane	59–60°/0.2 mm			150–151°		18
9-Methyl[6][2.3]pyridinophane	73–75°/0.05 mm			160–161°		18
[8][2.3]Pyridinophane	m.p. 21.8–23.4°	1.5370^{26}	0.9726	154–155°		446
[13][2.3]Pyridinophane	Oil	1.5282		134.5–136°	Picrolonate 158–159°	450

Compound				Ref.
11,13-Dimethyl[9](3.5)-pyridinophane	Liquid	No constants given		448
[7](2,6)Pyridinophane	70–73°/3 mm			447
[10](2,6)Pyridinophane	m.p. 15.5–16.6°	1.5241^{26}		446
2-Methyl[10](2,6)pyridinophane racemic (Muscopyridine)	m.p. 163–166° (decomp.)	1.5206^{25}	165–166°	446
Resolved $[\alpha]_D^{25}$ + 13.31° (±0.22°)			Picrolonate 163–166° (decomp.)	446
2,2-Dimethyl[10](2,6)pyridinophane	Oil		Picrolonate 170–172° (decomp.)	446
13-Methyl[10](2,6)pyridinophane	Oil			449
[8](2.5)Pyridinophane	70–75°/0.01 mm			60
[9](2.5)Pyridinophane	80–81°/0.04 mm			60
(+)-[9](2.5)Pyridinophane	$[\alpha]_D + 152°$; $[\alpha]_{546} + 195°$			60
[10](2.5)Pyridinophane	75–80°/0.01 mm			60
[11](2.5)Pyridinophane	90–95°/0.03 mm			60
[12](2.5)Pyridinophane	100°/0.01 mm			60
[2.2](3.5)Pyridinophane	263–265°			453
[2.2.2](3.5)Pyridinophane	259–260°			453
[2.2.2.2](3.5)Pyridinophane	277–278°			453
[9,9](2.5)Pyridinophane	94.5–97.5°			60
[2.2]Metacyclo(2,6)pyridinophane	No constants given			451
[2.2.2](2,6)Pyridinophane	205–206°			454
[2.2.2.2.2](2,6)Pyridinophane	160–161°			454
[2.2](2,6)Pyridinophane-1,9-diene	127.5–128°			455

TABLE V-8. Aralkylpyridine

Compound	M.p.	B.p.	n_D^{25}	Picrate (m.p.)	Ref.
2-Benzylpyridine		145–147°/15 mm		141–142°	360
3-Benzylpyridine		159–160°/17 mm		118°	360
4-Benzylpyridine		156–158°/16 mm		140.5–141°	360
2-Benzyl-6-methylpyridine		150–153°/17 mm		141–142°	360
2,6-Dibenzylpyridine		164–168°/0.2 mm		175–176°	360
2-Benzyl-3,6-dimethyl-4-phenylpyridine		179–181°/1 mm			142
4-(Phenethyl)pyridine		160°/25 mm	1.5786[20]	168°	63
4-(Phenylpropyl)pyridine		152°/20 mm	1.5681[20]	146°	63, 394
2-Methyl-3-phenethylpyridine		130–132°/2 mm	1.5672[27]		59
4-Methyl-3-phenethylpyridine		140–142°/3 mm	1.5711[26]		59
o-Chlorophenethylpyridine		135–136°/1 mm	1.5805[26]		59
m-Chlorophenethylpyridine		145–147.5°/1.2 mm	1.5806		57
p-Chlorophenethylpyridine		174.5°/5.5 mm	1.5793		57
2-Cyano-3-phenethylpyridine	36.5–40°	183–185°/3 mm	1.5763		57
2-Cyano-5-phenethylpyridine	55–58°				57
3-(o-Chlorophenethyl)-2-cyanopyridine	84–88°				57
3-(m-Chlorophenethyl)-2-cyanopyridine	72–74°				57
3-(p-Chlorophenethyl)-2-cyanopyridine	73–74°				57
5-(m-Chlorophenethyl)-2-cyanopyridine	47–49°				57
5-(p-Chlorophenethyl)-2-cyanopyridine	116–118°				57
1-Phenyl-3-(4-pyridyl)propane		149–149.5°	1.5627[20]		394
2-(p-Chlorophenyl)-1-(2-pyridyl)propane		134–140°/1 mm	1.5676[26]		469
3-(p-Chlorophenyl)-4-(2-pyridyl)hexane		154–158°/2 mm	1.5535		469
2-(p-Methoxyphenyl)-1-(2-pyridyl)propane		128–132°/0.5 mm	1.5540[27]		469
3-Phenyl-4-(2-pyridyl)hexane		122–125°/2 mm	1.5425		469
4-(2-Pyridyl)-3-p-tolylhexane		149–153°/5 mm	1.5395[24]		469
2-(p-Methoxyphenyl)-3-(2-pyridyl)butane		147–153°/4 mm	1.5508		469
2-(p-Methoxyphenyl)-3-(2-pyridyl)pentane		165–170°/4 mm	1.5520[20]		469
Isomer I[a]	50–51°				
Isomer II		147–152°/1 mm	1.5478		469

294

Compound	mp (°)	bp	n	
2-(p-Hydroxyphenyl)-3-(2-pyridyl)pentane				
Isomer I[a]	163–165°			
Isomer II	155–157°			
3-(p-Methoxyphenyl)-2-(2-pyridyl)pentane		149–154°/1 mm	1.5517^{24}	469
3-(p-Methoxyphenyl)-4-(2-pyridyl)hexane		145–150°/1 mm	1.5466^{24}	469
Isomer I HCl[a]	237–239°			469
Isomer II		155–160°/3 mm		469
3-(p-Hydroxyphenyl)-4-(2-pyridyl)hexane				
Isomer I HCl[a]	130–132°			469
Isomer II	253–256°			469
	126–128°			
3-(p-Methoxyphenyl)-4-(2-pyridyl)-2-methylhexane		161–165°/1 mm	1.5473^{24}	469
3-(o-Methoxyphenyl)-4-(2-pyridyl)hexane		142–143°/1 mm	1.5504^{24}	469
3-(p-Methoxyphenyl)-4-(2-pyridyl)-1-(diethylamino)hexane		185–188°/1 mm	1.5333	
3-(p-Diethylaminoethoxyphenyl)-4-(2-pyridyl)hexane		200–204°/1 mm	1.5328^{26}	469
2-(p-Methoxyphenyl)-3-(3-pyridyl)pentane		155–156°/1 mm	1.5557^{24}	469
3-(p-Methoxyphenyl)-4-(3-pyridyl)hexane	78–80°	164–165°/2 mm	1.5531^{21}	469
3-(p-Hydroxyphenyl)-4-(3-pyridyl)hexane	185–187°	167–169°/1 mm		469
3-(p-Ethoxyphenyl)-4-(3-pyridyl)hexane		175–178°/2 mm		469
3-(p-Methoxyphenyl)-4-(4-pyridyl)hexane				
Isomer I HCl[a]	205–206°			469
Isomer II HCl	190–193°			469
3-(p-Methoxyphenyl)-2-methyl-2-(4-pyridyl)pentane		169–170°/2 mm		469
5-(Dimethylamino)-2-(p-methoxyphenyl)-3-phenyl-3-(2-pyridyl)pentane		207–208°/1 mm		469
1-Dimethylamino-4-(p-methoxyphenyl)-3-phenyl-3-(2-pyridyl)hexane		210–217°/1 mm		469
2-(p-Methoxyphenyl)-1-(2-pyridyl)phenylpropane				
Isomer I[a]	76–80°			469
	106–110°			469

Table V-8 (*Continued*)

Compound	M.p.	B.p.	n_D^{25}	Picrate (m.p.)	Ref.
2-(p-Methoxyphenyl)-1-(2-pyridyl)-1-ethylphenylpropane		206–208°/1 mm			469
p-Methoxyphenyl-2-(2-pyridyl)cyclohexane		190–192/3 mm	1.5766		99
1-(2-Pyridyl)-2,3-diethyl-6-methoxyindane		185–190°/1 mm	1.5696^{26}		99
2,4-Dimethyl-5-[2-(4-pyridyl)ethyl]pyrrole	79–80°				432
2,4-Dimethyl-3-ethyl-5-[2-(4-pyridyl)ethyl]pyrrole	97–98°				432
2,4-Dimethyl-5-[2-(3-methyl-4-pyridyl)ethyl]pyrrole	No constants given				432
2,4-Dimethyl-5-[2-(2-ethyl-4-pyridyl)ethyl]pyrrole	No constants given				432
1,2,4-Trimethyl-5-[2-(2-pyridyl)ethyl]pyrrole	No constants given				432
1-Phenethyl-2,4-dimethyl-5-[2-(4-pyridyl)ethyl]pyrrole	No constants given				432
2,5-Bis[2-(3-pyridyl)ethyl]pyrrole	No constants given				611
2,5-Bis[2-(4-pyridyl)ethyl]pyrrole	123–124°				611
2,5-Bis[2-(5-ethyl-2-pyridyl)ethyl]pyrrole	No constants given				611
1-Methyl-2,5-bis[2-(4-pyridyl)ethyl]pyrrole	125–126°				611
1-Phenethyl-3-ethyl-4-methyl-2,5-bis-[2-(4-pyridyl)ethyl]pyrrole	No constants given				611
2-[2-(2-Pyridyl)ethyl]thiophene	100–101°				117
3-Methyl-2-[2-(2-pyridyl)ethyl]thiophene	56–59°	124°/0.2 mm			117
2-(3-Indolylmethyl)-4-methylpyridine	130–131°				98
2-(3-Indolylethyl)pyridine	118–120°	170–185°/0.3 mm		HCl 157–159°	608, 610
4-(3-Indolylethyl)pyridine	149–151°			HCl 260–262°	607, 608, 610
4-(1-Indolylethyl)pyridine	41–45°	160–165°/0.1 mm		Methobromide 210–211° HCl 206–208°	608

296

Compound	M.p. (°)	B.p. (mm)	n_D	Other (°)	Salt, m.p. (°)	References
2-(1-Methyl-3-indolylethyl)pyridine		170–185°/0.5 mm	1.6140		HCl 152–153°	610
4-(1-Methyl-3-indolylethyl)pyridine	96–98°				HCl 164–166° (decomp.)	608, 610
2-(3-Indolylethyl)-5-ethylpyridine	112–113°				HCl 159–160°	610
2-(1-Methyl-3-indolylethyl)-5-ethylpyridine		170–185°/0.5 mm	1.6140			608
2-(1-Methyl-3-indolylethyl)-5-ethylpyridine		175–200°/0.5 mm	1.5957^{24}			608, 610
4-(2-Methyl-3-indolylethyl)pyridine	153–154°				HCl 242–243°	608
4-(3-Methyl-1-indolylethyl)pyridine					HCl 210–212°	608
4-(1-Ethyl-3-indolylethyl)pyridine	47–50°				HCl 167–169°	607, 608
4-(1-Benzyl-3-indolylethyl)pyridine					HCl 199–200°	608, 610
3-(5-Methoxy-1-indolylethyl)pyridine		225–230°/0.3 mm				608
4-(1,3-Dimethyl-2-indolylethyl)pyridine	No constants given					608
2-1-(1-Methyl-3-indolyl)-2-propylpyridine	No constants given					608
4-(3,3'-Diindolylmethyl)pyridine	No constants given					607
2-(1-Benzimidazolylethyl)pyridine	152–155°				HCl 205–207°	607
4-(1-Benzimidazolylethyl)pyridine	97–98°				HCl 208–211°	607, 608
2-(1-Benzotriazolylethyl)pyridine					HCl 161–163°	607, 608
2-(2-Benzotriazolylethyl)pyridine	59–62°	150–158°/0.4 mm			HCl 183–185°	607, 608
4-(1-Benzotriazolylethyl)pyridine	100–102°				HCl 200–202°	607, 608
2-(2-Benzotriazolylethyl)pyridine	92–95°				HCl 197–200°	607, 608
2-(Indenylethyl)pyridine		145–155°/0.5 mm	1.5987		HCl 147–149.5° (decomp.)	608, 610
4-(Indenylethyl)pyridine	96–97°				HCl 223–225°	608, 610
Di-(2-Pyridyl)methane		176–186°/2 mm				431
(6-Methyl-2-pyridyl)(2-pyridyl)methane		107–133°/1.2 mm		196°		431
1,2-Di-(2-pyridyl)ethane	48–49°	114–117°/1.7 mm				95, 428
1,2-Di-(6-methyl-2-pyridyl)ethane	52–53°	140–153°/2.5 mm				95
1,2-Di-(4,6-dimethyl-2-pyridyl)ethane	84–85°	170–200°/4 mm				95
1,2-Di-(2-pyridyl)propane		87–88°/0.05 mm	1.5565	199–200°		434
1,3-Di-(2-pyridyl)propane		173–176°/18 mm	1.5620^{27}	196–197°		428, 433
2-Methyl-1,3-di-(2-pyridyl)propane		102–103°/0.05 mm	1.5593^{23}	175–176°		434
1,3-Di-(2-pyridyl)butane		96–97°/0.08 mm	1.5540^{19}	190–191°		433, 434
1,4-Di-(2-pyridyl)butane	43–45°	110°/0.15 mm		226–227°	Dimethiodide 260–261°	58, 76, 428
1,4-Di-(4-pyridyl)butane	119–121°			220–221°		76

Table V-8 (*Continued*)

Compound	M.p.	B.p.	n_D^{25}	Picrate (m.p.)	Ref.
1-(2-Pyridyl)-4-(4-pyridyl)butane		122–124°/0.05 mm		192–193°	58
2,4-Di-(2-pyridyl)pentane		102–103°/0.05 mm	1.5500[20]	211–213°	433
1,5-Di-(2-pyridyl)pentane		135–137°/0.8 mm		138–139°	428
1,6-Di-(2-pyridyl)hexane		172–173°/1.2 mm		195–196°	428
1,6-Di-(3-pyridyl)hexane	33–34°	165–166°/1.5 mm	1.5421[20]	Dimethiodide 243–245°	429
1,6-Di-(4-pyridyl)hexane		176–178°/1 mm	1.5402[20]	Dimethiodide 173–175°	429
1-(2-Pyridyl)-6-(3-pyridyl)hexane		153–157°/0.2 mm			429
1,7-Di-(2-pyridyl)heptane		160–161°/0.4 mm		Dipicrolonate 185–186°	428
1,7-Di-(4-pyridyl)heptane	28–30°	170–175°/0.1 mm			429
1,8-Di-(2-pyridyl)octane		167–169°/0.4 mm		Dipicrolonate 192–193°	428
1,8-Di-(4-pyridyl)octane	36–37°	185–190°/0.5 mm			429
2,7-Di-(4-pyridyl)octane		167–170°/0.1 mm	1.5349[20]		429
1,11-Di-(2-pyridyl)undecane	29–30°	208–211°/0.6 mm		178–180°	428
1,12-Di-(2-pyridyl)dodecane	42–43°	215–217°/0.6 mm		238–241°	428
6,7-Di-(2-pyridyl)dodecane		215–217°/13 mm	1.5163	148–150°	96
1,13-Di-(2-pyridyl)tridecane		206–210°/1 mm		Dipicrolonate 165–166°	430
1,14-Di-(2-pyridyl)tetradecane		185–190°/0.8 mm		112–113°	430
1,15-Di-(2-pyridyl)pentadecane		220–223°/0.4 mm		92–94°	430
1,16-Di-(2-pyridyl)hexadecane		231–234°/0.6 mm		126–128°	430
1,17-Di-(2-pyridyl)heptadecane		248–250°/0.8 mm		86–87°	430
1,18-Di-(2-pyridyl)octadecane		256–259°/0.8 mm		106–108°	430
Diphenyl-n-octyl-(2-pyridyl)methane	41–42.5°	190–195°/0.1 mm		178–179°	437
Diphenyl-n-octyl-(3-pyridyl)methane		187–190°/0.1 mm		117–118°	437
Diphenyl-n-octyl-(4-pyridyl)methane		200–202°/0.1 mm		139–141°	437
p-Chlorobenzyldiphenyl-(2-pyridyl)methane	148–151.5°				437

Compound	mp/bp	Ref.
p-Chlorobenzyldiphenyl-(3-pyridyl)methane	174–175°	437
p-Chlorobenzyldiphenyl-(4-pyridyl)methane	157–160°/0.1 mm	437
3,3-Dimethylbutyldiphenyl-(2-pyridyl)methane	230–240°/0.2–0.3 mm	437
3,3-Dimethylbutyldiphenyl-(4-pyridyl)methane	70.5–71.5°	437
Diphenyl-(1-imidazolyl)-(2-pyridyl)methane	92–93.5°	439
Diphenyl-(1-imidazolyl)-(3-pyridyl)methane	222–224°	439
Diphenyl-(1-imidazolyl)-(4-pyridyl)methane	208–210°	439
o-Fluorophenyl-(1-imidazolyl)-phenyl-(2-pyridyl)methane	217–218°	439
o-Fluorophenyl-(1-imidazolyl)-phenyl-(3-pyridyl)methane	193–194°	439
p-Fluorophenyl-(1-imidazolyl)-phenyl-(2-pyridyl)methane	172–173°	439
p-Fluorophenyl-(1-imidazolyl)-phenyl-(4-pyridyl)methane	162–164°	439
o-Chlorophenyl-(1-imidazolyl)-phenyl-(2-pyridyl)methane	145–146°	439
o-Chlorophenyl-(1-imidazolyl)-phenyl-(3-pyridyl)methane	145–149°	439
o-Chlorophenyl-(1-imidazolyl)-phenyl-(4-pyridyl)methane	116–118°	439
p-Chlorophenyl-(1-imidazolyl)-phenyl-(2-pyridyl)methane	72–75°	439
	138–140°	439

Table V-8 (*Continued*)

Compound	M.p.	B.p.	n_D^{25}	Picrate (m.p.)	Ref.
m-Chlorophenyl-(1-imidazolyl)-phenyl-(4-pyridyl)methane	130°				439
p-Chlorophenyl-(1-imidazolyl)-phenyl-(4-pyridyl)methane	157–158°				439
p-Bromophenyl-(1-imidazolyl)-phenyl-(2-pyridyl)methane	133°				439
p-Bromophenyl-(1-imidazolyl)-phenyl-(4-pyridyl)methane	136–139°				439
Phenyl-m-trifluorophenyl-(1-imidazolyl)-(2-pyridyl)methane	94–96°				439
Phenyl-m-trifluorophenyl-(1-imidazolyl)-(4-pyridyl)methane	110–112°				439
p-Methylthiophenyl-(1-imidazolyl)-phenyl-(2-pyridyl)methane	150–152°				439
Diphenyl-(2-methyl-1-imidazolyl)-(4-pyridyl)methane	175–178°				439

[a] The isomers I and II refer to stereoisomers (*erythro* and *threo*) separated by fractional crystallization. Definitive conformational assignments were not made.

TABLE V-9. pK_a of Pyridines at 25°C

Base	pK_a
Pyridine	5.25
2-Methylpyridine	5.94
2-Amylpyridine	6.00
2-Hexylpyridine	5.95
2-Benzylpyridine	5.13
2-Vinylpyridine	4.98

found to be the most reactive, while cobalt was selective and catalyzed exchange α to the N atom. The positions *ortho* to a methyl group were deactivated, and protons flanked by two methyl groups did not undergo exchange (154).

Pyridines containing a C-2–CD_3 group have properties markedly different from those of the C-3 and C-4–CD_3 substituted compounds. Thus, the C-2 isomer shows a significant heat of reaction with BF_3 (155), and enhanced rates of reaction with methyl, ethyl, and isopropyl iodides (156), when compared with the C-3 and C-4 isomers. Brown and co-workers have shown that this secondary isotope effect is a steric phenomenon rather than due to an inductive or hyperconjugative effect (155).

The association of pyrrole with pyridine and methylpyridines was studied by means of the dielectric polarization of the complexes using infrared measurements. The low dipole moment of the 1:1 complexes was explained by hydrogen bonding between both the N-atom and the π-electron system of pyridine ring with the N–H of the pyrrole (157). A study of the comparative merits of various thermodynamic parameters for establishing the stability of complexes of the picolines and cobalt, nickel, zinc, and cadmium tetra-fluoroborates was undertaken (158).

Many Werner complexes of alkylpyridines with metal salts were made and found to be useful as oxidation accelerators for paints, polymerization catalysts, anticorrosion agents, insecticides, fungicides, and oil-soluble metal carriers (159, 160), and as selective clathrate agents for separating organic compounds such as the xylenes (160). Molybdates of pyridine and 2-, 3-, and 4-methylpyridine were studied (161), as were the carbonyl complexes obtained from the action of $Fe(CO)_5$ and $Co(CO)_4$ on various pyridine bases (162, 163). Poly(ethylene sulfide) precipitated many pyridine derivatives as coloured complexes (164). Aluminium alkyls containing heteroaryl rings such as pyridine were patented as polymerization catalysts to prepare hydrocarbon polymers (165). Decaborane adducts of pyridine and 2,4,6-trimethylpyridine were prepared and used to effect vulcanization of natural rubber (166). The picryl-methylpyridinium chlorides were described

(167). Fisel and Franchevici studied the complex salts of various pyridines with sodium palmitate and stearate. 3- And 4-methylpyridines formed complex salts but 2-methyl-, 2,4-dimethyl-, and 2,4,6-trimethylpyridines do not form such salts due to steric effects (168).

Certain pyridine salts have found application in chemical synthesis. N-Alkyl-N,N-diarylamines undergo selective dealkylation with hydrochlorides and hydrobromides of pyridine homologs. The C-alkyl or alkoxy groups are not affected by this reaction (169). The pyridine-cuprous chloride complex was found to be an effective autoxidation catalyst for converting primary aromatic amines to azo compounds (170). Beckmann rearrangements were catalyzed by pyridine hydrochloride (171). Clark obtained rate constants and other thermodynamic data for the decomposition of malonic acid in pyridine and the three picolines (172). The presence of alkyl substituted pyridines also prevents the thermal decomposition of t-alkyl chlorides and bromides (173).

Pyridine, 2-, 3-, and 4-methylpyridines and 2,6-dimethylpyridine all reacted with calcium in the presence of liquid ammonia to form unstable colored compounds (174).

The polarographic behavior of cobalt (II), nickel (II), and zinc (II) in pyridine and 4-methylpyridine was quite similar but differed markedly when 2-methylpyridine was used as the solvent, possibly as a consequence of a chemical reaction coupled with the charge-transfer process (175).

The reversible photochromism shown by nitrobenzylpyridines due to the tautomeric photoreaction V-93 to V-94 has been further investigated (176).

A new color test for pyridine and its derivatives was described by Lisboa. When pyridine and its derivatives (0.05 to 5% in a phosphate buffer) reacts with BrCN (10%) in the presence of barbituric acid (1%), colors are produced—red-orange for pyridine and 2-methylpyridine and weak violet for 2,4-lutidine. The colors for various other pyridine derivatives and by using other barbiturates were also described (177).

The use of pyridine, its homologues and derivatives in agriculture and medicine has been reviewed (178).

B. Properties—Spectroscopic

Ultraviolet, infrared, nuclear magnetic resonance, and mass spectroscopy are becoming increasingly important for determining the structure of pyridines and for establishing the isomer distribution in reaction products. Reviews containing sections on the electronic absorption (179a), infrared (179b), and nuclear magnetic resonance (179c) spectra of pyridines have been published.

The ultraviolet spectra of purified pyridine, 2-, 3-, and 4-alkyl (methyl to *n*-heptyl) pyridines, 2,3-, 2,5-, 2,6-, and 3,5-dimethyl pyridines were recorded (146). The Mills-Nixon effect in pyridines was investigated by examining the ultraviolet and infrared spectra of 2,3-dimethylpyridine, 2,3-cyclo-penteno-, 2,3-cyclohexeno-, and 2,3-cycloheptenopyridines. The ultraviolet absorption spectra showed differences indicative of ring strain in the pyridine nucleus and evidence of steric hindrance of the C-4 proton in 2,3-cyclo-heptenopyridine was shown from its infrared spectrum (180).

The infrared spectra of purified pyridine, 2-, 3-, and 4-alkyl-(methyl to *n*-heptyl) pyridines, and all the lutidine isomers were recorded (146, 147). In a compilation of the infrared spectra of commercially available organic compounds numerous pyridine and condensed pyridine compounds are included (181). Infrared spectroscopy provides a convenient method of distinguishing between isomers and establishing the orientation of mono- and disubstituted pyridines since each isomer exhibits typical absorption patterns. Table V-10 summarizes the relevant pyridine infrared absorption bands.

Complete analysis of the proton magnetic resonance spectra of pyridine (AB_2X_2) (191), 2,3-dimethylpyridine, and 2,6-dimethylpyridine (AB_2) (192) were reported. A double resonance method was employed to measure the ^{14}N chemical shifts of pyridine and the pyridinium ion (193). The ^{13}C nuclear magnetic resonance spectroscopy of substituted pyridines has been reviewed (855).

The mass spectrometric fragmentation pattern of an alkylpyridine is profoundly affected by the position of substitution and the nature of the alkyl group. The mass spectrum of 3-alkyl substituted pyridines (**V-95**; R—CH$_2$— at C-3) has an intense peak at *m/e* 92, corresponding to the fragment $C_5H_4NCH_2^+$ (**V-96**), which is absent in the mass spectrum of the 2-isomer and is of medium intensity in the 4-isomer (194, 195).

V-95   **V-96**

TABLE V-10. Infrared Absorption Bands of Pyridine and Alkyl Substituted Pyridines (see 179b)[a]

Compound			Region (cm^{-1})		
	3,100–3,000	2,000–1,600	1,600–1,500	1,300–950	950–700
Pyridine	3,083, 3,054, 3,036 (182)			1,218, 1,148, 1,085, 1,068, 1,029, 992	700 (182)
2-Substituted		1,960 and 1,755 (87)		1,279, 1,147, 1,093, 1,048, 994 (187)	794–781, 752–746 (183)
3-Substituted		1,930 to 1,905 (185)		1,202–1,182, 1,124, 1,103, 1,038, 1,025 (188)	810–789, 715–712 (183)
4-Substituted	Two bands at	1,950 to 1,920 (87, 186)		1,232–1,208, 1,067, 993 (189)	820–794 (183)
2,3-Disubstituted	3,095–3,060		1,588–1,578		800–787, 741–725 (190)
2,4-Disubstituted	and	(110)	1,605–1,599, 1,136–1,099, 998–988		
2,5-Disubstituted	3,055–3,010		1,605–1,599 (2,5-dialkyl alone absorb at 1,028) (183)		826–813, 735–725 (190)
2,6-Disubstituted	(183, 184)[b]		1,588–1,578		813–769, 752–725 (190)
3,4-Disubstituted			1,605–1,599		
Trisubstituted				1,044–1,032, 1,001–996 (183)	

[a] Most of these data are taken from the article by A. R. Katritzky and A. P. Ambler (179b). We thank the authors and Academic Press for permission to utilize these data. Further data on the infrared spectra of pyridines are presented in "Physical Methods in Heterocyclic Chemistry," Vol. 4. Academic, New York. 1971.
[b] Numbers in parentheses are reference numbers.

304

Pyridines with a C-2 propyl or higher alkyl group (**V-97**), undergo a McLafferty rearrangement with the elimination of an equivalent amount of olefin. This type of rearrangement does not occur to any appreciable extent with the 3- or 4-isomers (194, 195).

V-97 → V-98

C. Purification and Separation of Isomers

Most of the present methods used for the separation and purification of various pyridines are based on:

1. Azeotropic distillation with water.
2. Crystallization of salts or complexes.
3. Conversion to the pyridine-1-oxides, purification of the *N*-oxides followed by deoxygenation to the pyridine (see Chapter IV and Section I.2.B.f).

Occasionally countercurrent or chromatographic methods have been employed.

Azeotropic distillation with water was used to separate and purify 2- and 4-methylpyridines (196), pyridine and 3-methylpyridine (197), 2,4-dimethylpyridine (198), 3- and 4-methylpyridines (199–201), and 2,6-dimethylpyridine of sufficient purity for organic synthesis and physiochemical investigation (201), and other alkylpyridines (202, 203). This process was convenient for separating 5-ethyl-2-methylpyridine from 2-methyl-5-vinylpyridine (204) (see also Chapter II).

Various alkylpyridines were isolated or purified as complexes with copper (II) chloride or sulfate (205–211), manganese thiocyanate (212), Werner complexes (213), lithium halides (214) and cobalt (II), iron (II and III), nickel (II), and zinc (II) chlorides (141, 215–221). A study of the solubility and stability of these last salt complexes of pyridine, the picolines, and 2,6-dimethylpyridine in water led to a method of preparing 3- and 4-methylpyridines (81%) free from 2,6-dimethylpyridine. Oxidation of these compounds provided nicotinic and isonicotinic acids (216). The use of calcium chloride gave high purity (99.2%) 4-methylpyridine (222). Complexes of 3- and 4-methylpyridines with borate esters (223) and of formamide with 2,4,6-trimethylpyridine (224) were used for purifying these compounds. Phosphate (225, 226), nitrate (227), hydrochloride (209), oxalate (41, 228–231), picrate (35, 41), phthalate and salicylate (232), dichromate (40), and o-cresol (233, 234) salts were also utilized for the purification of various pyridines.

Fractionation of N-oxides obtained from alkylpyridine mixtures, followed by deoxygenation of the separated N-oxides provides pure alkylpyridines (235–238).

Crude pyridine and 3-methylpyridine were purified by treating them with chlorine or bromine at room temperature (239). Separations involving countercurrent methods (240–242), preferential adsorption on molecular sieves (243, 244), and chromatography over alumina impregnated with metal salts (245) were reported.

The paper chromatography of alkylpyridines was investigated (246). Sorbate–sorbent interactions in the gas chromatography of pyridine compounds were examined (247) and the gas chromatography of alkylpyridines on silicone and bentone-silicone phases has been described (248).

The isolation and purification of the alkylpyridines are also reviewed in detail in Chapter II.

4. Reactions

Three reviews have appeared covering various aspects of the reactions of pyridines. Eisch and Gilman have compared the chemistry of aza-hetero-aromatic compounds from a theoretical point of view (249). Abramovitch and Saha have considered the electrophilic, nucleophilic, and homolytic substitution of pyridines with respect to the effect of substituents on these reactions (100), and Pfleiderer has discussed the reactivity and hetero-aromatic character of six-membered nitrogen heterocycles from the point of view of the energy distribution within the molecule in the transition state (250).

A. Isomerization

The isomerization of polyfluoroalkylpyridines by fluoride ion was described. The structure of the isomers was established by NMR spectral analysis as **V-99** and **V-100**. Since **V-100** rearranged to the symmetrical isomer **V-99** on treatment with KF, **V-100** is the product of kinetic and **V-99** of thermodynamic control (251).

V-99 V-100

Pentakis(pentafluoroethyl)pyridine (**V-101**) was prepared from the reaction of pentafluoropyridine with tetrafluoroethylene in DMF in the presence of CsF as catalyst. The Dewar type pyridine, pentakis(pentafluoroethyl)-1-azabicyclo[2.2.0]hexa-2,5-diene (**V-102**) was obtained from **V-101** on photolysis (< 270 nm), while irradiation (> 200 nm) gave the prismane, pentakis-(pentafluoroethyl)-1-azatetracyclo[2.2.0.0^{2,6}.0^{3,5}]hexane (**V-103**) (252).

V-101 **V-102** **V-103**

B. Dealkylation

The hydrodealkylation of alkylpyridines using nickel (253–255), nickel on alumina (256–259), vanadium pentoxide (260), and silver and platinum (261) catalyst was reported. 5-Ethyl-2-methylpyridine was selectively demethylated (254, 256, 262), and this method is recommended as an economic way of preparing 3-ethylpyridine (34 to 41 % conversion, 93 to 95 % purity) (256).

TABLE V-11. Products of Dealkylation of Butylpyridines (%) (263)

Product	2-Butylpyridine (temp.)			3-Butylpyridine (temp.)		
	470°	570°	600°	470°	570°	600°
Starting material	88	30	23			
Pyridine		5	17	1	5	
2-Methylpyridine	9	20	19	3	5	
3-Methylpyridine				19	28	
2-Ethylpyridine	3	17	17	5		
3-Ethylpyridine					10	
2-Propylpyridine			2			
Indolizine			1–10			1–10
Unidentified				5	8	

The complex reactions involved in the demethylation of 2-methylpyridine has been studied (258).

The products obtained by passing 2- and 3-butylpyridines separately over a 20% chromia-alumina catalyst at various temperatures were identified. These results are summarized in Table V-11 (263). The pyrolysis of 2-phenylethylpyridine (700°) was studied. Among the products identified were pyridine (5%), 2-methylpyridine (22.1%), 2-vinylpyridine (1.3%), and α-stilbazole (2.3%). A mechanism for this decomposition was proposed (264).

C. Oxidation

Pyridine aldehydes, ketones, acids, and cyanides are obtained by the oxidation of alkylpyridines. Oxidizing agents that have been used include ozone, selenium dioxide, potassium permanganate, sodium dichromate, nitric acid and nitrates, sulfur, and iodine, and vapor phase air oxidations. Electrochemical oxidations are also described as well as the use of a microbiological method.

The reaction of ozone with pyridines is of some value for their structural evaluation (see Chapter V, 1st ed., p. 184). The products isolated from substituted pyridines were indicative of initial attack at the 2,3- and/or 5,6-bonds in the ring, that is, bonds with the most double bond character (265). There is also evidence that attack by ozone can lead to pyridine-1-oxides (266, 267).

Heating 2-methylpyridine in an autoclave with an aqueous dichromate solution in the presence of phosphoric acid at 225 to 350° gave pyridine-2-carboxylic acid (80%). By this method 2,5-dimethylpyridine formed pyridine-2,5-dicarboxylic acid (75%) (268). Permanganate oxidation of 3-methylpyridine gave pyridine-3-carboxylic acid (60%) (269); and of 5-ethyl-2-methylpyridine produced isocinchomeronic acid (60 to 65%) (270, 271). Vaculik and Kuthan have shown that aqueous permanganate can be used to oxidize 3,5-dimethyl-4-alkylpyridines (V-104) to the 4-alkyl-3-methyl-pyridine-5-carboxylic acid (V-105) or 4-alkylpyridine-3,5-dicarboxylic acid (V-106; R = CH_3, C_2H_5, C_4H_9, and $PhCH_2$) (272).

V-106   ⟵   V-104   ⟶   V-105

An interesting modification of this method is described by Prostakov and co-workers. 4-Benzyl-2,5-dimethylpyridine **(V-107)** was oxidized with permanganate in the presence of magnesium nitrate to give 4-benzoyl-2,5-dimethylpyridine **(V-108)** which, on further oxidation, gave 4-benzoyl-pyridine-2,5-dicarboxylic acid **(V-109)** (273).

Selenium dioxide converted methyl- and dimethylpyridines to mixtures containing the aldehydes and acids. 3-Methylpyridine was more difficult to oxidize than the 2- and 4-isomers, which gave the aldehydes (about 25%) and the acids (about 50%) (274). The use of selenium dioxide in the presence of NO_2 in an inert solvent such as 1,2,4-trichlorobenzene converted 3-methylpyridine to nicotinic acid (about 85%) and 4-methylpyridine to isonicotinic acid (about 88%) (275).

Nitric acid is a convenient reagent for oxidizing alkylpyridines to the carboxylic acids. 4-Methyl- and ethylpyridines, on heating in an autoclave with aqueous nitric acid, formed isonicotinic acid (75 to 85%), the inclusion of air decreasing the amount of nitric acid required (276–278). 5-Alkyl-2-methylpyridines **(V-110)** gave 6-methylnicotinic acid **(V-111)** at 250 to 325°F and 30 to 650 psi pressure (279). By changing the conditions (250° and 750 psi and in the presence of ferric nitrate), both alkyl groups were oxidized, to give pyridine-2,5-dicarboxylic acid **(V-112)** (98%) (280, 281). More drastic conditions resulted in the loss of the thermally more labile C-2 carboxyl group to give pyridine-3-carboxylic acid **(V-113)** (270, 282–284). The use of

cupric nitrate and air in place of nitric acid gave the copper salt of iso-cinchomeronic acid **(V-112)** (270, 285, 286). The optimum conditions for this latter reaction were established (270). Titova and co-workers have determined the optimum conditions for the nitric acid oxidation—30% molar excess of nitric acid and a reaction temperature of 180 to 190°—and have discussed the mechanism of this reaction (284).

The vapor phase oxidation of alkylpyridines to produce aldehydes and acids has continued to invoke interest because it is a convenient route to these derivatives. The catalyst usually employed is vanadium pentoxide with added oxides of molybdenum, chromium, or tin. The three pyridine aldehydes (287, 288), 6-methylpyridine-2-carboxaldehyde (288), and 5-methylpyridine-3-carboxaldehyde (289) were obtained by this method. Vapor phase oxidation to the pyridinecarboxylic acids has been reported (290–294). The 3-methylpyridine in a 3-picoline fraction could be concentrated by selectively oxidizing the 2- and 4-methyl groups present in the other components of this fraction and removing the acids formed (290). Another application based on the preferential rapid oxidation of the methyl groups in the 2- and 4-position is the preparation of 3-ethylpyridine from 5-ethyl-2-methylpyridine. Oxidation over a nickel–alumina catalyst produces 5-ethylpyridine-2-carboxylic acid which undergoes decarboxylation (295).

Vapor phase oxidations in the presence of ammonia-amoxidations-producing cyanopyridines were used extensively (296–308). Cyanides were also obtained by treating alkylpyridines with NO in the presence of a rhenium catalyst (309).

An interesting reaction patented by Cislak is the vapor phase reaction of an alkylpyridine and an acid such as acetic or propionic acids to form the acetyl or propionylpyridine. Thus 2-, 3-, or 4-methylpyridines with acetic acid gave the 2-, 3-, or 4-acetylpyridines respectively; 5-ethyl-2-methyl-pyridine gave 2,5-diacetylpyridine and 4-methylpyridine gave 4-propionyl-pyridine with propionic acid. No yields were reported (310).

The Willgerodt reaction of methylpyridines with sulfur and amines to form the pyridinethiocarboxamides has been investigated (311–315). A C-2 methyl group is more reactive than a C-4 methyl group, although 4-methylpyridine also undergoes this reaction (311). By-products from the reaction of 2- and 4-methylpyridine with sulfur and aniline were the N,N'-diphenylpicolin- and isonicotin-amidines and the 2-(2- or 4-pyridyl)benzo-thiazoles, respectively (313) (see Chapter V, 1st ed., p. 186). The reaction of 4-picoline with sulfur gave a mixture of 1,2-bis(4-pyridyl)ethane and 1,2-bis(4-pyridyl)ethylene (316) (see also Chapter V, 1st ed., p. 186).

Electrochemical oxidations of alkylpyridines to the carboxylic acids were investigated in detail (317–319).

The microbiological oxidation of 3-methylpyridine to nicotinic acid was accomplished with the aid of microorganisms such as *Mycobacterium* in a nutrient broth at 29 to 30° (320).

Godin and Graham found that heating 5-ethyl-2-methylpyridine and a catalytic amount of a peroxide in an oxygen atmosphere produced 5-acetyl-2-methylpyridine in up to 40% yield (321).

A convenient synthesis of pyridine-2-aldehydes (V-117) was discovered by Markovac and co-workers. Heating 2-methylpyridines (V-114) with iodine in dimethylsulfoxide (DMSO) gave the 2-aldehydes (V-117) *via* an oxy-dimethylsulfonium halide intermediate (V-116) (322).

D. Nuclear Reduction

The catalytic hydrogenation of various pyridines using platinum metal catalysts and various solvents (acidic, neutral, and basic) was reviewed (323). Rhodium, although little used for the reduction of pyridines, is a particularly effective catalyst under mild conditions (323). There appears to be a "temperature threshold" for the reduction of pyridines. Thus, with palladium in acetic acid many pyridines were reduced smoothly to the piperidines in high yields only when the hydrogenations were run at 70 to 80°C (324). Nickel catalysts are also effective for reducing substituted pyridines to the piperidines (325–329).

A review has also appeared on the reduction of nitrogen heterocycles, including pyridines and pyridinium salts with complex metal hydrides. The products of these reductions are usually dihydropyridines and tetrahydropyridines (330). Of particular interest is the reduction of pyridine with lithium aluminium hydride, a reaction studied extensively by Lansbury and Peterson. The aged solution reduces ketones and aldehydes but not carboxylic acids (331–333), and was used by Giam and Abbott (106) for the preparation of 3-substituted pyridines (see Section 2.C.c).

In a chapter on the 3-Pirideines (1,2,3,6-tetrahydropyridines), Ferles and Pliml reviewed the literature on the reduction of pyridines using sodium and boiling alcohols (the Ladenburg reduction), aluminium hydride and complex hydrides, electrolytic reductions, and the Lukeš reduction (pyridine salts with formic acid) (334). 2,4,6-Trimethylpyridine was reduced electrolytically, with sodium and alcohol, and sodium and butanol and the ratio of products formed from each reaction established (335).

E. Nuclear Substitution

Alkylpyridines were halogenated in oleum (336, 337), sulfuryl chloride, and phosphorus oxychloride (338). An ionic mechanism was proposed for these reactions (337). Chlorination and bromination of pyridine and methyl-pyridines was also affected by reaction with the appropriate halogen in presence of aluminium chloride (339). The halogenation of pyridines is discussed in detail in Chapter VI.

The perfluoroalkylation of pyridines usually produces mixtures of isomers which can be separated as a result of the difference in basicities of the isomers (340, 341), and the structures established by ^{19}F NMR spectroscopy (342). The reaction is conveniently performed by treating pyridine or its homologues with perfluoroalkyl iodide at 170 to 185° in the presence of excess base (340, 341) or sodium acetate (343), or by heating pentafluoropyridine or 3,5-dichlorotrifluoropyridine and perfluoroalkenes at 110° with catalytic amounts of cesium or potassium fluoride in DMF (342) (also see Section 3.A).

Nitration of 2,4-dimethylpyridine gave a mixture of the isomeric 2,4-dimethyl-3- and 5-nitropyridines which were separated (344). Pyridine underwent N-nitration and nitrosation when treated with nitronium and nitrosonium tetrafluoroborate in nitromethane, acetonitrile, tetramethylene sulfone, and liquid sulfur dioxide. There was no evidence of N to C migration, even on heating (345, 856). The nitration of pyridines is discussed in Chapter VIII.

The sulfonation of the methylpyridines using oleum and mercuric sulfate is reported to form 2-methylpyridine-5-sulfonic acid (62%), 3-methyl-pyridine-5-sulfonic acid (61%), and 4-methylpyridine-5-sulfonic acid (60%) (346). 2,6-Di-t-butylpyridine can be sulfonated at C-3, since the steric effect of the two t-butyl groups results in a decrease in the basicity of the ring. This reaction does not occur with 2,6-diisopropylpyridine (347, 348). Chapter XV also discusses the sulfonation of pyridines.

The Chichibabin reaction, in which a metal amide reacts with a pyridine to form 2- or 4-aminopyridines, is reviewed in detail in Chapters II and IX. The reaction of 4-methylpyridine with sodium hydrazide to form 2-hydrazino-4-methylpyridine (V-118) (349) probably proceeds by a similar route. The

V-118

mechanistically similar alkylation and arylation of pyridines using organo-lithium compounds is discussed in Sections I.2.C.c and IV.1.A and Chapter VII (also see Ref. 100).

The heterogeneous bimolecular alkylation of pyridines, the Emmert reaction, was studied extensively by Bachman and co-workers (350, 351). This reaction involves the reaction of a pyridine with a carbonyl compound in the presence of magnesium amalgam, when the 2-pyridyl alcohol was obtained, or aluminum, when a 4:1 ratio of the 2- and 4-pyridyl alcohols resulted. This reaction is suggested to proceed *via* an intermediate such as **V-119**. In the case of aluminium, the greater size of the aluminium atom permits a possible transannular bridge between the nitrogen on the 4-position as in **V-120** and thus permits formation of the 4-isomer (350). Bachman and

V-119 M = Mg or Al   **V-120**

Schisla found that substitution of the ketone or aldehyde component by an acid derivative (ester, amide, or cyanide) gave acylpyridines (351). Abramo-vitch and Vinutha studied the reaction of cyclohexanone and 2-methyl-cyclohexanone with pyridine and 3-methylpyridine and found that the C-3 methyl group directed attack to C-6. The conformation of the products of the reaction of 4-*t*-butylcyclohexanone with pyridine was established by NMR spectroscopy, the ratio of *cis* to *trans* alcohol being 54.2:45.8 showing that there is little stereoselectivity in this reaction. The mechanism was discussed (352).

F. Side-Chain Halogenation and Other Substitution

The chlorination of methylpyridines has, until recently, been useful only for the preparation of polychloromethylpyridines such as **V-122** to **V-124** (1, 2, 353, 354) (see also Chapter V).

V-121   **V-122**   **V-123**   **V-124**

Kutney and co-workers found that N-bromosuccinimide (NBS) with benzoyl peroxide converted 4-ethylpyridine to 4-(1,1-dibromoethyl)pyridine. HBr and 4-methylpyridine to 4-tribromomethylpyridine·HBr. In the latter case, the mono- and dibromomethylpyridines were not obtained. 3-Methylpyridine did not undergo bromination under these conditions. The order of reactivity toward NBS was 4-methyl > 2-methyl > 3-methyl (355). On treatment with chlorine in fuming sulfuric acid in the presence of a free-radical initiator, or on irradiation, 2,6-dimethylpyridine gave 2,6-bis-(chloromethyl)pyridine and 2-chloromethyl-6-methylpyridine; 3-methylpyridine gave 3-chloromethylpyridine (356). Mathes and Schüly found that chlorination of 2-methylpyridine with chlorine (2.5 equivalents) in carbon tetrachloride in the presence of anhydrous sodium carbonate gave 2-chloromethylpyridine (V-212) (65%) which was purified by extraction into $2N$ hydrochloric acid. The 2-dichloromethylpyridine (5%) coformed and other products remained in the carbon tetrachloride layer. 2,6-Dimethylpyridine gave 2-chloromethyl-6-methylpyridine (57%), 2,4,6-trimethylpyridine gave 2-chloromethyl-4,6-dimethylpyridine (55%), and 2,3-dimethylpyridine gave 2-chloromethyl-3-methylpyridine (64%) by this process. The properties, reactions, and uses of the monohalomethylpyridines were discussed (357, 358). Chlorine is reported to react with 5-ethyl-2-methylpyridine in concentrated hydrochloric acid to form 5-(α-chloroethyl)-2-methylpyridine from which 2-methyl-5-vinylpyridine could be obtained (359).

The Friedel-Crafts reaction fails in the case of the pyridines and the possible reasons for this lack of activity in pyridines was discussed in Chapter I (1st ed.). Although the pyridines do not undergo a Friedel-Crafts reaction with benzyl halides, the reverse reaction in which a chloromethylpyridine-HCl reacts with benzene in presence of aluminum chloride produces isomer-free benzylpyridines in high yields (see Table V-12) (360).

TABLE V-12. Benzylpyridines from the Friedel-Crafts Reaction Using Chloromethylpyridine·HCl (360)[a]

Reactant	Product	Yield (%)	b.p./mm	m.p. (picrate)
2-Chloromethylpyridine·HCl	2-Benzylpyridine	80	145–147°/15	141–142
3-Chloromethylpyridine·HCl	3-Benzylpyridine	79	159–160°/17	118
4-Chloromethylpyridine·HCl	4-Benzylpyridine	75	156–158°/16	140.5–144
2-Chloromethyl-6-methylpyridine·HCl	2-Benzyl-6-methylpyridine	73	150–153°/17	141–142
2,6-Bis-(chloromethyl)-pyridine·HCl	2,6-Dibenzylpyridine	76	164–168°/0.2	175–176

[a] We thank *Verlag Chemie* for permission to use this table.

2-Phenacylpyridine was prepared by heating benzoyl chloride and 2-methylpyridine (361). A similar reaction was observed on treating 4-methylpyridine and 4-ethylpyridine with benzenesulfonyl chloride, the acyl group migrating from the N atom to the C-4 α-carbon (362).

G. Side-Chain Metalation

The lateral metalation of the 2-, 3-, or 4-methyl group in substituted pyridines provides a convenient method of introducing various substituents into the α-position. The metalation can be accomplished by treating the picoline with sodium or potassium amide in a solvent, usually liquid ammonia, or by reacting the picoline with an organolithium compound in a solvent such as ether or tetrahydrofuran in an inert atmosphere. These metalated pyridines react with alkyl halides to provide the homologues of the picolines (146, 363) (see also Section I.2.C.c.), with halo-substituted polycyclic alkadienes such as 1,2-dichloro-2,5-norbornadiene to form polyhalo-substituted polycyclic pyridine compounds useful as insecticides (364), and with ω-halogenoacetylenes (V-126) to form alkynylpyridines (V-127) (365)

$$R\text{—}\underset{N}{\bigcirc}\text{—}CH_2M \;+\; X(CH_2)_nC\!\equiv\!CH \;\longrightarrow\; R\text{—}\underset{N}{\bigcirc}\text{—}(CH_2)_{n+1}C\!\equiv\!CH$$

V-125 M = Li, Na, K   V-126   V-127

The laterally metalated (Li and Na) 3- and 4-methyl-, 2-ethyl, and 2-isobutylpyridines (V-125) reacted with esters (V-128) to form the picolyl ketones (V-129) (366–368). When organolithium compounds were employed for the metalation, the products of addition to the azomethine bond were also present (367). 2-Picolyllithium reacted with ethyl 2-pyridylacetate or 2-pyridylacetonitrile in the same way to form 1,3-di-(2-pyridyl)acetone (369). Forcing conditions using excess picolyllithium favored further reaction with the ketone to produce the carbinol V-130. These compounds had useful

V-125 + R^{1}COOR^2 $\longrightarrow$

V-128

$$R\text{—}\underset{N}{\bigcirc}\text{—}CH_2COR^1 \;\xrightarrow{R\text{—}\underset{N}{\bigcirc}\text{—}CH_2Li}\; \left[R\text{—}\underset{N}{\bigcirc}\text{—}CH_2\text{—}C\underset{R^1}{\overset{OH}{|}} \right]_2$$

V-129   V130

pharmacological properties. They diminished the tremors induced by Tremorine and inhibited enzymatic oxidation reactions of the liver enzymes (370).

Reaction with aldehydes and ketones gave the *sec*- and *t*-carbinols respectively (371–373), while formaldehyde gave primary alcohols (374). Kauffmann and co-workers found that 1,2-bis(6-methyl-2-pyridyl)ethane (V-131) underwent selective monolithiation with butyllithium at the methyl group and *not* the methylene group (373).

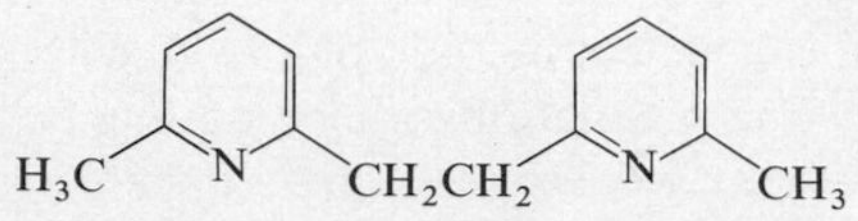

V-131

The picolyllithium compounds on treatment with carbon dioxide gave the pyridylacetic acids which were isolated as their esters (375, 376). Chloromethoxymethane produced the β-methoxyethylpyridine (377) and chloroalkylnitrile gave dicyanoalkanes **V-132** (378, 379).

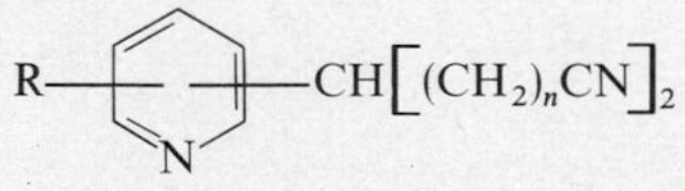

V-132

The nitrosation of 2-, 3-, and 4-methyl-, 2- and 4-ethyl-, 2- and 4-phenethyl-, 2- and 4-benzyl-, and 2,4- and 2,6-dimethylpyridines was accomplished by treating these compounds with sodium or potassium amide in liquid ammonia and then with an alkyl nitrite. The products were the respective aldoxime or ketoxime (380–382).

Treating 1,2-dimethylpyridinium iodide or bromide with sodium methoxide in methanol and butyl nitrite gave 1-methylpyridine-2-aldoxime iodide or bromide (383, 384). Alkyl nitrates gave the nitro derivatives in high yields (70 to 90%) (385).

The addition of 2-picolyllithium to Schiff bases is a convenient route to *N*,1-disubstituted-2-(2-pyridyl)ethylamines (V-133) (386), and chloromethyltrimethylsilane or chloromethylphenyldimethylsilane gave the organosilicon substituted pyridines which were useful as solvents and lubricants (387, 388).

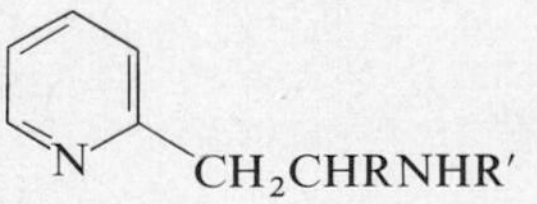

V-133

Grignard reagents prepared by treating methyl substituted pyridines with methylmagnesium bromide or isopropylmagnesium chloride gave the carbinols on reaction with ketones (389, 390).

H. Condensation Reactions

The side-chain alkylation and aralkylation of 2- and 4-alkylpyridines (391–394) has already been discussed (see Section I.2.C.a). The cyanoethylation of the methyl group of 2- and 4-methylpyridines using acrylonitrile and catalytic amounts of sodium was reported. By using the appropriate ratio of reactants, one or two cyanoethyl groups can be introduced (379).

Hydroxymethylation of the methyl group at the C-2 or C-4 position of the pyridine ring occurs on heating methyl substituted pyridines with formaldehyde or its polymers. The mono(hydroxymethyl)-compounds (V-134) are the usual products (395–400), although di- (V-135) and tri-(hydroxymethyl)-derivatives (V-136) can also be prepared (401–403).

V-134   V-135   V-136

The condensation of 2- or 4-alkylpyridines with carbonyl compounds to form unsaturated compounds continues to attract attention (404–410). The alkylpyridine methiodides also underwent condensation (408) and acetylation of the alkylpyridine prior to condensation was reported to improve the yield (404). The cis-isomer of α-stilbazole was separated from the known trans-isomer by crystallization of the trans-isomer. The liquid cis-isomer was purified by chromatography (410).

I. Synthesis of Condensed Ring Systems

2-Benzyl- and 2-ethylpyridines reacted with α-bromoketones to form indolizines (pyrrocolines). The reactivity of the C-1 and C-2 positions in the indolizine ring was explained by the increased electron densities at these positions (411).

Acetylenes react with pyridine and its homologues to give a variety of products depending on the solvent and reaction conditions employed. These reactions and the reaction mechanisms involved were the subject of a review by Acheson (412). An exothermic reaction took place on mixing pyridine, the picolines, and 3,5-dimethylpyridine with dimethyl acetylenedicarboxylate in ether. From these reactions "labile" 1:2 adducts (9aH-quinolizines) (**V-137**), "stable" 1:2 adducts (4H-quinolizines) (**V-138**) (RRI 1688), and "Kashimoto compounds" **V-139** were isolated. The "labile" adducts rapidly isomerized to the "stable" isomers (412). The 9aH-quinolizine could

            V-137   **V-138**   **V-139**

add dimethyl acetylenedicarboxylate to give a 1:3 adduct **V-140** (RRI 1688) (413). Jackman and co-workers isolated unstable products on treating pyridine or 3-methylpyridine with dimethyl acetylenedicarboxylate in ether at $-50°$ and suggested "ylid" structure **V-141** for these products (414), later changing it to the 1-pyridinium-1-(1,2-dicarbomethoxyvinyl)-β-carboxylate (**V-142**), the 1:1:1 adduct of pyridine, dimethyl acetylenedicarboxylate, and carbon dioxide (present in the cooling bath) (415). This latter structure was verified by Acheson and Plunkett (416).

            V-140   **V-141**

            V-142

In methanol as solvent the pyridines form indolizines (**V-143**) (RRI 1276) ($R^2 = COOCH_3$ and $CH(OCH_3)COOCH_3$) with dimethyl acetylene-dicarboxylate. A mechanism for the formation of these compounds was proposed (412). When pyridine, 3- and 4-methyl-, or 3,5-dimethylpyridine

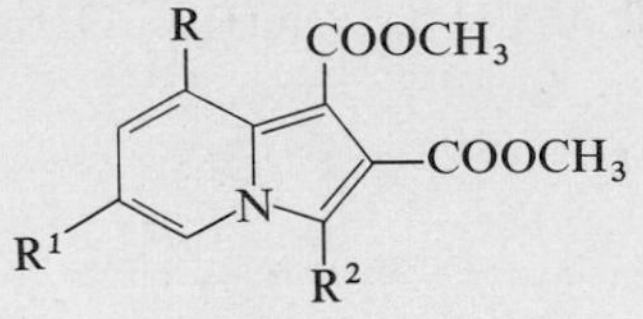

R COOCH₃

V-143

were treated with methyl propiolate in acetonitrile, the cycl[3.2.2]azines **V-144** (RRI 8333) were isolated (417). In the case of 3,5-dimethylpyridine the corresponding indolizine was also isolated and could be converted to the cyclazine on refluxing with methyl propiolate in toluene with palladium on charcoal (417, 418).

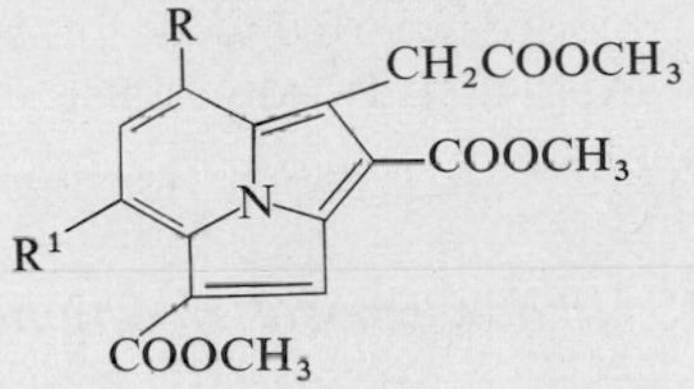

V-144

Thieno[2,3-*b*]pyridine (**V-145**) (RRI 1347) and thieno[3,2-*c*]pyridine (**V-146**) (RRI 1345) were obtained by treating pyridines substituted at the 3-position by ethyl, vinyl, α-hydroxyethyl, or acetyl groups with H_2S at 630° (419, 420).

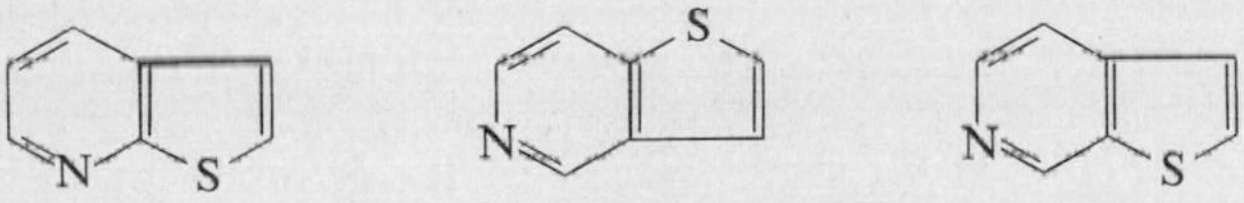

V-145   **V-146**   **V-147**

4-Ethylpyridine was converted to 4-ethyl-3-pyridinethiol, which was dehydrogenated to thieno[2,3-*c*]pyridine (**V-147**) (RRI 1346) (421).

2-Picolyllithium and 1,2-epoxyindan gave 1-(2-pyridylmethyl)-2-indanol (**V-148**), which underwent reductive cyclization to 5a,7,8,9,10,10a,11, 11a-octahydroindeno[1,2-*b*]indolizine (**V-149**) (422).

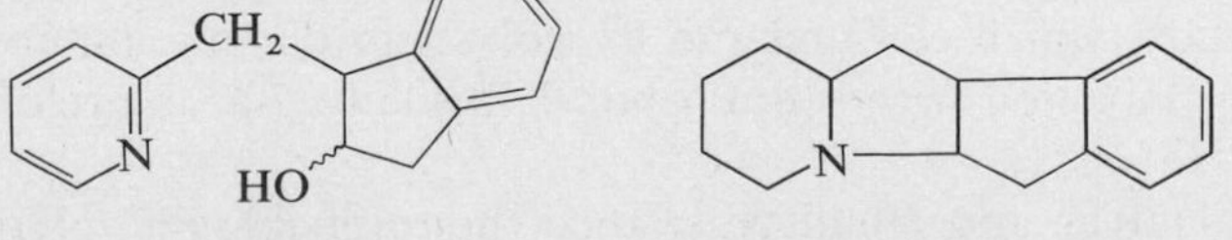

V-148   **V-149**

The 2-hydrazinopyridines, prepared by treating pyridine or 4-methyl-pyridine with sodium hydrazide, gave the cyclized products **V-150** (RRI

1094) with acetic acid (R = CH$_3$), urea (R = OH), and cyanogen bromide (R = NH$_2$) (349).

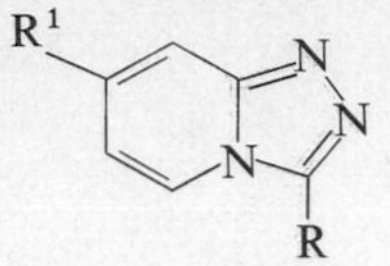

V-150

J. Ring-Opening Reactions

Small amounts of pyrrole are reported to be formed on treating pyridine with degassed Raney Nickel (423). Pyridines also undergo photochemical ring cleavage (see Section I.4.K).

K. Photochemical Reactions

The photochemical alkylation of pyridines and the photodimerization of pyridylalkenes have already been discussed (see Section I.2.C.e). Whitten and Lee found that irradiation of 1,2-bis(4-pyridyl)ethylene in methylcyclohexane or 2-propanol produces the photoreduction product 1,2-bis(4-pyridyl)ethane and the photoaddition products **V-151** (R = C$_7$H$_{13}$ and

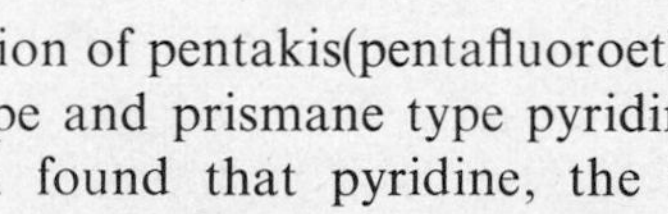

V-151

C(CH$_3$)$_2$OH, respectively). The mechanism of this reaction was discussed (424).

The photoisomerization of pentakis(pentafluoroethyl)pyridine gives thermally stable Dewar type and prismane type pyridines (see Section I.4.A). Wilzbach and Rausch found that pyridine, the picolines, and several lutidines (**V-152**) form thermally unstable Dewar type pyridines (**V-153**) on irradiation, which can undergo hydrolysis to olefinic amino aldehydes (**V-156**), or reduction with sodium borohydride to 2-azabicyclo[2.2.0]hex-5-enes (**V-155**) (425).

Joussot-Dubien and Houdard studied the reversible photolytic cleavage of pyridine in water or alcohol and proposed the formation of the unsaturated amino aldehyde (**V-156**) (R = R^1 = H), which was detected by ultraviolet absorption spectroscopy (426).

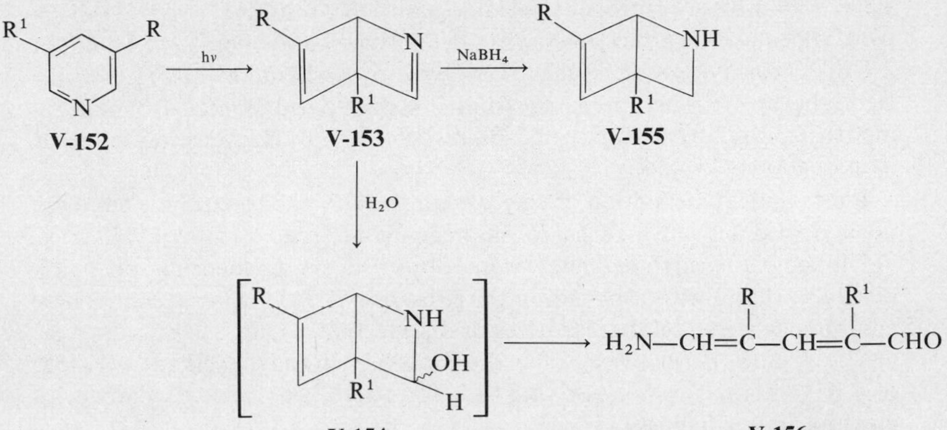

V-152 → V-153 → V-155

V-154 → V-156

The 3,5-dicarbalkoxy-substituted pyridines gave on photolysis a variety of products including pyrroles and dihydropyridines (427).

5. Heteroaralkylpyridines Including Di- and Polypyridylalkanes

The reaction of α,ω-dihaloalkanes with picolyllithium (428), picolylsodium (429), or pyridyl Grignard reagents (430) was used to prepare a series of di-(2-pyridyl)alkanes. Di-2-pyridylmethanes (V-157) were obtained from the reaction of pyridines with substituted 2-picolylsodium (431).

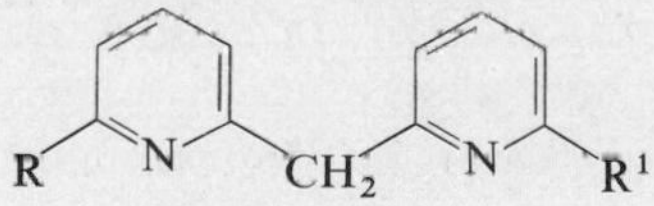

V-157

4-Vinylpyridine condensed with pyrroles in the presence of acid to form 2-[2-(4-pyridyl)ethyl]pyrroles (V-158). These compounds were found to possess hypotensive and antitremor activity (432). 2,4-Di(2-pyridyl)pentane

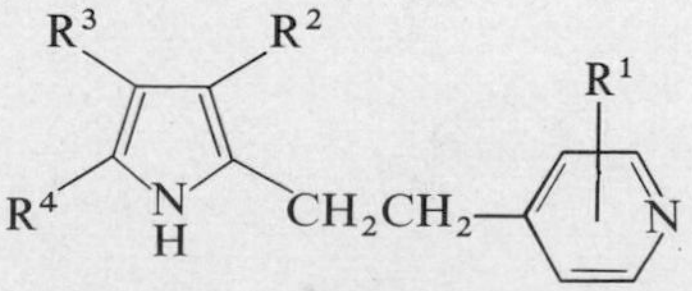

V-158

was prepared by refluxing 2-ethylpyridine and 2-vinylpyridine in the presence of potassium. Ethyl α-(2-pyridyl)propionate was condensed in the same way with 2-isopropenylpyridine, and the resulting ethyl 2,4-di-(2-pyridyl)pentane-2-carboxylate was hydrolyzed and decarboxylated to 2,4-di-(2-pyridyl)pentane (433). 1,3-Di-(2-pyridyl)butane and 1,2-di-(2-pyridyl)propane were prepared from 2,4-di-(2-pyridyl)butan-1-ol and 2-methyl-2,3-di-(2-pyridyl)propan-1-ol, respectively, by dehydration followed by reduction (433, 434).

Electrolytic dimerization of vinylpyridines (435) and electrolytic coupling of vinylpyridines with alkenes (75) has been used to prepare dipyridylalkanes.

Diphenyl(2-pyridyl)methanes were obtained on condensing phenyl(2-pyridyl)carbinol with a phenol in the presence of 94% sulfuric acid. These compounds were useful as laxatives or disinfectants (436).

The sodium derivatives of the three diphenylpyridyl methanes (V-159; R = H), obtained by reaction with sodamide in toluene, were alkylated with alkyl or aralkyl halides (437).

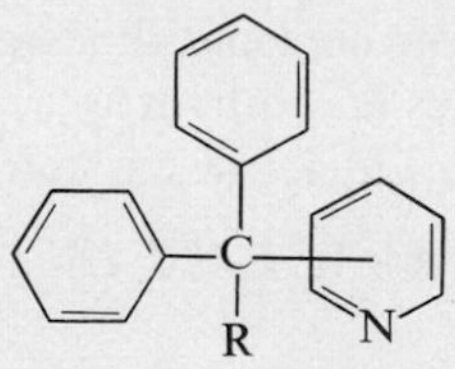

V-159

The sodium derivatives of various dipyridylalkanes gave, on treatment with ethylene oxide, bis-(β-hydroxyethyl)-dipyridylalkanes which could be used to enhance the dyeing properties of synthetic fibers to improve the quality of rubber and to protect steel (438).

N-(Diarylpyridylmethyl)imidazoles (V-160) with antimycotic properties (439) and 2-[α-(2-pyridyl)benzyl]imidazoline (V-161) with antidepressive activity (440) were prepared.

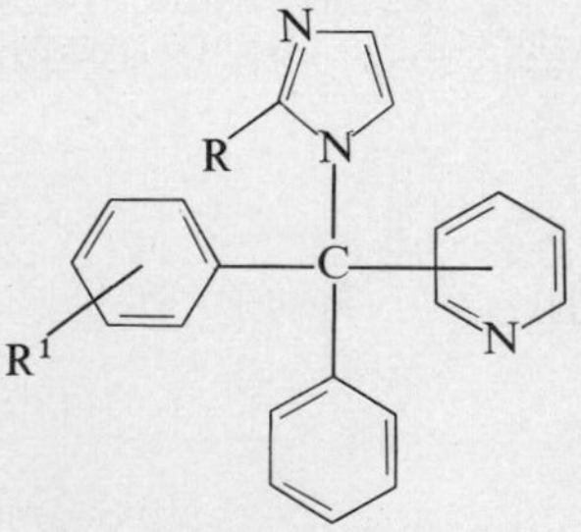

V-160

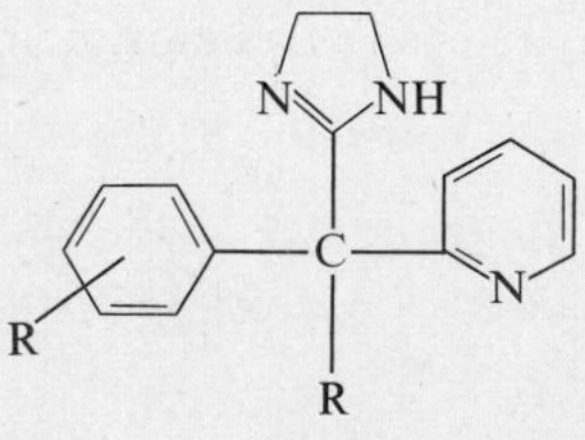

V-161

6. Pyridinophanes

The pyridinophanes comprise the group of pyridines bridged at the *ortho*-(2,3- or 3,4-), *meta*-(2,4- or 3,5- or 2,6-), or *para*-(2,5-) positions by hydrocarbon or substituted hydrocarbon residues. Substitution of the pyridine ring would increase the number of isomers possible. Since the bridge may itself carry one or more pyridine or other aryl or heteroaryl moieties, a large number of pyridinophanes are possible. A book by Smith covers the nomenclature, methods of synthesis, and properties of the bridged aromatic compounds (including the pyridinophanes) (441), and Vögtle and Neumann have recently proposed a systematic method of nomenclature for the phanes (442, 443). This section on the pyridinophanes is included because of the current interest in this area and because some of the pyridinophanes, due to steric interactions, possess properties (spectroscopic and chemical) quite different from those shown by the alkylpyridines (441, 444, 445). In Section I.1, some of the methods of synthesising *ortho*-bridged pyridinophanes were described.

The structure of muscopyridine, a base isolated from the perfume gland of the musk deer was established as 9-methyl-[10](2,5)pyridinophane (**V-162**; R = CH$_3$, R^1 = H), and this structure was confirmed by a multi-stage synthesis (446).

V-162   V-163

The three-step synthesis by Gerlach and Huber of a series of [*n*](2,5)-pyridinophanes (**V-163**; X = CH$_2$) (61) has been mentioned (Section I.2.B.d). Reduction of the lower members (*n* < 12) of the [*n*](2,5)pyridinophan-*n*-ones using lithium aluminium hydride gave diastereomeric [*n*](2,5)-pyridinophan-*n*-ols. [9](2,5)Pyridinophane was also resolved into its enantiomers **V-164** and **V-165** (61).

V-164   V-165

Nozaki, Fujita, and Mori prepared [7](2,6)pyridinophane (**V-166**) (44%) by the action of hydroxylamine hydrochloride (2.5 equivalents) on cyclo-dodecane-1,5-dione. This compound failed to form an N-oxide as a result of the steric effect of the bridge (447), while the corresponding decamethylene bridged compound formed an N-oxide readily (446).

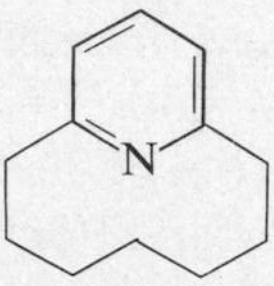

V-166

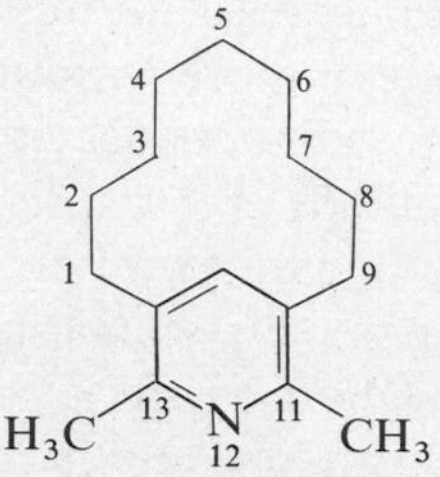

V-167

11,13-Dimethyl[9](3,5)pyridinophane (**V-167**) was prepared by Balaban from the corresponding pyrylium perchlorate (448). Georgi and Rétey employed a similar approach to prepare 13-methyl[10](2,6)pyridinophane (**V-162**; R = H, R^1 = CH_3) and the dimethyl[10.10](2,6)pyridinophane (**V-168**) (449).

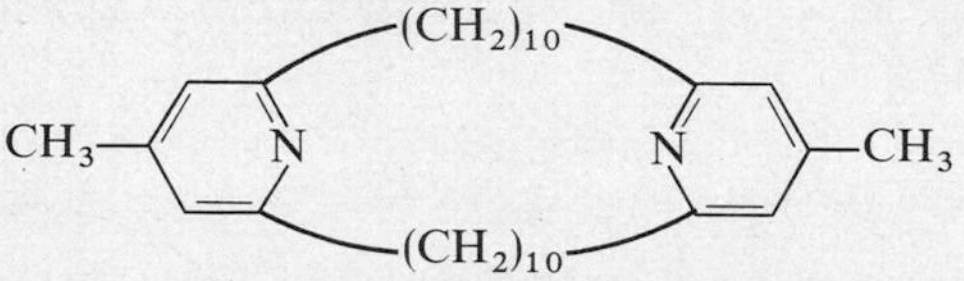

V-168

Prelog and Geyer used a multistage process for the synthesis of [13](2,3)-pyridinophane (450).

Sutherland and co-workers have prepared a series of [2.2]metacyclo-phanes (**V-169**; A = B = CH; A = CH, B = N; A = B = N) which were shown to have a non-planar chair conformation as shown. These structures

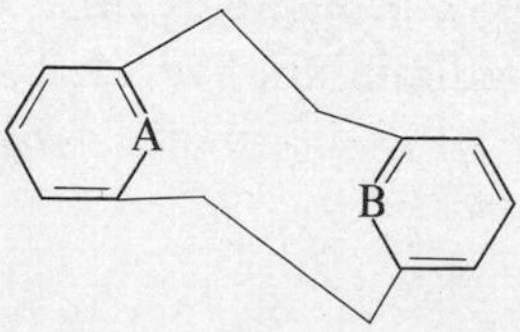

V-169

were conformationally rigid and ring inversion occurred only on heating, as shown by NMR spectroscopy. From this study the nonbonded inter-actions were found to be in the order N, N < N, CH < CH, CH (451, 452).

Jenny and Holzrichter subjected 3,5-bis(chloromethyl)pyridine to a Wurtz reaction with sodium and tetraphenylethylene and, by gradient elution chromatography of the product, isolated [2.2](3,5)pyridinophane (V-170) (5%), [2.2.2](3,5)pyridinophane (V-171) (4.2%), and [2.2.2.2](3,5)-pyridinophane (V-172; A = N, B = CH) (1.5%) (453). The same reaction using 1,2-bis(6-chloromethyl-2-pyridyl)ethane gave [2.2.2.2](2,6)pyridino-phane (V-172; A = CH, B = N) (4.2%) and [2.2.2.2.2.2](2,6)pyridinophane (2.1%). The octameric pyridinophane was also detected spectrometrically (454).

V-170   V-171

V-172

Starting with 2,6-bis(chloromethyl)pyridine Bockelheide and Lawson pre-pared [2.2](2,6)pyridinophane-1,9-diene (V-173) by an interesting sequence of reactions. This compound did not undergo spontaneous valence tautomer-ization to 15,16-dihydro-15,16-diazapyrene (V-174) as predicted by earlier

V-173   V-174

workers (455). Boekelheide and Pepperdine also prepared 4,6,8,12,14,16-hexamethyl[2.2](3,5)pyridinophane (V-175) by a Wurtz reaction on 3,4-bis(bromomethyl)-2,4,6-trimethylpyridine and 4,6,8,16-tetramethyl-5-aza-[2.2]metacyclophan-1-ene (V-176) which was dehydrogenated by ruthenium-on-alumina to trans-1,3,15,16-tetramethyl-2-azadihydropyrene (V-177). This latter compound, like other dihydropyrenes, undergoes photoisomerization to the metacyclophane-1,9-diene valence tautomer (V-178) (456).

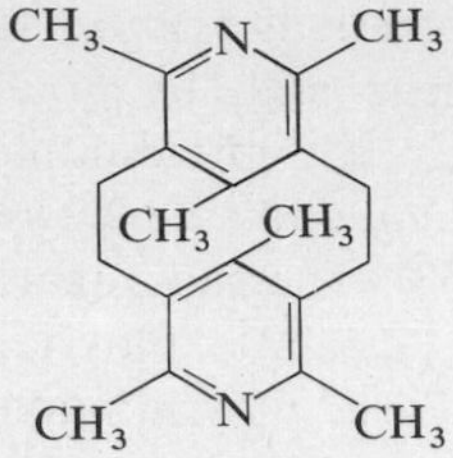

CH₃ N CH₃

V-175 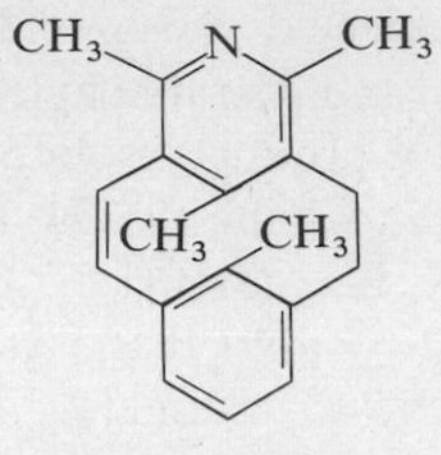   **V-176** 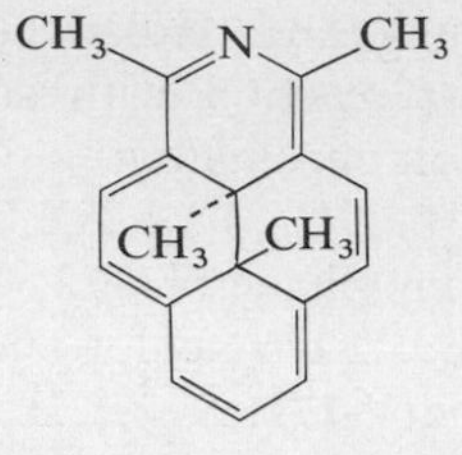   **V-177**

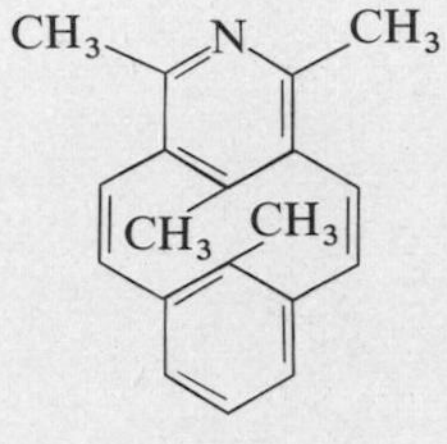

V-178

A novel hexadentate macrocyclic ligand Schiff base, 6,12,19,25-tetra-methyl-7,11,20,24-dinitrilodibenzo[*b*,*m*][1,4,12,15]tetraazacyclodocosine (**V-179**) was obtained by condensing 2,6-diacetylpyridine with *o*-phenylene-diamine (457).

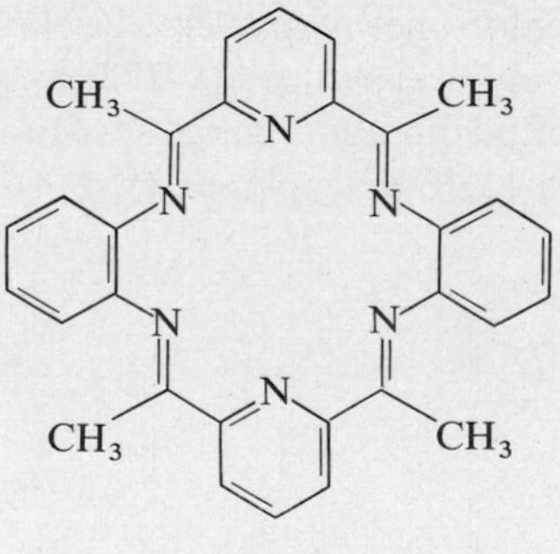

V-179

7. Pyridine Ylids

Certain reactions of 1-methylpyridinium halides are recognized as proceeding *via* one of two possible pyridine ylid intermediates **V-180** or **V-181**. Studies on the deuterium exchange of 1-methylpyridinium salts have indicated the formation of the ylid **V-180** as a reaction intermediate (458).

V-180   **V-181**

Kröhnke found, however, that 1-(2-hydroxy-2-phenylethyl)pyridinium bromide (**V-182**) was the product (over 80%) when 1-methylpyridinium bromide was condensed with benzaldehyde, the reaction proceeding through the ylid **V-181** (459).

Quast and co-workers have shown that the ylid **V-180** is conveniently obtained by the thermal decomposition of the betaine 1-methylpyridinium-2-carboxylate. The ylid can be trapped by electrophilic reagents (460–462).

Howe and Ratts have studied these reactions and found that thermolyzing 1-methylpyridinium-2-carboxylate hydrochloride in the presence of benz-aldehyde gave 2-(α-hydroxybenzyl)-1-methylpyridinium chloride (**V-183**)

V-182   **V-183**

(83%) (463). This novel synthetic approach merits further investigation since it has the obvious advantages of simplicity and high yields.

II. Alkenyl- and Aralkenylpyridines

Reviews have appeared on the synthesis and properties of 2-vinylpyridine (464, 465), 3-vinylpyridine and 4-vinylpyridine (465), and 5-ethyl-2-vinyl-pyridine (466).

1. Preparation

A. From Metallopyridine Compounds

Three different approaches have been employed for the synthesis of alkenyl and aralkenylpyridines from metallopyridines. Picolyllithium or picolyl-magnesium bromides on reaction with alkenyl halides form the alkenyl or aralkenylpyridines (467) (see also Chapter 1, 1st ed.).

The lithium derivatives of pyridines, picolines, or benzylpyridine react with carbonyl groups to form the respective carbinols, which are then dehydrated (468–471). Conversely, an organolithium compound can be treated with a pyridine aldehyde or ketone to give the carbinol which, on dehydration, produces a mixture of the *cis-* and *trans-*stilbazoles (472). *Trans-*1,2-di-(2-pyridyl)ethylene (**V-184**) was obtained in 60% yield (468), and Villani and co-workers prepared a number of stilbazoles by this method and examined their hypocholesteremic and estrogenic properties (469). The reaction of 4-picolyllithium with isobuteroyl chloride in tetrahydrofuran gave the carbinol (**V-185**) which was dehydrated to the olefin (**V-186**) (470).

V-184   V-185

V-186

A third approach utilizes the addition of a picolyllithium to a Schiff base. At low temperatures and with short reaction periods the product is a pyridylethylamine. On warming or with prolonged reaction times there is a rapid elimination of an amine with the formation of a stilbazole. 4-Stilbazole (36%) and 1,1-diphenyl-2-(4-pyridyl)ethylene (81.4%) were prepared by this method (473).

B. Condensation with Carbonyl Compounds

The condensation of aldehydes and ketones with 2- and 4-alkylpyridines containing two α-hydrogen atoms provides a convenient method of preparing alkenyl compounds (see Sections I.2.C.a and I.4.H and Chapter V, 1st ed.). In addition to the 2- and 4-alkylpyridines (466, 474–490), their quarternary salts (491–493), N-oxides (478), and N-ethoxypyridinium halides (494) also condense with carbonyl compounds. Some of the pyridine-1-oxides were reported to undergo concomitant deoxygenation (478). In these condensations the *trans*-stilbazole isomer predominates (474). Condensations employing higher aromatic ring aldehydes were successful (479). 1,2-Bis(2-pyridyl)ethylene was prepared by condensing pyridine-2-aldehyde with 2-methylpyridine using zinc chloride as the catalyst (490). Preliminary treatment of the alkylpyridine with an acid chloride followed by condensation with the carbonyl compound (404, 478), and the use of chlorocyclophosphazines (477) and piperidinium acetate (478) as catalysts have been described. Condensation with dialdehydes such as glyoxal or terephthaldehyde produces the dialkenyl compounds (476, 491).

The use of formaldehyde for the preparation of vinylpyridines *via* the β-hydroxyethylpyridine was investigated in detail (481–493). In all cases the intermediate carbinol could be isolated. 2-Picoline gave 2-vinylpyridine (485, 487, 488); 2-methyl-5-vinylpyridine gave 2,5-divinylpyridine (482, 483, 489); 2,6-dimethylpyridine gave 2-methyl-6-vinylpyridine (484) and 2,6-divinylpyridine (481, 484); 5-ethyl-2-methylpyridine gave 5-ethyl-2-vinyl-pyridine (466); and 2,4,6-trimethylpyridine gave 4,6-dimethyl-2-vinyl-pyridine, 4-methyl-2,6-divinylpyridine, and 6-methyl-2,4-divinylpyridine (486) by this method.

The pyridylacetic acids (including the 3-isomer) condensed with aldehydes and the resulting β-phenyl-α-pyridylacrylic acids (**V-187**) gave stilbazoles on decarboxylation (495–497).

V-187

3-(Pyridylmethylene)oxindoles (**V-188**) have been prepared by the condensation of isatin with picoline (857, 858). The Knoevenagel condensation of the pyridine aldehydes with oxindole is reported to be a versatile method for preparing these compounds, whose geometry has not been established (498). Pyridine-3-aldehyde condensed in the same way with fluorene to form 9-(3-pyridylmethylidene)fluorene (50 to 60%) (499).

V-188

The versatile Wittig reaction has only recently been applied to the synthesis of alkenylpyridines (456, 500–502). Baker and Gibson prepared a large number of 4-stilbazoles (**V-189**) by reacting pyridine-4-aldehyde (**V-190**) with benzyltriphenylphosphonium halide (**V-191**) or, conversely, by the action of aryl aldehydes (**V-192**) on picolyltriphenylphosphonium halides (**V-193**), in the presence of base. These stilbazoles were tested as inhibitors of choline acetyltransferase (500). These condensations gave mixtures of the *cis*- and

V-190   **V-191**   **V-189**

V-192   **V-193**

trans-stilbazoles which were separated and identified by their ultraviolet absorption spectra (500).

Bodalski and co-workers used *O,O*-diethyl 3-pyridylmethylphosphonate with potassium *t*-butoxide as the base to prepare alkenylpyridines (502).

The stilbazole **V-194**, obtained in 66% yield by the Wittig reaction, gave the novel 4-azahexahelicine (**V-195**) on photolytic cyclodehydrogenation in the presence of iodine (501).

V-194   **V-195**

Neobetenamine (**V-196**) was synthesized by the Vilsmeier-Haack reaction between 4-methylpyridine and *N*-formylindoline (503).

V-196

C. Condensation with Dienes

Alkyl-substituted pyridines containing at least one α-hydrogen atom undergo base-catalyzed condensations with dienes to give alkenylpyridines (504–509). Usually a mixture of products is obtained. Lithium, sodium, potassium, sodamide, and potassium *t*-butoxide were used as the catalyst. Pines and co-workers have shown that there is a marked difference in the products obtained depending on whether lithium, sodium, or potassium is employed. Potassium is the most reactive but the least selective, while lithium is the most selective but has the slowest addition rate. Thus in the condensation of 4-isopropylpyridine with isoprene, lithium gives the "tail" addition product **V-197** as the major product, while potassium provides an approximately 1:1 mixture of both "tail" and "head" (**V-198**) addition products. Diaddition products are also produced in these reactions (506).

V-197   V-198

Another complication that arises particularly when sodium is employed is the migration of double bonds (509).

These reactions usually employ excess of the alkylpyridine as the solvent. Pines and Stalick found that a homogeneous reaction is possible when potassium *t*-butoxide is used in dipolar aprotic solvents such as hexamethylphosphoramide and dimethylsulfoxide. Advantages claimed are short reaction times, high yields of products, and little side reaction. Alkylations and aralkylation also proceed smoothly in this system (507).

The Diels-Alder reaction between butadiene and acrylonitrile normally produces cyclohexenenitrile. However in the presence of a catalyst, 2-vinylpyridine is also formed. Janz and Duncan have studied this cycloaddition in detail (510, 511).

D. Dehydration of Carbinols

The conversion of the readily available pyridinecarbinols into alkenylpyridines may be brought about by direct dehydration or by conversion to the alkyl halide followed by dehydrohalogenation. The dehydration is usually accomplished in the presence of a basic or acidic catalyst or by thermolysis of the alcohol or the derived acetate.

The preparation of vinylpyridines from the hydroxyethylpyridines (see Section II.1.B) is usually accomplished by heating with aqueous sodium or potassium hydroxide (481–493, 512–514). A modification of this method in which 2-pyridylethanol is dehydrated and the product removed as it is formed, by steam distillation, gave an essentially quantitative yield of 2-vinylpyridine (514).

The dehydration of carbinols using acid catalysts is reported (515–519). Alkenylpyridines containing amino substituents in the pyridine ring (518) or in the side chain (515) were made by this process. Pyridylindenes (**V-200**) were made by the dehydration of the indanol (516) or by the cyclodehydration of 2-phenyl-3-pyridylbutan-2,3-diols (**V-199**) (517). An interesting synthesis

V-199   V-200

of 5-*n*-butyl-2-propenylpyridine is that of Hardegger and Nickles. 5-*n*-Butyl-2-propylpyridine was converted to its *N*-oxide, which with acetic anhydride gave 2-(α-acetoxy-*n*-propyl)-5-*n*-butylpyridine. This compound was hydrolyzed to the alcohol and dehydrated to the alkenylpyridine (519).

The thermolysis of α-hydroxalkyl- or α-acetoxyalkylpyridines is a

V-201   V-203   V-202   V-204

V-205

satisfactory method for the synthesis of vinylpyridines (520–522). 2-Vinyl-pyridine (67 %), 2-(1-butenyl)pyridine (34 %), and 4-vinylpyridine (83 %) were prepared by this route (522). Yoshida and co-workers studied the flash thermolysis of 1-arylbut-3-enyl acetates. They found that compounds such as 1-(4-pyridyl)but-3-enyl acetate (**V-201**) gave the diene **V-202** and the Cope reaction products **V-203** and **V-204** (RRI 1708). A totally unexpected product **V-205** (RRI 1707) was also found and its formation is explained by an azacyclodecapentene intermediate (523).

Heating 2-methyl- and 4-methylpyridine-1-oxides with benzyl alcohol in the presence of potassium hydroxide gave the 2- and 4-stilbazoles. 2-Pyridine-methanol, 4-pyridinemethanol, and 6-methyl-2-pyridinemethanol could be used instead of benzaldehyde, to prepare the *trans*-1,2-dipyridylethylenes (524).

E. Side-Chain Dehydrohalogenation or Dehalogenation

The method of dehydrohalogenation is seldom used because the required haloalkanylpyridines are usually obtained from the carbinols. 2-Methyl-5-vinylpyridine (41 to 43 %) was obtained by treating 5-(α-chloroethyl)-2-methylpyridinium hydrochloride with trialkylamine and the resulting quarternary ammonium salt with strong aqueous potassium hydroxide (525). The unstable 6-butenyl-2-methylpyridine was also obtained from the alcohol *via* the chloride (526).

The zinc debromination of 1,2-dibromo-2-(2,5-dimethylpyrid-4-yl)-1-phenylethane gave 2,5-dimethyl-4-styrylpyridine (**V-206**) (527).

V-206

F. Side-Chain Dehydrogenation

The vapor phase catalytic dehydrogenation of 5-ethyl-2-methylpyridine (MEP), usually in the presence of steam, to produce 2-methyl-5-vinyl-pyridine (MVP), was studied extensively (528–537). Yields of up to 80 % MVP at 70 to 80 % conversion of MEP were reported (529). Purification of the product by distillation presents problems because of the polymerization of MVP. Haskell and McKay effected purification by treatment with *n*-pentane, in which the polymer alone is insoluble (538). Dehydrogenation of ethylvinylpyridine to divinylpyridine (539) and 4-ethylpyridine to 4-vinyl-pyridine (536, 540, 541) were reported. A method of depolymerization leading to the recovery of 48.7 to 51.1 % of vinylpyridine was patented (542). The use of sulfur, carbon disulfide, thiourea, thioglycollic acid, or thioacetamide to effect dehydrogenation of alkylpyridines was described (543).

2. Properties

The physical properties of the alkenylpyridines are summarized in Table V-13, and the aralkenylpyridines in Table V-14.

The stilbazoles **V-207** and **V-208** (A = CH) and the 1,2-dipyridylethylenes **V-207** and **V-208** (A = N) exhibit geometric isomerism, and the *cis*-**V-208** and *trans*- **V-207** isomers of the 2-stilbazoles (474, 500, 544, 545), 1,2-bis(2-pyridyl)ethylene (490, 524, 546), and 1,2-bis(4-pyridyl)ethylene (524, 546, 547) were separated by column chromatography. The isomers so far isolated have usually been the *trans*-isomers because they are more stable and invariably have higher melting points than the *cis*-isomers.

V-207   V-208

The photolytic *cis-trans* isomerization of the 2-stilbazoles was studied. The methyl iodide or hydrochloride salt of 2-stilbazole on irradiation underwent dimerization (139, 545) (see Section I.2.C.e). In solution there was interconversion of the *cis*- and *trans*-isomers, the equilibrium being determined by the wavelength of the radiation. Spectroscopy was employed for the analysis of these mixtures (544, 545, 548).

The infrared spectra of several alkenylpyridines have been published (181). The infrared and ultraviolet spectra of *cis*-2-stilbazole (544) and *cis*- and *trans*-1,2-bis(4-pyridyl)ethylene (547) were reported. It was observed that the 1,2-disubstituted ethylenes show a strong *trans*-CH=CH band at 975 to 991 cm^{-1} in the infrared region (524). The alkenylpridines in methanol, hydrochloric, or sodium hydroxide solution show two absorption bands in the ultraviolet region. In the case of the vinylpyridines these bands are at about 282 and 236 mμ in both methanolic or basic solutions, while in acid solution they occur at 292 and 235 mμ (549). Ultraviolet spectroscopy

TABLE V-13. Alkenylpyridines

Compound	B.p.	n_{D}^{20}	Picrate (m.p.)	Salt (m.p.)	Ref.
2-Vinylpyridine	49–51°/11 mm	1.5476			522
4-Vinylpyridine	78–82°/23 mm	1.5450	189–190°		520
2-(1-Butenyl)pyridine	52–56°/4 mm	1.5422			522
2-(4-Butenyl)pyridine	74–75°/12 mm	1.5029			467
2-(3-Pentenyl)pyridine	68–75°/4 mm	1.5080			393
3-(3-Pentenyl)pyridine	97.6–97.8°/10 mm	1.5133			393
2-(6-Hexenyl)pyridine	103–104°/11 mm	1.5059			467
1-(3-Pyridyl)-2-methylprop-1-ene	84–86°/19 mm	1.5420²²	134–135°		502
2-(2,7-Nonadien-5-yl)pyridine	113–119°/5–6 mm	1.5118			393
3-(2,7-Nonadien-5-yl)pyridine	89–92°/1.0 mm	1.5206	64.5–65.5°		393, 505
3-(Tri-2-butenyl)methylpyridine	138–145°/1.0 mm	1.5290			505
2,5-Divinylpyridine	70°/4 mm	1.5945			483
2,6-Divinylpyridine	88–89°/16 mm	1.5710	140.5°	Picrolonate m.p. 250°	481, 484
2-Methyl-6-vinylpyridine	123–124°/0.5 mm	1.5620	162–163°	Picrolonate m.p. 228°	481, 484
4-Methyl-2-vinylpyridine	76–77°/15 mm		157–158°	Picrolonate m.p. 178–180°	395
5-Isopropenyl-2-methylpyridine	116–117°/50 mm	1.5360	191–192°	Methiodide m.p. 152–153°	851
6-Butenyl-2-methylpyridine	103.4–104.5°/20 mm	1.5370	110.5–111.5°		526
4,6-Dimethyl-2-vinylpyridine	81–82°/15 mm		155°		486
4-Methyl-2,6-divinylpyridine			165–166°		486
6-Methyl-2,4-divinylpyridine	91–92°/16 mm		143–144°		486
3,4-Dimethyl-5-isopropenylpyridine	222–223°		127–129°		849
3-Cyclohexylidenemethylpyridine	74–75°/0.3 mm	1.5636²⁵	134–136°		502

TABLE V-14. Aralkenylpyridines

Compound	M.p.	B.p.	n_D	Picrate (m.p.)	Ref.
Trans-2-stilbazole	91°	97°/0.1 mm			410, 613, 570
Cis-2-stilbazole	−50°	141°/10 mm			410
Trans-1-(3-methyl-2-pyridyl)-2-phenylethylene	54–55°	105°/0.1 mm			570
Trans-1-(4-methyl-2-pyridyl)-2-phenylethylene	66–67°	102°/0.1 mm			570
Trans-1-(5-methyl-2-pyridyl)-2-phenylethylene	75–76°	100°/0.1 mm			570
Trans-1-(6-methyl-2-pyridyl)-2-phenylethylene	43–44.5°	100°/0.1 mm			526, 570
1-(3,6-Dimethyl-2-pyridyl)-2-phenylethylene		115°/0.1 mm			570
1-(4,6-Dimethyl-2-pyridyl)-2-phenylethylene		113°/0.1 mm			570
Trans-1-(2-pyridyl)-2-*p*-tolylethylene	86–87°	110°/0.1 mm			613, 570
Trans-2-(4-methoxyphenyl)-1-(2-pyridyl)ethylene	74°				613
Trans-2-(4-chlorophenyl)-1-(2-pyridyl)ethylene	83°				613
Trans-2-(4-bromophenyl)-1-(2-pyridyl)ethylene	90°				613
Trans-2-(3-nitrophenyl)-1-(2-pyridyl)ethylene	125–126°				613
Trans-2-(4-nitrophenyl)-1-(2-pyridyl)ethylene	133–134°				613
1-(4-Butyl-2-pyridyl)-2-phenylethylene		220–204°/5 mm		220–222°	474
1-(4-Amyl-2-pyridyl)-2-phenylethylene		185–190°/3 mm		178°	474
1-(6-Butenyl-2-pyridyl)-2-phenylethylene		143.5°/0.2 mm		175–177.5° and 161–163°	526
2-(4-Methylthiophenyl)-1-phenyl-1-(2-pyridyl)propene					
Isomer I	75–77°				471
Isomer II	103–105°				471
1,2-Diphenyl-1-(2-pyridyl)propene	48–54°			HCl 165–167°	471
2-(4-Methoxyphenyl)-1-phenyl-1-(2-pyridyl)propene				HCl 137–139°	471
2-Phenyl-1-(2-pyridyl)butene				HCl 179–181°	469
2-(4-Chlorophenyl)-1-(2-pyridyl)propene	62–64°			HCl 183–186°	469
1-(2-Pyridyl)-2-*p*-tolypropene		163–169°/2 mm	1.6161^{27}		469
2-(4-Methoxyphenyl)-1-(2-pyridyl)propene	41–42°		1.5937^{27}		469
2-(4-Methoxyphenyl)-1-(2-pyridyl)butene		175–179°/2 mm			469
2-(4-Methylthiophenyl)-1-(2-pyridyl)propene	151–153°				469
2-Cyclopropyl-2-(4-methoxyphenyl)-1-(2-pyridyl)ethylene				HCl 255–256°	469

Compound	M.p. (°)	B.p. (°/mm)	n_D	Derivative	References
4-Phenyl-3-(2-pyridyl)-3-hexene		107–110°/1 mm	1.5660^{27}		469
3-(4-Methoxyphenyl)-2-(2-pyridyl)-2-butene		145–150°/1 mm	1.5827^{24}		469
2-(4-Methoxyphenyl)-3-(2-pyridyl)-2-pentene		172–174°/3 mm	1.5758^{24}		469
4-(4-Methoxyphenyl)-3-(2-pyridyl)-3-hexene		169–170°/1 mm	1.5665^{24}	HCl 154–156°	469
4-(4-Hydroxyphenyl)-3-(2-pyridyl)-3-hexene	144–146°				469
4-(2-Methoxyphenyl)-3-(2-pyridyl)-3-hexene		152–155°/2 mm	1.5688^{24}		469
4-(4-Methoxyphenyl)-3-(2-pyridyl)-5-methyl-3-hexene		147–150°/1 mm	1.5639^{25}		469
4-(4-Diethylaminoethoxyphenyl)-3-(2-pyridyl)-3-hexene		204–206°/1 mm	1.5452^{26}		469
Trans-3-stilbazole	80–81°	105°/0.1 mm		171–172°	502, 570
Cis-3-stilbazole	111°	120°/0.1 mm			570
Trans-1-(3-pyridyl)-2-*p*-tolylethylene	87°	115°/0.1 mm			570
Cis-1-(3-pyridyl)-2-*p*-tolylethylene	101°	125°/0.1 mm			570
Trans-2-(4-chlorophenyl)-1-(3-pyridyl)ethylene	153°	140°/0.1 mm			570
Cis-2-(4-chlorophenyl)-1-(3-pyridyl)ethylene	103°	125°/0.1 mm			570
Trans-2-(4-bromophenyl)-1-(3-pyridyl)ethylene	153°	135°/0.1 mm			570
Cis-2-(4-bromophenyl)-1-(3-pyridyl)ethylene	101–102°				570
Trans-2-(4-iodophenyl)-1-(3-pyridyl)ethylene	132–133°				570
Cis-2-(4-iodophenyl)-1-(3-pyridyl)ethylene	92–93°				570
Trans-2-(4-methoxyphenyl)-1-(3-pyridyl)ethylene	86–88°				570
Cis-2-(4-methoxyphenyl)-1-(3-pyridyl)ethylene	78–79°				570
Trans-2-(4-nitrophenyl)-1-(3-pyridyl)ethylene	154–155°				469
Cis-2-(4-nitrophenyl)-1-(3-pyridyl)ethylene	96–97°				502
4-(4-Methoxyphenyl)-3-(3-pyridyl)-3-hexene		155–158°/1 mm	1.5645^{26}	185–187°	475, 570
4-Phenyl-1-(3-pyridyl)buta-1,3-diene		105°/0.1 mm			570
Trans-4-stilbazole		103°/0.1 mm			570
Trans-1-(3-methyl-4-pyridyl)-2-phenylethylene		105°/0.1 mm		TsOH 201–202°	500, 570
Trans-1-(4-pyridyl)-2-*o*-tolylethylene		100°/0.1 mm			500, 570
Trans-1-(4-pyridyl)-2-*m*-tolylethylene		110°/0.1 mm			570
Trans-1-(4-pyridyl)-2-*p*-tolylethylene		120°/0.1 mm			570
Trans-1-(3-methyl-4-pyridyl)-2-*o*-tolylethylene		115°/0.1 mm			570
Trans-1-(3-methyl-4-pyridyl)-2-*m*-tolylethylene	104–106°	115°/0.1 mm			570
Trans-2-(4-chlorophenyl)-1-(4-pyridyl)ethylene	114–115°	120°/0.1 mm			570
Trans-2-(4-methoxyphenyl)-1-(4-pyridyl)ethylene	134–136°				500, 570
Trans-2-(4-nitrophenyl)-1-(4-pyridyl)ethylene	171–172°				475, 500, 570

Table V-14 (*Continued*)

Compound	M.p.	B.p.	n_D	Picrate (m.p.)	Ref.
2-(4-Methoxyphenyl)-1-(4-pyridyl)propene	128–129°	173–176°/1 mm	1.5984²⁶	HCl 134–136°	469, 475
2-(4-Aminophenyl)-1-(4-pyridyl)ethylene	279–281° (decomp.)				500
2-(4-Dimethylaminophenyl)-1-(4-pyridyl)ethylene	238–239°				475
2-(4-Diethylaminophenyl)-1-(4-pyridyl)ethylene	187–188°				475
2-(4-Acetylaminophenyl)-1-(4-pyridyl)ethylene	218–219°				500
2-(3-Chlorophenyl)-1-(4-pyridyl)ethylene	61–63°			HCl 229–230°	500
2-(3-Methoxyphenyl)-1-(4-pyridyl)ethylene	191–192°				500
2-(3-Aminophenyl)-1-(4-pyridyl)ethylene	148–149°				500
2-(3-Cyanophenyl)-1-(4-pyridyl)ethylene				TsOH 151–152°	500
2-(3-Nitrophenyl)-1-(4-pyridyl)ethylene				HCl 227–230° (decomp.)	500
2-(2-Chlorophenyl)-1-(4-pyridyl)ethylene					500
2-(2-Methoxyphenyl)-1-(4-pyridyl)ethylene	70–71°			168–170°	500
2-(2,4-Dichlorophenyl)-1-(4-pyridyl)ethylene					500
2-(3,4-Dichlorophenyl)-1-(4-pyridyl)ethylene	162–163°			TsOH 267–268°	500
2-(2-Naphthyl)-1-(4-pyridyl)ethylene	95–100°				500
2-(3,4-Methylenedioxyphenyl)-1-(4-pyridyl)ethylene					500
1-(4-Pyridyl)-2-(3,4,5-trimethoxyphenyl)ethylene	150–151°			247–248°	500
2-(4-Ethoxyphenyl)-1-(4-pyridyl)ethylene	184–185°				500
2-(4-Benzyloxyphenyl)-1-(4-pyridyl)ethylene	96–97°			TsOH 194–195°	500
2-(4-Phenylpropoxyphenyl)-1-(4-pyridyl)ethylene	105–106°				500
2-(4-Ethylphenyl)-1-(4-pyridyl)ethylene					500
2-(4-Benzylphenyl)-1-(4-pyridyl)ethylene	127–130°			HCl 265–266°	500
1-(4-Pyridyl)-2-(4-styrylphenyl)ethylene					500
2-(4-Phenylbutylphenyl)-1-(4-pyridyl)ethylene				TsOH 158–159°	500
2-(3-Ethoxyphenyl)-1-(4-pyridyl)ethylene	78–81°			HCl 193–194°	500
2-(3-Isoamyloxyphenyl)-1-(4-pyridyl)ethylene					500
2-(3-Benzyloxyphenyl)-1-(4-pyridyl)ethylene					500

Compound	M.p.	B.p.	n_D	Derivative	Ref.
2-(3-Phenoxypropoxyphenyl)-1-(4-pyridyl)ethylene				HCl 169–171°	500
2-(3-Benzylphenyl)-1-(4-pyridyl)ethylene	91–92°				500
2-(3-Phenylbutylphenyl)-1-(4-pyridyl)ethylene	51–54°				500
2-(3-Methoxy-5-methylphenyl)-1-(4-pyridyl)ethylene				HCl 235–238°	500
1-(2,5-Dimethyl-4-pyridyl)-2-phenylethylene	86–88°				527
1-(3-Ethyl,5-methyl-4-pyridyl)-2-phenylethylene		153°/0.02 mm	1.6144^{20}	182–183°	121
1-(4-Chlorophenyl)-2-(4-pyridyl)propene	106°				405
Trans-1-(4-nitrophenyl)-2-(4-pyridyl)propene	108–109°				475
Cis-1-(4-nitrophenyl)-2-(4-pyridyl)propene	75–77°				475
Trans-1-phenyl-2-(4-pyridyl)propene	68–70°				475
Trans-1-(4-methoxyphenyl)-2-(4-pyridyl)propene	76–78°				475
Trans-1-(4-dimethylaminophenyl)-2-(4-pyridyl)propene	163–164°[c]				475
Cis-1-(4-dimethylaminophenyl)-2-(4-pyridyl)propene	118–120°				475
Trans-1-(4-diethylaminophenyl)-2-(4-pyridyl)propene	115–116°				475
2-(4-Biphenylyl)-1-(4-pyridyl)ethylene	213–214°[c]			HCl 257–259°	479
2-(5-Acenaphthyl)-1-(4-pyridyl)ethylene	113–113.5°[c]			HCl > 300° (decomp.)	479
2-(9-Anthryl)-1-(4-pyridyl)ethylene	161–162°[c]				479
2-(3-Pyrenyl)-1-(4-pyridyl)ethylene	207.5–208°[c]				479
Trans-1,2-di-(2-pyridyl)ethylene	120–121°[c]				490
Cis-1,2-di-(2-pyridyl)ethylene	48–49°				490
Trans-1,2-di-(4-pyridyl)ethylene	149–150°				546
Trans-1-(2-pyridyl)-2-(4-pyridyl)ethylene	75–76°				546
1-(3-Pyridyl)-2-(4-pyridyl)ethylene	65–68°	155–165°/5 mm		234–236°	502
Trans-2-(6-methyl-2-pyridyl)-1-(2-pyridyl)ethylene	56.5–57.5°[c]	157–162°/6 mm			546
Trans-1,2-di-(6-Methyl-2-pyridyl)ethylene	112.3°	156–159°/6 mm			546
Trans-1-(6-methyl-2-pyridyl)-2-(4-pyridyl)ethylene	71°	170–173°/7 mm			546
Trans-1-(2-pyridyl)-2-(2-quinolyl)ethylene	95–96°				546
2,4-Di-(2-pyridyl)-1-butene		98–99°/0.05 mm	1.5820^{25}	161–162°	433

was used to distinguish between *cis-* and *trans-*isomers (500, 544, 547). Galiazzo and co-workers prepared the six isomeric styrylpyridines (*cis-* and *trans-*2-, 3- and 4-stilbazoles) and some of their substituted derivatives and investigated the effect of substitution and geometrical and nitrogen-positional isomerization on the dissociation constant, ultraviolet, and infrared spectra of these compounds. This work shows that spectroscopy can be used to distinguish between the isomers (550).

The biological activity of several stilbazole analogues were investigated. The preparation of 1-(5-nitro-2-furyl)-2-(2-pyridyl)ethylenes (**V-209**), of use as antibacterial and fungicidal agents was described (551). A number of

V-209

substituted stilbazoles and dihydrostilbazoles were tested for their hypocholesteremic and estrogenic activity (469), and forty-three 4-stilbazole analogs were examined for their activity in inhibiting choline acetyltransferase from rabbit brain (500).

The isolation of vinylpyridines from mixtures by distillation after acid treatment (552) or by precipitation as a copper chloride complex (553) were reported. The purification of MVP by gradual freezing (554) and of vinylpyridines by counter current exchange methods (555, 556) are also possible. Various compounds were recommended for inhibiting the polymerization of alkenylpyridines (557–569, 571, 572).

3. Reactions

A. Oxidation

The ozonolysis of vinylpyridines was investigated by Callighan and Wilt. Ozonolysis at low temperatures ($-40°C$) in methanol as solvent, with reduction of the active oxygen products by sodium sulfite, gave high yields of pyridine aldehydes. At room temperature ($+30°C$) pyridine carboxylic acids were isolated. Table V-15 summarizes the data (573).

Peracetic acid reacts with *trans*-α-stilbazole in ether to precipitate an unstable solid which slowly evolved acetic acid and produced α-stilbazole-1-oxide (574).

TABLE V-15. Ozonolysis of Vinylpyridines

	Low temperature ($-40°C$)		Room temperature ($+30°C$)	
Pyridine	Product	Yield (%)	Product	Yield (%)
2-Vinyl	2-Aldehyde	65	2-Carboxylic acid	62
4-Vinyl	4-Aldehyde	50	4-Carboxylic acid	61
2-Methyl-5-vinyl	2-Methyl-5-aldehyde	75	2-Methyl-5-carboxylic acid	40
6-Methyl-2-vinyl	6-Methyl-2-aldehyde	80	6-Methyl-2-carboxylic acid	62
5-Ethyl-2-vinyl	5-Ethyl-2-aldehyde	63	5-Ethyl-2-carboxylic acid	40

B. Reduction and Hydrogenation

6-Ethyl-2-methylpyridine was prepared in 92% yield by the hydrogenation of 2-methyl-6-vinylpyridine using Raney nickel as catalyst (575). 2-, 3-, And 4-amyl- or -nonylpyridine were also obtained by the catalytic hydrogenation of the appropriate alkenylpyridine (65, 576).

The electrolytic reductive dimerization of vinylpyridines was mentioned earlier (Section I.2.B.g) (75, 76).

C. Addition of Inorganic Molecules

The addition of Cl_3SiH to 2- or 4-vinylpyridines to form β-(2- or 4-pyridyl)-ethyltrichlorosilane occurred readily on heating to 150° in the absence of a catalyst (577–579). The compounds could be hydrolyzed to the pyridyl-ethylpolysiloxanes (577–579) useful as anion exchange resins (578), or treated with methyl magnesium bromide to form the β-trimethylsilylethylpyridine (577). The alkenylpyridines also added CH_3Cl_2SiH in a similar manner. Subsequent treatment with ethyl orthoformate gave the β-pyridylethyl methyldiethoxysilane (580).

Reaction of 2-vinylpyridine with decaborane in benzene gave a red thermoplastic powder recommended for use in rocketry (581).

The rate of the iodocyclization of 2-(3-butenyl)pyridine (**V-210**) was studied, the product being **V-211** (582).

V-210   V-211

Phosphites or thiophosphites add to 2-vinylpyridine on heating. Lower reaction temperatures are possible when sodium ethoxide is used as a catalyst, yields of the order of 60% being realized (583).

Hydrogen sulfide adds to 2-, 3-, and 4-vinylpyridine. Hansch and Carpenter were able to identify only di-(4-pyridylethyl)sulfide (V-212) from the

$$\text{N}\!\!\diagdown\!\!\diagup\!\!-\text{CH}_2\text{CH}_2\text{SCH}_2\text{CH}_2\!\!-\!\!\diagup\!\!\diagdown\!\!\text{N}$$

V-212

reaction with 4-vinylpyridine (584). Klemm and co-workers, however, have shown that thieno[3,2-b]pyridine (V-213) (RRI 1344) was formed on heating 2-vinylpyridine with hydrogen sulfide (585), thieno[2,3-b]pyridine (V-145), and thieno[3,2-c]pyridine (V-146) are obtained from 3-vinylpyridine (419) and thieno[2,3-c]pyridine (V-147) from 4-vinylpyridine (420). These bicyclic compounds are formed by dehydrocyclization of the pyridylethyl

V-213   V-145   V-146   V-147

mercaptans. Benzyl mercaptan could be used instead of hydrogen sulfide (420, 586). Other negatively substituted aliphatic mercaptans have also been added to vinylpyridines (75 to 90% yields) (65, 587).

D. Addition of Organic Molecules

The addition of amines (65, 399, 588–597), hydrazines and acid hydrazides (595, 598–601), and hydroxylamine (600, 602) to vinylpyridines was investigated. The reactions, usually acid-catalyzed, proceeded as expected in most cases to form the pyridylethyl derivatives. Diamines, aliphatic, and aromatic amines could be used. 2- Or 4-vinylpyridine gave excellent yields of the N,N-dipyridylethyloxyamines (V-124; R = H or CH$_2$C$_6$H$_5$) with hydroxylamine or O-benzylhydroxylamine (600). Phthalhydrazide gave N-[2-(2- or

V-214

V-215

4-pyridylethyl]phthalhydrazide (78 and 93%) **(V-215)** (RRI 1628) with 2- or 4-vinylpyridine (600). 2- And 4-vinylpyridine reacted with *N*-hydroxy-pyrrolidine and similar *N,N*-disubstituted compounds to form products such as **V-216** which do not contain oxygen (602).

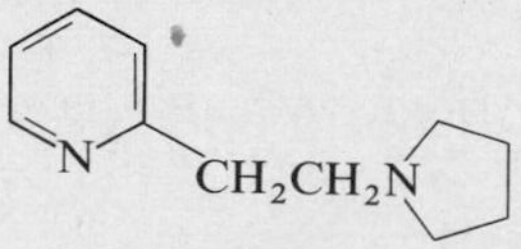

V-216

Further examples of the pyridylethylation of organic compounds containing active hydrogen atoms were reported (594, 603–612). 2- And 4-vinylpyridines added to alkyl and aralkylnitriles (594, 603), acid amides (594), and keto esters (604) in the presence of base (594, 604) or Triton B (603). Indoles (605–610), benzotriazole, benzimidazole (607), indenes (610), and pyrroles (611, 612) were pyridylethylated by vinylpyridines. The use of the sodium or potassium salt of indoles resulted in the formation of the *N*-pyridylethyl derivative **V-217** (606, 607, 609), while in an acid medium addition occurred at C-3 to give **V-218** (607–610).

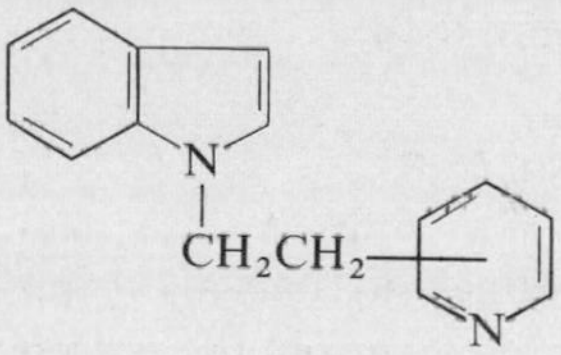

V-217

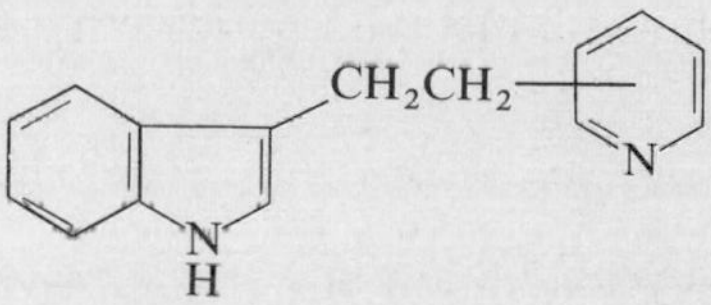

V-218

In an acid medium the pyrroles underwent pyridylethylation at both the C-2- and C-5-positions to give the 2,5-bis(pyridylethyl)pyrroles **(V-219)** (432, 611).

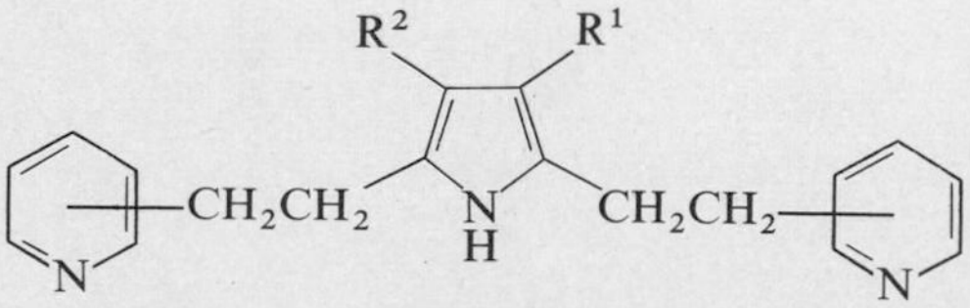

V-219

Carbethoxycyclopropylpyridines were obtained from the reaction of vinylpyridines with ethyl diazoacetate. The product was shown to consist of a mixture of the *cis-* **(V-220)** and *trans-* **(V-221)** isomers (612).

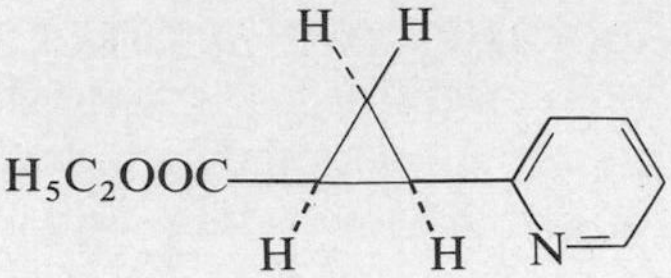

V-220 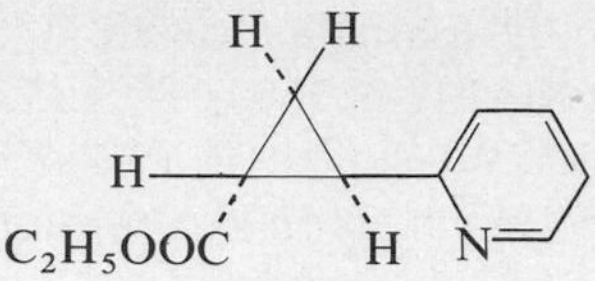   V-221

The reaction of 2-vinylpyridine with aryldiazonium chlorides with added copper (II) chloride gave **V-222** which could be dehydrochlorinated to 2-stilbazoles (613).

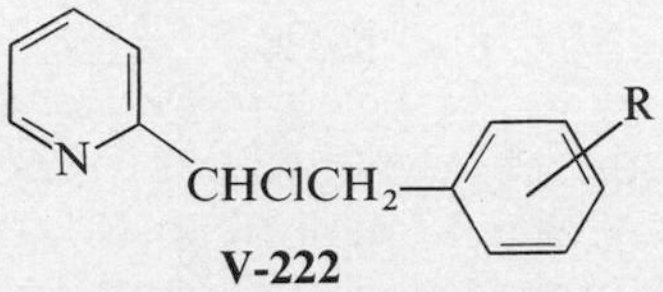

V-222

E. Willegerodt Reaction

The Willgerodt or Willgerodt-Kindler reaction was used to prepare 2-pyridylacetamide from 2-vinylpyridine (389) and 2-methylpyridyl-5-acetic acid from 2-methyl-5-vinylpyridine (614). Suminov has shown that in the Willgerodt-Kindler reaction using 2-methyl-5-vinylpyridine the vinyl group is more reactive and is attacked preferentially (614, 615). A by-product of the reaction was 2-methyl-5-pyridylethyl mercaptan (614).

F. Diels-Alder Reaction

The cycloaddition of 2-, 3-, or 4-vinylpyridines or their alkyl derivatives (**V-223**) to substituted oxazoles (**V-11**) in a polar solvent in the presence of an acid produces 4-pyridylpyridines (**V-225**) *via* the adduct **V-224** (616, 617).

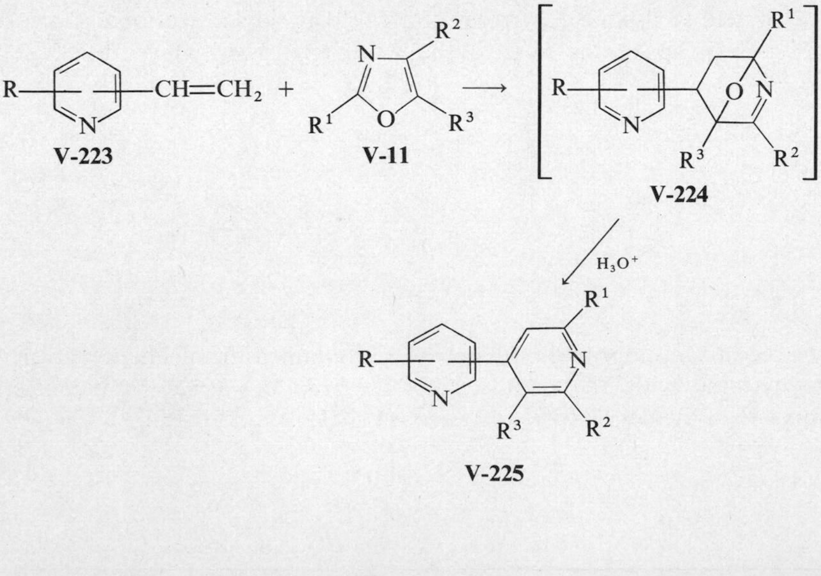

V-223   V-11   V-224

H_3O^+

V-225

Vinylpyridines also condensed with isoprene and piperylene to form 1-methyl-4-pyridylcyclohexene and its isomers (618).

The Diels-Alder reaction of vinylpyridines with anthracene produces 11-pyridyl-*cis*-9,10-dihydro-9,10-*endo*-ethanoanthracenes (**V-226**) which have potential use as antihistaminic and diuretic agents (619).

V-226

G. Synthesis of Condensed Ring Systems

The synthesis of the thienopyridines by the addition of hydrogen sulfide to the vinylpyridines has already been discussed (Section II.3.c).

Acheson and Feinberg have studied the condensation of *cis*- and *trans*-stilbazole with dimethyl acetylenedicarboxylate. The initial "labile" adduct was shown to be tetramethyl 6-styryl-9*aH*-quinolizine-1,2,3,4-tetracarboxylate (**V-227**) (RRI 1688), which on heating was converted to the "stable" 4*H*-isomer **V-228** (RRI 1689), and finally to tetramethyl 1,2-dihydrocycl-[3,3,2]azine-2a,3,4,5-tetracarboxylate (**V-229**) (620) (see Section I.4.J).

V-227   **V-228**   **V-229**

The photocyclization of aralkenylpyridines provides a convenient route to many condensed ring systems which are comparatively inaccessible by other methods. Loader and co-workers have cyclized 2-, 3-, and 4-stilbazoles and 1,2-di-(2- and 4-pyridyl)ethylenes by this method (621). Bortolus and co-workers extended this reaction to the four *cis*-stilbazoles substituted with a methyl group in the pyridine ring (622) and have shown that 3-styrylpyridine can photocyclize at the C-2 or C-4 position of the pyridine ring to give the two isomers, 5,6-benzoquinoline (**V-230**) (RRI 3524) and 7,8-benzoquinoline

V-230   V-231

(**V-231**) (RRI 3526), respectively (623). Kumler and Dybas have examined the effect of various parameters on this photocyclization reaction (624).

The synthesis of an azahelicine by this method has been reported (501) (see Section II.1.B).

H. Polymers and Their Properties

Numerous publications have appeared on the polymerization and copolymerization of alkenyl- and aralkenylpyridines (467, 488, 625–638). The mechanism of the polymerization of vinylpyridines (625) and the kinetics of the solution polymerization of 4-vinylpyridine (626) were studied. Polymers of 2-vinylpyridine (628) and copolymers of vinylpyridines with styrene, dichlorostyrene, and α-methylstyrene (627) were recommended as ion-exchange resins. Polymers containing heavy-metal salts had water-proofing, fungicidal, and insecticidal properties (629), and polymers with metal acetylacetonate incorporated were examined for their thermal stability (636). Acrylonitrile graft polymers with vinylpyridines can be spun into fibers from an N,N-dimethylacetamide solution and can be dyed with acid dyes (632). General purpose rubbers with excellent physical properties were made (623, 634). Copolymers of trans-1,2-di-(2-pyridyl)ethylene with butadiene and styrene have been made (467).

I. Photochemical Reactions

The photochemical reactions of the aralkenylpyridines have been studied extensively (see Sections I.2.C.e, II.2., and II.3.G). The aralkenylpyridines and their salts undergo photodimerization in the solid state (139). In dilute solutions, the cis- or trans-stilbazoles undergo an initial cis-trans isomerization followed by a dehydrocyclization to condensed ring systems, the reaction being catalyzed by iodine. In concentrated solutions dimerization also occurs.

III. Alkynyl- and Aralkynylpyridines

1. Preparation

The dehydrohalogenation of dihaloalkyl- and aralkylpyridines was used to prepare 2-, 3-, and 4-ethynylpyridines (639, 640), 5-ethynyl-2-methyl-pyridine (641, 642), 6-butynyl-2-phenylethynylpyridine, and 2-methyl-6-phenylethynylpyridine (526) and di-(2-pyridyl)-, (2-pyridyl)(3-pyridyl)-, and (2-pyridyl)(4-pyridyl)acetylenes (643). The halo compounds were prepared by brominating the alkene (526, 641–643) or the alkane (641) or by treating a pyridyl ketone with phosphorus pentachloride (639, 640). When an alcohol was used as the solvent for the dehydrohalogenation step, the α-alkoxy-alkenylpyridine was also obtained as a by-product (526, 641, 642).

Alkynylpyridines have also been prepared from organometallic compounds. Alkali metal acetylides V-233 form 2-ethynylpyridine (V-234) with the quaternary pyridinium salts of pyridine-1-oxides (V-232) (644).

2-, 3-, Or 4-alkynylpyridines containing terminal triple bonds (V-237) were prepared by treating the appropriate picolyllithium with α-bromo-ω-alkoxy-alkanes, when the ω-alkoxypyridylalkane (V-235) was formed. This compound gave the ω-bromopyridylalkane (V-236) with concentrated HBr, and this was treated with sodium acetylide to form V-237 (645).

Disubstituted alkynes containing a pyridine ring (V-239) have been obtained by treating a sodium or lithium derivative of an alkynylpyridine (V-238) with an alkyl halide, or by the action of a chloroalkyne such as

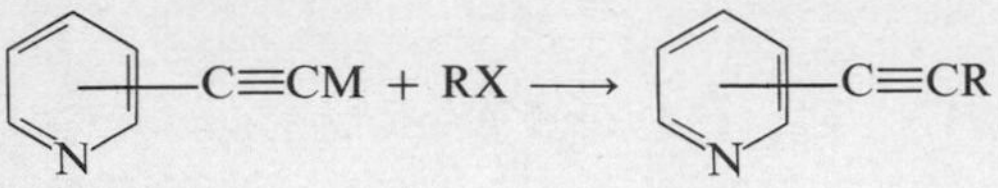

$$\text{(pyridyl)}-C{\equiv}CM + RX \longrightarrow \text{(pyridyl)}-C{\equiv}CR$$

M = Li or Na   **V-239**

V-238

7-chloro-5-heptyne (**V-240**) on a picolyllithium or sodium, the product in this instance being 8-pyridyl-5-octyne (**V-241**) (40%) (646). In these latter

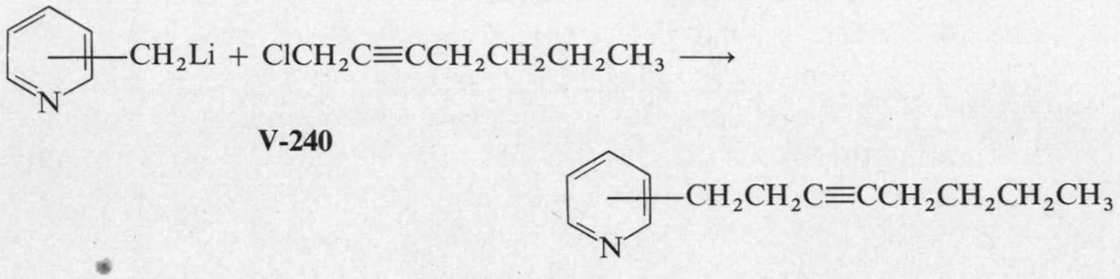

$$\text{(pyridyl)}-CH_2Li + ClCH_2C{\equiv}CCH_2CH_2CH_2CH_3 \longrightarrow$$

V-240

$$\text{(pyridyl)}-CH_2CH_2C{\equiv}CCH_2CH_2CH_2CH_3$$

V-241

reactions a double substitution of the methyl group of 4-picolylsodium

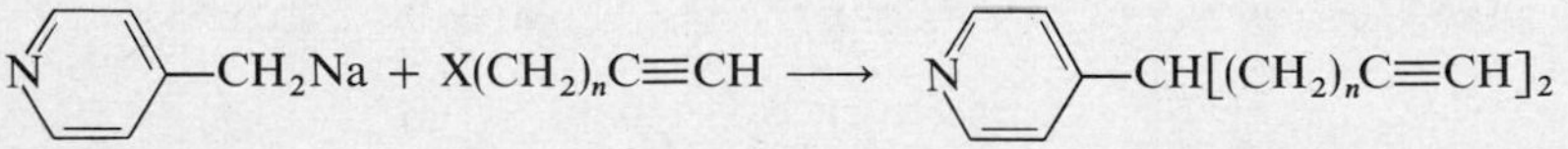

$$\text{(pyridyl)}-CH_2Na + X(CH_2)_nC{\equiv}CH \longrightarrow \text{(pyridyl)}-CH[(CH_2)_nC{\equiv}CH]_2$$

V-242

sometimes occurs. With 3-picolylsodium, hydrogen-sodium exchange took place and 3-methylpyridine was recovered (365).

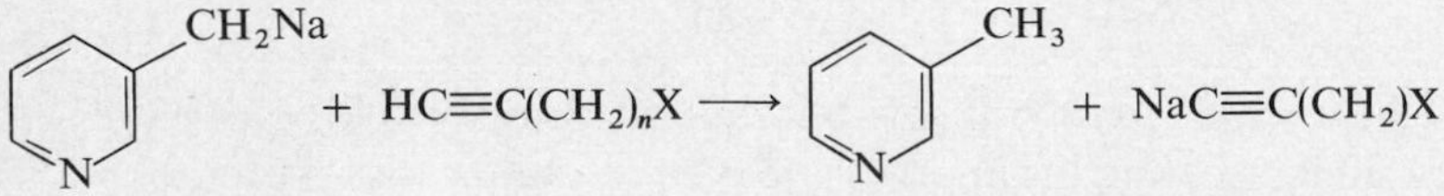

$$\text{(pyridyl)}{-}CH_2Na + HC{\equiv}C(CH_2)_nX \longrightarrow \text{(pyridyl)}{-}CH_3 + NaC{\equiv}C(CH_2)X$$

1,4-Bis(2-methyl-5-pyridyl)-1,3-butadiyne (**V-244**), a pyridyl polyacetylene, was prepared in 88% yield by shaking 5-ethynyl-2-methylpyridine (**V-243**) with copper (I) chloride in pyridine in an oxygen atmosphere (647).

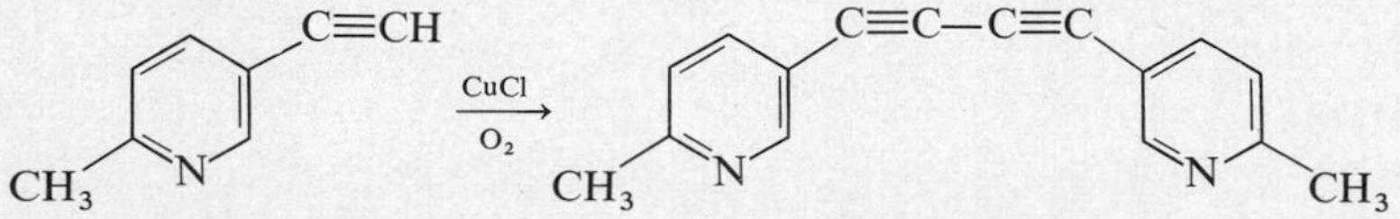

$$\overset{C{\equiv}CH}{\underset{CH_3}{\text{(pyridyl)}}} \xrightarrow[O_2]{CuCl} \overset{C{\equiv}C-C{\equiv}C}{\underset{CH_3 \qquad CH_3}{\text{(bis-pyridyl)}}}$$

V-243   **V-244**

2-Phenylethynylpyridine (**V-247**) (88%) and 2-(p-anisylbutadiynyl)pyridine (77%) were obtained by the action of 2-iodopyridine (**V-245**) on phenyl-acetylene (**V-246**) or p-anisyldiacetylene, respectively, in the presence of copper, using dimethylformamide or pyridine as solvent (648, 649).

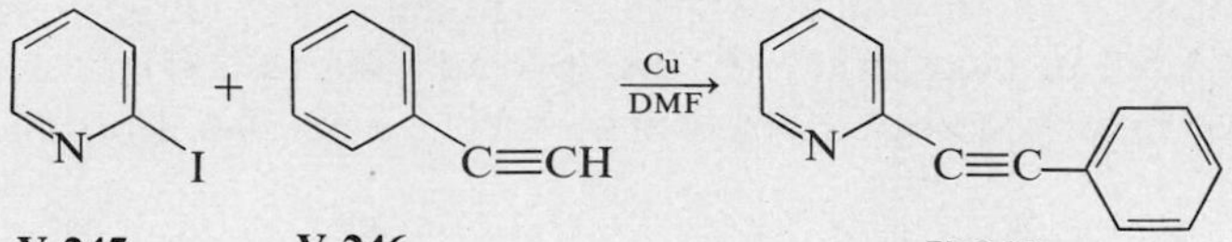

V-245   **V-246**   **V-247**

2. Reactions

Dipyridylacetylenes were reduced to the dipyridylethanes by using palladium on charcoal, while hydrogenation using Lindlar catalysts in methanol gave the *cis*-dipyridylethylenes (643). Nickel–aluminium alloy with sodium hydroxide also reduced alkynylpyridines to the alkylpyridines (650), as did hydrogenation over Raney nickel (651).

A general reaction for the preparation of pyridylacetylenic *t*-carbinols (**V-249**) involves treating a ketone with the lithium or sodium derivative of a pyridylacetylene (**V-248**). This same reaction can be brought about by

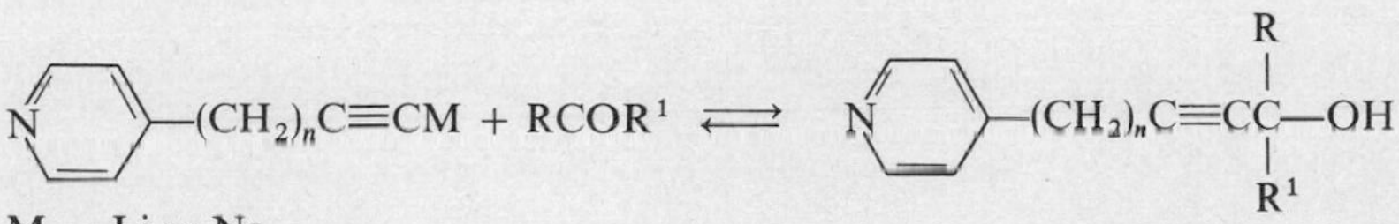

M = Li or Na

V-248

V-249

condensing the ketone with the pyridylacetylene using powdered potassium hydroxide (652–654).

These carbinols were also prepared *via* Grignard compounds. Reacting 5-ethynyl-2-methylpyridine with ethylmagnesium bromide followed by acetone gave 1-(2-methyl-5-pyridyl)-3-methylbutyn-3-ol (44.1%) (655).

These condensations are reversible, since the pyridylacetylenes (**V-248**) were regenerated on treating the carbinols with concentrated aqueous potassium hydroxide. The rates of decomposition in this system was determined (653).

Mannich reactions involving the acetylenic proton of 2- and 4-pyridylalkynes to form **V-250** were reported (639, 642, 645, 656).

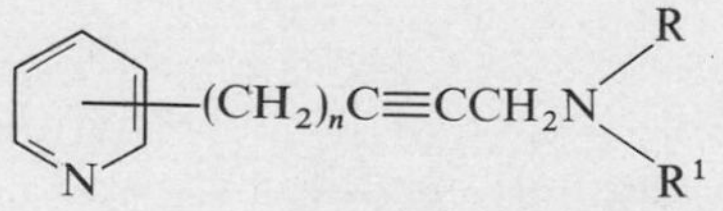

V-250

Alkylation of 2-pyridylalkynes with a terminal acetylenic group produced the ω-alkylated product, while the 4-pyridylalkynes (**V-251**) gave a mixture (3:1) of the ω-alkylated product **V-252** and the α-alkylated product **V-253**.

$$\text{N} \overset{}{\underset{}{\bigcirc}} -(CH_2)_2C\equiv CH \xrightarrow{\text{RX}}$$

V-251

$$\text{N} \overset{}{\underset{}{\bigcirc}} -(CH_2)_2C\equiv CR \quad + \quad \text{N} \overset{}{\underset{}{\bigcirc}} -\overset{R}{\underset{}{C}}HCH_2C\equiv CH$$

V-252   **V-253**

The use of two molar equivalents of the alkyl halide gave α,ω-dialkylated 4-pyridylalkynes (656).

The hydration of the acetylenes using sulfuric acid in the presence of mercury (II) sulfate was a convenient route to pyridylketones (526, 642, 645, 651).

Treatment of 5-ethynyl-2-methylpyridine with an alcohol gave the 5-(2-alkoxyethenyl)-2-methylpyridine (**V-254**; R = OR^1) and with amines, the 5-(2-aminoethenyl)-2-methylpyridine (**V-254**; R = NR^{1}R^2) (642).

$$\underset{CH_3}{\overset{CH=CHR}{\bigcirc}}$$

V-254

Migration of the triple bond is base-catalyzed. Thus heating 6-(2-pyridyl)-1-hexyne with potassium hydroxide gave a near quantitative yield of 6-(2-pyridyl)-2-hexyne. Similar high yield migrations of the acetylenic bond from the α to the β position was observed with 6-(4-pyridyl)-1-hexyne and 7-(4-pyridyl)-1-heptyne (651).

Pyridylacetylenic acids were prepared by treating the sodium pyridyl-acetylides with carbon dioxide (657).

1,3-Dipolar cycloaddition of diazomethane to 3-ethynylpyridines (**V-255**) leading to a series of 3-(3-pyrazolyl)pyridines (**V-256**) was reported (658).

$$\underset{R}{\overset{C\equiv CH}{\bigcirc}} + CH_2N_2 \longrightarrow \underset{R}{\bigcirc}\overset{}{\underset{}{\overset{N}{\underset{N}{\diagdown}}}}NR^1 \qquad R^1 = H, CH_3$$

V-255   **V-256**

The physical properties of the alkynyl- and aralkynylpyridines are summarized in Table V-16.

TABLE V-16. Alkynyl- and Aralkynylpyridines

Compound	M.p.	B.p.	n_D	Picrate (m.p.)	Ref.
2-Ethynylpyridine		86–88°/14 mm	1.5534[21]		640, 645
4-Ethynylpyridine	45°				640
5-Ethynyl-2-methylpyridine	51–52°	64–66°/5 mm			642
2-Methyl-5-(3-methyl-3β-cyanoethoxy-1-butynyl)pyridine		162°/10 mm		140–142°	654
2-Methyl-5-(3-methyl-3-buten-1-ynyl)pyridine		120–122°/20 mm	1.5345[20]		654
5-(2-Pyridyl)-1-pentyne		103–104°/16 mm		106°	645
5-(3-Pyridyl)-1-pentyne		124–125°/18 mm		89–90°	645
5-(4-Pyridyl)-1-pentyne		120–121°/16 mm		91°	645
6-(3-Pyridyl)-1-hexyne		131–132°/14 mm		71–72°	645
7-(2-Pyridyl)-4-heptyne		100–101°/1 mm		106.5°	646
8-(2-Pyridyl)-5-octyne		154°/20 mm		69°	646
8-(4-Pyridyl)-5-octyne		149–150°/13 mm		100°	646
10-(2-Pyridyl)-5-decyne		91–94°/15 mm			646
10-(4-Pyridyl)-5-decyne		145°/1.4 mm		77.5°	646
11-(2-Pyridyl)-5-undecyne		183–184°/14 mm			646
11-(3-Pyridyl)-6-undecyne		141–142°/0.5 mm		60°	646
12-(2-Pyridyl)-6-dodecyne		152–153°/0.9 mm			646
12-(2-Pyridyl)-9-dodecyne		155°/1.25 mm			646
12-(3-Pyridyl)-6-dodecyne		146°/0.7 mm		54°	646
12-(4-Pyridyl)-6-dodecyne				79.5°	646
15-(2-Pyridyl)-4-pentadecyne		169–170°/0.5 mm		53°	646
15-(3-Pyridyl)-4-pentadecyne		186–188°/0.7 mm		87°	646
1-(2-Methyl-5-pyridyl)-1,3-butadiyne	94°				647
1-(2-Methyl-5-pyridyl)-4-phenyl-1,3-butadiyne	121–122°			191–193°	647
6-Butynyl-2-phenylethynyl pyridine				166.5–167.5°	526
6-(2-Ethoxybutenyl)-2-phenylethynylpyridine				134.5–135°	526
2-Methyl-6-phenylethynyl-pyridine	45–47°	124.5–126.5°/0.5 mm			526
Di-(2-pyridyl)acetylene	69–70°				643
(2-Pyridyl)(3-pyridyl)acetylene	41–42°				643
(2-Pyridyl)(4-pyridyl)acetylene	64–65°				643
1,4-Di-(2-pyridyl)butadiyne	119–119.5° (decomp.)				649
1,4-Di-(2-methyl-5-pyridyl)-1,3-butadiyne	192–192.5°			229–230°	647
1,8-Di-(2-methyl-5-pyridyl)-1,3,5,7-octatetrayne	175° (decomp.)				647

IV. Arylpyridines

1. Preparation

Many of the methods used for the preparation of the alkylpyridines can also be applied to the arylpyridines. Most of the published work on the synthetic applications of organometallic compounds and free radical reactions have been reviewed already in Sections I.2.C.c and I.2.C.d; hence only new material of particular relevance to the arylpyridines is considered in the following sections.

A. Arylations Involving Organometallic Derivations

The arylation of pyridines by the addition of aryllithium compounds to the azomethine bond of pyridines followed by aromatization of the dihydro-pyridine and the effect of substituents in the C-3 position on the orientation of the products has already been reviewed (Section I.2.C.c).

Prostakov and co-workers have described another approach using aryllithium compounds which introduces the aryl group into the C-4 position. 2,5-Dimethyl-4-piperidone (V-257) reacts with phenyllithium to form 2,5-dimethyl-4-hydroxy-4-phenylpiperidine (V-258) which is dehydrated to the $\Delta^{4,5}$-compound V-259 and dehydrogenated to 2,5-dimethyl-4-phenylpyridine (V-260) (659, see also 660). The 2,5-dimethyl-4-phenyl-pyridine (V-260) thus obtained could be further treated with phenyllithium to give 3,6-dimethyl-2,4-diphenylpyridine (**V-261**). *p*-Tolyllithium behaved similarly (661).

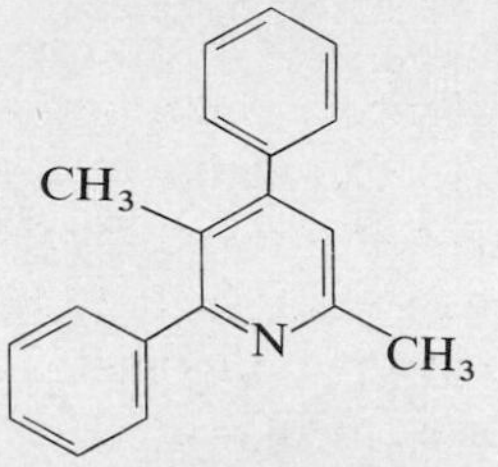

V-261

Pyridyllithium compounds (**V-262**) react with cyclohexanone to give 1-pyridylcyclohexanols (**V-263**) which are aromatized on treatment with concentrated sulfuric acid and glacial acetic acid to the phenylpyridines (**V-264**). Using this method, 3-methyl-5-phenylpyridine, 3-methyl-4-phenyl-pyridine, and various tolylpyridines were prepared (662, 859).

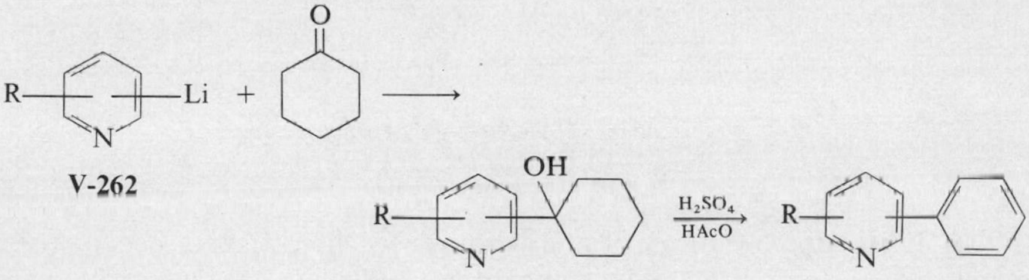

V-262

V-263   **V-264**

Grashey and Huisgen have shown that 1-methylpyridinium iodide reacts with phenyllithium to form the unstable 1-methyl-2-phenyl-1,2-dihydro-pyridine (663).

B. Free Radical Arylations

Some of the work on the homolytic arylation of the pyridines has already been summarized (see Section I.2.C.d. and Ref. 100).

The free radical phenylation of pyridine (664), 2-methylpyridine (665, 666), 3- and 4-methylpyridines (666), and the dimethylpyridines (666, 667) were studied, the phenyl radical being generated by the thermal decomposition of benzoyl peroxide. A quantitative study of the isomer distribution and the total rate ratios for the homogeneous Gomberg-Hey phenylation of 3- and 4-methylpyridines was reported (662). Lynch and co-workers found that the phenylation of pyridinium ions gave high yields (about 85% of the product) of 2-phenylpyridine when the phenyl radical was generated from benzoyl peroxide (664) or from benzenediazonium tetrafluoroborate (668). These results are in agreement with molecular-orbital calculations (128). Bonnier and co-workers measured the rate constants for the phenylation of the dimethylpyridines in neutral media and found that the effects of the methyl groups were additive (667). The radical phenylation of pyridine using phenylazotriphenylmethane was studied by Huisgen and Nakaten (669) and a by-product (7%) was shown to be 2-phenyl-5-tritylpyridine (663). The free radical phenylation of pyridine-metal complexes using N-nitroso-sym-diphenylurea was reported (670).

Fields and Meyerson have studied the high temperature (600°C) arylation of pyridine using nitrobenzene and nitrobenzene-d_5. The nitrobenzene dissociates at these high temperatures into the phenyl and NO_2 radicals. The phenyl radical then reacts with the pyridine to form the phenylpyridines, whose isomer distribution was similar to that obtained with other phenyl radical sources. Other products identified were hydroxypyridine, bipyridines, and biphenylpyridines. Mechanisms for the formation of all these products were suggested (671).

C. Miscellaneous Methods

The Diels-Alder synthesis of arylpyridines by the action of benzonitrile on dienes or tetraphenylcyclopentadienone (672) has been considered in Section I.1 and the synthesis of 4-pyridylpyridines by the action of vinylpyridine on oxazoles in Section II.3.F.

A novel and convenient synthesis of 3,4-diarylpyridines is that of Westphal, Feix, and Joos. 6,7-Disubstituted thiazolo[3,2-a]-pyridinium salts (V-265) (673) are converted to 3,4-diarylpyridines (V-266) (Ar = C_6H_5, 2-pyridyl, 2-furyl) on heating with Raney nickel in 90% ethanol (674). Alternatively, 3,4-diarylquinolinium salts (V-267) (RRI 1689) are partially hydrogenated to the tetrahydro derivative, oxidized to the carboxylic acid and decarboxylated to the 3,4-diarylpyridine (V-266; R = C_6H_5) (674).

2,6-Diphenylpyridine (55%) and 2,6-diphenyl-3-methylpyridine (5%)

V-265   V-266   V-267

were obtained by the thermal rearrangement of acetophenone *N,N,N*-trimethylhydrazonium tetrafluoroborate and 2,6-diphenyl-3,5-dimethylpyridine (56%) from propiophenone *N,N,N*-trimethylhydrazonium tetrafluoroborate. The mechanism of this interesting rearrangement was discussed (675).

The vapor-phase reaction of allyl alcohol, aromatic ketones, and ammonia was reported to form phenylpyridines in high yields. 2-Phenylpyridine, 3-methyl-2-phenylpyridine and 2-methyl-3-phenylpyridine were made by this method (676).

3,4-Diphenylpyridine was obtained by treating 2,3-diphenylglutarimide with phosphorus pentachloride and reducing the 2,4,6-trichloro-3,4-diphenylpyridine (677). Cyclization of **V-268** with sulfuric acid gave 2,5-diphenyl-6-pyridone (**V-269**) which was treated with phosphorus oxychloride and the product reduced to give 2,5-diphenylpyridine (**V-270**) (678). Various

6-arylpyridines were synthesized by the decarboxylation of 6-arylnicotinic acids (144).

2,4,6-Triarylpyridines (**V-272**) were obtained by treating benzylidene di-p-methylacetophenone (**V-271**; R = CH$_3$) with hydroxylamine (680) or from the pyryllium salts **V-273** (681).

Brown and England found that certain pyrimidine compounds gave the 3,5-diarylpyridines on treatment with ammonia (682).

Many other methods, particularly those suited to the synthesis of heteroarylpyridines, have been used. These synthetic approaches are described in the appropriate sections.

2. Properties

The physical properties of the new arylpyridines are summarized in Table V-17, of the 5-membered heteroarylpyridines in Table V-18, and those of the 6-membered heteroarylpyridines in Table V-19.

The infrared spectra of the phenylpyridines have been published (181). The infrared and nuclear magnetic resonance spectra of 3-methyl-2-phenylpyridine (112) and various dimethylphenylpyridines (667) and the ultraviolet absorption curves for 3-methyl-2-phenylpyridine, 5-methyl-2-phenylpyridine, 2-phenylnicotine, and 6-phenylnicotine (110) were reported.

Triphenylpyridine forms a black paramagnetic adduct μ_{eff} 0.161 μB and a violet diamagnetic adduct μ_{eff} 0.58 μB with lithium in tetrahydrofuran under anaerobic conditions (684).

Paris, Garmaise, and Komlossy have studied the quaternization of 2-p-dimethylaminophenylpyridine using methyl iodide and obtained both the 1-methyl-2-p-dimethylaminophenylpyridinium iodide (**V-274**) and 2-p-trimethylammoniumphenylpyridine iodide (**V-275**) (685).

V-274   V-275

A series of p-chlorophenylpyridines were made and found to have antiinflammatory, analgesic and antipyretic activity, and fibrinogen and cholesterol lowering activity in the blood (686).

7- And 12-(2-, 3-, and 4-pyridyl)benz[a]anthracenes **V-276** (RRI 5253)

V-276   V-277

TABLE V-17. Arylpyridines

Compound	M.p.	B.p.	n_D^{20}	d_{20}	Picric (m.p.)	Ref.
2-Phenylpyridine		146–147°/16 mm	1.6182^{25}		175–175.5°	676
2-Methyl-3-phenylpyridine		136–137°/15 mm	1.5910^{25}		135–136.5°	676
2-Methyl-6-phenylpyridine		125–126°/5 mm	1.6057	1.0598	141–142°	847, 144
2-Methyl-6-(p-tolyl)pyridine	22°	139–140°/3 mm	1.6030	1.0372	151°	847
6-(p-Ethylphenyl)-2-methylpyridine		145–146°/2 mm	1.5812	1.0085	146–147°	847
6-(p-Chlorophenyl)-2-methylpyridine	62.5–63.5°	142–152°/8 mm				144
3-Methyl-2-phenylpyridine		154–156°/18 mm	1.6001		165–166°	112, 676
3-Methyl-6-phenylpyridine		162–164°/20 mm			181–183°	112
3-Ethyl-2-phenylpyridine		156–158°/18 mm			104°	111
5-Ethyl-2-phenylpyridine		179–180°/22 mm			155–156°	111
2-Phenyl-3-isopropylpyridine		152–154°/22 mm				111
2-Phenyl-5-isopropylpyridine	53–53.5°					111
3-t-Butyl-2-phenylpyridine		140°/18 mm				111
5-t-Butyl-2-phenylpyridine		130°/0.72 mm				111
2,6-Diphenylpyridine	82°					675
3,4-Diphenylpyridine	114°					675
2,4-Dimethyl-6-phenylpyridine		100–120°/1.5 mm				95
2,6-Dimethyl-4-phenylpyridine					243° (decomp.)	30
2,5-Dimethyl-4-(m-tolyl)pyridine	40–41°	138–146°/5 mm				660
2,6-Diphenyl-3-methylpyridine		150–160°/0.5 mm				675
4-(4-Dimethylaminophenyl)-2,6-diphenylpyridine	138°					681
2,6-Di-(p-tolyl)-4-phenylpyridine	159–160°					680
2,4,6-Tri-(p-tolyl)pyridine	173–175°					680
3,6-Dimethyl-2,4-diphenylpyridine		167–169°/2 mm				661
3,6-Dimethyl-4-phenyl-2-(p-tolyl)pyridine		186°/2 mm				661

357

TABLE V-18. Five-membered Heteroarylpyridines

Compound	M.p.	B.p.	Salt m.p.	Ref.
4-(3-Furyl)pyridine	Unstable solid		Methiodide 215–217°	690
3,4-Di-(2-furyl)pyridine	59–60°			674
4-(2-Thienyl)pyridine			Methiodide 181° (decomp.)	690
4-(3-Thienyl)pyridine			Methiodide 176–177°	690, 698
4-(2-Ethoxycarbonyl-5-methyl-3-pyrrolyl)pyridine	165–166°			690
4-(2-Carboxy-5-methyl-3-pyrrolyl)pyridine	214° (decomp.)			690
4-(5-Methyl-3-pyrrolyl)pyridine	228–229°		Methiodide 210–211° (decomp.)	690
2,5-Di-(2-pyridyl)pyrrole	96°			702
3-(4-Pyridyl)indole	210–212°			704
1-(β-Dimethylaminoethyl)-3-(4-pyridyl)indole			HCl 258–260° (decomp.)	704
1-(β-Diethylaminoethyl)-3-(4-pyridyl)indole			HCl 225° (decomp.)	704
1-(β-Morpholinoethyl)-3-(4-pyridyl)indole			HCl 245° (decomp.)	704
1-[2-(N'-Methylpiperazino)ethyl]-3-(4-pyridyl)indole			HCl 260° (decomp.)	704
1-(β-Aminoethyl)-3-(4-pyridyl)indole		204–207°/0.4 mm		704
1-(Cyanomethyl)-3-(4-pyridyl)indole	190–192°			704
1-(Cyanoethyl)-3-(4-pyridyl)indole	126–128°			704
1-(γ-Aminopropyl)-3-(4-pyridyl)indole		209–212°/0.4 mm		704
1-(γ-Dimethylaminopropyl)-3-(4-pyridyl)indole		184–186°/0.1 mm		704
1-(β-Dimethylaminoethyl)-3-(4-pyridyl)indole			HCl 258–260° (decomp.)	704
2-(5-Methyl-3-pyrazolyl)pyridine	115–116°		Picrate 172–174°	689, 712
2-(4-Bromo-3-methyl-5-pyrazolyl)pyridine	165–166°			712
2-(5-Methyl-3-Δ^2-pyrazolinyl)-pyridine		105–110°/0.1 mm		712
2-(4-Formyl-3-methyl-5-pyrazolyl)pyridine	167–168°			712
2-(4-Carboxy-3-methyl-5-pyrazolyl)pyridine	229–231°			712
3-(5-Methyl-3-pyrazolyl)pyridine	137°			689
4-(3-Pyrazolyl)pyridine	157–158°			689
4-(4-Pyrazolyl)pyridine	198–200°			689
4-(5-Methyl-3-pyrazolyl)pyridine	180–183°		Methiodide 252–253°	689

Table V-18 (*Continued*)

Compound	M.p.	B.p.	Salt m.p.	Ref.
4-(4-Methyl-3-pyrazolyl)pyridine	Oil		Methiodide 213–214°	689
4-(5-Ethyl-3-pyrazolyl)pyridine	116–117°		Methiodide 213–214°	689
4-(5-Isobutyl-3-pyrazolyl)pyridine	156–157°		Methiodide 206–207°	689
4-(5-*n*-Hexyl-3-pyrazolyl)pyridine	111–112°		Methiodide 78–79°	689
4-(5-Cyclopropyl-3-pyrazolyl)pyridine	126–127°		Methochloride 259–261°	689
4-(5-Trifluoromethyl-3-pyrazolyl)pyridine	184–185°		Methochloride 254°	689
4-(5-Carbethoxy-3-pyrazolyl)pyridine	209–210°		Methochloride 201–202°	689
4-(4,5-Cyclohexeno-3-pyrazolyl)pyridine	198–199°			689
4-(5-Phenyl-3-pyrazolyl)pyridine	207–208°		Methiodide 211–212°	689
4-(5-Benzyl-3-pyrazolyl)pyridine	136–137°			689
3-Methyl-4-(5-methyl-3-pyrazolyl)pyridine	136–138°		Methochloride 263–265°	689
2-(3-Methyl-1-phenyl-5-pyrazolyl)pyridine		136–140°/0.2 mm	HCl 169–171°	706
3-(3-Methyl-1-phenyl-5-pyrazolyl)pyridine		160–170°/2 mm	HNO₃ 153–155°	706
4-(1-*o*-Chlorophenyl-3-methyl-5-pyrazolyl)pyridine	113–114°		HCl 227–229°	706
2-(1-*m*-Chlorophenyl-3-methyl-5-pyrazolyl)pyridine		156–162°/0.2 mm	HCl 154–156°	706
3-(1-*m*-Chlorophenyl-3-methyl-5-pyrazolyl)pyridine		156–162°/0.2 mm	HNO₃ 144–145°	706
4-(1-*m*-Chlorophenyl-3-methyl-5-pyrazolyl)pyridine	75–76°		HCl 241–244°	706
4-(1-*m*-Chlorophenyl-5-methyl-3-pyrazolyl)pyridine	75–76°		HCl 208–210°	706
2(1-*p*-Chlorophenyl-3-methyl-5-pyrazolyl)pyridine	94–96°		HCl 187–189°	706
4-(1-*p*-Chlorophenyl-3-methyl-5-pyrazolyl)pyridine	117–118°		HCl 254–257°	706
4-(1-*p*-Chlorophenyl-5-methyl-3-pyrazolyl)pyridine	90–92°		HCl 260–263°	706
4-(1-*p*-Bromophenyl-3-methyl-5-pyrazolyl)pyridine	124–126°		HCl 249–253°	706
4-[1-(2,4-Dichlorophenyl)-3-methyl-5-pyrazolyl]pyridine	135–136°			706
4-[1-(2,4-Dichlorophenyl)-5-methyl-3-pyrazolyl]pyridine	133–134°		HCl 244–249°	706
4-[1-(2,5-Dichlorophenyl)-3-methyl-5-pyrazolyl]pyridine	132–134°		HCl 246–249°	706

Table V-18 (*Continued*)

Compound	M.p.	B.p.	Salt m.p.	Ref.
4-[1-(3-Chloro-2-methylphenyl)-3-methyl-5-pyrazolyl]pyridine	Oil		HCl 213–215°	706
4-[1-(3-Chloro-2-methylphenyl)-5-methyl-3-pyrazolyl]pyridine	Oil		HCl 284–290°	706
4-[1-(5-Chloro-2-methoxyphenyl)-3-methyl-5-pyrazolyl]pyridine	158–159°			706
4-[1-(5-Chloro-2-methoxyphenyl)-5-methyl-3-pyrazolyl]pyridine	140–141°			706
3-(1-*m*-Methoxyphenyl-3-methyl-5-pyrazolyl)pyridine		165–170°/0.5 mm	HCl 168–170°	706
4-(1-*m*-Methoxyphenyl-3-methyl-5-pyrazolyl)pyridine			HNO$_3$ 136–139°	706
3-(1-*m*-Hydroxyphenyl-3-methyl-5-pyrazolyl)pyridine	130–132°		HNO$_3$ 126° (decomp.)	706
4-(1-*m*-Hydroxyphenyl-3-methyl-5-pyrazolyl)pyridine			HBr 251–253°	706
2-Methyl-5-(1-*p*-tolyl-3-pyrazol-5-one)pyridine	216–218°			707
4-(4-Methyl-1-phenyl-3-pyrazol-5-one)pyridine	174–174.5°		Picrate 213–214°	708
4-(4-Ethyl-1-phenyl-3-pyrazol-5-one)pyridine	174–176°		Picrate 190–192°	708
4-(1-Phenyl-3-pyrazol-5-one)pyridine	229–230° (decomp.)		Picrate 112°	708
4-(4-Methyl-3-pyrazol-5-one)pyridine	303° (decomp.)		Picrate 242–244°	708
4-(4-Ethyl-3-pyrazol-5-one)pyridine	282° (decomp.)		Picrate 228°	708
4-(1-Benzyl-4-methyl-3-pyrazol-5-one)pyridine	203–205°			708
4-(1-Benzyl-4-ethyl-3-pyrazol-5-one)pyridine	188–190°			708
4-(5-Isoxazolyl)pyridine	100–102°		Methochloride 182.3° (decomp.)	692, 693
4-(3-Methyl-5-isoxazolyl)pyridine	67–68°		Methochloride 250° (decomp.)	692
4-(5-Methyl-3-isoxazolyl)pyridine			Methochloride 221.2°	692
4-(3-Ethyl-5-isoxazolyl)pyridine	48–49°		Methochloride 205–206° (decomp.)	692, 693
4-(5-Trifluoromethyl-3-isoxazolyl)pyridine	82–83°		Methochloride 230° (decomp.)	692, 693
3-Methyl-4-(3-methyl-5-isoxazolyl)pyridine	87–88°		Methochloride 246–247° (decomp.)	692, 693
4-(2-Imidazolyl)pyridine	211–212°		Methochloride 277–278° (decomp.)	694

Table V-18 (*Continued*)

Compound	M.p.	B.p.	Salt m.p.	Ref.
4-(2-Benzimidazolyl)pyridine	214°			723
4-(2-Oxazolyl)pyridine			Methochloride 244° (decomp.)	695
4-(2-Methyl-5-oxazolyl)pyridine	79–81°			695
4-(5-Methyl-2-oxazolyl)pyridine	99–100°		Methiodide 213–214°	695
4-(2-Methyl-4-thiazolyl)pyridine	79–80°		Methochloride 228–231°	696
4-(2-Cyclopropyl-4-thiazolyl)pyridine		100°/0.1 mm	Methiodide 233–234°	696
4-(2-Methyl-5-thiazolyl)pyridine	<30°		Methiodide 253–255° (decomp.)	696
4-(5-Methyl-2-thiazolyl)pyridine	88–90°		Methochloride 242–244° (decomp.)	696
4-(4,5-Dimethyl-2-thiazolyl)pyridine	101–102°			696
2-(1,2,4-Oxadiazol-3-yl)pyridine	109–110°			729
2-(1,3,4-Oxadiazol-2-yl)pyridine	117.5–118.5°			729
2-(5-Methyl-1,2,4-oxadiazol-3-yl)pyridine	87–88.5°			729
4-(5-Methyl-1,3,4-oxadiazol-2-yl)pyridine			Methiodide 275–277° (decomp.)	694
2-(5-*o*-Chlorophenyl-1,2,4-oxadiazol-3-yl)pyridine	93–95°			727, 728
3-(5-*o*-Chlorophenyl-1,2,4-oxadiazol-3-yl)pyridine	85°			727, 728
4-(5-*o*-Chlorophenyl-1,2,4-oxadiazol-3-yl)pyridine	138–140°		Methiodide 231–232°	727, 728
4-(5-Methyl-1,2,4-oxadiazol-3-yl)pyridine	97°			727, 728
4-(5-*p*-Chlorophenyl-1,2,4-oxadiazol-3-yl)pyridine	168–170°			727, 728
3-(3-*o*-Ethoxyphenyl-1,2,4-oxadiazol-5-yl)pyridine	121–122°			727, 728
4-(3-Styryl-1,2,4-oxadiazol-5-yl)pyridine	115°			727, 728
3-(5-*p*-Aminophenyl-1,2,4-oxadiazol-3-yl)pyridine	217°			727, 728
3-(5-*p*-Chlorophenoxymethyl-1,2,4-oxadiazol-3-yl)pyridine	135–138°			727, 728
3-[5-(3-Pyridyl)-1,2,4-oxadiazol-3-yl]pyridine	134°			727, 728
4-[5-(4-Pyridyl)-1,2,4-oxadiazol-3-yl]pyridine	164°			727, 728
2-Ethyl-4-(5-methyl-1,2,4-oxadiazol-3-yl)pyridine			HCl 221°	727, 728
4-(5-*o*-Chlorophenyl-1,2,4-oxadiazol-3-yl)-2-ethylpyridine	66°			727, 728

Table V-18 (*Continued*)

Compound	M.p.	B.p.	Salt m.p.	Ref.
2-Ethyl-4-[5-(2-pyridyl)-1,2,4-oxadiazol-3-yl]pyridine			HCl 230°	727, 728
2-Ethyl-4-[5-(*o*-hydroxyphenyl)-1,2,4-oxadiazol-3-yl]pyridine	103°			727, 728
2-Ethyl-4-[5-(4-pyridyl)-1,2,4-oxadiazol-3-yl]pyridine	67°			727, 728
2-Ethyl-4-[5-(2-ethyl-4-pyridyl)-1,2,4-oxadiazol-3-yl]pyridine			HCl 253°	727, 728
4-[5-(*p*-Chlorobenzyl)-1,2,4-oxadiazol-3-yl]-2-ethylpyridine			HCl 185–187°	727, 728
4-[5-(*p*-Chlorophenoxymethyl)-1,2,4-oxadiazol-3-yl)pyridine	146–147°			727, 728
4-(5-Methyl-1,3,4-thiadiazol-2-yl)pyridine			Methiodide 226–227° (decomp.)	694
4-(5-Methyl-1,2,4-triazol-3-yl)pyridine			Methochloride 281–284°	694
2-(5-Methyl-1,2,4-triazol-3-yl)pyridine	167–168°			733
2-(3-Phenyl-1,2,4-triazol-5-yl)pyridine	212–213°			733
2-[3-(2-Pyridyl)-1,2,4-triazol-5-yl]pyridine	208–209°			733
4-Methyl-2-(3-phenyl-1,2,4-triazol-5-yl)pyridine	221–222°			733
4-Phenyl-2-(3-phenyl-1,2,4-triazol-5-yl)pyridine	236–237°			733
2-(5-Tetrazolyl)pyridine	215–216° (decomp.)			737, 738
3-(5-Tetrazolyl)pyridine	238° (decomp.)			737, 738
4-(5-Tetrazolyl)pyridine	263° (decomp.)		Methochloride 235° (decomp.)	694, 737
4-Methyl-3-(5-tetrazolyl)pyridine	225–227° (decomp.)			737
5-Methyl-3-(5-tetrazolyl)pyridine	223° (decomp.)			737
3-(5-Tetrazolylmethyl)pyridine	192–193° (decomp.)		HCl 192–193° (decomp.)	737, 738
3-(1-Methyl-5-tetrazolyl)pyridine	78–80°			738
3-(2-Methyl-5-tetrazolyl)pyridine	127.5–129°			738
2-Pyridylferrocene	87–89°			739
1,1′-Bis(2-pyridyl)ferrocene	179–180°			739

TABLE V-19. Six-Membered Heteroarylpyridines

Compound	M.p.	B.p.	Picrate (m.p.)	Salt (m.p.)	Ref.
2,4'-Bipyridyl	61.5°	148–150°/11 mm	218°	Methiodide 188–190°	780, 781
				Styphnate 205° (decomp.)	
3,3'-Bipyridyl	64°	160°/9 mm		Methiodide 168–169°	785
				HCl 290–295°	
3,4'-Bipyridyl	61–62°	144–146°/15 mm	208–209°	Styphnate 230°	777, 780
4-Methyl-2,3'-bipyridyl		165–175°/1 mm	199.5–201°	Picrolonate 219.5–221°	784
4-Methyl-3,3'-bipyridyl	48.5–50°		205–207°	Picrolonate 256–258°	784
				(decomp.)	
6-Methyl-4,2'-bipyridyl		165–170°/11 mm	209°	Perchlorate 164–165°	780
6-Methyl-4,3'-bipyridyl	58–59°	155–160°/12 mm	199–200°		780
4-Ethyl-2,3'-bipyridyl			160–161°	Picrolonate 211–212.5°	784
4-Ethyl-3,3'-bipyridyl		195–210°/1 mm	204–205°	Picrolonate 245–246.5°	784
				(decomp.)	
2,6-Dimethyl-2,4'-bipyridyl		180–183°/27 mm	202°	Methiodide 232–234°	781
2,6-Dimethyl-4,4'-bipyridyl		178°/17 mm		Methiodide 310°	781
				(decomp.)	
2,6-Dimethyl-4,2'-bipyridyl	81–83°	113–115°/0.2 mm			780
2,2'-Dimethyl-4,4'-bipyridyl	170–172°				750
4,4'-Dimethyl-2,2'-bipyridyl	114.5–115°				679
5,5'-Dimethyl-2,2'-bipyridyl	88.5–89.5°				679
6,6'-Dimethyl-2,2'-bipyridyl					679
4,4'-Dimethyl-2,2'-bipyridyl		130°/0.2 mm	153–154°		679
			(decomp.)		

363

Table V-19 (*Continued*)

Compound	M.p.	B.p.	Picrate (m.p.)	Salt (m.p.)	Ref.
4,4'-Distyryl-2,2'-bipyridyl	268.5–269.5°				679
4,4'-Diphenethyl-2,2'-bipyridyl	147.5–148°				679
4,4'-Diphenyl-2,2'-bipyridyl	187–188°				679
5,5'-Diethyl-4,4'-dimethyl-2,2'-bipyridyl	141.5–142°				679
5,5'-Diethyl-4,4'-distyryl-2,2'-bipyridyl	182–183°				679
5,5'-Diethyl-4,4'-diphenethyl-2,2'-bipyridyl	133–135°				679
2,2'.6,6'-Tetramethyl-4,4'-bipyridyl	150–151°	147°/2 mm			750
4,4',6,6'-Tetramethyl-2,2'-bipyridyl	142.5–143°				679
2,4-Di-(3-pyridyl)pyridine (nicotelline)	147–148°		216–217°		782, 783
2,6-Di-(3-pyridyl)pyridine	82–83°		263–268°	HgCl$_2$ 235–236°	783
3,4-Di-(2-pyridyl)pyridine	101°				674
2,5-Bis-(2-pyridyl)pyrazine	226–227°				831
2,4,6-Tris-(2-pyridyl)-1,3,5-triazine	Hydrate 244–245°				831
2,4,6-Tris-(4-methyl-2-pyridyl)-1,3,5-triazine	Hydrate 213–214°				831
2,4,6-Tris-(4-phenyl-2-pyridyl)-1,3,5-triazine	244–245°				831
3-(2-Pyridyl)-1,2,4-triazine	86–87°				733
5-Phenyl-3-(2-pyridyl)-1,2,4-triazine	140.5–141°				830
5,6-Dimethyl-3-(2-pyridyl)-1,2,4-triazine	92–93°				733
3-(4-Methyl-2-pyridyl)-1,2,4-triazine	106–107°				733
5,6-Dimethyl-3-(4-methyl-2-pyridyl)-1,2,4-triazine	147–148°				733
5,6-Dimethyl-3-(4-phenyl-2-pyridyl)-1,2,4-triazine	112–113°				733
2,4,6-Tris-(4-ethyl-2-pyridyl)-1,3,5-triazine	105–106°				831

364

and **V-277** were made and screened for possible carcinogenic activity (687, 688).

The properties and reactions of the heteroarylpyridines are discussed in the appropriate sections.

V. Heteroarylpyridines

The heteroarylpyridines in which the pyridine ring is directly attached to a five- or six-membered heteroaryl moiety form an important class of compounds. Some of these compounds or their reduced forms occur

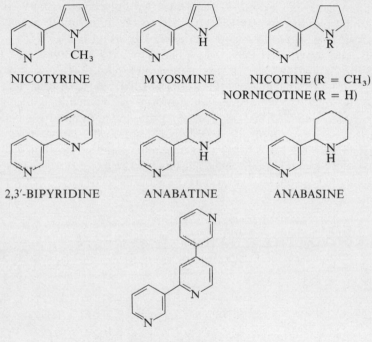

NICOTYRINE   MYOSMINE   NICOTINE (R = CH₃)

NORNICOTINE (R = H)

2,3'-BIPYRIDINE   ANABATINE   ANABASINE

NICOTELLINE

naturally as, for example, the tobacco alkaloids. Other derivatives such as "Diquat" find application in agriculture as herbicides. Still other heteroarylpyridines have potential value in medicine.

2 Br⁻

DIQUAT

1. Five-Membered Heteroarylpyridines

During a general screening program for oral hypoglycemic activity, 1-methyl-4-[5(3)-methyl-3(5)-pyrazolyl]pyridinium iodide (**V-278**) was found to possess excellent activity. The isomeric 2- and 3-pyridinium analogues and the isomer in which the 4-pyridine was attached to the C-4 position of the pyrazole moiety were completely devoid of activity (689). Utilizing this lead, Wiegand, Bauer, and co-workers synthesized a series of 4-azolyl-pyridinium salts and examined their oral hypoglycemic activity. In this series, the 4-pyridine moiety was attached to the following five-membered heteroaryl rings: 3-furyl, 2- and 3-thienyl, 3-pyrrolyl (690, 691), 3-indolyl (691), 3- and 4-pyrazolyl (689), 5-isoxazolyl (692, 693), 2-imidazolyl (694), 2- and 5-oxazolyl (695), 2-, 4-, and 5-thiazolyl (696), 3-(1,2,4-oxadiazolyl) (697), 2-(1,3,4-oxadiazolyl), 2-(1,3,4-thiadiazolyl), 4-(1,2,4-triazolyl), and 5-tetrazolyl (694). In all these compounds, the five-membered ring was synthesized by standard methods from a pyridine having a suitable substituent in the C-4 position. Of all the 4-azolylpyridinium salts prepared, only the 1,2,4-triazolyl-, 1,3,4-thiadiazolyl-, imidazolyl-, and tetrazolylpyridinium salts showed no oral hypoglycemic activity.

V-278

A. Furyl, Thienyl, and Pyrrolyl Pyridines

4-(3-Furyl)pyridine (690), 4-(2-thienyl)pyridine (698, 699), 4-(3-thienyl)-pyridine (698), and 4-(2-methyl-4-pyrrolyl)pyridine (690) were prepared by standard methods. Hellmann and Dieterich have described a convenient synthesis of nicotine, myosmine, 5-(2-pyridyl)proline, and 5-(3-pyridyl)-proline starting from nicotinoyl acetic ester, paraformaldehyde, and acetaminomalonic ester (700).

The structure of anhydroproferrorosamine B, a metabolite of *Pseudomonas roseus fluorescens* was found to be 2-(2-pyridyl)-5-carboxypyrrole-3-acetic acid (**V-279**) (701).

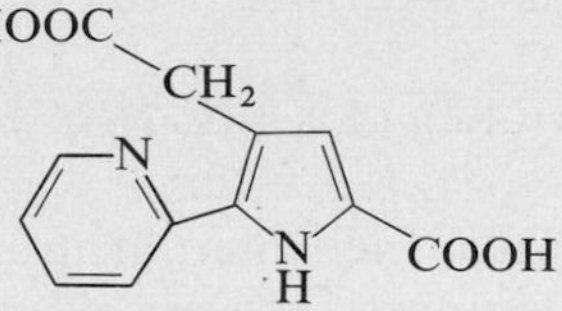

V-279

The multistage synthesis of 2,5-bis-(2-pyridyl)pyrrole, which is a versatile ligand for the preparation of many metal inner complexes (**V-280**), starting from 2-picoline was reported (702).

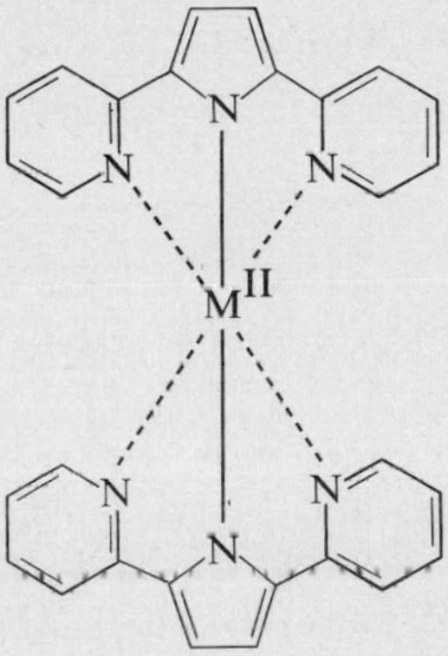

V-280

M = Fe, Co, Ni, Zn, Cd, Cu

The high pressure (1000 psi) hydrogenation of 4-methyl-4-nitro-3-phenyl-1-(3-pyridyl)-1-pentanone (**V-281**) over Raney nickel produces 5,5-dimethyl-4-phenyl-2-(3-pyridyl)-Δ^1-pyrroline (**V-282**) as the major product. The

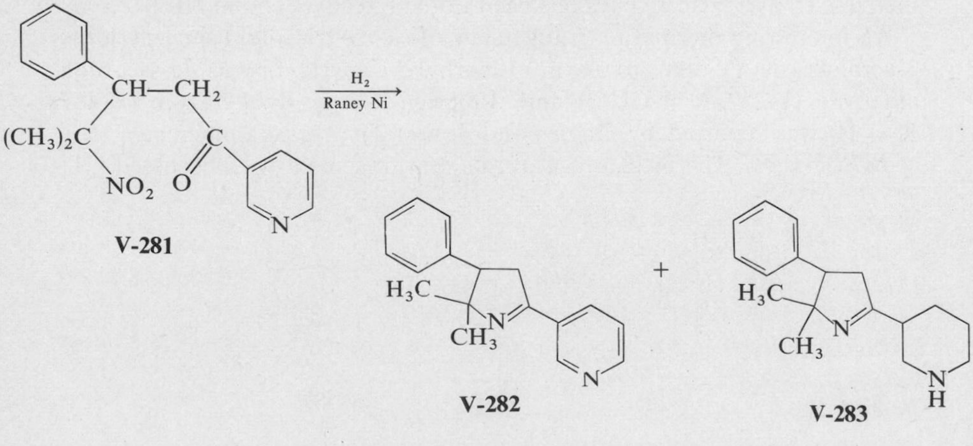

V-281

V-282

+

V-283

minor product was assigned structure **V-283** on the basis of spectroscopic evidence (703).

1-*t*-Aminoalkyl-3-(4-pyridyl)indoles (**V-284**) were found to possess anti-inflammatory and antiprotozal properties and, in addition, to stimulate the biosynthesis of steroids (704). Indolyltetrahydropyridines **V-285** with anti-histaminic and antiserotonine activity were also patented (705).

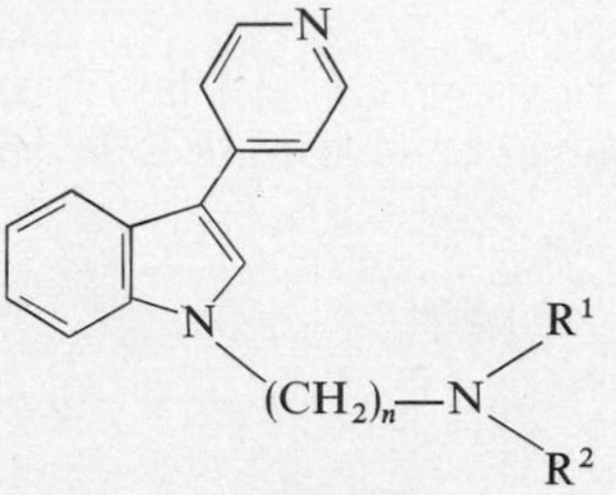

V-284

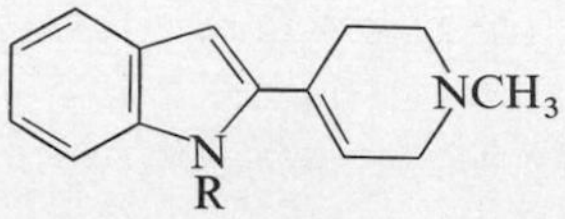

V-285

B. *1,2-Azolyl Pyridines*

The reaction of arylhydrazines with 4-pyridoylacetone gave 1-aryl-3-methyl-5-(4-pyridyl)pyrazoles and 1-aryl-5-methyl-3-(4-pyridyl)pyrazoles (689, 706), and the reaction with pyridoylacetates (or their *N*-oxides) gave the 1-aryl-3-pyridylpyrazol-5-ones (707, 708). The pyridylpyrazolones had analgesic and antispasmodic activity (707). 3-Pyridyl β-styryl ketone on treatment with phenylhydrazine gave 1,5-diphenyl-3-(3-pyridyl)-Δ^2-pyrazoline (709). 4-(2-Hydroxyphenyl)-5-(4-pyridyl)pyrazoles which are useful against adenovirus, parainfluenza virus, and rhino virus have been reported (710).

The preparation of pyrazolylpyridines by the cycloaddition of diazomethane to pyridylacetylenes has been noted already (Section III.2).

An interesting thermal rearrangement of *cis-vic*-triazolo[1,5-*a*]-pyridine-3-acraldehyde (**V-286**) to form 3-methyl-5-(2-pyridyl)pyrazole-4-carboxaldehyde (**V-287**; R = CHO) and 3-methyl-5-(2-pyridyl)pyrazole (**V-287**; R = H) was reported by Davies and Jones. The *trans*- isomer gave only **V-287** (R = H). The mechanism of this rearrangement was discussed (711, 712).

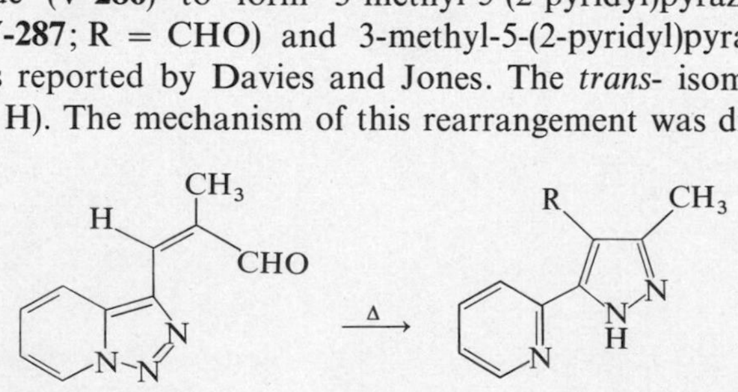

V-286　　　　　　　　V-287

Bowie found that the reaction of hydrazine with 5-chloro-1,8-naph-thyridines (**V-288**) gave 2-amino-3-(5-pyrazolyl)pyridines (**V-289**) which could be cyclized with orthoesters to pyrazolo[1,5-c]pyrido[3,2-e]-pyrimidines (**V-290**) (713).

A comparison of the experimental and calculated dipole moments has shown that 1-pyridylpyrazoles and 5-amino-1-(3- or 4-pyridyl)pyrazoles have nonplanar configurations. In the case of 5-amino-1-(2-pyridyl)-pyrazoles, intramolecular hydrogen bonding produces a planar *trans* configuration. The dihedral angles between the pyrazole and pyridine rings for various compounds was calculated (714).

Hydroxylamine reacted with 1-(4-pyridyl)-1,3-alkanediones to form the 4-isoxazolylpyridines (692, 693) and with pyridoylacetonitrile to give 5-amino-3-pyridylisoxazoles (715).

C. 1,3-Azolyl Pyridines

4-(2-Thiazolyl)pyridines (**V-291**) were prepared by the action of thio-isonicotinamide on α-bromoketones (696, 716), or by treating isonicotin-amidoacetone with phosphorus pentasulfide (696). Bromoacetylpyridine reacted with thioamides, including thiourea, to form the 4-(4-thiazolyl)-pyridines (**V-292**) (696, 717), and 4-(2-methyl-5-thiazolyl)pyridine (**V-293**) was prepared by the action of phosphorus pentasulfide on 4-acetylamino-acetylpyridine (696).

The "Diquat" thiazole analogue **V-294** has herbicidal properties (718), and the thiazolinylpyridyl phosphorothioates **V-295** have insecticidal, nematocidal, and acaricidal properties (719).

V-294   **V-295**

Pyridylbenzothiazolines (**V-296**) were synthesized by condensing pyridyl-aldehydes or ketones with substituted *o*-aminobenzenethiols or their hydro-chlorides in alcohol. The compounds **V-296** (R = R^1 = H) were readily oxidized by manganese dioxide to the benzothiazoles (**V-297**) (RRI 1152), but attempts to methylate them using methyl iodide in methanolic sodium hydroxide resulted in ring cleavage to **V-298**. These compounds possessed *in vitro* activity against *Mycobacterium tuberculosis* (H 37Rv) but only slight *in vivo* activity (720).

V-298

V-296   **V-297**

4-(2-Oxazolyl)pyridine (721), 4-(5-methyl-2-oxazolyl)pyridine and 4-(2-methyl-5-oxazolyl)pyridine (695), and 4-(2-imidazolyl)pyridine (694) have been prepared by standard methods. Pyridylimidazolines (**V-299**) were synthesized by condensing a pyridine aldehyde with the appropriate primary-secondary diamine. Reaction of the aldehyde with *N*-methyl-aminoethanol gave the pyridyloxazolidine derivatives (**V-300**). These nicotine analogues were devoid of pharmacological activity (722).

V-299   **V-300**

Heating a mixture of isonicotinic acid and *o*-phenylenediamine gave 2-(4-pyridyl)benzimidazole (723). 5-Benzyl- and 5-phenyl-5-(3-pyridyl)-hydrantoins **V-301** had little anticonvulsant activity on rats (724).

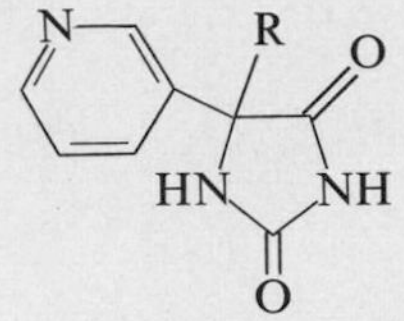

V-301

D. Oxadiazolyl, Thiadiazolyl, and Triazolyl Pyridines

4-(5-Methyl-1,3,4-oxadiazol-2-yl)pyridine (**V-302**) (R = CH$_3$) (725) 2-(1,3,4-oxadiazol-2-yl-5-phenyl)pyridine (**V-302**) (R = C$_6$H$_5$) (726), and a number of pyridyl-1,2,4-oxadiazoles (**V-303**) (R and/or R^1 = 2-, 3-, or 4-pyridyl) (727, 728) were synthesized, and the 1,2,4-oxadiazoles were reported to have antitussive, spasmolytic, local anaesthetic, and coronary dilating properties (727).

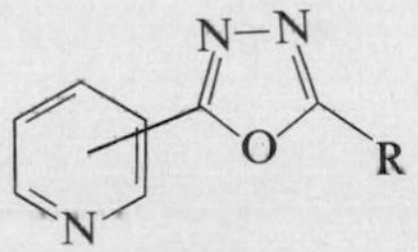

V-302

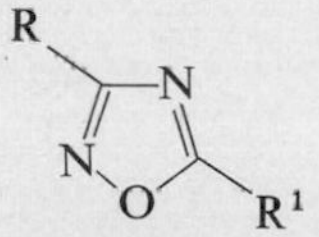

V-303

An interesting reaction is the nitrosation of imidazo[1,5-*a*]pyridine (**V-304**) which produces 3-(2-pyridyl)-1,2,4-oxadiazole (**V-303**; R = 2-pyridyl, R^1 = H) in 70% yield (729).

4-(5-Methyl-1,3,4-thiadiazol-2-yl)pyridine (725), 2- and 3-(5-alkylthio-1,3,4-thiadiazol-2-yl)pyridines (730), and many pyridyl-1,2,4-triazoles (726, 731, 732) have been reported. Heating an ethanolic solution of pyridine-2-carboxaldehyde and acetamidazone produced 2-methyl-5-(2-pyridyl)-1,2,4-triazoline (**V-305**) (81%) (733).

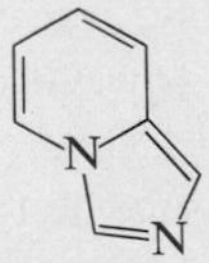

V-304

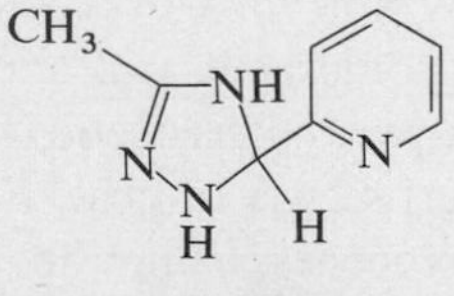

V-305

The oxidation of 1-(2-pyridyl)benzotriazole (**V-306**) (RRI 1097) using peracetic acid was studied to find out which nitrogen is the most susceptible to attack. The product was identified as 1-(2-pyridyl)benzotriazole-3-oxide (**V-307**) (734).

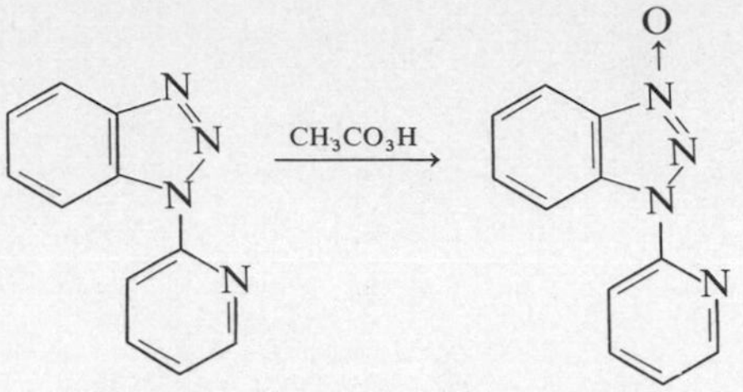

V-306   V-307

Blackman and Polya prepared a series of 1,2,4-triazolinethiones and studied their spectroscopic properties. 4-(4-Pyridyl)-1,2,4-triazoline-5-thione (**V-308**) alone behaved anomalously and this was explained as due to the extended conjugation in the compound (735).

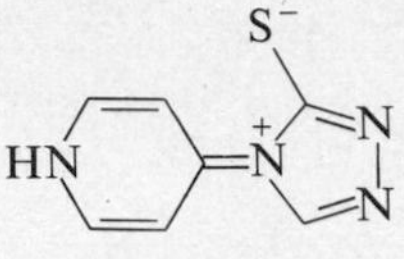

V-308 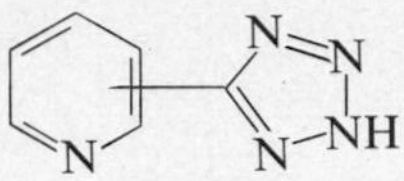   V-309

E. Tetrazolyl Pyridines

2-, 3-, And 4-cyanopyridines react with ammonium azide in dimethylformamide to give the 5-pyridyltetrazoles (**V-309**) (736–738). Of the series of compounds prepared, 5-(3-pyridyl)tetrazole was found to be the most active lipolysis inhibitor (737, 738).

F. Pyridyl Ferrocene

Nesmeyanov and co-workers found that on heating a mixture of 1,1'-ferrocenyldiboronic acid and copper acetate in pyridine under a nitrogen atmosphere, 2-pyridylferrocene (**V-311**) was formed in low yields, probably by way of a ferrocenyl radical. Ferrocenyllithium (a mixture of lithio- and 1,1'-dilithioferrocenes from the reaction of butyllithium with ferrocene), however, on heating with pyridine for 3 hr in a nitrogen atmosphere gave 2-pyridylferrocene (**V-311**) (24%) and 1,1'-bis(2-pyridyl)ferrocene (**V-312**) (3%) (739).

2-Pyridylferrocene was oxidized by permanganate to pyridine-2-carboxylic acid and gave a red hydrochloride (739). Lithiation of 2-pyridylferrocene followed by reaction with tributyl borate gave 2-(2-pyridyl)ferroceneboronic

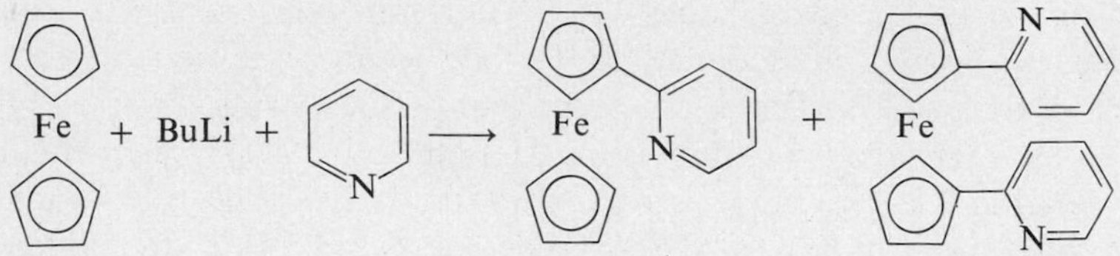

V-310   **V-311**   **V-312**

acid (740, 741). Photolysis of 2-(2-pyridyl)ferrocene methiodide (**V-313**) gave
1-methyl-2-cyclopentadienylidene-1,2-dihydropyridine (**V-314**), cyclopentadiene, and iron (742).

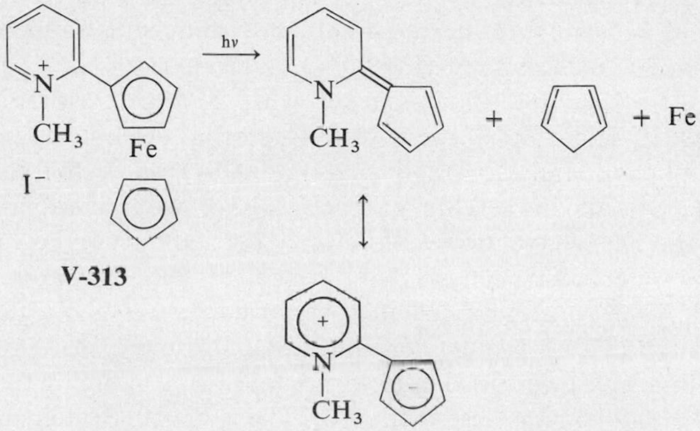

V-313

V-314

2. Six-Membered Heteroarylpyridines

A. Bi- and Polypyridines

a. PREPARATION. The synthesis of bipyridines by the reaction of pyridines
with metals such as sodium and magnesium has been further investigated
(743–761). 4,4′-Bipyridyl is the major product of this reaction when pyridine
is used (743, 748–753, 755). 2,2′-Bipyridyl and 2,4′-bipyridyl are also formed.
It is usual to use excess pyridine as the solvent, but dialkylarylamines, when
used as diluents, are reported to give improved yields (743). A continuous
process for the preparation of bipyridyls using magnesium and following the
reaction by electrical conductance measurements was reported (748). The
optimum conditions for preparing bipyridyls from pyridine and sodium
followed by oxidation was established (750). Oxidants that have been used

are bromine (755), nitroalkanes (751, 758), nitrobenzene (752), and air. 2,2'-Bipyridyl (65%) is preferentially formed when pyridine is heated with zinc chloride, cadmium chloride, or mercury (II) chloride (759), or salts of the Group VIII metals (760), or iron (III) chloride (761). Using this latter catalyst with 2,2'-bipyridyl and 2,2'-bipyridyl hydrochloride, 2,2',2'',2'''-tetrapyridyl was synthesized (761). When methyl chloride was added to a reaction mixture containing pyridine and sodium amide in liquid ammonia the product was N,N'-dimethyl-4,4'-bipyridylium chloride (65%) (744).

2,2'-, 3,3'-, 4,4'-, And 2,3'-bipyridyls were prepared by means of the Ullman and Grignard reactions (762–767). The 3-halopyridines were found to be less active than the 2- and 4-halopyridines in both reactions (762). A mixture of 2-bromopyridine and 4-chloropyridine with copper powder in nitrobenzene gave 2,2'-bipyridyl (57.5%), 4,4'-bipyridyl (12.5%), and 2,4'-bipyridyl (14.5%) (766). Raney nickel with 2-chloropyridine gave 2,2'-bipyridyl (763). The effect of the solvent—tetralin, decalin, pseudocumene, decane, diphenyl, and dimethylformamide (764)—and the form of the copper catalyst (765) in the Ullman reaction have been investigated.

Badger and Sasse have reviewed the work carried out on the dimerization of pyridines using metal catalysts (768); hence only recent work in this area is considered. Raney nickel has usually been used; however, Ziegler catalysts (769), rhenium (770), rhodium, osmium, and iridium (771), and platinum and ruthenium (772) were also used. A method for prolonging the life of the Raney nickel catalyst was described (773). The vapor phase dehydrogenation of 4-(4-pyridyl)piperidine gave 4,4'-bipyridyl (774). The optimum reaction conditions for the preparation of 2,2'-bipyridyl were established (775). Radiolysis of pyridine using ^{60}Co gamma rays gave 2,2'-bipyridyl (776).

Heating pyridine-3-carboxaldehyde with acetaldehyde, ammonia, and ammonium acetate gave 3,4'-bipyridyl (**V-315**) (32%) (777). Bipyridyls were prepared in a similar way by the vapor phase reaction of substituted pyridines with ammonia (778). The liquid phase ammonolysis of 3-(4-pyridyl)-1,5-pentanediol gave 4,4'-bipyridyl (779).

V-315

2,4'-, 3,4'-Bipyridyls, 4-(3-pyridyl)picoline, and 4-(2-pyridyl)picoline were synthesized unambiguously by the decarboxylation of the respective acids (780, 781). A new synthesis of nicotelline, 2,4-di-(3-pyridyl)pyridine *via* a pyridone has been reported (782, 783).

The Gomberg reaction using pyridine or alkyl-substituted pyridines and pyridine-3-diazonium chloride gave bipyridyls (784, 785).

Bipyridyls have also been obtained from bipyrylium salts (786). The dimerization of pyridinium salts by cyanide is reported to yield dihydro-bipyridyls (787).

b. PROPERTIES AND REACTIONS. 4,4'-Bipyridyl, which is used in the preparation of herbicides, is separated from 2,4'-bipyridyl and 2,2'-bipyridyl which are also usually present in reaction products, in the form of an insoluble dihydrate (788–790). 2,2'-Bipyridyl can be removed as an iron (II) sulfate complex (791). It is also possible to separate the 4,4'-pyridyl from the highly toxic contaminant 2,2',6',2"-terpyridyl by means of the iron (II) complex of the terpyridyl (792). A sulfonic acid cation-exchange resin was used to separate bipyridinium salts from the pyridinium salts (793).

Complexes of 2,2'-bipyridyl with vanadium (794), cobalt, rhodium, and iridium (795) were studied. The peroxyvanadate of 2,2'-bipyridyl was obtained by adding 2,2'-bipyridyl to a solution of ammonium vanadate in 5% hydrogen peroxide (796). Beck and Halmos have determined the stability constants of copper (II) complexes of 2,2'-dipyridyl, 2-(2-pyridyl)piperidine, and 2,2'-dipiperidyl (797). The magnetic properties of 2,2'-dipyridylcobalt dicyanide (798) and the exchange and electron-transfer reactions of tris-(2,2'-dipyridyl)cobalt (III) ions (799) were studied. 2,2'-Dipyridyl reacted with titanium tetrachloride and diethylaluminium chloride to form a violet complex (V-316) (800). The stability constant of bis(2,2',2"-terpyridyl) iron (II) complex was measured (801).

V-316

Perchlorobipyridyls were obtained in high yields by heating a mixture of the bipyridyl and phosphorus pentachloride in an autoclave. These compounds can be used as fungicides, herbicides, insecticides, and bactericides (802). 5,5'-Dibromo-2,2'-bipyridyl (**V-317**; $R = R^1 = Br$) (4%) and 5-bromo-2,2'-bipyridyl (**V-317**; $R = H, R^1 = Br$) (11.4%) were obtained by treating 2,2'-bipyridyl hydrobromide with bromine at 250°. 3,3'-Bipyridyl gave 5-bromo-3,3'-bipyridyl (**V-318**; $R = Br$) (803).

R

V-317   V-318

The sulfonation of 2,2′-bipyridyl at 300° yields 2,2′-bipyridyl-5-sulfonic acid (**V-317**; R = H, R^1 = SO$_3$H) and 2,2′-bipyridyl-5,5′-disulfonic acid (**V-317**; R = R^1 = SO$_3$H) (804). 3,3′-Bipyridyl on sulfonation at 300° gives the 5-sulfonic acid (**V-318**; R = SO$_3$H) (805). 3,4′-Bipyridyl gave 3,4′-bipyridyl-3′-sulfonic acid (**V-319**) (806). 4,4′-Bipyridyl with concentrated

SO$_3$H

V-319

sulfuric acid at 300° gave 4,4′-bipyridyltetrasulfonic acid (807). With SO$_3$ at room temperature, 4,4′-bipyridyl formed a crystalline complex which on heating to 200° gave 4,4′-bipyridyl-3-sulfonic acid (808).

4,4′-Dimethyl-2,2′-bipyridyl was oxidized with selenium dioxide to 4-aldehydo-2,2′-bipyridyl-4′-carboxylic acid (9%) (809).

Hydrogenation of 4,4′-bipyridyl using Raney nickel (810) and platinum oxide (811) produces 4-(4-pyridyl)piperidine. The zinc dust reduction of 2,2′-bipyridyl dimethiodide gave a fairly stable radical cation **V-320** (812).

CH$_3$

CH$_3$

V-320

The reduction of 3,3′-bipyridyl using sodium in isoamyl alcohol produced all possible hydrogenation products from the di- to the decahydrobipyridyls. The products were separated by chromatography on alumina (813). Reduction of 4,4′-bipyridyl over platinum catalysts and by means of sodium in ethanol, or of anabasine over a Raney nickel catalyst also gave a mixture of all the possible hydrogenation products, which were separated over alumina (814).

Quarternary bipyridyl salts have been investigated for their herbicidal action (815–818). Black and Summers have shown that the molecular size is important in determining the herbicidal activity of Diquat analogues (818).

The stereochemistry of the bridged salts of 2,2′-dipyridyl were studied by ultraviolet spectroscopy and by the measurement of the redox potentials (819). Chloroacetic acid reacted with 4,4′-bipyridyl to give N,N′-bis(acetic acid)-4,4′-bipyridinium chloride (820).

1-(2-, 3- Or 4-pyridyl)-3,4-dihydroisoquinolines (V-321), tetrahydro-isoquinolines, and tetrahydroquinolines were prepared (820–827) and are reported to have analgesic, antipyretic, spasmolytic, and sedative activity.

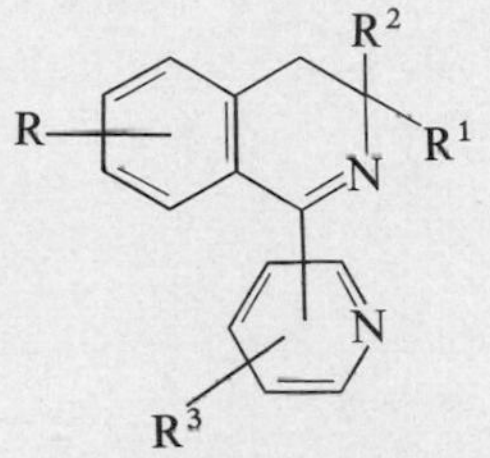

V-321

B. Diazinyl- and Triazinylpyridines

The reported work in this area has been extremely limited. 5-(4-Pyridyl)-2-(5-nitro-2-furyl)pyrimidine (V-322) was found to be a potent inhibitor of *Trichomonas vaginalis* and also to possess antibacterial activity (829).

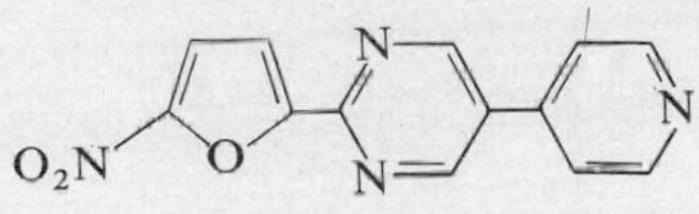

V-322

3-(2-Pyridyl)-5-(substituted phenyl)-*as*-triazines (V-323) were prepared by treating 2-pyridylhydrazine with phenylglyoxals. Compounds V-324

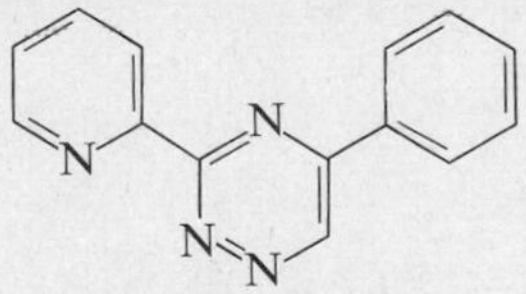

V-323 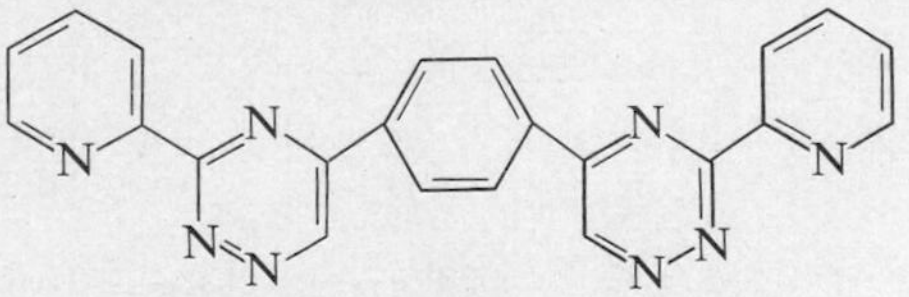   V-324

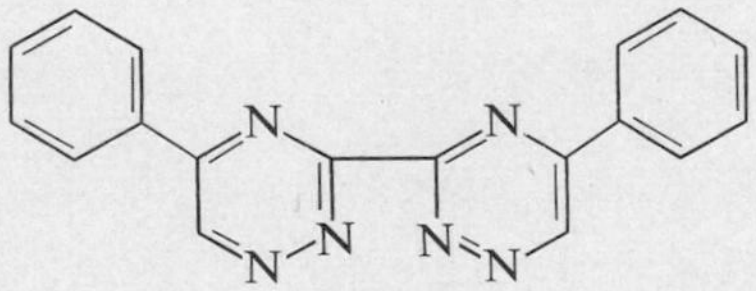

V-325

and **V-325** were also prepared. These compounds have corrosion inhibiting properties (830).

2-Cyanopyridine was trimerized by heating it with sodium hydride to form 2,4,6-tris-(2-pyridyl)-1,3,5-triazine **(V-326)**. 2,5-Bis-(2-pyridyl)pyrazine **(V-327)** was synthesized by the dimerization of 2-pyridyl aminomethyl ketone, followed by dehydrogenation (831).

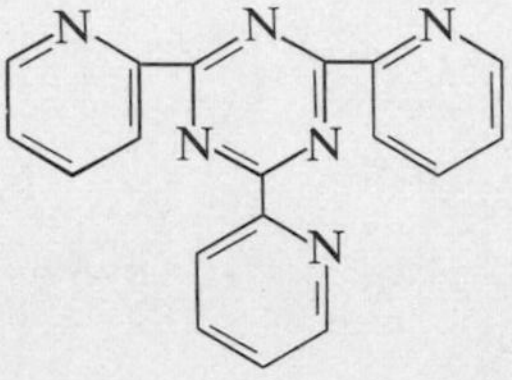

V-327

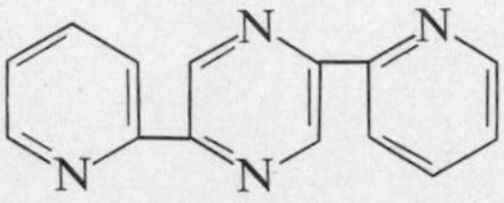

V-326

VI. Cyclopentadienylidenedihydropyridines

Sesquifulvalene **(V-328)** is a highly unstable compound with the properties of a reactive polyolefin. The nitrogen-containing analogues, the cyclopentadienylidenedihydropyridines, **V-329** and **V-330**, by contrast are relatively stable and many of these compounds have been prepared and characterized (832–843).

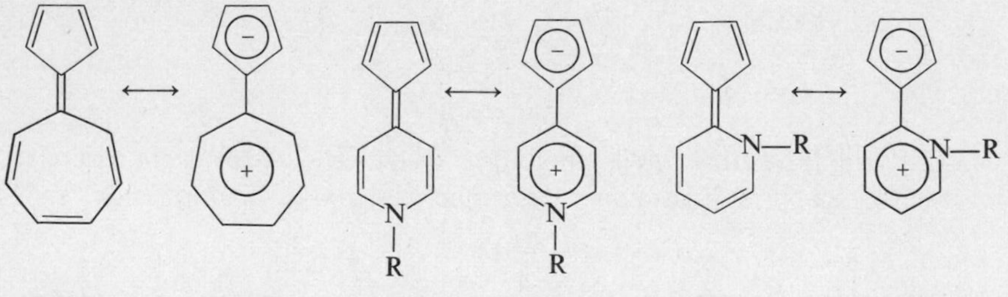

V-328a   V-328b   V-329a   V-329b   V-330a   V-330b

The synthesis of N-methyl-4-cyclopentadienylidene-1,4-dihydropyridine by the photolysis of N-methylpyridiniumferrocene iodide has already been reported (Section IV.3.).

Four other general methods of synthesizing these sesquifulvalene analogues have been employed.

The 4-cyclopentadienylidenedihydropyridines **(V-329**; R = $CH_2C_6H_3$-Cl_2-2,6 and $CH_2C_6H_5$) were made by the dehydrogenative addition of cyclopentadiene to a pyridinium salt (832, 833, 838).

Sodium cyclopentadienyl (**V-331**) reacts with 1-alkyl-4-bromopyridinium bromide (**V-332**; X = Br) or 1-alkyl-4-methoxy- or phenoxypyridinium iodide (**V-332**; X = OR¹) to form compounds of the 4-pyridine series (**V-329**) (835, 841, 842). The 2-halopyridinium salts (**V-333**) produce by this same reaction the 2-pyridinium compounds (**V-330**) (837, 840, 842, 843).

Another unambiguous synthesis of these compounds was reported by Berson and co-workers. 2-Pyridyllithium reacts with cyclopentanone to form the *tert*-alcohol **V-334**, which was dehydrated *via* its methosulfate to **V-330**. Using this same route, 2- and 4-pyridyllithium reacted with fluorenone to form the benzo derivatives of **V-329** and **V-330** (834).

V-334

Heterosequifulvalenes (**V-335**) containing oxygen as the heteroatom undergo a near quantitative nucleophilic O/N–R exchange when treated with aliphatic amines, to form the aza analogues **V-336** (836, 839).

Table V-20 summarizes the physical data on these compounds.

The color of the anhydro base (**V-330**) and its derivatives is reversibly discharged on addition of a proton acid as a result of conversion to the conjugate acid forms **V-337** and **V-338**. Berson and co-workers have shown from a study of the nuclear magnetic resonance spectra of the conjugate acids that

TABLE V-20. Cyclopentadienylidenedihydropyridines

Compound	M.p.	Salt (m.p.)	Ref.
2-Cyclopentadienylidene-1-methyl-1,2-dihydropyridine	Dimorphous 56–57° and 74.5–75°	Perchlorate 80° (decomp.)	742, 834, 843
1-Benzyl-2-cyclopentadienylidene-1,2-dihydropyridine	141–142°		834
4-Cyclopentadienylidene-1-methyl-1,4-dihydropyridine	235° (decomp.)		834
1-Butyl-4-cyclopentadienylidene-1,4-dihydropyridine	170°		838
1-Benzyl-4-cyclopentadienylidene-1,4-dihydropyridine	205–209°		834
4-Cyclopentadienylidene-1,4-dihydro-1,2-dimethylpyridine	208° (decomp.)		838
4-Cyclopentadienylidene-1,4-dihydro-1,3-dimethylpyridine	132–133°		838
1-Butyl-4-cyclopentadienylidene-1,4-dihydro-2-methylpyridine	170–173° (decomp.)		838
1-Butyl-4-cyclopentadienylidene-1,4-dihydro-3-methylpyridine	139° (decomp.)		838
1-Butyl-4-cyclopentadienylidene-1,4-dihydro-2,3-dimethylpyridine	156° (decomp.)		838
1-Butyl-4-cyclopentadienylidene-1,4-dihydro-2,5-dimethylpyridine	118° (decomp.)		838
2-Fluorenylidene-1-methyl-1,2-dihydropyridine	171–172° (decomp.)		834, 837
4-Fluorenylidene-1-methyl-1,4-dihydropyridine	187–188°		834
1-Benzyl-4-fluorenylidene-1,4-dihydropyridine	209–210°		834
2-Indenylidene-1-methyl-1,2-dihydropyridine	161–162° (decomp.)		834, 837
4-Indenylidene-1-methyl-1,4-dihydropyridine	215° (decomp.)		834
1-Butyl-4-indenylidene-1,4-dihydropyridine	102–103° (decomp.)	Perchlorate 102.5–104° Picrate 112–113°	834
1,2-Dimethyl-4-indenylidene-1,4-dihydropyridine	205–207° (decomp.)		838
1-Butyl-4-indenylidene-2-methyl-1,4-dihydropyridine	139.5° (decomp.)		838
1-Butyl-4-indenylidene-3-methyl-1,4-dihydropyridine	135–136° (decomp.)		838

V-337   V-338   V-339

the other possible form **V-339** is not present (834). The 4-pyridine compounds **V-329** behave in a similar way (see also Ref. 835).

The ultraviolet-visible spectra of the 2-pyridine compounds (**V-330**) show four maxima at about 210, 272, 360, and 420 mμ, shifting to two maxima at 230 to 260 and 342 mμ in acid solution (834). The 4-pyridine compounds (**V-329**) show maxima at 242, 335, 428, and 440 (inflection) mμ, shifting to 249, 300 (inflection), and 348 mμ in acid solution (835). A theoretical correlation of the excited-state properties of the 2- and 4- series has been made (844).

The pK_a values for the conjugate acids was measured (834), and a theoretical correlation of ground-state properties such as dipole moments and acidities reported (845).

The nuclear magnetic resonance spectra of several members in this series of compounds were reported (835, 838).

Compounds of the 2- series (**V-330**) react rapidly with electrophilic reagents such as bromine, nitrosyl chloride, acid chlorides, and acetic anhydride, although no pure products have as yet been isolated or identified from these reactions (834). Boyd and co-workers used deuteration experiments to show that electrophilic attack should take place at the C-2 and C-5 positions of the cyclopentadiene ring (838). These workers have also discussed the internal rotation and aromaticity of these compounds (838).

References

1. K. Schofield, Heteroaromatic Nitrogen Compounds: Pyrroles and Pyridines, Butterworth, London, 1967.
2. R. A. Abramovitch, "Pyridine and Pyridine Derivatives," in "Kirk-Othmer Encyclopedia of Chemical Technology," Vol. 16, 2nd ed., Wiley, New York, 1968.
3. L. I. Vereshchagin and I. L. Kotlyarevskii, *Russ. Chem. Rev.*, **30**, 426 (1961).
4. A. Paris, *Ind. Chim.* (France), **56**, 9, 46 (1969); *Chem. Abstr.*, **72**, 100380d (1970).
5. A. N. Neuz and M. Pieroni, *Hydrocarbon Process*, **47**, (11) 139 (1968); *Chem. Abstr.*, **73**, 109586d (1970).
6. E. A. Coulson and J. B. Ditcham, *J. Chem. Soc.*, 356 (1957).
7. F. N. Stepanov, N. L. Dovgan, and A. G. Yurchenko, *Zh. Obshch. Khim.*, **6**, 1823 (1970); *Chem. Abstr.*, **73**, 130850v (1970).

8. A. M. van Leusen and J. C. Jagt, *Tetrahedron Lett.*, 971 (1970).
9. T. Jaworski and S. Kwiatkowski, *Rocz. Chem.*, **44**, 555 (1970); *Chem. Abstr.*, **73**, 130845x (1970).
10. M. Ya Karpeiskii and U. L. Florent'ev, *Russ. Chem. Rev.*, **38**, 540 (1969).
11. G. Ya. Kondrat'eva and H. Chih-hêng, *Dokl. Akad. Nauk SSSR*, **164**, 816 (1965).
12. Takeda Industries Ltd., French Patent 1,400,843, 1965; *Chem. Abstr.*, **63**, 9922d (1965).
13. J. Falke and F. Korte, *Brennstoff-Chem.*, **46** (9), 276 (1965); *Chem. Abstr.*, **64**, 3467g (1966).
14. A. G. Mohan, *J. Org. Chem.*, **35**, 3982 (1970).
15. J. Epsztajn, W. E. Hahn, and B. K. Boleslaw, *Rocz. Chem.*, **44**, 431 (1970); *Chem. Abstr.*, **72**, 132478h (1970).
16. H. L. Lochte and A. G. Pittman, *J. Org. Chem.*, **25**, 1462 (1960).
17. H. L. Lochte and A. G. Pittman, *J. Amer. Chem. Soc.*, **82**, 469 (1960).
18. E. Breitmaier and E. Bayer, *Tetrahedron Lett.*, 3291 (1970).
19. E. Breitmaier and E. Bayer, *Angew. Chem. Inter. Edit.*, **8**, 765 (1969).
20. E. Breitmaier, S. Gassenmann, and E. Bayer, *Tetrahedron*, **26**, 5907 (1970).
21. M. Ohashi, H. Kamachi, H. Kakisawa, and G. Stork, *J. Amer. Chem. Soc.*, **89**, 5460 (1967).
22. M. Ohashi, *Nippon Kagaku Zasshi*, **91**, (1) 12 (1970); *Chem. Abstr.*, **72**, 100556r (1970).
23. I. Adachi, *Chem. Pharm. Bull. Jap.*, **17**, (11), 2209 (1969).
24. W. Dittmar, J. Sauer, and A. Steigel, *Tetrahedron Lett.*, 5171 (1969).
25. R. L. Jones and C. W. Rees, *J. Chem. Soc.*, C, 2249 (1969).
26. R. L. Jones and C. W. Rees, *J. Chem. Soc.*, C, 2255 (1969).
27. R. Nicoletti and M. L. Forcellese, *Tetrahedron Lett.*, 153 (1965).
28. W. E. Parham, R. W. Davenport, and J. B. Biasotti, *J. Org. Chem.*, **35**, 3775 (1970).
29. Yu. I. Chumakov and V. P. Shertstyuk, *Tetrahedron Lett.*, 129 (1965).
30. P. F. G. Praill and A. L. Whitear, *Proc. Chem. Soc.*, 312 (1959).
31. G. N. Dorofeenko, Yu, A. Zhdanov, and L. N. Etmetchenko, *Khim. Geterotsikl. Soedin.*, 781, (1969); *Chem. Abstr.*, **72**, 111223f (1970); G. N. Dorofeenko and V. U. Kransnikov, *J. Org. Chem. USSR*, **8**, 2674 (1972).
32. L. Fedotova and G. Vanags, *Latvijas PSR Zinātnu Akad. Vēstis*, (7) 81 (1958); *Chem. Abstr.*, **53**, 11367d (1959).
33. L. Fedotova and G. Vanags, *Trudy Inst. Khim. Akad. Nauk Latv. S.S.R.*, **2**, 53 (1958); *Chem. Abstr.*, **53**, 21941j (1959).
34. L. Fedotova and G. Vanags, *Latvijas PSR Zinātnu Akad. Vēstis*, (5) 93 (1958); *Chem. Abstr.*, **53**, 1348a (1959).
35. L. Fedotova and G. Vanags, *Latvijas PSR Zinātnu Akad. Vēstis*, (2) 75 (1959); *Chem. Abstr.*, **54**, 526f (1960).
36. C. H. Jarboe and C. J. Rosene, *J. Chem. Soc.*, 2455 (1961).
37. Y. Fushizaki, *Technol. Repts. Osaka Univ.*, **1**, 309 (1951); *Chem. Abstr.*, **46**, 7746i (1952).
38. K. Fujinaga, Japanese Patent 176,238, 1948; *Chem. Abstr.*, **45**, 7603e (1951).
39. Vojtěch Štajgr., Czechoslovakian Patent 85,147, 1956; *Chem. Abstr.*, **51**, 8808b (1957).
40. E. Profft and F. Melichar, *J. Prakt. Chem.* **2**, [4] 87 (1955).
41. Tsu-Hang Yeh and I. V. Kalechits, *Jan Liao Hsüeh Pao*, **2**, 146 (1957); *Chem. Abstr.*, **52**, 20145b (1958).
42. P. Arnall, *J. Chem. Soc.*, 1702 (1958).
43. H. Suzumura, *Bull. Chem. Soc. Jap.*, **34**, 1097 (1962).
44. S. Heřmanek, *Collect. Czech. Chem. Commun.*, **24**, 2748 (1959).
45. Y. Sawa and R. Maeda, Japanese Patent 2148, 1965; *Chem. Abstr.*, **65**, 14635e (1965).
46. A. N. Wilson and S. A. Harris, *J. Amer. Chem. Soc.*, **73**, 2388 (1951).
47. J. M. Bobbitt and D. A. Scola, *J. Org. Chem.*, **25**, 560 (1960).

48. N. Sperber, M. Sherlock, D. Papa, and D. Kender, *J. Amer. Chem. Soc.*, **81**, 704 (1959).
49. W. Hoefling and G. Reckling, British Patent 974,113, 1964; *Chem. Abstr.*, **62**, 7734d (1965).
50. T. Naito and K. Ueno, *Yakugaku Zasshi*, **79**, 1277 (1959); *Chem. Abstr.*, **54**, 4566c (1960).
51. R. Lukeš and M. Pergál, *Chem. Listy*, **52**, 68 (1958).
52. F. Šorm, *Collect. Czech. Chem. Commun*, **13**, 57 (1948).
53. M. Ferles and M. Prystǎs, *Collect. Czech. Chem. Commun.*, **24**, 3326 (1959).
54. M. Ferles and A. Tesařova, *Collect. Czech. Chem. Commun.*, **32**, 1631 (1967).
55. H. Lund, *Acta Chem. Scand.*, **17**, 972 (1963).
56. T. Nakashima, *Yakugaku Zasshi*, **77**, 698 (1957); *Chem. Abstr.*, **51**, 16462g (1957).
57. W-C. Liu and L. Wang, *Yao Hsueh Hsueh Pao*, **12**, (1), 56 (1965); *Chem. Abstr.*, **62**, 16186b (1965).
58. J. Michalski, C. Piechucki, and H. Zajac, *Bull. Acad. Pol. Sci.*, *Ser. Sci. Chim.*, **17**, (6) 347 (1969); *Chem. Abstr.*, **72**, 31552x (1970).
59. F. J. Villani, P. J. L. Daniels, C. A. Ellis, T. A. Mann, and K-C. Wang, *J. Heterocycl. Chem.*, **8**, 73 (1971).
60. H. Gerlach and E. Huber, *Helv. Chim. Acta*, **51**, 2027 (1968).
61. G. Lock, *Osterr. Chem. -Ztg.*, **50**, 5 (1949); *Chem. Abstr.*, **44**, 1928f (1950); *idem, Monatsh Chem.*, **85**, 802 (1954).
62. R. Lukeš and P. Vaculík, *Chem. Listy*, **51**, 1510 (1957).
63. K. B. Prasad, H. N. Al-Jallo, and K. S. Al-Dulaimi, *J. Chem. Soc.*, *C*, 2134 (1969).
64. Yu. I. Chumakov and V. M. Ledovskikh, *Metody Poluch. Khim. Reaktivov Prep.*, (17) 7 (1967); *Chem. Abstr.*, **71**, 61162s (1969).
65. E. Profft and H. W. Linke, *Chem. Ber.*, **93**, 2591 (1960).
66. R. A. Abramovitch, M. Saha, E. M. Smith, and R. T. Coutts, *J. Amer. Chem. Soc.*, **89**, 1537 (1967); R. A. Abramovitch, E. M. Smith, E. E. Knaus, and M. Saha, *J. Org. Chem.* **37**, 1690 (1972).
67. F. A. Daniher and B. E. Hackley, Jr., U.S. Patent 3,467,659, 1969; *Chem. Abstr.*, **71**, 124246v (1969).
68. H. Alper and J. T. Edward, *Can. J. Chem.*, **48**, 1543 (1970).
69. S. Miyano, *Chem. Pharm. Bull. Jap.*, **14**, 663 (1966).
70. L. Horner and H. Röder, *Chem. Ber.*, **101**, 4179 (1968).
71. M. E. C. Biffin, J. Miller, and D. B. Paul, *Tetrahedron Lett.*, 1015 (1969).
72. V. J. Traynelis and K. Yamauchi, *Tetrahedron Lett.*, 3619 (1969).
73. T. Koenig and J. Wieczorek, *J. Org. Chem.*, **35**, 508 (1970).
74. M. Jankovsky and M. Ferles, *Collect. Czech. Chem. Commun.*, **35**, 2802 (1970).
75. M. M. Baizer, U.S. Patent 3,218,245, 1965; *Chem. Abstr.*, **64**, 17554f (1966).
76. J. D. Anderson, M. M. Baizer, and E. J. Prill, *J. Org. Chem.*, **30**, 1645 (1965).
77. Ciba Ltd., British Patent 898,322, 1962; *Chem. Abstr.*, **57**, 13740f (1962).
78. K. K. Moll and H. J. Uebel, German (East) Patent 54,006, 1967; *Chem. Abstr.*, **68**, 59436d (1968).
79. A. D. Dariev, V. N. Gudz, and V. V. Vampilova, *Tr. Buryat. Kompleks. Nauch.-Issled. Inst. Akad. Nauk. SSSR, Sib. Otd.*, (20) 3 (1966); *Chem. Abstr.*, **68**, 78091d (1968).
80. V. N. Gudz, A. D. Dariev, and S. A. Trifonova, *Tr. Buryat. Kompleks. Nauk.-Issled. Inst. Akad. Nauk SSSR, Sib. Otd.*, (20) 13 (1966); *Chem. Abstr.*, **68**, 104924z (1968).
81. A. D. Dariev, V. N. Gudz, and N. S. Kozlov, *Izv. Sib. Otd. Akad. Nauk SSR, Ser. Khim. Nauk*, (3) 105 (1966); *Chem. Abstr.*, **67**, 64193f (1967).
82. R. C. Myerly and K. Weinberg, U.S. Patent 3,428,641, 1969; *Chem. Abstr.*, **71**, 91324p (1969).
83. K. K. Moll and H. J. Uebel, German (East) Patent 58,740, 1967; *Chem. Abstr.*, **69**, 106564b (1968).

84. F. E. Cislak and W. R. Wheeler, U.S. Patent 2,786,846, 1957; *Chem. Abstr.*, **51**, 13941d (1957).
85. H. Pines and S. V. Kannan, *Chem. Commun.*, 1360 (1969).
86. H. Pines and B. Notari, *Amer. Chem. Soc. Div. Petrol. Chem. Preprints*, **4** (4) B47 (1959); *Chem. Abstr.*, **57**, 11151g (1962); *idem.*, *J. Amer. Chem. Soc.*, **82**, 2209 (1960).
87. H. Pines and D. Wunderlich, *J. Amer. Chem. Soc.*, **81**, 2568 (1959).
88. W. K. R. Musgrave, R. D. Chambers, and R. A. Storey, British Patent 1,195,692, 1970; *Chem. Abstr.*, **73**, 77061k (1970).
89. D. Taub, R. D. Hoffsommer, C. H. Kuo, and N. L. Wendler, *J. Org. Chem.*, **30**, 3229 (1965).
90. N. Gospodinov, Sh. Levi, and El. Mutafchieva, *Farmatsiya*, **17** (4) 12 (1967); *Chem. Abstr.*, **68**, 104929e (1068).
91. J. P. Kutney and T. Tabata, *Can. J. Chem.*, **41**, 695 (1963).
92. C. G. Screttas, J. F. Estham, and C. W. Kamienski, *Chimia*, **24**, 109 (1970).
93. C. G. Screttas, French Patent 1,585,052, 1970; *Chem. Abstr.*, **73**, 130895p (1970).
94. A. M. Jones, C. A. Russell, and S. Skidmore, *J. Chem. Soc. C*, 2245 (1969).
95. A. M. Jones and C. A. Russell, *J. Chem. Soc.*, *C*, 2246 (1969).
96. J. T. Edward, *Can. J. Chem.*, **42**, 965 (1964).
97. I. L. Kotlyarevskii, L. N. Korolenok, L. G. Stadnikova, and T. G. Shishmakova, *Izv. Akad. Nauk SSSR, Ser. Khim.*, 1224 (1966).
98. T. Shavel, Jr. and G. C. Morrison, U.S. Patent 3,359,273, 1967; *Chem. Abstr.*, **69**, 2872j (1968).
99. F. J. Villani and C. A. Ellis. *J. Med. Chem.*, **13**, 1245 (1970).
100. R. A. Abramovitch and J. G. Saha, in "Advances in Heterocyclic Chemistry," A. R. Katritzky and A. J. Boulton, Eds., Vol. 6, Academic, New York, 1966, p. 229.
101. R. A. Abramovitch and G. A. Poulton, *Chem. Commun.*, 274 (1967).
102. R. A. Abramovitch, C. S. Giam, and G. A. Poulton, *J. Chem. Soc. C*, 128 (1970).
103. G. Fraenkel and J. C. Cooper, *Tetrahedron Lett.*, 1825 (1968).
104. C. S. Giam and J. L. Stout, *Chem. Commun.*, 142 (1969).
105. C. S. Giam and J. L. Stout, *Chem. Commun.*, 478 (1970).
106. C. S. Giam and S. D. Abbott, *J. Amer. Chem. Soc.*, **93**, 1294 (1971).
107. R. Levine and W. M. Kadunce, *Chem. Commun.*, 921 (1970).
108. R. A. Abramovitch and C. S. Giam, *Can. J. Chem.*, **42**, 1627 (1964).
109. R. A. Abramovitch and G. A. Poulton, *J. Chem. Soc.*, *B*, 901 (1969).
110. R. A. Abramovitch, C. S. Giam, and A. D. Notation, *Can. J. Chem.*, **38**, 761 (1960).
111. R. A. Abramovitch and C. S. Giam, *Can. J. Chem.*, **40**, 213 (1962).
112. R. A. Abramovitch, A. D. Notation, and C. S. Giam, *Tetrahedron Lett.*, 1 (1959).
113. F. Haglid, *Acta Chem. Scand.*, **21**, 329 (1967).
114. F. Haglid, *Acta Chem. Scand.*, **21**, 335 (1967).
115. Yu. I. Chumakov and V. M. Ledovskikh, U.S.S.R. Patent 158,574, 1963; *Chem. Abstr.*, **60**, 10656b (1964).
116. A. P. Gray, W. L. Archer, E. E. Spinner, and C. J. Cavallito, *J. Amer. Chem. Soc.*, **79**, 3805 (1957).
117. P. N. Gordon and W. C. Austin, South African Patent 67 04,638, 1968; *Chem. Abstr.*, **70**, 57658v (1969).
118. D. E. Ames and J. L. Archibald, *J. Chem. Soc.*, 1475 (1962).
119. N. Collignon and P. Pastour, *C. R. Acad. Sci. Paris, Ser. C.*, **269**, (15), 857 (1969); *Chem. Abstr.*, **72**, 43765g (1970).
120. D. Bryce-Smith, P. J. Morris, and B. J. Wakefield, *Chem. Ind.* (London), 495 (1964).

121. J. P. Kutney and T. Tabata, *Can. J. Chem.*, **40**, 1140 (1962).
122. A. F. Vompe and N. V. Monich, U.S.S.R. Patent 136,377, 1961; *Chem. Abstr.*, **56**, 3464a (1962).
123. D. Korbonits, K. Harsanyi, and G. Leszkovszky, Hungarian Patent 151,257, 1964; *Chem. Abstr.*, **60**, 14482c (1964).
124. O. Červinka, *Chem. Ind.* (London), 1482 (1960).
125. G. Fraenkel, J. W. Cooper, and C. M. Fink, *Angew. Chem. Inter. Edit.*, **9**, 523 (1970).
126. R. D. Brown, *J. Chem. Soc.*, 272 (1956).
127. C. A. Coulson and C. Longuet-Higgins, *Proc. Roy. Soc.*, **A192**, 16 (1947).
128. R. D. Brown and M. L. Heffernan, *Aust. J. Chem.*, **9**, 83 (1956).
129. W. H. Rieger, U.S. Patent 2,502,174, 1950; *Chem. Abstr.*, **44**, 5396 (1950).
130. S. Goldschmidt and M. Minsinger, *Chem. Ber.*, **87**, 956 (1954).
131. K. Schwetlick and R. Lungwitz, *Z. Chem.*, **4**, 458 (1964).
132. R. A. Abramovitch and K. Kenaschuk, *Can. J. Chem.*, **45**, 509 (1967).
133. K. C. Bass and P. Nababsing, *J. Chem. Soc.*, *C*, 2169 (1970).
134. K. C. Bass and P. Nababsing, *J. Chem. Soc.*, *C*, 388 (1969).
135. E. Hardegger and E. Nikles, *Helv. Chim. Acta*, **40**, 2421 (1957).
136. K. K. Chiu and H. H. Huang, *J. Chem. Soc.*, *C*, 2758 (1969).
137. E. F. Travecedo and V. I. Stenberg, *Chem. Commun.*, 609 (1970).
138. S. Caplain, J. P. Catteau, and A. La Blache-Combier, *Chem. Commun.*, 1475 (1970).
139. J. L. R. Williams, *J. Org. Chem.*, **25**, 1839 (1960).
140. J. Kuthan, N. V. Koshmina, J. Palecek, and V. Skala, *Collect. Czech. Chem. Commun.*, **35**, 2787 (1970).
141. Yu. I. Chumakov and V. F. Novikova, *Khim. Prom. Ukr.*, (2) 47 (1968); *Chem. Abstr.*, **69**, 43743m (1968).
142. N. S. Prostakov, S. S. Moiz, V. P. Zvolinskii, and G. I. Cherenkova, *Khim. Geterotsikl. Soedin.*, (6) 779 (1970); *Chem. Abstr.*, **73**, 109635u (1970).
143. C. Mercier and J. P. Dubose, *Bull. Soc. Chim. Fr.*, (12) 4425 (1969).
144. N. K. Kochetkov, E. D. Khomutova, and L. M. Likhosherstov, *Zh. Obshch. Khim.*, **29**, 1657 (1959); *Chem. Abstr.*, **54**, 8815f (1960).
145. J. A. Zoltewicz, C. L. Smith, and J. D. Meyer, *Tetrahedron*, **24**, 2269, 2273 (1968).
146. C. T. Kyte, G. H. Jeffery, and A. I. Vogel, *J. Chem. Soc.*, 4454 (1960).
147. E. A. Coulson, J. D. Cox, E. F. G. Herington, and J. F. Martin, *J. Chem. Soc.*, 1934 (1959).
148. R. V. Helm, W. J. Lanum, G. L. Cook, and J. S. Ball, *Amer. Chem. Soc. Div. Petrol. Chem. Preprints*, **2**, (4) A17 (1957); *Chem. Abstr.*, **54**, 22,634C (1960).
149. J. Ploquin, *Bull. Soc. Pharm. Bordeaux*, **95**, 177 (1956); *Chem. Abstr.*, **51**, 17911b (1957).
150. R. H. Linnell, *J. Org. Chem.*, **25**, 290 (1960).
151. A. Albert, in "Physical Methods in Heterocyclic Chemistry," Vol. 1, A. R. Katritzky, Ed., Academic, New York, 1963, p. 67.
152. R. Bhattacharya, *Proc. Natl. Inst. Sci. India*, **24A**, 304 (1958); *Chem. Abstr.*, **53**, 11371a (1959).
153. T. I. Abramovitch, I. P. Gragerov, and V. V. Perekalin, *Dokl. Akad. Nauk S.S.S.R.*, **121**, 295 (1958).
154. G. E. Calf and J. L. Garnett, *Aust. J. Chem.*, **21**, 1221 (1968).
155. H. C. Brown, M. E. Azzaro, J. G. Koelling, and G. J. McDonald, *J. Amer. Chem. Soc.*, **88**, 2520 (1966).
156. H. C. Brown and G. J. McDonald, *J. Amer. Chem. Soc.*, **88**, 2514 (1966).
157. M. Gomel and H. Lumbroso, *C. R. Acad. Sci. Paris*, *Ser. C.*, **252**, 3039 (1961).

158. G. C. Bhattacharya and H. C. Dasgupta, *J. Ind. Chem. Soc.*, **47**, (2) 157 (1970); *Chem. Abstr.*, **72**, 121306z (1970).

159. W. S. Dorsey and W. D. Schaeffer, U.S. Patent 2,827,462, 1958; *Chem. Abstr.*, **54**, 21681e (1960).

160. W. D. Schaeffer, U.S. Patent 2,827,463, 1958; *Chem. Abstr.*, **54**, 21681g (1960).

161. R. G. Beiles and E. M. Beiles, *Zh. Neorg. Khim.*, **12**, (4) 884 (1967); *Chem. Abstr.*, **67**, 82060g (1967).

162. W. Hieber and N. Kahlen, *Chem. Ber.*, **91**, 2234 (1958).

163. W. Hieber and R. Wiesboeck, *Chem. Ber.*, **91**, 1146 (1958).

164. F. Wolf and K. Ziegenhorn, *Z. Chem.*, **8**, (5) 192 (1968).

165. A. P. Kottenhahn, U.S. Patent 3,154,528, 1964; *Chem. Abstr.*, **62**, 1847e (1965).

166. V. D. Aftandilian, U.S. Patent 2,961,444 (1960); *Chem. Abstr.*, **55**, 7440e (1961).

167. K. Okoń, *Bull. Acad. Polon. Sci. Sér. Sci. Chim. Géol. Géograph.*, **6**, 331 (1958); *Chem. Abstr.*, **52**, 20153d (1958).

168. S. Fisel and H. Franchevici, *Acad. Rep. Populare Romaine, Filiala Iasi, Studii Cercetari Stunt. Chim.*, **10**, 41 (1959); *Chem. Abstr.*, **57**, 2191e (1962).

169. N. P. Buu-Hoi, G. Saint-Ruf, and B. Lobert, *Bull. Soc. Chim. Fr.*, (5) 1769 (1969).

170. A. P. Terent'ev and Ya. D. Mogilyanskii, *Zh. Obshch. Khim.*, **28**, 1959 (1958); *Chem. Abstr.*, **53**, 1327 (1959).

171. R. Royer, P. Demerseman, G. Colin, and A. Cheutin, *Bull. Soc. Chim. Fr.*, (10) 4090 (1968).

172. L. W. Clark, *J. Phys. Chem.*, **60**, 1583 (1956).

173. C. W. Kruse and R. F. Kleinschmidt, U.S. Patent 3,227,766, 1966; *Chem. Abstr.*, **64**, 8054f (1966).

174. A. R. Utke and R. T. Sanderson, *J. Org. Chem.*, **28**, 3582 (1963).

175. G. H. Aylward, R. W. Lee, and E. C. Watton, *Chem. Commun.*, 1594 (1970).

176. A. L. Bluhm, J. Weinstein, and J. A. Sousa, *J. Org. Chem.*, **28**, 1989 (1963).

177. B. P. Lisboa, *Naturwissenschaften*, **44**, 618 (1957).

178. P. Arnall, *Chem. Prods.*, **24**, 451 (1961); *Chem. Abstr.*, **56**, 4872f (1962).

179. "Physical Methods in Heterocyclic Chemistry," Vol. 2, A. R. Katritzky, Ed., Academic, New York, 1963; (a) S. F. Mason, p. 20, 35; (b) A. R. Katritzky and A. P. Ambler, p. 274; (c) R. F. M. White, p. 141.

180. E. Godar and R. P. Mariella, *J. Amer. Chem. Soc.*, **79**, 1402 (1957).

181. C. J. Pouchert, "The Aldrich Library of Infrared Spectra," Aldrich Chemical Co. Inc., United States, 1970.

182. J. K. Wilmshurst and H. J. Bernstein, *Can. J. Chem.*, **35**, 1183 (1957).

183. G. L. Cook and F. M. Church, *J. Phys. Chem.*, **61**, 458 (1957).

184. D. Heinert and A. E. Martell, *J. Amer. Chem. Soc.*, **81**, 3933 (1959).

185. C. R. Eddy and A. Eisner, *Anal. Chem.*, **26**, 1428 (1954).

186. H. Shindo, *Pharm. Bull. Jap.*, **5**, 472 (1957).

187. A. R. Katritzky and A. R. Hands, *J. Chem. Soc.*, 2202 (1958).

188. A. R. Katritzky, A. R. Hands, and R. A. Jones, *J. Chem. Soc.*, 3165 (1958).

189. A. R. Katritzky and J. N. Gardner, *J. Chem. Soc.*, 2198 (1958).

190. H. E. Podall, *Anal. Chem.*, **29**, 1423 (1957).

191. W. G. Schneider, H. J. Bernstein, and J. A. Pople, *Can. J. Chem.*, **35**, 1487 (1957).

192. H. J. Bernstein, J. A. Pople, and W. G. Schneider, *Can. J. Chem.*, **35**, 65 (1957).

193. J. D. Baldeschwieler and E. W. Randall, *Proc. Chem. Soc.*, 303 (1961).

194. K. Bieman, "Mass Spectrometry," McGraw-Hill, New York, 1962, pp. 130, 134.

195. G. Spiteller, in "Advances in Heterocyclic Chemistry," Vol. 7, A. R. Katritzky and A. J. Boulton, Eds., Academic, New York, 1966, p. 317.

196. E. C. Craven, British Patent 766,387, 1957; *Chem. Abstr.*, **51**, 12151h (1957).

197. Japan Gas Chem. Co., British Patent 1,108,686, 1968; *Chem Abstr.*, **69**, 43808m (1968).

198. L. D. Gluzman, Yu. A. Slachinskii, and V. P. Kostochka, *Koks Khim.* (1) 42 (1970); *Chem. Abstr.*, **72**, 78824c (1970).

199. V. E. Privalov, L. D. Gluzman, V. M. Efimenko, and Yu. A. Slachinskii, *Koks Khim*, (5) 38 (1970); *Chem. Abstr.*, **73**, 25252d (1970).

200. S. Neuhaeuser and F. Wolf, *Brennst.-Chem.*, **49**, 355 (1968); *Chem. Abstr.*, **70**, 114965u (1969).

201. J. Swiderski, A. Szuchnik, and J. Wasiak, *Rocz. Chem.*, **38**, (7/8) 1145 (1964); *Chem. Abstr.*, **61**, 16046g (1964).

202. S. Neuhaeuser, F. Wolf, and E. Brand, German (East) Patent 57,852, 1967; *Chem. Abstr.*, **69**, 59106s (1968).

203. Polska Akademia Nauk Instytut Chemii Fizycznej, French Patent 1,484,583, 1967; *Chem. Abstr.*, **68**, 87185q (1968).

204. M. N. Dilbert, U.S. Patent 2,995,500, 1961; *Chem. Abstr.*, **56**, 2430f (1962).

205. M. O. Shrader and H. L. Dimond, U.S. Patent 2,767,187, 1956; *Chem. Abstr.*, **51**, 9710c (1957).

206. E. M. Gepshtein, *Nauchn. Tr. Vost. Nauchn.-Issled. Uglekhim. Inst.*, **16** (1) 26 (1963); *Chem. Abstr.*, **61**, 13275c (1964).

207. Yu. I. Chumakov and L. P. Lugovskaya, U.S.S.R. Patent 179,318 (1966); *Chem. Abstr.*, **65**, 2232f (1966).

208. Yu. I. Chumakov and L. P. Lugovskaya, U.S.S.R. Patent 177,890, 1964; *Chem. Abstr.*, **65**, 694h (1966).

209. F. Wolf and S. Neuhaeuser, *Brennst.-Chem.*, **50**, (3) 83 (1969); *Chem. Abstr.*, **70**, 114965u (1969).

210. J. Vymetal and P. Tvaruzek, Czechoslovakian Patent 133,170, 1969; *Chem. Abstr.*, **73**, 77069u (1970).

211. F. Wolf and S. Neuhaeuser, German (East) Patent 65,923 (1969); *Chem. Abstr.*, **72**, 21614e (1970).

212. Yu. I. Chumakov, U.S.S.R. Patent 101,691, 1955; *Chem. Abstr.*, **51**, 12986i (1957).

213. W. D. Schaffer, U.S. Patent 3,095,420, 1963; *Chem. Abstr.*, **59**, 12767b (1963).

214. P. Fotis and E. K. Fields, U.S. Patent 3,112,322 (1963); *Chem. Abstr.*, **60**, 10657b (1964).

215. Yu. I. Chumakov and V. F. Vivlkova, U.S.S.R. Patent 201,407, 1967; *Chem. Abstr.*, **69**, 19026j (1968).

216. Yu. I. Chumakov, *Med. Prom. S.S.S.R.*, **12** (12) 13 (1958); *Chem. Abstr.*, **53**, 21932i (1959).

217. Yu. I. Chumakov and L. P. Lugovskaya, U.S.S.R. Patent 178,377 (1966); *Chem. Abstr.*, **64**, 19570f (1966).

218. Yu. I. Chumakov, L. P. Lugovskaya, and V. F. Novikova, U.S.S.R. Patent 177,891, 1966; *Chem. Abstr.*, **64**, 19569a (1966).

219. Yu. I. Chumakov and N. I. Chumakova, U.S.S.R. Patent 106,597, 1957; *Chem. Abstr.*, **52**, 2088i (1958).

220. Yu. I. Chumakov, U.S.S.R. Patent 105,811, 1957; *Chem. Abstr.*, **51**, 14829e (1957).

221. Yu. I. Chumakov and E. V. Lugovskoi, U.S.S.R. Patent 180,600, 1966; *Chem. Abstr.*, **65**, 12176h (1966).

222. J. Baron, J. Szewczyk, A. Karafiol, W. Smalur, and H. Bartyla, Polish Patent 59,225, 1970; *Chem. Abstr.*, **73**, 45363u (1970).

223. W. C. Von Dohlen and W. F. Tully, U.S. Patent 2,924,602, 1960; *Chem. Abstr.*, **54**, 13148d (1960).

224. H. Pfeiffer, German Patent 931,473, 1955; *Chem. Abstr.*, **52**, 18473a (1958).

225. E. M. Gepshtein, *Nauchû. Tr. Vost. Nauchû.-Issled. Uglekhim. Inst.*, **16** (1) 36 (1963); *Chem. Abstr.*, **61**, 13275g (1964).
226. E. M. Gepshtein, U.S.S.R. Patent 149,784, 1962; *Chem. Abstr.*, **58**, 11335c (1963).
227. L. Achremowicz and Z. Skrowaczewska, Polish Patent 52,039, 1966; *Chem. Abstr.*, **68**, 95698z (1968).
228. N. E. Podklenov, *Soobshcheniya Sakhalin. Kompleks. Nauch.-Issledovatel. Inst.*, **3**, 131 (1957); *Chem. Abstr.*, **54**, 21084h (1960).
229. R. P. Williams and G. D. Sammons, *Ind. Eng. Chem., Chem. Eng. Data Ser.*, **2**, 76 (1957); *Chem. Abstr.*, **52**, 15516h (1958).
230. H. L. Dimond, U.S. Patent 2,818,411, 1957; *Chem. Abstr.*, **52**, 4964h (1958).
231. N. E. Podkletnov and D. F. Podkopaeva, *Soobshch. Sakhalinsk. Kompleksn. Nauchn.-Issled. Inst. Akad. Nauk SSSR.*, (8) 121 (1959); *Chem. Abstr.*, **57**, 11152h (1962).
232. L. Kuczyński and A. Nawojski, *Przem. Chem.*, **34**, 190 (1955); *Chem. Abstr.*, **53**, 3220h (1959).
233. H. Wille, R. Oberkobusch, and L. Rappen, German Patent 1,245,963, 1967; *Chem. Abstr.*, **68**, 87192q (1968).
234. H. Wille, R. Oberkobusch, and L. Rappen, German Patent 1,245,962, 1967; *Chem. Abstr.*, **68**, 87173j (1968).
235. D. Jerchel and H. Hippchen, German Patent 1,019,653, 1957; *Chem. Abstr.*, **54**, 1557i (1960).
236. D. Jerchel and W. Melloh, *Ann. Chem.*, **613**, 144 (1958).
237. A. Albert, *Chem. Ind.* (*London*), 582 (1958).
238. R. F. Evans and W. Kynaston, *J. Chem. Soc.*, 5556 (1961).
239. T. Akira and K. Yoshitaka, Japanese Patent 6820,187, 1968; *Chem. Abstr.*, **70**, 57666w (1969).
240. W. Waddington, British Patent 943,193, 1963; *Chem. Abstr.*, **60**, 4115c (1964).
241. S. Neuhaeuser and F. Wolf, German (East) Patent 63,492, 1968; *Chem. Abstr.*, **70**, 77809y (1969).
242. A. Yamamoto, Japanese Patent 1517. 1956; *Chem. Abstr.*, **51**, 8809d (1957).
243. R. N. Fleck and C. G. Wight, U.S. Patent 3,029,242, 1962; *Chem. Abstr.*, **57**, 12445b (1962).
244. R. N. Fleck and C. G. Wight, U.S. Patent 3,064,002, 1962; *Chem. Abstr.*, **58**, 9032h (1963).
245. Yu. I. Chumakov and E. V. Lugovskoi, U.S.S.R. Patent 180,601, 1966; *Chem. Abstr.*, **65**, 12177a (1966).
246. V. Galík and S. Landa, *Collect. Czech. Chem. Commun.*, **29**, 2562 (1964).
247. C. S. Giam, S. D. Abbott, and W. B. Davis, *J. Chromatog.*, **42**, 457 (1969).
248. L. H. Klemm, J. Shabtai, and F. H. W. Lee, *J. Chromatog.*, **51**, 433 (1970).
249. J. Eisch and H. Gilman, *Chem. Rev.*, **57**, 525 (1957).
250. W. Pfleiderer, in "Topics in Heterocyclic Chemistry," R. N. Castle, Ed., Interscience, 1969, p. 56.
251. R. D. Chambers, R. P. Corbally, J. A. Jackson, and W. K. R. Musgrave, *Chem. Commun.*, 127 (1969).
252. M. G. Barlow, J. G. Dingwall, and R. N. Haszeldine, *Chem. Commun.*, 1580 (1970).
253. M. Matsumoto, M. Naito, K. Mizushiro, and M. Morita, *Aromatikkusu*, **19**, (5) 209 (1967); *Chem. Abstr.*, **70**, 11512q (1969).
254. R. C. Myerly and K. Weinberg, U.S. Patent 3,334,101, 1967; *Chem. Abstr.*, **68**, 87176n (1968).
255. L. V. Zykova, *Uchenye Zapiski, Saratov. Gosudarst. Univ. N.G. Chernyshevskogo*, **43**, 89 (1956); *Chem. Abstr.*, **54**, 21079e (1960).

256. J. Herzenberg, R. Covini, M. Pieroni, and A. Neuz, *Ind. Eng. Chem. Prod. Res. Develop.*, **6** (3) 195 (1967); *Chem. Abstr.*, **67**, 90628w (1967).

257. A. A. Balandin, L. I. Sovalov, and T. A. Slovokhotova, *Dokl. Akad. Nauk S.S.S.R.*, **110**, 79 (1956).

258. L. I. Zamyshlyaeva, A. A. Balandin, and T. A. Slovokhotova, *Vestn. Mosk. Univ. Ser II, Khim*, **20**, (1) 38 (1965); *Chem. Abstr.*, **62**, 14619d (1965).

259. L. I. Zamyshlyaeva, A. A. Balandin, and T. A. Slovokhotova, *Izv. Akad. Nauk SSSR, Ser. Khim.*, 330 (1965); *Chem. Abstr.*, **62**, 14622e (1965).

260. K. K. Moll, K. Pelzing, H. Baltz, and E. Leibnitz, German (East) Patent 59,568, 1968; *Chem. Abstr.*, **70**, 47311n (1969).

261. F. Mensch, *Erdoel Kohle, Erdgas, Petrochem*, **22**, (2) 67 (1969); *Chem. Abstr.*, **70**, 87480y (1969).

262. A. Neuz, R. Covini, and M. Pieroni, *Chim. Ind. (Milan)*, **49**, (3), 259 (1967); *Chem. Abstr.*, **67**, 32550d (1967).

263. N. I. Shuikin, L. A. Erivanskaya, G. S. Korosteleva, and A. P. Polyakov, *Izv. Akad. Nauk SSSR. Ser. Khim.*, 2216 (1965); *Chem. Abstr.*, **64**, 9674e (1966).

264. J. W. Sweeting and J. F. K. Wilshire, *Aust. J. Chem.*, **15**, 800 (1962).

265. C. Longuet-Higgins and C. A. Coulson, *Trans. Faraday Soc.*, **43**, 87 (1947).

266. G. Slomp Jr., *J. Org. Chem.*, **22**, 1277 (1957).

267. G. Slomp Jr. and J. L. Johnson, *J. Amer. Chem. Soc.*, **80**, 915 (1958).

268. L. H. Beck, U.S. Patent 3,154,549, 1964; *Chem. Abstr.*, **62**, 1673d (1965).

269. M. V. Rubtsov, L. N. Yakhontov, and S. V. Yatsenko, *Zhur. Priklad. Khim.*, **36**, 315 (1957); *Chem. Abstr.*, **51**, 12908a (1957).

270. L. O. Shnaidman and M. I. Siling, *Tr. Vses. Nauchn.-Issled. Vitamin, Inst.*, **7**, 18 (1961); *Chem. Abstr.*, **60**, 7988b (1964).

271. M. I. Farberov, B. F. Ustavshchikov, and T. S. Titov, *Metody Polucheniya Khim. Reaktivov Preparatov, Gos. Kom. Sov. Min. SSSR. Khim.*, (II) 58 (1964); *Chem. Abstr.*, **64**, 15832d (1966).

272. P. Vaculik and J. Kuthan, *Collect. Czech. Chem. Commun.*, **25**, 1591 (1960).

273. N. S. Prostakov, L. A. Shakhparonova, and L. M. Kirillova, *Zh. Obshch. Khim.*, **34** (10) 3231 (1964).

274. T. Slebodzinski, H. Kielczewska, and W. Biernacki, *Przem. Chem.*, **48** (2) 90 (1969); *Chem. Abstr.*, **71**, 38751z (1969).

275. Ruetgerswerke and Teerverwertung, A.-G. British Patent 1,132,746, 1968; *Chem. Abstr.*, **70**, 37658s (1969).

276. M. Levi and A. Georgiev, *Farmatsiya*, (3) 27 (1955), *Chem. Abstr.*, **52**, 7315e (1958).

277. Lonza Elektrizitaetswerke and Chemische Fabriken A.-G., Swiss Patent 339,625, 1955; *Chem. Abstr.*, **56**, 2430e (1962).

278. R. Stager, Swiss Patent 335,519, 1959; *Chem. Abstr.*, **54**, 18558f (1960).

279. J. E. Mahan and R. P. Williams, U.S. Patent 2,993,904, 1957; *Chem. Abstr.*, **56**, 1434a (1962).

280. M. E. Abraham and G. Wilbert, U.S. Patent 2,884,415, 1959; *Chem. Abstr.*, **54**, 18558d (1960).

281. L. O. Shnaidman and M. I. Siling, *Tr. Vses. Nauchn.-Issled. Vitamin. Inst.*, **8**, 5 (1961); *Chem. Abstr.*, **57**, 16551h (1962).

282. R. Aries, French Patent 1,509,049, 1968; *Chem. Abstr.*, **70**, 37657r (1969).

283. A. Stocker, O. Marti, T. Pfammatter, and G. Schreiner, German Patent 1,956,117, 1970; *Chem. Abstr.*, **73**, 45352q (1970).

284. T. S. Titova, B. F. Ustavshchikov. M. I. Farberov, and E. V. Degtyarev, *Zh. Prikl. Khim. (Leningrad)*, **42**, (4) 910 (1969); *Chem. Abstr.*, **71**, 38750y (1969).

390   Alkylpyridines and Arylpyridines

285. T. Kato, *Bull. Chem. Soc. Jap.*, **34**, 636 (1961).
286. T. I. Baranova, L. F. Titova, and A. M. Kut'in, *Khim. Prom.*, **43** (3) 204 (1967); *Chem. Abstr.*, **67**, 21792h (1967).
287. L. Leitis and M. V. Shimanskaya, *Khim. Geterotsikl. Soedin.*, 507 (1967); *Chem. Abstr.*, **68**, 29553n (1968).
288. L. Leitis and M. V. Shimanskaya, *Khim. Geterotsikl. Soedin.*, 281 (1969); *Chem. Abstr.*, **71**, 22002v (1969).
289. W. Mathes and W. Sauermilch, *Chem. Ber.*, **93**, 286 (1960).
290. M. V. Shimanskaya and S. Hillers, *Khim. Tekhnol. Primenenie Proizvodnykh Piridina Khinolina, Materialy Soveshchaniya, Inst. Khim., Akad. Nauk Latv. S.S.R. Riga*, 185 (1957); *Chem. Abstr.*, **55**, 16545f (1961).
291. S. K. Bhattacharyya and A. K. Kar, *Indian J. Appl. Chem.*, **30**, (1–2) 35 (1967); *Chem. Abstr.*, **68**, 21807n (1968).
292. S. K. Bhattacharyya and A. K. Kar, *Indian J. Appl. Chem.*, **30**, (1–2) 42 (1967), *Chem. Abstr.*, **68**, 21801f (1968).
293. W. Schwarze, German Patent 1,071,085, 1959; *Chem. Abstr.*, **57**, 11172g (1962).
294. B. V. Suvorov, A. D. Kagarlitskii, T. A. Afanas'eva, and I. I. Kan, *Khim. Geterotsikl. Soedin.*, 1129 (1969); *Chem. Abstr.*, **72**, 121320z (1970).
295. Societa Edison S.p.A.-Settore Chimico, Netherlands Patent Appl. 65,09,249, 1966; *Chem. Abstr.*, **65**, 5447f (1966).
296. B. V. Suvorov, A. D. Kagarlitskii, T. A. Afanas'eva, O. B. Lebedeva, A. I. Loiko, and V. A. Serazetdinova, *Khim. Geterotsikl. Soedin.* 1024 (1969); *Chem. Abstr.*, **72**, 132456x (1970).
297. M. Yamada, K. Iwakiri, and Y. Kimura, Japanese Patent 70,25,289, 1970; *Chem. Abstr.*, **73**, 98803u (1970).
298. T. A. Afanas'eva, A. D. Kagarlitskii, I. I. Kan, and B. V. Suvorov, *Khim. Geterotsikl. Soedin.*, 675 (1969); *Chem. Abstr.*, **72**, 43366c (1970).
299. M. Takeishi, K. Fujii, and T. Matsui, Japanese Patent 7013,572, 1970; *Chem. Abstr.*, **73**, 77059r (1970).
300. Farbenfabriken Bayer A.-G., French Patent 2,012,674, 1970; *Chem. Abstr.*, **74**, 3522k (1971).
301. D. J. Hadley and B. Wood, U.S. Patent 2,839,535, 1958; *Chem. Abstr.*, **52**, 13805b (1958).
302. B. Lipka, B. Buszynska, and E. Treszezanowicz, *Przem. Chem.*, **49**, (1) 34 (1970); *Chem. Abstr.*, **72**, 111231g (1970).
303. V. I. Trubnikov, E. S. Zhdanovich, and N. A. Preobrazhenskii, *Khim.-Farm. Zh.*, **3** (12) 42 (1969); *Chem. Abstr.*, **72**, 111241k (1970).
304. B. Lipka and B. Buszynska, *Przem. Chem.*, **49** (5) 275 (1970); *Chem. Abstr.*, **73**, 98761d (1970).
305. K. Smeykal, K. K. Moll, L. Bruesehaber, I. Krafft, K. Pelzing, and U. Schrattenholz, German (East) Patent 62,055, 1968; *Chem. Abstr.*, **70**, 87586n (1969).
306. Yu. N. Solntsev, A. I. Loiko, A. D. Kagarlitskii, D. Kh. Sembaev, and B. V. Suvorov, *Izv. Akad. Nauk Kaz. SSR, Ser. Khim.*, **19** (6) 53 (1969); *Chem. Abstr.*, **72**, 55191e (1970).
307. B. Lipka, E. Treszczanowicz, I. Jaworska, and A. Jurewicz, *Przem. Chem.*, **37**, 484 (1958); *Chem. Abstr.*, **53**, 5261i (1959).
308. A. F. D'Alessandro, U.S. Patent 2,861,999 (1958); *Chem. Abstr.*, **53**, 9258h (1959).
309. E. Fischer, *Chimia*, **22**, (11) 437 (1968), *Chem. Abstr.*, **70**, 37624c (1969).
310. F. E. Cislak, U.S. Patent 3,118,899, 1964; *Chem. Abstr.*, **60**, 9250f (1964).
311. B. Emmert, *Chem. Ber.*, **91**, 1388 (1958).
312. A. N. Kost, P. B. Terent'ev, and L. A. Golovleva, *Vestn. Mosk. Univ. Ser II, Khim.*, **19** (6) 56 (1964); *Chem. Abstr.*, **62**, 9099h (1965)
313. P. E. Miller, G. L. Oliver, J. R. Dann, and J. W. Gates, Jr., *J. Org. Chem.*, **22**, 664 (1957).

References 391

314. H. Najer, P. Chabrier, R. Giudicelli, and E. Joannic-Voisinet, *C. R. Acad. Sci., Paris, Ser. C*, **244**, 2935 (1957).
315. T. P. Sycheva and M. N. Shchukina, *Biol. Aktivn. Soedin. Akad. Nauk. SSSR.*, 42 (1965); *Chem. Abstr.*, **64**, 6607h (1966).
316. T. L. Tolbert and J. Preston, *J. Heterocycl. Chem.*, **6**, 963 (1969).
317. S. S. Kruglikov, V. G. Khomyakov, and L. I. Kazakova, *Tr. Mosk. Khim.-Tekhnol. Inst.*, (32) 201 (1961); *Chem. Abstr.*, **57**, 16543c (1962).
318. S. S. Kruglikov and V. G. Khomyakov, *Tr. Mosk. Khim.-Tekhnol. Inst.*, (32) 194 (1961); *Chem. Abstr.*, **57**, 16542i (1962).
319. V. G. Khomyakov, S. S. Kruglikov, and L. I. Kazakova, *Tr. Mosk. Khim.-Tekhnol. Inst.*, (32) 189 (1961); *Chem. Abstr.*, **57**, 15065b (1962).
320. L. A. Golovleva, G. K. Skryabin, A. N. Kost, and P. B. Terent'ev, U.S.S.R. Patent 228,688, 1968; *Chem. Abstr.*, **70**, 77797t (1969).
321. G. W. Godin and A. R. Graham, U.S. Patent 3,075,989, 1963; British Patent 839522, 1960; *Chem. Abstr.*, **55**, 5542g (1961).
322. A. Markovac, C. L. Stevens, A. B. Ash, and B. E. Hackley, Jr., *J. Org. Chem.*, **35**, 841 (1970).
323. P. N. Rylander, "Catalytic Hydrogenation over Platinum Metals," Academic, New York, 1967, p. 375.
324. G. N. Walker, *J. Org. Chem.*, **27**, 2966 (1962).
325. K. Smeykal and K. K. Moll, British Patent 1,062,900, 1967; *Chem. Abstr.*, **67**, 100008k (1967).
326. A. S. Safaev, S. Abidova, S. Akhundzhanov, and A. S. Sultanov, *Katal. Pererab. Uglevodorodn. Syr'ya*, (1) 63 (1967); *Chem. Abstr.*, **71**, 12963v (1969).
327. N. I. Shuikin and V. M. Brusnikina, *Izvest. Akad. Nauk S.S.S.R. Otdel. Khim. Nauk*, 1288 (1959); *Chem. Abstr.*, **54**, 1516f (1960).
328. M. Maruoka, *Nippon Kagaku Zasshi*, **82**, 1257 (1961); *Chem. Abstr.*, **58**, 11324d (1963).
329. H. Yasui and II. Echi, Japanese Patent 6814,467, 1968; *Chem. Abstr.*, **70**, 77796s (1969).
330. R. E. Lyle and P. S. Anderson, in "Advances in Heterocyclic Chemistry," Vol. 6, A. R. Katritzky and A. J. Boulton, Eds., Academic, New York, 1966, p. 45.
331. P. T. Lansbury and J. O. Peterson, *J. Amer. Chem. Soc.*, **85**, 2236 (1963).
332. P. T. Lansbury and J. O. Peterson, *J. Amer. Chem. Soc.*, **84**, 1756 (1962).
333. P. T. Lansbury and J. O. Peterson, *J. Amer. Chem. Soc.*, **83**, 3537 (1961).
334. M. Ferles and J. Pliml, in "Advances in Heterocyclic Chemistry," Vol. 12, A. R. Katritzky and A. J. Boulton, Eds., Academic, New York, 1970, p. 59.
335. A. Silhankova, D. Doskocilova, and M. Ferles, *Collect. Czech. Chem. Commun.*, **34**, (7) 1985 (1969).
336. W. Drzenick and P. Tomasik, *Rocz. Chem.*, **44**, (4) 779 (1970); *Chem. Abstr.*, **74**, 3475x (1971).
337. L. van der Does and H. J. Hertog, *Rec. Trav. Chim.*, **84**, (7) 951 (1965).
338. T. Batkowski, D. Tomasik, and P. Tomasik, *Rocz. Chem.*, **41**, (2) 2101 (1967).
339. D. E. Pearson, W. W. Stargrove, J. K. T. Chow, and B. R. Suthers, *J. Org. Chem.*, **26**, 789 (1961).
340. L. M. Yagupol'skii, A. G. Galushko, and M. A. Rzhavinskaya, *Zh. Obshch. Khim.*, **38** (3) 668 (1968); *Chem. Abstr.*, **69**, 59064b (1968).
341. L. M. Yagupol'skii and A. K. Galushko, U.S.S.R. Patent 232,221, 1968; *Chem. Abstr.*, **70**, 96632b (1969).
342. H. C. Fielding, British Patent 1,133,492, 1968; *Chem. Abstr.*, **70**, 37660m (1969).
343. L. M. Yagupol'skii and A. G. Galushko, *Zh. Obshch. Khim.*, **39** (9) 2087 (1969); *Chem. Abstr.*, **72**, 31676r (1970).

344. L. Achremowicz, T. Batkowski, and Z. Skrovaczewska, *Rocz. Chem.*, **38** (9) 1317 (1964); *Chem. Abstr.*, **66**, 1630b (1965).
345. G. A. Olah, J. A. Olah, and N. A. Overchuk, *J. Org. Chem.*, **30**, 3373 (1965).
346. J. Delarge, *Farmaco (Pavia)*, *Ed. Sci.*, **20** (9) 629 (1965); *Chem. Abstr.*, **64**, 3467e (1966).
347. H. C. van der Plas and H. J. den Hertog, *Chem. Weekblad*, **53**, 560 (1957); *Chem. Abstr.*, **52**, 10074i (1958).
348. H. C. van der Plas and T. H. Crawford, *J. Org. Chem.*, **26**, 2611 (1961).
349. Th. Kauffmann, H. Hacker, Ch. Kosel, and K. Vogt, *Z. Naturforsch*, **14b**, 601 (1959), *Chem. Abstr.*, **54**, 4584e (1960).
350. G. B. Bachman, M. Hamer, E. Dunning, and R. M. Schisla, *J. Org. Chem.*, **22**, 1296 (1957).
351. G. B. Bachman and R. M. Schisla, *J. Org. Chem.*, **22**, 1302 (1957).
353. R. A. Abramovitch and A. R. Vinutha, *J. Chem. Soc.*, *C*, 2104 (1969).
353. H. Johnston, U.S. Patent 3,317,549, 1967; *Chem. Abstr.*, **68**, 29611q (1968).
354. W. H. Taplin III, U.S. Patent 3,424,754, 1969; *Chem. Abstr.*, **70**, 77806v (1969).
355. J. P. Kutney, W. Cretney, T. Tabata, and M. Frank, *Can. J. Chem.*, **42** (3) 698 (1964).
356. J. Kollonitsch, French Patent 1,394,362, 1965; *Chem. Abstr.*, **63**, 8326e (1965).
357. W. Mathes and H. Schüly, *Angew. Chem. Inter. Edit.*, **2**, 144 (1963).
358. W. Mathes and H. Schüly, U.S. Patent 3,123,608, 1964; *Chem. Abstr.*, **64**, 2063e (1966).
359. R. H. Hall and I. D. Fleming, British Patent 866,380, 1961; *Chem. Abstr.*, **55**, 24795f (1961).
360. D. Jerchel, S. Noetzei, and K. Thomas, *Chem. Ber.*, **93**, 2966 (1960).
361. Yu. V. Kurbatov, A. S. Kurbatov, M. A. Solekhova, O. S. Otroshchenko, and A. S. Sadykov, *Tr. Samarkand. Gos. Univ.*, (167) 192 (1969); *Chem. Abstr.*, **73**, 109642u (1970).
362. Z. Foldi, *Chem. Ind. (London)*, 684 (1958).
363. H. C. Brown and B. Kanner, U.S. Patent 2,780,626, 1957; *Chem. Abstr.*, **51**, 9710e (1957).
364. J. P. Luvisi, U.S. Patent 2,889,329, 1959; *Chem. Abstr.*, **54**, 2364g (1960).
365. J. A. Gautier and M. Miocque, *C. R. Acad. Sci.*, *Paris, Ser. C.*, **249**, 2785 (1959).
366. C. Osuch and R. Levine, *J. Org. Chem.*, **21**, 1099 (1956).
367. C. Osuch and R. Levine, *J. Org. Chem.*, **22**, 939 (1957).
368. S. Raynolds and R. Levine, *J. Amer. Chem. Soc.*, **82**, 472 (1960).
369. R. Bodalski, J. Michalski, and B. Mlotkowska, *Rocz. Chem.*, **43** (4) 677 (1969); *Chem. Abstr.*, **71**, 61159w (1969).
370. B. Brust, R. I. Fryer, and L. H. Sternbach, U.S. Patent 3,400,131, 1968; *Chem. Abstr.*, **69**, 10652z (1968).
371. A. D. Miller and R. Levine, *J. Org. Chem.*, **24**, 1364 (1959).
372. N. S. Postakov and N. M. Mikhailova, *Khim. Geterotsikl. Soedin*, 1086 (1970); *Chem. Abstr.*, **74**, 64161v (1971).
373. Th. Kauffmann, G. Beissmer, W. Sahm, and A. Woltermann, *Angew. Chem. Inter. Ed.*, **9**, 808 (1970).
374. F. E. Cislak, U.S. Patent 2,891,959, 1959; *Chem. Abstr.*, **53**, 20096a (1959).
375. R. J. Dummel, W. Wrinkle, and H. S. Mosher, *J. Amer. Chem. Soc.*, **78**, 1936 (1956).
376. H. Zimmer and D. K. George, *Chem. Ber.*, **89**, 2285 (1956).
377. P. Arnall, British Patent 918,179, 1963; *Chem. Abstr.*, **58**, 13923a (1963).
378. F. E. Cislak, U.S. Patent 2,868,795, 1959; *Chem. Abstr.*, **53**, 10256a (1959).
379. F. E. Cislak, U.S. Patent 2,868,794, 1959; *Chem. Abstr.*, **53**, 10255d (1959).
380. T. Kato and Y. Goto, *Chem. Pharm. Bull. (Tokyo)*, **11**, 461 (1963).
381. S. E. Forman, *J. Org. Chem.*, **29**, 3323 (1964).
382. T. Kato and Y. Goto, *Yakugaku Zasshi*, **85**, (5) 451 (1965); *Chem. Abstr.*, **63**, 5596f (1965).

383. Z. Binenfeld and J. Vorkapic, *Farm. Glas.*, **25** (6–7) 257 (1969); *Chem. Abstr.*, **71**, 112763y (1969).
384. R. B. Margerison, French Patent 1,448,775, 1966; *Chem. Abstr.*, **67**, 43683 (1967).
385. H. Feuer and J. P. Lawrence, *J. Amer. Chem. Soc.*, **91**, 1856 (1969).
386. R. Schuman and E. D. Amstutz, *Rec. Trav. Chim. Pays-Bas*, **84** (4) 441 (1965).
387. L. H. Sommer, U.S. Patent 2,838,515, 1958; *Chem. Abstr.*, **52**, 13805b (1958).
388. Midland Silicones Ltd., British Patent 757,855, 1956; *Chem. Abstr.*, **51**, 15600e (1957).
389. E. Profft and F. Schneider, *J. Prakt. Chin.*, [4] **2**, 316 (1955).
390. T. Kato, H. Yamanaka, and T. Adachi, *Yakugaku Zasshi*, **85** (7) 611 (1965); *Chem. Abstr.*, **63**, 9911c (1965).
391. B. Notari and H. Pines, *J. Amer. Chem. Soc.*, **82**, 2945 (1960).
392. H. Pines and N. E. Sartoris, *J. Org. Chem.*, **34**, 2113 (1969).
393. Yu. I. Chumakov and V. M. Ledovskikh, *Metody Polucheniya Khim. Reaktivov Preparatov, Gos. Kom. Sov. Min. SSSR po Khim*, (7) 38 (1963); *Chem. Abstr.*, **61**, 10653a (1964).
394. V. M. Ledovskikh and Yu. I. Chumakov, U.S.S.R. Patent 154,278, 1963; *Chem. Abstr.*, **60**, 5466g (1964).
395. J. Michalski and K. Studniarski, *Chem. Tech.* (Berlin), **9**, 96 (1957); *Chem. Abstr.*, **51**, 10530c (1957).
396. M. I. Farberov, B. F. Ustavshchikov, A. M. Kut'in, and V. A. Bukhareva, *Metody Polucheniya Khim. Reaktivov Preparatov, Gos. Kom. Sov. Min. SSSR po Khim*, (11) 108 (1964); *Chem. Abstr.*, **64**, 15830c (1966).
397. I. N. Azerbaev, A. A. Nikolaeva, and A. I. Teterevkov, *Tr. Khim.-Met. Inst. Akad. Nauk Kaz. SSR*, **2**, 151 (1968); *Chem. Abstr.*, **70**, 68084f (1969).
398. M. Baba, Y. Kawi, and M. Dotani, Japanese 7025,095, 1970; *Chem. Abstr.*, **73**, 109703q (1970).
399. E. Profft, *Chem. Ber.*, **91**, 957 (1958).
400. N. Yamaguchi, K. Takei, A. Nishi, and K. Ozawa, Japanese Patent 18,285, 1960; *Chem. Abstr.*, **56**, 2429d (1962).
401. P. A. Gangrskii, E. C. Chvyeva, and Yu. I. Chumakov, *Med. Prom. SSSR*, **13**, (3) 13 (1959); *Chem. Abstr.*, **54**, 18512d (1960).
402. R. Lukeš and V. Galik, *Chem. Listy*, **51**, 2319 (1957).
403. R. Lukeš, V. Galik, and J. Jizba, *Collect. Czech. Chem. Commun.*, **26**, 2727 (1961).
404. A. K. Sheinkman, U.S.S.R. Patent 158,576, 1963; *Chem. Abstr.*, **60**, 10655f (1964).
405. T. Zsolnai, G. Lugosi, and G. Csermely, Hungarian Patent 149,812, 1962; *Chem. Abstr.*, **60**, 9252a (1964).
406. N. S. Prostakov, K. J. Mathew, and D. Pkhal'gumani, *Zh. Org. Khim.*, **1** (6) 1128 (1965); *Chem. Abstr.*, **63**, 16300c (1965).
407. R. Palmerio, *Corriere Farm.*, **21**, (17) 445 (1966); *Chem. Abstr.*, **67**, 73499n (1967).
408. N. S. Prostakov and V. A. Kurichev, *Khim. Geterotsikl. Soedin.*, 124 (1968); *Chem. Abstr.*, **69**, 106486c (1968).
409. C. F. Boehringer and Soehne G.m.b.H., Belgian Patent 615,319, 1962; *Chem. Abstr.*, **58**, 11333h (1963).
410. T. Katsumoto, *Bull. Chem. Soc. Jap.*, **33**, 242 (1960).
411. F. N. Stepanov and N. A. Aldanova, *Khim. Tekhnol. Promenenic Proizvodnykh Piridina Khinolina, Materialy Soveshchaniya, Inst. Khim. Akad. Nauk. Latv. SSR Rega*, 203 (1957) (Publ. 1960); *Chem. Abstr.*, **55**, 23529e (1961).
412. R. M. Acheson, in "Advances in Heterocyclic Chemistry," Vol. 1, A. R. Katritzky, Ed., Academic, New York, 1963, p. 125.

413. R. M. Acheson, M. W. Foxton, and A. R. Hands, *J. Chem. Soc., C*, 387 (1968).
414. L. M. Jackson, A. W. Johnson, and J. C. Tebby, *J. Chem. Soc.*, 1579 (1960).
415. A. W. Johnson and J. C. Tebby, *J. Chem. Soc.*, 2126 (1961).
416. R. M. Acheson and A. O. Plunkett, *J. Chem. Soc.*, 2676 (1964).
417. R. M. Acheson and D. A. Robinson, *Chem. Commun.*, 175 (1967).
418. A. Galbraith, T. Small, R. A. Barnes, and V. Boekelheide, *J. Amer. Chem. Soc.*, **83**, 453 (1961).
419. L. H. Klemm and D. R. McCoy, *J. Heterocycl. Chem.*, **6**, 73 (1969).
420. L. H. Klemm, D. R. McCoy, J. Shabtai, and W. K. T. Kiang, *J. Heterocycl. Chem.*, **6**, 813 (1969).
421. C. Hansch, W. Carpenter, and J. Todd, *J. Org. Chem.*, **23**, 1924 (1958).
422. J. Sam, J. N. Plampin, and D. W. Alwani, *J. Org. Chem.*, **27**, 4543 (1962).
423. A. M. Sargeson and W. H. F. Sasse, *Proc. Chem. Soc.*, 150 (1958).
424. D. G. Whitten and Y. J. Lee, *J. Amer. Chem. Soc.*, **92**, 416 (1970).
425. K. E. Wilzbach and D. J. Rausch, *J. Amer. Chem. Soc.*, **92**, 2178 (1970).
426. J. Joussot-Dubien and J. Houdard, *Tetrahedron Lett.*, 4389 (1967).
427. R. M. Kellogg, T. J. van Bergen, and H. Wynberg, *Tetrahedron Lett.*, 5211 (1969).
428. G. Bianchetti, *Farmaco (Pavia) Ed. Sci.*, **11**, 346 (1956); *Chem. Abstr.*, **53**, 9209f (1959).
429. D. E. Ames and J. L. Archibald, *J. Chem. Soc.*, 1475 (1962).
430. G. Bianchetti, *Farmaco (Pavia) Ed. Sci.*, **12**, 441 (1957); *Chem. Abstr.*, **51**, 17914b (1957).
431. K. Thiele, A. Gross, K. Posselt, and W. Schuler, *Chim. Ther.*, **2** (5) 366 (1967); *Chem. Abstr.*, **69**, 51949y (1968).
432. J. L. Archibald, U.S. Patent 3,429,885, 1969; *Chem. Abstr.*, **70**, 96643f (1969).
433. R. Bodalski and J. Michalski, *Rocz. Chem.*, **41** (3) 549 (1967); *Chem. Abstr.*, **67**, 73492e (1967).
434. R. Bodalski and J. Michalski, *Rocz. Chem.*, **41** (5) 939 (1967); *Chem. Abstr.*, **67**, 73493f (1967).
435. M. M. Baizer and E. J. Prill, U.S. Patent 3,218,246, 1965; *Chem. Abstr.*, **64**, 17554h (1966).
436. Aktieselskabet Pharmacia, Danish Patent 86,820, 1959; *Chem. Abstr.*, **53**, 18966d (1959).
437. O. Martensson and E. Nilsson, *Acta Chem. Scand.*, **15**, 1021 (1961).
438. F. E. Cislak, C. K. McGill, and G. W. Campbell, Jr., U.S. Patent 3,317,550, 1967; *Chem. Abstr.*, **67**, 21842z (1967).
439. W. Draber, K. Buechel, and M. Plempel, South African Patent 6904,723, 1970; *Chem. Abstr.*, **73**, 35370e (1970).
440. L. A. Walter, German Patent 1,905,353, 1969; *Chem. Abstr.*, **72**, 31790y (1970).
441. B. H. Smith, "Bridged Aromatic Compounds," Academic, New York, 1964.
442. F. Vögtle and P. Neumann, *Tetrahedron Lett.*, 5329 (1969).
443. F. Vögtle and P. Neumann, *Tetrahedron*, **26**, 5847 (1970).
444. F. Vögtle and P. Neumann, *Tetrahedron*, **26**, 5299 (1970).
445. V. Prelog, in "Perspectives in Organic Chemistry," Interscience, New York, 1956, p. 96.
446. K. Biemann, G. Büchi, and B. H. Walker, *J. Amer. Chem. Soc.*, **79**, 5558 (1957).
447. H. Nozaki, S. Fujita, and T. Mori, *Bull. Chem. Soc. Jap.*, **42**, 1163 (1969).
448. A. T. Balaban, *Tetrahedron Lett.*, 4643 (1968).
449. U. K. Georgi and J. Rétey, *Chem. Commun.*, 32 (1971).
450. V. Prelog and U. Geyer, *Helv. Chim. Acta*, **28**, 1677 (1945).
451. J. R. Fletcher and I. O. Sutherland, *Chem. Commun.*, 1504 (1969).
452. I. Gault, B. J. Price, and I. O. Sutherland, *Chem. Commun.*, 540 (1967).
453. W. Jenny and H. Holzrichter, *Chimia*, **22** (3) 139 (1968); *Chem. Abstr.*, **69**, 59065c (1968).
454. W. Jenny and H. Holzrichter, *Experientia*, **22**, 306 (1968).
455. V. Boekelheide and J. A. Lawson, *Chem. Commun.*, 1558 (1970).

456. V. Boekelheide and W. Pepperdine, *J. Amer. Chem. Soc.*, **92**, 3686 (1970).
457. R. W. Stotz and R. C. Stoufer, *Chem. Commun.*, 1682 (1970).
458. J. A. Zoltewicz and C. L. Smith, *J. Amer. Chem. Soc.*, **89**, 3358 (1967).
459. F. Kröhnke, *Angew. Chem.*, **65**, 605 (1953).
460. H. Quast and E. Frankenfeld, *Angew. Chem. Inter. Edit.*, **4**, 691 (1965).
461. H. Quast and E. Schmitt, *Ann. Chem.*, **732**, 43 (1970).
462. H. Quast and E. Schmitt, *Ann. Chem.*, **732**, 64 (1970).
463. R. K. Howe and K. W. Ratts, *Tetrahedron Lett.*, 4745 (1967).
464. E. R. Wallsgrove, *Manuf. Chemist*, **30**, 206 (1959); *Chem. Abstr.*, **53**, 14577f (1959).
465. M. Yoshida, *Yûki Gôsei Kagaku Kyôkai Shi*, **16**, 571 (1958); *Chem. Abstr.*, **53**, 1342c (1959).
466. E. Profft, *Chemiker-Ztg.*, **81**, 427 (1957); *Chem. Abstr.*, **52**, 2011h (1958).
467. E. Profft and R. Stumpf, *J. Prakt. Chem.*, **19**, (5–6) 266 (1963).
468. C. S. Marvel, A. T. Tweedie, and J. Economy, *J. Org. Chem.*, **21**, 1420 (1956).
469. F. J. Villani, C. A. Ellis, R. F. Tavares, M. Steinberg, and S. Tolksdorf, *J. Med. Chem.*, **13**, 359 (1970).
470. B. Brust, R. I. Fryer, and L. H. Sternbach, British Patent 1,106,028, 1968; *Chem. Abstr.*, **69**, 35962s (1968).
471. J. H. Burckhalter, U.S. Patent 3,124,585, 1964; *Chem. Abstr.*, **60**, 15841f (1964).
472. F. Zymalkowski and E. Reimann, *Ann. Chem.*, **715**, 98 (1968).
473. M. E. Derieg, I. Douvan, and R. I. Fryer, *J. Org. Chem.*, **33**, 1290 (1968).
474. T. Nakashima, *Yakugaku Zasshi*, **77**, 698 (1957); *Chem. Abstr.*, **51**, 16462g (1957).
475. A. K. Sheinkman, A. N. Kost, V. I. Sheichenko, and A. N. Rozenberg, *Ukr. Khim. Zh.*, **33** (9) 941 (1967); *Chem. Abstr.*, **68**, 104919b (1968).
476. H. Pommer and W. Sarnecki, German Patent 1,903,191, 1970; *Chem. Abstr.*, **73**, 77068t (1970).
477. B. I. Stepanov, G. I. Migachev, and N. S. Petukhova, U.S.S.R. Patent 358, 312, 1969; *Chem. Abstr.*, **72**, 132532w (1970).
478. E. D. Parker and A. Furst, *J. Org. Chem.*, **23**, 201 (1958).
479. G. Drefahl, K. Ponsold, and E. Gerlach, *Chem. Ber.*, **93**, 481 (1960).
480. A. N. Kost and A. K. Sheinkman, U.S.S.R. Patent 170,986 (1965); *Chem. Abstr.*, **63**, 9922a (1965).
481. B. M. Kuindzhi, M. A. Zepalova, L. D. Gluzman, A. K. Val'kova, R. M. Tsiu, I. V. Zaitseva, and A. A. Rok, *Metody Poluch Khim. Reaktivov Prep.*, (15) 93 (1967); *Chem. Abstr.*, **68**, 114378p (1968).
482. B. M. Kuindzhi, M. A. Zepalova-Parnitskaya, L. A. Matyushenko-Ugoltseva, I. D. Pavlova-Kravtsova, A. K. Va'lkova, I. V. Zaitseva, L. N. Nikolenko, and A. B. Pashkov, French Patent 1,525,475, 1968; *Chem. Abstr.*, **71**, 101715h, (1969).
483. J. Vahldieck and G. Buchmann, German (East) Patent 54,361, 1967; *Chem. Abstr.*, **67**, 100009m (1967).
484. R. Bodalski, J. Michalski, and K. Studnearski, *Rocz. Chem.*, **38** (9) 1337 (1964); *Chem. Abstr.*, **62**, 1627c (1965).
485. L. D. Gluzman, R. M. Tsin, and A. A. Rok, *Koks i Khim.*, (11) 48 (1961); *Chem. Abstr.*, **57**, 8537a (1962).
486. R. Bodalski and J. Michalski, *Bull. Acad. Polon. sci.*, *Sér. Sci. Chim.*, **8**, 217 (1960); *Chem. Abstr.*, **55**, 18721h (1961).
487. B. M. Kuindzhi, L. D. Gluzman, M. A. Zepalova, R. M. Tsip, A. K. Val'kova, and A. A. Rok, U.S.S.R. Patent 135,488, 1961; *Chem. Abstr.*, **55**, 16571b (1961).
488. M. M. Koton, *Khim. Tekhnol. Primenenie Proizvodnykh Piridina Khinolina, Materialy Soveshchaniya, Inst. Khim. Akad. Nauk Latv. SSR Riga*, 119 (1957); *Chem. Abstr.*, **55**, 16546g (1961).

489. B. M. Kuindzhi, M. A. Zepalova, L. A. Matyushenko, I. D. Pavlova, A. K. Val'kova, I. V. Zaitseva, L. N. Nikolenko, and A. B. Pashkov, British Patent 1,156,964, 1969; *Chem. Abstr.*, **71**, 61236u (1969).
490. T. Katsumoto, *Bull. Chem. Soc. Jap.*, **32**, 1019 (1959).
491. T. M. B. Wilson, British Patent 1,189,330, 1970; *Chem. Abstr.*, **73**, 14710d (1970).
492. R. N. Castle and C. W. Whittle, *J. Org. Chem.*, **24**, 1189 (1959).
493. A. P. Phillips and R. B. Burrows, U.S. Patent 3,376,297, 1968; *Chem. Abstr.*, **69**, 43805h (1968).
494. C. v. d. M. Brink and P. J. de Jager, *Tydskr. Natuurwetenskappe*, **3**, 74 (1963); *Chem. Abstr.*, **60**, 5447e (1964).
495. J. A. T. Beard and A. R. Katritzky, *Rec. Trav. Chim. Pays-Bas*, **78**, 592 (1959).
496. F. A. Al-Tai, G. Y. Sarkis, and F. A. Al-Najjar, *Bull. Coll. Sci. Univ. Baghdad*, **10**, 81 (1967); *Chem. Abstr.*, **72**, 43377g (1970).
497. H. Erdtman and A. Rosengren, *Acta Chem. Scand.*, **22**, 1475 (1968).
498. G. N. Walker, R. T. Smith, and B. N. Weaver, *J. Med. Chem.*, **8**, 626 (1965).
499. Societe Rocal, French Patent M 3339, 1965; *Chem. Abstr.*, **63**, 11515h (1965).
500. B. R. Baker and R. E. Gibson, *J. Med. Chem.*, **14**, 315 (1971).
501. R. H. Martin and M. Deblecker, *Tetrahedron Lett.*, 3597 (1969).
502. R. Bodalski, A. Malkiewicz, and J. Michalski, *Bull. Acad. Polon. Sci. Ser. Sci. Chim.*, **13** (3) 139 (1965); *Chem. Abstr.*, **63**, 8310e (1965).
503. B. Badgett, I. Parikh, and A. S. Dreiding, *Helv. Chim. Acta.*, **53**, 433 (1970).
504. Yu. I. Chumakov and V. M. Ledovskikh, U.S.S.R. Patent 158,575, 1963; *Chem. Abstr.*, **60**, 10655d (1964).
505. Yu. I. Chumakov and V. M. Ledovskikh, *Tetrahedron*, **21**, 937 (1965).
506. W. M. Stalick and H. Pines, *J. Org. Chem.*, **35**, 1712 (1970).
507. H. Pines and W. M. Stalick, *Tetrahedron Lett.*, 3723 (1968).
508. H. Pines and N. C. Sih, *J. Org. Chem.*, **30**, 280 (1965).
509. H. Pines, N. C. Sih, and E. Lewicki, *J. Org. Chem.*, **30**, 1457 (1965).
510. G. J. Janz and N. E. Duncan, *J. Amer. Chem. Soc.*, **75**, 5389 (1953).
511. G. J. Janz and N. E. Duncan, *J. Amer. Chem. Soc.*, **75**, 5391 (1953).
512. Y. Kawai and M. Dotani, Japanese Patent 6932,787, 1969; *Chem. Abstr.*, **72**, 78889c (1970).
513. J. G. Carey, British Patent 1,117,384, 1968; *Chem. Abstr.*, **69**, 43804g (1968).
514. H. L. Dimond, L. J. Fleckenstein, and M. O. Shrader, U.S. Patent 2,848,456, 1958; *Chem. Abstr.*, **53**, 1384d (1959).
515. G. de Stevens and C. F. Huebner, U.S. Patent 3,120,519, 1964; *Chem. Abstr.*, **60**, 9253a, (1964).
516. J. A. Berson and E. M. Evleth, *Chem. Ind.* (London), 901 (1959).
517. M. J. Allen, U.S. Patent 3,118,895, 1964; *Chem. Abstr.*, **60**, 10656e (1964).
518. F. E. Cislak, U.S. Patent 2,868,796, 1959; *Chem. Abstr.*, **53**, 10256i (1959).
519. E. Hardegger and E. Nikles, *Helv. Chim. Acta.*, **40**, 1016 (1957).
520. Yu. I. Chumakov and Yu. P. Shapovalova, *Metody Polucheniya Khim. Reaktivov Preparatov, Gos. Kom. Sov. Min. SSSR po Khim*, 43 (1964); *Chem. Abstr.*, **64**, 15832g (1966).
521. Yu. I. Chumakov and Yu. P. Shapovalova, U.S.S.R. Patent 162,534, 1964; *Chem. Abstr.*, **61**, 10660g (1964).
522. Yu. I. Chumakov and Yu. P. Shapovalova, *Zh. Org. Khim.*, **1** (5) 940 (1965); *Chem. Abstr.*, **63**, 6960h (1965).
523. M. Yoshida, H. Sugihara, S. Tsushima, and T. Miki, *Chem. Commun.*, 1223 (1969).
524. S. Miyano, A. Uno, and N. Abe, *Chem. Pharm. Bull.* (Tokyo), **15** (4) 515 (1967).
525. M. Maruoka, K. Isagawa, and Y. Fushizaki, *Kobunshi Kagaku*, **18**, 751 (1961); *Chem. Abstr.*, **58**, 5630g (1963).

526. V. Hněvsová and I. Ernest. *Collect. Czech. Chem. Commun.*, **25**, 748 (1960).
527. N. S. Prostakov and V. A. Kurichev, *Khim. Geterotsikl. Soedin.*, 679 (1967); *Chem. Abstr.*, **68**, 104932a (1968).
528. G. Cevidalli, J. Herzenberg, and A. Neuz, Italian Patent 596,924, 1959; *Chem. Abstr.*, **57**, 12443g (1962).
529. S. Yoshioka, T. Ohmae, and N. Hasegawa, *Kogyo Kagaku Zasshi*, **65**, 1995 (1962); *Chem. Abstr.*, **59**, 2762g (1963).
530. S. Yoshioka, T. Ohmae, and N. Hasegawa, Japanese Patent 6536, 1964; *Chem. Abstr.* **61**, 13289c (1964).
531. T. Oga, *Koru Taru*, **14**, 380 (1962); *Chem. Abstr.*, **58**, 5631 (1963).
532. E. Hesse and P. Müller, German (East) Patent 16,727 (1959); *Chem. Abstr.*, **54**, 18558h (1960).
533. Y. Matsuda, Y. Nakahara, R. Kato, T. Yasuda, Y. Moriyama, and Y. Ito, Japanese Patent 7728 (1959) 7729 (1959); *Chem. Abstr.*, **54**, 15406c (1960).
534. H. Ichinokawa, *Tokyo Kogyo Shikensho Hokoku*, **61** (12) 514 (1966); *Chem. Abstr.*, **67**, 11407x (1967).
535. R. W. Sudhoff, U.S. Patent 3,344,143, 1967; *Chem. Abstr.*, **68**, 87168m (1968).
536. F. Runge, G. Naumann, and M. Morgner, British Patent 828,205, 1960; *Chem. Abstr.*, **54**, 13149b (1960).
537. E. W. Pitzer, U.S. Patent 2,866,790, 1958; *Chem. Abstr.*, **53**, 7466e (1959).
538. D. M. Haskell and D. L. McKay, U.S. Patent 2,768,169, 1956; *Chem. Abstr.*, **51**, 11393e (1957).
539. B. M. Kuindzhi, A. K. Val'kova, M. A. Zepalova, L. A. Matyushenko, A. B. Pashkov, L. N. Nikolenko, and T. A. Ivanova, U.S.S.R. Patent 182,160, 1966; *Chem. Abstr.*, **65**, 16949f (1966).
540. A. A. Artamonov, A. A. Balandin, G. M. Marukyan, and M. I. Kotelenets, *Khim. Geterotsikl. Soedin.*, 512 (1967); *Chem. Abstr.*, **64**, 21798k (1968).
541. R. N. Lacey and B. Yeomans, British Patent 852,129, 1960; *Chem. Abstr.*, **55**, 13452h (1961).
542. S. D. Turk and B. D. Simpson, U.S. Patent 2,996,509, 1959; *Chem. Abstr.*, **56**, 4740f (1962).
543. K. Saruto and H. Maekawa, Japanese Patent 10,696, 1962; *Chem. Abstr.*, **59**, 3898h (1963).
544. T. Katsumoto and A. Honda, *Nippon Kagaku Zasshi*, **8** (6) 527 (1963); *Chem. Abstr.*, **59**, 15254a (1963).
545. J. L. R. Williams, S. K. Webster, and J. A. Van Allan, *J. Org. Chem.*, **26**, 4893 (1961).
546. S. Miyano and N. Abe, *Chem. Pharm. Bull.* (Tokyo), **15** (4) 511 (1967).
547. T. Katsumoto, *Bull. Chem. Soc. Jap.*, **33**, 1376 (1960).
548. J. L. R. Williams, *J. Amer. Chem. Soc.*, **84**, 1323 (1962).
549. G. B. Gechele and S. Pietra, *J. Org. Chem.*, **26**, 4412 (1961).
550. G. Cauzzo, G. Galiazzo, M. Mazzucato, and N. Mongiat, *Tetrahedron*, **22**, 689 (1966).
551. S. Minami, K. Fujimoto, Y. Takase, and M. Nakata, Japanese Patent 7011,496, 1970; *Chem. Abstr.*, **73**, 45364v (1970).
552. J. A. Price, U.S. Patent 2,810,714, 1957; *Chem. Abstr.*, **52**, 3402g (1958).
553. A. A. Artamonov, A. A. Balandin, G. M. Marukyan, and M. I. Kotelenets, *Dokl. Akad. Nauk SSSR*, **163** (2) 359 (1965).
554. P-T. Li, Y-T. Liu, C-C. Chiang, H-F. Kao, and C-Y. Kung, *Chung-Kuo K'o Hsueh Yuan Ying Yung Hua Hsueh Yen Chin So Chi K'an*, (7) 12 (1963); *Chem. Abstr.*, **63**, 13200g (1965).
555. R. A. Findlay, U.S. Patent 2,799,677, 1957; *Chem. Abstr.*, **52**, 450g (1958).
556. E. N. Pennington, U.S. Patent 2,879,272, 1959; *Chem. Abstr.*, **53**, 17152e (1959).
557. S. D. Turk and M. F. Potts, U.S. Patent 2,965,645, 1960; *Chem. Abstr.*, **55**, 11440i (1961).

558. W. L. Smith and M. F. Potts, U.S. Patent 2,824,105, 1958; *Chem. Abstr.*, **52**, 10178g (1958).

559. A. M. Schnitzer and R. E. Reusser, U.S. Patent 2,857,389, 1958; *Chem. Abstr.*, **53**, 5296f (1959).

560. C. W. Mertz, U.S. Patent 2,860,140, 1958; *Chem. Abstr.*, **53**, 6260h (1959).

561. C. W. Mertz, U.S. Patent 2,866,789, 1958; *Chem. Abstr.*, **53**, 6687b (1959).

562. K. Saruwatari and M. Matsushima, Japanese Patent 5872, 1959; *Chem. Abstr.*, **54**, 14274d (1960).

563. P. F. Warner, U.S. Patent 2,861,997, 1958; *Chem. Abstr.*, **53**, 11414c (1959).

564. W. B. Reynolds and R. E. Reusser, U.S. Patent 2,861,998 (1958); *Chem. Abstr.*, **53**, 11308e (1959).

565. H. R. Snyder, U.S. Patent 2,862,927, 1958; *Chem. Abstr.*, **53**, 8168c (1959).

566. C. W. Mertz, U.S. Patent 2,842,551 (1958); *Chem. Abstr.*, **52**, 19253g (1958).

567. J. J. Costolow, U.S. Patent 3,239,433 (1966); *Chem. Abstr.*, **64**, 15850d (1966).

568. R. E. Reusser and A. M. Schnitzer, U.S. Patent 2,812,329, 1957; *Chem. Abstr.*, **52**, 3402d (1958).

569. C. W. Mertz, U.S. Patent 2,745,834, 1956; *Chem. Abstr.*, **51**, 1295b (1957).

570. G. Galiazzo, *Gazz. Chim. Ital.*, **95** (11) 1322 (1965).

571. J. E. Mahan, U.S. Patent 2,893,995 (1959); *Chem. Abstr.*, **53**, 19459a (1959).

572. W. L. Smith, M. F. Potts, and P. S. Hudson, U.S. Patent 2,874,159, 1959; *Chem. Abstr.*, **53**, 12310a (1959).

573. R. H. Cullighan and M. H. Wilt, *J. Org. Chem.*, **26**, 4912 (1961).

574. V. I. Pansevich-Kolyada, *Zh. Obshch. Khim.*, **37** (3) 745 (1967); *Chem. Abstr.*, **67**, 54009m (1967).

575. Yu. I. Chumakov, V. S. Oleinik, and V. M. Ledovskikh, *Metody Polucheniya Khim. Reaktivov Preparatov*, Gos. Kom. Sov. Min. SSSR po Khim., (11) 77 (1964); *Chem. Abstr.*, **64**, 15832b (1966).

576. Yu. I. Chumakov and V. M. Ledovskikh, U.S.S.R. Patent 168,702, 1965; *Chem. Abstr.*, **63**, 1771e (1965).

577. S. Nozakura, *Bull. Chem. Soc. Jap.*, **29**, 784 (1956).

578. J. F. Brown Jr., U.S. Patent 2,924,601, 1960; *Chem. Abstr.*, **54**, 15406g (1960).

579. F. E. Cislak, U.S. Patent 2,854,455, 1958; *Chem. Abstr.*, **53**, 5292i (1959).

580. B. A. Bluestein, U.S. Patent 3,071,561, 1963; *Chem. Abstr.*, **59**, 1682a (1963).

581. J. Green and B. Groten, U.S. Patent 3,190,863, 1965; *Chem. Abstr.*, **63**, 5440d (1965).

582. Yu. A. Serguchev and E. A. Shilov, *Khim. Str. Svoistva Reaktivnost Org. Soedin.*, 23, (1969); *Chem. Abstr.*, **73**, 44556x (1970).

583. E. Maruszewska-Wieczorkowska and J. Michalski, *Bull. Acad. Polon. Sci.*, *Sér. Sci. Chim. géol. Géograph.*, **6**, 19 (1958); *Chem. Abstr.*, **52**, 16349g (1958).

584. C. Hansch and W. Carpenter, *J. Org. Chem.*, **22**, 936 (1957).

585. L. H. Klemm and D. Reed, *J. Org. Chem.*, **25**, 1816 (1960).

586. L. H. Klemm, J. Shabtai, D. R. McCoy, and W. K. T. Kiang, *J. Heterocycl. Chem.*, **5**, 883 (1968).

587. I. E. Il'ichev, V. S. Etlis, and A. P. Terent'ev, *Dokl. Akad. Nauk. SSSR*, **185** (4) 832 (1969).

588. F. F. Flicke, U.S. Patent 2,792,403, 1957; *Chem. Abstr.*, **51**, 13941f (1957).

589. S. I. Suminov and A. N. Kost, *Zh. Obshch. Khim.*, **34** (7) 2421 (1964); *Chem. Abstr.*, **61**, 11966b (1964).

590. O. B. Nielsen, V. Tvaermore, P. W. Feit, and H. H. Frey, German Patent 2,013,179, 1970; *Chem. Abstr.*, **74**, 22704n (1971).

591. C. M. McCloskey, U.S. Patent 3,410,861, 1968; *Chem. Abstr.*, **70**, 77812u (1969).

592. E. Profft and S. Lojack, *Rev. Chim. Acad. Rep. Populaire Roumaine*, **7** (1) 405 (1962), *Chem. Abstr.*, **59**, 8696b (1963).

593. A. T. Phillip, A. T. Casey, and C. R. Thompson, *Aust. J. Chem.* **23**, 491 (1970); *Chem. Abstr.*, **72**, 111230f (1970).

594. G. Magnus and R. Levine, *J. Amer. Chem. Soc.*, **78**, 4127 (1956).

595. A. N. Kost, A. P. Terent'ev, E. V. Vinogradova, P. B. Terent'ev, and V. V. Ershov, *Zh. Obshch. Khim.*, **30**, 2556 (1960); *Chem. Abstr.*, **55**, 13423b (1961).

596. N. F. Kazarinova, N. V. Dzhigirei, and N. A. Shabaeva, *Metody Poluch. Khim. Reaktivov Prep.*, (14) 38 (1966); *Chem. Abstr.*, **67**, 64187g (1967).

597. E. Profft and G. Busse, *Z. Chem.*, 19 (1961).

598. S. I. Suminov and A. N. Kost, *Zh. Org. Khim.*, **1** (11) 2055 (1965); *Chem. Abstr.*, **64**, 9677e (1966).

599. A. N. Kost, S. I. Suminov, and V. I. Vysotskii, *Zh. Org. Khim.*, **1** (11) 2071 (1965); *Chem. Abstr.*, **64**, 9678d (1966).

600. L. Bauer, A. Shoeb, and V. C. Agwada, *J. Org. Chem.*, **27**, 3153 (1962).

601. E. Profft and R. Kaden, *Arch. Pharm.*, **297** (11) 673 (1964); *Chem. Abstr.*, **62**, 11772c (1965).

602. L. A. Paquette, *J. Org. Chem.*, **27**, 2870 (1962).

603. E. Profft and W. Steinke, *Chem. Ber.*, **94**, 2267 (1961).

604. A. M. Parsons, U.S. Patent 3,019,229, 1962; British Patent 847051, 1960; *Chem. Abstr.*, **55**, 7441c (1961).

605. E. V. Vinogradova, A. N. Kost, V. N. Mitropol'skaya, and A. P. Terent'ev, *Zh. Obshch. Khim.*, **33** (5) 1556 (1963); *Chem. Abstr.*, **59**, 13938h (1963).

606. M. E. Freed and L. M. Rice, U.S. Patent 3,154,556, 1964; *Chem. Abstr.*, **61**, 16050h (1964).

607. A. P. Gray, H. Kraus, and D. E. Heitmeier, *J. Org. Chem.*, **25**, 1939 (1960).

608. C. J. Cavallito and A. P. Gray, U.S. Patent 3,300,506, 1967; *Chem. Abstr.*, **67**, 32595x (1967).

609. A. P. Gray and W. L. Archer, *J. Amer. Chem. Soc.*, **79**, 3554 (1957).

610. Irwin, Neisler and Co., British Patent 842,996, 1960; *Chem. Abstr.*, **55**, 7439g (1961).

611. J. L. Archibald, U.S. Patent 3,431,268, 1969; *Chem. Abstr.*, **70**, 106395t (1969).

612. A. P. Gray and H. Kraus, *J. Org. Chem.*, **31**, 399 (1966).

613. K. G. Tashchuk, A. V. Dombrovskii, and V. S. Fedorov, *Ukr. Khim. Zh.*, **30** (5) 496 (1964); *Chem. Abstr.*, **61**, 5606f (1964).

614. S. I. Suminov, U.S.S.R. Patent 242,897, 1969; *Chem. Abstr.*, **71**, 12462z (1969).

615. S. I. Suminov, *Khim. Geterotsikl. Soedin.*, 375 (1968); *Chem. Abstr.*, **69**, 86920j (1968).

616. P. B. Terent'ev, M. Islam, A. A. Zaitsev, and A. N. Kost, *Vestn. Mosk. Univ. Khim.*, **24** (2) 123 (1969); *Chem. Abstr.*, **71**, 12970v (1969).

617. A. N. Kost, P. B. Terent'ev, M. Islam, and A. A. Zaitsev, U.S.S.R. Patent 250, 142 (1969); *Chem. Abstr.*, **72**, 78883w (1970).

618. P. B. Terent'ev, M. Islam, and I. V. Chaikovskii, *Khim. Geterotsikl. Soedin.*, 1659 (1970); *Chem. Abstr.*, **74**, 53463n (1971).

619. F. G. Fallon, E. M. Roberts, G. P. Claxton, and E. L. Schumann, U.S. Patent 2,991,287, 1961; *Chem. Abstr.*, **56**, 3464e (1962).

620. R. M. Acheson and R. S. Feinberg, *J. Chem. Soc. C*, 351 (1968).

621. C. E. Loader, M. V. Sargent, and C. J. Timmons, *Chem. Commun.*, 127 (1965).

622. P. Bortolus, G. Cauzzo, and G. Galiazzo, *Tetrahedron Lett.*, 239 (1966).

623. G. Galiazzo, P. Bortolus, and G. Cauzzo, *Tetrahedron Lett.*, 3717 (1966).

624. P. L. Kumler and R. A. Dybas, *J. Org. Chem.*, **35**, 125 (1970).

625. R. L. Dannley, J. A. Schufle, I. Cohen, and J. R. Chambers, *J. Polymer Sci.*, **19**, 285 (1956); *Chem. Abstr.*, **52**, 2009h (1958).

626. R. M. Fuoss, M. Watanabe, and B. D. Coleman, *Mezhdunarsol Simpozium po Makromol. Khim. Doklady, Moscow*, **3**, 134 (1960); *Chem. Abstr.*, **55**, 7411c (1961).

627. Y. Murata, U.S. Patent 2,828,270, 1958; *Chem. Abstr.*, **52**, 11320a (1958).

628. C. E. Adams and C. N. Kimberlin Jr., U.S. Patent 2,899,396, 1959; *Chem. Abstr.*, **54**, 1770b (1960).
629. J. E. Pritchard, U.S. Patent 2,861,902, 1958; *Chem. Abstr.*, **53**, 8526h (1959).
630. H. B. Irwin, U.S. Patent 2,830,975, 1958; *Chem. Abstr.*, **52**, 13323c (1958).
631. J. E. Pritchard, U.S. Patent 2,862,902, 1958; *Chem. Abstr.*, **53**, 7670d (1959).
632. D. B. Capps, U.S. Patent 2,850,479, 1958, *Chem. Abstr.*, **53**, 4823c (1959).
633. R. W. Jones, U.S. Patent 2,860,123, 1958; *Chem. Abstr.*, **53**, 4797f (1959).
634. H. E. Railsback and C. C. Biard, *Ind. Eng. Chem.*, **49**, 1043 (1957).
635. M. M. Koton and O. K. Surnina, *Dokl. Akad. Nauk SSSR.*, **113**, 1063 (1957).
636. K. V. Martin, *J. Amer. Chem. Soc.*, **80**, 233 (1958).
637. I. Utsumi, T. Ida, S. Takahashi, and N. Sugimoto, *J. Pharm. Sci.*, **50**, 592 (1961); *Chem. Abstr.*, **55**, 21487f (1961).
638. A. Neuz and G. B. Gechele, *Chim. Ind.* (Milan), **43**, 142 (1961); *Chem. Abstr.*, **55**, 19921f (1961).
639. I. L. Kotlyarevskii, L. G. Fedenok, and L. N. Korolenok, *Izv. Sib. Otd. Akad. Nauk SSSR*, *Ser. Khim. Nauk*, (2) 111 (1969); *Chem. Abstr.*, **71**, 81101z (1969).
640. U. Haug and H. Fürst, *Chem. Ber.*, **93**, 593 (1960).
641. A. N. Kost, P. B. Terent'ev, and L. N. Moshentseva, *Metody Polucheniya Khim. Reaktivov Preparatov, Gos. Kom. Sov. Min. SSSR Khim.*, (11) 73 (1964), *Chem. Abstr.*, **64**, 15830g (1966).
642. A. N. Kost, P. B. Terent'ev, and T. Zavada, *Dokl. Akad. Nauk. SSSR.*, **130**, 326 (1960).
643. D. Jerchel and W. Melloh, *Ann. Chem.*, **622**, 53 (1959).
644. C. K. McGill, U.S. Patent 3,045,022, 1962; *Chem. Abstr.*, **57**, 15077g (1962).
645. M. Miocque and J. A. Gautier, *C. R. Acad. Sci., Paris, Ser. C.*, **252**, 2416 (1961).
646. M. Miocque, *Bull. Soc. Chim. Fr.*, **2**, 326 (1960).
647. I. L. Kotlyarevskii and L. I. Vereshchagin, *Izv. Akad. Nauk SSSR, Otd. Khim. Nauk*, 162 (1962); *Chem. Abstr.*, **57**, 11155b (1962).
648. M. S. Shvartsberg, V. N. Andrievskii, and I. L. Kotlyarevskii, *Izv. Akad. Nauk SSSR, Ser. Khim.*, (11) 2665 (1968); *Chem. Abstr.*, **70**, 68085g (1969).
649. M. S. Shvartsberg, I. L. Kotlyarevskii, A. N. Kozhevnikova, and V. N. Andrievskii, *Izv. Akad. Nauk SSSR, Ser. Khim.*, (5) 1144 (1970), *Chem. Abstr.*, **73**, 130851w (1970).
650. A. N. Kost, P. B. Terent'ev, and L. A. Moshentseva, *Zh. Obshch. Khim.*, **34** (9) 3035 (1964); *Chem. Abstr.*, **61**, 14632h (1964).
651. M. Miocque, *Bull. Soc. Chim. Fr.*, **2**, 332 (1960).
652. M. Miocque, *Bull. Soc. Chim. Fr.*, **2**, 330 (1960).
653. I. L. Kotlyarevskii, L. I. Vereshchagin, and O. G. Yashina, *Izv. Sibirsk. Otd. Akad. Nauk SSSR*, (11) 148 (1962); *Chem. Abstr.*, **59**, 3885c (1963).
654. A. N. Kost, P. B. Terent'ev, and A. A. Sochegolev, *Zh. Obshch. Khim.*, **32**, 2606 (1962); *Chem. Abstr.*, **58**, 9020b (1962).
655. I. L. Kotlyarevskii, L. I. Vereshchagin, O. G. Yashima, E. K. Vasil'ev, and Yu. M. Faershtein, *Izv. Sibirsk. Otd. Akad. Nauk SSSR.*, (9) 80 (1962); *Chem. Abstr.*, **59**, 1584a (1963).
656. J. A. Gautier and M. Miocque, *C. R. Acad. Sci., Paris, Ser. C.*, **250**, 719 (1960).
657. J. A. Gautier, I. M. Olomucki, and M. Miocque, *C. R. Acad. Sci., Paris, Ser. C.*, **251**, 562 (1960).
658. T. B. Terent'ev, T. P. Moskvina, L. V. Moshentseva, and A. N. Kost, *Khim. Geterotsikl. Soedin.*, 498 (1970); *Chem. Abstr.*, **73**, 45406k (1970).
659. N. S. Prostakov and N. N. Mikhailova, *Med. Prom. SSSR.*, **14** (2) 11 (1960); *Chem. Abstr.*, **54**, 22637b (1960).
660. N. S. Prostakov, L. A. Gaivoronskaya, L. M. Kirillova, and M. K. S. Mokhomon, *Khim. Geterotsikl. Soedin.*, 782 (1970); *Chem. Abstr.*, **73**, 120467m (1970).

661. N. S. Prostakov, N. M. Mikhailova, and Yu. M. Talanov, *Khim. Geterotsikl. Soedin.*, 1359 (1970); *Chem. Abstr.*, **74**, 53448m (1971).
662. R. A. Abramovitch and M. Saha, *Can. J. Chem.*, **44**, 1765 (1966).
663. R. Grashey and R. Huisgen, *Chem. Ber.*, **92**, 2641 (1959).
664. H. J. M. Dou and B. M. Lynch, *Tetrahedron Lett.*, 896 (1965).
665. J. M. Bonnier, J. Court, and T. Fay, *Bull. Soc. Chim. Fr.*, 1204 (1967).
666. H. J. M. Dou, G. Vernin, and J. Metzger, *Tetrahedron Lett.*, 953 (1968).
667. J. M. Bonnier, J. Court, and M. Gelus, *Bull. Soc. Chim. Fr.*, 139, 142 (1970).
668. B. M. Lynch and H. S. Chang, *Tetrahedron Lett.*, 2965 (1964).
669. R. Huisgen and H. Nakaten, *Ann. Chem.*, **586**, 70 (1954).
670. R. J. Gritter and A. W. Godfrey, *J. Amer. Chem. Soc.*, **86**, 4724 (1964).
671. E. K. Fields and S. Meyerson, *J. Org. Chem.*, **35**, 62 (1970).
672. S. B. Needleman and M. C. Chang Kuo, *Chem. Rev.*, **62**, 405 (1962).
673. O. Westphal and A. Joos, *Angew. Chem. Inter. Edit.*, **8**, 73 (1969).
674. O. Westphal, G. Felix, and A. Joos, *Angew. Chem. Inter. Edit.*, **8**, 74 (1969).
675. G. R. Newkome and D. L. Fishel, *Chem. Commun.*, 916 (1970).
676. Y. Ishiguro, Y. Morita, and K. Ikushima, *Yakugaku Zasshi*, **78**, 220 (1958); *Chem. Abstr.*, **54**, 24712d (1960).
677. P. I. Mortimer, *Aust. J. Chem.*, **21** (2) 467 (1968).
678. D. Muenzner, H. Lattau, and H. Schubert, *Z. Chem.*, **7** (7) 278 (1967).
679. W. H. F. Sasse and C. P. Whittle, *J. Chem. Soc.*, 1347 (1961).
680. E. K. Fields, U.S. Patent 3,150,146, 1964; *Chem. Abstr.*, **64**, 16051a (1964).
681. S. V. Krivun, G. N. Dorofunko, and A. S. Kovalevskii, *Khim. Geterotsikl. Soedin.*, 733 (1970); *Chem. Abstr.*, **73**, 98769n (1970).
682. D. J. Brown and B. T. England, *Aust. J. Chem.*, **23** (3) 625 (1970).
683. T. Ishiguro, Y. Morita, and K. Ikushima, *Yagaku Zasshi*, **78**, 268 (1958); *Chem. Abstr.*, **52**, 11847b (1958).
684. K. Luchder and H. Langanke, *Z. Chem.*, **10** (2) 74 (1970).
685. G. Y. Paris, D. L. Garmaise, and J. Komlossy, *J. Heterocycl. Chem.*, **8**, 169 (1971).
686. P. Doyle and G. J. Stacey, South African Patent 6706,809, 1969; *Chem. Abstr.*, **72**, 12592w (1970).
687. F. A. Vingiello, E. B. Ellerbe, T. J. Delia, and J. Yanez, *J. Med. Chem.*, **7**, 121 (1964).
688. F. A. Vingiello and T. J. Delia, *J. Org. Chem.*, **29**, 2180 (1964).
689. V. J. Bauer, H. P. Dalalian, W. J. Fanshawe, S. R. Safir, E. C. Tocus, and C. R. Boshart, *J. Med. Chem.*, **11**, 981 (1968).
690. G. E. Wiegand, V. J. Bauer, S. R. Safir, D. A. Blickens, and S. J. Riggi, *J. Med. Chem.*, **14**, 214 (1971).
691. W. J. Fanshawe, G. E. Wiegand, V. J. Bauer, and S. R. Safir, German Patent 1,933,853, 1970; *Chem. Abstr.*, **73**, 3807m (1970).
692. V. J. Bauer, W. J. Fanshawe, H. P. Dalalian, and S. R. Safir, *J. Med. Chem.*, **11**, 984 (1968).
693. V. J. Bauer and S. R. Safir, British Patent 1,178,604, 1970; *Chem. Abstr.*, **72**, 79017d (1970).
694. V. J. Bauer, G. E. Wiegand, W. J. Fanshawe, and S. R. Safir, *J. Med. Chem.*, **12**, 944 (1969).
695. G. E. Wiegand, V. J. Bauer, S. R. Safir, D. A. Blickens, and S. J. Riggi, *J. Med. Chem.*, **12**, 943 (1969).
696. G. E. Wiegand, V. J. Bauer, S. R. Safir, D. A. Blickens, and S. J. Riggi, *J. Med. Chem.*, **12**, 891 (1969).
697. W. J. Fanshawe, V. J. Bauer, S. R. Safir, D. A. Blickens, and S. J. Riggi, *J. Med. Chem.*, **12**, 381 (1969).
698. H. Wynberg, T. J. van Bergen, and R. M. Kellogg, *J. Org. Chem.*, **34**, 3175 (1969).

699. K. Kahmann, H. Sigel, and H. Erlenmeyer, *Helv. Chim. Acta*, **47**, 1754 (1964).
700. H. Hellmann and D. Dieterich, *Angew. Chem. Inter. Edit.*, **1**, 53 (1962).
701. M. Pouteau-Thouvenot, M. Choussy, A. Gaudemer, and M. Barbier, *Bull. Soc. Chim. Biol.*, **52** (1) 51 (1970); *Chem. Abstr.*, **73**, 55934v (1970).
702. F. Hein and U. Beierlein, *Pharm. Z.*, **96** (8–9) 401 (1957); *Chem. Abstr.*, **52**, 14608d (1958).
703. M. C. Kloetzel, F. L. Chubb and J. L. Pinkus, *J. Amer. Chem. Soc.*, **80**, 5773 (1958).
704. D. Beck and K. Schenker, Netherlands Patent 6,514,606, 1966; *Chem. Abstr.*, **65**, 12175h (1966).
705. R. N. Schut and F. E. Ward, German Patent 1,901,637, 1969; *Chem. Abstr.*, **72**, 66828u (1970).
706. W. B. Wright, Jr., H. J. Brabander, R. A. Hardy, Jr., and W. Fulmor, *J. Amer. Chem. Soc.*, **81**, 5637 (1959).
707. M. Kawai, Japanese Patent 7,010,148, 1970; *Chem. Abstr.*, **73**, 45506t (1970).
708. M. Bieganowska, *Acta Pol. Pharm.*, **27** (4) 369 (1970); *Chem. Abstr.*, **74**, 13055z (1971).
709. M. C. Kloetzel and F. L. Chubb, *J. Amer. Chem. Soc.*, **79**, 4226 (1957).
710. F. Binon and J. M. Beiler, German Patent 1,949,813, 1970; *Chem. Abstr.*, **73**, 14847a (1970).
711. L. S. Davies and G. Jones, *Tetrahedron Lett.*, 1049 (1970).
712. L. S. Davies and G. Jones, *J. Chem. Soc. C.*, 759 (1971).
713. R. A. Bowie, *Chem. Commun.*, 565 (1970).
714. S. A. Alieva, Yu. V. Kolodyazhuyi, A. D. Garnovskii, O. A. Osipov, I. I. Grandberg, and N. F. Krokhina, *Khim. Geterotsikl. Soedin.*, 45 (1970); *Chem. Abstr.*, **72**, 110632b (1970).
715. Ciba Ltd., French Patent M 6700, 1969; *Chem. Abstr.*, **74**, 53762j (1971).
716. K. Wallenfels and M. Gellrich, *Ann. Chem.*, **621**, 210 (1959).
717. A. Taurens and A. Blaga, *J. Heterocycl. Chem.*, **7**, 1137 (1970).
718. J. E. Downes, British Patent 1,171,524, 1969; *Chem. Abstr.*, **72**, 90519q (1970).
719. K. Gubler and O. Kristiansen, German Patent 2,014,886, 1970; *Chem. Abstr.*, **74**, 53766p (1971).
720. P. J. Palmer, R. B. Trigg, and J. V. Warrington, *J. Med. Chem.*, **14**, 248 (1971).
721. M. Dadkah and B. Prijns, *Helv. Chim. Acta.*, **45**, 375 (1962).
722. R. N. Castle, *J. Org. Chem.*, **23**, 69 (1958).
723. B. L. Bastić, *Bull. Soc. Chim. Belgrade*, **16**, 141 (1951); *Chem. Abstr.*, **48**, 5860c (1954).
724. A. Novelli and J. R. Barrio, *Biol. Soc. Quim. Peru*, **35**, (4) 118 (1969); *Chem. Abstr.*, **73**, 25361p (1970).
725. F. H. McMillan, F. Leonard, R. I. Meltzer, and J. A. King, *J. Amer. Pharm. Assoc.*, **42**, 457 (1953).
726. P. M. Hergenrother, *J. Heterocycl. Chem.*, **6**, 965 (1969).
727. K. Harsanyi, J. Reiter, D. Korbonits, K. Takacs, E. Bako, G. Leszkovszky, L. Tardos, and C. Vertesy, German Patent 1,920,037, 1970; *Chem. Abstr.*, **74**, 13156h (1971).
728. K. Harsanyi, J. Reiter, D. Korbonits, C. Gonczi, K. Takacs, E. Bako, G. Leszkovszky, L. Tardos, and C. Vertesy, Hungarian Patent 156,976, 1970; *Chem. Abstr.*, **72**, 100719w (1970).
729. W. W. Paudler and J. E. Kuder, *J. Org. Chem.*, **32**, 2430 (1967).
730. S. Kubota, T. Okitsu, and Y. Koida, *Yakugaku Zasshi*, **90** (7) 841 (1970); *Chem. Abstr.*, **73**, 77147t (1970).
731. R. Huisgen, K. v. Fraunberg, and H. J. Sturm, *Tetrahedron Lett.*, **30**, 2589 (1969).
732. B. Eisert and E. Endres, *Ann. Chem.*, **734**, 56 (1970).
733. F. H. Case, *J. Heterocycl. Chem.*, **7**, 1001 (1970).
734. A. J. Hubert and G. Anthoine, *Bull. Soc. Chim. Belg.*, **78**, (9–10) 553 (1969).
735. A. J. Blackman and J. B. Polya, *J. Chem. Soc. C.*, 1016 (1971).
736. D. D. Libman and R. Slack, *J. Chem. Soc.*, 2253 (1956).

737. G. F. Holland, Netherlands Patent 6,516,322, 1966: *Chem. Abstr.*, **67**, 54165j (1967).
738. G. F. Holland and J. N. Pereira, *J. Med. Chem.*, **10**, 149 (1967).
739. A. N. Nesmeyanov, V. A. Sazonova, and A. V. Gerasimenko, *Dokl. Akad. Nauk SSSR*, **147**, 634 (1962).
740. A. N. Nesmeyanov, V. A. Sazonova, and V. E. Fedorov, *Izv. Akad. Nauk SSSR, Ser. Khim.*, (9) 2133 (1970); *Chem. Abstr.*, **74**, 31837k (1971).
741. D. J. Booth, G. Marr, B. W. Rockett, and A. Rushworth, *J. Chem. Soc. C.*, 2701 (1969).
742. A. N. Nesmeyanov, V. A. Sazonova, A. V. Gerasimenko, and N. S. Sazonova, *Dokl. Akad. Nauk SSSR*, **149**, 1354 (1963).
743. Imperial Chemical Industries Ltd., Belgian Patent 620,301, 1963; *Chem. Abstr.*, **59**, 8712f (1963).
744. Imperical Chemical Industries Ltd., Netherlands Patent 6,607,292, 1966; *Chem. Abstr.*, **67**, 73530r (1967).
745. W. E. Kramer and L. A. Joo, U.S. Patent 3,127,359, 1964; *Chem. Abstr.*, **61**, 792b (1964).
746. R. H. Linnell, *J. Org. Chem.*, **22**, 1691 (1957).
747. F. E. Cislak, U.S. Patent 2,992,224, 1961; *Chem. Abstr.*, **56**, 461f (1962).
748. R. S. Fanshawe and A. W. Olleveant, Belgian Patent 638,139, 1964; *Chem. Abstr.*, **63**, 584f (1965).
749. K. Holowiecki, J. Horak, and M. Rozmarynowicz, *Lodz. Towarz. Nauk. Wydzial III, Acta. Chim.*, **9**, 177 (1964); *Chem. Abstr.*, **62**, 9100c (1965).
750. Imperial Chemical Industries Ltd., Belgian Patent 617,852, 1962; *Chem. Abstr.*, **65**, 5447b (1966).
751. F. R. Bradbury and A. Campbell, British Patent 1,016,548, 1966; *Chem. Abstr.*, **65**, 693f (1966).
752. F. R. Bradbury and A. Campbell, British Patent 957,098, 1964; *Chem. Abstr.*, **61**, 13286c (1964).
753. Imperial Chemical Industries Ltd., Belgian Patent 617,748, 1962; *Chem. Abstr.*, **58**, 13922h (1963).
754. A. Baines and A. Campbell, Belgian Patent 617,852, 1962; *Chem. Abstr.*, **60**, 2907f (1964).
755. R. Setton, *C. R. Acad. Sci., Paris, Ser. C.*, **244**, 1205 (1957).
756. Imperial Chemical Industries Ltd., Belgian Patent 629,552, 1963; *Chem. Abstr.*, **60**, 14482e (1964).
757. Imperial Chemical Industries Ltd., French Patent 1,341,585, 1963; *Chem. Abstr.*, **60**, 14482f (1964).
758. P. B. Dransfield and M. H. Watson, French Patent 1,380,806, 1964; *Chem. Abstr.*, **62**, 10417h (1965).
759. A. H. Jubb, British Patent 869,955, 1961; *Chem. Abstr.*, **56**, 461i (1962).
760. G. H. Lang, British Patent 1,014,076, 1965, *Chem. Abstr.*, **64**, 9691e (1966).
761. A. H. Jubb, British Patent 869,954, 1961; *Chem. Abstr.*, **56**, 3465e (1962).
762. Yu. V. Kurbatov, O. S. Otroshchenko, A. S. Sadykov, and M. Goshaev, *Tr. Samarkand. Gos. Univ.*, (167) 85 (1969), *Chem. Abstr.*, **73**, 98760c (1970).
763. C. Gardner, British Patent 829,838, 1960; *Chem. Abstr.*, **54**, 15406b (1960).
764. W. Karminski and Z. Kulichi, *Chem. Stosowana, Ser. A.*, **9** (1) 129 (1965); *Chem. Abstr.*, **63**, 18018d (1965).
765. Z. Kulicki and W. Karminski, *Zeszyty Nauk. Politech. Slask. Chem.*, **16** (85) 11 (1963), *Chem. Abstr.*, **62**, 4001c (1965).
766. M. Goshaev, O. S. Otroshchenko, A. S. Sadykov, and N. Kuznetsova, *Izv. Akad. Nauk Turkm. SSR, Ser. Fiz.-Tekh. Khim. Geol. Nauk*, (5) 114 (1970), *Chem. Abstr.*, **74**, 3480v (1971).

767. O. S. Otroshchenko, M. Goshaev, A. S. Sadykov, and N. V. Kuznetsova, U.S.S.R. Patent 253,066, 1969; *Chem. Abstr.*, **72**, 121372t (1970).
768. G. M. Badger and W. H. F. Sasse, in "Advances in Heterocyclic Chemistry," Vol. 2, A. R. Katritzky, Ed., Academic, New York, 1963, p. 179.
769. D. Y. Waddan and D. Williams, German Patent 1,950,074, 1970; *Chem. Abstr.*, **73**, 3799k (1970).
770. G. H. Lang, British Patent 1,026,822, 1966; *Chem. Abstr.*, **65**, 2232b (1966).
771. G. H. Lang, U.S. Patent 3,173,920, 1965; British Patent 981,353, 1965; *Chem. Abstr.*, **63**, 14826d (1965).
772. G. H. Lang and R. G. A. New, U.S. Patent 3,210,366, 1965; British Patent 1,000,656, 1965; *Chem. Abstr.*, **63**, 16315a (1965).
773. G. L. Varcoe, German Patent 1,445,079, 1969; *Chem. Abstr.*, **72**, 21615f (1970).
774. R. D. Bowden, German Patent 2,022,928, 1970; *Chem. Abstr.*, **74**, 42285e (1971).
775. L. D. Gluzman, L. P. Stolyarenko, and S. N. Ol'shanskaya, *Sb. Nauch. Tr. Ukr. Nauch-Issled. Uglekhim. Inst.*, (20) 160 (1967); *Chem. Abstr.*, **70**, 3770g (1969).
776. F. Antoine, *C. R. Acad. Sci., Paris, Ser. C.*, **258** (19) 4742 (1964).
777. O. S. Otroshchenko, A. A. Ziyaev, and A. S. Sadykov, *Khim. Geterotsikl. Soedin.*, 365 (1969); *Chem. Abstr.*, **71**, 22001u (1969).
778. R. D. Bowden, German Patent 1,913,732, 1970; *Chem. Abstr.*, **72**, 132535z (1970).
779. R. D. Bowden, German Patent 2,033,958, 1971; *Chem. Abstr.*, **74**, 64208j (1971).
780. F. Kuffner and F. Straberger, *Monatsh. Chem.*, **88**, 793 (1957).
781. R. F. Homer, *J. Chem. Soc.*, 1574 (1958).
782. J. Thesing and A. Müller, *Angew. Chem.*, **68**, 577 (1956).
783. J. Thesing and A. Müller. *Chem. Ber.*, **90**, 711 (1957).
784. R. L. Frank and J. V. Crawford, *Bull. Soc. Chim. Fr.*, 419 (1958).
785. R. Lukeš and Jiŕé Pliml, *Chem. Listy*, **52**, 759 (1958).
786. R. D. Bowden, German Patent 1,913,150, 1969; *Chem. Abstr.*, **72**, 43449g (1970).
787. L. J. Winters, N. G. Smith, and M. I. Cohen, *Chem. Commun.*, 642 (1970).
788. Imperial Chemical Industries Ltd., Belgian Patent 628,926, 1963; *Chem. Abstr.*, **61**, 6995e (1964).
789. R. S. Fanshawe and C. Shepherd, Belgian Patent 628,926, 1963; *Chem. Abstr.*, **61**, 6995e (1964).
790. R. S. Fanshawe, Belgian Patent 628,927, 1963; *Chem. Abstr.*, **60**, 15671e (1964).
791. G. I. Mikhailov and L. I. Mizrakh, U.S.S.R. Patent 123,946, 1959; *Chem. Abstr.*, **54**, 9963f (1960).
792. Imperial Chemical Industries Ltd., Netherlands Patent 6,401,239, 1964; *Chem. Abstr.*, **62**, 4011f (1965).
793. J. E. Colchester and J. H. Entwistle, British Patent 1,074,977, 1967; *Chem. Abstr.*, **68**, 95697y (1968).
794. S. Herzog, *Z. Anorg. u. Allgem. Chem.*, **294**, 155 (1958); *Chem. Abstr.*, **52**, 13721f (1958).
795. B. Martin and G. M. Waind, *Proc. Chem. Soc.*, 169 (1958).
796. R. G. Beiles and V. V. Malysheva, U.S.S.R. Patent 172,784, 1965; *Chem. Abstr.*, **64**, 708d (1966).
797. M. T. Beck and M. Halmos, *Nature*, **191**, 1090 (1961).
798. L. Cambi and E. Paglia, *Atti Accad. Nazl. Lincei Rend. Classe Sci. Fis. Mat. Nat.*, **21**, 372 (1956); *Chem. Abstr.*, **51**, 12093h (1957).
799. P. Ellis, R. G. Wilkins, and M. J. G. Williams, *J. Chem. Soc.*, 4456 (1957).
800. A. N. Plyusnin, B. A. Uvarov, V. I. Tsvetkova, and N. M. Chirkov, *Izv. Akad. Nauk SSSR, Ser. Khim.*, (10) 2324 (1969); *Chem. Abstr.*, **72**, 43774j (1970).
801. W. W. Brandt and J. P. Wright, *J. Amer. Chem. Soc.*, **76**, 3082 (1954).

802. R. D. Chambers, D. Lomas, and W. K. R. Musgrave, British Patent 1,163,472, 1969; *Chem. Abstr.*, **71**, 124269e (1969).
803. Yu. V. Kurbatov, V. K. Kiryukhin, O. S. Otroschchenko, and A. S. Sadykov, *Nauch. Tr. Tashkent. Gos. Univ.*, (286) 92 (1966); *Chem. Abstr.*, **67**, 82059w (1967).
804. O. S. Otroshchenko, Yu. V. Kurbatov, and A. S. Sadykov, *Nauchn. Tr. Tashkentsk. Gos. Univ.*, (263) 27 (1964), *Chem. Abstr.*, **63**, 4248c (1965).
805. O. S. Otroshchenko, Yu. V. Kurbatov, A. S. Sadykov, and F. Pirnazarova, *Nauchn. Tr. Tashkentsk. Gos. Univ.*, (263) 33 (1964), *Chem. Abstr.*, **63**, 4248e (1965).
806. O. S. Otroshchenko, A. A. Ziyaev, and A. S. Sadykov, *Nauch. Tr. Tashkent. Gos. Univ.*, (286) 85 (1966); *Chem. Abstr.*, **67**, 82057u (1967).
807. O. S. Otroshchenko, A. S. Sadykov, and A. A. Ziyaev, *Zh. Obshch. Khim.*, **31**, 678 (1961); *Chem. Abstr.*, **55**, 23527b (1961).
808. A. A. Ziyaev, O. S. Otroshchenko, and A. S. Sadykov, *Khim. Geterotsikl. Soedin. Akad. Nauk Latv. SSR*, (1) 90 (1965); *Chem. Abstr.*, **63**, 5595b (1965).
809. M. Seyhan and W. C. Fernelius, *Chem. Ber.*, **91**, 469 (1958).
810. D. C. Allport, British Patent 1,130,551, 1968; *Chem. Abstr.*, **70**, 28832s (1969).
811. D. C. Allport, British Patent 1,129,511, 1968; *Chem. Abstr.*, **70**, 28833t (1969).
812. L. A. Summers, *Naturwissenschaften*, **54**, (18) 491 (1967).
813. Yu. N. Forostyan and E. I. Efimova, *Zh. Obshch. Khim.*, **39** (9) 2122 (1969); *Chem. Abstr.*, **72**, 31561z (1970).
814. Yu. N. Forostyan, E. I. Efimova, and V. Novilov, *Zh. Obshch. Khim.*, **38** (4) 839 (1968); *Chem. Abstr.*, **69**, 77076m (1968).
815. R. F. Homer and J. E. Downes, British Patent 1,070,504, 1967; *Chem. Abstr.*, **67**, 73529x (1967).
816. R. C. Brian, G. W. Driver, R. F. Homer, and R. L. Jones, British Patent 813,532, 1959; *Chem. Abstr.*, **55**, 4541h (1961).
817. R. F. Homer, U.S. Patent 3,202,500, 1965; *Chem. Abstr.*, **63**, 16368b (1965).
818. A. L. Black and L. A. Summers, *J. Heterocycl. Chem.*, **8**, 29 (1971).
819. R. F. Homer and T. E. Tomlinson, *J. Chem. Soc.*, 2498 (1960).
820. N. F. Kazarinova, K. A. Solomko, and M. N. Kotelenets, *Metody Poluch. Khim. Reaktivov Prep.*, (14) 22 (1966); *Chem. Abstr.*, **67**, 2975d (1967).
821. K. Thomae, G.m.b.H., French Patent 1,536,543, 1968; *Chem. Abstr.*, **72**, 31643c (1970).
822. K. Thomae, G.m.b.H., French Patent 2,002,366, 1969; *Chem. Abstr.*, **72**, 66846y (1970).
823. K. Thomae, G.m.b.H., French Patent 93,006, 1969; *Chem. Abstr.*, **69**, 106574e (1970).
824. K. Thomae, G.m.b.H., French Patent 1,555,417, 1969; *Chem. Abstr.*, **72**, 31638e (1970).
825. K. Thomae, G.m.b.H., British Patent 1,174,385, 1969; *Chem. Abstr.*, **72**, 66845x (1970).
826. K. Matsumori, A. Ide, and H. Watanabe, *Nippon Kagaku Zasshi*, **91** (6) 575 (1970), *Chem. Abstr.*, **73**, 109646y (1970).
827. T. Tokuda, K. Matsumori, A. Ide, and H. Watanabe, *Nippon Kagaku Zasshi*, **91** (6) 572 (1970); *Chem. Abstr.*, **73**, 109645x (1970).
828. W. Engel, E. Seeger, H. Teufel, and H. Machleidt, *Chem. Ber.*, **104**, 248 (1971).
829. R. Albrecht, K. Gutsche, H. J. Kessler, and E. Schroeder, *J. Med. Chem.*, **13**, 733 (1970).
830. B. M. Culbertson, U.S. Patent 3,498,981, 1970; *Chem. Abstr.*, **73**, 35416z (1970).
831. F. H. Case and E. Koft, *J. Amer. Chem. Soc.*, **81**, 905 (1959).
832. F. Kröhnke, K. Ellegast, and E. Bertram, *Ann. Chem.*, **600**, 176 (1956).
833. D. N. Kursanov, N. K. Baranetskaja, and V. N. Setkina, *Proc. Acad. Sci. USSR, Chem. Sect.*, **113**, 191 (1957).
834. J. A. Berson, E. M. Evleth, Jr., and Z. Hamlet, *J. Amer. Chem. Soc.*, **87**, 2887 (1965).
835. G. V. Boyd and L. M. Jackman, *J. Chem. Soc.*, 548 (1963).
836. G. Seitz and H. Mönnighoff, *Ann. Chem.*, **732**, 11 (1970).

837. J. A. Berson and E. M. Evleth, Jr., *Chem. Ind.* (London), 901 (1959).
838. G. V. Boyd, A. W. Ellis, and M. D. Harms, *J. Chem. Soc. C*, 800 (1970).
839. G. Seitz, *Angew. Chem. Inter. Edit.*, **8,** 478 (1969).
840. E. M. Evleth, Jr., J. A. Berson, and S. L. Manatt, *Tetrahedron Lett.*, 3087 (1964).
841. G. V. Boyd, *Proc. Chem. Soc.*, 253 (1960).
842. J. A. Berson and E. M. Evleth, Jr., *Chem. Ind.* (London), 1362 (1961).
843. J. A. Berson, E. M. Evleth, Jr., and Z. Hamlet, *J. Amer. Chem. Soc.*, **82,** 3793 (1960).
844. E. M. Evleth, Jr., J. A. Berson, and S. L. Manatt, *J. Amer. Chem. Soc.*, **87,** 2908 (1965).
845. J. A. Berson, E. M. Evleth, Jr., and S. L. Manatt, *J. Amer. Chem. Soc.*, **87,** 2901 (1965).
846. T. Ishiguro, Y. Morita, and K. Ikushima, *Yakugaku Zasshi*, **77,** 660 (1957); *Chem. Abstr.*, **51,** 16463c (1957).
847. I. L. Kotlyarewskii, E. K. Vasilev, and L. I. Vereshchagin, *Izvest. Sibir. Otdel. Akad. Nauk SSSR*, (9) 52 (1959); *Chem. Abstr.*, **54,** 9915h (1960).
848. A. S. Bailey and J. S. A. Brunskill, *J. Chem. Soc.*, 2554 (1959).
849. J. P. Kutney and R. C. Selby, *J. Org. Chem.*, **26,** 2733 (1961).
850. N. S. Prostakov, S. Ya. Govor, N. N. Mikeeva, and P. K. P. Franko, *Khim. Geterotsikl. Soedin.*, 1018 (1969); *Chem. Abstr.*, **73,** 35181u (1970).
851. J. E. Mahan, S. D. Turk, A. M. Schnitzer, R. P. Williams, and G. D. Sammons, *Ind. Eng. Chem. Eng. Data, Ser. 2*, 76 (1957).
852. P. Doyle and R. R. J. Yates, *Tetrahedron Lett.*, 3371 (1970).
853. R. A. Abramovitch and B. Vig, *Can. J. Chem.*, **41,** 1961 (1963).
854. R. A. Abramovitch, W. C. Marsh, and J. G. Saha, *Can. J. Chem.*, **43,** 2631 (1965).
855. J. B. Stothers, "Carbon-13 NMR Spectroscopy," Academic, New York, 1971.
856. J. Jones and J. Jones, *Tetrahedron Lett.*, 2117 (1964).
857. A. M. Akkermann and H. Veldstra, *Rec. Trav. Chim. Pays-Bas*, **73,** 629 (1954); A. M. Akkermann, D. K. De Jongh, and H. Veldstra, *Rec. Trav. Chim. Pays-Bas*, **70,** 899 (1951).
858. R. A. Abramovitch and D. H. Hey, *J. Chem. Soc.*, 1697 (1954).
859. R. A. Abramovitch and J. G. Saha, *J. Chem. Soc.*, 2175 (1964).

CHAPTER VI

Halopyridines

MAX M. BOUDAKIAN

Olin Corporation, Rochester, New York

I. Introduction

The number of new compounds listed in Tables VI-1 through VI-10 attest to the preparation of many new halopyridines and molecular addition compounds synthesized since the previous survey (1). Particularly noteworthy has been the description of improved routes to known compounds such as pentachloropyridine and pentafluoropyridine, which has generated considerable interest in their properties and conversion to new halopyridines.

II. Nuclear Halogen Derivatives

1. Preparation

A. Direct Halogenation

a. FLUORINATION.

The vapor phase (150 to 280°) fluorination of pyridine (2) or 2-fluoropyridine (3) provides only 0.2 to 3.0% yield of perfluoro-(N-fluoropiperidine). The pyridine electrochemical fluorination technique developed by Simons (4, 5) gave perfluoropiperidine in 13% yield (6–8); this method was extended to the preparation of the corresponding perfluoropiperidine derived from 2-, 3-, and 4-picoline (9) and 4-n- and isopropylpyridine (9). Perfluoropiperidines constitute useful precursors to pentafluoropyridine (7, 8, 10) and perfluoropicolines (9) by aromatization at 560 to 610° with iron or nickel (see Section III.1.G.).

b. CHLORINATION

(1) *Pyridine and Alkylpyridines.* Vapor phase chlorination routes for the preparation of chloropyridines in the presence of numerous diluents have recently been described (11–16). Scission of the trichloromethyl group during the vapor phase chlorination of polychloro-2-(trichloromethyl)-pyridine gave the tri-, tetra-, and pentachloropyridines (17). Saturated heterocyclics such as piperidine undergo simultaneous gas phase (580°) chlorination and aromatization to give pentachloropyridine in 32% yield (18a). This product was also obtained in 75% yield from the related chlorination of N-substituted piperidines, for example, N-chlorocarbonylpiperidine (18b, c). N-Acylated 2,3,4,5-tetrachloro-6-aminopyridines are obtained from a similar halogenation of cyanoethylated-piperidine (18d).

Boudakian and co-workers described a convenient route to 2-chloropyridine (**VI-2**), based on the photochemical chlorination of pyridine (**VI-1**) in refluxing carbon tetrachloride (19). A similar transformation has been

$$\text{VI-1} + Cl_2 \xrightarrow[\substack{CCl_4 \\ (62-78\%)}]{h\nu} \text{VI-2}$$

VI-1   VI-2

effected by reaction of pyridine hydrochloride with cupric chloride or cupric chloride–chlorine (20). The "swamping catalyst" technique developed by

Pearson (21), based on the chlorination of pyridine-aluminum chloride complex (VI-3), presents a useful route to 3-chloropyridine (VI-5). The contrasting inertness of pyridine hydrochloride–aluminum complex to chlorination has been attributed to immobilization of the electrons on nitrogen and deactivation of the ring toward electrophilic attack. Excess aluminum chloride in the "swamping catalyst" system may also serve to provide a medium of high dielectric constant and to transform chlorine into an active electrophile.

High yield routes to polychlorinated pyridines by liquid phase chlorination techniques have also been disclosed. The photochemical chlorination of 2-chloropyridine at 150° gave a 95% yield of 2,6-dichloropyridine (21a). Johnston obtained a 68% yield of 2,3,4,5-tetrachloropyridine by chlorination of 2-chloropyridine hydrochloride in hexachlorobutadiene as solvent at 95 to 100° (21b). Likewise, Smith prepared 2,3,5,6-tetrachloropyridine or pentachloropyridine in 85 to 96% yield by the ferric chloride-catalyzed chlorination of 2,6-dichloropyridine at 180° (21c).

Phosphorus pentachloride has been found to be a useful chlorinating agent. An improved preparation of 4-chloropyridine (VI-7) involves the reaction of N-(4-pyridyl)pyridinium chloride hydrochloride (VI-6) with this reagent (22, 23). The technique was also applied to the preparation of 4-chloro-3-picoline in 67% yield. Higher yields (97%) of pentachloropyridine from phosphorus pentachloride and pyridine were realized by modification of Sell and Dootson's initial conditions (24) by use of elevated temperatures (350°) and larger proportions of halogenating agent (25, 26). This technique was extended to the preparation in 87% yield of 2,2'-octachlorobipyridyl from 2,2'-bipyridyl (27).

(2) *Pyridine-1-oxides.* 2- And 4-chloropyridine were obtained on passage of chlorine-sulfur dioxide into a chloroform solution of pyridine-1-oxide (28). Abramovitch and Knaus (29) recently prepared 2,6-dichloro-3,4-lutidine-1-oxide (**VI-9**) by the successive metallation of 3,4-lutidine-1-oxide (**VI-8**), and chlorination.

VI-8   **VI-9**

Phosphorus oxychloride effects the simultaneous deoxygenation-chlorination (α- or γ-) of pyridine-1-oxides. While 2-chloropyridine-1-oxide (**VI-10**) gave 2,6-dichloropyridine (**VI-11**) (30), 2-picoline-1-oxide provided 4-chloro-2-picoline (**VI-12**) (31). 2-Chloro-3-chloromethylpyridine (32), 2-chloro-3-cyanomethylpyridine (33), and 4-chloro-2,6-lutidine (34) were also prepared

VI-12   **VI-10**   **VI-11**

by deoxygenation-chlorination of the *N*-oxides of 3-chloromethylpyridine, 3-cyanomethylpyridine, and 2,6-lutidine, respectively.

Taylor and Crovetti described the one-step deoxygenation-chlorination-dehydration of nicotinamide-1-oxide to 2-chloronicotinonitrile by use of phosphorus oxychloride-phosphorus pentachloride (35).

While 3,5-dibromopyridine-1-oxide undergoes deoxygenation-chlorination with sulfuryl chloride to give 2- and 4-chloro-3,5-dibromopyridine, this reagent effects α- and γ-chlorination of 3,5-diethoxypyridine-1-oxide without loss of the *N*-oxide function (36).

The conversion of 4-nitropyridine-1-oxide to 4-chloropyridine-1-oxide with acetyl chloride (37) was extended to other substrates such as the 4-nitro derivatives of 2- and 3-picoline-1-oxide (38, 42), nicotinic acid-1-oxide (38), nicotinamide-1-oxide (43), 3,5-lutidine-1-oxide (39), and 2,3,5,6-tetramethyl-pyridine-1-oxide (39). In addition to the product (**VI-14**) expected from the reaction of 4-nitro-2-picoline-1-oxide (**VI-13**) and acetyl chloride, nitrosation (with the acetyl nitrite formed in the reaction) of the methyl group occurs followed by subsequent formation of the oxime (**VI-15**) and/or nitrile (**VI-16**) (44). Displacement of the nitro group in 4-nitropyridine-1-oxides can

VI-13   **VI-14**   **VI-15**   **VI-16**

also be effected by phosphorous trichloride, hydrochloric acid, or hydrogen chloride gas (45–48).

Brown demonstrated that 2-nitropyridine-1-oxides undergo similar displacement with acetyl chloride to give 2-chloropyridine-1-oxides (49); however, phosphorous trichloride effected displacement-deoxygenation of 3-hydroxy-2-nitropyridine-1-oxide to give 2-chloro-3-hydroxypyridine (50).

3-Nitropyridine-1-oxide undergoes nuclear chlorination-deoxygenation without displacement of the nitro group: phosphorus oxychloride converted 3-nitropyridine-1-oxide to 2- and 6-chloro-3-nitropyridine (51), while 3,5-dinitropyridine-1-oxide was transformed into 2-chloro-3,5-dinitropyridine (52a). 3-Chloropyridines have been obtained as by-products in the direct acylamination of pyridine-1-oxides with imidoyl chlorides (52b).

(3) *Substituted Pyridines.* Application of the hydrogen chloride-hydrogen peroxide halogenation technique to 2-aminopyridine (53) or 6-amino-2-picoline (54) gave 2-amino-5-chloropyridine and 6-amino-3,5-dichloro-2-picoline, respectively. A convenient two-step process to penta-chloropyridine (**VI-19**) involves successive chlorination of 2,6-diamino-pyridine (**VI-17**) in hydrochloric acid to give 3,4,5-trichloro-6-hydroxy-2-pyridone (**VI-18**), followed by halogenation with phosphorus pentachloride-phosphorus oxychloride (55). While 4-cyanopyridine is converted to 4-cyanotetrachloropyridine with phosphorus pentachloride at 350°, both 2- and 3-cyanopyridine form pentachloropyridine under the same conditions (56). In contrast, gas phase chlorination of 2-, 3-, or 4-cyanopyridine can

VI-17   **VI-18**   **VI-19**

be controlled to give the corresponding mono-, di-, or tetrachlorocyano-pyridine (15, 57, 57a, b).

t-Butyl hypochlorite effected chlorination of 2-pyridone to give 5-chloro-2-pyridone (58).

Reinvestigation of the isonicotinic acid-thionyl chloride reaction gave only 2-chloroisonicotinic acid (59); none of the previously reported 3-chloro isomer was found (60).

c. BROMINATION

(1) *Pyridines and Alkylpyridines.* Facile bromination of pyridine can be accomplished in fuming sulfuric acid (65% SO_3) to give high yields of 3-bromopyridine (61, 62). [Negligible bromination occurred in 90% sulfuric acid or in the presence of silver sulfate (62)]. Sulfur trioxide serves to form a neutral adduct with pyridine, thereby causing less ring deactivation than that in the pyridinium ion (i.e., in 90% sulfuric acid). In addition, higher concentrations of the active electrophile (bromine cation?) are present in the fuming sulfuric acid system. This promising β-bromination technique was extended to alkylpyridines such as 2-, 3-, and 4-picoline and 2,6-lutidine (61, 63). The comparable "swamping catalyst" method, involving aluminum chloride instead of sulfur trioxide, has been applied to the liquid phase β-bromination of 2-, 3-, and 4-picoline (21, 64).

Garcia and his co-workers effected the β-bromination of pyridine in boiling thionyl chloride or sulfur monochloride to give a 40% yield of 3,5-dibromopyridine (**VI-20**) (65). 5-Bromonicotinic acid has been prepared in

75 % yield by the method above. Sulfuryl chloride or phosphorus oxychloride can also be employed in the bromination procedure above (66).

Giam and Abbott described a novel β-halogenation technique involving bromination of lithium tetrakis(N-dihydropyridyl)aluminate (prepared from pyridine and lithium aluminum hydride) to give a 41 % yield of 3-bromopyridine (66a).

The hydroxyl group in 3-hydroxypyridine and in the corresponding 1-oxide exhibits a powerful directional influence toward 2-substitution during alkaline bromination (in the presence of 10 % sodium hydroxide) to give the corresponding 2-bromopyridine (67). Bromination of 2,4-dihydroxypyridine, 2-ethoxy-4-hydroxypyridine, and 4-ethoxy-2-hydroxypyridine resulted in bromination at the 3-position (68). These observations have been interpreted in terms of the bromination of the pyridone forms. In contrast,

VI-21

2,4-diethoxypyridine undergoes substitution at the 5-position. The hydrogen bromide-hydrogen peroxide technique was applied to 2-aminopyridine to give 2-amino-5-bromopyridine in 70 % yield (53).

den Hertog observed the strong directing influence of reactor packing in the vapor phase bromination of 2,6-dibromopyridine above 450°: with ferrous bromide-pumice, 3- and 5-substitution occur to give 2,3,6-tri- and 2,3,5,6-tetrabromopyridine; with pumice packing, 4-substitution results to give 2,4,6-tribromopyridine (69). Iodine enhanced the rate of 4-bromination. Boudakian and co-workers observed predominant α-halogenation during the gas phase reaction of pyridine with bromine chloride (carbon tetrachloride diluent) to give 2-bromopyridine (**VI-22**) and 2-chloropyridine (**VI-23**) in 75 and 21 % yields, respectively (70a, b). The altered product

orientation for the bromination of **VI-1** with increasing temperature has been generally attributed to a change from electrophilic (300°) to a radical-type bromination (500°) (71). Several investigators have recently suggested that the relative thermodynamic stability of 2- and 3-bromopyridine may play a significant role in determining such orientation at 500° (72). The gas phase (600°) reaction of 2-, 3-, or 4-cyanopyridine with excess bromine gave as new products the corresponding tetrabromocyanopyridines (72a).

A new route to 4-bromopyridine features the reaction of N-(4-pyridyl)-pyridinium chloride hydrochloride and phosphorus pentabromide (22).

Carbon tetrachloride-soluble N-bromohydantoins exhibit comparable bromination activity as the less-soluble N-bromosuccinimide and have thus been recommended for the halogenation of substrates such as pyridine (73).

(2) *Pyridine-1-oxides.* Bromination of pyridine-1-oxide can be effected in acetic acid-sodium acetate-chloroform to give 3,5-dibromopyridine-1-oxide (35% yield) (74) or in 90% sulfuric acid-silver sulfate to form 2- and 4-bromopyridine-1-oxide (10% yield, 1:2 ratio) (75). Bromination is resisted in the presence of iron powder (76) or 90% sulfuric acid alone (75). Fuming sulfuric acid alters the orientation to give 3-bromopyridine-1-oxide; this has been attributed to the formation of the stable sulfur trioxide-pyridine-1-oxide adduct which is deactivated at the 2-, 4-, and 6-positions toward electrophilic attack (77). Bromine water converted 1-hydroxy-2-pyridone to 3,5-dibromo-1-hydroxy-2-pyridone (78).

Ring bromination, as well as coupling, result from the reaction of pyridine-1-oxide (**VI-24**) with n-butyllithium and bromine (29).

4-Nitropyridine-1-oxides are converted to the corresponding 4-bromopyridine with phosphorous tribromide (45, 64, 78) or to the N-oxide with acetyl bromide (42, 79, 80). 4-Nitropyridine-1-oxide follows different paths when heated with hydrobromic acid in aqueous or in acetic acid media, giving 3,5-dibromo-4-pyridone and 4-bromopyridine-1-oxide, respectively (**VI-25**) (81, 82). Phosphorus oxybromide effected nuclear bromination-

deoxygenation of 4-nitropyridine-1-oxide to give 2-bromo-4-nitropyridine (83).

d. IODINATION. The iodination of 2-, 3-, and 4-pyridinol in aqueous sodium carbonate gave 5-iodo-2-pyridone, 2-iodo-3-pyridinol, and 3-iodo-4-pyridone (VI-26) (84). Above 255°, VI-26 disproportionates to give 3,5-diiodo-4-pyridone (VI-27).

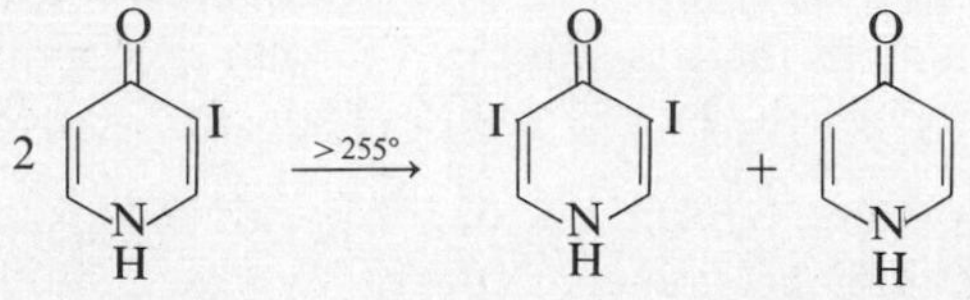

VI-26   VI-27

B. Halogen Molecular Addition Compounds

a. BONDING. Mulliken's charge transfer theory has been applied to bonding in halogen molecular addition compounds (85, 86). Solutions of pyridine and iodine gives rise to the following equilibria (VI-28):

$$Py + I_2 \rightleftharpoons \underset{\substack{\text{"outer"}\\ \text{complex"}}}{Py \cdot I_2} \rightleftharpoons \underset{\substack{\text{"inner"}\\ \text{complex"}}}{(PyI)^+ I^-} \rightleftharpoons (PyI)^+ + I^-$$

VI-28

X-ray studies (87) suggest the following structures for the 2:1 and 1:1 complexes from 4-picoline and iodine (88): The "outer complex" (i.e., undis-

2:1 complex

VI-29

1:1 complex

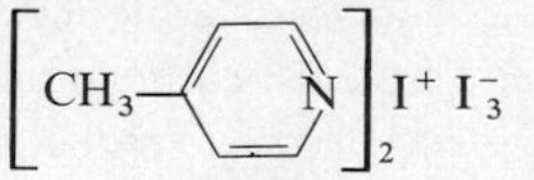

VI-30

sociated) structures are arbitrarily employed in the text and Tables VI-1, VI-2, and VI-3.

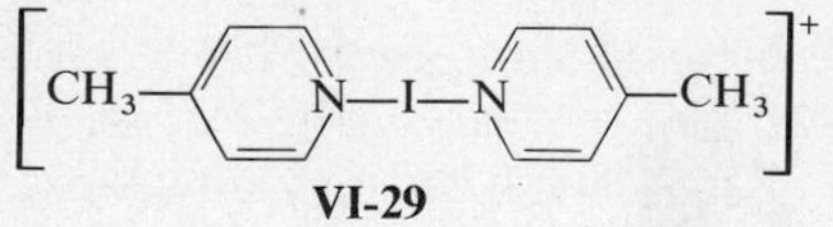

b. FLUORINE ADDITION COMPOUNDS. Fluorine and pyridine react in trichlorofluoromethane at $-80°$ to give a solid adduct possessing ionic character; violent decomposition occurs above $-2°$ to give 2-fluoropyridine (89). The adduct also serves as a fluorinating agent in the above solvent at $-10°$ to effect 1,2-difluoro addition to chlorinated olefins.

TABLE VI-1. Pyridine-Halogen Molecular Addition Compounds[a]

Compound	M.p. (°C)	Ref.
(Pyridine)$_2$·BrF	Melts with dec.	95
Pyridine·2IBr	53	92
Pyridine·2ICl	35	92, 269
Pyridine·ICl$_3$	175–177	101
Pyridine·2ICl$_3$	140	93
(Pyridine)$_2$·ICl$_3$	69[c]	93
Pyridine·IF	80–97 (dec.); 110 (dec.)	94, 96c
(Pyridine)$_2$·IF	112–132 (dec.)	95
Pyridine·IF$_3$[b]	166–168	96–98
(Pyridine)$_2$·IF$_3$	165	96b
Pyridine·IF$_5$	9 (est.)	91a, b

[a] Arbitrarily shown as undissociated complex.

[b] A product, m.p. 166°, designated as $[I(Py)_2]^+[IF_6]^-$, dipyridineiodine (I) hexafluoroiodate, has also been reported (97a, b).

[c] Incongruent.

TABLE VI-2. Molecular Addition Compounds of Substituted Pyridines[a]

Compound	M.p. (°C)	Ref.
3-Iodo-2,6-lutidine·Cl$_2$	240	285
3-Iodo-sym-collidine·Cl$_2$	251	285
3-Picoline·Br$_2$	41	270
4-Picoline·Br$_2$	73	270
4-n-Amylpyridine·Br$_2$	85–87.5	102
4,4'-Bipyridyl·2Br$_2$	Not given	103
(2-BrC$_5$H$_4$NH)$^+$Br$^-$·(2-BrC$_5$H$_4$N·Br$_2$)	87–91 (dec.)	105
(2-ClC$_5$H$_4$NH)$^+$Br$^-$·(2-ClC$_5$H$_4$N·Br$_2$)	74–83 (dec.)	105
2-Picoline·I$_2$	38	270
3-Picoline·I$_2$	62	270
4-Picoline·I$_2$	83.2–83.4	88, 270
(4-Picoline)$_2$·I$_2$	223–224	88
2-Fluoropyridine·I$_2$	Not given	91
4,4'-Bipyridyl·I$_2$	Not given	103
4-Picoline·BrCl	89–90	432
4-n-Amylpyridine·BrCl	105.0–105.5	102
2,4-Lutidine·BrCl	96–98	432
2,5-Lutidine·BrCl	94–96	432
2,6-Lutidine·BrCl	99–100	432
3,4-Lutidine·BrCl	93–94	432
3,5-Lutidine·BrCl	106–108	432
2,3,6-Collidine·BrCl	102–104	432
2-Chloropyridine·IBr	44–45	91
2-Fluoropyridine·IBr	44–45	91
3-Bromopyridine·IBr	77–78	91

Table VI-2 (*Continued*)

Compound	M.p. (°C)	Ref.
3-Chloropyridine·IBr	47	91
3-Fluoropyridine·IBr	70–72	91
4-Chloropyridine·IBr	193–195	91
2-Picoline·IBr	67–68	90
4-Picoline·IBr	78.0–78.5	92
2,6-Lutidine·IBr	106–108	90
4-*n*-Amylpyridine·IBr	100.5–101.0	102
4,4′-Bipyridyl·IBr	260–265	103
2-Chloropyridine·ICl	80–82	91
2-Fluoropyridine·ICl	56	91
3-Bromopyridine·ICl	90–92	91
3-Chloropyridine·ICl	56	91
3-Fluoropyridine·ICl	95–97	91
4-Chloropyridine·ICl	224–226	91
2-Picoline·ICl	77.5	90
3-Picoline·ICl	55–56	440
4-Picoline·ICl	107–108	101
2,6-Lutidine·ICl	112–113	90
2,2′-Bipyridyl·2ICl	95; 140 (starts to decompose at 100°)	100, 104
4,4′-Bipyridyl·2ICl	Decomposes at 230°	103
4,4′-Dimethyl-2,2′-bipyridyl·2ICl	130–140	100
2-Fluoropyridine·ICl$_3$	56–65	91
2-Picoline·ICl$_3$	131–133	101
4-Picoline·ICl$_3$	133–134	101
2,6-Lutidine·ICl$_3$	103–105	101
2,2′-Bipyridyl·IF	110 (dec.)	96c
2,2′-Bipyridyl·IF$_3$	142 (dec.)	96d
2-Fluoropyridine·IF$_5$	Not given	91

[a] Arbitrarily shown as undissociated complex.

TABLE VI-3. Salts of Pyridine Molecular Addition Compounds

Compound	M.p. (°C)	Ref.
Pyridine·HICl$_2$·ICl	85–86	99
2,6-Lutidine·HICl$_2$·ICl	Oil	99
2,2′-Bipyridyl·HCl·ICl	139–140	100, 104
2,2′-Bipyridyl·HICl$_2$·ICl	65–67	99
4,4′-Bipyridyl·2HCl·2ICl	191–194	103
4,4′-Bipyridyl·2HBr·2IBr	Not given	103
4,4′-Dimethyl-2,2′-bipyridyl·HICl$_2$	153	99
4,4′-Dimethyl-2,2′-bipyridyl·HICl$_2$·ICl	57–59	99

c. INTERHALOGEN ADDITION COMPOUNDS. Spectrometric studies by Popov and Rygg indicate that the stability of interhalogen pyridine complexes in carbon tetrachloride is associated with the acid strength of the interhalogen (ICl > IBr > I_2) (90). The strengths of pyridine·ICl complex (in nonpolar solvents) are also dependent on the location of substituents in the pyridine ring. Thus, the stability of chloropyridine-interhalogen complexes decreased in the order 4- > 3- > 2-chloro, as would be expected from the increasing magnitude of the inductive effect as chlorine approaches nitrogen (91). Likewise, increased melting points of 1:1 interhalogen-pyridine adducts (ICl_3 > ICl > IBr) have been associated with increase in electron acceptor strength of the interhalogen (92, 93). Polar solvents (e.g., acetonitrile) promote ionic dissociation of these adducts (90).

$$2\ C_5H_5N\cdot ICl \xrightarrow{CH_3CN} (C_5H_5N)_2I^+ + ICl_2^- \qquad \text{VI-31}$$

Complexes of iodine fluorides with pyridine have also attracted attention. Both pyridine·IF (94, 96c) and (pyridine)$_2$·IF (95, 96a) have been synthesized. Pyridine·IF_3 has been prepared from pyridine and iodine trifluoride in trichlorofluoromethane (96a, b), or by electrolysis of (pyridine)$_2$·IF in acetonitrile (98). [Recent studies by Meinert suggest the possible existence of dipyridineiodine (I) hexafluoroiodate, $[I(Py)_2]^+[IF_6]^-$ (97a, b)]. Phase diagrams indicate that the 1;1 iodine pentafluoride-pyridine complex is not stable (91a, b). A solid of unknown composition was formed from the violent reaction of iodine heptafluoride and pyridine (98a).

While (pyridine)$_2$·BrF was prepared from pyridine, bromine, and silver fluoride in acetonitrile (95), no stable pyridine-bromine trifluoride complex could be formed at temperatures as low as $-42°$ (91). However, the thermally sensitive pyridine-bromine pentafluoride adduct dissolved in acetonitrile at -30 to $-40°$ without decomposition (97c).

d. ACID SALTS OF ADDITION COMPOUNDS. Yagi and Popov isolated a new class of polyhalide complex ions (suggested to be $I_2Cl_3^-$) from salts of interhalogen addition complexes (99, 100).

$$C_5H_5N\cdot HCl\cdot ICl + ICl \xrightarrow{C_2H_4Cl_2} C_5H_5N\cdot HICl_2\cdot ICl \qquad \text{VI-31A}$$

e. ADDITION COMPOUNDS OF SUBSTITUTED PYRIDINES. Table VI-2 summarizes novel halogen addition compounds, based on halopyridines (91), alkylpyridines (88, 90, 92, 101, 102), and bipyridyls (101, 103, 104). An unusual amine-halogen complex (**VI-32**) precipitated from chloroform solutions of 2-halopyridines and bromine (105). (Hydrogen bromide in the 2-halopyridinium bromide may arise from bromination of chloroform).

$$(XC_5H_4NH)^+Br^-\cdot XC_5H_4N\cdot Br_2$$
$$X = Cl,\ Br$$

VI-32

C. From Pyridinols, Pyridones, and Related Starting Materials

2- (**VI-2**) And 3-chloropyridine (**VI-5**) were prepared in 84 to 88% yield by the thermal decomposition of mixed tetraaryloxyphosphorus mono-halides (**VI-33**) (ArOH = 2- or 3-hydroxypyridine) (106).

$$\text{ArOH} \xrightarrow{\text{PCl}_5} (\text{ArO})_3\text{PCl}_2 \xrightarrow{\text{Ar'OH}} (\text{Ar'O})_3(\text{Ar'O})\text{PCl} \xrightarrow{250°} \textbf{VI-2} + (\text{ArO})_3\text{PO} \quad \textbf{VI-34}$$

or

VI-33   **VI-5**

During the conversion of 1-hydroxy-4-pyridones to 4-halopyridines, the proper choice of halogenating agent can also permit nuclear chlorination: 4-chloropyridine (phosphorus oxychloride/trace phosphorus pentachloride) or 2,4-dichloropyridine (sulfuryl chloride) from 1-hydroxy-4-pyridone; 3,4-dichloropyridine (phosphorus oxychloride) or 2,3,4-trichloropyridine (sulfuryl chloride) from 3-chloro-1-hydroxy-4-pyridone (107). [Hydroxyl groups can be protected during simultaneous nuclear chlorination-deoxygenation of hydroxypyridine-1-oxides by phosphorus oxychloride (108)]. Substituted 1-hydroxy-2-pyridones also undergo bromination-deoxygenation with phosphorus oxybromide: 2,3,5-tribromopyridine from 3,5-dibromo-1-hydroxy-2-pyridone (78), 2-bromo-3,5-dinitropyridine from 1-hydroxy-3,5-dinitro-2-pyridone (78), and 2-bromo-5-nitropyridine from 1-hydroxy-5-nitro-2-pyridone (109). 4-Methoxynicotinamide-1-oxide (**VI-35**) gave 2,4-dichloronicotinonitrile (**VI-36**) through a sequence involving displacement of the alkoxy group, nuclear chlorination-deoxygenation, and dehydration by the phosphorus oxychloride-phosphorus pentachloride reagent (38).

VI-35   **VI-36**

Cyanuric fluoride has served as a fluorinating agent for the preparation of 2-fluoropyridine from 2-pyridone (110). The phosphorus pentachloride-phosphorus oxychloride combination has been extended to substrates such as 1-methyl-3,4,5,6-tetrachloro-2-pyridone (111), 3,4-dichloro-5,6-dimethyl-2-pyridone (112), and 6-methyl (or phenyl)-4-trifluoromethyl-2-pyridone (113) to give pentachloropyridine, 2,3,4-trichloro-5,6-dimethylpyridine, and 6-methyl (or phenyl)-4-trifluoromethyl-2-chloropyridine, respectively. Phosphorus pentachloride was employed to convert 3,4,5-trichloro-2-pyridone to 2,3,4,5-tetrachloropyridine (114). 2,3-Dibromopyridine, previously iso-lated as a minor product from the gas phase bromination of pyridine (115), can be prepared from 3-bromo-2-pyridone and phosphorus oxybromide in 65% yield (116, 117). 4-Pyridone has been successively converted to 3,5-dibromo-4-pyridone (with bromine), 2,3,5,6-tetrabromo-4-pyridone (bromi-nation in 70% oleum), and pentabromopyridine (phosphorus pentabromide) (118).

Reinvestigation of the glutarimide (**VI-37**)-phosphorus pentachloride reaction gave a mixture of di-, tri-, tetra-, and pentachloropyridine (119).

VI-37   (or tautomer)

$$C_5H_2Cl_3N;$$
$$C_5HCl_4N;$$
$$C_5Cl_5N$$

The formation of pentachloropyridine becomes important when the reaction is conducted at 275°C for 24 hr.

D. From Aminopyridines

Finger and co-workers prepared 2,3-difluoropyridine and 2,6-difluoro-pyridine by the Schiemann reaction on 3-amino-2-fluoropyridine and 6-amino-2-fluoropyridine, respectively (120, 121). Earlier warnings con-cerning the thermal instability of 3-pyridyldiazonium tetrafluoroborate (122) were again confirmed by a recent detonation involving this salt (123). An alternate diazotization route to 2-fluoropyridine featured decomposition of 2-pyridyldiazonium hexafluorophosphate (124). The *in situ* decomposition of these unstable salts yields significant amounts of 2-pyridones, presumably arising from the acid-catalyzed hydrolysis of the corresponding 2-fluoro-pyridine (124a).

Talik has prepared numerous substituted 4-halopyridines by diazotization of the corresponding 4-aminopyridine (125–128). Substituted 4-iodo-pyridines have been synthesized by successive diazotization of the 4-amino-pyridine-1-oxide, addition of potassium iodide, and deoxygenation with phosphorous trichloride (129). Diazotization-bromination (cuprous bromide in hydrobromic acid) of 4-aminotetrafluoropyridine to 4-bromotetrafluoro-pyridine could be successfully accomplished in aqueous hydrofluoric acid medium (130). (The latter serves to offset loss of fluoride ion during diazotization of negatively substituted amines).

Numerous halopyridines have been obtained by replacement of the nitramino group in 2-, 3-, or 4-(nitramino)pyridines by halogen derived from phosphorus halides (131a–c).

$$\text{VI-38} \quad \underset{\text{N}}{\bigcirc}\text{NHNO}_2 \quad \xrightarrow[(76\%)]{\text{PCl}_5} \quad \underset{\text{N}}{\bigcirc}\text{Cl} + \text{N}_2\text{O} + \text{HCl} + \text{POCl}_3$$

E. Halogen Exchange

The halogen-exchange technique (132) for the preparation of fluoronitropyridines from halonitropyridines and potassium fluoride in polar solvents was extended to the preparation of fluorocyanopyridines (133). This technique was also applied to halopyridines devoid of nitro or cyano activating groups, thereby permitting the synthesis of new fluoropyridines or providing alternate routes to known compounds. With polar solvents such as tetramethylene sulfone ("sulfolane"), dimethyl sulfone or N-methyl-2-pyrrolidone, exchange fluorination of α- and/or γ-halogens was effected: 2-chloro- and 2-bromopyridine (134); 2,3-, 2,5-, and 2,6-dichloropyridine (134, 135); 5-bromo-2-chloropyridine (134); 2,3,5-trichloropyridine (134); 2,3,5,6-tetrachloropyridine, as well as mixtures of isomeric tetrachloropyridines (25); and pentachloropyridine (25, 26, 136, 137). While β-halogens resisted fluorination in the lower chlorinated pyridines, β-fluorination of the tetra- and pentachloropyridines was sluggish.

The halogen-exchange technique was further modified by carrying out the fluorination in the absence of solvent with selected substrates. Thus pentachloropyridine (VI-39) was readily converted by potassium fluoride to pentafluoropyridine (VI-40) in 69 to 83 % yield at 480 to 500° in an autoclave (25, 26, 136, 137). The solvent-free fluorination technique was also applied to

$$\underset{\text{VI-39}}{\overset{\text{Cl}}{\underset{\text{Cl}}{\overset{\text{Cl}}{\bigcirc}}}\text{Cl}} \quad \xrightarrow[480-500°]{\text{KF}} \quad \underset{\text{VI-40}}{\overset{\text{F}}{\underset{\text{F}}{\overset{\text{F}}{\bigcirc}}}\text{F}}$$

completely halogenated substrates such as 4-cyanotetrachloropyridine (56) and octachloro-2,2'-bipyridyl (27) to give the corresponding perfluorinated product.

This type of fluorination does not follow a consistent pattern with partially chlorinated pyridines: degradation occurred with 2-chloropyridine (138) or 2,3,4,6- and 2,3,5,6-tetrachloropyridines (above 420°) (25, 136) and partial halogen substitution was observed with the tetrachloropyridines above (below 400°) (25, 136) and complete halogen exchange with 2,6-dichloropyridine (139).

$$\underset{\textbf{VI-41}}{Cl\text{—}\fbox{N}\text{—}Cl \quad \xrightarrow[\substack{400° \\ (80\%)}]{KF} \quad F\text{—}\fbox{N}\text{—}F}$$

The solvent-free fluorination technique was recently modified by Boudakian (138) through the use of potassium bifluoride. While charring occurred with 2-chloropyridine-potassium fluoride at 315°, a 74% yield of 2-fluoropyridine (**VI-42**) was obtained with potassium bifluoride at 315° (4 hr). This is to be contrasted with the 21-day period required for the conversion of 2-chloropyridine to **VI-42** with potassium fluoride in polar solvents (134).

$$\underset{\textbf{VI-42}}{\fbox{N}\text{—}Cl \;+\; KHF_2 \;\xrightarrow[(74\%)]{315°}\; \fbox{N}\text{—}F \;+\; KCl \;+\; HF}$$

An atmospheric pressure solvent-free fluorination technique was recently described by Fielding and co-workers (140, 141) who passed 3,5-dichloro-trifluoropyridine and other hetero-pentahalopyridines at 750 to 770° through a fused salt melt (45% potassium fluoride:55% potassium chloride).

Other halide ions can also serve as nucleophiles. For example, the reaction of pentafluoropyridine (**VI-40**) with sodium iodide in dimethyl formamide (142), or with potassium chloride at elevated temperatures (141) gave 4-iodo- or 4-chloro-tetrafluoropyridine (**VI-43**), respectively. Heterohalogenated pyridines, for example 3-chloro-2,4,5,6-tetrafluoropyridine, also exhibit similar lability at the 4-position towards halide ions (142). [While 4-substitution occurred when pentachloropyridine was treated with sodium bromide in dimethylformamide to give 4-bromo-tetrachloropyridine (142a), reaction of this substrate with potassium fluoride in tetramethylene sulfone gave 2-fluorotetrachloropyridine (142b)].

$$\textbf{VI-40} \;+\; NaI \;\xrightarrow{DMF}\; \underset{\substack{\textbf{VI-43} \\ X = I,\,Cl}}{F\text{—}\overset{X}{\underset{N}{\fbox{}}}\text{—}F} \;\xleftarrow[temp.]{high}\; KCl \;+\; \textbf{VI-40}$$

Musgrave recently observed that pentafluoropyridine and 3,5-dichloro-trifluoropyridine undergo α- and γ-halogen exchange reactions with hydrogen halides (HCl, HBr, and HI) in sulfolane or aluminum halides (142c, d).

F. Involving Ring Closure

Roedig and co-workers developed several ring closure routes to penta-chloropyridine (**VI-39**), based on amides of pentachloro-2,4-pentadienoic acid (**VI-44, VI-45**) (111, 143). The oxime of perchloropenta-1,3-dien-5-al

(**VI-46**) was cyclized to 2,3,4,5-tetrachloropyridine-1-oxide (**VI-47**), which was subsequently converted to pentachloropyridine (**VI-39**) with sulfuryl chloride (144). 2-Alkyl- or aryl-tetrachloropyridines (**VI-49**) were prepared from perchloropentadienonitrile (**VI-48**) and Grignard reagents (114).

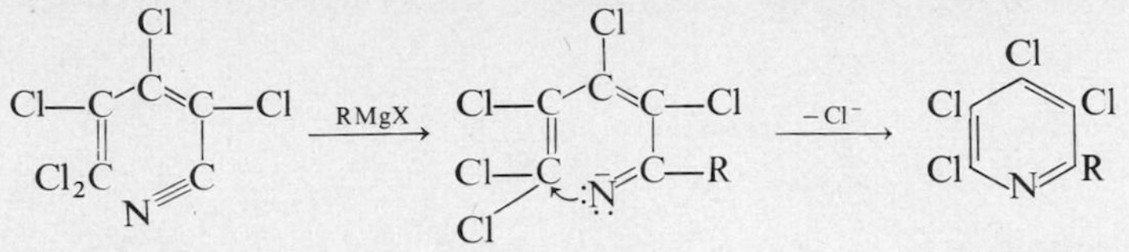

VI-48   **VI-49**

The reaction of **VI-48** and alkoxides gave 2,6-dialkoxy-3,4,5-trichloro-pyridines (145); 3,4,5-trichloro-6-hydroxy-2-pyridone was synthesized by successive cyclization of pentachloro-2,4-dienoic acid to give perchloro-2-pyrone, followed by amination (111).

Grohe and Roedig (112) transformed β-ketocarboxylic esters/trichloro-acrylonitrile to 2,3-disubstituted-3,4,5-trichloropyridine (**VI-50**) by a three-step sequence:

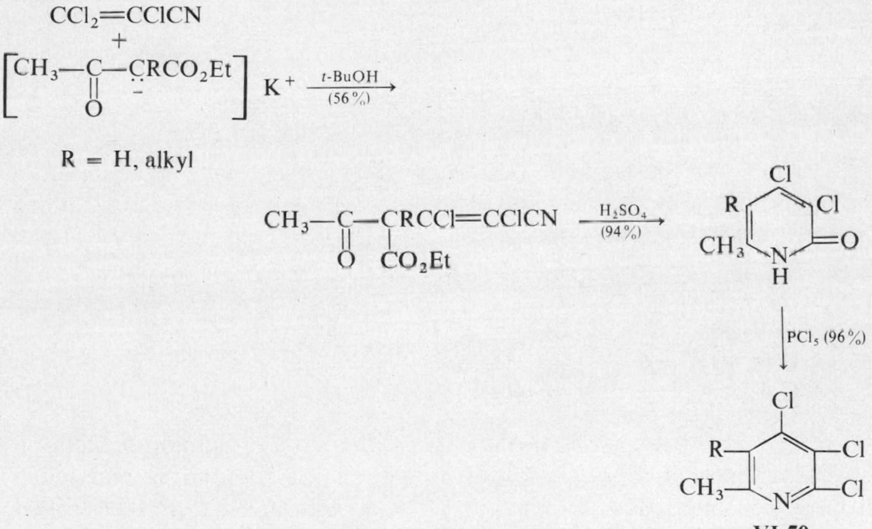

R = H, alkyl

VI-50

Another cyclization route involving nitriles features the recently dis-covered transformation of γ-cyanocarboxylic acids, $(NC)CH=CRCH_2COX$, by hydrogen chloride to give substituted 6-chloro-2-pyridones; the latter can then be converted to the corresponding 2,6-dichloropyridine with phosphorus oxychloride (111a).

The Diels-Alder reaction has been applied to the preparation of 4-chloro-and 5-chloro-2-picolinonitrile from cyanogen and chloroprene (146, 148), as well as 2-bromo-3,4,5,6-tetrafluoropyridine (VI-52) from cyanogen bromide and perfluorocyclohex-1,3-diene (VI-51) (149). Esters of 3,4,5-trichloro-6-benzoylpicolinic acid (VI-54) have been similarly prepared by

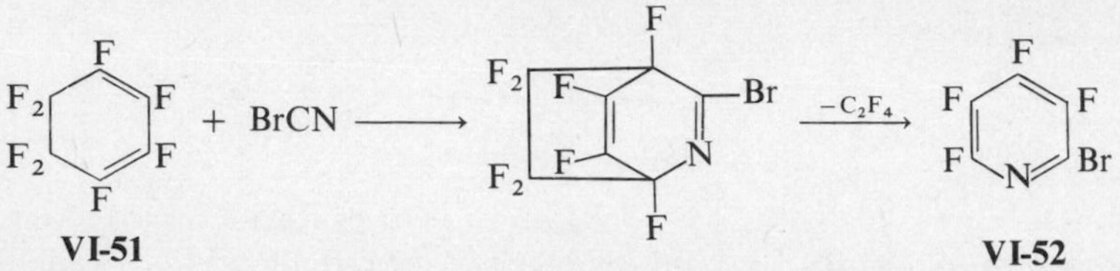

VI-51   VI-52

Jaworski from dialkyl ketals of 1,2,3,4-tetrachlorocyclopentadienone (VI-53) and benzoyl cyanide (150, 151).

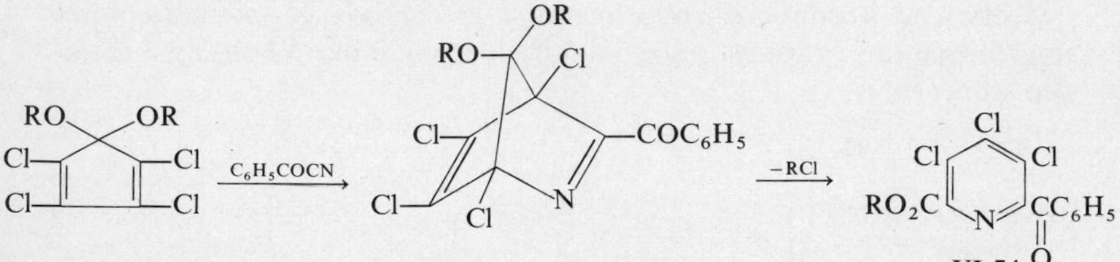

VI-54

Johnson and co-workers (152) described new routes to 2-amino-6-halopyridines (VI-57) which involve the treatment of 3-hydroxyglutaro-nitrile (VI-55) or glutacononitrile (VI-56) with hydrogen halides. Hydrogen

NCCH$_2$C(OH)RCH$_2$CN
VI-55
R = H, alkyl, aryl
X = Br, I

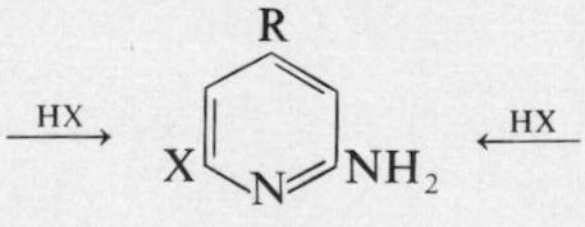

$\xrightarrow{HX}$

VI-57

$\xleftarrow{HX}$

NCCH=C(R)CH$_2$CN
VI-56

bromide converted malononitrile or its dimer to 2,4-diamino-6-bromo-5-cyanopyridine (VI-58) (153a, b). However, the corresponding reaction of isopropylbromide and malononitrile gave 6-bromo-3-cyano-2,4-bis(iso-propylamino)pyridine (153b).

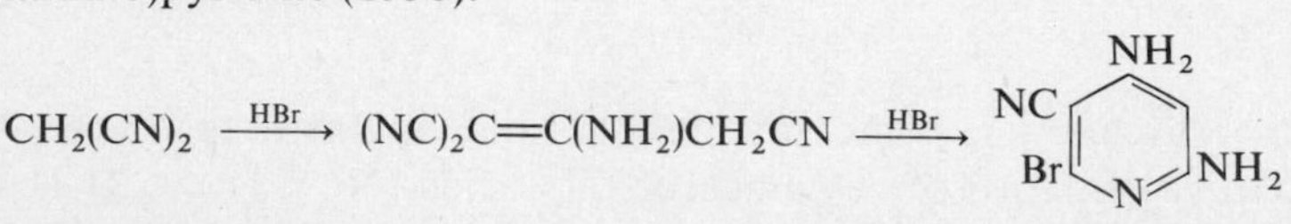

CH$_2$(CN)$_2$ $\xrightarrow{HBr}$ (NC)$_2$C=C(NH$_2$)CH$_2$CN $\xrightarrow{HBr}$

VI-58

Elvidge treated malonyl chloride with a variety of nitriles to give sub-stituted 2-chloropyridines; for example, 2-chloro-3-cyano-4-hydroxy-6-pyridone was obtained from malononitrile, while 2-chloro-3-(p-chloro-phenyl)-4-hydroxy-6-pyridone was formed from p-chlorophenylacetonitrile

(153c-e). Salts of substituted 1,1,3,3-tetracyanopropenes have likewise been transformed to 2-amino-6-halo-3,5-dicyanopyridines (**VI-59**) (154).

$$\text{Na}[(NC)_2C{=}C(R)C(CN)_2] \xrightarrow{HX}$$
$$R = H, OR, CN, Br, NH_2, C_6H_5$$
$$X = Cl, Br, I$$

VI-59

Successive ring scission and cyclization during the high temperature (600°C) chlorination of 1,2-dicyanocyclobutane provided high yields of the isomeric tetrachlorocyanopyridines (155). These products were also obtained from the vapor phase chlorination of adiponitrile or 1,4-dicyano-1-(or -2-)butene; n-valeronitrile gave pentachloropyridine (155a).

Treatment of glutaraldehydes or cyclic enol precursors with ammonium chloride and an oxidizing agent provides chlorinated pyridines: 3-chloro- and 3,5-dichloropyridine from glutaraldehyde or 2-methoxy-3,4-dihydropyran; and 3-chloro-4-picoline and 3,5-dichloro-4-picoline from β-methylglutaraldehyde or 2-methoxy-4-methyl-3,5-dihydropyran (156, 157). 5-Chloro-3-hydroxy-2-pyridone was obtained from furfural, chlorine, and sulfamic acid (158).

The high temperature (450°) reaction of carbon tetrachloride and ammonia gave a mixture of pentachloropyridine (27%) and cyanuric chloride (70%) (159).

G. Other Methods

a. RING ENLARGEMENT REACTIONS. Jones and Rees converted pyrroles to substituted 3-halopyridines in 55 to 90% yield by reaction with dichlorocarbene generated from sodium trichloroacetate in neutral aprotic solvents, such as 1,2-dimethoxyethane (160). For example, a 70% yield of 3-chloro-2,6-lutidine (**VI-61**) was realized from 2,5-dimethylpyrrole (**VI-60**) and dichlorocarbene by this technique, in contrast to the 9% yield obtained under basic conditions (chloroform/sodium ethoxide). Nicoletti converted

VI-60   VI-61   VI-62

2-dichloromethyl-2,5-dimethylpyrrolenine (**VI-62**) to **VI-61** with excess n-butyllithium (162).

b. DISPLACEMENT OF THE HALOMERCURI GROUP. The previously reported claim (163) that 4-bromopyridine-1-oxide represented the main product from the successive mercuration-bromination of pyridine-1-oxide was shown by Van Ammers and den Hertog to be incorrect; instead, 2- and 3-bromo- and 2,6-dibromopyridine-1-oxides were found to be the major products (164, 165). Halogenation (bromine and iodine) of 3-chloromercuri-4-aminopyridine or 3,5-bis(chloromercuri)-4-aminopyridine gave the corresponding 3-halo- or 3,5-dihalo-4-aminopyridine (166).

c. DEOXYGENATION OF PYRIDINE-1-OXIDES. Halopyridines have been prepared from the corresponding N-oxide by application of various reduction systems: Raney nickel/hydrogen (167); Urushibara nickel/hydrogen (168); palladium-on-carbon/hydrogen (169); ammonia (169); iron-acetic acid (36, 78, 164, 165); sodium borohydride-aluminum trichloride (170); nitric oxide-sulfuric acid (171); sulfur dioxide (172); phosphorous trichloride (45, 129, 173); phosphorous tribromide (45); phosphoryl bromide (83); benzenesulfenyl chloride (174); and ferrous sulfate-ammonium hydroxide (128).

While deoxygenation of 4-iodo-3-picoline-1-oxide with phosphorous trichloride in chloroform gave 4-iodo-3-picoline, the corresponding deoxygenation in chloroform-water furnished 4-chloro-3-picoline (129).

Half-wave potentials ($-E_{1/2}$, V) from the polarographic reduction of halopyridine-1-oxides at pH 5.0 have been determined: 3-bromo, 1.040; 3-chloro, 1.071; 3-fluoro, 1.144; 4-bromo, 1.094; 4-chloro, 1.174 (129a).

2. Physical Properties

Properties of new compounds synthesized since the previous review (1), well as the inclusion of new methods to selected known compounds have been compiled in the following tables:

    Halopyridines (Table VI-4)
    Alkylhalopyridines (Table VI-5)
    Arylhalopyridines (Table VI-6)
    Haloalkylhalopyridines (Table VI-7)

The base strength of more halogenated pyridines were measured by Fischer and co-workers (175): 4-chloropyridine (pK_a 3.83); 4-bromopyridine (3.75); 3,5-dichloropyridine (0.67); and 3,5-dibromopyridine (0.82). The pK_a of pentafluoropyridine has been estimated to be -11 by use of ^{19}F NMR (175a).

Cumper and Vogel measured the dipole moments of 15 isomeric mono-, di-, and tri-halogenated pyridines; these have been interpreted on the basis of resonance interaction of the halogen with the pyridine ring (176).

TABLE VI-4. Properties of the Halopyridines

Compound	Method of preparation	Physical properties	Ref.
2,3-Difluoropyridine	From Schiemann diazotization of 3-amino-2-fluoropyridine	b.p. 118°; n_D^{25} 1.4420	121
2,4-Difluoropyridine	From 2,4-dichloropyridine and KF in sulfolane (ethylene glycol initiator)	b.p. 104–105	463
2,6-Difluoropyridine	From Schiemann diazotization of 6-amino-2-fluoropyridine	b.p. 124.5°; n_D^{25} 1.4349	120, 121
2,4-Diiodopyridine	Diazotization of 4-amino-2-iodopyridine; 2-iodo-4-nitraminopyridine and PI₃	m.p. 74°; HgCl₂ salt, m.p. 239°	127, 131c
3,4-Diiodopyridine	Diazotization of 4-amino-3-iodopyridine; 3-iodo-4-nitraminopyridine and PI₃	m.p. 114–115°; picrate, m.p. 143°	128a, 131c
2,4,6-Trifluoropyridine	Catalytic reduction of 3,5-dichlorotrifluoropyridine; 2,4,6-trichloropyridine and KF in sulfolane (ethylene glycol initiator)	b.p. 94–95°	180, 463
2,5,6-Trifluoropyridine	By-product from catalytic reduction of 3-chlorotetrafluoropyridine	b.p. 115–116°	180
3,4,5-Trifluoropyridine	Pyrolysis of 3,3,4,4,5,5-hexafluoropiperidine over NaF	Not described	271
2,3,5,6-Tetrafluoropyridine	From pentafluoropyridine and LiAlH₄ or cat. reduction; decarboxylation of 2,3,5,6-tetrafluoropyridine-4-carboxylic acid; 2,3,5,6-tetrafluoro-4-hydrazinopyridine and aq. CuSO₄	b.p. 102°	9, 180, 181
2,4,5,6-Tetrafluoropyridine	Catalytic reduction of 3-chlorotetrafluoropyridine	b.p. 89–90°	180
3,4,5,6-Tetrafluoropyridine	By-product from catalytic reduction of pentafluoropyridine	b.p. 87–88°	180
3-Chloro-2-fluoropyridine	2,3-Dichloropyridine, KF, and dimethyl sulfone	b.p. 94–95°/100 mm; n_D^{25} 1.5020	134
4-Chloro-2-fluoropyridine	From 2,4-dichloropyridine and KF in sulfolane (ethylene glycol initiator)	b.p. 138°	463

Table VI-4 (*Continued*)

Compound	Method of preparation	Physical properties	Ref.
5-Chloro-2-fluoropyridine	2,5-Dichloropyridine, KF and dimethylsulfoxide or -sulfone; Schiemann diazotization of 2-amino-5-chloropyridine	b.p. 88°/100 mm; n_D^{25} 1.4970	134
6-Chloro-2-fluoropyridine	2,6-Dichloropyridine and KF	b.p. 81–82°/51 mm; m.p. 24–25°	139
2-Chloro-3-fluoropyridine	Modified Schiemann diazotization (hexafluoro-phosphoric acid) on 3-amino-2-chloropyridine	b.p. 84–85°/55 mm; n_D^{27} 1.5022	465
4-Chloro-3-fluoropyridine	From 3-fluoro-4-nitropyridine-1-oxide and PCl_3; diazotization of 4-amino-3-fluoropyridine	b.p. 138–139°; picrate, m.p. 134°	128b, 272
2-Chloro-4-fluoropyridine	From 2,4-dichloropyridine and KF in sulfolane (ethylene glycol initiator)	b.p. 140°	463
3-Bromo-2-fluoropyridine	From 3-bromo-2-chloropyridine, KF, and dimethylsulfone	b.p. 76°/20 mm; n_D^{20} 1.5380	462
5-Bromo-2-fluoropyridine	5-Bromo-2-chloropyridine, KF, and dimethyl sulfone; Schiemann diazotization of 2-amino-5-bromopyridine; 2,5-dibromopyridine and KHF_2	b.p. 63°/15 mm; 162–164°/750 mm; n_D^{25} 1.5300	134, 451
4-Bromo-3-fluoropyridine	From 3-fluoro-4-nitropyridine-1-oxide and PBr_3; diazotization of 4-amino-3-fluoropyridine	b.p. 160–163°; picrate, m.p. 114°	128b, 272
5-Bromo-3-fluoropyridine	Schiemann diazotization of 5-amino-3-bromopyridine	b.p. 78°/11 mm; m.p. 24.5–25.0°	273
2-Fluoro-4-iodopyridine	Diazotization of 4-amino-2-fluoropyridine	m.p. 58°	128b
3-Fluoro-4-iodopyridine	Diazotization of 4-amino-3-fluoropyridine	m.p. 87°; picrate, m.p. 140°	128b
4-Bromo-2-chloropyridine	Diazotization of 4-amino-2-chloropyridine; 2-chloro-4-nitraminopyridine and PBr_3 or PBr_5	m.p. 26–27°; $HgCl_2$ salt, m.p. 152°	125, 131c
6-Bromo-2-chloropyridine	Diazotization of 2-amino-6-bromopyridine	m.p. 67–68°	226
4-Bromo-3-chloropyridine	From 3-chloro-4-nitropyridine-1-oxide and PBr_3; diazotization of 4-amino-3-chloropyridine; 3-chloro-4-nitraminopyridine and PBr_3 or PBr_5	m.p. 71°; picrate, m.p. 126°	45, 128a, 131c

430

Compound	Method	Properties	References
2-Bromo-4-chloropyridine	2-Bromo-4-nitraminopyridine and PCl$_5$	b.p. 98°/15 mm; HgCl$_2$ salt, m.p. 163°	131c
3-Bromo-4-chloropyridine	3-Bromo-4-nitropyridine-1-oxide and PCl$_3$	b.p. 196–198°	45
2-Chloro-4-iodopyridine	Diazotization of 4-amino-2-chloropyridine; 2-chloro-4-nitraminopyridine and PI$_3$	m.p. 43°; HgCl$_2$ salt, m.p. 173°	125, 131c
3-Chloro-4-iodopyridine	Diazotization of 4-amino-3-chloropyridine; 3-chloro-4-nitraminopyridine and PI$_3$	m.p. 105–106°; picrate, m.p. 143°	128a, 131c
4-Chloro-2-iodopyridine	2-Iodo-4-nitraminopyridine and PCl$_5$	m.p. 29°; HgCl$_2$ salt, m.p. 158°	131c
4-Chloro-3-iodopyridine	From 3-iodo-4-nitropyridine-1-oxide and PCl$_3$; diazotization of 4-amino-3-iodopyridine; 3-iodo-4-nitraminopyridine and PCl$_3$	m.p. 79°; picrate, m.p. 136°	45, 128a, 131c
2-Bromo-4-iodopyridine	Diazotization of 4-amino-2-bromopyridine; 2-bromo-4-nitraminopyridine and PI$_3$	m.p. 61°; HgCl$_2$ salt, m.p. 151°	127, 131c
3-Bromo-4-iodopyridine	Diazotization of 4-amino-3-bromopyridine; 3-bromo-4-nitraminopyridine and PI$_3$	m.p. 112°; picrate, m.p. 154°	128a, 131c
3-Bromo-5-iodopyridine	Diazotization of 5-amino-3-bromopyridine	m.p. 132–134°	273
4-Bromo-2-iodopyridine	Diazotization of 4-amino-2-iodopyridine; 2-iodo-4-nitraminopyridine and PBr$_3$ or PBr$_5$	m.p. 46°; 55°; HgCl$_2$ salt, m.p. 152°	127, 131c
4-Bromo-3-iodopyridine	Diazotization of 4-amino-3-iodopyridine; 3-iodo-4-nitraminopyridine and PBr$_3$ or PBr$_5$	m.p. 77–78°; picrate, m.p. 131°	128, 131c
6-Chloro-2,4-difluoropyridine	From 2,4,6-trichloropyridine and KF in sulfolane (ethylene glycol initiator)	b.p. 135°	463
4-Chloro-2,6-difluoropyridine	From 2,4,6-trichloropyridine and KF in sulfolane (ethylene glycol initiator)	b.p. 135–137°	463
3,5-Dichloro-2-fluoropyridine	From 2,3,5-trichloropyridine, KF, and dimethyl sulfone; Schiemann diazotization of 2-amino-3,5-dichloropyridine	m.p. 42–43°	134
2-Bromo-3,5-dichloropyridine	Diazotization of 2-amino-3,5-dichloropyridine	m.p. 41.5°	274
4-Bromo-2,6-dichloropyridine	Diazotization of 4-amino-2,6-dichloropyridine	m.p. 95°	126
2,6-Dichloro-4-iodopyridine	Diazotization of 4-amino-2,6-dichloropyridine	m.p. 160°	126
4-Chloro-3,5-dibromopyridine	From 3,5-dibromo-4-pyridine and PCl$_5$	m.p. 101°	275

431

Table VI-4 (*Continued*)

Compound	Method of preparation	Physical properties	Ref.
3-Chloro-2,4,6-trifluoropyridine	From 2,3,4,6-tetrachloropyridine, KF in sulfolane	b.p. 128–129°	25
3-Chloro-2,5,6-trifluoropyridine	From 3-chloro-2,4,5,6-tetrachloropyridine and LiAlH$_4$	NMR, IR, and elemental analysis	26
3,5-Dichloro-2,6-difluoropyridine	From 2,3,5,6-tetrachloropyridine, KF in sulfolane or dimethyl sulfone	m.p. 45–46.3°	25, 134
3-Chloro-2,4,5,6-tetrafluoropyridine	Pentachloropyridine, KF in sulfolane or N-methyl-2-pyrrolidone	b.p. 119–121°; n_D^{20} 1.4307–1.4374	25, 26, 136, 137
4-Chloro-2,3,5,6-tetrafluoropyridine	Pentafluoropyridine and KCl	Not given	141
2-Bromo-3,4,5,6-tetrafluoropyridine	Perfluorocyclohexa-1,3-diene and BrCN	b.p. 140–142°	149
4-Bromo-2,3,5,6-tetrafluoropyridine	Diazotization of 4-amino-tetrafluoropyridine	b.p. 134–135°; n_D^{20} 1.4579	130
4-Iodo-2,3,5,6-tetrafluoropyridine	Pentafluoropyridine, NaI/DMF; 4-hydrazino-tetrafluoropyridine, Ag$_2$O, and CH$_3$I	b.p. 84°/40 mm; m.p. 47–48°	142
4-Bromo-2,3,5,6-tetrachloropyridine	Pentachloropyridine, NaBr/DMF; 2,3,5-tetrachloro-4-hydrazinopyridine with Br$_2$/dil. HCl	m.p. 146–148°; 150–152°	142a, 460
2-Fluoro-3,4,5,6-tetrachloropyridine	Pentachloropyridine, KF, and sulfolane	b.p. 234°	142b
4-Iodo-2,3,5,6-tetrachloropyridine	2,3,5,6-Tetrachloro-4-hydrazinopyridine and Ag$_2$O/CH$_3$I	m.p. 178–180°	460
3-Chloro-4-iodo-2,5,6-trifluoropyridine	3-Chloro-2,4,5,6-tetrafluoropyridine and NaI/DMF	b.p. 78–80°/10 mm; m.p. 39–41°	142
3,5-Dichloro-2,4,6-trifluoropyridine	Pentachloropyridine and KF in sulfolane or N-methyl-2-pyrrolidone	b.p. 150°; 159–160, $n_D^{20.5}$ 1.4804	25, 26, 136, 137
2,6-Difluoro-3,4,5-trichloropyridine	Pentachloropyridine, KF, and sulfolane	b.p. 196°	142b
3,5-Difluoro-2,4,6-tribromopyridine	Pentafluoropyridine, HBr, and sulfolane	m.p. 105.5–106.5°	142d
Perfluoro-2,3,4,5-tetrahydropyridine	Defluorination of perfluoro-(N-fluoropiperidine) with steel, ferrocene, mercury, or Mn(CO)$_4$H; 2,6,6-trichloro-3,3,4,4,5,5-hexafluoropiperideine and AgF	b.p. 43°; n_D^{22} 1.2800	3, 8, 276–278
2,3,4,5-Tetrachloro-2,3,4,5,6-penta-fluorotetrahydropyridine	Pentafluoropyridine and Cl$_2$/uv	b.p. 60°/8 mm	439

432

TABLE VI-5. Properties of Alkylhalopyridines

Compound	Method of preparation	Physical properties	Ref.
3-Fluoro-2-picoline	Schiemann diazotization of 3-amino-2-picoline	b.p. 120–122°; picrate, m.p. 141°	279
4-Fluoro-2-picoline	Schiemann diazotization of 4-amino-2-picoline	b.p. 128–132°; n_D^{20} 1.4713; picrate, m.p. 152°; methiodide, m.p. 238°	280
5-Fluoro-2-picoline	Schiemann diazotization of 5-amino-2-picoline	b.p. 120–121°; n_D^{20} 1.4696	411
6-Fluoro-2-picoline	Schiemann diazotization of 6-amino-2-picoline	b.p. 142°; n_D^{25} 1.4673	281
4-Fluoro-3-picoline	4-Nitramino-3-picoline and BF$_3$/acetic acid	b.p. 128°; picrate, m.p. 154°; HgCl, m.p. 143°	450
6-Fluoro-3-picoline	Schiemann diazotization of 6-amino-3-picoline	b.p. 155–156°	282
2-Fluoro-4-picoline	Schiemann diazotization of 2-amino-4-picoline	b.p. 157°; n_D^{25} 1.4690	281
3-n-Butyl-2-fluoropyridine	From 3-bromo-2-fluoropyridine and n-BuLi in tetrahydrofuran (−60°)	b.p. 70°/5 mm	462
3-Fluoro-2,6-lutidine	Schiemann diazotization of 3-amino-2,6-lutidine	b.p. 136–138°; 140°; picrate, m.p. 164°; 177°	444, 445
4-Fluoro-2,6-lutidine	4-Nitramino-2,6-lutidine and BF$_3$/acetic acid	b.p. 138°; picrate, m.p. 191°; HgCl$_2$, m.p. 159° (dec.)	450
2-Fluoro-4,6-lutidine	Schiemann diazotization of 2-amino-4,6-lutidine	b.p. 169–170°	424
4-Chloro-2-picoline	2-Picoline-1-oxide with Cl$_2$/SO$_2$ or POCl$_3$; 4-nitro-2-picoline-1-oxide and HCl: diazotization of 4-amino-2-picoline; PCl$_3$ or PCl$_5$ with 4-(nitramino)-2-picoline	b.p. 162–163°; b.p. 54–56°/15 mm; n_D^{20} 1.5240; picrate, m.p. 162–163°; HNO$_3$ salt, m.p. 156–157°; HgCl$_2$ salt, m.p. 168°	28, 31, 46, 126, 131b
4-Chloro-3-picoline	3-Picoline-1-oxide with Cl$_2$/SO$_2$, PCl$_3$ or POCl$_3$; N-[3-methyl-4-pyridyl-]pyridinium chloride hydrochloride and PCl$_5$; 4-nitro-3-picoline-1-oxide with CH$_3$COCl, then Fe/acetic acid; 4-(nitramino)-3-picoline and PCl$_3$ or PCl$_5$	b.p. 76–78°/24 mm; picrate, m.p. 150–151°; HgCl$_2$ salt, m.p. >260°	22, 28, 41, 131b, 283

433

Table VI-5 (*Continued*)

Compound	Method of preparation	Physical properties	Ref.
6-Chloro-3-picoline	5-Methyl-2-pyridone and POCl$_3$/PCl$_5$; 6-(nitramino)-3-picoline with PCl$_3$ or PCl$_5$	b.p. 56°/2.5 mm; HgCl$_2$ salt, m.p. 178°	41, 131b
2-Chloro-4-picoline	4-Picoline-1-oxide and Cl$_2$/SO$_2$; PCl$_3$ or PCl$_5$ and 2-(nitramino)-4-picoline	b.p. 84°/17 mm; 194°; picrate, m.p. 113–115°; HgCl$_2$ salt, m.p. 156°	28, 131b
3-Chloro-4-picoline	4-Picoline and Cl$_2$/excess AlCl$_3$	b.p. 101°/74 mm; n_D^{25} 1.5279	21
4-Chloro-2-ethylpyridine	2-Ethylpyridine-1-oxide with Cl$_2$/SO$_2$; diazotization of 4-amino-2-ethylpyridine	b.p. 75–77°/15 mm; picrate, m.p. 117–119°; HCl salt, m.p. 179–180°	28, 284
3-Chloro-2,6-lutidine	Diazotization of 3-amino-2,6-lutidine (HCl/Cu) (new route)	b.p. 75°/35 mm; 176–177°/757 mm; picrate, m.p. 148°	207, 285
6-t-Butyl-3-chloro-2-picoline	2-Methyl-5-t-butylpyrrole, CHCl$_3$, KOH/aq. EtOH	b.p. 116°/25 mm	161
6-t-Butyl-5-chloro-2-picoline	2-Methyl-5-t-butylpyrrole, CHCl$_3$, KOH/aq. EtOH	b.p. 106°/23 mm	161
4-Chloro-2,6-di-t-butylpyridine	2,6-Di-t-butyl-4-pyridone and PCl$_5$/POCl$_3$	b.p. 110–112°/12 mm; chloroaurate, m.p. 252–254°	265
3-Chloro-sym-collidine	Diazotization of 3-amino-sym-collidine	b.p. 190–192°; picrate, m.p. 150–151°	285
3-Chloro-2,4,5,6-tetramethylpyridine	Tetramethylpyrrole, CHCl$_3$, KOH/aq., EtOH, or CCl$_3$CO$_2$Na/1,2-dimethyoxyethane	m.p. 41–42°; picrate, m.p. 154°	160
6-Chloro-2-allylpyridine	2,6-Dichloropyridine and allylmagnesium chloride	b.p. 110–114°/27 mm; n_D^{20} 1.5338	428
3-Bromo-2-picoline	2-Picoline, Br$_2$, and fuming H$_2$SO$_4$	b.p. 72–73°/14 mm	61
4-Bromo-2-picoline	Diazotization of 4-amino-2-picoline; 4-nitro-2-picoline with acetyl bromide, 48% HBr, or urea/HBr; 1-hydroxy-2-methyl-4-pyridone and PBr$_3$; 4-(nitramino)-2-picoline with PBr$_3$ or PBr$_5$	b.p. 184°; 74–76°/14 mm; n_D^{20} 1.5542; picrate, m.p. 186°; methiodide, m.p. 214°; HgCl$_2$ salt, m.p. 183°	126, 131b, 280, 286

Compound	Method	Properties	References
4-Bromo-3-picoline	Diazotization of 4-amino-3-picoline (Craig); 4-nitro-3-picoline-1-oxide or CH_3COBr/Fe/acetic acid; 4-(nitramino)-3-picoline with PBr_3 or PBr_5	b.p. 108–110°/60 mm; picrate, m.p. 141°; $HgCl_2$ salt, m.p. 248° (dec.)	61, 131b, 257
5-Bromo-3-picoline	3-Picoline, Br_2, and fuming H_2SO_4 or excess $AlCl_3$; diazotization of 5-amino-3-picoline (Craig procedure)	b.p. 108–110°/25 mm; m.p. 16.5–17°; n_D^{20} 1.5604; picrate, m.p. 180–181°	61, 64
2-Bromo-4-picoline	Diazotization of 2-amino-4-picoline; 2-(nitramino)-4-picoline with PBr_3 or PBr_5	b.p. 213°; 67°/21 mm; n_D^{25} 1.5021; $HgCl_2$ salt, m.p. 149°	131b, 287
3-Bromo-4-picoline	4-Picoline, Br_2 with excess $AlCl_3$, or fuming H_2SO_4	b.p. 47–48°/0.6 mm; n_D^{25} 1.5613; picrate, m.p. 138–139°; $HgCl_2$ salt, m.p. 175–177°	21, 61, 63
4-Bromo-2-ethylpyridine	Diazotization of 4-amino-2-ethylpyridine (Craig method)	b.p. 97–98°/30 mm	284
4-Bromo-3-ethylpyridine	From 4-nitro-3-ethylpyridine-1-oxide and CH_3COBr, then Fe/acetic acid; Craig diazotization of 4-amino-3-ethylpyridine	b.p. 104°/24 mm; picrate, m.p. 133–134°	257, 442
6-Bromo-3-ethylpyridine	Diazotization of 6-amino-3-ethylpyridine	b.p. 120–122°/23 mm	442
4-Bromo-3-isopropylpyridine	From 4-nitro-3-isopropylpyridine-1-oxide and CH_3COBr, then Fe/acetic acid	picrate, m.p. 149–150°	257
2-Bromo-4-s-butylpyridine	Bromination of 4-s-butylpyridine	b.p. 104–107°/5 mm	288
3-Bromo-2,5-lutidine	2,5-Lutidine. Br_2 and fuming H_2SO_4	b.p. 87–88°/12 mm; picrate, m.p. 172–174°	79
4-Bromo-2,5-lutidine	From 2,5-dimethyl-4-nitropyridine-1-oxide and acetyl bromide, then Fe/acetic acid	b.p. 89–90°/14 mm; n_D^{20} 1.5501	79
3-Bromo-2,6-lutidine	2,6-Lutidine, Br_2, and fuming H_2SO_4; diazotization of 3-amino-2,6-lutidine	b.p. 194–196°; 82.5–83.0°/14 mm; n_D^{20} 1.5598; picrate, m.p. 150°	61, 285
4-Bromo-2,6-lutidine	Diazotization of 4-amino-2,6-lutidine; 4-(nitramino)-2,6-lutidine with PBr_3 or PBr_5	b.p. 194°; picrate, m.p. 178°; $HgCl_2$ salt, m.p. 192°	126, 131b, 249

Table VI-5 (*Continued*)

Compound	Method of preparation	Physical properties	Ref.
4-Bromo-5-ethyl-2-methylpyridine	From 5-ethyl-2-methyl-4-nitropyridine-1-oxide and PBr$_3$	b.p. 105–110°/20 mm; picrate, m.p. 141–142°; methiodide, m.p. 167–169°	78
3-Bromo-*sym*-collidine	Diazotization of 3-amino-*sym*-collidine; bromination of *sym*-collidine in 65% oleum	b.p. 219–221°; 84–87°/4 mm; picrate, m.p. 149–150°	285, 448
4-Iodo-2-picoline	Diazotization of 4-amino-2-picoline or the *N*-oxide; 4-(nitramino)-2-picoline and PI$_3$	b.p. 84–86°/14 mm: m.p. 42°; picrate, m.p. 209°; methiodide, m.p. 238°	126, 129, 131b, 280
2-(Nitramino)-3-picoline	2-(Nitramino)-3-picoline and PI$_3$	b.p. 105°/15 mm; HgCl$_2$ salt, m.p. 164°	131b
4-Iodo-3-picoline	Diazotization of 4-amino-3-picoline-1-oxide, then KI and PCl$_3$; PI$_3$ and 4-(nitramino)-3-picoline	m.p. 46–48°; picrate, m.p. 157–158°	129, 131b
6-Iodo-3-picoline	6-(Nitramino)-3-picoline and PI$_3$	m.p. 52°; HgCl$_2$ salt, m.p. 164°	131b
2-Iodo-4-picoline	2-(Nitramino)-4-picoline and PI$_3$	b.p. 112°/15 mm; HgCl$_2$ salt, m.p. 164°	131b
3-Iodo-2,6-lutidine	2,6-Lutidine, I$_2$, and 30% oleum; diazotization of 3-amino-2,6-lutidine	m.p. 30–31°; b.p. 105°/11 mm; picrate, m.p. 154°	289
4-Iodo-2,6-lutidine	4-(Nitramino)-2,6-lutidine and PI$_3$; diazotization of 4-amino-2,6-lutidine or the *N*-oxide	m.p. 101–102°; picrate, m.p. 189–190°; 197°	126, 129, 131b
3-Iodo-*sym*-collidine	*sym*-Collidine, I$_2$, and 65% oleum; diazotization of 3-amino-*sym*-collidine	b.p. 123°/12 mm; 109–110°/3 mm; m.p. 19°; picrate m.p. 132°	285, 449, 457
5-Ethyl-4-iodo-2-methylpyridine	Diazotization of 4-amino-5-ethyl-2-methylpyridine-1-oxide	m.p. 65–66°; picrate, m.p. 143–144°	129
3,5-Dichloro-4-picoline	4-Picoline, Cl$_2$, and excess AlCl$_3$	b.p. 125–127°/14 mm; m.p. 48–49°	21

436

Compound	Method	Properties	References
2,6-Dichloro-4-picoline	2,6-Dihydroxy-4-picoline and POCl₃ or C₆H₅POCl₂; from 6-chloro-4-methyl-2-pyridone and POCl₃	b.p. 122–123°/22.5 mm; m.p. 65°	111a, 290, 443
4,6-Dichloro-3-vinylpyridine	4,6-Dichloro-3-(2-chloroethyl)pyridine and NaOH/EtOH	b.p. 88–91°/2 mm; m.p. 15.5°	435
2,6-Dichloro-3,4-lutidine	From 2,6-dihydroxy-3,4-lutidine and POCl₃ or C₆H₅POCl₂	b.p. 137°/18 mm; m.p. 71–73°	290, 443
2,6-Dichloro-3,5-diethylpyridine	From α,α-diethylglutarimide and PCl₅	m.p. 43.5–44.5°	434
2,6-Dichloro-3,5-di-n-propylpyridine	From α,α-di-n-propylglutarimide and PCl₅	b.p. 145–147°/11 mm; m.p. 34.5–35.5°	434
2,6-Dichloro-3,5-di-n-butylpyridine	From α,α-di-n-butylglutarimide and PCl₅	b.p. 130–131.5°/2 mm; n_D^{20} 1.5214	434
2,6-Dichloro-3,5-di-n-hexylpyridine	From α,α-di-n-hexylglutarimide and PCl₅	b.p. 169–170°/2 mm; n_D^{20} 1.5078	434
3,5-Dichloro-2,4,6-collidine	Chlorination-decarboxylation of 2,4,6-trimethyl-3,5-pyridinecarboxylic acid (K salt)	m.p. 78–80°	63
2,6-Dichloro-3,4,5-trimethylpyridine	2,6-Dihydroxy-3,4,5-trimethylpyridine and POCl₃	m.p. 99°	291, 292
2,6-Dichloro-3,4-trimethylenepyridine	6-Chloro-3,4-trimethylene-2-pyridone and POCl₃	m.p. 36°	111a
2,6-Dichloro-3,4-dimethyl-5-ethylpyridine	2,6-Dihydroxy-3,4-dimethyl-5-ethylpyridine and POCl₃	m.p. 66°	291
2,6-Dichloro-3,5-diethyl-4-methylpyridine	2,6-Dihydroxy-3,5-diethyl-4-methylpyridine and POCl₃	b.p. 109°/0.05 mm; m.p. 19–20°; $n_D^{21.5}$ 1.5470	291
2,5-Dibromo-3-picoline	3-Picoline, Br₂, and fuming H₂SO₄ (65% SO₃); diazotization of 2-amino-5-bromo-3-picoline (Br₂/48% HBr)	m.p. 42–43°	61
5,6-Dibromo-3-picoline	Diazotization of 6-amino-5-bromo-3-picoline (Br₂/48% HBr)	m.p. 52–54°	466
2,5-Dibromo-4-picoline	Diazotization of 2-amino-5-bromo-4-picoline (48% HBr/Br₂)	m.p. 37–38°	61
3,5-Dibromo-4-picoline	4-Picoline, Br₂, and fuming H₂SO₄; from 3,5-dibromo-2-hydrazino-4-picoline and aq. CuSO₄	m.p. 105–107°; HgCl₂ salt, m.p. 219–221°	61, 63
3,5-Dibromo-2,4-lutidine	Bromination of 2,4-lutidine in 65% oleum	m.p. 32–33°; picrate, m.p. 145–147°	449

Table VI-5 (*Continued*)

Compound	Method of preparation	Physical properties	Ref.
3,5-Dibromo-2,6-lutidine	Bromination of 2,6-lutidine in 65% oleum	m.p. 64°	61, 456
3,5-Dibromo-*sym*-collidine	Bromination of *sym*-collidine in 65% oleum	m.p. 80.5–81°; picrate, m.p. 170–173°	448, 449
3,5-Diiodo-4-picoline	4-Picoline, I$_2$, and 20% oleum	m.p. 167°	293
3,5-Diiodo-*sym*-collidine	*sym*-Collidine, I$_2$, and 65% oleum	m.p. 113–114°; picrate, m.p. 174°	449, 457
5-Bromo-2-chloro-3-picoline	Diazotization of 2-amino-5-bromo-3-picoline	m.p. 40–41°	466
5-Bromo-6-chloro-3-picoline	Diazotization of 6-amino-5-bromo-3-picoline	m.p. 67–69°	465
5-Bromo-2-fluoro-3-picoline	Schiemann diazotization on 2-amino-5-bromo-3-picoline	m.p. 62–64°	466
5-Bromo-6-fluoro-3-picoline	Schiemann reaction or modification (diazonium hexafluorophosphate) on 6-amino-5-bromo-3-picoline	m.p. 54–55°	465
6-Bromo-5-fluoro-3-picoline	Modified Schiemann (diazonium hexafluorophosphate) on 5-amino-6-bromo-3-picoline	m.p. 32–33°	464
6-Chloro-5-fluoro-3-picoline	Modified Schiemann (diazonium hexafluorophosphate) on 5-amino-6-chloro-3-picoline	b.p. 90–92°/25 mm	464
4-Chloro-3-fluoro-2,6-lutidine	3-Fluoro-4-nitro-2,6-lutidine-1-oxide and PCl$_3$	b.p. 174–175°; picrate, m.p. 125°	445
4-Bromo-3-fluoro-2,6-lutidine	3-Fluoro-4-nitro-2,6-lutidine-1-oxide and PBr$_3$	m.p. 67°; picrate, m.p. 140°	445
3-Bromo-5-iodo-*sym*-collidine	Bromination of 3-iodo-*sym*-collidine or iodination of 3-bromo-*sym*-collidine in 65% oleum	m.p. 82–83°; picrate, m.p. 175–176°	449
3,5,6-Trifluoro-2,4-lutidine	Pentafluoropyridine and CH$_3$Li	b.p. 148–150°; n_D^{20} 1.4366; d^{20} 1.27	248
2-*n*-Butyl-4-methyl-3,5,6-trifluoropyridine	2,3,5,6-Tetrafluoro-4-picoline and *n*-BuLi	b.p. 206–207°; n_D^{20} 1.4453	248
2,4-Di-(1-propenyl)-3,5,6-trifluoropyridine	Pentafluoropyridine and propenyl-lithium	b.p. 91–92°/5 mm	181
3,5,6-Trichloro-2-picoline	3,5-Dichloro-2-methyl-6-pyridone and POCl$_3$; diazotization of 6-amino-3,5-dichloro-2-picoline	m.p. 71–72°	54
2,4,6-Trichloro-3-picoline	2-Chloro-4,6-dihydroxypyridine and POCl$_3$	m.p. 28.5°	153c
2,3,6-Trichloro-4-picoline	2,3,5,6-Tetrachloro-4-picoline and *n*-BuLi, followed by hydrolysis	Not described	236

438

4-Ethyl-2,3,6-trichloropyridine	4-Ethyl-2,3,5,6-tetrachloropyridine and n-BuLi, followed by hydrolysis	b.p. 180°/1.5 mm	236
4,5,6-Trichloro-2,3-lutidine	3,5-Dichloro-5,6-dimethyl-2-pyridone and PCl_5	m.p. 66.5–67.5°	112
3,4,5-Trichloro-2,6-lutidine	Pentachloropyridine-1-oxide and CH_3MgI, then PCl_3 deoxygenation	m.p. 71–72°	454
3,4,5-Trichloro-2,6-diethylpyridine	Pentachloropyridine-1-oxide and EtMgBr, then PCl_3 deoxygenation	b.p. 90°/1.5 mm	454
2,3,5-Tribromo-4-picoline	4-Picoline and fuming H_2SO_4; diazotization of 2-amino-3,5-dibromo-4-picoline (48 % HBr/Br_2)	m.p. 56–57°	61
2,3,5,6-Tetrafluoro-4-picoline	Pentafluoropyridine and CH_3Li	b.p. 130–131°; n_D^{20} 1.4109, d_4^{20} 1.44	248
4-(1-Propenyl)-2,3,5,6-tetrafluoropyridine	Pentafluoropyridine and propenyl lithium	b.p. 54°/10 mm	181
4-n-Butyl-2,3,5,6-tetrafluoropyridine	Pentafluoropyridine, n-BuLi	b.p. 185–186°; n_D^{20} 1.4249	248
3,4,5,6-Tetrachloro-2-picoline	Perchloropentadienenitrile and CH_3MgI; pentachloropyridine-1-oxide and CH_3MgI, then PCl_3 deoxygenation	m.p. 93–94°	114, 454
2,3,5,6-Tetrachloro-4-picoline	4-Methoxytetrachloropyridine and CH_3Li; tetrachloro-4-pyridyl-lithium and dimethyl sulfate; pyrolysis of 2,3,5,6-tetrachloropyridine-4-acetic acid	m.p. 88.5°; b.p. 90°/0.02 mm	236, 295, 426
4-Ethyl-2,3,5,6-tetrachloropyridine	4-Methoxytetrachloropyridine and C_2H_5MgBr; pentachloropyridine and C_2H_5MgBr; 4-tetrachloropyridylmagnesium chloride and EtI; 2,3,5,6-tetrachloro-4-picoline, n-BuLi and dimethyl sulfate	m.p. 67.5–68.5°; b.p. 93°/0.02 mm	225, 236, 295, 426
4-n-Propyl-2,3,5,6-tetrachloropyridine	Pentachloropyridine and C_3H_7MgBr	b.p. 94–95°/0.02 mm	426
4-Allyl-2,3,5,6-tetrachloropyridine	2,3,5,6-Tetrachloro-4-pyridylcopper and allyl bromide	b.p. 99–101°/0.1 mm	430, 446
4-n-Butyl-2,3,5,6-tetrachloropyridine	4-Methoxytetrachloropyridine and n-BuMgBr or n-BuLi; 4-methylmercapto-2,3,5,6-tetrachloropyridine and n-BuLi; pentachloropyridine and C_4H_9Br	b.p. 140–145°/3 mm; 100°/0.07 mm	236, 294, 295, 426
2-Ethynyl-3,4,5,6-tetrachloropyridine	Electrolytic reduction of heptachloro-2-vinylpyridine	m.p. 108–111°	459
3-Ethynyl-2,4,5,6-tetrachloropyridine	Electrolytic reduction of heptachloro-3-vinylpyridine	m.p. 85–88°	459

439

TABLE VI-6. Properties of the Arylhalopyridines

Compound	Method of preparation	Physical properties	Ref.
2-Chloro-3-phenylpyridine	3-Phenyl-2-pyridone and $POCl_3$; 2-chloro-3-N-nitrosoacetamidopyridine and benzene	b.p. 108°/0.1 mm; m.p. 55°	455, 458
4-Chloro-3-phenylpyridine	3-Phenyl-4-pyridone and $POCl_3$	Liquid; converted to 4-diethylaminopropylamino-3-phenylpyridine dioxalate, m.p. 155–157°	431
6-Chloro-3-phenylpyridine	6-Chloro-3-N-nitrosoacetamidopyridine and benzene	m.p. 65°	458
2-Chloro-3-(m-methoxyphenyl)pyridine	3-(m-Methoxyphenyl)-2-pyridone and $POCl_3$	b.p. 126–128°/0.5 mm	455
2-Chloro-3-(2,4,6-trichlorophenyl)pyridine	2-Chloropyridine-3-sulfonyl chloride, 1,3,5-trichlorobenzene, and CuCl	b.p. 144–145°/0.3 mm; m.p. 92–95°	429
2-Chloro-5-(2,4,6-trichlorophenyl)pyridine	2-Chloropyridine-5-sulfonyl chloride, 1,3,5-trichlorobenzene, and CuCl	m.p. 82–84°	429
2-Chloro-5-(2,3,4,5,6-pentachlorophenyl)-pyridine	2-Chloropyridine-5-sulfonyl chloride, pentachlorobenzene, and CuCl	b.p. 184°/0.6 mm; m.p. 105–110°	429
2-Chloro-5-(2,4,6-tribromophenyl)-pyridine	2-Chloropyridine-5-sulfonyl chloride, 1,3,5-tribromobenzene, and CuCl	b.p. 164°/0.15 mm; m.p. 78–80°	429
4-Chloro-3-(2,4,6-trichlorophenyl)pyridine	4-Chloropyridine-3-sulfonyl chloride, 1,3,5-trichlorobenzene, and CuCl	b.p. 134°/0.7 mm; m.p. 81°	429
2-Chloro-5,6-dimethyl-4-phenylpyridine	5,6-Dimethyl-4-phenyl-2-pyridone and $POCl_3$	m.p. 72–73°	296
2-Chloro-3,4-diphenylpyridine	3,4-Diphenyl-2-pyridone and $POCl_3$	m.p. 145–146°	297
2-Chloro-3,6-diphenylpyridine	2,5-Diphenyl-6-pyridone and $POCl_3$	m.p. 102–103°	298
3-Chloro-4,5-diphenylpyridine	Partial reduction (10% Pd–C/hydrogen) of 2,5,6-trichloro-3,4-diphenylpyridine	m.p. 106–107°	433
3-Chloro-4,5-di-(p-methoxyphenyl)pyridine	Partial reduction (10% Pd–C/hydrogen) of 2,5,6-trichloro-3,4-di-(p-methoxyphenyl)pyridine	m.p. 109°	433
3-Chloro-4,5-dimethyl-2,6-diphenylpyridine	3,4-Dimethyl-2,5-diphenyl-1H-pyrrole, CCl_3CO_2Na, and 1,2-dimethoxyethane	m.p. 148°; picrate, m.p. 155°	160
2-Bromo-3-(p-chlorophenyl)-6-picoline	3-(p-Chlorophenyl)-6-methyl-2-pyridone and PBr_3/DMF	m.p. 96–97°	436

440

Compound	Method	Properties	Ref.
2-Bromo-5-(2,4,6-tribromophenyl)pyridine	2-Bromopyridine-5-sulfonyl chloride, 1,3,5-tribromobenzene, and CuCl	b.p. 180–182°/0.1 mm; m.p. 73°	429
3-Bromo-2-phenylpyridine	Craig diazotization of 3-amino-2-phenylpyridine	b.p. 305–307°/720 mm	299
3-Bromo-6-phenylpyridine	Pyridine and phenyl-lithium, then bromine	m.p. 71–72.5°; picrate, m.p. 160–161°	424, 427
3-Bromo-5-(2,4,6-trichlorophenyl)pyridine	3-Bromopyridine-5-sulfonyl chloride, 1,3,5-trichlorobenzene, and CuCl	b.p. 178°/0.7 mm; m.p. 65°	429
2-Bromo-5,6-dimethyl-4-phenylpyridine	5,6-Dimethyl-4-phenyl-2-pyridone and $POBr_3$	m.p. 75–76°	296
2,6-Dichloro-4-phenylpyridine	6-Chloro-4-phenyl-2-pyridone and $POCl_3$	m.p. 54°	111a
2,6-Dichloro-4-methyl-3-phenylpyridine	2,6-Dihydroxy-4-methyl-3-phenylpyridine and $POCl_3$	m.p. 60–61°	300
3,5-Dichloro-2-phenylpyridine	2-Phenyl-3,4,5,6-tetrachloropyridine and HI/CH_3CO_2H	m.p. 48°	114
2,5-Dichloro-3,4-diphenylpyridine	Partial reduction (10% Pd–C/hydrogen) of 2,5,6-trichloro-3,4-diphenylpyridine	m.p. 160°	433
3,5-Dichloro-2,4,6-triphenylpyridine	Tetrachloro-4-methoxypyridine and phenyllithium	m.p. 157–158°	438
2,3,6-Trichloro-4-phenylpyridine	Tetrachloro-4-phenylpyridine, n-BuLi, and ether, then H_2O	m.p. 60–61°	438
2,5,6-Trichloro-3-phenylpyridine	Phenylglutarimide and PCl_5	m.p. 91–92°	433
2,3,6-Trichloro-4-(p-methoxyphenyl)-pyridine	Tetrachloro-4-(p-methoxyphenyl)pyridine, n-BuLi, and ether, then H_2O	m.p. 136–137°	438
4-(p-Dimethylaminophenyl)-2,3,6-trichloropyridine	Tetrachloro-4-(p-dimethylaminophenyl)pyridine, n-BuLi, and ether, then H_2O	m.p. 182–183°	438
2,3,6-Trichloro-4-(p-trifluoromethylphenyl)pyridine	Tetrachloro-4-(p-trifluoromethylphenyl)pyridine, n-BuLi, and ether, then H_2O	m.p. 75–76°	438
4-(p-Methoxyphenyl)-5-methyl-2,3,6-trichloropyridine	Tetrachloro-4-(p-methoxyphenyl)pyridine, n-BuLi, and ether, then dimethyl sulfate	m.p. 138–139°	438
4,6-Diphenyl-2,3,5-trichloropyridine	Tetrachloro-4-methoxypyridine and phenyllithium	m.p. 102–103°	438
3,4-Diphenyl-2,5,6-trichloropyridine	2,3-Diphenylglutarimide and PCl_5	m.p. 149–151°	433
2,6-Diphenyl-3,4,5-trichloropyridine	Pentachloropyridine-1-oxide and C_6H_5MgBr, then PCl_3 deoxygenation	m.p. 166°	454
3,4-Di-(p-methoxyphenyl)-2,5,6-trichloropyridine	2,3-Bis(p-methoxyphenyl)glutarimide and PCl_5	m.p. 132°	433
2,3,5-Trichloro-4,6-(p-trifluoromethylphenyl)pyridine	Tetrachloro-4-methoxypyridine and p-trifluoromethylphenyl lithium	m.p. 156–157°	438

Table VI-6 (*Continued*)

Compound	Method of preparation	Physical properties	Ref.
4-Phenyl-2,3,5,6-tetrafluoropyridine	Pentafluoropyridine and phenyl-lithium; 4-hydrazino-tetrafluoropyridine, bleaching powder, and benzene	m.p. 106–107°	181, 192
4-(Pentafluorophenyl)-2,3,5,6-tetrafluoropyridine	Pentafluoropyridine and C_6F_5MgBr	m.p. 98.5–99.5°	130
2-(Pentafluorophenyl)-3,4,5,6-tetrafluoropyridine	Perfluorocyclohexa-1,3-diene and C_6F_5CN	b.p. 200°	149
4-Benzyl-2,3,5,6-tetrachloropyridine	Pentachloropyridine and benzylmagnesium chloride	m.p. 107–108°; b.p. 165°/15 mm	225, 426
4-(p-Methoxyphenyl)-2,3,5,6-tetrachloropyridine	Tetrachloro-4-methoxypyridine and p-methoxyphenyl-lithium	m.p. 143–144°	438
4-(p-Dimethylaminophenyl)-2,3,5,6-tetrachloropyridine	Tetrachloro-4-methoxypyridine and p-dimethyl-aminophenyl-lithium	m.p. 257–259°	438
4-Phenyl-2,3,5,6-tetrachloropyridine	Tetrachloro-4-methoxypyridine and C_6H_5Li; 2,3,5,6-tetrachloro-4-pyridylcopper and iodobenzene; 2,3,5,6-tetrachloro-4-hydrazinopyridine with Ag_2O/benzene	m.p. 137–138°; 140–141°	236, 438, 446, 460
4-(p-Trifluoromethylphenyl)-2,3,5,6-tetrachloropyridine	Tetrachloro-4-methoxypyridine and p-trifluoromethylphenyl-lithium	m.p. 155–156°	438
2-(p-Fluorophenyl)-3,4,5,6-tetrachloropyridine	Perchloropentadienonitrile and p-FC_6H_4MgBr	m.p. 141–142°	114
2-(α-Naphthyl)-3,4,5,6-tetrachloropyridine	Perchloropentadienonitrile and α-$C_{10}H_7MgBr$	m.p. 141.5–142.5°	114
2-Phenyl-3,4,5,6-tetrachloropyridine	Perchloropentadienonitrile and C_6H_5MgBr; pentachloropyridine-1-oxide and C_6H_5MgBr, then PCl_3 deoxygenation	m.p. 113.5–114°	114, 454
2-(p-Tolyl)-3,4,5,6-tetrachloropyridine	Perchloropentadienonitrile and p-$CH_3C_6H_4MgBr$	m.p. 107–107.5°	114
3,5-Dichloro-2,6-difluoro-4-phenylpyridine	3,5-Dichloro-2,6-difluoro-4-hydrazinopyridine, bleaching powder, and benzene	m.p. 48.5–49.0°	142b

442

TABLE VI-7. Properties of the Haloalkylhalopyridines

Compound	Method of preparation	Physical properties	Ref.
4-Chloro-2-(chloromethyl)pyridine	4-Chloro-2-picoline-1-oxide and *p*-toluenesulfonyl chloride	b.p. 51°/0.9 mm; n_D^{25} 1.5225; picrate, m.p. 123–125°	368
5-Chloro-2-(chloromethyl)pyridine	5-Chloro-2-picoline-1-oxide and *p*-toluenesulfonyl chloride	b.p. 48°/0.5 mm; n_D^{25} 1.5293; picrate, m.p. 96–98°	368
2-Chloro-3-(chloromethyl)pyridine	3-Chloromethylpyridine-1-oxide and phosphorus oxychloride	b.p. 115°/13 mm	32
4-(Chloromethyl)-2,3,5,6-tetrafluoropyridine	2,3,5,6-Tetrafluoro-4-picoline and Cl_2/hv or SO_2Cl_2	b.p. 167.5°; n_D^{20} 1.4512	248
4,6-Dichloro-3-(2-chloroethyl)pyridine	4,6-Dihydroxy-3-(2-chloroethyl)pyridine and $POCl_3$	b.p. 123–125°/1.5 mm	435
2,4,6-Trichloro-3-(2-chloroethyl)pyridine	4-Chloro-5-methyl-6-oxo-2,3-dihydro-5-azabenzofuran and $POCl_3$	m.p. 55°	435
2-(Bromomethyl)-6-fluoropyridine	6-Fluoro-2-picoline, NBS/benzoyl peroxide	HBr salt, m.p. 108–112° (dec.)	452
3-(Bromomethyl)-2-fluoropyridine	2-Fluoro-3-picoline, NBS/benzoyl peroxide	HBr salt, m.p. 105–120° (dec.)	452
3-(Bromomethyl)-6-fluoropyridine	6-Fluoro-3-picoline, NBS/benzoyl peroxide	HBr salt, m.p. 114–120° (dec.)	452
4-(Bromomethyl)-2-fluoropyridine	2-Fluoro-4-picoline, NBS/benzoyl peroxide	HBr salt, m.p. 93–105° (dec.)	452
2-(Bromomethyl)-6-chloropyridine	6-Chloro-2-picoline, NBS/benzoyl peroxide	HBr salt, m.p. 95–111° (dec.)	453
3-(Bromomethyl)-2-chloropyridine	2-Chloro-3-picoline, NBS/benzoyl peroxide	HBr salt, m.p. 98–110° (dec.)	453
3-(Bromomethyl)-6-chloropyridine	6-Chloro-3-picoline, NBS/benzoyl peroxide	HBr salt, m.p. 92–98° (dec.)	453
4-(Bromomethyl)-2-chloropyridine	2-Chloro-4-picoline, NBS/benzoyl peroxide	HBr salt, m.p. 113–119° (dec.)	453
6-Bromo-2-(bromomethyl)pyridine	6-Bromo-2-picoline, NBS/benzoyl peroxide	HBr salt, m.p. 150–157° (dec.)	453
2-Bromo-3-(bromomethyl)pyridine	2-Bromo-3-picoline, NBS/benzoyl peroxide	HBr salt, m.p. 135–143° (dec.)	453
6-Bromo-3-(bromomethyl)pyridine	6-Bromo-3-picoline, NBS/benzoyl peroxide	HBr salt, m.p. 160–170° (dec.)	453
2-Bromo-4-(bromomethyl)pyridine	2-Bromo-4-picoline, NBS/benzoyl peroxide	HBr salt, m.p. 120–124° (dec.)	453
4-(Bromomethyl)-2,3,5,6-tetrafluoropyridine	2,3,5,6-Tetrafluoro-4-picoline and Br_2/hv or NBS/peroxide	b.p. 187–188°; n_D^{20} 1.4802	248
2-(Dichloromethyl)-3,4,5-trichloropyridine	Chlorination of 2-picoline·HCl	b.p. 100–110°/1 mm; m.p. 62–63°	390, 391

Table VI-7 (*Continued*)

Compound	Method of preparation	Physical properties	Ref.
4-(Dichloromethyl)-2,3,5,6-tetrafluoropyridine	2,3,5,6-Tetrafluoro-4-picoline and $Cl_2/h\nu$ or SO_2Cl_2	b.p. 179°; n_D^{20} 1.4684	248
4-(Dibromomethyl)-2,3,5,6-tetrafluoropyridine	2,3,5,6-Tetrafluoro-4-picoline and $Br_2/h\nu$	b.p. 208–209°; n_D^{20} 1.5185	248
2-Chloro-6-methyl-4-trifluoromethyl-pyridine	6-Methyl-4-trifluoromethyl-2(1H)-pyridone and $POCl_3/PCl_5$	b.p. 162.0–162.5°; 78–79°/25 mm; n_D^{25} 1.45084	113
2-Chloro-6-phenyl-4-trifluoromethyl-pyridine	6-Phenyl-4-trifluoromethyl-2(1H)-pyridone and $POCl_3/PCl_5$	m.p. 49.6–50.6°	113
Perfluoro-(2-picoline)	Defluorination of perfluoro-(N-fluoro-2-methyl-piperidine); CF_3CN and perfluorocyclohex-1,3-diene	b.p. 102–103°	9, 149
Perfluoro-(3-picoline)	Defluorination of perfluoro-(N-fluoro-3-methylpiperidine)	b.p. 102–103°	9
Perfluoro-(4-picoline)	Defluorination of perfluoro-(N-fluoro-4-methylpiperidine)	b.p. 102–103°	9
Perfluoro-(4-isopropylpyridine)	Pentafluoropyridine, perfluoropropene, KF, and sulfolane	b.p. 128–129°	324
3-Chloro-2-trichloromethylpyridine	Aqueous chlorination of 2-picoline	b.p. 100–104°/2 mm	392
6-Chloro-2-trichloromethylpyridine	Chlorination of 2-picoline·HCl or 2-(trichloromethyl)-pyridine	b.p. 136–137.5°/11 mm; m.p. 62.5–62.9°	392–394
4-Chloro-6-methyl-2-trichloromethyl-pyridine	Minor product from 2,6-dimethyl-4-pyridone and $POCl_3/PCl_5$	m.p. 74–75°	395
4,6-Difluoro-2-trifluoromethylpyridine	4,6-Dichloro-2-trifluoromethylpyridine and KF (350°)	Not described	461
3,6-Dichloro-2-trichloromethylpyridine	Chlorination of 3-chloro-2-trichloromethylpyridine	m.p. 47–48°	393, 396
4,5-Dichloro-2-trichloromethylpyridine	Aqueous chlorination of 2-picoline	b.p. 100–102°/1 mm	437

444

Compound	Preparation	Properties	Ref.
4,6-Dichloro-2-trichloromethylpyridine	2,4-Dichloro-6-dichloromethylpyridine, Cl_2/benzoyl peroxide; Cl_2 and 4-chloro-2-trichloromethylpyridine	b.p. 100–103°/1.5 mm	390, 393
5,6-Dichloro-2-trichloromethylpyridine	Chlorination of 5-chloro-2-trichloromethylpyridine	m.p. 38°	393
2,6-Dichloro-4-trichloromethylpyridine	Chlorination (480°) of γ-picoline	m.p. 56–58°	15
4,5,6-Trichloro-2-trichloromethylpyridine	Chlorination of 4,5-dichloro-2-trichloromethylpyridine	m.p. 60.2°	393
3,5,6-Trichloro-2-trichloromethylpyridine	Chlorination of 3,5-dichloro-2-trichloromethylpyridine	b.p. 120°/1 mm; m.p. 58–59°	393
3,4,6-Trichloro-2-trichloromethylpyridine	Chlorination of 3,4-dichloro-2-trichloromethylpyridine	Not described	393
Mixture of 3,4,5-trifluoro-5-chloro-2-trifluoromethylpyridine and 4,5,6-trifluoro-3-chloro-2-trifluoromethyl-pyridine	3,4,5,6-Tetrachloro-2-trifluoromethylpyridine and KF	b.p. 41°/23 mm	461
4,6-Difluoro-3,5-dichloro-2-trifluoromethylpyridine	3,4,5,6-Tetrachloro-2-trifluoromethylpyridine and KF	b.p. 60°/23 mm	461
2,3,5,6-Tetrafluoro-4-trichloromethyl-pyridine	2,3,5,6-Tetrafluoro-4-picoline and Cl_2/hv or SO_2Cl_2	b.p. 197–198°; n_D^{20} 1.4820	248
3,4,5,6-Tetrachloro-2-trichloromethyl-pyridine	Chlorination of 2-picoline·HCl or 3,4,5-trichloro-2-trichloromethylpyridine	m.p. 58–60°	391, 393
2,3,5,6-Tetrafluoro-4-(1-chlorotetra-fluoroethyl)pyridine	Pentafluoropyridine, chlorotrifluoroethylene, and KF	b.p. 140°	324
5-Chloro-2,4-bis(trifluoromethyl)pyridine	5-Chloro-2,4-bis(trichloromethyl)pyridine and HF	b.p. 85–86°/100 mm; n_D^{25} 1.4113	301, 302
Perfluoro-(2,4-diisopropylpyridine)	Pentafluoropyridine, perfluoropropene, KF, and sulfolane	b.p. 158–160°	324
Perfluoro-(2,4,5- and 2,4,6-tri-isopropylpyridine)(3:1 mixture)	Perfluoro-(4-isopropylpyridine), perfluoropropene, KF, and sulfolane	b.p. 190–192°	324
Trans-β-3,4,5,6-pentachloro-2-vinylpyridine	Electrolytic reduction of heptachloro-2-vinylpyridine	m.p. 117–121°	459

3. Reactions

A. Hydrogenolysis, Reduction, and Coupling

Comparative hydrogenolysis studies with Raney nickel-potassium hydroxide-methanol disclosed that 2-, 3-, and 4-bromopyridine were reduced within 0.5 to 0.75 hr; longer periods were required for the chloropyridines: 2- (2 hr) < 4- (3 hr) ≪ 3- (16 hr) (177a, b). Polarographic studies revealed the following order of ease of halopyridine reduction: I > Br > Cl and 4 > 2 > 3 (178). The preparation of pyridine 2-*d*, 3-*d*, 4-*d* by treatment of 2-, 3-, or 4-halopyridines with zinc-2*N* D_2SO_4 has been described (179).

Preferential catalytic hydrogenolysis of chlorofluoropyridines has provided several novel compounds, for example 2,4,6-trifluoropyridine from 3,5-dichlorotrifluoropyridine (180). Replacement of the 4-fluorine atom in pentafluoropyridine by hydrogen to give 2,3,5,6-tetrafluoropyridine occurs under free radial (catalytic hydrogenation) or nucleophilic (lithium aluminum hydride) conditions (180, 181). Substituents in the 4-position are more susceptible to displacement by hydride ion (from $LiAlH_4$) than other ring substituents; for example, 3-chloro-2,4,5,6-tetrafluoropyridine was reduced to 3-chloro-2,5,6-trifluoropyridine (26). While pentahalopyridines generally undergo multiple nucleophilic displacement at positions 2, 4, or, hydride ion can effect displacement at position 3 in pentachloropyridine (55). The latter was converted by lithium aluminum hydride under ambient conditions to 2,3,6-trichloropyridine in 90% yield (181a, b).

Hydrogenolysis of 2-amino-5-chloropyridine to 2-aminopyridine without the formation of coupling products was effected through use of the hydrazine-palladium charcoal-ethanol system (182).

In an evaluation of solvent systems for the Ullmann coupling reaction, dimethylformamide gave the highest yields of 2,2'-bipyridyl from 2-bromopyridine (183). Ullmann coupling (at the 4-position) was demonstrated for 4-bromo- and 4-iodotetrafluoropyridine (130, 142). Perfluoro-(β,β'-pyridylenes) were prepared from 3-chlorotetrafluoropyridine and 3,5-dichlorotrifluoropyridine with copper (184), while poly(β,β'-pyridylenes) were obtained from 3,5-dibromopyridine and potassium in dioxane (185) or 3,5-dichloropyridine and lithium in tetrahydrofuran (186).

B. Hydrolysis

While the alkaline hydrolysis (4*M* KOH) of 2- or 4-halopyridines (X = Cl, Br, I) gave 2- or 4-pyridone, 3-halopyridines formed mixtures of 3-hydroxypyridine and 4-pyridone (186a). Zoltewicz and Sale have suggested competing direct substitution and elimination-addition mechanisms. The latter would involve 3,4-pyridyne (**VI-63b**) as an intermediate.

Pentafluoropyridine, 3-chlorotetrafluoropyridine, 3,5-dichlorotrifluoro-pyridine, and pentachloropyridine gave the corresponding tetrahalo-4-pyridone upon reaction with potassium or sodium hydroxide in water or aqueous ethanol (107, 181, 187). The use of t-butyl alcohol solvent permits appreciable 2-substitution; this has been interpreted in terms of the steric requirements of the solvent and ring substituents (187). Application of this system to heterohalogenated pyridines, for examples 4-bromo- or 4-iodo-tetrafluoropyridine, results in preferential displacement of 2-fluorine to give the tetrahalo-2-pyridone (130, 142). Suschitzky observed that 2-substitution preferentially occurred from the reaction of pentachloropyridine with a mixture of acetic and sulfuric acids to give tetrachloro-2-pyridone; this has been cited as an example of nucleophilic substitution upon pentachloro-pyridinium ion (187a).

Tetrahalo-4-pyridinethiones have been prepared from pentachloro- or pentafluoropyridine and potassium hydrosulfide (188–190a).

C. Alcoholysis

Abramovitch and co-workers (190b) found that the rates of reaction of the 2-halopyridines toward methoxide ion in methanol were in the order $F \gg Br > Cl$. Other kinetic studies (in methanol and in dimethyl sulfoxide) showed that the rates were in the sequence 2-halopyridine > 2-halo-3-picoline > 2-halo-5-picoline and were dependent upon E_{act} when the halogen was bromine but upon $\Delta S^{\ddagger}$ when it was chlorine. [The accelerating role of the *ortho*-methyl group in the bromopicolines has been interpreted in light of the combined effects of London forces and ion-dipole interactions (117, 190b)]. In contrast to the rearrangement products arising from the alkaline hydrolysis of 3-halopyridines (186a), the corresponding alcoholysis (sodium methoxide-methanol) gave only direct substitution (190c).

While attack at the 4-position represents the main substitution path for the reaction of alkoxides with pentachloropyridine (191), a more detailed study by Flowers, Haszeldine, and Majid (55) disclosed that a mixture of 2- and 4-alkoxytetrachloropyridines is obtained. Larger nucleophiles lead to increased 2-substitution, presumably due to less steric hindrance.

Alkoxide (solvent)	Ratio of 4-:2-substitution
$CH_3OK(CH_3OH)$	85:15
$C_2H_5OK(C_2H_5OH)$	65:35
$n\text{-}C_4H_9OK(n\text{-}C_4H_9OH)$	57:43

Lability of the 4-fluorine atom in pentafluoropyridine was demonstrated by shaking the latter with sodium methoxide/methanol under ambient conditions (181, 192). Prolonged heating gave 3,5-difluoro-2,4,6-trimethoxy-pyridine. While the nitro group is displaced in the reaction of 4-nitrotetra-fluoropyridine and sodium methoxide (193), fluorine is removed in penta-fluoronitrobenzene or 2,3,5,6-tetrafluoronitrobenzene. The striking activation of the 4-position in polyfluoropyridines has been attributed to the stabilization of a transition state in which the high electron density can be placed on the ring-nitrogen (194). [The preferential displacement of fluorine by methoxide in 4-bromo- or 4-iodotetrafluoropyridine to give 2-methoxy-3,5,6-trifluoro-4-halopyridine would require another rationalization (130, 142)].

D. Aminolysis

Since Levine and Leake's (195) initial observation concerning the *in situ* generation of 3,4-pyridyne (**VI-63b**) from 3-bromopyridine, acetophenone, and sodium amide to give 4-aminopyridine and 4-phenacylpyridine, considerable interest has been shown in the pyridine counterpart of benzyne (196, 197). Kaufmann studied the isomer distribution from the reaction of 2-, 3-, and 4-halopyridines with lithium piperidide and other secondary amines (198–201). Both 3-bromo- and 3-chloropyridine provided nearly equal amounts of 3- and 4-piperidinopyridine, while 3-fluoropyridine gave a 96:4 isomer ratio. (Direct substitution predominated with 2-fluoro- and 4-chloropyridine).

The generation of 3,4-pyridyne (**VI-63b**) was also demonstrated when 3-bromo-, 3-chloro-, and 3-iodopyridine were treated with potassium amide-liquid ammonia to give 3- and 4-aminopyridine (1:2 ratio) (203, 204). Since Zoltewicz observed a more rapid exchange of deuterium in 3-chloro-pyridine-4-*d*, as compared with 3-chloropyridine-2-*d* with sodium amide-ammonia, these pyridyne-generated reactions by use of amide ion probably involve formation of 3-chloro-4-pyridyl anion (**VI-63a**), followed by elimination of halide ion to give (**VI-63b**) (208, 209). In contrast, 2-halopyridines

VI-63a   VI-63b

(202–204) and 4-fluoropyridine (205) undergo direct substitution under similar amination conditions, while 3-fluoropyridine is converted to a mixture of 2,4'- and 4,4'-bipyridyls (206).

Den Hertog and co-workers observed a surprising ring transformation in which amination of 2,6-dibromopyridine provided 4-amino-2-methyl-pyrimidine (**VI-64**) as the sole product (205). In contrast, amination of

2,6-difluoropyridine gave 2,6-diaminopyridine. Amination of 2,3-, 2,4-, 2,5-, 3,4-, and 3,5-dibromopyridine gave mixtures of diamino and amino-bromopyridines which were generated from a bromo-3,4-pyridyne.

Substituted halopyridines undergo ring contraction under selected amination conditions. Potassium amide-liquid ammonia converted 3-amino-2-bromopyridine to 3-cyanopyrrole (210). 2-Bromo-3-hydroxypyridine and 2,6-dibromo-3-hydroxypyridine (**VI-65**) were transformed to pyrrole-2-carbonamide and 5-bromopyrrole-2-carbonamide (**VI-66**), respectively (211).

Ring scission to give 1,3-dicyanopropene, $NCCH{=}CH{-}CH_2CN$, occurred in the reaction of 2-bromo-6-nitropyridine or 6-amino-2-bromo-pyridine with excess potassium amide in liquid ammonia (212).

Steric considerations play an important role in the reaction of nitrogen-containing nucleophiles with pentachloropyridine: 4-substitution is favored with ammonia (213, 214), sodium amide-ammonia-ether (143), and hydrazine hydrate-ether (143), while 2-substitution predominates with secondary amines such as dimethyl amine, piperidine, morpholine, and pyrrolidine in benzene (215, 216). Haszeldine and his co-workers conducted a more detailed study on the amounts of 2- and 4-substitution with change in nucleophile (ethanol solvent); larger nucleophiles favor 2-substitution because of decreased steric hindrance (55).

Amine	Ratio of 4- : 2- substitution
NH_3	70:30
$n\text{-}C_4H_9NH_2$	25:75
$(CH_3)_2NH$	20:80
$(C_2H_5)_2NH$	1:99

4-Substitution predominates when pentafluoropyridine, 3-chlorotetra-fluoropyridine, or 3,5-dichlorotrifluoropyridine react with ammonia, hydrazine, or dimethyl amine (181, 187, 192). These studies associated greater reactivity with increasing substrate chlorine content (187). A 2-fluoro substituent was displaced exclusively in the reaction of 4-bromo- or 4-iodotetrafluoropyridine with aqueous ammonia (130, 142). Nitro group displacement was noted during the amination of 4-nitrotetrafluoropyridine (193).

No 2-(N-β-hydroxyethylanilino)pyridine (**VI-68a**) was obtained from the reaction of 2-bromopyridine (**VI-67**) and N-phenylethanolamine; instead, rearrangement to give N-(β-anilinoethyl)-2-pyridone (**VI-68b**) occurred (217).

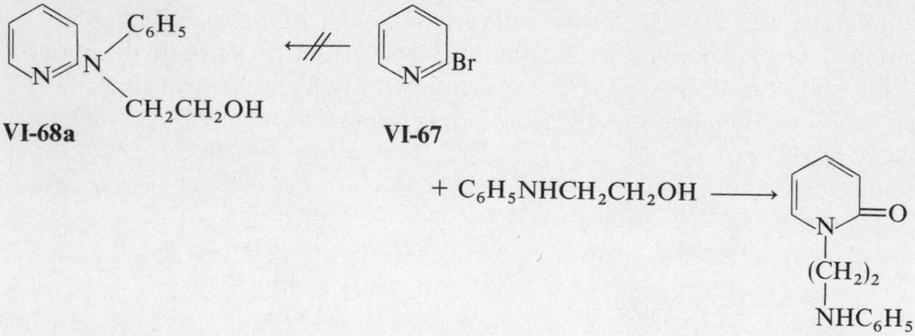

VI-68a   **VI-67**

$$+ C_6H_5NHCH_2CH_2OH \longrightarrow$$

VI-68b

Heinisch recently demonstrated the synthesis of 3-amino-s-triazolo-[4,3-a]pyridines (**VI-69**) from the reaction of 2-bromopyridine (**VI-67**) and thiosemicarbazide (217a).

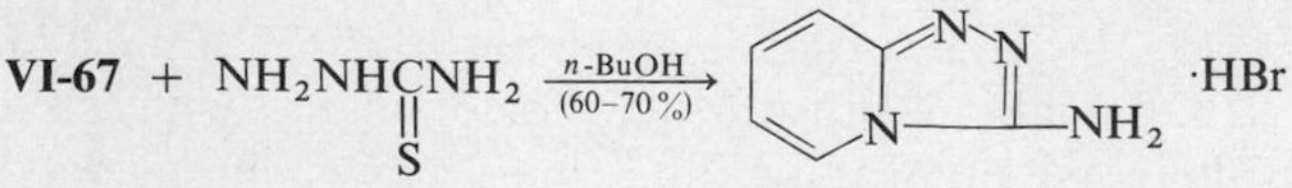

$$\textbf{VI-67} + NH_2NHCNH_2 \xrightarrow[(60-70\%)]{n\text{-BuOH}}$$

·HBr   **VI-69**

E. Displacement by Cyano and Similar Groups

2-Cyanopyridine was obtained from the gas phase reaction of 2-chloropyridine and hydrogen cyanide over nickel oxide-alumina catalyst at 580° (218). The conversion of 3-hydroxy-2-iodo- or 3-iodo-4-pyridone to the corresponding nitrile with cuprous cyanide (in xylene or isoamyl alcohol) simplified the former multistep routes to 3-hydroxypicolinic and 4-hydroxynicotinic acids (219). While hexafluorobenzene is inert to metal cyanides in aprotic solvents (220), pentafluoropyridine reacts with sodium cyanide in dimethyl formamide at low temperatures to give 4-cyanotetrafluoropyridine (56).

The solvent-free reaction of cuprous cyanide with the isomeric 2-bromo-3-, 5-, and 6-picolines gave good yields of the corresponding nitrile; in contrast, 2-bromo-4-picoline provided only small amounts of the desired nitrile (221).

F. Formation of Organometallic Compounds

A new route to pyridylmagnesium bromides or iodides (VI-70) features the reaction of 2-, 3-, or 4-halopyridines with phenylmagnesium halides in tetrahydrofuran or diethyl ether (222).

$$X = Cl, Br \qquad Y = Br, I \qquad \textbf{VI-70}$$

The mode of exchange with n-butyl-lithium and pentachloropyridine is solvent-dependent: tetrachloro-2-pyridyllithium is the major product in hydrocarbon solvents, while metal-halogen exchange at the 4-position predominates in diethyl ether (223, 224). Substitution at the 4-position is also favored for the pentabromopyridine/n-butyl-lithium system in diethyl ether (224a). Pentachloropyridine undergoes metalation with alkyl- and arylsilyllithium reagents to give 4-(alkyl- or arylsilyl)tetrachloropyridines (225). Heterohalogenated pyridines undergo selective metal-halogen exchange: n-butyl-lithium converted 2-bromo-6-chloropyridine and 4-iodotetrafluoropyridine to 6-chloro-2-pyridyl-lithium (226), and 2,3,5,6-tetrafluoro-4-pyridyl-lithium (142), respectively, while 4-iodotetrachloropyridine and lithium dimethylcopper gave 2,3,5,6-tetrachloro-4-pyridylcopper (227).

Pentachloropyridine (223, 224, 228, 229), pentabromopyridine (224a), and 4-bromo- and 4-iodotetrafluoropyridine (130, 142, 230) form Grignard reagents at the 4-position. Tetrachloro-4-pyridylmagnesium halides or the corresponding lithium derivative have been coupled with copper (I) halides (227), chlorotrimethylsilane (229), or mercuric chloride (224) to give 2,3,5,6-tetrachloro-4-pyridylcopper, trimethyl-(2,3,5,6-tetrachloro-4-pyridyl)silane, or bis(2,3,5,6-tetrachloro-4-pyridyl)mercury, respectively. Bis(2,3,5,6-tetrafluoro-4-pyridyl)mercury was prepared from tetrafluoro-4-pyridylmagnesium bromide and mercuric chloride or from 4-bromotetrafluoropyridine and lithium amalgam (*via* tetrafluoro-4-pyridyl-lithium) (230). (See also Ch. VII.)

Metal carbonyl anions react with pentafluoropyridine or 4-bromo-tetrafluoropyridine to give 4-substituted transition metal complexes (**VI-71**) (231–234a, b). Nuclear hydrogen atoms in tetrahalopyridines are sufficiently

$M = Mn(CO)_5$; $Re(CO)_5$; $(\pi\text{-}C_5H_5)Ru(CO)_2$;
$Mn(CO)_4PPh_3$; $(\pi\text{-}C_5H_5)Fe(CO)_2$

VI-71

acidic to undergo displacement by metals: for example, 2,3,4,6-tetrafluoro-pyridine can be converted to 2,3,4,6-tetrafluoro-5-pyridyl-lithium by *n*-butyl-lithium (180), while lithium dimethylcopper transformed 2,3,5,6-tetrachloropyridine to 2,3,5,6-tetrachloro-4-pyridylcopper (227).

2,3-Halopyridynes have been generated (through the substituted 3-pyridyl-lithium) by the reaction of *n*-butyl-lithium with 2,3,5,6-tetrachloro-4-(1-piperidyl)pyridine (235) or 2,3,5,6-tetrachloro-4-(trimethylsilyl)pyridine (225) and from *t*-butyl-lithium with 2,3,5,6-tetrachloro-4-methoxypyridine (236). [The formation of 4,5,6-trichloro-2,3-pyridyne and 2,5,6-trichloro-3,4-pyridyne from tetrachloropyridyl-lithium precursors has been intensively studied (237, 238)]. Lithium amalgam has permitted the generation of 2,3-pyridynes (**VI-72**) from 2-chloro-3-bromopyridine (79, 202, 239) or 4-, 5-, or 6-ethoxy-2,3-dibromopyridine (240); the pyridyne can be trapped with furan and converted to quinoline derivatives. 3,4-Pyridyne (**VI-63b**) has

VI-72

been generated from 3-bromo-4-chloropyridine using lithium amalgam (241); this method could not be applied successfully to the formation of 2,5,6-trifluoro-3,4-pyridyne from 4-bromotetrafluoropyridine (230).

G. C-Alkylation Reactions

2-Ethynylpyridines are formed from 2-iodopyridines and cuprous acetylides (242). 2-Bromopyridine has been widely employed in C-alkylation reactions with substituted phenylacetonitriles and sodium amide (243–246). α-Alkylation occurs preferentially with 2,5-dibromopyridine in the latter system (243).

4-Chloropyridine and Grignard reagents provide 4-alkylpyridines (247). Pentachloropyridine (**VI-19**) undergoes 4-substitution with Grignard reagents (225), or ethyl sodiomalonate (143). 4-Alkylation is also favored in

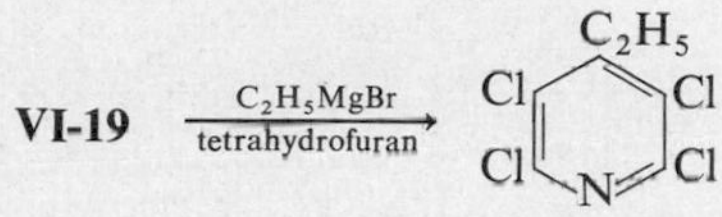

$$\textbf{VI-19} \xrightarrow[\text{tetrahydrofuran}]{C_2H_5MgBr}$$

VI-73

the reaction of pentafluoropyridine with alkyl- or aryl-lithium reagents (181, 192, 248). The alkylation site can be altered by changing the substrate. Pentachloropyridine-1-oxide undergoes 2- and 2,6-substitution with Grignard reagents; subsequent deoxygenation provides the corresponding alkylpyridine (454a,b). 2-Substitution results with tetrafluoro-4-picoline and organic lithium reagents (248).

H. Other Metathetical Reactions

The order of replacement of halogen by sulfite ion in the monochloropyridines was 4-Cl > 2-Cl > 3-Cl (249, 250). Preferential 4-substitution occurred in the reaction of pentafluoropyridine with sulfur nucleophiles such as the sulfite, thiophenoxide, and benzenesulfinate ions in aprotic solvents (250a).

O-Pyridyl ketoximes (**VI-74**) have been prepared from 2- or 4-bromopyridine and acetone oxime (251).

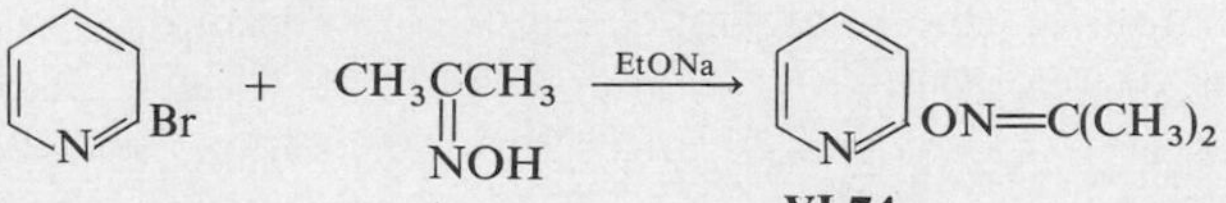

VI-74

2-Bromopyridine undergoes nucleophilic attack by pyridine-1-oxide (or picoline-, quinoline-, or isoquinoline-1-oxides), followed by an intramolecular nucleophilic attack to give N-(2-pyridyl)-2-pyridone (**VI-75**) (252–257) (see also Chapter IV). In contrast, 4-bromo- or 4-chloropyridine

VI-75

follows a different reaction path with these N-oxides, that is, the predominant formation of N-(4-pyridyl)-4-pyridone presumably arises by hydrolysis of the dimer of the 4-halopyridine (258, 259).

The Arbuzov rearrangement can be extended to heterocyclics; for example, pentachloropyridine (**VI-19**) and triethyl phosphite react to give diethyl 2,3,5,6-tetrachloro-4-pyridylphosphonate (**VI-76**) (142a).

I. Polymerization

2-Halopyridines do not self-condense under ambient conditions. 2-Fluoropyridine dimerizes at 130 to 140° to give N-(2-pyridyl)-2-fluoropyridinium fluoride (260, 261); 2-bromopyridine undergoes similar self-condensation at 250 to 300° (262). Prolonged heating of 2-chloro-, 2-bromo-, or 2-iodopyridine gave resins of uncertain structure (261).

Both 3-chloro- and 3-bromopyridine resist polymerization because of the low reactivity of the 3-halogen (262). During the preparation of 3-bromopyridine, a black solid with a polypyridylene structure was isolated (263).

Wibaut established the following order of conversion of 4-halopyridines to N-(4-pyridyl)-halopyridinium halide: F > Br > Cl > I (261). The self-condensation of 4-chloro- or 4-bromopyridine is acid-catalyzed; treatment of the glass equipment with alkali avoids this problem (247). 2,6-Di-*t*-butyl-4-chloropyridine exhibits greater storage stability than does 4-chloropyridine (265). The polymeric electrolytes previously obtained by the spontaneous polymerization of 4-chloro- or 4-bromopyridine (258, 259) are examples of "onium polymerization" (262, 266–268). The growth of the polymer chain takes place through the unshared electrons of the hetero atoms involving charge transfer complexes (**a**), followed by dimer formation (**b**), and so on (**VI-77**). The polymers exhibited electrical conductivity (10^{-3} to 10^{-7} Ω/cm) and were paramagnetic. Polypyridinium salts such as **VI-77** also show nematocidal properties (264).

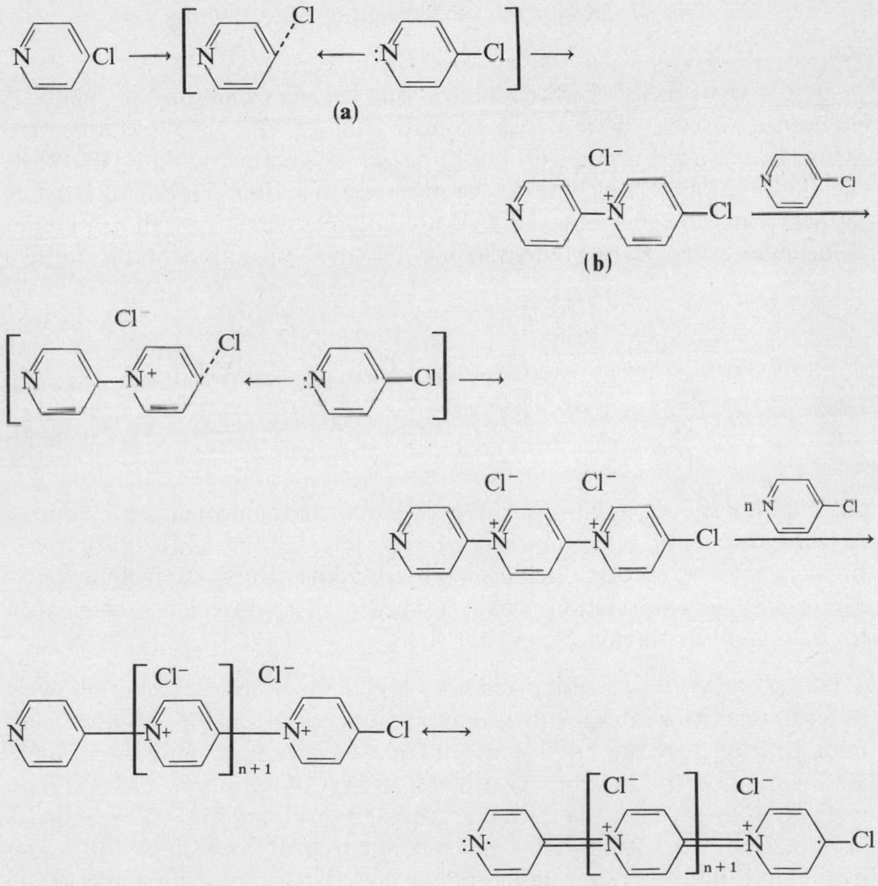

(**a**)

(**b**)

VI-77

III. Side-Chain Halogen Derivatives

1. Preparation

Many new techniques have been also described for the preparation of side-chain halogen derivatives, especially perfluoroalkyl pyridines. For example, the classical method for the preparation of such compounds featured successive side-chain chlorination of an alkyl pyridine, followed by halogen exchange with hydrogen fluoride (301, 302).

A. Halogenation of Saturated Side Chains

a. CHLORINATION. Earlier studies on the chlorination of picolines proceeded directly to the trichloromethyl stage (301–305). Mathes and Schuely developed a one-step route to 2-chloromethylpyridine (**VI-78**) by the chlorination of 2-picoline in the presence of sodium carbonate (carbon tetrachloride solvent) (306, 307). With polymethylated pyridines, monochlorination occurred predominantly at the 2-position. Other direct

$$2 \quad \underset{\underset{N}{\bigcirc}}{CH_3} \; + \; 2Cl_2 \; \xrightarrow[CCl_4]{Na_2CO_3} \; \underset{\underset{N}{\bigcirc}}{CH_2Cl} \; + \; \underset{\underset{N}{\bigcirc}}{CHCl_2} \; + \; \underset{\underset{N}{\bigcirc}}{CCl_3}$$

VI-78　　　　　　5%　　　　　　　<5%
65%

chlorination routes to chloromethylpyridines include chlorination conducted in sulfuric acid and in the presence of free-radical initiators [azo(isobutyronitrile)], for example, 3-chloromethylpyridine from 3-picoline (308). The gas-phase chlorination (steam as diluent) has provided a convenient route to 4-chloroalkylpyridines (309, 310).

b. BROMINATION.　Kutney and co-workers (311) observed the following order of reactivity rates for the isomeric picolines toward N-bromosuccinimide/benzoyl peroxide: 4 > 2 ≫ 3. (The 4-isomer gave only 4-tribromomethylpyridine, the 2-isomer furnished a mixture of the mono- and dibromomethylpyridines, and the 3-isomer resisted bromination.) The sluggish bromination of 2,3,5,6-tetrafluoro-4-picoline to give low yields of 4-bromomethyl-tetrafluoropyridine has been interpreted in terms of steric interaction of the fluorine atoms in the 3- and 5-positions (248).

B. Displacement of Oxygen

Chloromethylpyridines can be prepared from trichloroacetyl chloride with the corresponding 2-, 3-, or 4-hydroxymethylpyridine (315).

Substituted 2-chloromethylpyridines have been synthesized by the action of *p*-toluenesulfonyl chloride upon 2,6-lutidine-1-oxide (316, 317) and 5-ethyl-2-methylpyridine-1-oxide (316). Trichloroacetyl chloride also effected the conversion of 2-picoline-1-oxide into 2-chloromethylpyridine in 70 to 80 % yield (318). The transformation of picoline-1-oxides by reactive halides into chloromethylpyridines may involve a mechanism suggestive of nucleophilic attack by chloride ion on an exocyclic methylene carbon (**VI-79**) (316, 319).

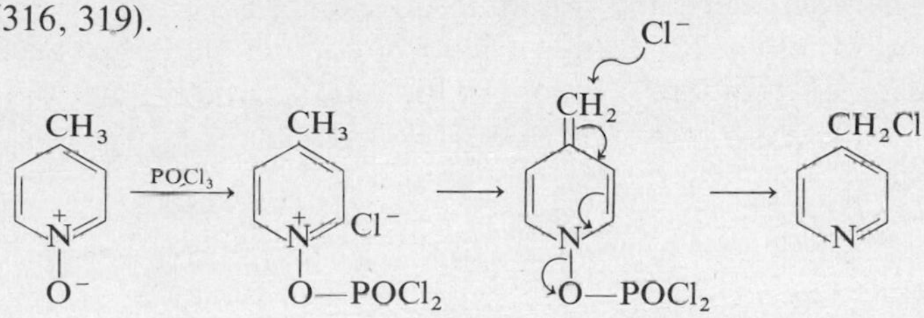

VI-79

Pyridinecarboxylic acids are converted to the corresponding trifluoromethylpyridine (**VI-80**) by sulfur tetrafluoride-hydrogen fluoride involving the *in situ* formation of the acid fluoride (320–322).

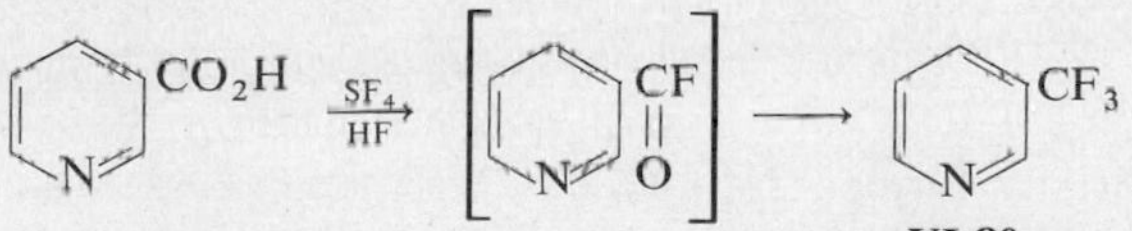

VI-80

C. Introduction of a Halogenated Side Chain

a. FREE RADICAL PERFLUOROALKYLATION OF PYRIDINE. Mixtures of α-, β-, and γ-perfluoroalkylpyridines (**VI-81**) are formed by the reaction of perfluoroalkyl iodides with pyridine (323). The α-isomer can be isolated from this mixture by treatment with 20 % hydrochloric acid.

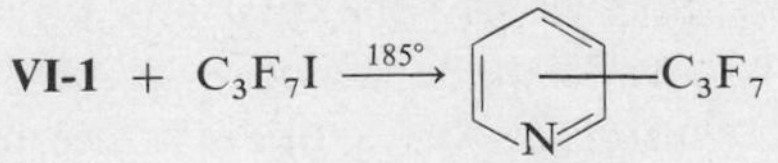

$$\textbf{VI-1} \ + \ C_3F_7I \ \xrightarrow{185°} \ \text{(pyridine)}-C_3F_7$$

VI-81

b. PERFLUOROALKYLATION WITH POLYFLUOROCARBANIONS. A technique described as the nucleophilic equivalent of the Friedel-Crafts reaction involving polyfluorocarbanions and perfluorinated substrates (e.g., pentafluoropyridine) gave displacement products such as **VI-82** (324a, b). Chambers and co-workers also noted that higher fluoro-olefin pressures favored

$$CF_3CF{=}CF_2 \xrightarrow[\text{sulfolane}]{KF} (CF_3)_2\bar{C}F \overset{+}{K} \xrightarrow[(94\%)]{C_5F_5N}$$

VI-82

polysubstitution, for example, perfluoro(pentaethylpyridine), $C_5(C_2F_5)_5N$, from tetrafluoroethylene and pentafluoropyridine. Heterohalogenated substrates, for example 3,5-dichlorotrifluoropyridine, undergo nucleophilic displacement at positions 2 and 4 with tetrafluoroethylene and cesium fluoride in dimethyl formamide **(VI-83)** (325).

VI-83

Steric factors influence the type of trisubstitution in the reaction of polyfluoroalkylcarbanions with pentafluoropyridine: with $CF_3CF_2^-$, **VI-84** (a product of kinetic control) is formed exclusively; the more bulky $(CF_3)_3C^-$ provides only **VI-85** (the isomer favored by thermodynamic control) (326, 326a, 327). With $(CF_3)_2CF^-$, a nucleophile of intermediate steric requirements, the initially formed 2,4,5-trialkyl isomer **(VI-84)** was subsequently rearranged to **VI-85** by heating with fluoride ion.

VI-84   **VI-85**

c. PYRIDINE-1-OXIDE ALKYLATION REARRANGEMENT. A novel approach to 2-polyfluoroalkylpyridines involves alkylation of pyridine- or picoline-1-oxides with terminally unsaturated perfluoroalkenes, which may involve rearrangement of a postulated isooxazolidine intermediate **VI-86** (328). For example, pyridine-1-oxide and hexafluoropropylene gave 2-(1,2,2,2-tetrafluoroethyl)pyridine **VI-87**.

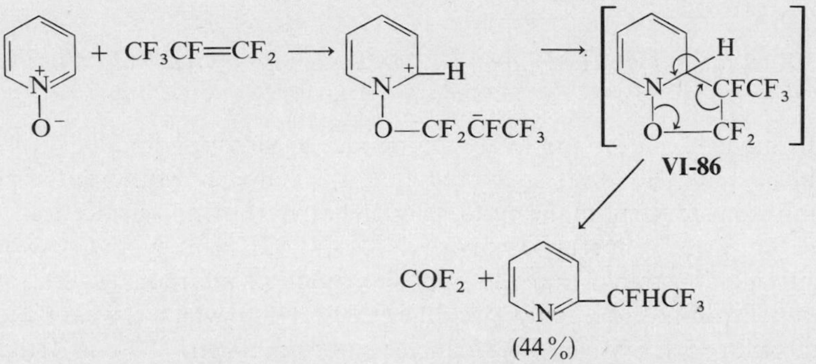

$$\text{(pyridine N-oxide)} + CF_3CF{=}CF_2 \longrightarrow \overset{+}{N}{-}H \longrightarrow \left[\begin{array}{c} H \\ \text{VI-86} \end{array} \right]$$

$$COF_2 + \text{(pyridine)}{-}CFHCF_3$$

(44%)

VI-87

d. PYRIDINE ANION ROUTES. 2-Pyridyllithium and alkylpyridine anions react with fluorinated-olefins (329), -esters (330), or -ketones (331–333) to provide pyridines bonded to fluorinated functional groups. Fluorinated

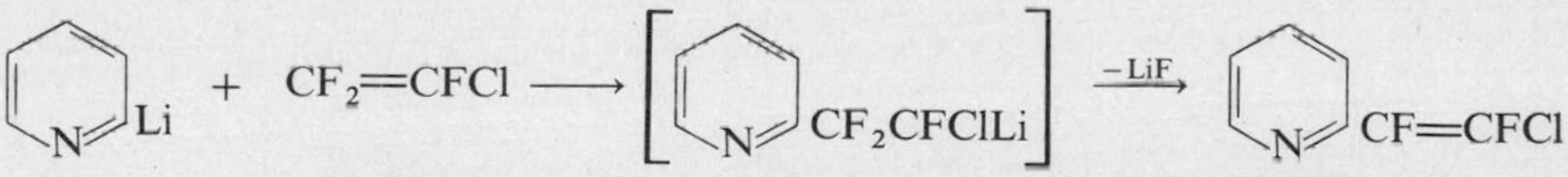

$$\text{(pyridine)}{-}Li + CF_2{=}CFCl \longrightarrow \left[\text{(pyridine)}{-}CF_2CFClLi \right] \xrightarrow{-LiF} \text{(pyridine)}{-}CF{=}CFCl$$

VI-88

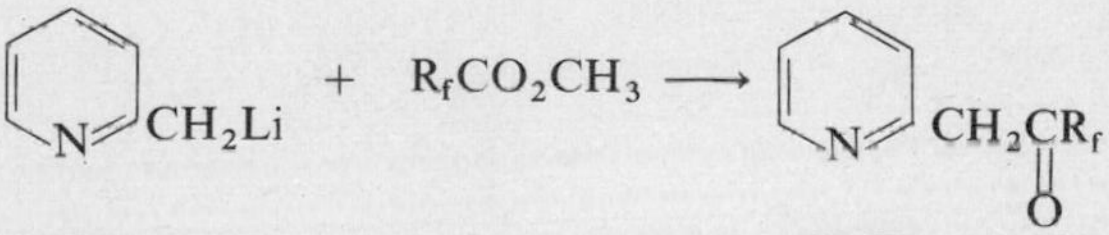

$$\text{(pyridine)}{-}CH_2Li + R_fCO_2CH_3 \longrightarrow \text{(pyridine)}{-}CH_2\underset{O}{\overset{\|}{C}}R_f$$

VI-89

$$\text{(pyridine)}{-}CH(C_6H_5)Li + C_6H_5\underset{O}{\overset{\|}{C}}CF_3 \longrightarrow \text{(pyridine)}{-}CH(C_6H_5)C(OH)(C_6H_5)CF_3$$

$$\xrightarrow{-H_2O} \text{(pyridine)}{-}C(C_6H_5){=}C(C_6H_5)CF_3$$

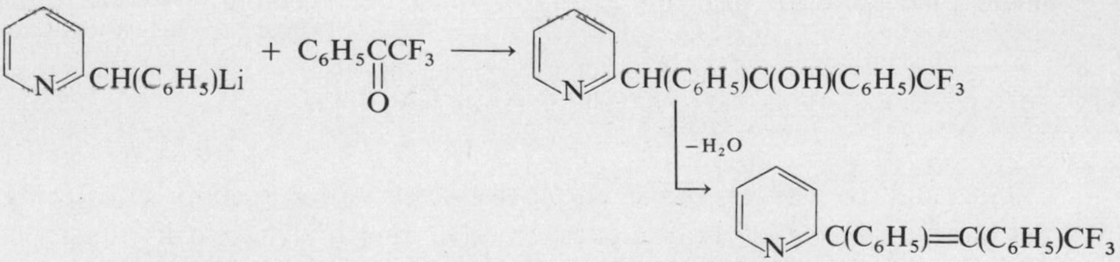

VI-90

heterocyclic β-diketones have been synthesized from the base-catalyzed reactions of acetylpyridines and fluorinated esters (334).

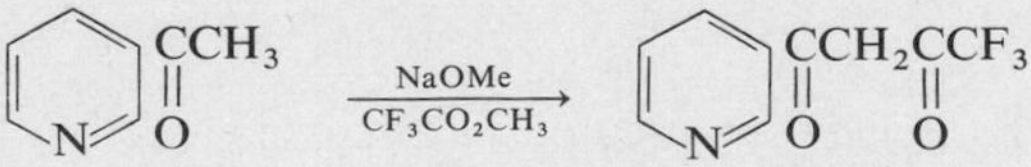

$$\text{(pyridine N-oxide)}{-}\underset{O}{\overset{\|}{C}}CH_3 \xrightarrow[CF_3CO_2CH_3]{NaOMe} \text{(pyridine N-oxide)}{-}\underset{O}{\overset{\|}{C}}CH_2\underset{O}{\overset{\|}{C}}CF_3$$

VI-91

D. *Addition of Hydrogen Halide to Unsaturated Side Chains*

The mode of addition of hydrogen halides to vinylpyridines has not been established. It had been suggested that the halogen would enter at the carbon atom adjacent to the pyridine nucleus (Markownikoff addition) (335). However, for 3,5-dimethyl-4-vinylpyridine (**VI-92**), Kutney and co-workers recently demonstrated that the opposite mode of addition occurs to give 3,5-dimethyl-4-(2-bromoethyl)pyridine (**VI-93**) (336) (which is what would be expected for such a Michael-type nucleophilic addition). These investigators postulated that vinylpyridines such as **VI-92** should react according to

resonance structure **VI-92a**; the initial step would be expected to involve nucleophilic addition of bromide ion to the terminal carbon in **VI-92a**.

E. *Involving Ring Closure*

Janz and co-workers have employed Diels-Alder reactions featuring perfluorinated nitriles and dienes to give thermodynamically unstable dihydropyridines (**VI-94**), which then undergo spontaneous aromatization to form 2-(perfluoroalkyl)pyridines (**VI-95**) (337–342). An analogous

transformation involves the previously discussed (Section II.1.F) reaction of perfluorocyclohex-1,3-diene with perfluorinated nitriles, the intermediate undergoing aromatization to give 2-(perfluoroalkyl)pyridines (149). The high reactivity for both types of ring closure reactions has been attributed to the enhanced dienophilic reactivity of the nitrile group imparted by the electronegative perfluoroalkyl group.

Another cyclization route involves the reaction of fluorinated 1,3-dicarbonyl compounds with cyanoacetamide to give trifluoromethyl-2(1H)-pyridones (**VI-96**) (343).

$$R = CH_3;\ CF_3;\ C_6H_5$$

The condensation of malonyl chloride with β-chloropropionitrile permitted the formation of 2-chloro-3-chloromethyl-4-hydroxy-6-pyridone (**VI-97**) in 40% yield (153d).

F. Ring Enlargement Reactions

Haloalkylpyridines have been generated in small quantities by ring enlargement reactions involving the treatment of aziridino-pyrrolines (**VI-98**) with hydrochloric acid: **VI-98** gave primarily 2-chloromethyl-2,5-dimethylpyrrolenine (**VI-99**) and lesser amounts of 2-chloromethyl-5-methylpyridine (**VI-100**) (344, 345).

G. *Aromatization Processes*

Perfluoro-(2-, 3-, and 4-picolines), for example **VI-103**, have been obtained by successive electrochemical fluorination of the specific picoline (e.g., **VI-101**), followed by defluorination of the corresponding perfluoro-(*N*-fluoromethylpiperidines) **(VI-102)** with iron at elevated temperatures (580°/2 mm) (9). One preparative disadvantage is the low yield (2 to 4%) during the electrochemical fluorination step.

VI-101   VI-102   VI-103

2. Properties

The stability of the haloalkylpyridines is in the order: $F > Cl > Br > I$. Increasing halogen substitution is associated with greater haloalkylpyridine stability: $CX_3 > CHX_2 > CH_2X$ (306). While 2-chloromethylpyridine can be distilled, both 3- and 4-chloromethylpyridine, as well as 2-bromomethylpyridine, undergo decomposition. Monochloroalkylpyridines exhibit instability under ambient conditions, which may be reduced by storage at low temperatures or in hydrocarbon solvents up to 20% concentration (310). 3-Chloromethylpyridine, upon standing at room temperature, develops heat until vigorous detonation occurs (306). Distillation of large quantities of 4-(3-chloropropyl)pyridine occasionally leads to violent, exothermic polymerization (346). However, monohaloalkylpyridines exhibit sufficient basicity to form stable hydrogen halide salts.

Dihaloalkylpyridines are relatively unstable and resinify on standing, but can form hydrochloride and picrate salts. In contrast, trihaloalkylpyridines are appreciably stable; however, their weak basicity precludes formation of the salts above.

Haloalkylpyridines cause intense inflammation of the skin. Vapors of solutions of the free bases produce strong irritation of the eyes and mucous membranes (306). Some of these compounds, for example, 4-bromomethylpyridine and its hydrobromide salt, are severe vesicants (347).

TABLE VI-8. Properties of the Side-Chain Halopyridines

Compound	Method of preparation	Physical properties	Ref.
2-Chloromethylpyridine	2-Hydroxymethylpyridine and $SOCl_2$; 2-picoline with Cl_2/Na_2CO_3 or Cl_2/steam (new routes)	b.p. 45–47°/1.5 mm; n_D^{20} 1.5365; picrate, m.p. 149–150°; HCl salt, m.p. 128°	306, 309, 373
3-Chloromethylpyridine	3-Picoline with Cl_2/H_2SO_4/azobis(isobutyronitrile) or Cl_2/steam (new routes)	Unstable oil; picrate m.p. 130–131°	308, 309
4-Chloromethylpyridine	4-Picoline-1-oxide and $POCl_3$; 4-picoline with Cl_2/steam (new routes)	b.p. 39–40°/1.5 mm	309, 319
2-Chloromethyl-3-picoline	2,3-Lutidine and Cl_2/Na_2CO_3	b.p. 59–61°/1.4 mm; n_D^{20} 1.5407; picrate, m.p. 146–146.5°; HCl salt, m.p. 157°	306
2-Chloromethyl-4-picoline	2,4-Lutidine and Cl_2/Na_2CO_3	b.p. 51–52°/0.9 mm; n_D^{20} 1.5326; picrate, m.p. 159–160°	306
6-Chloromethyl-3-picoline	6-Hydroxymethyl-3-picoline with $SOCl_2$ or $POCl_3$	b.p. 80°/5 mm; picrate, m.p. 160–161°	345, 397
6-Chloromethyl-2-picoline	2,6-Lutidine-1-oxide and $POCl_3$; 2,6-lutidine with Cl_2/Na_2CO_3 or $PCl_5/POCl_3$; 6-hydroxymethyl-2-picoline and $SOCl_2$ (new routes)	b.p. 81°/12 mm; n_D^{20} 1.5315; picrate, m.p. 162–163°; HCl salt, m.p. 155–156°	34, 306, 307, 316, 375
2-Chloromethyl-4,6-lutidine	2,4,6-Collidine and Cl_2/Na_2CO_3	b.p. 68–70°/1.4 mm; n_D^{20} 1.5290; picrate, m.p. 126–128°	306, 307
2-Chloromethyl-5-ethylpyridine	5-Ethyl-2-methylpyridine-1-oxide with $POCl_3$ or CH_3COCl; 5-ethyl-2-methylpyridine with Cl_2/steam	b.p. 90°/3 mm; picrate, m.p. 121°	308, 309, 316, 398
3-Chloromethyl-2-picoline	3-Hydroxymethyl-2-picoline and $SOCl_2$	Picrate, m.p. 140–141° (of 3-methoxymethyl-2-picoline derivative)	399

463

Table VI-8 (*Continued*)

Compound	Method of preparation	Physical properties	Ref.
3-Chloromethyl-4-picoline	3-Hydroxymethyl-4-picoline and SOCl$_2$	b.p. 81–85°/35 mm; picrate, m.p. 161°	400
5-Chloromethyl-2-picoline	5-Hydroxymethyl-2-picoline and SOCl$_2$ or POCl$_3$	b.p. 65–71°/3 mm; picrate, m.p. 160–161°	397, 401
2-(1-Chloroethyl)pyridine	2-Ethylpyridine and Cl$_2$/steam	b.p. 61–62°/4.5 mm; picrate, m.p. 98–99.5°	309, 310
4-(1-Chloroethyl)pyridine	4-Ethylpyridine and Cl$_2$/steam; 4-(1-hydroxyethyl)-pyridine and SOCl$_2$	b.p. 74–77°/4 mm; picrate, m.p. 100–101°	309, 310, 382
3-(2-Chloroethyl)pyridine	3-(2-Hydroxyethyl)pyridine and SOCl$_2$	HCl salt, m.p. 154°	402
6-(2-Chloroethyl)-3-picoline	6-(2-Hydroxyethyl)-3-picoline and SOCl$_2$	b.p. 85–86°/5 mm; picrate, m.p. 129°	382
6-(1-Chloroethyl)-3-picoline	6-(1-Hydroxyethyl)-3-picoline and SOCl$_2$	b.p. 77–80°/7 mm; picrate, m.p. 108–109°	382
2-(1-Chloropropyl)pyridine	2-(1-Hydroxypropyl)pyridine and POCl$_3$	b.p. 80–90°/4 mm; picrate, m.p. 138–140°	356
4-(1-Chloropropyl)pyridine	4-(1-Hydroxypropyl)pyridine and POCl$_3$	b.p. 90–95°/3 mm; picrate, m.p. 111–113°	356
4-(3-Chloropropyl)pyridine	4-(3-Hydroxypropyl)pyridine and SOCl$_2$	b.p. 90–91°/1.5 mm; n_D^{25} 1.5230	346, 403
2-(1-Methyl-1-chloroethyl)pyridine	2-iso-Propylpyridine, Cl$_2$/steam	b.p. 69–70°/7 mm	309
4-(4-Chlorobutyl)pyridine	4-(4-Hydroxybutyl)pyridine and SOCl$_2$	b.p. 96–115°/0.7 mm	403
2-Bromomethylpyridine	2-Hydroxymethylpyridine and PBr$_3$ or HBr	HBr salt, m.p. 145–146°; lachrymator	373, 404, 405
6-Bromomethyl-2-picoline	6-Hydroxymethyl-2-picoline and HBr; 2,6-lutidine, NBS/benzoyl peroxide, and hν	Picrate, m.p. 142–143°; dermatic and lachrymatory	312, 375
2-(1-Bromoethyl)pyridine	2-Ethylpyridine, NBS, and benzoyl peroxide	b.p. 89.5–90.5/7 mm; n_D^{25} 1.5550–1.5561	313
4-(1-Bromoethyl)pyridine	4-Ethylpyridine, NBS, and benzoyl peroxide	HBr salt, m.p. 151.0–151.5°	336
4-(2-Bromoethyl)-3,5-dimethylpyridine	4-(1-Hydroxyethyl)-3,5-dimethylpyridine and HBr; 3,5-dimethyl-4-vinylpyridine and HBr	HBr salt, m.p. 208–209°	336

464

Compound	Preparation	Properties	Ref.
3-(3-Bromopropyl)pyridine	3-(3-Ethoxypropyl)pyridine and HBr	HBr salt, m.p. 108°; picrate, m.p. 136°	406
3-(1-Bromobutyl)pyridine	3-Butylpyridine, N-bromosuccinimide, benzoyl peroxide	Oil	407
3-(4-Bromobutyl)pyridine	3-(4-Ethoxybutyl)pyridine and HBr	HBr salt, m.p. 110–111°; picrate, m.p. 87°	406
2-(5-Bromopentyl)pyridine	2-(5-Phenoxypentyl)pyridine and HBr	Reacted with $(CH_3)_2NH$ and characterized as 2-(5-dimethylaminopentyl)pyridine, b.p. 100–108°/2 mm	408
2-(6-Bromohexyl)pyridine	2-(6-Acetoxyhexyl)pyridine and HBr/CH_3CO_2H	Picrate, m.p. 155–159°	409
2-Dichloromethylpyridine	2-Picoline, Cl_2, and Na_2CO_3	b.p. 62–64°/1.2 mm; n_D^{20} 1.5530; picrate, m.p. 117–118°	306, 307
2-(1,2-Dichloroethyl)pyridine	2-Vinylpyridine·HCl and Cl_2	HCl salt, m.p. 125–126.5°	410
2-(1,2-Dibromoethyl)pyridine	2-Vinylpyridine·HCl and Br_2	HCl salt, m.p. 143–145.5°; picrate, m.p. 118–119°; free base polymerizes on standing	410, 412
4-(1,1-Dibromoethyl)pyridine	4-Ethylpyridine, NBS/benzoyl peroxide	HBr salt, m.p. 158–160°	311
2-Trifluoromethylpyridine	CF_3CN and butadiene; picolinic acid and SF_4/HF	b.p. 143°/745 mm; n_D^{25} 1.4155	320, 337
3-Trifluoromethylpyridine	Nicotinic acid and SF_4/HF	b.p. 113–115°; n_D^{25} 1.4150	320
4-Trifluoromethylpyridine	Isonicotinic acid and SF_4/HF	b.p. 108–110°; n_D^{25} 1.4155	320
2-Difluoromonochloromethylpyridine	$ClCF_2CN$ with butadiene; HF and 2-trichloromethyl-pyridine	b.p. 162°; 103.5–106.0°/100 mm; n_D^{25} 1.4646	301, 341
3-Methyl-2-trifluoromethylpyridine	CF_3CN and pentadiene	b.p. 174°; n_D^{20} 1.466	342
4-Methyl-2-trifluoromethylpyridine	CF_3CN and isoprene	b.p. 170°; n_D^{20} 1.4298	342
5-Methyl-2-trifluoromethylpyridine	CF_3CN and isoprene	b.p. 171°; n_D^{20} 1.4298	342
6-Methyl-2-trifluoromethylpyridine	CF_3CN and pentadiene	b.p. 154°; n_D^{20} 1.4278	342
6-Methyl-3-trifluoromethylpyridine	6-Methylnicotinic acid and SF_4/HF	b.p. 129–130°; m.p. 24–25.5°	411
4-Tribromomethylpyridine	4-Picoline, NBS/benzoyl peroxide	m.p. 71–72°; picrate, m.p. 162.5–167.0°	311
2-(1,1,2-Trichloroethyl)pyridine	2-(α-Chlorovinyl)pyridine·HCl and Cl_2	HCl salt, m.p. 128–129°	410

Table VI-8 (*Continued*)

Compound	Method of preparation	Physical properties	Ref.
2-(1,2,2,2-Tetrafluoroethyl)pyridine	Pyridine-1-oxide and hexafluoropropylene	b.p. 142°; n_D^{28} 1.4109	328
6-(1,2,2,2-Tetrafluoroethyl)-2-picoline	2-Picoline-1-oxide and hexafluoropropylene	b.p. 154°; n_D^{25} 1.4189	328
2- and 6-(1,2,2,2-Tetrafluoroethyl)-3-picoline (80:20 mixture)	3-Picoline-1-oxide and hexafluoropropylene	b.p. 166°; n_D^{26} 1.4248	328
2-(1,2,2,2-Tetrafluoroethyl)-4-picoline	4-Picoline-1-oxide and hexafluoropropylene	b.p. 166°; n_D^{26} 1.4248	328
2-Pentafluoroethylpyridine	CF_3CF_2CN and butadiene	b.p. 151°; n_D^{25} 1.3949	341
2-(2-Iodotetrafluoroethyl)pyridine	Iodine and silver 3-(2-pyridyl)perfluoropropionate	b.p. 129–131°/45 mm; m.p. 20.0–20.5°; n_D^{18} 1.5081	441
2-Heptafluoropropylpyridine	$CF_3CF_2CF_2CN$ and butadiene; C_3F_7I and pyridine; decarboxylation (400°) of 2-(heptafluoropropyl)-isonicotinic acid	b.p. 162°; n_D^{25} 1.3814 d_{23} 1.438	323a,b, 341
3- and 4-Heptafluoropropylpyridine (mixture)	C_3F_7I and pyridine	b.p. 137–138°; n_D^{20} 1.3762	323
2-(Heptafluoropropyl)-4-picoline	$CF_3CF_2CF_2CN$ and isoprene; 4-picoline and C_3F_7I	b.p. 176°; 107–108°/80 mm; n_D^{25} 1.4040	323b, 342
3-(Heptafluoropropyl)-4-picoline	4-Picoline and C_3F_7I	b.p. 87°/90 mm; n_D^{18} 1.3928 d_{18} 1.4300	323b
2-(Heptafluoropropyl)-5-picoline	$CF_3CF_2CF_2CN$ and isoprene	b.p. 181°; n_D^{25} 1.3934	342
2-(3,4-Dichloro-heptafluorobutyl)pyridine	Pyridine and $ICF_2CF_2CFClCF_2Cl$	b.p. 124–125°/25 mm; n_D^{20} 1.4298	323
3- and 4-(3,4-Dichloroheptafluorobutyl)-pyridine (mixture)	Pyridine and $ICF_2CF_2CFClCF_2Cl$	n_D^{20} 1.4304	323
2,3-Bis(chloromethyl)pyridine	2,3-Bis(hydroxymethyl)pyridine/$SOCl_2$	Picrate, m.p. 143–144°	413
2,6-Bis(chloromethyl)pyridine	2,6-Bis(hydroxymethyl)pyridine/$SOCl_2$; 2,6-lutidine, H_2SO_4, Cl_2, and azobis(isobutyronitrile)	b.p. 95–110°/3 mm; m.p. 74–75°; lachrymator	308, 375
3,5-Bis(chloromethyl)pyridine	3,5-Bis(hydroxymethyl)pyridine/$SOCl_2$	m.p. 86–87°	413
3,4-Bis(chloromethyl)-6-methylpyridine	3,4-Bis(hydroxymethyl)-6-methylpyridine/$SOCl_2$	HCl salt, m.p. 175–177°	414
3,5-Bis(chloromethyl)-2,6-lutidine	3,5-Bis(hydroxymethyl)-2,6-lutidine and $SOCl_2$	m.p. 107–108.5°	363
2,6-Bis(chloromethyl)-3,4,5-trimethyl-pyridine	2,6-Bis(hydroxymethyl)-3,4,5-trimethylpyridine and $POCl_3$	m.p. 96–96.3°	415

466

Compound	Method	Properties	Ref.
3,5-Bis(1-bromoethyl)pyridine	From 3,5-bis(1-hydroxyethyl)pyridine and HBr	HBr salt, m.p. 185°	416
2,6-Bis(1-bromoethyl)pyridine	From 2,6-bis(1-hydroxyethyl)pyridine and HBr	m.p. 84–89°; dermatic and lachrymatory	375
2,3-Bis(trifluoromethyl)pyridine	2,3-Pyridinedicarboxylic acid with SF_4/HF	b.p. 158°; 35°/6 mm; m.p. −2°	321, 322
2,4-Bis(trifluoromethyl)pyridine	2,4-Pyridinedicarboxylic acid with SF_4/HF	b.p. 121–122°	322
2,5-Bis(trifluoromethyl)pyridine	2,5-Pyridinedicarboxylic acid with SF_4/HF	b.p. 130–131°; m.p. 27–28°	322
2,6-Bis(trifluoromethyl)pyridine	2,6-Pyridinedicarboxylic acid with SF_4/HF	b.p. 152–154°; m.p. 55–56°	322
3,4-Bis(trifluoromethyl)pyridine	3,4-Pyridinedicarboxylic acid with SF_4/HF	b.p. 132–133°	322
3,5-Bis(trifluoromethyl)pyridine	3,5-Pyridinedicarboxylic acid with SF_4/HF	b.p. 117–118°; m.p. 35–36°	322
2,6-Bis(trichloromethyl)pyridine	2,6-Lutidine and $POCl_3/PCl_5$	m.p. 82–83°	395
Perfluoro(pentaethylpyridine)	Pentafluoropyridine, tetrafluoroethylene, KF, and sulfolane	m.p. 77–78°	324
3-(Heptafluoro-n-propyl)-4-dichloromethylpyridine	3-(Heptafluoro-n-propyl)-4-methylpyridine and Cl_2/CCl_4/benzoyl peroxide	b.p. 128–130°/85 mm; m.p. 31–32°	323b
2-(Heptafluoro-n-propyl)-4-trichloromethylpyridine	2-(Heptafluoro-n-propyl)-4-methylpyridine and Cl_2/CCl_4/benzoyl peroxide	b.p. 117–118°/22 mm; d_{13} 1.6535; n_D^{13} 1.4400	323b
2-(1-Chlorovinyl)pyridine	From 2-(1,2-dichloroethyl)pyridine and Et_3N or EtOH	HCl salt, m.p. 139–143°	410
2-(4-Chloro-3-pentenyl)pyridine	2-Picolyl-lithium and $ClCH_2CH=CClCH_3$	b.p. 118–121°/9 mm; n_D^{20} 1.5261; d_{20} 1.0702; picrate, m.p. 98–99°	417
2-(1-Bromovinyl)pyridine	From 2-(1,2-dibromoethyl)pyridine and Et_3N or EtOH	b.p. 98°/15 mm; HBr salt, m.p. 150–153°; picrate, m.p. 134°	410, 412
5-(1-Bromovinyl)-2-picoline	From 5-vinyl-2-picoline and Br_2/acetic acid	b.p. 103–104°/10 mm; m.p. 8°; n_D^{20} 1.5820; d_{20} 1.4154; picrate, m.p. 167°	418
2-(1-Chloro-1,2-difluorovinyl)pyridine	2-Pyridyl-lithium and chlorotrifluoroethylene	b.p. 115°/16 mm	329
2-(Perfluorovinyl)pyridine	Pyrolysis of sodium 3-(2-pyridyl)perfluoropropionate; dehydrofluorination of 2-(1,2,2,2-tetrafluoroethyl)pyridine	b.p. 42°/7 mm; n_D^{19} 1.4708	441, 447
4-(2-Pyridyl)heptafluoro-1-butene	2-(3,4-Dichloroheptafluorobutyl)pyridine and zinc dust	b.p. 100–102°/46 mm; n_D^{21} 1.4074; d^{21} 1.495	441

467

TABLE VI-9. Properties of the Arylhaloalkyl Pyridines and Related Compounds

Compound	Method of preparation	Physical properties	Ref.
2-Chloromethyl-6-(p-chlorophenyl)-pyridine	6-(p-Chlorophenyl)-2-picoline-1-oxide and acetyl chloride	m.p. 110–111°	436
3-Chloromethyl-6-phenylpyridine	5-Hydroxymethyl-2-phenylpyridine and SOCl$_2$	b.p. 100–110°/0.02 mm; m.p. 80–81°; HCl salt, m.p. 193–195°	419
3-Chloromethyl-6-(p-chlorophenyl)-2-picoline	3-Hydroxymethyl-6-(p-chlorophenyl)-2-picoline and SOCl$_2$	HCl salt, m.p. 171–172°	436
4-Chloromethyl-2-phenylpyridine	4-Hydroxymethyl-2-phenylpyridine and SOCl$_2$	HCl salt, m.p. 203–205°	419
2-(α-Chlorobenzyl)pyridine	2-(α-Hydroxybenzyl)pyridine and SOCl$_2$	b.p. 126–131°/0.3 mm	420
3-(α-Chlorobenzyl)pyridine	3-(α-Hydroxybenzyl)pyridine and SOCl$_2$	Not described	421
4-(α-Chlorobenzyl)-2,6-lutidine	4-(α-Hydroxybenzyl)-2,6-lutidine and SOCl$_2$	b.p. 140°/0.07 mm; picrate, m.p. 155–157°	404
1,2-Bis(6-bromomethyl-2-pyridyl)ethane	1,2-Bis(6-hydroxymethyl-2-pyridyl)ethane and HBr/CH$_3$CO$_2$H	m.p. 118–120° (dec.)	375

TABLE VI-10. Properties of Halostilbazoles and Related Compounds

Compound	Method of preparation	Physical properties	Ref.
1,2-Bis(4-chloro-3-pyridyl)ethane	4,4'-Dichloro-3,3'-dipicolylether-1,1'-dioxide with NH_3 or H_2/Pd-C	m.p. 115–115.5°; 2HCl salt, m.p. 291° (dec.); picrate, m.p. 228°	169
1,2-Dibromo-bis(6-methyl-2-pyridyl)ethane	1,2-Bis(6-methyl-2-pyridyl)ethane and $Br_2/CHCl_3$	m.p. 194–196° (dec.)	375
o-Bromostilbazole dibromide	o-Bromostilbazole and Br_2	m.p. 166–169°	422
1-Phenyl-2-(2-pyridyl)-1-trifluoromethylethane	Reduction of 1-phenyl-2-(4-pyridyl)-1-trifluoromethylethylene	m.p. 70–74°	333
1,1,2,2-Tetrachloro-bis(2,3,5,6-tetrafluoro-4-pyridyl)ethane	2,3,5,6-Tetrafluoro-4-picoline with Cl_2/hv or SO_2Cl_2	m.p. 288–289°	248
1,1,2,2-Tetrachloro-bis[2-(heptafluoro-n-propyl)-4-pyridyl]ethane	2-(Heptafluoro-n-propyl)-4-(trichloromethyl)pyridine and copper	m.p. 148–150°; N-methyl perchlorate, m.p. 174–175°	323b
1,2-Bis(3-pyridyl)hexafluoropropane	3-Iodopyridine, 1,2-diodohexafluoropropane, Cu, and DMF	m.p. 63°	423
Perfluoro-1,3-bis(2-pyridyl)propane	Perfluorocyclohexa-1,3-diene and 1,2-dicyano-hexafluoropropane	b.p. 45–55°/ca. 10^{-1} mm	149
1,2-Difluoro-1,2-bis(2-pyridyl)ethylene	2-Pyridyl-lithium and $CF_2{=}CF_2$	m.p. 91°	329
1-Phenyl-2-(4-pyridyl)-1-trifluoromethyl-ethylene	Dehydration of 1-phenyl-2-(4-pyridyl)-1-trifluoromethyl-ethanol	m.p. 158–171°	333
1,2-Diphenyl-2-(2-pyridyl)-1-trifluoromethyl-1,2-ethylene	Dehydration of 1,2-diphenyl-2-(2-pyridyl)-1-trifluoromethyl-ethanol	m.p. 87.0–88.5°	331
2-(2-Phenyltetrafluoroethyl)-4-picoline	4-Picoline, 2-phenyl-1-iodotetrafluoroethane, and sodium acetate	m.p. 52–53°; N-methyl perchlorate, m.p. 153–154°	441

469

Properties of new compounds synthesized since the previous review (1) are listed as follows: side-chain halopyridines (Table VI-8); arylhaloalkyl pyridines and related compounds (Table VI-9); and halostilbazoles and related compounds (Table VI-10).

3. Reactions

A. Hydrolysis

Hydrolysis of 4-chloromethyl-tetrafluoropyridine (VI-104) could not be effected by acid, while aqueous alkali carbonates gave polyethers (VI-106), arising through nucleophilic attack by alkoxide derived from VI-105 at position 2 of another molecule (248).

CH$_2$Cl — VI-104 → (Na$_2$CO$_3$) → [CH$_2$OH — VI-105] → (Na$_2$CO$_3$) →

CH$_2$O — ... CH$_2$O — ... CH$_2$OH

VI-106 $n = 1 - 3$

An attractive route to pyridyl ketones, for example, 4-acetylpyridine, features hydrolysis of 4-(1,1-dibromoethyl)pyridine with aqueous sodium carbonate (311). Trifluoromethyl groups bound to pyridine are hydrolyzed to the carboxylic acid by oleum [for example, tetrafluoropyridine-4-carboxylic acid from perfluoro(4-picoline) (9)]. The cyano group in ethyl 5-cyano-6-methyl-2-trifluoromethyl-3-pyridinecarboxylate is preferentially hydrolyzed by concentrated sulfuric acid (348).

B. Metathesis

a. SULFUR-CONTAINING NUCLEOPHILES. Numerous sulfur-containing pyridines (VI-107) have been prepared from 2- or 4-chloromethylpyridine with sulfur nucleophiles (319, 349–352) (see also Chapter XV). Trilithium phosphorothioate and 2-chloromethylpyridine react at pH 9 to form dilithium-S-(2-pyridyl)methylphosphorothioate (353). Biologically active

$$\text{(pyridine ring)}-CH_2Y$$

VI-107

$Y = SH; SCN; CSNH_2; SC(=NH)NH_2\cdot HCl; S\text{-alkyl}; SO_2\text{-alkyl}; S\text{-aryl}; SO_2\text{-aryl};$

$SO_2C_6H_4NH_2\text{-}p$ (and N-acetyl)

compounds of the type **VI-108** are formed from the reaction of halomethyl-pyridines and $RR'P(:X)YH(X, Y = O$ or $S)$ (354).

$$\text{(pyridine ring)}-CH_2YP(X)RR'$$

VI-108

The first recorded preparation of a pyridylmethanesulfonic acid was accomplished by the reaction of ethyl 2-bromomethylnicotinate and aqueous sodium sulfite (314).

b. ALKOXIDE. Chloromethylpyridines undergo exchange reactions with alkoxides derived from alcohols (355, 356), glycols (357), cellulose (358), and phenol (356). Nuclear halogens are displaced in preference to side-chain halogen in perfluorinated alkylpyridines (325).

$$\underset{\text{CF(CF}_3)_2}{\text{F-pyridine-F}} + NaOCH_3 \longrightarrow \underset{\text{CF(CF}_3)_2}{\text{F-pyridine-OCH}_3} \quad \textbf{VI-109}$$

c. AMINATION. Preferential displacement of side-chain halogen in 4-bromomethyltetrafluoropyridine with aqueous ammonia occurs under ambient conditions to give only 4-aminomethyltetrafluoropyridine (248). Displacement of side chain halogen by a wide variety of amines have been reported: iso- and n-propylamine (359, 360), N,N-diethylethylenediamine (361), N-(4-hydroxybutyl)aminoethanol (362), ethyleneimine (363), and theophylline (364). The isomeric N-(2-, 3-, and 4-pyridylmethyl)-1,2-diaminoethanes prepared from ethylene diamine exhibit chelating properties (365). Aniline hydrochloride and 2-chloromethylpyridine form N-(2-picolyl)aniline **VI-110** at 135 to 140°; at higher temperatures (225 to 230°), rearrangement occurs to give 2-(p-aminobenzyl)pyridine (**VI-111**) (366).

In contrast to benzotrifluoride, the trifluoromethyl group in 2-(trifluoro-methyl)pyridine can be displaced by treatment with sodium amide-ammonia to give 2-aminopyridine in 88% yield (367).

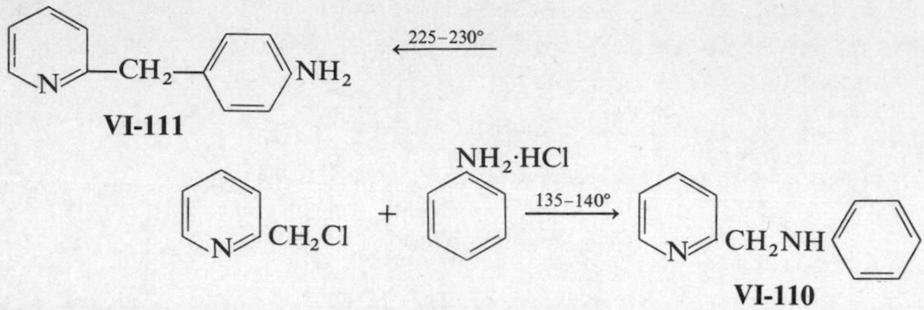

VI-111

VI-110

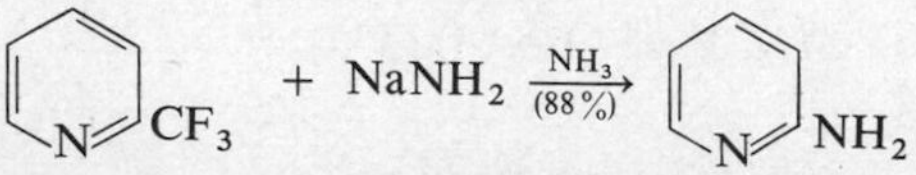

VI-112

2-Pyridine aldoximes were prepared by Danaher and co-workers from 2-chloromethylpyridines and hydroxylamine hydrochloride in aqueous ethanol (368). The initial nucleophilic displacement provides 2-pyridylmethylhydroxylamine **VI-113**, followed by protonation-elimination to give the aldimine **VI-114** and subsequent attack by excess hydroxylamine to form 2-pyridine aldoxime **VI-115**.

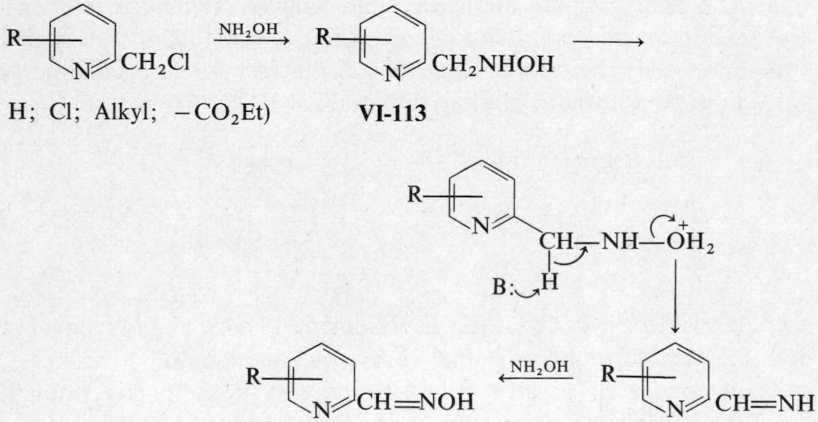

(R = H; Cl; Alkyl; −CO₂Et)   **VI-113**

VI-115   **VI-114**

d. PHOSPHORUS NUCLEOPHILES. Numerous pyridylmethylphosphonates **(VI-116)** have been synthesized from 2-, 3-, or 4-chloromethylpyridine with di- or trialkyl phosphites and diphenylphosphine oxide (369–372).

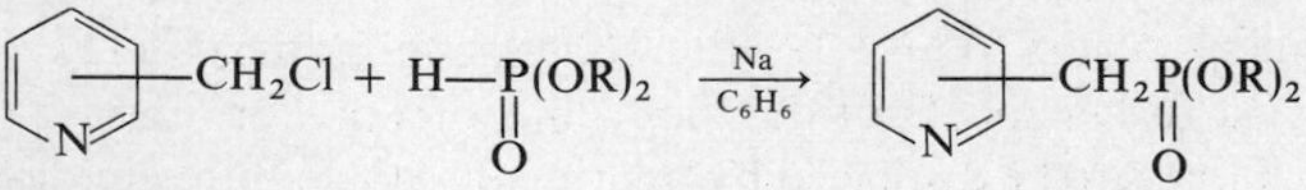

VI-116

e. CYANIDES. 2- And 4-chloromethylpyridine, 2-chloromethyl-6-methyl-pyridine, and 2,6-bis(chloromethyl)pyridine undergo typical exchange reactions with alkali cyanides (373–375).

C. C-Alkylation and Coupling Reactions

Isomer-free benzylpyridines can be prepared from the alkylation of benzene by 2-, 3-, or 4-chloromethylpyridine, 2-chloromethyl-6-methyl-pyridine or 2,6-bis(chloromethyl)pyridine (376).

VI-117

Nitriles containing active hydrogen atoms, e.g., phenylacetonitrile (317), diphenylacetonitrile (377), or α-(1-naphthyl)acetonitrile (378) undergo C-alkylation reactions with 2-, 3-, and 4-chloromethylpyridine (377–380), 2-chloromethyl-4-methylpyridine (317), and 2-chloromethyl-6-methylpyri-dine (377). Haloalkylpyridines alkylate the sodium salts of diethyl acetamido-malonate (347), diethyl alkylmalonate (381), and diethyl malonate (382).

4-Chloromethylpyridine undergoes addition with acrylate esters or acrylonitrile to give 4-cyclopropylpyridines (**VI-118**) (383).

$(X = CO_2Et, CO_2Me, CN)$ **VI-118**

Wurtz coupling with sodium in xylene was successfully applied for the conversion of 2-bromomethyl-6-methylpyridine to 1,2-bis(6-methyl-2-pyridyl)ethane (375) and the reaction of 2-chloromethylpyridine with bis(2-chloroethyl)methyl amine to give bis[3-(2-pyridyl)propyl]methyl amine (384). The synthesis of m-pyridinophanes (**VI-119**) from the reaction of 3,5-bis(chloromethyl)pyridine and sodium in tetraphenylethylene has been described (385, 386). Extension of Wurtz conditions to prepare di-(pyridine-2,6-methylene) (**VI-122**) from 2,6-bis(bromomethyl)pyridine (**VI-120**) or 1,2-bis(6-bromomethyl-2-pyridyl)ethane (**VI-121**) failed. However, the de-sired transformation was effected starting from **VI-121** with n-butyl-lithium in ether (375).

VI-119

$$n = 0, 1, 2.$$

The most convenient method for the preparation of the parent compound in the dipyrido[1,2-α:1',2'-c]imidazol-10-ium series (**VI-123**) is accomplished by the reaction of 2-bromomethylpyridine and 2-bromopyridine (387).

VI-123
(X$^-$ = perchlorate, picrate)

D. Polymerization

Both 2-bromomethylpyridine (388) and 2-(β-bromoethyl)pyridine (425) form dimers. In contrast, 4-bromomethylpyridine undergoes intermolecular quaternization to give a polymer (388). The faster rate of quaternization of the 4-isomer relative to the 2-isomer has been attributed to steric factors (347). More recently, Berlin has studied the "onium polymerization" of 4-chloromethylpyridine; these polymers do not contain paramagnetic centers, as found in the corresponding polymers from 4-chloro- or 4-bromopyridine (262).

$$\text{N}\text{CH}_2\text{Cl} \longrightarrow \text{N}\text{CH}_2 - \left[\overset{+}{\text{N}} - \text{CH}_2 \right]_n \overset{+}{\text{N}}\text{CH}_2\text{Cl}$$

Cl⁻ Cl⁻

VI-124

Polymers arising from intermolecular quaternization during the preparation of 2,6-bis(bromomethyl)pyridine have been reported (312). Polyethynylpyridines have been prepared from the dehydrohalogenation-polymerization of 2- or 4-(1,2-dibromoethyl)pyridine and 5-(1,2-dibromoethyl)-2-methylpyridine by calcium oxide at 200° (389).

2-(Trifluorovinyl)pyridine has been claimed to impart dye and pigment receptivity to polytetrafluoroethylene (447).

E. Isomerization

Haszeldine (389a) isolated the first stable valence-bond isomers of a six-membered-ring heterocycle by the ultraviolet irradiation of pentakis-(pentafluoroethyl)pyridine (**VI-125**) in perfluoro-*n*-pentane to give quantitative yields of either the *para*-bonded valence isomer, pentakis(pentafluoroethyl)-1-azabicyclo[2,2,0]hexa-2,5-diene (**VI-126**), or the corresponding prismane, pentakis(pentafluoroethyl)-1-azatetracyclo[2,2,0,0^{2,6}, 0^{3,5}]hexane (**VI-127**).

$$\text{VI-125} \quad \text{VI-126} \quad \text{VI-127}$$

References

1. H. E. Mertel, "Halopyridines," in "Pyridine and Its Derivatives," E. Klingsberg, Ed., Part II, Interscience, New York, 1961.
2. R. N. Haszeldine, *J. Chem. Soc.*, 1966 (1950).
3. R. E. Banks and G. E. Williamson, *J. Chem. Soc.*, 815 (1965).
4. J. H. Simons, *J. Electrochem. Soc.*, **45**, 47 (1949).
5. J. H. Simons, U.S. Patent 2,490,098, 1949; *Chem. Abstr.*, **44**, 6443 (1950).
6. T. C. Simmons, F. W. Hoffman, R. B. Beck, H. V. Holler, T. Katz, R. S. Koshar, E. R. Larsen, J. E. Mulvaney, K. E. Paulson, F. E. Rogers, B. Singleton, and R. E. Sparks, *J. Amer. Chem. Soc.*, **79**, 3429 (1957).
7. R. E. Banks, A. E. Ginsberg, and R. N. Haszeldine, *J. Chem. Soc.*, 1740 (1961).
8. R. E. Banks, W. M. Cheng, and R. N. Haszeldine, *J. Chem. Soc.*, 3407 (1962).

9. R. E. Banks, J. E. Burgess, and R. N. Haszeldine, *J. Chem. Soc.*, 2720 (1965).
10. J. Burdon, D. Gilman, C. R. Patrick, M. Stacey, and J. C. Tatlow, *Nature*, **186,** 231 (1960).
11. M. O. Schrader, H. L. Dimond, and J. C. E. Schult, U.S. Patent 2,839,534, 1958; *Chem. Abstr.*, **52,** 17293 (1958).
12. D. A. Shermer, U.S. Patent 2,820,791, 1958; *Chem. Abstr.*, **52,** 9221 (1958).
13. J. A. Zaslowsky, U.S. Patent 3,153,044, 1964; *Chem. Abstr.*, **62,** 534 (1965).
14. W. H. Taplin, U.S. Patent 3,251,848, 1966; *Chem. Abstr.*, **65,** 2231 (1966).
15. W. H. Taplin, U.S. Patent 3,420,833, 1969; *Chem. Abstr.*, **71,** 3279 (1969).
16. Belgian Patent 659,475, 1965; *Chem. Abstr.*, **64,** 2064 (1966).
17. British Patent 991,526, 1965; *Chem. Abstr.*, **63,** 9921 (1965).
18. (a) H. Johnston and S. H. Ruetman, U.S. Patent 3,583,988, 1971; *Chem. Abstr.*, **75,** 63621 (1971); (b) H. Holtschmidt and W. Zecher, Belgian Patent 622,382, 1962; *Chem. Abstr.*, **59,** 11534 (1963); (c) H. Holtschmidt and H. Tarnow, *Ind. Eng. Belg.*, (32) 68 (1967); (d) G. Beck, H. Holtschmidt, and H. Heitzer, *Ann. Chem.*, **731,** 45 (1970).
19. (a) M. M. Boudakian, F. F. Frulla, D. F. Gavin, and J. A. Zaslowsky, *J. Heterocycl. Chem.*, **4,** 375 (1967); (b) M. M. Boudakian and J. A. Zaslowsky, U.S. Patent 3,297,556, 1967; *Chem. Abstr.*, **66,** 75915 (1967).
20. W. G. Tucker, Ph.D. Dissertation, University of Oklahoma, 1962; *Diss. Abstr.*, **23,** 1522 (1963).
21. D. E. Pearson, W. W. Hargrove, J. K. T. Chow, and B. R. Suthers, *J. Org. Chem.*, **26,** 789 (1961); (a) R. R. Stringham and F. E. Torba, U.S. Patent 3,557,124, 1971; *Chem. Abstr.*, **74,** 141552 (1971); (b) H. Johnston, U.S. Patent 3,555,032, 1971; *Chem. Abstr.*, **75,** 5727 (1971); (c) E. Smith, U.S. Patent 3,538,100, 1970; *Chem. Abstr.*, **74,** 53542 (1971).
22. D. Jerchel, H. Fischer, and K. Thomas, *Chem. Ber.*, **89,** 2921 (1956).
23. K. Thomas and D. Jerchel, in "Newer Methods of Preparative Organic Chemistry," Vol. III, W. Foerst, Ed., Academic, New York, 1964, p. 53.
24. W. J. Sell and F. W. Dootson, *J. Chem. Soc.*, **73,** 432 (1898).
25. R. D. Chambers, J. Hutchinson, and W. K. R. Musgrave, *J. Chem. Soc.*, 3573 (1964).
26. R. E. Banks, R. N. Haszeldine, J. V. Latham, and I. M. Young, *J. Chem. Soc.*, 594 (1965).
27. R. D. Chambers, D. Lomas, and W. K. R. Musgrave, *Tetrahedron*, **24,** 5633 (1968).
28. W. Hofling, D. Eilhauer, G. Reckling, and F. Andreas, British Patent 958,877, 1964; *Chem. Abstr.*, **61,** 8262 (1964).
29. R. A. Abramovitch and E. E. Knaus, *J. Heterocycl. Chem.*, **6,** 989 (1969); R. A. Abramovitch, J. Campbell, E. E. Knaus, and A. Silhankova, *ibid*, **9,** 1367 (1972).
30. M. Hamana and M. Yamazaki, *Yakugaku Zasshi*, **81,** 574 (1961); *Chem. Abstr.*, **55,** 24743 (1961).
31. T. Kato, *J. Pharm. Soc. Jap.*, **75,** 1239 (1955); *Chem. Abstr.*, **50,** 8665 (1956).
32. F. Gadient, E. Jucker, A. Lindenmann, and M. Taeschler, *Helv. Chim. Acta*, **45,** 1860 (1962).
33. S. Okuda and M. M. Robison, *J. Amer. Chem. Soc.*, **81,** 740 (1959).
34. T. Kato, *J. Pharm. Soc. Jap.*, **75,** 1236 (1955); *Chem. Abstr.*, **50,** 8665 (1956).
35. E. C. Taylor and A. J. Crovetti, *Org. Synth.*, *Coll. Vol. IV*, 166 (1963).
36. H. J. den Hertog and C. Hoogzand, *Rec. Trav. Chim. Pays-Bas*, **76,** 261 (1957).
37. E. Ochiai, *J. Org. Chem.*, **18,** 534 (1953).
38. E. C. Taylor and A. J. Crovetti, *J. Amer. Chem. Soc.*, **78,** 214 (1956).
39. J. M. Essery and K. Schofield, *J. Chem. Soc.*, 4953 (1960).
40. J. Delarge, *J. Pharm. Belg.*, **22,** 257 (1967); *Chem. Abstr.*, **68,** 104913 (1968).
41. W. Herz and D. R. K. Murty, *J. Org. Chem.*, **26,** 122 (1961).
42. I. Suzuki, *J. Pharm. Soc. Jap.*, **68,** 126 (1948); *Chem. Abstr.*, **47,** 8074 (1953).
43. T. Wieland and H. Biener, *Chem. Ber.*, **96,** 266 (1963).

44. T. Kato and H. Hayashi, *Yakugaku Zasshi*, **83**, 352 (1963); *Chem. Abstr.*, **59**, 7473 (1963).
45. T. Talik and Z. Talik, *Rocz. Chem.*, **36**, 417 (1962); *Chem. Abstr.*, **58**, 5627 (1963).
46. E. Profft and W. Rolle, *Wiss. Z. Tech. Hochsch. Chem. Leuna-Merseburg*, **2**, 187 (1959–1960); *Chem. Abstr.*, **55**, 1609 (1961).
47. E. Profft and G. Schulz, *Arch. Pharm.* (Weinheim), **294**, 292 (1961).
48. H. W. Krause and W. Langenbeck, *Chem. Ber.*, **92**, 155 (1959).
49. E. V. Brown, *J. Amer. Chem. Soc.*, **79**, 3565 (1957).
50. K. Lewicka and E. Plazek, *Rec. Trav. Chim. Pays-Bas*, **78**, 644 (1959).
51. E. C. Taylor and J. S. Driscoll, *J. Org. Chem.*, **25**, 1716 (1960).
52. (a) E. Ochiai and C. Kaneko, *Chem. Pharm. Bull.* (Tokyo), **8**, 28 (1960); (b) R. A. Abramovitch and G. M. Singer, *J. Amer. Chem. Soc.*, **91**, 5672 (1969); R. A. Abramovitch and R. B. Rogers, *Tetrahedron Lett.*, 1951 (1971).
53. F. Friedrich and R. Pohloudek-Fabini, *Pharmazie*, **19**, 677 (1964); *Chem. Abstr.*, **62**, 7720 (1965).
54. (a) S. D. Moschitskii, Y. N. Ivaschenko, L. S. Sologub, and A. A. Kisilenko, *Ukr. Khim. Zh.*, **35**, 524 (1969); *Chem. Abstr.*, **71**, 49727 (1969); (b) S. D. Moschitskii, L. S. Sologub, and Y. N. Ivaschenko, *Khim. Geterosikl. Soedin*, (6) 1068 (1968); *Chem. Abstr.*, **70**, 77737 (1969).
55. W. T. Flowers, R. N. Haszeldine, and S. A. Majid, *Tetrahedron Lett.*, 2503 (1967).
56. R. E. Banks, R. N. Haszeldine, and I. M. Young, *J. Chem. Soc.*, *C*, 2089 (1967).
57. R. M. Bimber, U.S. Patent 3,325,503, 1967; *Chem. Abstr.*, **68**, 68896 (1968); (a) W. H. Taplin and S. H. Ruetman, U.S. Patent 3,591,597, 1971; *Chem. Abstr.*, **75**, 76619 (1971); (b) W. H. Taplin, U.S. Patent 3,575,992, 1971; *Chem. Abstr.*, **75**, 35767 (1971).
58. Y. N. Ivashchenko, O. N. Yaklovleva, S. D. Moschitskii, and A. F. Pavlenko, U.S.S.R. Patent 194,823, 1967; *Chem. Abstr.*, **69**, 27253 (1968).
59. H. H. Fox and J. T. Gibas, *J. Org. Chem.*, **23**, 64 (1958).
60. H. Meyer and R. Graf, *Ber.* **61B**, 2213 (1928).
61. L. van der Does and H. J. den Hertog, *Rec. Trav. Chim. Pays-Bas*, **84**, 951 (1965).
62. H. J. den Hertog, L. van der Does, and C. A. Landheer, *Rec. Trav. Chim. Pays-Bas*, **81**, 864 (1962).
63. K. Palat, L. Novacek, and M. Celadnik, *Collect. Czech. Chem. Comm.*, **32**, 1191 (1967).
64. R. A. Abramovitch and M. Saha, *Can. J. Chem.*, **44**, 1765 (1966).
65. E. E. Garcia, C. V. Greco, and I. M. Hunsberger, *J. Amer. Chem. Soc.*, **82**, 4430 (1960).
66. T. Batkowski, D. Tomasik, and P. Tomasik, *Rocz. Chem.*, **41**, 2101 (1967); (a) C. S. Giam and S. D. Abbott, *J. Amer. Chem. Soc.*, **93**, 1294 (1971).
67. K. Lewicka and E. Plazek, *Rocz. Chem.*, **40**, 405 (1966); *Chem. Abstr.*, **65**, 7134 (1966).
68. C. R. Kolder and H. J. den Hertog, *Rec. Trav. Chim. Pays-Bas*, **79**, 474 (1960).
69. H. J. den Hertog, W. P. Combé, and C. R. Kolder, *Rec. Trav. Chim. Pays-Bas*, **66**, 77 (1958).
70. (a) M. M. Boudakian, D. F. Gavin, and R. J. Polak, *J. Heterocycl. Chem.*, **4**, 377 (1967); (b) R. M. Thomas, U.S. Patent 3,153,044, 1964; *Chem. Abstr.*, **62**, 534 (1965).
71. G. W. Wheland, "Resonance in Organic Chemistry," Wiley, New York, 1955, p. 485; Calculations based on a molecular orbital treatment are found in Wheland, *J. Amer. Chem. Soc.*, **64**, 900 (1942).
72. See footnote 21, Ref. 70a. Another recent treatment of this subject is provided by E. C. Kooyman, *Adv. Free-Radical Chem.*, Vol. 1, G. H. Williams, Ed., Logos Press, London, 1965, p. 137; (a) S. H. Ruetman, U.S. Patent 3,598,868, 1971; *Chem. Abstr.*, **75**, 98453 (1971).
73. R. A. Corral, O. O. Orazi, and J. D. Bonafede, *Anales Asoc. Quim. Arg.*, **45**, 151 (1957); *Chem. Abstr.*, **53**, 342 (1959).

74. M. Hamana and M. Yamazaki, *Chem. Pharm. Bull.* (Tokyo), **9**, 414 (1961); *Chem. Abstr.*, **55**, 24749 (1961).

75. H. C. van der Plas, H. J. den Hertog, M. van Ammers, and B. Haase, *Tetrahedron Lett.*, **32**, (1961).

76. H. S. Mosher and F. J. Welch, *J. Amer. Chem. Soc.*, **77**, 2902 (1955).

77. M. van Ammers, H. J. den Hertog, and B. Haase, *Tetrahedron*, **18**, 227 (1962).

78. T. B. Lee and G. A. Swan, *J. Chem. Soc.*, 771 (1956).

79. R. J. Martens and H. J. den Hertog, *Rec. Trav. Chim. Pays-Bas*, **86**, 655 (1967).

80. J. W. Streef and H. J. den Hertog, *Rec. Trav. Chim. Pays-Bas*, **85**, 803 (1966).

81. H. J. den Hertog and W. P. Combé, *Rec. Trav. Chim. Pays-Bas*, **70**, 581 (1951).

82. E. Ochiai, T. Ito, and S. Okuda, *J. Pharm. Soc. Jap.*, **71**, 591 (1951); *Chem. Abstr.*, **46**, 980 (1952).

83. M. Hamana, Y. Hoshide, and K. Kaneda, *J. Pharm. Soc. Jap.*, **76**, 1337 (1956); *Chem. Abstr.*, **51**, 6639 (1957).

84. F. W. Broekman and H. J. C. Tendeloo, *Rec. Trav. Chim. Pays-Bas*, **81**, 107 (1962).

85. R. S. Mulliken, *J. Amer. Chem. Soc.*, **74**, 811 (1952).

86. C. Reid and R. S. Mulliken, *J. Amer. Chem. Soc.*, **76**, 3869 (1954).

87. O. Hassel, C. Rømming, and T. Tufte, *Acta Chem. Scand.*, **15**, 967 (1967).

88. D. L. Glusker and A. Miller, *J. Chem. Phys.*, **26**, 331 (1957).

89. H. Meinert, *Z. Chem.*, **5**, 64 (1965).

90. A. I. Popov and R. H. Rygg, *J. Amer. Chem. Soc.*, **79**, 4622 (1957).

91. (a) M. T. Rogers and W. K. Meyer, *J. Phys. Chem.*, **66**, 1397 (1962); (b) W. K. Meyer, Ph.D. Dissertation, Michigan State University, 1958, p. 88.

92. R. D. Whitaker, *J. Inorg. Nucl. Chem.*, **26**, 1405 (1964).

93. R. D. Whitaker and J. R. Ambrose, *J. Inorg. Nucl. Chem.*, **24**, 285 (1962).

94. R. A. Zingaro and W. E. Tolberg, *J. Amer. Chem. Soc.*, **81**, 1353 (1959).

95. H. Schmidt and H. Meinert, *Angew. Chem.*, **71**, 126 (1959).

96. (a) M. Schmeisser and E. Scharf, *Angew. Chem.*, **72**, 324 (1960); (b) M. Schmeisser, W. Ludovici, D. Naumann, P. Sartori, and E. Scharf, *Chem. Ber.*, **101**, 4214 (1968); (c) M. Schmeisser, P. Sartori, and D. Naumann, *Chem. Ber.*, **103**, 880 (1970); (d) M. Schmeisser, P. Sartori, and D. Naumann, *Chem. Ber.*, **103**, 590 (1970).

97. (a) H. Meinert and D. Jahn, *Z. Chem.*, **7**, 195 (1967); (b) H. Meinert and H. Klamm, *Z. Chem.*, **8**, 195 (1968); (c) H. Meinert and U. Gross, *Z. Chem.*, **9**, 190 (1969).

98. H. Schmidt and H. Meinert, *Angew. Chem.*, **72**, 109 (1960); (a) O. Ruff and R. Keim, *Z. Anorg. Allg. Chem.*, **193**, 176 (1930).

99. Y. Yagi and A. I. Popov, *Inorg. Nucl. Chem. Lett.*, **1**, 21 (1965).

100. Y. Yagi and A. I. Popov, *J. Amer. Chem. Soc.*, **87**, 3577 (1965).

101. R. D. Whitaker, J. R. Ambrose, and C. M. Hickam, *J. Inorg. Nucl. Chem.*, **17**, 254 (1960).

102. R. A. Zingaro and W. B. Witmer, *J. Phys. Chem.*, **64**, 1705 (1960).

103. A. I. Popov, J. C. Marshall, F. B. Stute, and W. B. Person, *J. Amer. Chem. Soc.*, **83**, 3586 (1961).

104. A. I. Popov and R. T. Pflaum, *J. Amer. Chem. Soc.*, **79**, 570 (1957).

105. W. B. Witmer and R. A. Zingaro, *J. Inorg. Nucl. Chem.*, **15**, 82 (1960).

106. D. G. Coe, H. N. Rydon, and B. L. Tonge, *J. Chem. Soc.*, 323 (1957).

107. H. J. den Hertog, J. Maas, C. R. Kolder, and W. P. Combé, *Rec. Trav. Chim. Pays-Bas*, **74**, 59 (1955).

108. G. R. Redford, A. R. Katritzky, and H. M. Wuest, *J. Chem. Soc.*, 4600 (1963).

109. M. van Ammers and H. J. den Hertog, *Rec. Trav. Chim. Pays-Bas*, **74**, 1160 (1955).

110. A. Dorlars, British Patent 845,062, 1961; *Chem. Abstr.*, **55**, 5544 (1961).

111. A. Roedig and G. Märkl, *Ann. Chem.*, **636**, 1 (1960); (a) G. Simchen, *Chem Ber.*, **103**, 389 (1970).

112. K. Grohe and A. Roedig, *Chem. Ber.*, **100**, 2953 (1967).

113. S. Portnoy, *J. Heterocycl. Chem.*, **6**, 223 (1969).

114. A. Roedig, K. Grohe, D. Klatt, and H.-G. Kleppe, *Chem. Ber.*, **99**, 2813 (1966).

115. H. J. den Hertog, *Rec. Trav. Chim. Pays-Bas*, **64**, 85 (1945).

116. J. W. Streef and H. J. den Hertog, *Rec. Trav. Chim. Pays-Bas*, **85**, 803 (1966).

117. R. A. Abramovitch, F. Helmer, and M. Liveris, *J. Org. Chem.*, **34**, 1730 (1969).

118. H. Pfanz and H. Dorn, *Arch. Pharm.*(Weinheim), **289**, 651 (1956); *Chem. Abstr.*, **51**, 8738 (1957).

119. R. W. Meikle and E. A. Williams, *Nature*, **210**, 523 (1966).

120. G. C. Finger and L. D. Starr, *Nature*, **191**, 595 (1961).

121. G. C. Finger, L. D. Starr, A. Roe, and W. J. Link, *J. Org. Chem.*, **27**, 3965 (1962).

122. A. Roe and G. F. Hawkins, *J. Amer. Chem. Soc.*, **69**, 2443 (1947).

123. (a) *Chem. Eng. News*, 44 (October 16, 1967); (b) *Chem. Eng. News*, 8 (December 18, 1967).

124. K. B. Rutherford, W. Redmond, and J. Rigamont, *J. Org. Chem.*, **26**, 5149 (1961); (a) F. L. Setliff, *Ark. Acad. Sci. Proceedings*, **22**, 88 (1968); *Chem. Abstr.*, **70**, 77723 (1969).

125. T. Talik and E. Plazek, *Rocz. Chem.*, **29**, 1019 (1955); *Chem. Abstr.*, **50**, 12045 (1956).

126. T. Talik and E. Plazek, *Rocz. Chem.*, **33**, 387 (1959); *Chem. Abstr.*, **53**, 18954 (1959).

127. T. Talik, *Rocz. Chem.*, **31**, 569 (1957); *Chem. Abstr.*, **52**, 5407 (1958).

128. (a) T. Talik, *Rocz. Chem.*, **36**, 1049 (1962); *Chem. Abstr.*, **59**, 7480 (1963); (b) T. Talik and Z. Talik, *Rocz. Chem.*, **42**, 1861 (1968).

129. Y. Suzuki, *Yakugaku Zasshi*, **81**, 1206 (1961); *Chem. Abstr.*, **56**, 3446 (1962); (a) T. Kubota and H. Miyazaki, *Bull. Chem. Soc. Jap.*, **39**, 2057 (1966).

130. R. D. Chambers, J. Hutchinson, and W. K. R. Musgrave, *J. Chem. Soc.*, 5040 (1965).

131. (a) A. Pusznynski and T. Talik, *Rocz. Chem.*, **41**, 917 (1967); (b) T. Talik and Z. Talik, *Rocz. Chem.*, **42**, 2061 (1968); (c) T. Talik and Z. Talik, *Rocz. Chem.*, **43**, 489 (1969).

132. G. C. Finger and L. D. Starr, *J. Amer. Chem. Soc.*, **81**, 2674 (1959).

133. G. C. Finger, D. R. Dickerson, T. Adl, and T. Hodgins, *Chem. Comm.*, 430 (1965).

134. G. C. Finger, L. D. Starr, D. R. Dickerson, H. S. Gutowsky, and J. Hamer, *J. Org. Chem.*, **28**, 1666 (1963).

135. J. Hamer, W. J. Link, A. Jurevich, and T. L. Vigo, *Rec. Trav. Chim. Pays-Bas*, **81**, 1058 (1962).

136. R. D. Chambers, J. Hutchinson, and W. K. R. Musgrave, *Proc. Chem. Soc.*, **83**, (1964).

137. R. E. Banks, R. N. Haszeldine, J. V. Latham, and I. M. Young, *Chem. Ind.*(London), 835 (1964).

138. (a) M. M. Boudakian, *J. Heterocycl. Chem.*, **4**, 381 (1967); (b) U.S. Patent 3,296,269, 1967; *Chem. Abstr.*, **66**, 65397 (1967).

139. M. M. Boudakian, *J. Heterocycl. Chem.*, **5**, 683 (1968).

140. H. C. Fielding, L. P. Gallimore, H. L. Roberts, and B. Tittle, *J. Chem. Soc.*, C, 2142 (1966).

141. N. R. Thompson and B. Tittle, "Halogenation and Halogen Exchange in Fused Salt Media," in "Halogen Chemistry," Vol. 2, V. Guttmann, Ed., Academic, New York, 1967, p. 381.

142. R. E. Banks, R. N. Haszeldine, E. Phillips, and I. M. Young, *J. Chem. Soc.*, C, 2091 (1967); (a) Y. N. Ivashchenko, L. S. Sologub, S. Moshchitskii, and A. V. Kirsanov, *J. Gen. Chem. U.S.S.R.*, **39**, 1662 (1969); (b) C. D. S. Tomlin, J. W. Slater, and D. Hartley, British Patent 1,161,492, 1969; *Chem. Abstr.*, **71**, 91313 (1969); (c) W. K. R. Musgrave, *Chem. Ind.* (London), 943 (1969); (d) R. D. Chambers, M. Hole, W. K. R. Musgrave, and J. G. Thorpe, *J. Chem. Soc.*, C, 61 (1971).

143. A. Roedig and K. Grohe, *Chem. Ber.*, **98**, 923 (1965).

144. A. Roedig, R. Kohlhaupt, and G. Maerkl, *Chem. Ber.*, **99**, 698 (1966).

145. A. Roedig, K. Grohe, and D. Klatt, *Chem. Ber.*, **99**, 2818 (1966).

146. I. M. Robinson and G. J. Janz, U.S. Patent 2,494,204, 1950; *Chem. Abstr.*, **44**, 7353 (1950).

147. G. J. Janz and A. G. Keenan, *Can. J. Res.*, **25B**, 283 (1947).
148. G. J. Janz and W. J. G. McCulloch, *J. Amer. Chem. Soc.*, **77**, 3143 (1955).
149. (a) L. P. Anderson, W. J. Feast, and W. K. R. Musgrave, *Chem. Comm.*, 1433 (1968); (b) *J. Chem. Soc.*, *C*, 2559 (1969).
150. T. Jaworski and W. Poloczkowa, *Rocz. Chem.*, **34**, 887 (1960); *Chem. Abstr.*, **55**, 8407 (1961).
151. T. Jaworski, *Rocz. Chem.*, **41**, 1521 (1967); *Chem. Abstr.*, **69**, 2822 (1968).
152. F. Johnson, J. P. Panella, A. A. Carlson, and D. H. Hunneman, *J. Org. Chem.*, **27**, 2473 (1962).
153. (a) R. A. Carboni, D. D. Coffman, and E. G. Howard, *J. Amer. Chem. Soc.*, **80**, 2828 (1958); (b) P. Boldt, W. Thielicke and J. Oberdoerfer, *Angew. Chem.*, *Inter. Ed. Engl.*, **9**, 377 (1970); (c) S. J. Davis, J. A. Elvidge, and A. B. Foster, *J. Chem. Soc.*, 3638 (1962); (d) J. A. Elvidge and N. A. Zaidi, *J. Chem. Soc.*, *C*, 2188 (1968); (e) T. Stensrud, E. Bernatek, and M. Johnsgaard, *Acta Chem. Scand.*, **25**, 523 (1971).
154. E. L. Little, W. J. Middleton, D. D. Coffman, V. A. Engelhardt, and G. N. Sausen, *J. Amer. Chem. Soc.*, **80**, 2832 (1958).
155. M. J. Marinak, U.S. Patent, 3,532,701, 1970; *Chem. Abstr.*, **74**, 53558 (1971); (a) H. Johnston, M. J. Marinak, and S. H. Ruetman, U.S. Patent 3,592,817, 1971; *Chem. Abstr.*, **75**, 88492 (1971).
156. W. Reppe, H. Pasedach, and M. Seefelder, German Patent 946,802, 1956; *Chem. Abstr.*, **53**, 6260 (1959).
157. J. E. Colchester, British Patent 1,102,261, 1968; *Chem. Abstr.*, **69**, 27266 (1968).
158. French Patent 1,538,729, 1968; *Chem. Abstr.*, **71**, 38809 (1969).
159. R. N. Haszeldine, R. E. Banks, and J. M. Birchall, U.S. Patent 3,359,267, 1967; *Chem. Abstr.*, **68**, 105015 (1968).
160. R. L. Jones and C. W. Rees, *J. Chem. Soc.*, *C*, 2249 (1969).
161. R. Nicoletti, M. L. Forcellese, and C. Germani, *Gazz. Chim. Ital.*, **97**, 685 (1967).
162. R. Nicoletti and M. L. Forcellese, *Tetrahedron Lett.*, 3033 (1965).
163. T. Ukai, Y. Yamamoto, and S. Hirano, *J. Pharm. Soc. Jap.*, **73**, 823 (1953); *Chem. Abstr.*, **48**, 9946 (1954).
164. M. van Ammers and H. J. den Hertog, *Rec. Trav. Chim. Pays-Bas*, **77**, 340 (1958).
165. M. van Ammers and H. J. den Hertog, *Rec. Trav. Chim. Pays-Bas*, **81**, 124 (1962).
166. E. Profft and K. H. Otto, *J. Prakt. Chem.*, **8**, 156 (1959); *Chem. Abstr.*, **54**, 5640 (1960).
167. E. Hayashi, H. Yamanaka, and K. Shimizu, *Chem. Pharm. Bull.* (Tokyo), **7**, 141 (1959); *Chem. Abstr.*, **54**, 22665 (1960).
168. E. Hayashi, H. Yamanaka, and C. Iijima, *Yakugaku Zasshi*, **80**, 839 (1960); *Chem. Abstr.*, **54**, 24730 (1960).
169. E. C. Taylor, A. J. Crovetti, and N. E. Boyer, *J. Amer. Chem. Soc.*, **79**, 3549 (1957).
170. R. F. Evans and H. C. Brown, *J. Org. Chem.*, **27**, 1665 (1962).
171. F. Kröhnke and H. Schäfer, *Chem. Ber.*, **95**, 1098 (1962).
172. F. A. Daniher and B. E. Hackley, *J. Org. Chem.*, **31**, 4267 (1966).
173. M. Hamana, *J. Pharm. Soc. Jap.*, **71**, 213 (1951); *Chem. Abstr.*, **46**, 4542 (1952).
174. S. Furakawa, *Pharm. Bull.* (Tokyo), **3**, 230 (1955); *Chem. Abstr.*, **50**, 8638 (1956).
175. A. Fischer, W. J. Galloway, and J. Vaughan, *J. Chem. Soc.*, 3591 (1964); (a) S. L. Bell, R. D. Chambers, W. K. R. Musgrave, and J. G. Thorpe, *J. Fluorine Chem.*, **1**, 51 (1971–1972).
176. C. W. N. Cumper and A. I. Vogel, *J. Chem. Soc.*, 4723 (1960).
177. (a) H. Kämmerer, L. Horner, and H. Beck, *Chem. Ber.*, **91**, 1376 (1958); (b) L. Horner, L. Schläfer, and H. Kämmerer, *Chem. Ber.*, **92**, 1700 (1959).
178. J. Holubek and J. Volke, *Collect. Czech. Chem. Comm.*, **27**, 680 (1962).
179. B. Bak, *J. Org. Chem.*, **21**, 797 (1956).

180. R. D. Chambers, F. G. Drakesmith, and W. K. R. Musgrave, *J. Chem. Soc.*, 5045 (1965).
181. R. E. Banks, J. E. Burgess, W. M. Cheng, and R. N. Haszeldine, *J. Chem. Soc.*, 575 (1965); (a) F. Binns, S. M. Roberts, and H. Suschitzky, *Chem. Comm.*, 1261 (1969); (b) F. Binns; S. M. Roberts, and H. Suschitzky, *J. Chem. Soc.*, C, 1375 (1970).
182. W. L. Mosby, *Chem. Ind.* (London), 1348 (1959).
183. W. Karminski and Z. Kulichi, *Chem. Stosowana*, *Ser. A*, **9**, 129 (1965); *Chem. Abstr.*, **63**, 18018 (1965).
184. H. C. Fielding, British Patent 1,085,882, 1967; *Chem. Abstr.*, **67**, 117564 (1967).
185. F. F. Cheshko, V. F. Sakhenko, and R. A. Parubova, *J. Gen. Chem. U.S.S.R.*, **36**, 1189 (1966).
186. P. R. Bloomfield and K. Parvin, U.S. Patent 3,159,589, 1964; *Chem. Abstr.*, **59**, 10303 (1963); (a) J. A. Zoltewicz and A. A. Sale, *J. Org. Chem.*, **36**, 1455 (1971).
187. R. D. Chambers, J. Hutchinson, and W. K. R. Musgrave, *J. Chem. Soc.*, 5634 (1964); (a) H. Suschitzky and G. E. Chivers, *J. Chem. Soc.*, C, 2867 (1971).
188. L. S. Kobrina, G. G. Furin, and G. G. Yakobson, *J. Gen. Chem. U.S.S.R.*, **38**, 505 (1968).
189. Netherlands Patent 6,516,409, 1966; *Chem. Abstr.*, **65**, 18564 (1966).
190. (a) H. Johnston, U.S. Patent 3,364,223, 1968; *Chem. Abstr.*, **69**, 27254 (1968); (b) R. A. Abramovitch, F. Helmer and M. Liveris, *J. Chem. Soc.*, B, 492 (1968); (c) J. A. Zoltewicz and A. A. Sale, *J. Org. Chem.*, **35**, 3462 (1970).
191. H. Johnston and M. S. Tomita, U.S. Patent 3,291,804, 1966; *Chem. Abstr.*, **66**, 115612 (1967).
192. R. D. Chambers, J. Hutchinson, and W. K. R. Musgrave, *J. Chem. Soc.*, 3736 (1964).
193. R. D. Chambers, J. Hutchinson, and W. K. R. Musgrave, *J. Chem. Soc.*, C, 220 (1966).
194. R. D. Chambers, J. A. H. MacBride, and W. K. R. Musgrave, *J. Chem. Soc.*, C, 2116 (1968).
195. R. Levine and W. W. Leake, *Science*, 780 (1955).
196. T. Kaufmann, *Angew. Chem. Inter. Ed. Engl.*, **4**, 543 (1965).
197. H. J. den Hertog and H. C. van der Plas, in "Advances in Heterocyclic Chemistry," Vol. 4, A. R. Katritzky, Ed., Academic, New York, 1965, p. 121.
198. T. Kaufmann and F. P. Boettcher, *Angew. Chem.*, **73**, 65 (1961).
199. T. Kaufmann and F. P. Boettcher, *Chem. Ber.*, **95**, 1528 (1962).
200. T. Kaufmann and R. Nürnberg, *Chem. Ber.*, **100**, 3427 (1967).
201. T. Kaufmann, H. Fischer, R. Nürnberg, M. Vestweber, and R. Wirthwein, *Tetrahedron Lett.*, 2911; 2917 (1967).
202. R. J. Martens and H. J. den Hertog, *Rec. Trav. Chim. Pays-Bas*, **83**, 621 (1964).
203. M. J. Pieterse and H. J. den Hertog, *Rec. Trav. Chim. Pays-Bas*, **80**, 1376 (1961).
204. M. J. Pieterse and H. J. den Hertog, *Rec. Trav. Chim. Pays-Bas*, **81**, 855 (1962).
205. H. J. den Hertog, H. C. van der Plas, M. J. Pieterse, and J. W. Streef, *Rec. Trav. Chim. Pays-Bas*, **84**, 1569 (1965).
206. R. J. Martens, H. J. den Hertog, and M. van Ammers, *Tetrahedron Lett.*, 3207 (1964).
207. T. Kato and T. Niitsuma, *Chem. Pharm. Bull.* (Tokyo), **13**, 963 (1965); *Chem. Abstr.*, **63**, 14805 (1965).
208. J. A. Zoltewicz and C. L. Smith, *J. Amer. Chem. Soc.*, **88**, 4766 (1966).
209. J. A. Zoltewicz and C. L. Smith, *Tetrahedron*, **25**, 4331 (1969).
210. H. J. den Hertog, R. J. Martens, H. C. van der Plas, and J. Bon, *Tetrahedron Lett.*, 4325 (1966).
211. W. A. Roelfsema and H. J. den Hertog, *Tetrahedron Lett.*, 5089 (1967).
212. J. W. Streef and H. J. den Hertog, *Tetrahedron Lett.*, 5945 (1968).
213. W. J. Sell and F. W. Dootson, *J. Chem. Soc.*, **73**, 777 (1898).
214. W. J. Sell and F. W. Dootson, *J. Chem. Soc.*, **77**, 771 (1900).

215. S. M. Roberts and H. Suschitzky, *Chem. Comm.*, 893 (1967).
216. S. M. Roberts and H. Suschitzky, *J. Chem. Soc.*, C, 1537 (1968).
217. R. G. Hiskey and J. Hollander, *J. Org. Chem.*, **29**, 3687 (1964); (a) L. Heinisch, *Z. Chem.*, **10**, 188 (1970).
218. A. V. Wilett and J. R. Pailthorp, U.S. Patent 2,716,646, 1955; *Chem. Abstr.*, **50**, 1914 (1956).
219. F. W. Broekman, A. van Veldhuizen, and H. Janssen, *Rec. Trav. Chim. Pays-Bas*, **81**, 792 (1962).
220. E. Felstead, H. C. Fielding, and B. J. Wakefield, *J. Chem. Soc.*, C, 708 (1966).
221. T. M. Moynehan, K. Schofield, R. A. Y. Jones, and A. R. Katritzky, *J. Chem. Soc. C*, 2637 (1962).
222. H. H. Paradies and M. Gorbing, *Angew. Chem.*, **81**, 293 (1969).
223. J. D. Cook, B. J. Wakefield, and C. J. Clayton, *Chem. Comm.*, 150 (1967).
224. J. D. Cook and B. J. Wakefield, *J. Organometal. Chem.*, **13**, 2342 (1969); (a) D. J. Berry and B. J. Wakefield, *J. Chem. Soc.*, C, 55 (1969).
225. S. S. Dua and H. Gilman, *J. Organometal. Chem.*, **12**, 299 (1968).
226. R. W. Meikle and P. M. Hamilton, *J. Agr. Food Chem.*, **13**, 377 (1965).
227. S. S. Dua, A. E. Jukes, and H. Gilman, *J. Organometal. Chem.*, **12**, P24 (1968).
228. I. F. Mikhailova and V. A. Barkash, *J. Gen. Chem. U.S.S.R.*, **37**, 2792 (1967); *Chem. Abstr.*, **69**, 59056 (1968).
229. S. S. Dua and H. Gilman, *J. Organometal. Chem.*, **12**, 234 (1968).
230. R. D. Chambers, F. G. Drakesmith, J. Hutchinson, and W. K. R. Musgrave, *Tetrahedron Lett.*, 1705 (1967).
231. (a) B. L. Booth, R. N. Haszeldine, and M. B. Taylor, *J. Organometal. Chem.*, **6**, 570 (1966); (b) B. L. Booth, R. N. Haszeldine, and M. B. Taylor, *J. Chem. Soc.*, A, 1974 (1970).
232. J. Cooke, M. Green, and F. G. A. Stone, *Inorg. Nucl. Chem. Lett.*, **3**, 47 (1967).
233. J. Cooke, M. Green, and F. G. A. Stone, *J. Chem. Soc.*, A, 173 (1968).
234. (a) M. Green, A. Taunton-Rigby, and F. G. A. Stone, *J. Chem. Soc.*, A, 2762 (1968); (b) T. Blackmore, M. I. Bruce and F. G. A. Stone, *J. Chem. Soc.*, A, 2158 (1968).
235. J. D. Cook and B. J. Wakefield, *Chem. Comm.*, 297 (1968).
236. R. A. Fernandez, H. Heaney, J. M. Jablonski, K. G. Mason, and T. J. Ward, *J. Chem. Soc.*, C, 1908 (1969).
237. J. D. Cook, B. J. Wakefield, H. Heaney, and J. M. Jablonski, *J. Chem. Soc.*, C, 2727 (1968).
238. J. D. Cook and B. J. Wakefield, *Tetrahedron Lett.*, 2535 (1967).
239. R. J. Martens and H. J. den Hertog, *Tetrahedron Lett.*, 643 (1962).
240. H. N. M. van der Lans and H. J. den Hertog, *Rec. Trav. Chim. Pays-Bas*, **87**, 549 (1968).
241. T. Kaufmann and F. P. Boettcher, *Chem. Ber.*, **95**, 949 (1962).
242. I. L. Kotlyarevski, V. N. Andrievskii, and M. S. Shvartsberg, *Khim. Geterotsikl. Soedin*, 308 (1967); *Chem. Abstr.*, **67**, 108533 (1967).
243. G. Erhart and W. Bestian, German Patent 922,824, 1955; *Chem. Abstr.*, **52**, 2932 (1958).
244. J. Klosa, *Arch. Pharm.* (Weinheim), **286**, 433 (1953); *Chem. Abstr.*, **49**, 8273 (1955).
245. T. Chokushigawara, Japanese Patent 2489, 1953; *Chem. Abstr.*, **49**, 3267 (1955).
246. Spanish Patent 236,259, 1957; *Chem. Abstr.*, **52**, 15596 (1958).
247. Ref. 23, pp. 74–75.
248. R. D. Chambers, B. Iddon, W. K. R. Musgrave, and L. Chadwick, *Tetrahedron*, **24**, 877 (1968).
249. R. F. Evans and H. C. Brown, *J. Org. Chem.*, **27**, 1329 (1962).
250. E. Ochiai and I. Suzuki, *Pharm. Bull.* (Tokyo), **2**, 247, (1954); *Chem. Abstr.*, **50**, 1015 (1956); (a) R. E. Banks, R. N. Haszeldine, D. R. Karsa, F. E. Rickett, and I. M. Young, *J. Chem. Soc.*, C, 1660 (1969).

251. L. A. Paquette, U.S. Patent 3,218,329, 1965; *Chem. Abstr.*, **64**, 3494 (1966).
252. K. Takeda, K. Hamamoto, and H. Tone, *J. Pharm. Soc. Jap.*, **72**, 1427 (1952); *Chem. Abstr.*, **47**, 8071 (1953).
253. K. Takeda and K. Hamamoto, *J. Pharm. Soc. Jap.*, **73**, 1158 (1953); *Chem. Abstr.*, **48**, 12748 (1954).
254. F. Ramirez and P. W. von Ostwalden, *J. Amer. Chem. Soc.*, **81**, 156 (1959).
255. S. Kajihara, *Nippon Kagaku Zasshi*, **85**, 672 (1964); *Chem. Abstr.*, **62**, 14624 (1965).
256. S. Kajihara, *Nippon Kagaku Zasshi*, **86**, 93 (1965); *Chem. Abstr.*, **63**, 577 (1965).
257. S. Kajihara, *Nippon Kagaku Zasshi*, **86**, 1060 (1965); *Chem. Abstr.*, **65**, 16936 (1966).
258. J. P. Wibaut and F. W. Broekman, *Rec. Trav. Chim. Pays-Bas*, **58**, 885 (1939).
259. J. P. Wibaut and F. W. Broekman, *Rec. Trav. Chim. Pays-Bas*, **78**, 593 (1959).
260. P. A. de Villiers and H. J. den Hertog, *Rec. Trav. Chim. Pays-Bas*, **75**, 1303 (1956).
261. J. P. Wibaut and W. J. Holmes-Kamminga, *Bull. Soc. Chim. Fr.*, 424 (1958); *Chem. Abstr.*, **52**, 20151 (1958).
262. A. A. Berlin and E. F. Razvodovski, *J. Polym. Sci.*, Part C, 369 (1967).
263. S. M. McElvain and M. A. Goese, *J. Amer. Chem. Soc.*, **65**, 2227 (1943).
264. Netherlands Patent 6,601,804, 1967; *Chem. Abstr.*, **68**, 68896 (1968).
265. H. C. van der Plas and H. J. den Hertog, *Rec. Trav. Chim. Pays-Bas*, **81**, 841 (1962).
266. A. A. Berlin and E. F. Razvodovskii, *Proc. Acad. Sci. U.S.S.R.*, *Chem. Sec.*, **140**, 925 (1961).
267. A. A. Berlin, L. V. Zherebtsova, and E. F. Razvodovski, *Polymer Sci. (U.S.S.R.)*, **6**, 67 (1964).
268. A. A. Berlin, E. F. Razvodovskii, and G. V. Korolev, *Polymer Sci. (U.S.S.R.)*, **6**, 2034 (1964).
269. Y. A. Fialkov and I. D. Muzika, *J. Gen. Chem. U.S.S.R.*, **18**, 1205 (1948).
270. J. Saxena and M. R. Gelra, *Indian Chem. J.*, **6**, 562 (1968); *Chem. Abstr.*, **70**, 68082 (1969).
271. M. T. Chaudrey, G. A. Powers, R. Stephens, and J. C. Tatlow, *J. Chem. Soc.*, 874 (1964).
272. T. Talik and Z. Talik, *Rocz. Chem.*, **38**, 777 (1964); *Chem. Abstr.*, **61**, 10653 (1964).
273. C. Zwart and J. P. Wibaut, *Rec. Trav. Chim. Pays-Bas*, **74**, 1062 (1955).
274. T. Maki and T. Takahashi, Japanese Patent 19,070, 1966; *Chem. Abstr.*, **66**, 37936 (1967).
275. H. Wuhrmann, French Patent 1,451,373, 1966; *Chem. Abstr.*, **66**, 94911 (1967).
276. R. E. Banks, R. N. Haszeldine, and R. Hatton, *J. Chem. Soc.*, C, 427 (1967).
277. H. Ulrich, E. Kober, H. Schroeder, R. F. W. Rätz, and C. Grundmann, *J. Org. Chem.*, **27**, 2585 (1962).
278. R. A. Mitsch, *J. Amer. Chem. Soc.*, **87**, 328 (1965).
279. Z. Talik and B. Brekiesz, *Rocz. Chem.*, **41**, 279 (1967); *Chem. Abstr.*, **67**, 73491 (1967).
280. E. Profft and H. Richter, *J. Prakt. Chem.*, **9**, 164 (1959); *Chem. Abstr.*, **54**, 8808 (1960).
281. A. Roe, P. H. Cheek, and G. F. Hawkins, *J. Amer. Chem. Soc.*, **71**, 4152 (1949).
282. J. T. Minor, G. F. Hawkins, C. A. VanderWerf, and A. Roe, *J. Amer. Chem. Soc.*, **71**, 1125 (1949).
283. I. Suzuki, *Pharm. Bull.* (Tokyo), **5**, 78 (1957); *Chem. Abstr.*, **52**, 1167 (1958).
284. N. F. Kucherova, R. M. Khomutov, E. I. Budovski, V. P. Evdakov, and N. K. Kochetkov, *J. Gen. Chem. U.S.S.R.*, **29**, 898 (1959).
285. T. Batkowski and E. Plazek, *Rocz. Chem.*, **36**, 51 (1962); *Chem. Abstr.*, **57**, 15066 (1962).
286. E. Ochiai and I. Suzuki, *Pharm. Bull.* (Tokyo), **2**, 147 (1954); *Chem. Abstr.*, **50**, 1015 (1954).
287. K. B. Wiberg, T. M. Shryne, and R. R. Kintner, *J. Amer. Chem. Soc.*, **79**, 3160 (1957).
288. T. Nakashima, *Yakugaku Zasshi*, **77**, 698 (1957); *Chem. Abstr.*, **51**, 16462 (1957).
289. T. Batkowski and E. Plazek, *Rocz. Chem.*, **25**, 251 (1951); *Chem. Abstr.*, **46**, 6127 (1952).
290. S. Sawa and R. Maeda, Japanese Patent 3689, 1964; *Chem. Abstr.*, **61**, 3075 (1964).

291. A. S. Bailey and J. S. A. Brunskill, *J. Chem. Soc.*, 2554 (1959).
292. K. Tsuda, H. Mishima, and M. Maruyama, *Pharm. Bull.* (Tokyo), **1**, 283 (1953); *Chem. Abstr.*, **49**, 8277 (1955).
293. K. Palat, M. Celadnik, L. Novacek, and R. Urbancik, *Rozhledy Tuberk.*, **19**, 716 (1959); *Chem. Abstr.*, **54**, 11016 (1960).
294. E. Ager, B. Iddon, and H. Suschitzky, *Tetrahedron Lett.*, 1507 (1969); *J. Chem. Soc. C*, 193 (1970).
295. J. D. Cook and B. J. Wakefield, *J. Chem. Soc.*, *C*, 1973 (1969).
296. T. Kametani, K. Ogasawara, and M. Shio, *Yagugaku Zasshi*, **86**, 809 (1066); *Chem. Abstr.*, **65**, 20092 (1962).
297. S. Masamune and K. Fukomoto, *Tetrahedron Lett.*, 4647 (1965).
298. D. Muenzner, H. Lattau, and H. Schubert, *Z. Chem.*, **7**, 278 (1967).
299. R. A. Abramovitch, K. S. Ahmed, and C. S. Giam, *Can. J. Chem.*, **41**, 1752 (1963).
300. J. N. Chatterjea and K. Prasad, *Chem. Ber.*, **93**, 1740 (1960).
301. E. T. McBee, H. B. Hass, and E. M. Hodnett, *Ind. Eng. Chem.*, **39**, 389 (1947).
302. E. T. McBee and E. M. Hodnett, U.S. Patent 2,516,402, 1950; *Chem. Abstr.*, **45**, 670 (1951).
303. B. R. Brown, D. L. Hammick, and B. H. Thewlis, *J. Chem. Soc.*, 1145 (1951).
304. W. J. Sell, *J, Chem. Soc.*, **87**, 799 (1905).
305. P. Dyson and D. L. Hammick, *J. Chem. Soc.*, 781 (1939).
306. W. Mathes and H. Schuely, *Angew. Chem. Inter. Ed. Engl.*, **2**, 144 (1963).
307. W. Mathes and H. Schuely, German Patent 1,204,231, 1965; *Chem. Abstr.*, **64**, 2603 (1966).
308. J. Kollonitsch, French Patent 1,394,362, 1965; *Chem. Abstr.*, **63**, 8326 (1965).
309. Belgian Patent 667,939, 1966; *Chem. Abstr.*, **65**, 5447 (1966).
310. P. Arnall and N. R. Clark, *Mfg. Chem. Aerosol News*, **37**, 39 (1966); *Chem. Abstr.*, **65**, 12163 (1966).
311. J. P. Kutney, W. Cretney, T. Tabata, and M. Frank, *Can. J. Chem.*, **42**, 698 (1964).
312. R. A. Barnes and H. M. Fales, *J. Amer. Chem. Soc.*, **75**, 3830 (1953).
313. B. H. Walker, *J. Org. Chem.*, **25**, 1047 (1960).
314. J. Hurst and D. G. Wibberley, *J. Chem. Soc.*, 119 (1962).
315. T. Koenig and J. S. Wieczorek, *J. Org. Chem.*, **33**, 1530 (1968).
316. E. Matsumura, T. Hirooka, and K. Imagawa, *Nippon Kagaku Zasshi*, **82**, 616 (1961); *Chem. Abstr.*, **57**, 12466 (1962).
317. T. Kato, *J. Pharm. Soc. Jap.*, **75**, 1236 (1955); *Chem. Abstr.*, **50**, 8665 (1956).
318. J. F. Vozza, *J. Org. Chem.*, **27**, 3856 (1962).
319. L. Bauer and L. A. Gardella, *J. Org. Chem.*, **28**, 1323 (1963).
320. M. S. Raasch, *J. Org. Chem.*, **27**, 1406 (1962).
321. S. Ng and C. H. Senderholm, *J. Chem. Phys.*, **40**, 2090 (1964).
322. Y. Kobayashi and E. Chinen, *Chem. Pharm. Bull.* (Tokyo), **15**, 1896 (1967).
323. (a) L. M. Yagupol'ski, A. G. Galushko, and M. A. Rzhavinskya, *J. Gen. Chem. U.S.S.R.*, **38**, 644 (1968); (b) L. M. Yagupol'ski, A. G. Galushko and V. I. Troitskaya, *J. Gen. Chem. U.S.S.R.*, **38**, 1692 (1968).
324. (a) R. D. Chambers, R. A. Storey, and W. K. R. Musgrave, *Chem. Comm.*, 384 (1966); (b) R. D. Chambers, J. A. Jackson, W. K. R. Musgrave, and R. A. Storey, *J. Chem. Soc.*, *C*, 2221 (1968); (c) W. K. R. Musgrave, R. D. Chambers, and R. A. Storey, British Patent 1,195,692, 1970; *Chem Abstr.*, **73**, 77061 (1970).
325. H. C. Fielding, British Patent 1,133,492, 1968; *Chem. Abstr.*, **70**, 37660 (1969).
326. R. D. Chambers, R. P. Corbally, J. A. Jackson, and W. K. R. Musgrave, *Chem. Comm.*, 127 (1969); (a) R. D. Chambers, R. P. Corbally, M. Y. Gribble, and W. K. R. Musgrave, *Chem. Comm.*, 1345 (1971).

327. C. J. Drayton, W. T. Flowers, and R. N. Haszeldine, *Chem. Comm.*, 662 (1970).
328. E. M. Mailey and L. R. Ocone, *J. Org. Chem.*, **33**, 3343 (1968).
329. S. Dixon, *J. Org. Chem.*, **21**, 400 (1956).
330. T. F. McGrath and R. Levine, *J. Amer. Chem. Soc.*, **77**, 3656 (1955).
331. J. R. Dice, L. Scheinman, and K. W. Berrodin, *J. Med. Chem.*, **9**, 176 (1966).
332. F. J. McCartey, C. H. Tilford, and M. G. van Campen, *J. Amer. Chem. Soc.*, **79**, 472 (1957).
333. Netherlands Patent 6,511,532, 1966; *Chem. Abstr.*, **65**, 3847 (1966).
334. R. Levine and J. K. Sneed, *J. Amer. Chem. Soc.*, **73**, 4478 (1951).
335. Ref. 1, p. 364.
336. J. P. Kutney and T. Tabata, *Can. J. Chem.*, **41**, 695 (1963).
337. J. M. S. Jarvie, W. E. Fitzgerald, and G. J. Janz, *J. Amer. Chem. Soc.*, **78**, 978 (1956).
338. J. M. S. Jarvie and G. J. Janz, *J. Phys. Chem.*, **60**, 1430 (1956).
339. G. J. Janz and M. A. de Crescente, *J. Org. Chem.*, **23**, 765 (1958).
340. P. J. Hawkins and G. J. Janz, *J. Chem. Soc.*, 1479 (1949).
341. G. J. Janz and A. R. Monahan, *J. Org. Chem.*, **29**, 569 (1964).
342. G. J. Janz and A. R. Monahan, *J. Org. Chem.*, **30**, 1249 (1965).
343. S. Portnoy, *J. Org. Chem.*, **30**, 3377 (1965).
344. R. Nicoletti and M. L. Forcellese, *Tetrahedron Lett.*, 153 (1965).
345. R. Nicoletti and M. L. Forcellese, *Gazz. Chim. Ital.*, **97**, 148 (1967).
346. K. C. Kennard and D. M. Burness, *J. Org. Chem.*, **24**, 464 (1959).
347. R. L. Bixler and C. Niemann, *J. Org. Chem.*, **23**, 575 (1958).
348. R. G. Jones, *J. Amer. Chem. Soc.*, **74**, 1489 (1952).
349. E. Maruszewska-Wieczorkowska, J. Michalski, and A. Skowronska, *Rocz. Chem.*, **31**, 543 (1957); *Chem. Abstr.*, **52**, 5406 (1958).
350. W. Mathes and A. Wolf, U.S. Patent 2,848,457, 1959; *Chem. Abstr.*, **53**, 1384 (1959).
351. K. Kahmann, F. Class, and H. Erlenmeyer, *Experientia*, **20**, 297 (1964).
352. M. W. Goldberg and S. Teitel, U.S. Patent 2,761,865, 1956; *Chem. Abstr.*, **51**, 3670 (1957).
353. S. Akerfeldt, *Acta Chem. Scand.*, **19**, 518 (1965).
354. Netherlands Patent 6,409,877, 1965; *Chem. Abstr.*, **63**, 8325 (1965).
355. T. Naito and K. Ueno, *Yakugaku Zasshi*, **79**, 1277 (1959); *Chem. Abstr.*, **54**, 4566 (1960).
356. I. Suzuki, *Pharm. Bull.* (Tokyo), **4**, 211 (1956); *Chem. Abstr.*, **51**, 7372 (1957).
357. M. Rink and H. W. Eich, *Arch. Pharm.* (Weinheim), **293**, 74 (1960); *Chem. Abstr.*, **60**, 15873 (1964).
358. L. S. Galbraikh, T. I. Tikhonenko, Z. V. Pushkareva, and Z. A. Rogorin, *Vysokolmolekul, Soedin Tsellyulozo i ee Proizvodnye, Sb. Statei*, 3 (1963); *Chem. Abstr.*, **60**, 14328 (1964).
359. G. Rey-Bellet, Swiss Patent 326,012, 1958; *Chem. Abstr.*, **53**, 9257 (1959).
360. G. Rey-Bellet and H. Spiegelberg, Swiss Patent 326,362, 1958; *Chem. Abstr.*, **53**, 9258 (1959).
361. V. Carelli, M. Cardellini, and F. Liberatore, *Farmaco, Ed. Sci.*, **15**, 803 (1960); *Chem. Abstr.*, **55**, 21109 (1961).
362. K. Saruto, Japanese Patent 26,852, 1964; *Chem. Abstr.*, **62**, 10418 (1965).
363. B. M. Mikhailov and T. K. Kozminskaya, *J. Gen. Chem. U.S.S.R.*, **26**, 2275 (1956).
364. E. Jucker, E. Rissi, and R. Sueess, German Patent 1,130,813, 1962; *Chem. Abstr.*, **58**, 4583 (1963).
365. E. Hoyer, *Chem. Ber.*, **93**, 2475 (1960).
366. M. Hamana and K. Funakoshi, *Yakagaku Zasshi*, **82**, 523 (1962); *Chem. Abstr.*, **58**, 3385 (1963).
367. Y. Kobayashi, I. Kumadaki, S. Taguchi, and Y. Hanzawa, *Tetrahedron Lett.*, 3901 (1970).
368. F. A. Daniher, B. E. Hackley, and A. B. Ash, *J. Org. Chem.*, **31**, 2709 (1966).
369. E. Maruszewska-Wieczorkowska, J. Michalski, and A. Skowronska, *Rocz. Chem.*, **30**, 1197 (1956); *Chem. Abstr.*, **51**, 11347 (1957).

370. P. Bednarek, R. Bodalski, J. Michalski, and S. Musierowicz, *Bull. Acad. Polon. Sci., Ser. Sci. Chim.*, **11**, 507 (1963); *Chem. Abstr.*, **60**, 5546 (1964).
371. R. Bodalski, A. Malkiewicz, and J. Michalski, *Bull. Acad. Polon. Sci., Ser. Sci. Chim.*, **13**, 139 (1965); *Chem. Abstr.*, **63**, 8310 (1965).
372. E. Maruszewska-Wieczorkowska and J. Michalski, *Rocz. Chem.*, **38**, 625 (1964); *Chem. Abstr.*, **61**, 10702 (1964).
373. K. Winterfeld and K. Flick, *Arch. Pharm.* (Weinheim), **26**, 448 (1956); *Chem. Abstr.*, **51**, 11346 (1959).
374. W. Schulze, *J. Prakt. Chem.*, **19**, 91 (1962).
375. W. Baker, K. M. Buggle, J. F. W. McOmie, and D. A. M. Watkins, *J. Chem. Soc.*, 3594 (1958).
376. D. Jerchel, S. Naetzel, and K. Thomas, *Chem. Ber.*, **93**, 2966 (1960).
377. R. W. Temple and L. F. Wiggins, British Patent 930,459, 1963; *Chem. Abstr.*, **59**, 13956 (1963).
378. E. Szarvasi, French Patent 82,715, 1964; *Chem. Abstr.*, **61**, 8281 (1964); addition to French Patent 1,342,159, 1969; *Chem. Abstr.*, **60**, 6827 (1964).
379. E. Szarvasi and L. Fontaine, French Patent M2765, 1964; *Chem. Abstr.*, **62**, 4014 (1965).
380. E. Szarvasi, French Patent 83,531, 1964; *Chem. Abstr.*, **62**, 535 (1965); addition to French Patent 1,342,159, 1969; *Chem. Abstr.*, **60**, 6827 (1964).
381. T. Kato, F. Hamaguchi, and T. Oiwa, *Yakugaku Zasshi*, **80**, 1293 (1960); *Chem. Abstr.*, **55**, 3601 (1961).
382. Y. Arata, *Yakugaku Zasshi*, **80**, 709 (1960); *Chem. Abstr.*, **54**, 21084 (1960).
383. A. P. Gray and H. Kraus, *J. Org. Chem.*, **31**, 399 (1966).
384. V. Carelli, M. Cardellini, and F. Liberatore, *Ann. Chim.* (Rome), **49**, 1761 (1959); *Chem. Abstr.*, **54**, 15380 (1960).
385. W. Jenny and H. Holzrichter, *Chimia*, **22**, 139 (1968).
386. W. Jenny and H. Holzrichter, *Chimia*, **21**, 509 (1967).
387. B. R. Brown and J. Humphreys, *J. Chem. Soc.*, 2040 (1959).
388. F. Sorm and L. Sedivy, *Collect. Czech. Chem. Comm.*, **13**, 289 (1948).
389. Y. M. Paushkin, S. A. Nisova, and V. D. Stytsenko, *Tr. Mosk. Inst. Neftekhim. Gaz. Prom.*, (51) 38 (1964); *Chem. Abstr.*, **63**, 740 (1965); (a) M. G. Barlow, J. G. Dingwall, and R. N. Haszeldine, *Chem. Comm.*, 1580 (1970).
390. H. Johnston, F. H. Norton, and M. S. Tomita, U.S. Patent 3,173,919, 1965; *Chem. Abstr.*, **62**, 14,638 (1965).
391. F. H. Norton and W. H. Taplin, U.S. Patent 3,256,167, 1966; *Chem. Abstr.*, **65**, 8882 (1966).
392. H. Johnston, Belgian Patent 644,105, 1964; *Chem. Abstr.*, **63**, 5609 (1965).
393. H. Johnston, M. S. Tomita, F. H. Norton, and W. H. Taplin, Belgian Patent 624,800 (1963); *Chem. Abstr.*, **61**, 1841 (1964).
394. W. H. Taplin, U.S. Patent 3,424,754, 1969; *Chem. Abstr.*, **70**, 77806 (1969).
395. T. Kato, H. Hayashi, and T. Anzai, *Yakugaku Zasshi*, **87**, 387 (1967); *Chem. Abstr.*, **67**, 64211 (1967).
396. H. Johnston, M. S. Tomita, F. H. Norton, and W. H. Taplin, Belgian Patent 624,800, 1963; *Chem. Abstr.*, **61**, 1841 (1964).
397. Y. Arata and K. Achiwa, *Yakugaku Zasshi*, **79**, 108 (1959); *Chem. Abstr.*, **53**, 10211 (1959).
398. E. Kobayashi and K. Matsumoto, Japanese Patent 7481, 1962; *Chem. Abstr.*, **59**, 1603 (1963).
399. Y. Sato, *Chem. Pharm. Bull.* (Tokyo), **7**, 241 (1959); *Chem. Abstr.*, **54**, 24726 (1960).
400. Y. Sawa and R. Maeda, Japanese Patent 2148, 1965; *Chem. Abstr.*, **62**, 14635 (1965).
401. T. Nakashima, *Yakugaku Zasshi*, **78**, 661 (1958); *Chem. Abstr.*, **52**, 18399 (1958).

402. D. Wagler and E. Hoyer, *Chem. Ber.*, **98**, 1073 (1965).
403. G. M. Singerman, R. Kimura, J. L. Riebsomer, and R. N. Castle, *J. Heterocycl. Chem.*, **3**, 74 (1966).
404. T. Kato, *J. Pharm. Soc. Jap.*, **75**, 1228 (1955); *Chem. Abstr.*, **50**, 8664 (1956).
405. M. Hamana, B. Umezawa, Y. Gota, and K. Noda, *Chem. Pharm. Bull.* (Tokyo), **8**, 692 (1960); *Chem. Abstr.*, **55**, 18723 (1961).
406. M. Miocque and J.-A. Gautier, *C. R. Acad. Sci.*, *Paris*, *Ser. C.*, **252**, 2416 (1961).
407. K. Steiner, U. Graf, and E. Hardegger, *Helv. Chim. Acta*, **46**, 690 (1963).
408. J. Krapcho and W. A. Lott, U.S. Patent 2,918,470, 1959; *Chem. Abstr.*, **54**, 6761 (1960).
409. T. R. Govindachari, N. S. Narasimhan, and S. Rajadurai, *J. Chem. Soc.*, 560 (1957).
410. J. E. Cochran, G. B. Kline, Q. F. Soper, and W. N. Cannon, U.S. Patent 3,180,872, 1965; *Chem. Abstr.*, **63**, 1773 (1965).
411. E. J. Blanz, F. A. French, J. R. DoAmaral, and D. A. French, *J. Med. Chem.*, **13**, 1124 (1970).
412. D. Leaver, W. K. Gibson, and J. D. R. Vass, *J. Chem. Soc.*, 6053 (1963).
413. K. Tsuda, N. Ikekewa, R. Takasaki, and Y. Yamakawa, *Pharm. Bull.* (Tokyo), **1**, 142 (1953); *Chem. Abstr.*, **50**, 13895 (1956).
414. H. Tani, *Yakugaku Zasshi*, **81**, 182 (1961); *Chem. Abstr.*, **55**, 14450 (1961).
415. K. Tsuda, H. Mishima, and M. Maruyama, *Pharm. Bull.* (Tokyo), **1**, 283 (1953); *Chem. Abstr.*, **49**, 8277 (1955).
416. F. Micheel and H. Dralle, *Ann. Chem.*, **670**, 57 (1963).
417. M. Hudlicky and F. Mares, *Chem. Listy*, **51**, 1875 (1957); *Chem. Abstr.*, **52**, 4628 (1958).
418. A. N. Kost, P. B. Ternetev, and T. Zavada, *Dokl. Akad. Nauk U.S.S.R.*, **130**, 326 (1960).
419. F. Bordin, F. Baccichetti, and G. Fattori, *Ann. Chim.* (Rome), **58**, 882 (1965).
420. British Patent 752,331, 1956; *Chem. Abstr.*, **51**, 9717 (1957).
421. British Patent 766,397, 1957; *Chem. Abstr.*, **51**, 15597 (1957).
422. R. L. Letsinger and A. J. Wysocki, *J. Org. Chem.*, **28**, 3199 (1963).
423. French Patent 1,538,366, 1968; *Chem. Abstr.*, **71**, 91046 (1969).
424. C. S. Giam, private communication, July 22, 1970.
425. V. Boekelheide and W. Feely, *J. Amer. Chem. Soc.*, **80**, 2217 (1958).
426. Y. N. Ivashchenko, S. D. Moschitskii, and A. K. Eliseeva, *Khim. Geterotsikl. Soedin* (1) 58 (1970); *Chem. Abstr.*, **72**, 100451 (1970).
427. C. S. Giam and J. L. Stout, *Chem. Comm.*, 478 (1970).
428. R. Koppang, A. C. Ranade, and H. Gilman, *J. Organometal. Chem.*, **22**, 1 (1970).
429. F. Long and K. N. Ayad, British Patent 1,147,438, 1969; *Chem. Abstr.*, **71**, 3280 (1969).
430. S. S. Dua, A. E. Jukes, and H. Gilman, *Org. Prep. Proc.*, **1**, 187 (1969).
431. W. G. Duncan and D. W. Henry, *J. Med. Chem.*, **11**, 909 (1968).
432. T. Surles and A. I. Popov, *Inorg. Chem.*, **8**, 2049 (1969).
433. P. I. Mortimer, *Aust. J. Chem.*, **21**, 467 (1968).
434. S. Hermanek, *Collect. Czech. Chem. Comm.*, **24**, 2748 (1959).
435. L. N. Yakhtanov, M. Y. Uritskaya, E. I. Lapan, and M. V. Rubtsov, *Khim. Geterotsikl. Soedin.*, 18 (1968); *Chem. Abstr.*, **69**, 106589 (1968).
436. P. Doyle and G. J. Stacey, South African Patent 6,706,809, 1969; *Chem. Abstr.*, **72**, 12592 (1970).
437. C. H. Brett and E. M. Hodnett, U.S. Patent 2,679,453, 1954; *Chem. Abstr.*, **48**, 9011 (1954).
438. J. D. Cook and B. J. Wakefield, *J. Chem. Soc.*, *C.* 2376 (1969).
439. R. E. Banks, W. M. Cheng, R. N. Haszeldine and G. Shaw, *J. Chem. Soc.*, *C*, 55 (1970).
440. J. Yarwood and W. B. Person, *J. Amer. Chem. Soc.*, **90**, 3930 (1968).
441. L. M. Yagupol'ski and A. G. Galushko, *J. Gen. Chem. U.S.S.R.*, **39**, 2041 (1969).

442. L. N. Yakhtanov, E. I. Lapan, and M. V. Rubtsov, *Khim. Geterotsikl. Soedin.*, 1063 (1967); *Chem. Abstr.*, **69**, 86786 (1968).
443. M. M. Robison, *J. Amer. Chem. Soc.*, **80**, 5481 (1958).
444. M. Bellas and H. Suschitzky, *J. Chem. Soc.*, 2096 (1965).
445. Z. Talik, T. Talik, and A. Puszynski, *Rocz. Chem.*, **39**, 601 (1965).
446. A. E. Jukes, S. S. Dua, and H. Gilman, *J. Organometal. Chem.*, **24**, 791 (1970).
447. E. A. Mailey, U.S. Patent 3,541,102 (1970); *Chem. Abstr.*, **74**, 31693 (1971).
448. P. Tomasik and W. Drzeniek, *Rocz. Chem.*, **43**, 569 (1969).
449. W. Drzeniek and P. Tomasik, *Rocz. Chem.*, **44**, 779 (1970).
450. T. Talik and Z. Talik, *Rocz. Chem.*, **44**, 1249 (1970).
451. J. Wielgat, *Rocz. Chem.*, **45**, 931 (1971).
452. P. T. Sullivan, C. B. Sullivan, and S. J. Norton, *J. Med. Chem.*, **14**, 211 (1971).
453. P. T. Sullivan and S. J. Norton, *J. Med. Chem.*, **14**, 557 (1971).
454. (a) F. Binns and H. Suschitzky, *J. Chem. Soc.*, C, 1223 (1971); (b) *Chem. Comm.*, 750 (1970).
455. F. Eloy and A. Deryckere, *Chim. Ther.*, **5**, 416 (1970).
456. W. Drzeniek and P. Tomasik, *Rocz. Chem.*, **43**, 1865 (1969).
457. E. Plazek, K. Dohaniuk, and Z. Grzyb, *Rocz. Chem.*, **26**, 106 (1952).
458. W. J. Adams, P. H. Hey, P. Mamalis, and R. E. Parker, *J. Chem. Soc.*, 3181 (1949).
459. J. N. Seiber, *J. Org. Chem.*, **36**, 2000 (1971).
460. I. Collins, S. M. Roberts, and H. Suschitzky, *J. Chem. Soc.*, C, 167 (1971).
461. F. E. Torba, U.S. Patent 3,609,158, 1971; *Chem. Abstr.*, **76**, 3699 (1972).
462. M. Mallet, G. Queguiner, and P. Pastour, *C. R. Acad. Sci., Paris, Ser. C*, **274**, 719 (1972).
463. A. Nicolson, German Patent 2,128,540, 1971; *Chem. Abstr.*, **76**, 59469 (1972).
464. F. L. Setliff, *Org. Prep. Proc. Int.*, **3**, 217 (1971).
465. W. J. Link, R. F. Borne, and F. L. Setliff, *J. Heterocycl. Chem.*, **4**, 641 (1967).
466. F. L. Setliff, *J. Chem. Eng. Data*, **15**, 590 (1970).

CHAPTER VII

Organometallic Compounds of Pyridine

HARRY L. YALE

Squibb Institute for Medical Research, New Brunswick, New Jersey

The preparation and reactions of pyridine derivatives where a metal atom is joined directly or by a carbon chain to the pyridine heterocycle comprise a major portion of this chapter. These pyridylmetallic compounds include derivatives of lithium, sodium, potassium, magnesium, copper, mercury, boron, silicon, tin, phosphorus, arsenic, and chromium.

A minor portion of this chapter discusses two areas of pyridine chemistry interrelated with organometallic compounds; (a) the addition of organometallic compounds and the alkali metal amides to the pyridine heterocycle; and (b) the generation of the highly reactive pyridynes. Both subjects represent major portions of the literature on the more recent aspects of pyridine chemistry, the first being concerned with the larger area of aromatic substitution and the second with the reactions of the pyridynes.

I. Pyridyllithium Compounds

1. Preparation

The metal-halogen exchange between a 2-, 3-, or 4-bromopyridine and n-butyllithium, usually in a mixture of diethyl ether and hexane, at temperatures of −30 to −70°, has resulted in the formation of 2-, 3-, and 4-pyridyllithium derivatives; this synthesis has been widely employed since it was first reported in 1940; competing reactions, e.g., addition to the azomethine linkage or coupling to form dipyridyl derivatives have been kept to a minimum with the bromopyridines by employing low temperatures and short reaction periods (1st ed., Part 2, pp. 422 ff) (1–16). With 3-fluoropyridine, proton abstraction is the predominant reaction even at −40°, as determined by subsequent reaction with 3-pentanone. As examples, 3-fluoropyridine and n-butyllithium give a mixture of 3-fluoro-2-pyridyl- and 3-fluoro-4-pyridyllithium, 2-fluoropyridine gives 2-fluoro-3-pyridyllithium (see below), and 2-chloro-3-fluoropyridine gives 2-chloro-3-fluoro-4-pyridyllithium (215). It should be noted, also, that 2-fluoropyridine and 2-thienyllithium are reported to yield only 2-(2-thienyl)pyridine (201a). 2-, 3-, and 4-Chloropyridines are reported to give the 6-, 4- and 2-, and 3-lithio derivatives, respectively,* along with concomittant formation of 6-butyl-2-chloro-, 2-butyl-3-chloro-, and 2-butyl-4-chloropyridine (215); again, 2,6-dichloropyridine and 2-thienyllithium are reported to give only 6-chloro-2-(2-thienyl)pyridine (201a). When a perhalopyridine and n-butyllithium react at low temperatures, the normal reaction occurs, but not exclusively with a single halogen (see also Chapter VI). Thus pentachloropyridine and n-butyllithium in diethyl ether at −60 to −40° give, in 70% yield, two products:

* The products were trapped with diethyl ketone to give the tertiary alcohols. Hydrogen-abstraction from the α-positions is remarkable and goes contrary to all other reports. It seems that the products may arise by a different mechanism (Ed.).

2,3,5,6-tetrachloro-4-pyridyllithium (**VII-1**) and 2,4,5,6-tetrachloro-3-pyrid-yllithium (**VII-2**), in a molar ratio of about 4:1 while in methylcyclohexane at room temperature there are obtained three products in 43 % yield, the major one being 3,4,5,6-tetrachloro-2-pyridyllithium (**VII-3**), along with lesser amounts of **VII-1** and **VII-2** in the molar ratio of 68:16:16 (6, 17, 208). These tetrachloropyridyllithium compounds are stable for short periods of time, even at 80°. When, however, **VII-1** and **VII-2** are subjected to prolonged heating in dilute solution in benzene, mesitylene, or durene, there occurs elimination of lithium chloride and the formation of 2,5,6-trichloro-3-pyridine, and this product reacts with durene to give the adduct **VII-4** (6, 17, 18) (see also p. 579).

VII-4

Prolonged heating of 2,5,6-trichloro-4-phenyl-3-pyridyllithium in the presence of furan leads to the formation of only trace amounts of the pyridyne adduct, 2,3-dichloro-5,8-epoxy-5,8-dihydro-4-phenylquinoline; the major product is 2,3,6-trichloro-4-phenylpyridine. Replacement of the 4-phenyl group by a *p*-anisyl, *p*-dimethylaminophenyl, or α,α,α-trifluoro-*p*-tolyl group, followed by heating with furan, or other trapping agents, gives none of the pyridyne adduct (19).

When pentachloropyridine is treated with methyllithium at −70° in tetrahydrofuran and then trimethylchlorosilane, a heptachlorobipyridyl (**VII-5**), of unknown structure is obtained in 55 % yield. The substitution of phenyllithium for the methyllithium has as a consequence the formation, in 32 % yield, of octachloro-4,4′-bipyridyl. None of the desired tetrachloro-pyridyltrimethylsilane is formed in either reaction (17). While no structure is proposed for **VII-5**, it is possible that this product is either one or both of the isomers **VII-6**, these being again derived from a trichloropyridyne (**VII-7**). The formation of the octachloro derivative may be explained by the interaction of the pentachloropyridine with 2,3,5,6-tetrachloro-4-pyridyllithium. It is not clear, however, why the reactions involved should be so radically different when the only modification involved is a change from an alkyl to an aryllithium compound.

Another anomalous behavior is found in the reaction between penta-fluoropyridine and propenyllithium in ether at −70°, since no halogen-metal

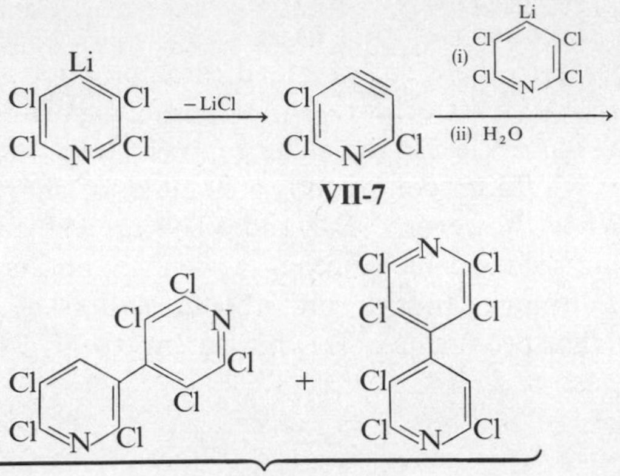

VII-7

VII-6

interconversion occurs. The product of the reaction is either 4-propenyl-2,3,5,6-tetrafluoropyridine or 2,4-di-(propenyl)-3,5,6-trifluoropyridine, depending on whether one or two moles of the propenyllithium are employed (20); a similar pattern is seen also when phenyllithium replaces the propenyllithium since the only product is 4-phenyl-2,3,5,6-tetrafluoropyridine (21).

Tetrachloro-4-pyridyllithium and benzonitrile, in boiling ether give a complex mixture of products, including 5,6,8-trichloro-2,4-diphenylpyrido-[3,4-d]pyrimidine (**VII-7a**), 4-(N-benzoylbenzimidoyl)tetrachloropyridine (**VII-7b**), and 4-benzimidoyltetrachloropyridine (**VII-7c**) in 11, 31, and 11 % yields, respectively; the same reactants kept at room temperature and then treated with boiling dilute hydrochloric acid, gave 4-benzoyltetrachloropyridine in 49 % yield. On boiling in ether tetrachloro-2-pyridyllithium and benzonitrile give a 9 % yield of 6,7,8-trichloro-2,4-diphenylpyrido[3,2-d]-pyrimidine (**VII-7d**), while at room temperature, the same reactants give a 12 % yield of 2-benzoyltetrachloropyridine. 4-(1-Piperidino)-2,5,6-trichloro-3-pyridyllithium and benzonitrile also give a complex mixture of products (208, 209).

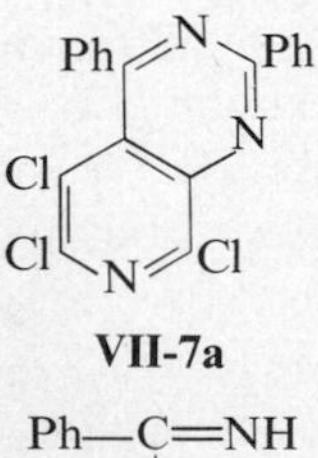

VII-7a

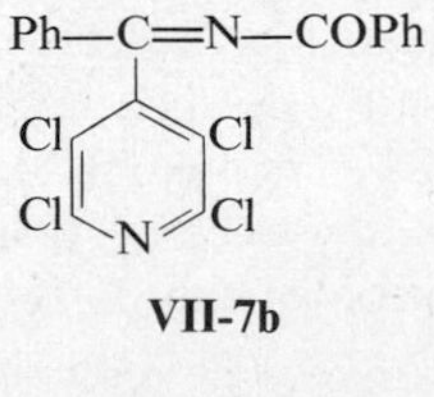

VII-7b

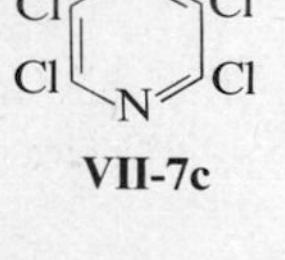

VII-7c

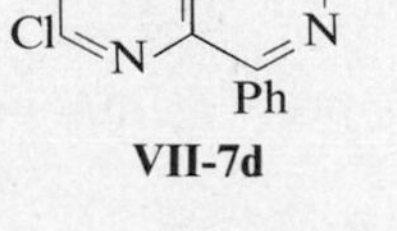

VII-7d

2,3,5,6-Tetrachloro-4-methoxypyridine, with a 12 molar excess of an aryl- or substituted aryllithium compound, undergoes three stepwise reactions (**VII-7e**) that are temperature dependent (19).

VII-7e

A series of related reactions have been reported to occur with **VII-7f** and **VII-7g** (18).

The lone protons in 2,4,5,6-tetrafluoropyridine and 2,3,5,6-tetrafluoropyridine and the two protons in 2,4,6-trifluoropyridine are strongly influenced by the electronegative fluorine atoms and are sufficiently acidic to undergo direct replacement by metal to give pyridyllithium compounds (**VII-8** to **VII-11**). Diethyl ether is a satisfactory solvent to use for the preparation of the first three compounds since these separate as they form; tetrahydrofuran, in which **VII-10** is soluble, will permit the second metalation to occur to give **VII-11** (15, 22, 23). Again, the lone proton in 2,4,5,6-tetrachloropyridine is also metalated by *n*-butyllithium to give the 3-lithio derivative (229).

4-Bromo-2,3,5,6-tetrafluoropyridine reacts with lithium amalgam to form 2,3,5,6-tetrafluoro-4-pyridyllithium. The organolithium derivative is then treated with mercuric chloride to give di-(2,3,5,6-tetrafluoropyridyl)-mercury (103).

The initial formation of an intermediate pyridyllithium compound is postulated in the reaction between 3-chloropyridine and lithium piperidide. Elimination of lithium chloride occurs next to give the pyridyne, and the last step involves addition of piperidine to give 1-(3-pyridyl)- and 1-(4-pyridyl) piperidine (**VII-12**) (24–26).

VII-12

The enhanced acidity of the proton at position-2 in pyridine-1-oxides has been utilized to prepare the unusual 2-pyridyllithium-1-oxide derivative (**VII-13**) by the reaction between 4-chloro-3-picoline-1-oxide and *n*-butyllithium in diethyl ether at −65°; **VII-13** reacts normally with cyclo-hexanone to give the carbinol in 38% yield and with carbon dioxide to give the carboxylic acid in 24% yield. The chlorine atom at position 4 does not influence the metalation reaction since 3,4-lutidine-1-oxide readily forms a lithium derivative **(VII-14)** and the latter, with cyclohexanone, gives the carbinol in 84% yield. Replacing diethyl ether with tetrahydrofuran, a

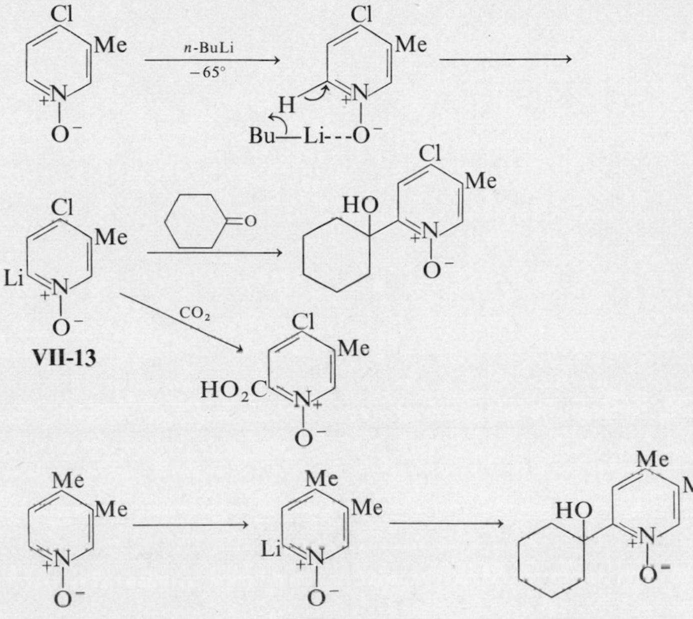

VII-13

VII-14

better solvent for those pyridine-1-oxides too insoluble in ether to undergo the reaction, and employing derivatives lacking a substituent at position 3, may result in metalation at both positions 2 and 6 **(VII-15)** (27, 225). Recently, pyridine-1-oxide has been reported to react with lithium hydride in diglyme at 60° to give 2-pyridyllithium-1-oxide (197).

The 2-pyridyllithium-1-oxides react with either oxygen or sulfur to give 1-hydroxy-2-pyridinones and -thiones, respectively; with chlorine or bromine, the lithium derivatives give a variety of dihalogenated derivatives (28, 197, 227). Several examples of these reactions are shown in **VII-15a**.

A contrasting reaction involves the adduct of pyridine and butyllithium-tetramethylethylenediamine complex **(VII-15b)**; the adduct reacts with benzophenone to give four products **(VII-15c)** (29).

2. Reactions

In the previous edition (Part 2, pp. 445ff) are enumerated the typical reactions of pyridyllithium compounds with active halogen compounds,

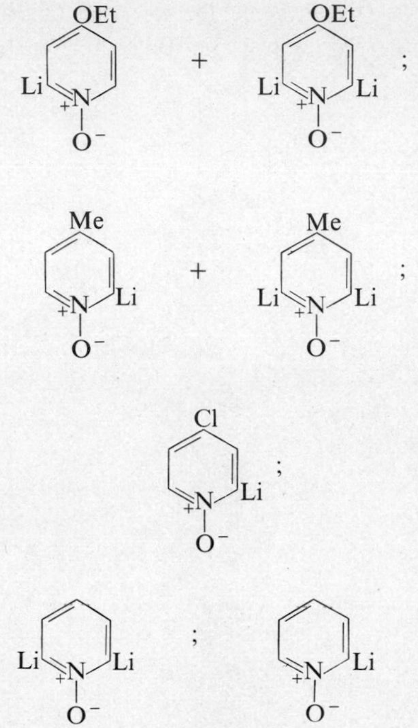

VII-15

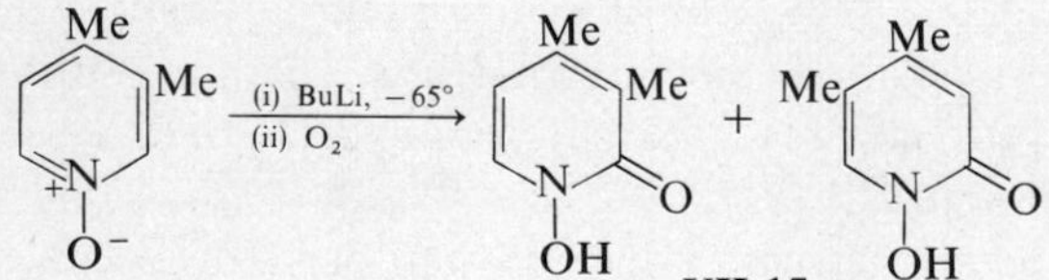

VII-15a

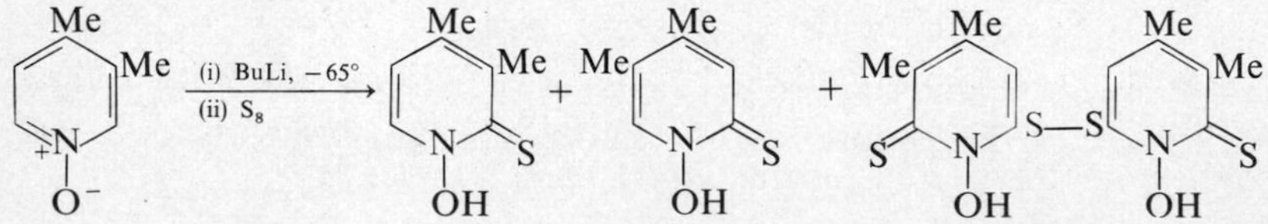

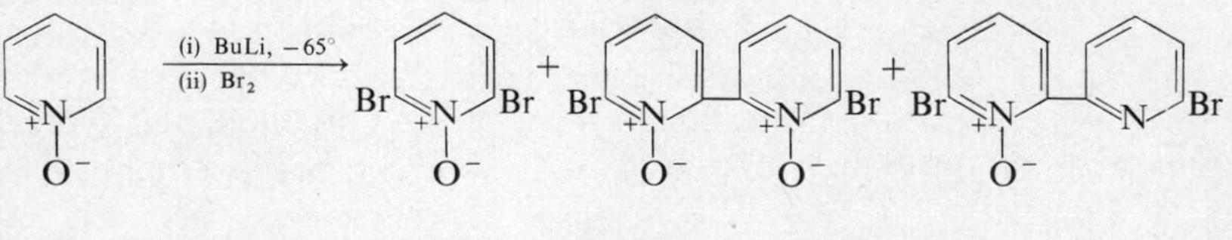

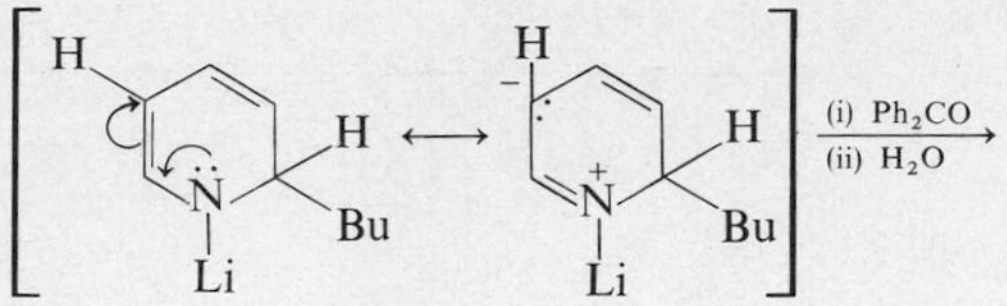

TMEDA
VII-15b

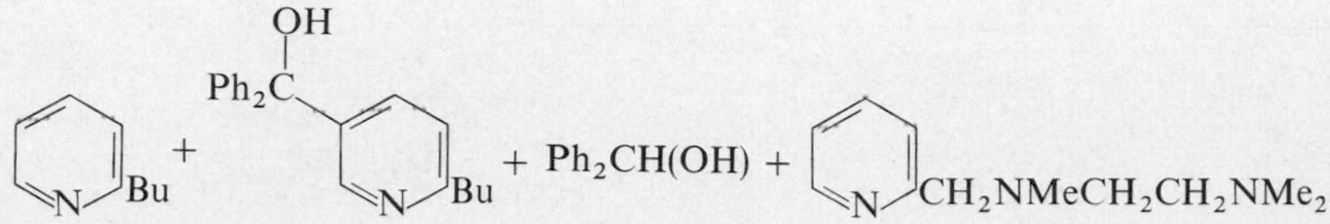

VII-15c

aldehydes, ketones, esters, nitriles, and carbon dioxide. A substantial number of similar reactions are summarized in Tables VII-7 to VII-10. These tables also include the typical reactions of pyridyllithium compounds with halides of several metallic and nonmetallic elements to give the corresponding pyridyl derivatives. One somewhat unusual reaction involves the cleavage of a disulfide by 2-pyridyllithium (**VII-16**) (9).

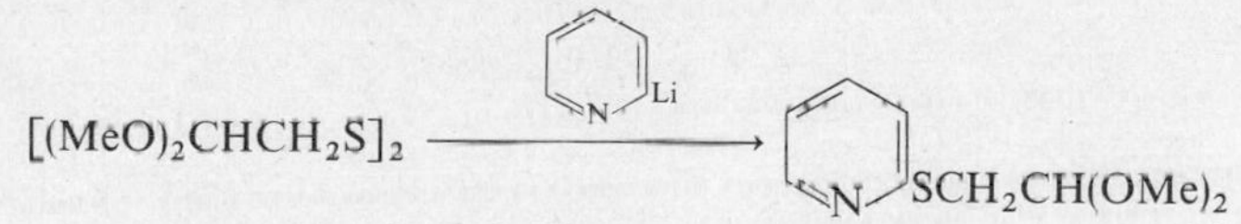

VII-16

II. Pyridylsodium and Pyridylpotassium Compounds

The enhanced acidity of the protons at position-2 in pyridine-1-oxides has led to the preparation of the corresponding 2-pyridylsodium and 2-pyridylpotassium compounds. The bases employed have been sodium hydride, potassium hydride, and potassium t-butoxide in solvents such as tetrahydrofuran, N,N-dimethylformamide, glyme, diglyme, and dimethylsulfoxide. These derivatives have been treated with sulfur to give the alkali metal salts of 2-mercaptopyridine-1-oxides, and with sulfur chloride to give the 2,2′-dithio-1,1′-dioxides. It is noteworthy that 3-methylpyridine-1-oxide and sodium hydride in glyme, after treatment with sulfur give the sodium salt of 5-methyl-2-mercaptopyridine-1-oxide (197). (See, however ref. 197a).

The reactions of the pyridylsodium and pyridylpotassium compounds are summarized in Table VII-4a.

TABLE VII-1. Reactions of 2-Pyridyllithium with Aldehydes and Ketones

Substituents in pyridine	R	R^1	Ref.
None	Me	$CH_2CH_2N(CH_2)_5$	4
None	Et	$CHMeCH_2NMe_2$	4
None	Pr	$CHEtCH_2NMe_2$	4
None	Me	$CHMeCH_2NMe_2$	4
None	Me	$CHMeCH_2N(CH_2)_5$	4
None	Me_3C	$CH_2CH_2N(CH_2)_5$	4
None	Me	$CH(CH_2NEt_2)CH_2NEt_2$	4
None	Me	$CH[CH_2N(CH_2)_5]CH_2N(CH_2)_5$	4
None	$Me_2N(CH_2)_2$	$(CH_2)_2NMe_2$	4
None	Me_2NCH_2CHMe	$CHMeCH_2NMe_2$	4
None	Me_2NCH_2CHMe	$(CH_2)_2NMe_2$	4
None	Me_2NCH_2CHEt	$(CH_2)_2NMe_2$	4
None	Et	EtCHPh	13
None	Et	$EtCHC_6H_4Cl$-p	13
None	Et	$EtCHC_6H_4[O(CH_2)_2NEt_2]p$	13
None			147
None		H	148
None	p-Chlorophenyl	Et	2
None	Ph	2-Pr	2
None	p-Chlorophenyl	2-Pr	2
None	p-Chlorophenyl	NEtMe	1
None	Ph	$CHMeCH_2N(CH_2)_5$	2
None	p-Chlorophenyl	$CHMeCH_2N(CH_2)_5$	2
None	Ph	$CHMeCH_2N(CH_2)_4$	2
None	p-Chlorophenyl	$CHMeCH_2N(CH_2)_4$	2
None	Ph	$CHPhCH_2N(CH_2)_5$	2

Table VII-1 (*Continued*)

Substituents in pyridine	R	R^1	Ref.
None	p-Chlorophenyl	CHPhCH$_2$N(CH$_2$)$_5$	2
None	Ph	CH$_2$CHMeN(CH$_2$)$_5$	2
None	Ph	CH$_2$CHMeN(CH$_2$)$_4$	2
None	p-Chlorophenyl	CH$_2$CHMeN(CH$_2$)$_4$	2
None	Ph	CH$_2$CHPhN(CH$_2$)$_5$	2
None	——————— Indanone ———————		3
None	p-Fluorophenyl	(CH$_2$)$_2$N(CH$_2$)$_4$	1
None	p-Bromophenyl	(CH$_2$)$_2$N(CH$_2$)$_4$	1
None	p-Methoxyphenyl	(CH$_2$)$_2$N(CH$_2$)$_4$	1
None	5-Chloro-2-thienyl	(CH$_2$)$_2$N(CH$_2$)$_4$	1
None	2-Thienyl	H	149
None	——————— N-methyl-2-pyrrolidone ———————		8
None	——— (MeO-substituted 1-oxo-2-phenyl-tetrahydronaphthalene structure, with O, Ph, MeO) ———		12
None	EtCHC$_6$H$_4$OMe-p	Et	13, 150
None	MeCHC$_6$H$_4$OMe-p	Et	13, 150
None	MeCHC$_6$H$_4$OMe-p	Me	13, 150
None	EtCHC$_6$H$_4$OMe-m	Et	13, 150
None	EtCHC$_6$H$_4$OMe-o	Et	13, 150
None	EtCHC$_6$H$_4$Me-p	Et	13, 150
None	MeCHC$_6$H$_4$(CHMe$_2$)-p	Et	13, 150
3-Me	Ph	CH$_2$CH$_2$N(CH$_2$)$_5$	2
3-Me	p-Chlorophenyl	CH$_2$CH$_2$N(CH$_2$)$_4$	2
4-Chloro-5 methyl-1-oxide		Cyclohexanone	27
4,5-Dimethyl-1-oxide		Cyclohexanone	27
4-Ethoxy-1-oxide		Cyclohexanone	27
4-Methyl-1-oxide		Cyclohexanone	27
None	Ph	Ph	164, 213
6-Phenyl	Ph	Ph	164
None	2-(p-methoxyphenyl)-cyclohexanone		220
None	1-Benzyl-3-pyrrolidinone		204
None	1-Ethoxycarbonyl-3-pyrrolidinone		204
None	Cyclohexyl	Cyclohexyl	213
	4-Methoxybenzaldehyde		214
2-Fluoro	Et	Et	215
3-Fluoro	Et	Et	215
4-Fluoro	Et	Et	215

499

TABLE VII-2. Reactions of 2-Pyridyllithium Compounds with Miscellaneous Reactants

Substituents in pyridine	Reactant	Product (yield if reported)	Ref.
None	$[(MeO)_2CHCH_2S]_2$	$2\text{-PySCH}_2CH(OMe)_2$	9
None	$(p\text{-MeOC}_6H_4)_2C\text{:}CHC_6H_4COCl\text{-}p$	$(p\text{-MeOC}_6H_4)_2C\text{:}CHC_6H_4(COPy\text{-}2)\text{-}p$ (34%)	9
None	$Ph_2C\text{:}CHC_6H_4CN\text{-}p$	$Ph_2C\text{:}CHC_6H_4(COPy\text{-}2)\text{-}p$ (73%)	11
None	$(p\text{-MeOC}_6H_4)_2C\text{:}CHC_6H_4CN\text{-}p$	$(p\text{-MeOC}_6H_4)_2C\text{:}CHC_6H_4(COPy\text{-}2)\text{-}p$ (34%)	11
None	CO_2	2-PyCO_2H	27
None	PCl_3	$(2\text{-Py})_3P$	10
None	$AsCl_3$	$(2\text{-Py})_3As$ (25%)	10
3,4,5,6-Tetrachloro[a]	H_2O		17, 208
3,4,5,6-Tetrafluoro	CO_2	3,4,5,6-Tetrafluoropicolinic acid	22, 208
4-Chloro-3-methyl-1-oxide	CO_2	4-Chloro-5-methylpicolinic acid-1-oxide (24%)	27, 227
4-Chloro-1-oxide	CO_2	4-Chloropicolinic acid-1-oxide (49%)	27, 227
4-Methyl-1-oxide	CO_2	4-Methylpicolinic acid-1-oxide	27
1-Oxide	O_2	No	28
1-Oxide	S_8	, 8%	28

4-Chloro-3-methyl-1-oxide	S_8	, 12%		28
4-Methyl-1-oxide	O_2	, 14%		28
4-Methyl-1-oxide	S_8	, 39%		28
3,4-Dimethyl-1-oxide	O_2	, 14%	, 10%	28
3,4-Dimethyl-1-oxide	S_8	24%	13%	37%	28

(Orientation of Me groups uncertain)

Table VII-2 (*Continued*)

Substituents in pyridine	Reactant	Product (yield if reported)	Ref.
3-Fluoro	Br_2	2-Bromo-3-fluoropyridine	215
4-Methyl-1-oxide	CO_2	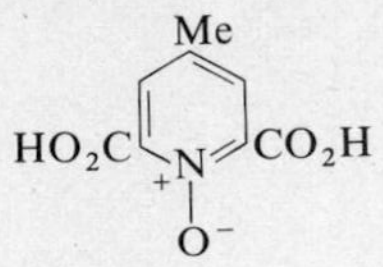	227
4-*n*-Propyl-1-oxide	S_8	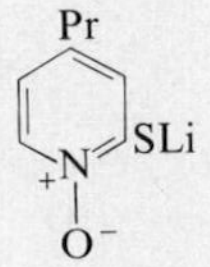	196
3-Chloro	Et_2CO	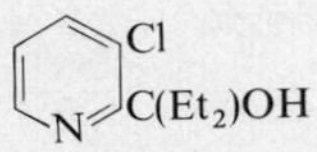	215
6-Chloro	Et_2CO	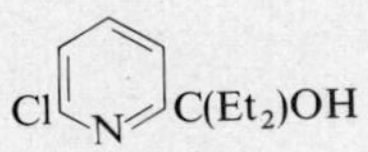 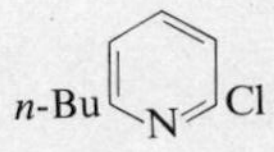	215
4-Chloro-3-fluoro	Et_2CO	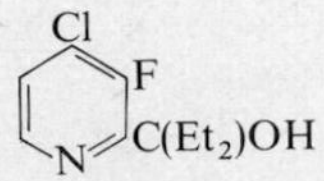	215

Reactant	Reagent	Product	Ref.
Cl,Cl,Cl,Cl-pyridinyl-Li	PhCN	Cl,Cl,Cl,Cl-pyridinyl-COPh	209
Me-pyridinium-N-oxide–Li	CO_2	Me-pyridinium-N-oxide–CO_2H (14%)	227
Me, Me, HO-cyclohexyl-pyridinium-N-oxide–Li	CO_2	Me, Me, HO-cyclohexyl-pyridinium-N-oxide–CO_2H (7%)	227
Me, Me-pyridinium-N-oxide–Li	CO_2	Me, Me-pyridinium-N-oxide–CO_2H (18%)	227
Me, Me-pyridinium-N-oxide–Li	$MeCO_2Et$	Me, Me-pyridinium-N-oxide–COMe (65%)	227
Me, Me, Me(Li)-pyridinium-N-oxide	$MeCO_2Et$	Me, Me, COMe-pyridinium-N-oxide (6%); Me, Me, COMe-pyridinium-N-oxide (20%)	227

Table VII-2 (Continued)

Substituents in pyridine	Reactant	Product (yield if reported)	Ref.
Me, Me pyridinium Li N-oxide (N—O⁻)	MeCONMe₂	MeCO / Me / Me pyridine N-oxide biaryl (13%)	227
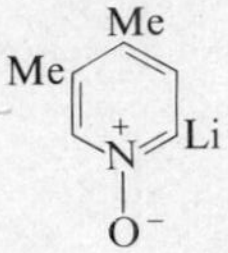 Me, Me pyridinium Li N-oxide (N—O⁻)	 morpholine NCOMe	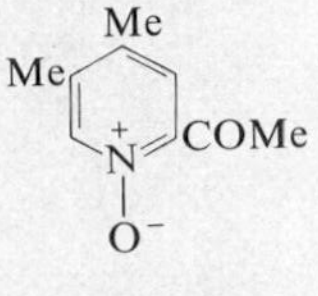 Me, Me, COMe pyridine N-oxide (3%)	227
EtO pyridinium (Li) N—O⁻	PrCO₂Et	EtO / PrCO pyridine N-oxide (19%); OEt / PrCO pyridine N-oxide COPr (17%)	227
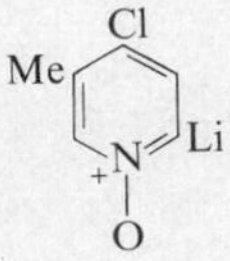 Cl, Me pyridinium Li N—O	PhCN	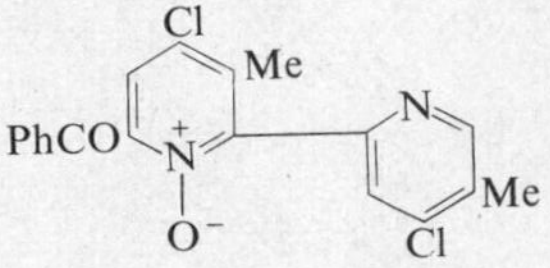 Me, Cl / PhCO pyridine N-oxide biaryl (12%)	227
None	D-(−)MeCH(OMe)CO₂Et	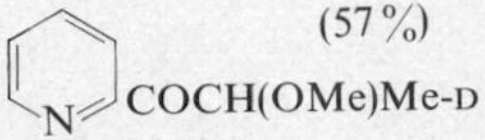 pyridine COCH(OMe)Me-D (57%)	227
None	PrCO₂Et	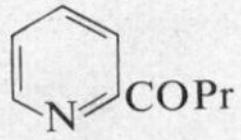 pyridine COPr (55%)	227

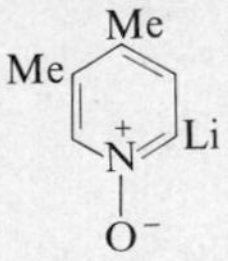

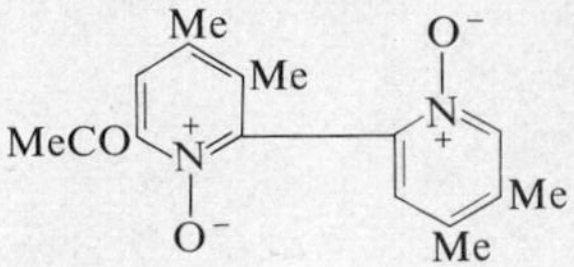

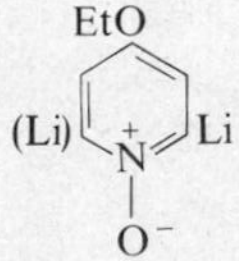

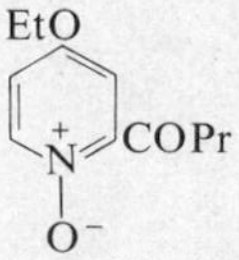

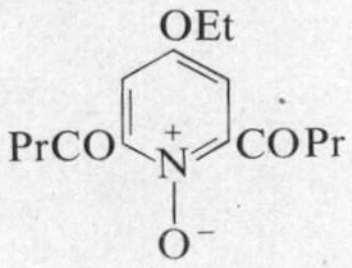

504

Substrate	Reagent	Products			Ref.
1-Oxide	Br_2	3%	8%	6%	28
4-Methyl-1-oxide	Br_2	5%	13%	18%	28
3,4-Dimethyl-1-oxide	Br_2	13%	4%		28
1-Oxide	Cl_2	5%			28
3,4-Dimethyl-1-oxide	Cl_2	9%			28

505

[a] The principal product when methylcyclohexane is the solvent in the reaction between 2,3,4,5,6-pentachloropyridine and *n*-butyllithium.

TABLE VII-3. Reactions of 3-Pyridyllithium Compounds with Miscellaneous Reactants

Substituents in pyridine	Reactant	Product (yield if reported)	Ref.
None	Ph$_2$C:CHC$_6$H$_4$CN-p	Ph$_2$C:CHC$_6$H$_4$CO(Py-3)-p (56%)	11
None	[6-MeO-2-Ph-1-tetralone structure: MeO, O, Ph]	[tetralinol structure: HO, Ph, (Py-3), MeO]	12
None	EtCH(C$_6$H$_4$OMe-p)COEt	EtCH(C$_6$H$_4$OMe-p)C(OH)Et(Py-3)	13, 150
None	(p-MeC$_6$H$_4$)$_2$CO	(p-MeC$_6$H$_4$)$_2$C(OH)(Py-3)	5
None	[thiophene-CHO structure: S, CHO]	3-PyCH(OH)[thienyl]	149
None	[thiophene-CHO structure: S, CHO]	3-PyCH(OH)[thienyl]	149
2-Cl	[dihydrofuran structure: O]	[bicyclic furo-pyridine structure: O, N], 31%	151
2-Cl	H$_2$O	2-PyCl (52%)	151
6-OH	EtCH(C$_6$H$_4$Me-p)COEt	EtCH(C$_6$H$_4$Me-p)C(OH)Et[pyridinone structure: N, H, O]	13, 150

506

Substrate	Reagent	Product	Ref.
piperidino-pyridine-Li (a)	H_2O	, 57%	101, 151
piperidino-pyridine-Li	CO_2	CO_2H , 31%	151
2,4,5,6-Tetrafluoro	HCONHMe	CHO , 40%	23
pyrrolidino-pyridine-Li	H_2O	, 39%	151
NMe_2 pyridine-Li	H_2O	NMe_2 , 45% and Me_2N , 6%	151

507

Table VII-3 (*Continued*)

Substituents in pyridine	Reactant	Product (yield if reported)	Ref.
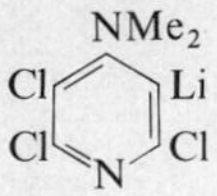	(i) BuLi (ii) H_2O	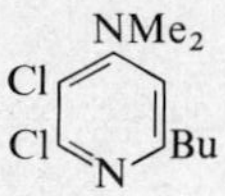 , 59%	151
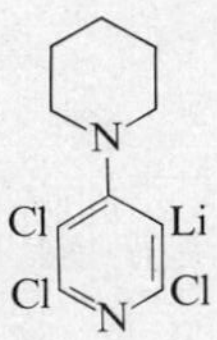		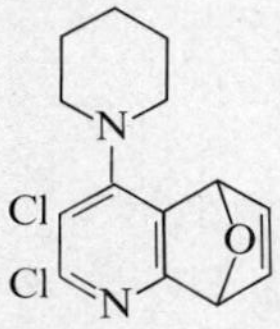 , 50%	151
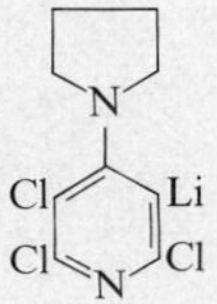		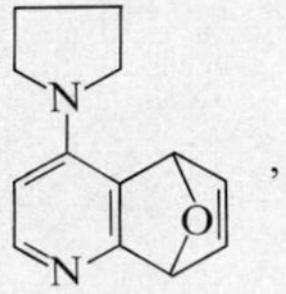 , 22%	151
4-Chloro	Et_2CO	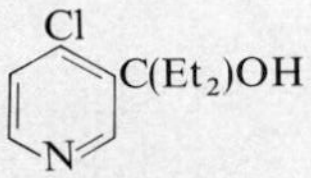 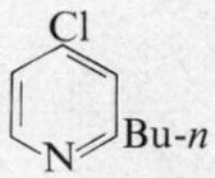	215

508

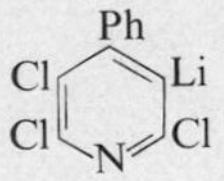	H_2O	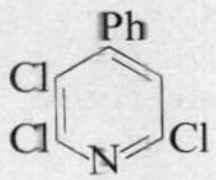 (67%)	233
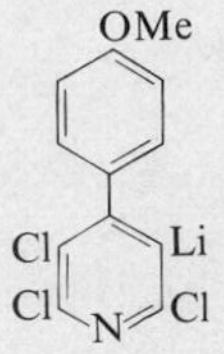	H_2O	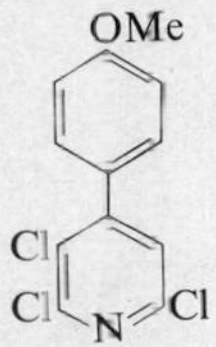 (44%)	233
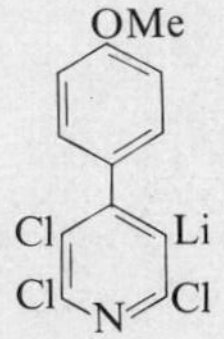	CO_2	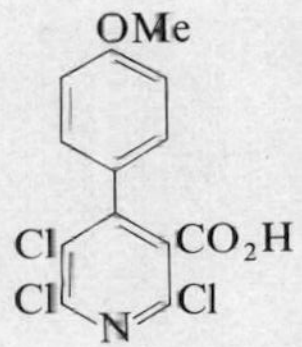 (75%)	233
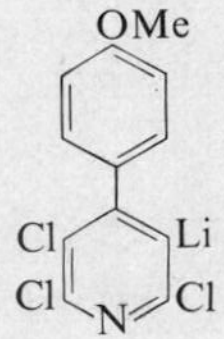	Me_2SO_4	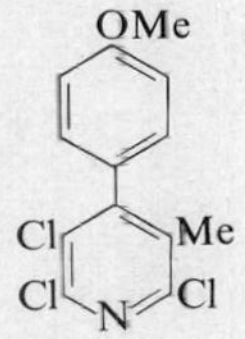 (94%)	233
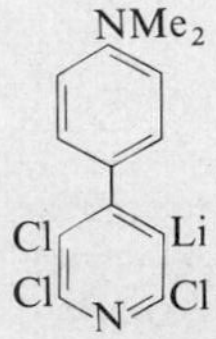	H_2O	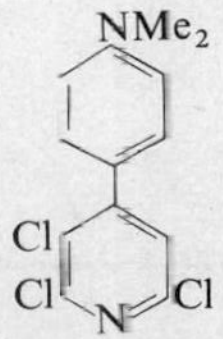 (36%)	233

Table VII-3 (*Continued*)

Substituents in pyridine	Reactant	Product (yield if reported)	Ref.
	H_2O	(80 %)	233
	H_2O	(32 %)	48
		(12 %)	48
	H_2O	(38 %)	48
	H_2O	(41 %)	48

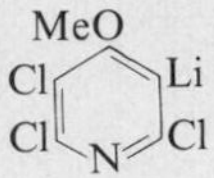

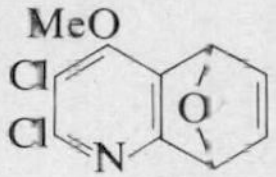

MeO (35%)

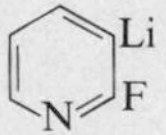

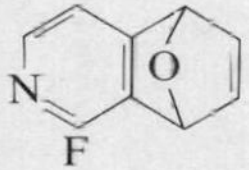

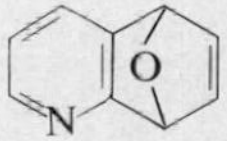

H₂O

OMe (26%)

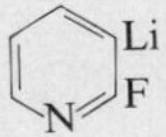

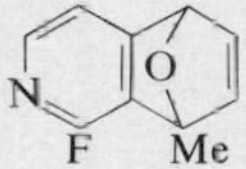

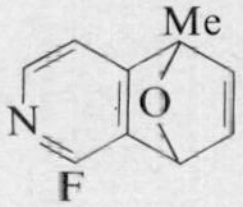

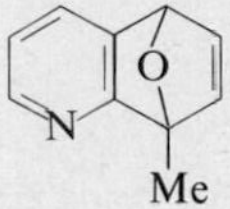

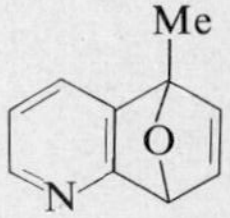

232

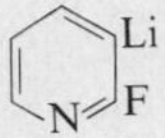

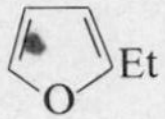

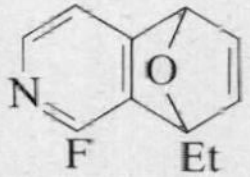

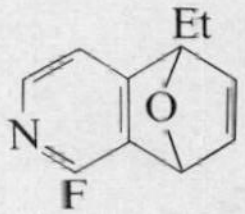

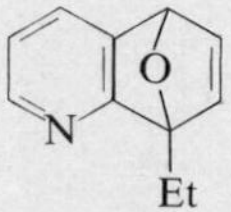

224

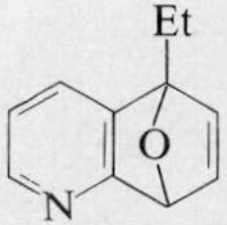

224

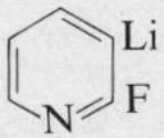

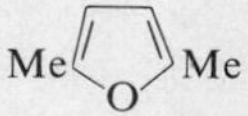

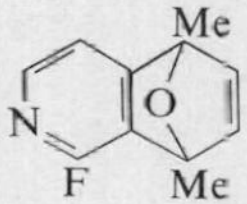

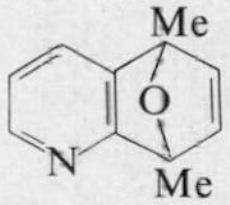

224

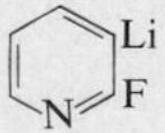

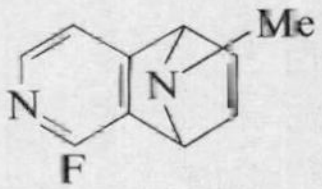

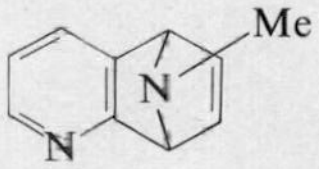

224

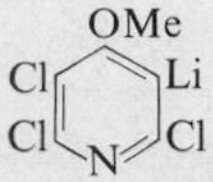

224

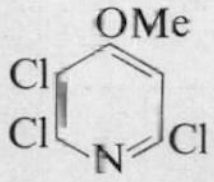

232

Table VII-3 (*Continued*)

Substituents in pyridine	Reactant	Product (yield if reported)	Ref.
(3,4,6-trichloro-5-methyl-2-lithiopyridine)	(1) CO₂ (2) CH₂N₂	(methyl 3,5,6-trichloro-4-methylpyridine-2-carboxylate) CO₂Me	232
(3,4,6-trichloro-5-ethyl-2-lithiopyridine)	H₂O	(2,3,5-trichloro-4-ethylpyridine) (51%)	232
(trichloro-lithiopyridine)	—	—	229
(piperidinyl-dichloro-lithiopyridine)	PhCN	Ph—C:NH (22%); and Ph—C:N—C:N—Ph (75%)	209
(NMe₂-trichloro-lithiopyridine)	(furan)	NMe₂ (43%)	151

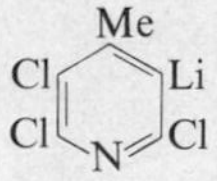

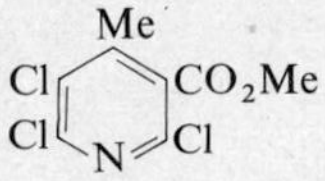

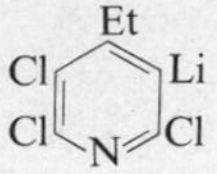

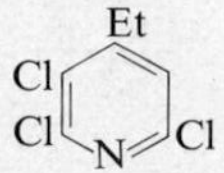

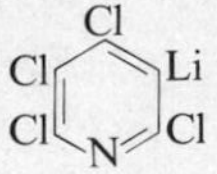

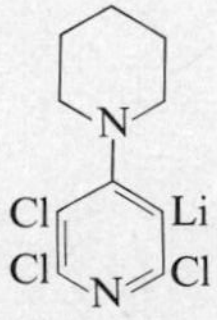

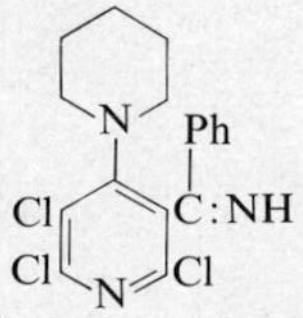

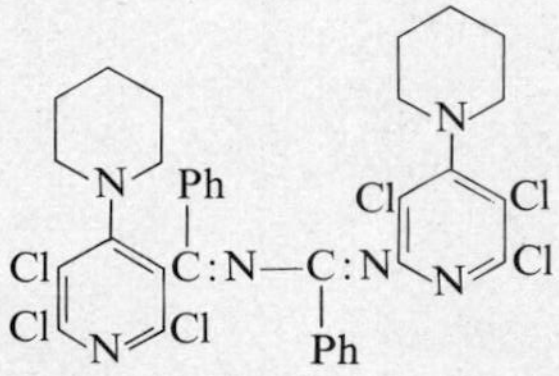

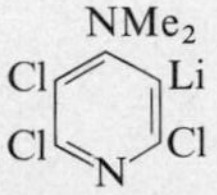

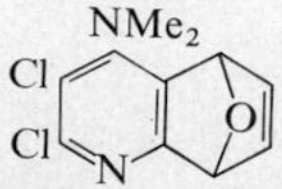

512

		, 42%	151
		, 45%	101
2,4,5,6-Tetrachloro[b]	H_2O		17
2,4,5,6-Tetrafluoro	CO_2	, 62%	15, 22
2,4,6-Trifluoro	CO_2	, 65%	15, 22
2,4,6-Trifluoro	CO_2		15, 22

[a] Obtained from the tetrachloro derivative and n-butyllithium in diethyl ether.

[b] Obtained from 2,3,4,5,6-pentachloropyridine and n-butyllithium in diethyl ether as a mixture with 2,3,5,6-tetrachloro-4-pyridyllithium.

[c] Obtained by the addition of BuLi to 3-bromo-2-fluoropyridine in ether at $-60°$, followed by the addition of the diene after *more than* 5 minutes, and warming.

[d] Obtained by the addition of BuLi to 3-bromo-2-fluoropyridine in ether at $-60°$, followed by the addition of the diene after *less than* 5 minutes, and warming. These results are remarkable and await an explanation.

513

TABLE VII-4. Reactions of 4-Pyridyllithium Compounds with Miscellaneous Reactants

Substituents in pyridine	Reactant	Product (yield if reported)	Ref.
None			12
None	EtCH(C$_6$H$_4$OMe-p)COEt	EtCH(C$_6$H$_4$OMe-p)C(OH)-(Py-4)Et	13, 150
None	(MeO)$_3$B	4-PyB(OMe)$_2$	30
None			16, 100
None			149
None			149
None	Ph$_2$CO	4-PyCPh$_2$(OH)	213
3-Me	Cyclohexanone	3-Me-4-PyCH(OH)(CH$_2$)$_5$	152
3-Fluoro[b]	Et$_2$CO	3-F-4-PyC(Et$_2$)OH 3-F-2-PyC(Et$_2$)OH	215
3-Fluoro	Br$_2$		215
3-Fluoro	HCONMe$_2$		215
2,3,5,6-Tetrachloro[a]	Me$_2$SO$_4$	, 64%	151
2,3,5,6-Tetrachloro[a]	CO$_2$		17

514

Table VII-4 (*Continued*)

Substituents in pyridine	Reactant	Product (yield if reported)	Ref.
2,3,5,6-Tetrachloro[a]	H_2O		17
2,3,5,6-Tetrachloro	Benzene		102
2,3,5,6-Tetrachloro	*p*-Xylene		102
2,3,5,6-Tetrachloro	*p*-Diisopropylbenzene		102
2,3,5,6-Tetrachloro	Mesitylene		102
2,3,5,6-Tetrachloro	Durene		6, 102

Table VII-4 (*Continued*)

Substituents in pyridine	Reactant	Product (yield if reported)	Ref.
			222
	H₂O	(88%)	48
		(32%)	48
	H₂O	(84%)	48
		(58%)	48
	H₂O	(77%)	48
		(65%)	48
	H₂O	(77%)	48
		(58%)	48

Table VII-4 (*Continued*)

Substituents in pyridine	Reactant	Product (yield if reported)	Ref.
![Cl-substituted pyridyllithium structure with Li at top, Cl at 2,3,5,6 positions]		![4,4'-bipyridine octachloro structure with Cl Cl / Cl Cl on rings]	226
2,3,5,6-Tetrachloro	1,3-Diphenylisobenzofuran	![bridged oxanorbornene product with Ph, Cl, O, N]	102
2,3,5,6-Tetrafluoro	CO_2	![tetrafluoropyridine with CO_2H], 50%	15, 22
2,3,5,6-Tetrafluoro	HCONHMe	![tetrafluoropyridine with CHO], 66%	23
2,3,5,6-Tetrafluoro	![tetrafluoropyridine-CHO]	![(tetrafluoropyridyl)$_2$—CH(OH)], 66%	23
2,3,5,6-Tetrafluoro	![(tetrafluoropyridyl)$_2$—CO]	![(tetrafluoropyridyl)$_3$—COH], 70%	23
2,3,5,6-Tetrafluoro	![tetrafluoropyridine-CO_2Et]	![(tetrafluoropyridyl)$_3$—COH], 77%	23
2,3,5,6-Tetrafluoro	$HgCl_2$	![(tetrafluoropyridyl)$_2$—Hg]	103
2,3,5,6-Tetrachloro	Mesitylene	Not identified	6

[a] Obtained as the principal product from reaction between 2,3,4,5,6-pentachloropyridine and *n*-butyllithium in diethyl ether; lesser amounts of 2,4,5,6-tetrachloro-3-pyridyllithium are also formed.

[b] From 3-fluoropyridine and butyl-lithium, followed by 3-pentanone. Again, the remarkable proton-abstraction from C-2 is reported (215).

TABLE VII-4A Reactions of 2-Pyridylsodium- and 2-Pyridylpotassium-1-Oxides with Miscellaneous Reagents

Substituent in pyridine-1-oxide	Reactant	Product	Ref.
None	S_8		197
None	S_8		197
None	S_2Cl_2		197
3-Me	S_8		197

III. Pyridylmagnesium Compounds

The pyridylmagnesium halides have been prepared most frequently by the entrainment procedure in diethyl ether (See 1st ed., Part 2, p. 437), employing ethyl bromide as the means of maintaining a clean magnesium surface. A recent recommendation is the use of dibromoethane instead of ethyl bromide, to be added at the same time as 3-bromopyridine for the synthesis of 3-pyridylmagnesium bromide (30). It is not known whether the substitution of tetrahydrofuran for diethyl ether would eliminate the need for an entraining agent. The polyfluorohalopyridines, however, have been converted to the Grignard reagents in good yield in tetrahydrofuran without these agents, and usually at -28 to $0°$. It is assumed that the highly electronegative fluorine atoms promote Grignard reagent formation. For example, in tetrahydrofuran, 2,3,5,6-tetrafluoro-4-pyridylmagnesium bromide, 2,4,6-trifluoro-5-chloro-3-pyridylmagnesium chloride, 3,5,6-trifluoro-2-methoxy-4-pyridylmagnesium bromide, 3,5-difluoro-2,6-dimethoxy-4-pyridylmagnesium bromide (31, 32), and 2,3,5,6-tetrachloro-4-pyridylmagnesium chloride (33, 34) have been prepared and have given the normal products when treated with active halogen compounds, aldehydes, ketones, and water.

In an unusual reaction which may involve a pyridylmagnesium compound, heating 145 g of pyridine, 6 g of magnesium, and 0.3 g of sodium in xylene under reflux followed by treatment of the mixture with air has produced 4,4'-bipyridyl in 55% yield (35).

The observation that 2- and 4-bromopyridine, but not 3-bromopyridine, react with methylmagnesium iodide to give 10 and 25% yields, respectively, of 2- and 4-picoline has suggested (36) an addition-elimination sequence involving **VII-16a**. Since, in the presence of CoCl$_2$, all three isomers react with methylmagnesium iodide to give 55, 26, and 60% of 2-, 3-, and 4-picoline, respectively, the function of the intermediate **VII-16a** is not clear.

VII-16a

The reactions of the pyridylmagnesium compounds are listed in Table VII-5.

IV. Pyridylalkylmetallic Compounds

1. Pyridylalkyllithium Compounds

Lateral metalation (1st ed., Part 2, p. 423) by means of phenyllithium or *n*-butyllithium remains the method of choice for the preparation of pyridylalkyllithium compounds from the corresponding alkylpyridines; metalation has also been effected with 2-furyllithium, 2-thienyllithium, or 1-methyl-2-pyrryllithium (213). The solvents for the reaction include diethyl ether, tetrahydrofuran, benzene, and hexane. A consequence of the commercial availability of hexane solutions of phenyllithium and *n*-butyllithium has been that a mixture of hexane and one of the other solvents is used frequently. The metalation reaction is generally performed at −20°.

Competitive reactions between equimolar amounts of 2,4-lutidine and phenyllithium have shown that metalation occurs predominantly at the 2-position (**VII-17**) (37–39). This preference, therefore, is contrary to that

TABLE VII-5. Reactions of Pyridylmagnesium Halides

Position of Mg X	X	Substituents in pyridine	Reactant	Product (yield if reported)	Ref.
2	Cl	None	4-MeOC$_6$H$_4$CHO	4-MeOC$_6$H$_4$CH(OH)Py-2	214
3	Br	None	(BuO)$_3$B	3-PyB(OBu)$_2$	30
4	Br	2,3,5,6-Tetrafluoro	CO$_2$	2,3,5,6-Tetrafluoro-4-pyridinecarboxylic acid	31, 32
4	Br	2,3,5,6-Tetrafluoro	MeCOEt	Ethyl methyl 4-(2,3,5,6-tetrafluoro)pyridyl carbinol	31, 32
4	Br	2,3,5,6-Tetrafluoro	2,3,4,5,6-Pentafluoropyridine	Octafluoro-4,4'-bipyridyl	31, 32
3	Cl	2,4,6-Trifluoro-5-chloro	BrCH$_2$CH$_2$Br	3-Chloro-2,4,6-trifluoropyridine	31, 32
3	Cl	2,4,6-Trifluoro-5-chloro	ClCH$_2$CH$_2$Br	3-Chloro-2,4,6-trifluoropyridine	31, 32
4	Br	2-Methoxy-3,5,6-trifluoro	H$_2$O	2-Methoxy-3,5,6-trifluoropyridine	31, 32
4	Br	3,5-Difluoro-2,6-dimethoxy	H$_2$O	3,5-Difluoro-2,6-dimethoxypyridine	31, 32
4	Cl	2,3,5,6-Tetrachloro	H$_2$O	2,3,5,6-Tetrachloropyridine (88 %)	31, 32, 151
4	Cl	2,3,5,6-Tetrachloro	MeCHO	1-(2,3,5,6-Tetrachloro-4-pyridyl)ethanol (30 %)	34
4	Cl	2,3,5,6-Tetrachloro	CO$_2$	2,3,5,6-Tetrachloroisonicotinic acid (35—40 %)	34
4	Cl	2,3,5,6-Tetrachloro	Pentachloropyridine	4,4'-Octachlorobipyridyl	33
4	Br	2,3,5,6-Tetrafluoro	HgCl$_2$	Bis(2,3,5,6-tetrafluoropyridyl)mercury (72%)	103

520

reported with sodium amide or potassium amide in liquid ammonia where metalation occurs preferentially at the 4-position (40, 41). It should be emphasized at this time that a competing reaction in these lateral metalations is the addition of phenyllithium, sodium amide, or potassium amide to the azomethine system of the pyridine heterocycle (see p. 31 and 1st ed., Part 2, p. 435).

VII-17

The previous edition (Part 2, Tables VII-6 to VII-10, pp. 449 ff) gives a list of the reactions of pyridylalkyllithium compounds with reactive halogen compounds, aldehydes, ketones, acid chlorides, acid anhydrides, esters, alkylene oxides, nitriles, oxygen, carbon dioxide, and the halogens. Related reactions reported since that compilation have been listed in Tables VII-6 to VII-8. The present discussion is limited to reactions of pyridylalkyllithium compounds not included in the previous edition.

The addition of 2-pyridylmethyllithium (**VII-18**) to a variety of N-(arylidene)anilines has given good to excellent yields of N,1-diaryl-2-(2-pyridyl)-ethylamines (**VII-19**). These reactions have taken place in ether under reflux,

VII-18   **VII-19**

and with the few exceptions noted below, the products are unaffected by these reaction conditions or even prolonged reaction periods at even higher temperatures (41). One exception to the general utility of this synthetic method is found in the reaction of **VII-18** with N-[p-(2-diethylaminoethoxy)-benzylidene]-4-(trifluoromethyl)aniline (**VII-20**), where the product is a black polymer; since **VII-18** and N-(p-methoxybenzylidene)-3-(trifluoromethyl) aniline give the normal product **VII-21**, direct attack at the 4-(trifluoromethyl) group is probably excluded. A likely mechanism would be

TABLE VII-6. Reactions of 2-Pyridylalkylmetallic Compounds

Organometallic compound	Reactant	Product (yield, if reported)	Ref.
a. Sodium and potassium compounds			
2-PyCH$_2$K	2-PrBr	2-PyCH$_2$CHMe$_2$ (83%)	41, 56
2-PyCHKMe	EtBr	2-PyCHEtMe (68%)	41, 56
2-PyCHKMe	2-PrBr	2-PyCHMeCHMe$_2$ (87%)	41, 56
2-PyCHKEt	EtBr	2-PyCHEt$_2$ (84%)	41, 56
2-PyCHKEt	2-PrBr	2-PyCHEt(CHMe$_2$) (45%)	41, 56
2-PyCKMe$_2$	EtBr	2-PyCEtMe$_2$ (70%)	41, 56
2-PyCKMeEt	EtBr	2-PyCEt$_2$Me (39%)	41, 56
2-PyCKMeCHMe$_2$	EtBr	No Reaction	41, 56
2-PyCKMeEt	2-PrBr	2-PyCEt(CHMe$_2$)Me (24%)	41, 56
2-PyCH$_2$Na	EtBr	2-PyPr-n (65%)	58
2-PyCH$_2$Na	2-PrBr	2-PyCH$_2$CHMe$_2$ (68%)	58
2-PyCHNaEt	EtBr	2-PyCHEt$_2$ (21%)	58
2-PyCHNaCHMe$_2$	2-PrBr	2-PyCH(CHMe$_2$)$_2$ (14%)	58
6-(Benzyloxy)-2-pyridylmethyl-potassium	Br(CH$_2$)$_4$Br	1,6-Bis(6-benzyloxy-2-pyridyl)hexane	46
6-(Benzyloxy)-2-pyridylmethyl-sodium	2(5-Bromopentyl)pyridine	2-[6-(6-Benzyloxy-2-pyridyl)hexyl]pyridine	46
6-Methoxy-2-pyridylmethyl-potassium	Br(CH$_2$)$_4$Br	1,6-Bis(6-methoxy-2-pyridyl)hexane	46
6-(Benzyloxy)-2-pyridylmethyl-potassium	CH$_3$(CH$_2$)$_5$Br	7-(6-Benzyloxy-2-pyridyl)heptane	61
6-(N,N-Diethylcarboxamido)-2-pyridylmethylpotassium	Me(CH$_2$)$_5$Br	N,N-Diethyl 6-heptylpicolinamide	61
6-(N,N-Diethylcarboxamido)-2-pyridylmethylpotassium	Me(CH$_2$)$_{10}$Br	N,N-Diethyl-6-dodecylpicolinamide	61
2-PyCNa(C$_6$H$_4$Cl-p)CONEt$_2$	Me$_2$N(CH$_2$)$_2$Cl	2-PyCH(C$_6$H$_4$Cl-p)CONEt$_2$ (44%) | (CH$_2$)$_2$NMe$_2$	132

2-PyCNaPhCON(CHMe$_2$)$_2$	Et$_2$N(CH$_2$)$_2$Cl	2-PyCPhCON(CHMe$_2$)$_2$ (44 %) —(CH$_2$)$_2$NEt$_2$	132
2-PyCNaPhCON(CH$_2$)$_5$	O(CH$_2$)$_4$N(CH$_2$)$_2$Cl	2-PyCPhCON(CH$_2$)$_5$ (32 %) —(CH$_2$)$_2$N(CH$_2$)$_4$O	132
2-PyCNaPhCON(CH$_2$)$_4$O	Et$_2$N(CH$_2$)$_2$Cl	2-PyCPhCON(CH$_2$)$_4$O (65 %) —(CH$_2$)$_2$NEt$_2$	132
2-PyCH$_2$Na	EtI	2-Py(CH$_2$)$_2$Me (62 %)	55
2-PyCH$_2$Na	EtBr	2-Py(CH$_2$)$_2$Me (64 %)	55
2-PyCHNaEt	EtBr	2-PyCHEt$_2$ (76 %)	55
2-PyCNaPh$_2$	n-C$_8$H$_{17}$Br	2-PyC(C$_8$H$_{17}$-n)Ph$_2$ (77 %)	62
2-PyCH$_2$Na	MeCO$_2$(CH$_2$)$_5$Cl	2-Py(CH$_2$)$_6$O$_2$CMe	153
2-PyCH$_2$Na-1-oxide	Me(CH$_2$)$_5$Br	2-Py(CH$_2$)$_6$Me-1-oxide	61
5-Methoxy-2-pyridylmethyl- potassium	MeCO$_2$(CH$_2$)$_5$Cl	MeO⟨pyridine⟩(CH$_2$)$_6$O$_2$CMe	153
3-Methyl-2-pyridylmethylsodium	MeI	2-Ethyl-3-methylpyridine	40
4-Methyl-2-pyridylmethylsodium	MeI	2-Methyl-4-(2-propyl)pyridine or 4-ethyl-2-methylpyridine	40
5-Methyl-2-pyridylmethylsodium	MeI	2-Ethyl-5-methylpyridine	40
6-Methyl-2-pyridylmethylsodium	MeI	2-Ethyl-6-methylpyridine and 2,6-diethylpyridine	40
2-PyCH$_2$K	2-(C$_4$H$_3$S)CH$_2$Cl	2-Py(CH$_2$)$_2$C$_4$H$_3$S)-2	47, 149, 154
2-PyCH$_2$Na	2-(4-Chlorobutoxy)tetrahydro-2H-pyran	2-[5-(Tetrahydro-2H-pyran-2-yloxy)pentyl]pyridine	46
2-PyCH$_2$Na	PhBr	2-PyCH$_2$Ph and 2-PyCHPh$_2$	53
2-PyCH$_2$K	Br(CH$_2$)$_{10}$CONMe$_2$	2-Py(CH$_2$)$_{11}$CONMe$_2$	61
2-PyCHNaCHMe$_2$	p-MeOC$_6$H$_4$CHBrEt	2-PyCHCHMe$_2$ CHEtC$_6$H$_4$OMe-p	13, 150

Table VII-6 (*Continued*)

Organometallic compound	Reactant	Product (yield, if reported)	Ref.
2-PyCHNa(CH₂)₂NEt₂	*p*-MeOC₆H₄CHBrEt	2-PyCH(CH₂)₂NEt₂ — CHEtC₆H₄OMe-*p*	13, 150
2-PyCHNaEt	*p*-MeC₆H₄CHBrEt	2-PyCHEt — CHEtC₆H₄Me-*p*	13, 150
2-PyCHNaEt	*p*-ClC₆H₄CHBrEt	2-PyCHEt — CHEtC₆H₄Cl-*p*	13, 150
(2-Py)₂CNaCON(CH₂)₄	PhCH₂CHMeNH(CH₂)₂Br	(2-Py)₂CCON(CH₂)₄ — (CH₂)₂NHCHMeCH₂Ph	94
Me-[Py-2]–CNa(Py-2)CON(CH₂)₄	PhCH₂CHMeNH(CH₂)₂Br	Me-[Py]–C(Py-2)CON(CH₂)₄ — (CH₂)₂NHCHMeCH₂Ph	94
(2-Py)₂CNaCON(CH₂)₄	*p*-MeC₆H₄CH₂CHMeNH(CH₂)₂Br	(2-Py)₂CCON(CH₂)₄ — (CH₂)₂NHCHMeCH₂C₆H₄Me-*p*	94
(2-Py)₂CNaCON(CH₂)₄	*p*-ClC₆H₄CH₂CHMeNH(CH₂)₂Br	(2-Py)₂CCON(CH₂)₄ — (CH₂)₂NHCHMeCH₂C₆H₄Cl-*p*	94
(2-Py)₂CNaCON(CH₂)₄	Ph(CH₂)₂CHMeNH(CH₂)₂Br	(2-Py)₂CCON(CH₂)₄ — (CH₂)₂NHCHMe(CH₂)₂Ph	94
(2-Py)₂CNaCON(CH₂)₄	PhCH₂CHMeNH(CH₂)₃Br	(2-Py)₂CCON(CH₂)₄ — (CH₂)₃NHCHMeCH₂Ph	94

(2-Py)₂CNaCON(CH₂)₄	p-ClC₆H₄CH₂CHMeNH(CH₂)₃Br	(2-Py)₂CCON(CH₂)₄—(CH₂)₃NHCHMeCH₂C₆H₄Cl-p	94
(2-Py)₂CNaCON(CH₂)₄	p-ClC₆H₄CH₂CMe₂NH(CH₂)₃Br	(2-Py)₂CCON(CH₂)₄—(CH₂)₃NHCMe₂CH₂C₆H₄Cl-p	94
(2-Py)₂CNaCON(CH₂)₄	Ph(CH₂)₂CHMeNH(CH₂)₃Br	(2-Py)₂CCON(CH₂)₄—(CH₂)₃NHCHMe(CH₂)₂Ph	94
2-PyCNaPh₂	p-ClC₆H₄CH₂Cl	2-PyC(CH₂C₆H₄Cl-p)Ph₂ (50 %)	62
2-PyCNaPh₂	Me₃C(CH₂)₂Br	2-PyC[(CH₂)₂CMe₃]Ph₂ (40 %)	62
2-PyCNaPhCON(CH₂)₄	Me₂N(CH₂)₂Cl	2-PyCPh[CON(CH₂)₄](CH₂)₂NMe₂ (40 %)	132
2-PyCNa(C₆H₄Cl-p)CONMe₂	p-ClC₆H₄CH₂Cl	2-PyC(C₆H₄Cl-p)(CONMe₂)CH₂C₆H₄Cl-p (63 %)	132
2-PyCNa(C₆H₄Cl-p)CONMe₂	O(CH₂)₄N(CH₂)₂Cl	2-PyC(C₆H₄Cl-p)(CONMe₂)(CH₂)₂N(CH₂)₄O (50 %)	132
2-PyCNa(C₆H₄Cl-p)CONMe₂	Me₂N(CH₂)₂Cl	2-PyC(C₆H₄Cl-p)CONMe₂—(CH₂)₂NMe₂	132
2-PyCNaPhCONEt₂	Me₂N(CH₂)₂Cl	2-PyCPhCONEt₂—(CH₂)₂NMe₂ (65 %); —(CH₂)₂NMe₂ (25 %)	132
(2-Py)₂CNaCON(CH₂)₄	[morpholino]N(CH₂)₂Cl	(2-Py)₂CCON(CH₂)₄—(CH₂)₂N(CH₂)₄O	94
(2-Py)₂CNaCONMe₂	Me₂N(CH₂)₂Cl	(2-Py)₂CCONMe₂—(CH₂)₂NMe₂	94
(2-Py)₂CNaCONMe₂	[pyrrolidino](CH₂)₂Cl	(2-Py)₂CCONMe₂—(CH₂)₂N(CH₂)₄	94

525

Table VII-6 (*Continued*)

Organometallic compound	Reactant	Product (yield, if reported)	Ref.
(2-Py)$_2$CNaCONMe$_2$	piperidine N-(CH$_2$)$_2$Cl	(2-Py)$_2$CCONMe$_2$ —(CH$_2$)$_2$N(CH$_2$)$_5$	94
(2-Py)$_2$CNaCONMe$_2$	morpholine N-(CH$_2$)$_2$Cl	(2-Py)$_2$CCONMe$_2$ —(CH$_2$)$_2$N(CH$_2$)$_4$O	94
(2-Py)$_2$CNaCONMe$_2$	piperidine N-CHMeCH$_2$Cl	(2-Py)$_2$CCONMe$_2$ —CHMeCH$_2$N(CH$_2$)$_5$	94
6-Methyl-2-pyridylmethylpotassium	CO(OEt)$_2$	Me—pyridine—CH$_2$CO$_2$Et (59–75 %)	198
(2-Py)$_2$CNaCONMe$_2$	morpholine N-CHMeCH$_2$Cl and morpholine N-CH$_2$CHMeCl	(2-Py)$_2$CCONMe$_2$ —CH$_2$CHMeN(CH$_2$)$_4$O and (2-Py)$_2$CCONMe$_2$ —CHMeCH$_2$N(CH$_2$)$_4$O	94
(2-Py)$_2$CNaCON(CH$_2$)$_4$	Me$_2$N(CH$_2$)$_2$Cl	(2-Py)$_2$CCON(CH$_2$)$_4$ —(CH$_2$)$_2$NMe$_2$	94

$\left[\text{Cl} \overset{\text{N}}{\underset{}{\bigcirc}} \right]_2 \text{CNaCON(CH}_2)_4$ → $\left[\text{Cl} \overset{\text{N}}{\underset{}{\bigcirc}} \right]_2 \overset{\displaystyle \mid}{\underset{\displaystyle (\text{CH}_2)_2\text{NMe}_2}{\text{C}}}\text{—CON(CH}_2)_4$

Reagent	Chloride	Product	Ref.
(2-Py)₂CNaCON(CH₂)₄	Me₂N(CH₂)₂Cl	(2-Py)₂CCON(CH₂)₄	94
(2-Py)₂CNaCON(CH₂)₄	(CH₂:CHCH₂)₂N(CH₂)₂Cl	—(CH₂)₂N(CH₂CH:CH₂)₂ / (2-Py)₂CCON(CH₂)₄	94
(2-Py)₂CNaCON(CH₂)₄	(CH₂:CHCH₂)₂NCHMeCH₂Cl	CH₂CHMeN(CH₂CH:CH₂)₂ / (2-Py)₂CCON(CH₂)₄ / —CHMeCH₂N(CH₂CH:CH₂)₂	
2-PyCHNaEt	p-MeOC₆H₄COEt	2-PyCHEt / C(OH)EtC₆H₄OMe-p	13, 150
2-PyCHNaEt	p-MeOC₆H₄COMe	2-PyCHEt / C(OH)MeC₆H₄OMe-p	13, 150
2-PyCHNaEt	p-ClC₆H₄COEt	2-PyCHEt / C(OH)EtC₆H₄Cl-p	13, 150
2-PyCHNaEt	p-BrC₆H₄COEt	2-PyCHEt / —C(CO)EtC₆H₄Br-p	13, 150
2-PyCHNaEt	3,4-(MeO)₂C₆H₃COMe	2-PyCHEt / —C(OH)MeC₆H₃(OMe)₂-3,4	13, 150
2-PyCHNaEt	p-[Me₂N(CH₂)₂O]C₆H₄COEt	2-PyCHEt / C(OH)C₆H₄[O(CH₂)₂NMe₂]-p	13, 150
2-PyCHNaEt	p-MeSC₆H₄COEt	2-PyCHEt / C(OH)C₆H₄SMe-p	13, 150

527

TABLE VII-6 (Continued)

Organometallic compound	Reactant	Product (yield, if reported)	Ref.
2-PyCHNaEt	p-MeC₆H₄COEt	2-PyCHEt—C(OH)C₆H₄Me-p	13, 150
2-PyCHNaEt	p-MeOC₆H₄COPy-2	2-PyCHEt—C(OH)(2-Py)C₆H₄OMe-p	13, 150
2-PyCNaEt₂	p-MeOC₆H₄COEt	2-PyCEt₂—C(OH)EtC₆H₄OMe-p	13, 150
2-PyCH₂Na	C₅H₅N	(2-Py)₂CH₂ (35%)	94
2-PyCH₂Na	2-PyMe	2-PyCH₂⟨ring⟩Me	94
(2-Py)₂CHNa	⟨N(CH₂)₄COCl⟩	(2-Py)₂CHCO[N(CH₂)₄]	94
NC ⟨pyridine⟩ CH₂K / KO	PhCO₂Me	⟨pyridinone⟩ CH₂COPh, 80%	67
NC ⟨pyridine, Ph⟩ CH₂K / KO	PhCO₂Me	⟨pyridinone, Ph⟩ CH₂COPh, 78%	67
NC ⟨pyridine⟩ CH₂K / KO	(CO₂Et)₂	⟨pyridinone⟩ CH:C(OH)CO₂Et, 46%	67

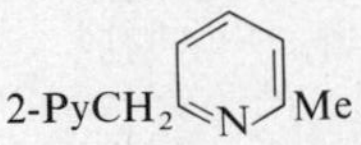

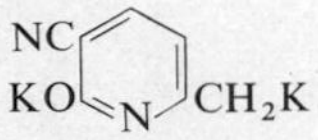

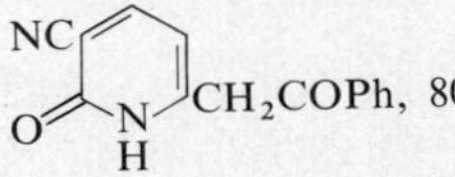

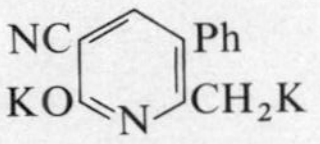

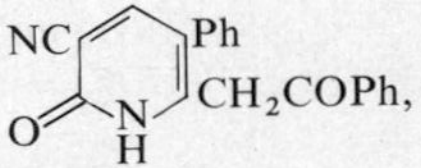

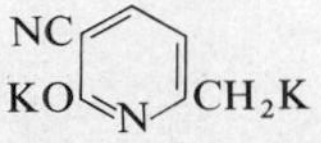

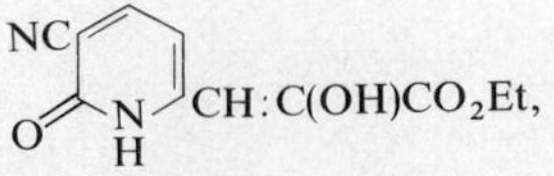

528

NC — KO⟩N⟨CH₂K	Ph₂CO	NC⟩N(H)⟨CH₂C(OH)Ph₂, 87%	67
NC — KO⟩N⟨CH₂K	PhCHO	NC⟩N(H)⟨CH₂CH(OH)Ph, 67%	67
KO⟩N⟨CH₂K	Ph₂CO	⟩N(H)⟨CH₂C(OH)Ph₂	67
NC — Ph ⟩N⟨CH₂K	PhCOCH:CHPh	NC — Ph ⟩N(H)⟨CH₂CHPhCH₂COPh, 87%	67
2-PyCH₂Na*	(ClCH₂CH₂)₂O	(tetrahydropyranyl-pyridine structure)	216
NC ⟩N⟨CH₂K	MeI	NC⟩N(H)⟨Et, 91%	68
NC ⟩N⟨CH₂K	PhCH₂Cl	NC⟩N(H)⟨CH₂CH₂Ph, 86%	68

* The intermediate 2-Py(CH$_2$)$_3$O(CH$_2$)$_2$Cl is treated further with NaNH$_2$.

Table VII-6 (*Continued*)

Organometallic compound	Reactant	Product (yield, if reported)	Ref.
(2-cyano-pyridone)CH$_2$K	Br(CH$_2$)$_3$Br	[NC—pyridone—CH$_2$CH$_2$CH$_2$]$_2$CH$_2$, 74%	68
(2-cyano-pyridone)CHKCH$_2$Ph	BuBr	NC—pyridone—CHBuCH$_2$Ph, 64%	68
(2-cyano-pyridone, Ph)CH$_2$K	PhCH$_2$Cl	NC—pyridone(Ph)—CH$_2$CH$_2$Ph,	68
b. Lithium compounds			
2-PyCH$_2$Li	C$_8$H$_{17}$Br	2-Nonylpyridine (38%)	155
		2-[(1-Octyl)nonyl]pyridine (2%)	
2-PyCH$_2$Li	C$_6$H$_{13}$CHMeBr	2-[(2-Methyl)octyl]pyridine (29%)	155
4-Methyl-2-pyridylmethyllithium	(EtO)$_2$CHCH$_2$Br	Me—pyridine—CH$_2$CH(OMe)$_2$	39
2-PyCH$_2$Li	Cl(CH$_2$)$_4$C⫶CC$_5$H$_{11}$-n	2-Py(CH$_2$)$_5$C⫶CC$_5$H$_{11}$-n (63%)	75
2-PyCH$_2$Li	ClCH$_2$C⫶CBu-n	2-Py(CH$_2$)$_2$C⫶CBu-n (53%)	75
2-PyCH$_2$Li	Cl(CH$_2$)$_3$C⫶CBu-n	2-Py(CH$_2$)$_4$C⫶CBu-n (63%)	75
2-PyCH$_2$Li	Cl(CH$_2$)$_9$C⫶CPr-n	2-Py(CH$_2$)$_{10}$C⫶CPr-n (66%)	75
2-PyCH$_2$Li	(p-ClC$_6$H$_4$)$_2$C⫶CHC$_6$H$_4$COCl-p	(p-ClC$_6$H$_4$)$_2$C⫶CHC$_6$H$_4$(COPy-2)-p (21%)	156
2-PyCH$_2$Li	Ph$_2$C⫶CHC$_6$H$_4$CN-p	Ph$_2$C⫶CHC$_6$H$_4$(COCH$_2$Py-2)-p (67%)	11
2-PyCH$_2$Li	(p-ClC$_6$H$_4$)$_2$C⫶CHC$_6$H$_4$CN-p	(p-ClC$_6$H$_4$)$_2$C⫶CHC$_6$H$_4$(COCH$_2$Py-2)-p (21%)	11

Organolithium reagent	Reactant	Product	Ref.
5-Methyl-4-phenyl-2-pyridyl-methyllithium	PhCH₂Cl	1-Phenyl-2-(5-methyl-4-phenyl-2-pyridyl)ethane + 1,3-Diphenyl-2-(5-methyl-4-phenyl-2-pyridyl)propane	157
2-PyCH₂Li	PhBr	2-PyCH₂Ph	38
2-PyCH₂Li	Cu₂Cl₂	2-PyCH₂CH₂Py-2	44
2-PyCH₂Li	Me₂AsI	2-PyCH₂AsMe₂	137
2-PyCH₂Li	PhCH:CHCH:CHCHO	PhCH:CHCH:CHCH:CHPy-2 (15%)	50
2-PyCH₂Li	p-Me₂NC₆H₄CHO	2-PyCH₂CH(OH)C₆H₄NMe₂-p	42
6-Methyl-2-pyridylmethyllithium	EtCHO	2-(2-Hydroxybutyl)-6-methylpyridine	158
2-PyCH₂Li	Ph₂CO	2-PyCH₂C(OH)Ph₂ (58%)	5
2-PyCH₂Li	PhCH₂COMe	2-PyCH₂CMe(OH)CH₂Ph	7
2-PyCH₂Li	p-MeOC₆H₄CH₂COMe	2-PyCH₂CMe(OH)CH₂C₆H₄OMe-p	7
2-PyCH₂Li	p-(p-MeOC₆H₄)C₆H₄COMe	2-PyCH₂CMe(OH)CH₂C₆H₄(C₆H₄OMe-p)p	7
2-PyCH₂Li	2-(2-Dimethylaminoethyl)indan-1-one	2-(2-Dimethylaminoethyl)-3-[(2-pyridyl)methyl]indene	159
2-PyCHLiMe	2-(2-Dimethylaminoethyl)indan-1-one	2-(2-Dimethylaminoethyl)-3-[1-(2-pyridyl)ethyl]indene	159
2-PyCHLiEt	2-(2-Dimethylaminoethyl)indan-1-one	2-(2-Dimethylaminoethyl)-3-[1-(2-pyridyl)propyl]indene	159
2-PyCLiMe₂	2-(2-Dimethylaminoethyl)indan-1-one	2-(2-Dimethylaminoethyl)-3-[dimethyl(2-pyridyl)methyl]indene	159
6-Methyl-2-pyridylmethyllithium	2-(2-Dimethylaminoethyl)indan-1-one	2-(3-Dimethylaminoethyl)-3-(6-methyl-2-pyridyl)methyl]indene	159
2-PyCHLiMe	2-(2-Diethylaminoethyl)indan-1-one	2-(2-Diethylaminoethyl)-3-[1-(2-pyridyl)ethyl]indene	159
2-PyCH₂Li	5-chloro-2-(2-dimethylaminoethyl)indan-1-one	5-chloro-2-(2-dimethylaminoethyl)-3-(pyridyl)methyl]indene	159
2-PhCHLiPh	2-(2-Dimethylaminoethyl)indan-1-one	2-(2-Dimethylaminoethyl)-3-[1-phenyl-1-(2-pyridyl)methyl]indene	160
2-PyCH₂Li	p-PhC₆H₄CO₂Me	2-PyCH₂COC₆H₄Ph-p (45%)	161
2-PyCH₂Li	⬡CO₂Et	2-PyCH₂COCH(CH₂)₅ (60%)	161

Table VII-6 (*Continued*)

Organometallic compound	Reactant	Product (yield, if reported)	Ref.
2-PyCH$_2$Li	PhCH$_2$CO$_2$Et	2-PyCH$_2$COCH$_2$Ph (57%)	161
2-PyCH$_2$Li	Ph(CH$_2$)$_2$CO$_2$Et	2-PyCH$_2$CO(CH$_2$)$_2$Ph	161
2-PyCH$_2$Li	p-O$_2$NC$_6$H$_4$CO$_2$Me	2-PyCH$_2$COC$_6$H$_4$NO$_2$-p	161
2-PyCH$_2$Li	Me–⟨thiophene⟩–CO$_2$Et	2-PyCH$_2$CO–⟨thiophene⟩–Me	47
2-PyCH$_2$Li	PhCS$_2$Me	2-PyCH$_2$CSPh (15%)	162
2-PyCH$_2$Li	PhCSOMe	2-PyCH$_2$COPh (18%)	162
2-PyCH$_2$Li	Me$_2$CHCSOMe	2-PyCH$_2$COCHMe$_2$ (45%)	162
2-PyCH$_2$Li	1-C$_{10}$H$_7$-CSOEt	2-PyCH$_2$CSC$_{10}$H$_7$-1 (50%)	163
2-PyCH$_2$Li	PhCSOEt	2-PyCH$_2$CSPh (15%)	163
2-PyCH$_2$Li	1-C$_{10}$H$_7$-CS$_2$Me	2-PyCH$_2$CSC$_{10}$H$_7$-1 (72%)	161
2-PyCH$_2$Li	⟨fluorene⟩CHC$_6$H$_4$CN-p	⟨fluorene⟩CHC$_6$H$_4$(COCH$_2$Py-2)-p	148
4-Methyl-2-pyridylmethyllithium	CH$_2$O	2-(4-Methyl-2-pyridyl)ethanol	37
4-Cyclohexyl-2-pyridylmethyl-lithium	CO$_2$	4-Cyclohexyl-2-pyridineacetic acid	164, 165
5-Cyclohexyl-2-pyridylmethyl-lithium	CO$_2$	5-Cyclohexyl-2-pyridineacetic acid	164, 165
6-Cyclohexyl-2-pyridylmethyl-lithium	CO$_2$	6-Cyclohexyl-2-pyridineacetic acid	164, 165
3-Methoxy-2-pyridylmethyl-lithium	MeCN	1-(3-Methoxy-2-pyridyl)acetone	166
3-Ethoxy-2-pyridylmethyl-lithium	MeCN	1-(3-Ethoxy-2-pyridyl)acetone	166

Substrate	Reagent	Product (yield)	Ref.
Me—N—CH₂CH₂—Me (pyridine)	Ph₂CO	Me—N—CH₂CH₂—N—CH₂C(OH)Ph₂ (61%)	199
Me—N—CH₂Li (pyridine)	Me—N—CH₂Cl	Me—N—CH₂CH₂—N—Me (60%)	199
(LiCH₂—N—)₂ (pyridine)	Me—N—CH₂Cl	CH₃—N—CH₂—(—N—CH₂—)₂—N—CH₃	199
LiCH₂—N—CH₂CH₂—N—CH₂CH₂—N—CH₂C(OH)Ph₂	Ph₂CO	[Ph₂(OH)CCH₂—N—CH₂—]₂CH₂	199
Me—N—CH₂Li (Me, Me)	MeI	Me—N—Et (40%)	217
Me—N—CH₂Li (Me, Me)	MeI	Me—N—Et (Me, Me) (60%)	217
Me—N—CH₂Li (Me, Me)	I(CH₂)₄I	Me(Me)—N—(CH₂)₆—N—Me(Me) (21%)	217

533

Table VII-6 (*Continued*)

Organometallic compound	Reactant	Product (yield, if reported)	Ref.
(Me, Me-substituted pyridine)–CH_2Li	$O_2(KMnO_4)$	$\left[\text{(Me, Me-pyridine)}-CH_2 \right]_2$	217–219
2-PyCHLiOPh	*o*-Toluonitrile	$2\text{-PyC}:C(NH_2)C_6H_4Me\text{-}o$ — OPh	167, 168
2-PyCHLiOPh	2,5-Dimethylbenzonitrile	$2\text{-PyC}:C(NH_2)C_6H_3Me_2\text{-}2,5$ — OPh	167, 168
2-PyCHLiOPh	*o*-(Methylthio)benzonitrile	$2\text{-PyC}{=}C(NH_2)C_6H_4SMe\text{-}o$ — OPh	168
2-PyCHLiOPh	*o*-Chlorobenzonitrile	$2\text{-PyC}:C(NH_2)C_6H_4Cl\text{-}o$ — OPh	167
2-PyCHLiOPh	*o*-(α-Naphthyl)benzonitrile	(naphthyl–phenyl structure) $2\text{-PyC}:C(NH_2)$ OPh	167
2-PyCHLiOC$_6$H$_4$Cl-*p*	*o*-Toluonitrile	$2\text{-PyC}:C(NH_2)C_6H_4Me\text{-}o$ — OC$_6$H$_4$Cl-*p*	167
2-PyCHLiOC$_6$H$_4$Cl-*p*	*o*-Chlorobenzonitrile	$2\text{-PyC}:C(NH_2)C_6H_4C_6H_4Cl\text{-}o$ — OC$_6$H$_4$Cl-*p*	167

2-PyCHLiOC$_6$H$_4$Me-o	o-Toluonitrile	2-PyC:C(NH$_2$)C$_6$H$_4$Me-o OC$_6$H$_4$Me-o	167
2-PyCHLiOC$_6$H$_4$Me-m	o-Toluonitrile	2-PyC:C(NH$_2$)C$_6$H$_4$Me-o OC$_6$H$_4$Me-m	167
2-PyCHLiOC$_6$H$_4$Me-m	o-Chlorobenzonitrile	2-PyC:C(NH$_2$)C$_6$H$_4$Cl-o OC$_6$H$_4$Me-m	167
2-PyCHLiOC$_6$H$_4$Me-p	o-Toluonitrile	2-PyC:C(NH$_2$)C$_6$H$_4$Me-o OC$_6$H$_4$Me-p	167
2-PyCHLiOC$_6$H$_4$OEt-p	o-Toluonitrile	2-PyC:C(NH$_2$)C$_6$H$_4$Me-o OC$_6$H$_4$OEt-p	167
2-PyCHLiOC$_6$H$_4$OEt-p	o-Chlorobenzonitrile	2-PyC:C(NH$_2$)C$_6$H$_4$Cl-o OC$_6$H$_4$OEt-p	167
2-PyCH$_2$Li	PhCH:NPy-2	2-PyCH$_2$CHPhNHPy-2 (78%)	42
2-PyCH$_2$Li	PhCH:NCH$_2$Ph	2-PyCH$_2$CHPhNHCH$_2$Ph (36%)	42
2-PyCH$_2$Li	(2-C$_4$H$_3$O)CH:NPh	2-PyCH$_2$CHi(C$_4$H$_3$O-2)NHPh (52%)	42
2-PyCH$_2$Li	(2-C$_4$H$_3$O)CH:N(CH$_2$)NEt$_2$	2-PyCH$_2$CHi(C$_4$H$_3$O-2)NH(CH$_2$)$_2$NEt$_2$ (55%)	42
2-PyCH$_2$Li	PhCH:N(CH$_2$)$_2$NEt$_2$	2-PyCH$_2$CHPhNH(CH$_2$)$_2$NEt$_2$ (61%)	42
2-PyCH$_2$Li	fluorenone *N*-phenylimine	55%	42
2-PyCH$_2$Li	xanthone *N*-phenylimine	63%	42

535

Table VII-6 (*Continued*)

Organometallic compound	Reactant	Product (yield, if reported)	Ref.
2-PyCH₂Li	PhCH:NPh	2-PyCH₂CHPhNHPh (33%)	42
2-PyCH₂Li	PhCMe:NPh	2-PyCH₂CMePhNHPh (47%)	42
2-PyCH₂Li	p-MeOC₆H₄CH:NC₆H₄OMe-p	2-PyCH₂CH(C₆H₄OMe-p)NHC₆H₄OMe-p (55%)	42
2-PyCH₂Li	p-Me₂NC₆H₄CH:NC₆H₄OMe-p	2-PyCH₂CH(C₆H₄NMe₂-p)NHC₆H₄OMe-p (83%)	42
2-PyCH₂Li	p-HOC₆H₄CH:NPh	2-PyCH₂CH (C₆H₄OH-p)NHPh (42%)	42
2-PyCH₂Li	p-ClC₆H₄CH:NC₆H₄OMe-p	2-PyCH₂CH(C₆H₄Cl-p)NHC₆H₄OMe-p (44%)	42
2-PyCH₂Li	p-ClC₆H₄CH:NC₆H₄OH-p	2-PyCH₂CH(C₆H₄Cl-p)NHC₆H₄OH-p (48%)	42
2-PyCH₂Li	p-MeOC₆H₄CH:NC₆H₄CF₃-m	2-PyCH₂CH(C₆H₄OMe-p)NHC₆H₄CF₃-m (45%)	42
2-PyCH₂Li	p-Et₂N(CH₂)₂OC₆H₄CH:NC₆H₄Cl-p	2-PyCH₂CH[(C₆H₄O(CH₂)₂NEt₂-p)NHC₆H₄Cl-p (59%)	42
2-PyCH₂Li	[3,4-(OCH₂O)C₆H₃]CH:NC₆H₄-NMe₂-p	2-PyCH₂CH[C₆H₃(OCH₂O)-3,4]NHC₆H₄Me₂-p (71%)	42
2-PyCH₂Li	p-Et₂N(CH₂)₂OC₆H₄CH:NC₆H₄-CF₃-p	Polymer	42
2-PyCH₂Li	CH:NPh	CH:CHPy-2	42
2-PyCH₂Li	CH:NC₆H₄Cl-p	CH:CHPy-2	42
	Ph(CH₂)₂Br	(CH₂)₃Ph, 79%	69

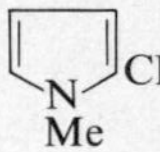

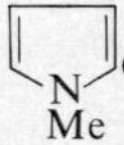

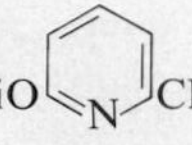

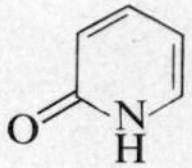

536

LiO–pyridine–CH₂Li	(ClCH₂)₂	[pyridone–(CH₂)₂]₂ , 60 %	69
LiO–pyridine–CH₂Li	Br(CH₂)₄Br	[pyridone–(CH₂)₃]₂ , 41 %	69
LiO–pyridine–CH₂Li	Cyclohexanone	pyridone–CH₂CH(OH)(CH₂)₅, 50 %	69
LiO–pyridine–CH₂Li	PhCHO	pyridone–CH₂CH(OH)Ph, 69 %	69
LiS–pyridine–CH₂Li	Ph(CH₂)₂Br	thiopyridone–(CH₂)₃Ph, 60 %	69
LiS–pyridine–CH₂Li	Ph₂CO	thiopyridone–CH₂C(OH)Ph₂, 80 %	69

537

Table VII-6 (*Continued*)

Organometallic compound	Reactant	Product (yield, if reported)	Ref.
c. Magnesium compounds			
2-PyCH$_2$MgBr	ClCH$_2$OMe	2-Py(CH$_2$)$_2$OMe (31%)	71
2-PyCH$_2$MgCl	ClCH$_2$OMe	2-Py(CH$_2$)$_2$OMe (6%)	71
2-PyCH$_2$MgBr	n-PrCHO	2-PyCH$_2$CH(OH)Pr-n (44%)	72
2-PyCH$_2$MgBr	iso-BuCHO	2-PyCH$_2$CH(OH)CHMe$_2$ (42%)	72
2-PyCH$_2$MgBr	n-BuCHO	2-PyCH$_2$CH(OH)Bu-n (54%)	72
2-PyCH$_2$MgBr	n-C$_5$H$_{11}$CHO	2-PyCH$_2$CH(OH)C$_5$H$_{11}$-n (44%)	72
2-PyCH$_2$MgBr	n-C$_6$H$_{13}$CHO	2-PyCH$_2$CH(OH)C$_6$H$_{13}$-n (43%)	72
2-PyCH$_2$MgBr	n-C$_7$H$_{15}$CHO	2-PyCH$_2$CH(OH)C$_7$H$_{15}$-n (46%)	72
2-PyCH$_2$MgBr	n-C$_8$H$_{17}$CHO	2-PyCH$_2$CH(OH)C$_8$H$_{17}$-n (37%)	72
2-PyCH$_2$MgBr	n-C$_9$H$_{19}$CHO	2-PyCH$_2$CH(OH)C$_9$H$_{19}$-n (41%)	72
2-PyCH$_2$MgBr	n-C$_{10}$H$_{21}$CHO	2-PyCH$_2$CH(OH)C$_{10}$H$_{21}$ (43%)	72
2-PyCH$_2$MgBr	n-C$_8$H$_{17}$CHMeCH$_2$CHO	2-PyCH$_2$CH(OH)CH$_2$CHMeC$_8$H$_{17}$-n (39%)	72
2-PyCH$_2$MgBr	CH$_2$:CHCHO	2-PyCH$_2$CH(OH)CH:CH$_2$ (28%)	72
2-PyCH$_2$MgBr	MeCH:CHCHO	2-PyCH$_2$CH(OH)CH:CHMe (44%)	72
2-PyCH$_2$MgBr	PhCH:CHCHO	2-PyCH$_2$CH(OH)CH:CHPh (53%)	72
2-PyCH$_2$MgBr	p-MeOC$_6$H$_4$CHO	2-PyCH$_2$CH(OH)C$_6$H$_4$OMe-p (38%)	72
2-PyCH$_2$MgBr	p-Me$_2$NC$_6$H$_4$CHO	2-PyCH$_2$CH(OH)C$_6$H$_4$NMe$_2$-p (48%)	72
2-PyCH$_2$MgBr	o-ClC$_6$H$_4$CHO	2-PyCH$_2$CH(OH)C$_6$H$_4$Cl-o (18%)	72
2-PyCH$_2$MgBr	p-BrC$_6$H$_4$CHO	2-PyCH$_2$CH(OH)C$_6$H$_4$Br-p (11%)	72
2-PyCH$_2$MgBr	MeCOPr-n	2-PyCH$_2$C(OH)MePr-n (52%)	72
2-PyCH$_2$MgBr	MeCOBu-n	2-PyCH$_2$C(OH)(n-Bu)Me (57%)	72
2-PyCH$_2$MgBr	MeCOC$_5$H$_{11}$-n	2-PyCH$_2$C(OH)Me(n-C$_5$H$_{11}$) (53%)	72
2-PyCH$_2$MgBr	MeCOC$_9$H$_{19}$-n	2-PyCH$_2$C(OH)Me(n-C$_9$H$_{19}$) (44%)	72
2-PyCH$_2$MgBr	Et$_2$CO	2-PyCH$_2$C(OH)Et$_2$ (45%)	72
2-PyCH$_2$MgBr	EtCOPr-n	2-PyCH$_2$C(OH)(n-Pr)Et (27%)	72
2-PyCH$_2$MgBr	n-PrCOBu-n	2-PyCH$_2$C(OH)(n-Bu)(n-Pr) (49%)	72
2-PyCH$_2$MgBr	(n-C$_5$H$_{11}$)$_2$CO	2-PyCH$_2$C(OH)(n-C$_5$H$_{11}$)$_2$ (37%)	72
2-PyCH$_2$MgBr	MeCOCMe$_3$	2-PyCH$_2$C(OH)(CMe$_3$)Me (56%)	72
2-PyCH$_2$MgBr	MeCOCH:CMe$_2$	2-PyCH$_2$C(OH)Me(CH:CMe$_2$) (44%)	72
2-PyCH$_2$MgBr	n-PrCH:CHCOMe	2-PyCH$_2$C(OH)Me(n-PrCH:CH) (46%)	72
2-PyCH$_2$MgBr	Cyclopentanone	2-PyCH$_2$C(OH)CH(CH$_2$)$_4$ (55%)	72
2-PyCH$_2$MgBr	Cyclohexanone	2-PyCH$_2$(OH)CH(CH$_2$)$_5$ (54%)	72

TABLE VII-7. Reactions of 3-Pyridylalkylmetallic Compounds

Organometallic compound	Reactant	Product (yield if reported)	Ref.
a. Sodium and potassium compounds			
$3\text{-PyCH}_2\text{Na}$	MeCl	3-PyEt (ca. 50%)^a	82
		3-PyMe	
		3-PyCHMe_2	
$3\text{-PyCH}_2\text{Na}$	EtBr	$3\text{-PyPr-}n$ (62%)	40, 58
$3\text{-PyCH}_2\text{Na}$	$\text{Br(CH}_2)_4\text{Br}$	$3\text{-Py(CH}_2)_6\text{Py-3}$	46
$3\text{-PyCH}_2\text{Na}$	$2\text{-Py(CH}_2)_5\text{Br}$	$3\text{-Py(CH}_2)_6\text{Py-3}$	46
$3\text{-PyCH}_2\text{Na}$	$\text{CH}_2\text{:CH(CH}_2)_3\text{Br}$	$3\text{-Py(CH}_2)_4\text{CH:CH}_2$ (90%)	211
$3\text{-PyCH}_2\text{Na}$	$\text{CH}_2\text{:CH(CH}_2)_4\text{Br}$	$3\text{-Py(CH}_2)_5\text{CH:CH}_2$ (69%)	211
$3\text{-PyCH}_2\text{Na}$	$\text{I(CH}_2)_9\text{C:CPr-}n$	$3\text{-Py(CH}_2)_{10}\text{:CPr-}n$ (54%)	75
$3\text{-PyCH}_2\text{Na}$	PhBr	$3\text{-PyCH}_2\text{Ph, 3-PyCHPh}_2$	53
$3\text{-PyCH}_2\text{K}$	$\text{MeCHBrPr-}n$	$3\text{-PyCH}_2\text{CHMePr-}n$ (86%)	57
3-PyCNaPh_2	$n\text{-C}_8\text{H}_{17}\text{Br}$	$3\text{-Py(C}_8\text{H}_{17}\text{-}n)\text{Ph}_2$ (64%)	62
$3\text{-PyCH}_2\text{K}$	$\text{Br(CH}_2)_{10}\text{CONMe}_2$	$3\text{-Py(CH}_2)_{11}\text{CONMe}_2$	61
$3\text{-PyCH}_2\text{K}$	Me_2CHCHO	$3\text{-PyCH}_2\text{CH(OH)CHMe}_2$ (6%)	57
$3\text{-PyCH}_2\text{K}$	PhCHO	$3\text{-PyCH}_2\text{CH(OH)Ph}$ (17%)	57
$3\text{-PyCH}_2\text{K}$	EtCOMe	$3\text{-PyCH}_2\text{C(OH)EtMe}$ (10%)	57
$3\text{-PyCH}_2\text{K}$	MeCOPr	$3\text{-PyCH}_2\text{C(OH)MePr}$ (9%)	57
$3\text{-PyCH}_2\text{K}$	[thiophene]—COMe	$3\text{-PyCH}_2\text{C(OH)MeC}_4\text{H}_3\text{S-2}$ (12%)	57
$3\text{-PyCH}_2\text{K}$	MeCOPh	$3\text{-PyCH}_2\text{C(OH)MePh}$ (11%)	57
$3\text{-PyCH}_2\text{K}$	Et_2CO	$3\text{-PyCH}_2\text{C(OH)Et}_2$ (10%)	57
$3\text{-PyCH}_2\text{K}$	EtCOPh	$3\text{-PyCH}_2\text{C(OH)EtPh}$ (14%)	57
$3\text{-PyCH}_2\text{K}$	$\text{PrCOPr-}n$	$3\text{-PyCH}_2\text{C(OH)PrPr-}n$ (10%)	57
$3\text{-PyCH}_2\text{K}$	Ph_2CO	$3\text{-PyCH}_2\text{C(OH)Ph}_2$ (27%)	57
$3\text{-PyCH}_2\text{K}$	$(p\text{-Me}_2\text{NC}_6\text{H}_4)_2\text{CO}$	$3\text{-PyCH}_2\text{C(OH)(C}_6\text{H}_4\text{NMe}_2p)_2$ (25%)	57

Table VII-7 (*Continued*)

Organometallic compound	Reactant	Product (yield if reported)	Ref.
3-PyCH$_2$Na	I(CH$_2$)$_4$C:CC$_5$H$_{11}$-n	3-Py(CH$_2$)$_5$C:CC$_5$H$_{11}$-n (48%)	57
3-PyCNaPh$_2$	p-ClC$_6$H$_4$CH$_2$Cl	3-PyC(CH$_2$C$_6$H$_4$Cl-p)Ph$_2$ (18%)	62
2-Phenyl-3-pyridylmethylsodium	MeCl	3-Ethyl-2-phenylpyridine (33%)	82
6-(Benzyloxy)-3-pyridylmethyl-potassium	Me(CH$_2$)$_5$Br	7-[6-(Benzyloxy)-3-pyridyl]heptane	61
2-(Benzyloxy)-3-pyridylmethyl-potassium	Me(CH$_2$)$_5$Br	7-[2-(Benzyloxy)-3-pyridyl]heptane	61
6-(Benzyloxy)-3-pyridylmethyl-potassium	Me(CH$_2$)$_{10}$Br	1-[6-(Benzyloxy)-3-pyridyl]dodecane	61
3-PyCK(C$_6$H$_4$Cl-p)$_2$	MeI	3-PyCMe(C$_6$H$_4$Cl-p)$_2$	52
3-PyCK(C$_6$H$_4$Cl-p)	PhCH$_2$Br	3-PyC(CH$_2$Ph)(C$_6$H$_4$Cl-p)$_2$	52
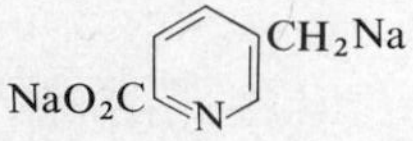	CH$_2$:CHCH$_2$Cl	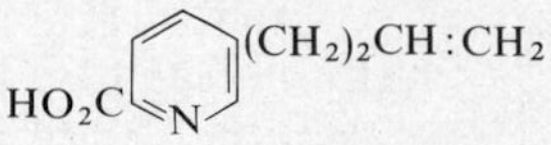 , 51%	65
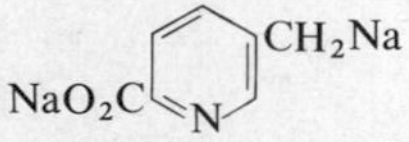	n-PrCl	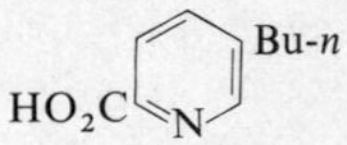 , 3%	65
(3-PyCOPh)$^=$Na$_2^{++}$	EtBr	3-PyC(OH)Et(Ph)	228
(3-PyCOPh)$^=$Na$_2^{++}$	isoPrBr	3-PyC(OH)Ph(Pr-iso)	228
(3-PyCOPh)$^=$Na$_2^{++}$	Me$_3$CBr	3-PyC(OH)Ph(CMe$_3$)	228
(3-PyCOPh)$^=$Na$_2^{++}$	PhCH$_2$Br	3-PyC(OH)Ph(CH$_2$Ph)	228
(3-PyCOPh)$^=$Na$_2^{++}$	Ph$_2$CO	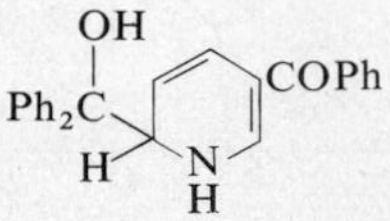	228

540

3-PyCHNaEt	p-MeOC$_6$H$_4$COEt	3-PyCHEt—EtC(OH)C$_6$H$_4$OMe-p	13, 150
3-PyCH$_2$Na	PhCO$_2$Et	3-PyCH$_2$COPh (78%)	54
3-PyCH$_2$Na	MeCO$_2$Et	3-PyCH$_2$COMe (47%)	54
3-PyCH$_2$Na	EtCO$_2$Et	3-PyCH$_2$COEt (61%)	54
3-PyCH$_2$Na	Me$_2$CHCO$_2$Et	3-PyCH$_2$COCHMe$_2$ (69%)	54
3-PyCH$_2$Na	Me$_3$CCO$_2$Et	3-PyCH$_2$COCMe$_3$ (79%)	54
3-PyCH$_2$Na^a	(ClCH$_2$CH$_2$)$_2$O		216
3-PyCNaPh$_2$	CO$_2$	3-PyC(CO$_2$H)Ph$_2$ (38%)	45
(3-Py)$_2$CNaPh	CO$_2$	(3-Py)$_2$C(CO$_2$H)Ph (40%)	45
3-PyCHNa(CH$_2$)$_2$NMe$_2$	PhCO$_2$Et	3-PyCHCOPh (64%)—(CH$_2$)$_2$NMe$_2$	54
3-PyCHNa(CH$_2$)$_2$NMe$_2$	Me$_2$CHCO$_2$Et	3-PyCHCOCHMe$_2$ (55%)—(CH$_2$)$_2$NMe$_2$	54

b. Lithium and Magnesium Compounds

5-Me-3-PyCH$_2$Li	Cyclohexanone		152
Mixt. 2-Me and 6-Me-3-PyCH$_2$Li	Cyclohexanone		169
3-PyCHLiMe	MeCl	3-PyCHMe$_2$ (ca. 55%)^b 3-PyEt	82
3-PyCLiMe$_2$	MeCl	3-PyCMe$_3$ (10%)	82

Table VII-7 (*Continued*)

Organometallic compound	Reactant	Product (yield if reported)	Ref.
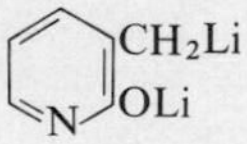 Pyridine–CH_2Li, OLi	$Ph(CH_2)_2Br$	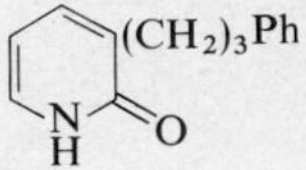 pyridinone, $(CH_2)_3Ph$, O, NH, 75%	69
 CH_2Li, SLi	$Ph(CH_2)_2Br$	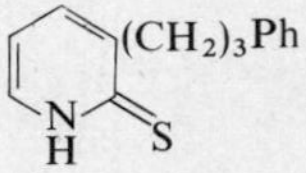 $(CH_2)_3Ph$, S, 19%	69
 CH_2Li, SLi	$p\text{-}ClC_6H_4(CH_2)_2Br$	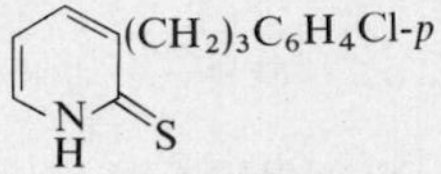 $(CH_2)_3C_6H_4Cl\text{-}p$, S, 70%	69
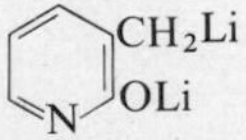 CH_2Li, OLi	Ph_2CO	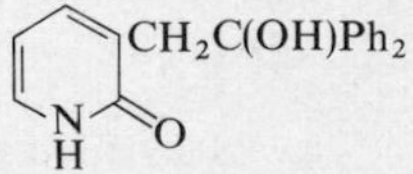 $CH_2C(OH)Ph_2$, O, 55%	69
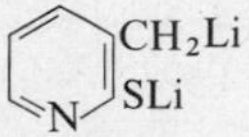 CH_2Li, SLi	Ph_2CO	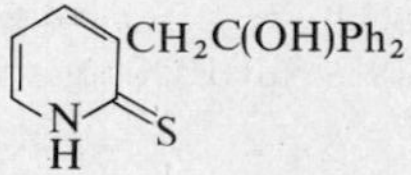 $CH_2C(OH)Ph_2$, S, 80%	69
5-Cyclohexyl-3-pyridylmethyl-lithium	CO_2	5-Cyclohexyl-3-pyridylacetic acid	164
6-Cyclohexyl-3-pyridylmethyl-lithium	CO_2	6-Cyclohexyl-3-pyridylacetic acid	164
3-(3-Pyridyl)propylmagnesium chloride	H_2O	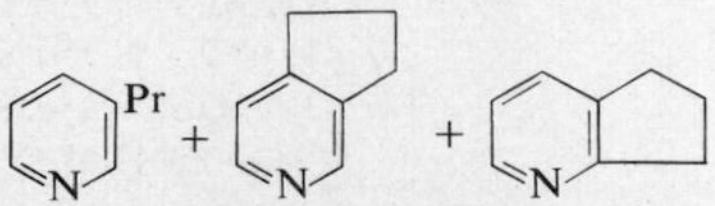 (cyclopenta-fused pyridines) + (3-Pr-pyridine)	73

[a] The intermediate $3\text{-}Py(CH_2)_3O(CH_2)_2Cl$ is treated further with one molar equivalent of $NaNH_2$.
[b] Mixture could not be separated by fractional distillation (compare Ref. 170).

542

TABLE VII-8. Reactions of 4-Pyridylalkylmetallic Compounds

Organometallic compound	Reactant	Product (yield if reported)	Ref.
a. Sodium and potassium compounds			
4-PyCH$_2$K	EtBr	4-PyC$_3$H$_7$-n (59 %)	41, 56
4-PyCH$_2$K	2-PrBr	4-PyCH$_2$CHMe$_2$ (48 %)	41, 56
4-PyCHKMe	EtBr	4-PyCHEMe (80 %)	41, 56
4-PyCHKMe	2-PrBr	4-PyCHMe(CHMe$_2$) (97 %)	41, 56
4-PyCHKEt	EtBr	4-PyCHEt$_2$ (81 %)	41, 56
4-PyCHKEt	2-PrBr	4-PyCHEt(CHMe$_2$) (68 %)	41, 56
4-PyCKMe$_2$	EtBr	4-PyCEtMe$_2$ (90 %)	41, 56
4-PyCKMeEt	EtBr	4-PyCEt$_2$Me (56 %)	41, 56
4-PyCKEt$_2$	EtBr	4-PyCEt$_3$ (25 %)	41, 56
4-PyCKMe(CHMe$_2$)	EtBr	No reaction	41, 56
4-PyCKMeEt	2-PrBr	4-PyCEtMe(CHMe$_2$) (50 %)	41, 56
4-PyCNaPh$_2$	n-C$_8$H$_{17}$Br	4-PyC(n-C$_8$H$_{17}$)Ph$_2$ (57 %)	62
4-PyCNaPh$_2$	p-ClC$_6$H$_4$CH$_2$Cl	4-PyC(CH$_2$C$_6$H$_4$Cl-p)Ph$_2$ (62 %)	62
4-PyCNaPh$_2$	Me$_3$CCH$_2$CH$_2$Br	4-PyC[(CH$_2$)$_2$CMe$_3$]Ph$_2$ (low yield)	62
3-Ethyl-4-pyridylmethylsodium	MeCl	3,4-Diethylpyridine (80 %)	64
4-PyCH$_2$Na	Br(CH$_2$)$_4$Br	4-Py(CH$_2$)$_6$Py-4	46
4-PyCH$_2$Na	Br(CH$_2$)$_5$Br	4-Py(CH$_2$)$_7$Py-4	46
4-PyCH$_2$Na	Br(CH$_2$)$_6$Br	4-Py(CH$_2$)$_8$Py-4	46
4-PyCH$_2$Na	BrCHMe(CH$_2$)$_4$Br	4-PyCH$_2$CHMe(CH$_2$)$_5$Py-4	46
4-PyCH$_2$Na	2-Py(CH$_2$)$_5$Br	4-Py(CH$_2$)$_6$Py-2	46
4-Pyridylmethylsodium-1-oxide	Br(CH$_2$)$_4$Br	1,6-Bis(1-oxido-4-pyridyl)hexane	46
2,6-Di(benzyloxy)-4-pyridylmethyl- sodium	Me(CH$_2$)$_5$Br	1-[2,6-Di(benzyloxy)-4-pyridyl]hexane	46
2,6-Di(benzyloxy)-4-pyridylmethyl- sodium	Br(CH$_2$)$_4$Br	1,6-Bis[2,6-Di(benzyloxy)-4-pyridyl]hexane	46
2-(Benzyloxy)-4-pyridylmethyl potassium	Br(CH$_2$)$_4$Br	1,6-Bis[2-(benzyloxy)-4-pyridyl]hexane	46

Table VII-8 (*Continued*)

Organometallic compound	Reactant	Product (yield if reported)	Ref.
2-Methoxy-4-pyridylmethylsodium	$Br(CH_2)_4Br$	1,6-Bis(2-methoxy-4-pyridyl)hexane	46
2-(Benzyloxy)-4-pyridylmethyl-potassium	$Me(CH_2)_5Br$	1-[2-(Benzyloxy)-4-pyridyl]heptane	61
2-(Benzyloxy)-4-pyridylmethyl-potassium	$PhCH_2Br$	1-[2-(Benzyloxy)-4-pyridyl]-2-phenylethane	61
2-(Benzyloxy)-4-pyridylmethyl-potassium	$Me(CH_2)_{10}Br$	1-[2-(Benzyloxy)-4-pyridyl]dodecane	61
4-$PyCH_2Na$	$PhBr$	4-$PyCH_2Ph$, 4-$PyCHPh_2$	53
4-PyCHNaEt	p-$MeOC_6H_4CHBrEt$	4-PyCHEt ∣ $CHEtC_6H_4OMe$-p	13, 150
4-PyCHNaEt	p-$MeOC_6H_4CMeBrEt$	4-PyCHEt ∣ $CEtMeC_6H_4OMe$-p	13, 150
4-$PyCH_2Na$	$ClCH_2C(OH)Ph_2$	4-$Py(CH_2)_2C(OH)Ph_2$ (50%)	171
2-(Phenylthio)-4-pyridylmethyl-potassium	$Me(CH_2)_5Br$	1-[2-(Phenylthio)-4-pyridyl]heptane	61
4-$PyCH_2K$	$Br(CH_2)_{10}CONMe_2$	4-$Py(CH_2)_{11}CONMe_2$	61
4-$PyCH_2Na$-1-oxide	$Me(CH_2)_5Br$	1-(1-Oxido-4-pyridyl)heptane	172
4-$PyCH_2Na$	Ph_2CO	4-$PyCH_2C(OH)Ph_2$ (61%)	172
4-PyCHNaEt	Ph_2CO	4-$PyCH(Et)C(OH)Ph_2$ (8%)	172
4-PyCHNaMe	Ph_2CO	4-$PyCH(Me)C(OH)Ph_2$ (7%)	172
4-$PyCH_2Na$	$Ph(p$-$MeC_6H_4)CO$	4-$PyCH_2C(OH)(C_6H_4Me$-$p)Ph$ (68%)	172
4-$PyCH_2Na$	$(p$-$MeC_6H_4)_2CO$	4-$PyCH_2C(OH)(C_6H_4Me$-$p)_2$ (42%)	172
4-$PyCH_2Na$	4-PyCOMe	4-$PyCH_2C(OH)(4$-$Py)Me$ (35%)	172
3-Methyl-4-pyridylmethylsodium	$Ph(p$-$MeC_6H_4)CO$	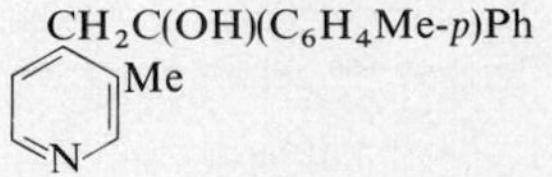 , 60%	172

544

2-Methyl-4-pyridylmethylsodium	Ph$_2$CO	, 85%	172
4-PyCH$_2$Na	6-Methoxy-1-tetralone	6-Methoxy-1-(4-pyridylmethylene)-1,2,3,4-tetrahydro-naphthalene and 3,4-dihydro-6-methoxy-1-(4-pyridyl-methyl)naphthalene (1:1)	49, 51
4-PyCHNaMe	6-Methoxy-1-tetralone	3,4-Dihydro-6-methoxy-1-[1-(4-pyridyl)ethyl]naphthalene (18%)	49, 51
4-PyCH$_2$Na	5-Methoxy-1-indanone	5-Methoxy-1-(4-pyridylmethylene)indane (47%)	49, 51
4-PyCH$_2$Na	5,6-Dimethoxy-1-indanone	5,6-Dimethoxy-1-(4-pyridylmethylene)indane (36%)	49, 51
4-PyCH$_2$Na	(4-PyCOPh	(4-PyCH$_2$)(4-Py)C(OH)Ph (20%)	172
4-PyCH$_2$Na	(3-PyCOPh	(4-PyCH$_2$)(3-Py)C(OH)Ph (39%)	172
4-PyCH$_2$Na	(2-PyCOPh	(4-PyCH$_2$)(2-Py)C(OH)Ph (51%)	172
4-PyCHNaEt	p-MeOC$_6$H$_4$COEt	4-PyCHEt | EtC(OH)C$_6$H$_4$OMe-p	13, 150
4-PyCHNaEt	p-MeOC$_6$H$_4$COMe	4-PyCHEt | MeC(OH)C$_6$H$_4$OMe-p	13, 150
4-PyCH$_2$Na	CH$_2$:CH(CH$_2$)$_3$Br	4-Py(CH$_2$)$_4$CH:CH$_2$ (88%)	211
4-PyCH$_2$Na	CH$_2$:CH(CH$_2$)$_4$Br	4-Py(CH$_2$)$_5$CH:CH$_2$ (89%)	211
4-PyCH$_2$Na*	(ClCH$_2$CH$_2$)$_2$O		216

* The intermediate 4-Py(CH$_2$)$_3$O(CH$_2$)$_2$Cl is further treated with NaNH$_2$.

Table VII-8 (*Continued*)

Organometallic compound	Reactant	Product (yield if reported)	Ref.
CH₂Na-pyridine-Me	MeI	Et-pyridine-Me (37%)	217
CH₂Na-pyridine-Me	EtI	Pr-*n*-pyridine-Me (27–37%)	217
CH₂Na-pyridine-Me	PrI	Bu-*n*-pyridine-Me (44%)	217
CH₂Na-pyridine-Me	PhCH₂Br	(CH₂)₂Ph-pyridine-Me (47%)	217
CH₂Na-pyridine-Me	I(CH₂)₄I	[(CH₂)₃-pyridine-Me]₂ (53%)	217
CH₂Na-pyridine-Me	Ph₂CO	CH₂CPh₂(OH)-pyridine-Me (47%)	217

CH₂Na (Me-pyridine)	$p\text{-MeOC}_6\text{H}_4\text{CHO}$	CH₂CH(OH)— (OMe-phenyl), CH₃-pyridine (60%)	217
CH₂Na (Me-pyridine)	fluorenone	HO, CH₂ fluorene, Me-pyridine	217
CH₂Na (Me-pyridine)	O_2	[CH₂ (Me-pyridine)]₂ ("Fair" yield)	217
CH₂Na (Me, Me-pyridine)	MeI	Et (Me, Me-pyridine) (47%)	217
CH₂Na (Me, Me-pyridine)	EtI	Pr-n (Me, Me-pyridine) (57%)	217
CH₂Na (Me, Me-pyridine)	PrI	Bu-n (Me, Me-pyridine) (42%)	217

547

Table VII-8 (*Continued*)

Organometallic compound	Reactant	Product (yield if reported)	Ref.
4-(CH₂Na)-2,6-Me₂-pyridine	PhCH₂Br	4-[(CH₂)₂Ph]-2,6-Me₂-pyridine (34%)	217
4-(CH₂Na)-2,6-Me₂-pyridine	Ph₂CO	4-[CH₂CPh₂(OH)]-2,6-Me₂-pyridine (96%)	217
4-(CH₂Na)-2,6-Me₂-pyridine	fluorenone	(24%)	217
2-Me-4-(CHNaEt)-pyridine	p-MeOC₆H₄COEt	4-PyCHEt—C(OH)C₆H₄OMe-p	13, 150
4-PyCH₂Na	PhCO₂Et	4-PyCH₂COPh (80%)	54
4-PyCH₂Na	MeCO₂Et	4-PyCH₂COMe (55%)	54
4-PyCH₂Na	EtCO₂Et	4-PyCH₂COEt (77%)	54
4-PyCH₂Na	Me₂CHCO₂Et	4-PyCH₂COCHMe₂ (78%)	54
4-PyCHNaPh	PhCO₂Et	4-PyCHPhCOPh (78%)	54
4-PyCHNaPh	EtCO₂Et	4-PyCHPhCOEt (79%)	54
4-PyCHNa(CH₂)₂NMe₂	PhCO₂Et	4-PyCHCOPh—(CH₂)₂NMe₂ (80%)	54

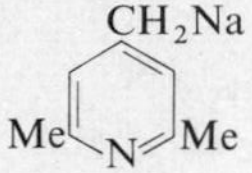

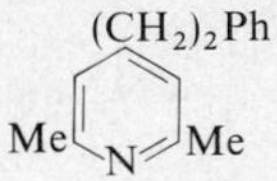

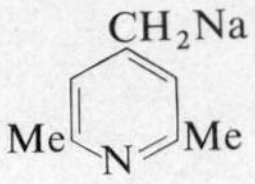

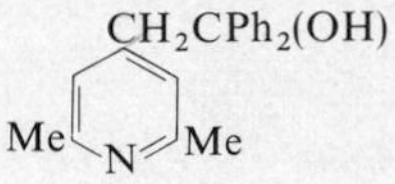

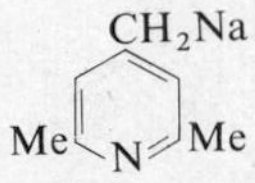

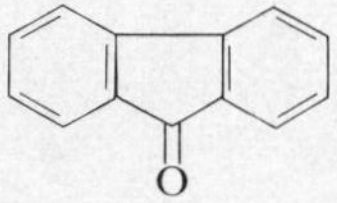

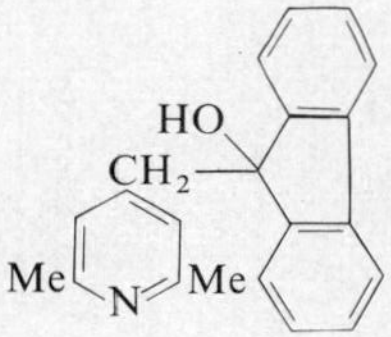

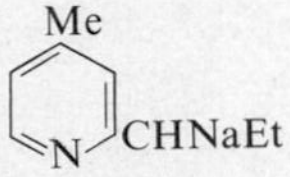

548

4-PyCHNa(CH₂)₂NMe₂	Me₂CHCO₂Et	4-PyCHCOCHMe₂ (78%) $\quad$(CH₂)₂NMe₂	54
4-PyCHNaMe	Ph₂C—CH₂O	4-PyCHMeCH₂C(OH)Ph₂ (50%)	171
[4-PyCHNa]₂CH₂	Ethylene oxide	[HO(CH₂)₂CHPy-4]₂CH₂	63
4-PyCHNaCH₂CHNaPy-3	Ethylene oxide	HO(CH₂)₂CH(Py-4)CH₂CH(Py-3)(CH₂)₂OH	63
4-PyCHNaCH₂CHNaPy-2	Ethylene oxide	HO(CH₂)₂CH(Py-4)CH₂CH(Py-2)(CH₂)₂OH	63
4-PyCHNaCH₂CHNa	Ethylene oxide	HOCH₂CH₂CHCH₂CHCH₂CH₂OH	63

CHNaCH₂CHNaPy-4 → Ethylene oxide → HOCH₂CH₂CHCH₂CHCH₂CH₂OH — 63

4-PyCHNa(CH₂)₂CHNaPy-4	Ethylene oxide	HOCH₂CH₂CH(CH₂)₂CHCH₂CH₂OH $\quad$Py-4$\qquad$Py-4	63
4-PyCHNa(CH₂)₆CHNaPy-4	Ethylene oxide	HOCH₂CH₂CH(CH₂)₆CHCH₂CH₂OH $\quad$Py-4$\qquad$Py-4	63
(4-PyCHNa)₂CH₂	Propylene oxide	(MeCH(OH)CH₂CHPy-4)₂CH₂	63

, 90% — 68

Table VII-8 (*Continued*)

Organometallic compound	Reactant	Product (yield if reported)	Ref.
CH₂K, NC, KO—Ph pyridine	Ph₂CO	CH₂C(OH)Ph₂ pyridone, 80%	67
CH₂K, NC, KO—Ph pyridine	PhCOCH:CHCOPh	CH₂CHPhCH₂COPh pyridone, 86%	67
b. Lithium compounds			
4-PyCH₂Li	ClCH₂C:CBu-n	4-Py(CH₂)₂C:CBu-n (40%)	75
4-PyCH₂Li	Cl(CH₂)₃C:CBu-n	4-Py(CH₂)₄C:CBu-n (47%)	75
4-PyCH₂Li	Cl(CH₂)₄C:CC₅H₁₁-n	4-Py(CH₂)₅C:CC₅H₁₁-n (45%)	75
CH₂Li pyridine, OLi	Ph(CH₂)₂Br	(CH₂)₃Ph, 37%	69
4-PyCH₂Li	Me₂CO	4-PyCH₂C(OH)Me₂	173, 174
4-PyCH₂Li	Et₂CO	4-PyCH₂C(OH)Et₂	173, 174
4-PyCH₂Li	Ph₂CO	4-PyCH₂C(OH)Ph₂	43
CH₂Li pyridine, OLi	Ph₂CO	CH₂C(OH)Ph₂ pyridone, 40%	69

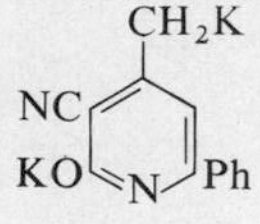

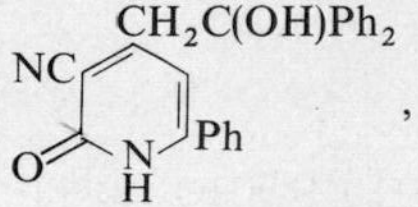

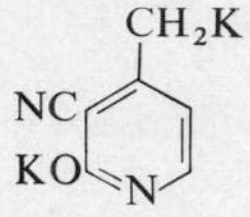

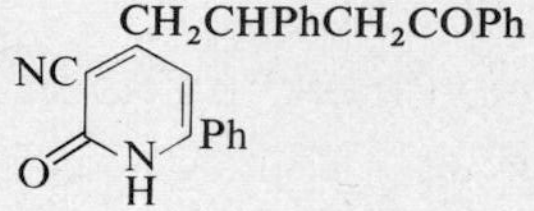

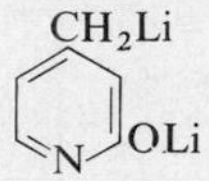

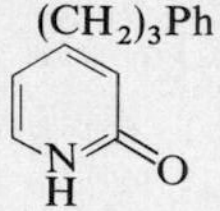

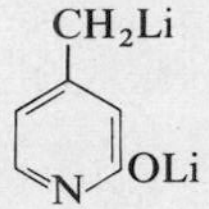

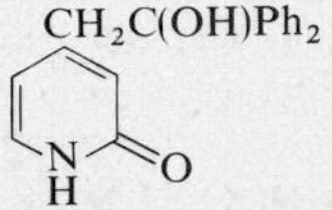

550

4-PyCH₂Li	PhCOC₄H₃S-2	4-PyCH₂CPh(OH)C₄H₃S-2
4-PyCH₂Li	p-(Me₂NC₆H₄)₂CO	4-PyCH₂C(OH)(C₆H₄NMe₂-p)₂
4-PyCH₂Li	PhOCH₂COMe	4-PyCH₂CMe(OH)CH₂OPh

173, 174
173, 174
173, 174

| 4-PyCH₂Li | PhCOCH₂—[piperidine NH] | 4-PyCH₂CPh(OH)CH₂—[piperidine NH] |

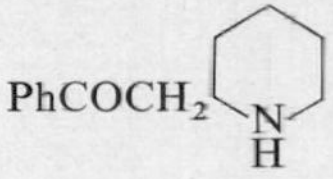

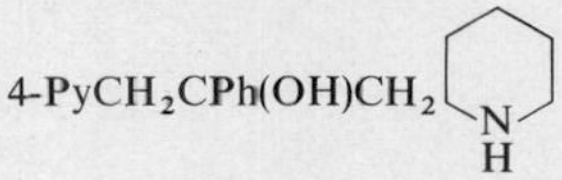

173, 174

| 4-PyCH₂Li | PhCOCH₂—[morpholine NH] | 4-PyCH₂CPh(OH)CH₂—[morpholine NH] |

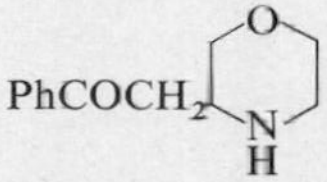

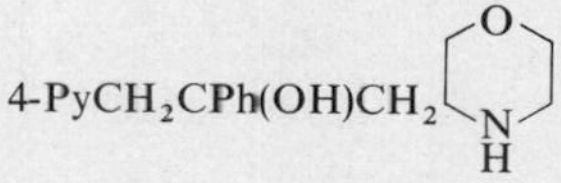

173, 174

| 4-PyCH₂Li | PhCO—[morpholine NH] | 4-PyCH₂CPh(OH)—[morpholine] |

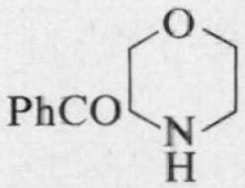

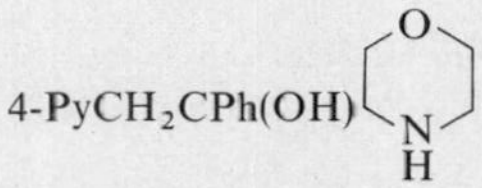

173, 174

4-PyCH₂Li	p-ClC₆H₄COMe	4-PyCH₂CMe(OH)C₆H₄Cl-p
Py-CH₂Li	PhCH:CHCOPy-4	4-PyCH₂C(CH:CHPh)(OH)Py-4
4-PyCH₂Li	Ph₂CHCH₂COCH₂Py-2	4-PyCH₂C(CH₂CHPh₂)(OH)Py-2
4-PyCH₂Li	PhCOCF₃	4-PyCH₂CPh(OH)CF₃
4-PyCH₂Li	Me₂CHCHEtCOMe	4-PyCH₂CMe(OH)CHEtCHMe₂
4-PyCH₂Li	Cyclohexanone	4-PyCH₂C(OH)(CH₂)₅
4-PyCH₂Li	1-Methyl-4-piperidone	4-PyCH₂C(OH)(CH₂)₄NMe

173,174
173, 174
173, 174
173, 174
173, 174
173, 174
173, 174

| 4-PyCH₂Li | PhCOCH₂—[piperidine NH] | 4-PyCH₂C(OH)PhCH₂—[piperidine NH] , 86% |

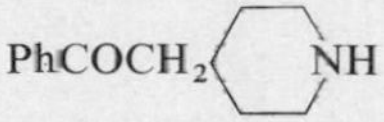

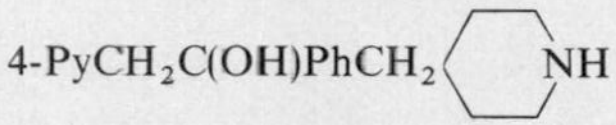

175, 176

4-PyCH₂Li	PhCOMe	4-PyCH₂CMe(OH)Ph
4-PyCH₂Li	o-ClC₆H₄COMe	4-PyCH₂CMe(OH)C₆H₄Cl-o
4-PyCH₂Li	4-PyCOMe	4-PyCH₂CMe(OH)Py-4

173, 174
173, 174
173, 174

| 4-PyCH₂Li | 3-Quinuclidone | [quinuclidine with OH, CH₂Py-4] |

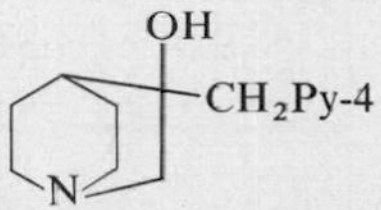

173, 174

551

Table VII-8 (*Continued*)

Organometallic compound	Reactant	Product (yield if reported)	Ref.
4-PyCH₂Li	Dibenzosuberone		173, 174
4-PyCH₂Li	2-(2-Dimethylaminoethyl)-indan-1-one	2-(2-Dimethylaminoethyl)-3-(4-pyridylmethyl)indene	159
4-PyCH₂Li	MeCOCl	(4-PyCH₂)₂C(OH)Me	176, 177
4-PyCH₂Li	Me₂CHCOCl	(4-PyCH₂)₂C(OH)CHMe₂	176, 177
4-PyCH₂Li	Me(CH₂)₆COCl	(4-PyCH₂)₂C(OH)[(CH₂)₆Me]	176, 177
4-PyCH₂Li	CH₂:CH(CH₂)₇COCl	(4-PyCH₂)₂C(OH)[(CH₂)₇CH:CH₂]	176, 177
4-PyCH₂Li	PhCOCl	4-PyCH₂COPh (44) + (4-PyCH₂)₂C(OH)Ph (53)	176–178
4-PyCH₂Li	PhCH₂COCl	(4-PyCH₂)₂C(OH)CH₂Ph	176, 177
4-PyCH₂Li			175, 179

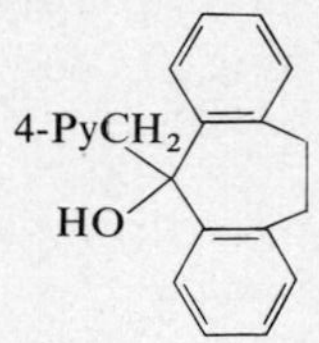

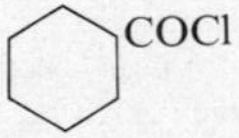

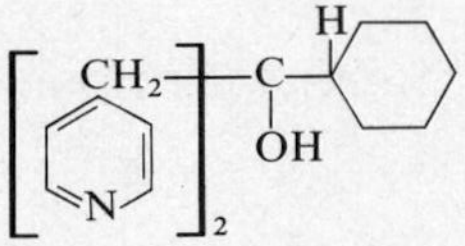

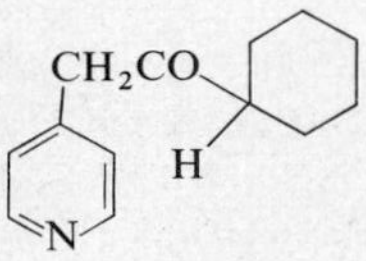

552

4-PyCH₂Li	(cyclopropanecarbonyl chloride structure)	(product structure)	179
4-PyCH₂Li	(cycloheptanecarbonyl chloride structure)	(product structure)	179
4-PyCH₂Li	HCO₂Et	(4-PyCH₂)₂C(OH)Et	176, 177
4-PyCH₂Li	Ethyl picolinate	(4-PyCH₂)₂C(OH)Py-2	176, 177
4-PyCH₂Li	Ethyl isonicotinate	(4-PyCH₂)C(OH)Py-4	176, 177
4-PyCH₂Li	Ethyl 2-furoate	(4-PyCH₂)₂C(OH)C₄H₃O-2	176, 177
4-PyCH₂Li	Ethyl 2-thiophenecarboxylate	(4-PyCH₂)₂C(OH)C₄H₃S-2	173, 177
4-PyCH₂Li	ClCO₂Et	(4-PyCH₂)₃COH, 32%	176, 177, 180
4-PyCH₂Li	CO₂	4-PyCH₂CO₂H	180
6-Cyclohexyl-4-pyridylmethyllithium	CO₂	6-Cyclohexyl-4-pyridylacetic acid	164
4-PyCH₂Li	PhCH:NPh	4-PyCH:CHPh 65%	43
4-PyCH₂Li	Ph₂C:NPh	4-PyCH₂CPh₂NHPh 71%	43
4-PyCH₂Li	4-PyCH₂CPh₂NHPh	PhC:NPh, 4-PyMe (97% total)	43
(trichloropyridine CH₂Li structure)	Me₂SO₄	(ethyl-trichloropyridine structure), 47%	151

553

Table VII-8 (*Continued*)

Organometallic compound	Reactant	Product (yield if reported)	Ref.

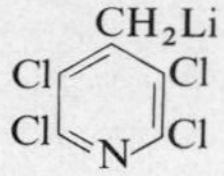

Pyridine bearing CH_2Li and Cl, Cl, Cl substituents / CO_2 / dichloropyridine bearing CH_2CO_2H, 73% / 151

4-(3-chloropropyl...)pyridine, $CH_2CH_2CH_2MgCl$ / H_2O / 4-Me-pyridine + 4-Pr-pyridine + 4-(cyclopropyl)pyridine (with H) + $CH_2{=}CH_2$ / 73

Pyridine bearing $C(Me)_2CH_2CH_2MgCl$ / H_2O / Pr-iso-pyridine + pyridine bearing $CH_2CH_2C(Me)Me$ + pyridine bearing Me, Me + $CH_2{=}CH_2$ / 181

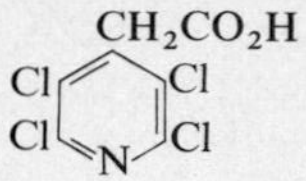

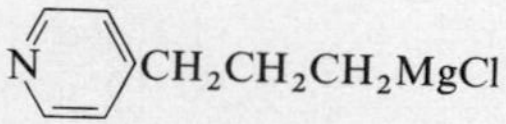

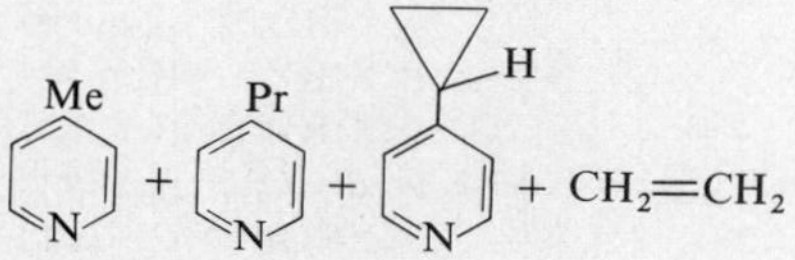

Pyridine bearing CH_2Li and Cl, Cl, Cl substituents / (i) CO_2 (ii) CH_2N_2 / dichloropyridine bearing CH_2CO_2Me / 232

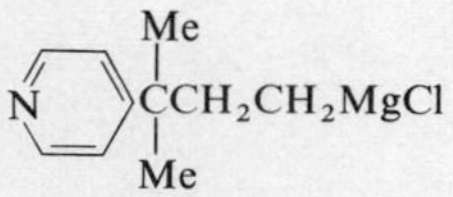

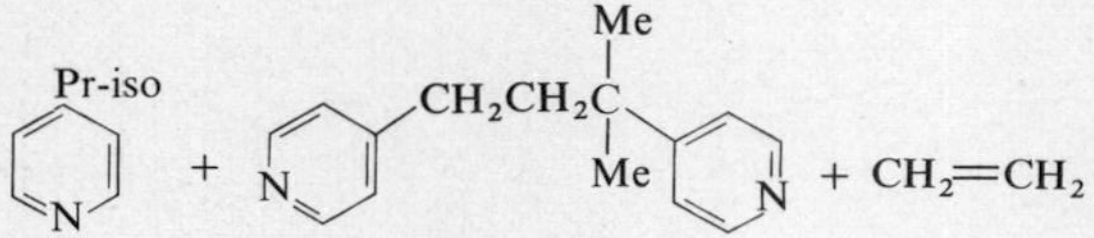

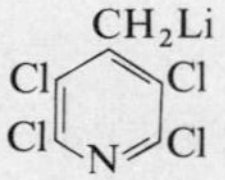

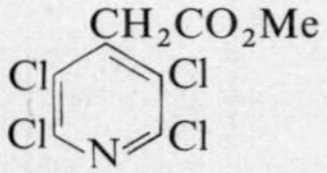

554

an initial attack of the 2-picolylcarbanion at the iminocarbon atom resulting in a displacement of fluoride ion and the formation of an α,α-difluoroquinone

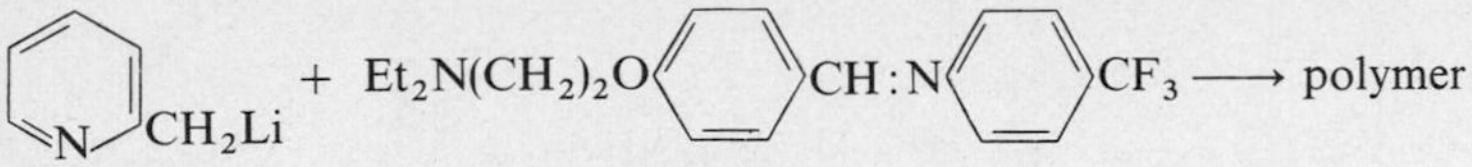

VII-20

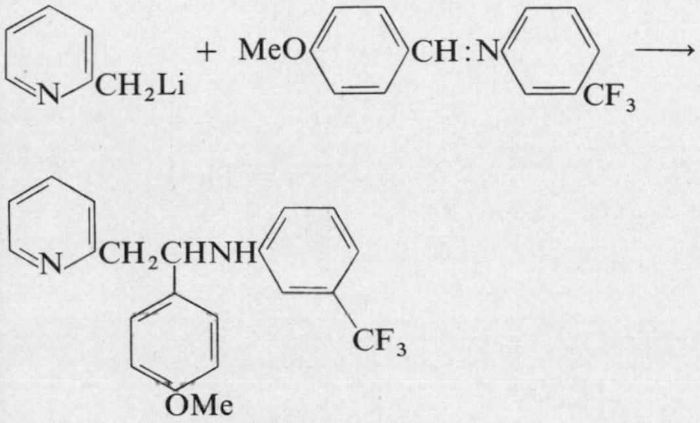

VII-21

imine, capable of polymerization (**VII-22**). Another example of elimination involves the reaction of *N*-(diphenylmethylene)aniline and **VII-18** to give

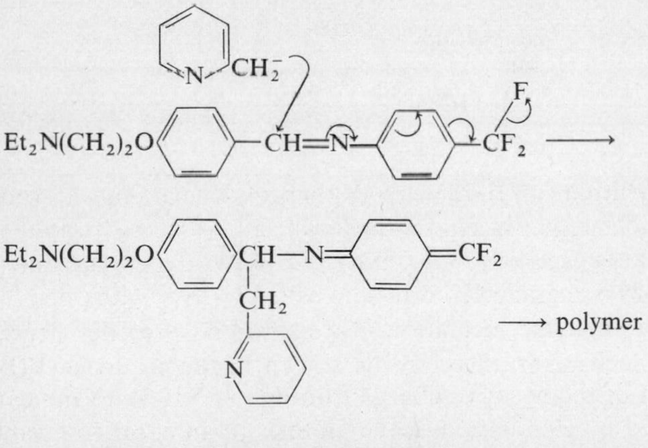

$\longrightarrow$ polymer **VII-22**

1,1-diphenyl-2-(2-pyridyl)ethanol **VII-23**; here, presumably, displacement of aniline by hydroxide ion is the mechanism involved; alternatively, a β-elimination can occur followed by the addition of hydroxide ion to the 2-vinylpyridine. Finally, there are the two examples of the recovery of only **VII-24** from the reaction of **VII-18** with either **VII-25** or **VII-26**.

The isolation of **VII-24** from these reactions has its counterpart in the isolation of 4-stilbazole **VII-27** as the only product from the reaction of

VII-18 + Ph$_2$C:NPh ⟶

$$\left[\underset{\text{not isolated}}{\left\lfloor \underset{N}{\bigcirc}\text{--CH}_2\text{CPh}_2\text{NHPh}\right\rfloor}\right] \xrightarrow[+\text{H}_2\text{O}]{-\text{PhNH}_2} \underset{N}{\bigcirc}\text{--CH}_2\text{C(OH)Ph}_2$$

$$\left[\underset{N}{\bigcirc}\text{--CH}{=}\text{CPh}_2\right]$$

VII-23

VII-18 +

$$\underset{\overset{|}{\text{Me}}}{\bigcirc_N}\text{--CH:NPh} \qquad \textbf{VII-25}$$

or

$$\underset{\overset{|}{\text{Me}}}{\bigcirc_N}\text{--CH:N}{\bigcirc}\text{Cl}$$

VII-26

⟶ $\underset{N}{\bigcirc}\text{--CH:CH}\underset{\overset{|}{\text{Me}}}{\bigcirc_N}$

VII-24

4-pyridylmethyllithium **VII-28** with *N*-(benzylidene)aniline at room temperature or under reflux in tetrahydrofuran; at $-25°$, with rapid workup, the same reactants give only *N*,1-diphenyl-2-(4-pyridyl)ethylamine (**VII-29**) (43). Since **VII-29** is degraded by warming with **VII-18** or with phenyllithium, it would appear that the formation of a carbanion precedes degradation. One possible mechanism, therefore, is shown in the sequence **VII-30** and involves a shift of the negative charge from anion **VII-31** to the carbanion **VII-32**, followed by elimination of the aniline. In an effort to establish the presence of the carbanion (**VII-32**), the amine **VII-29** isolated after a short reaction terminated by the addition of deuterium oxide was examined for deuterium incorporation; surprisingly, the PMR spectrum showed deuterium attached only to the nitrogen atom. Thus the carbanion **VII-32** is either extremely short-lived and undergoes immediate elimination of aniline, or **VII-24** and **VII-27** are formed by a concerted elimination mechanism (43).

 The reactions of **VII-28** with *N*-(diphenylmethylene)aniline show a

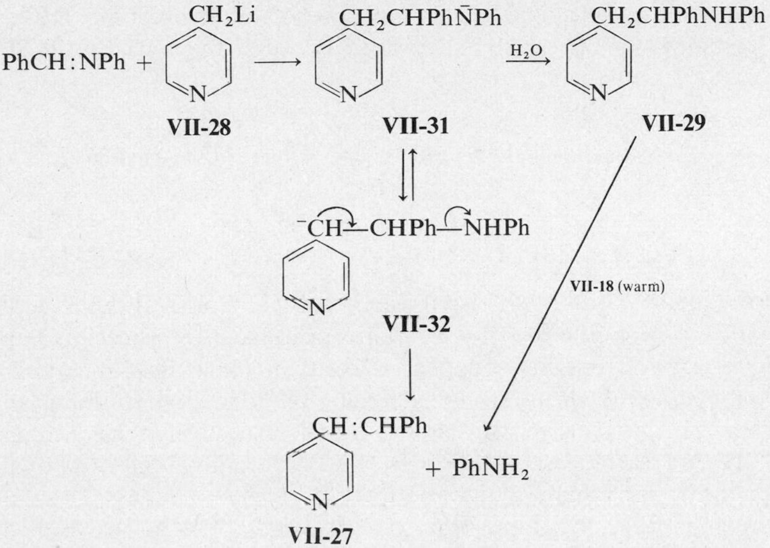

PhCH:NPh +

CH₂Li (VII-28) → CH₂CHPhN̄Ph (VII-31) —H₂O→ CH₂CHPhNHPh (VII-29)

VII-28   VII-31   VII-29

⇕

⁻CH⤸CHPh⤹NHPh

VII-32

VII-18 (warm)

VII-30

CH:CHPh   + PhNH₂

VII-27

similar pattern; thus at −30° with rapid workup, only **VII-33** is isolated, whereas after prolonged reflux, the product is the olefin **VII-34** (43). It should be noted that there is considerable similarity between these reactions

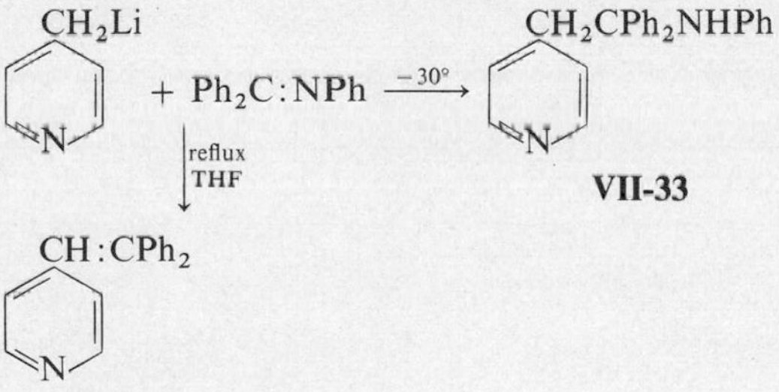

CH₂Li + Ph₂C:NPh —−30°→ CH₂CPh₂NHPh

VII-33

|reflux THF

CH:CPh₂

VII-34

and that involving formation of the carbinol (**VII-23**) in the reaction between 2-pyridylmethyllithium and N-(diphenylmethylene)aniline.

While an obvious conclusion might be that the mechanism of formation of **VII-34** is similar to that proposed for the formation of **VII-27**, this seems highly improbable since treatment of **VII-33** with **VII-28** or phenyllithium causes elimination of 4-picoline and the regeneration of N-(diphenylmethyl-ene)aniline in very high yield. It is also of interest that **VII-33** does not react with either sodium hydride or lithium hydride in tetrahydrofuran, indicating a striking decrease in acidity of the methylene protons. A possible explanation for the elimination of 4-picoline has been proposed and involves the formation of a complex (VII-35) which is degraded by water as shown in VII-36 (43).

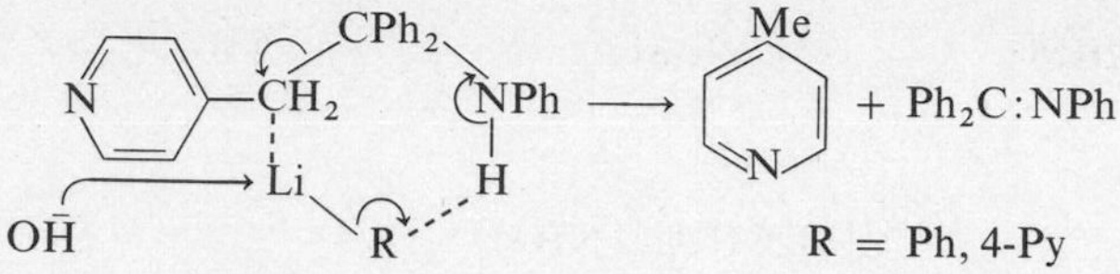

VII-35

The formation of 2,2′-bipyridyl from the reaction of 2-picolyllithium with cuprous chloride or cupric bromide has been postulated to proceed *via* an intermediate 2-pyridylmethyl copper derivative (44). Bis(6-methyl-2-pyridyl)ethane in tetrahydrofuran-*n*-hexane at −70° and *n*-butyllithium, in molar ratios of 1:1 or 1:2 give **VII-36a** or **VII-36b**, respectively; on warming to +20°, **VIII-36a** rearranges to **VII-36c**. Both **VII-36a** and **VII-36b** react with benzophenone or 6-methyl-2-pyridylmethyl chloride to give the expected products in good yield. Pyridophanes, *e.g.*, **VII-36d** and **VII-36e** are formed in low yield from **VII-36b** and cuprous chloride. A reaction of **VII-36f** is coupling (a) with cuprous chloride to give **VII-36g** or (b) with 2-pyridylmethyl chloride to give **VII-36h** (200, 201b).

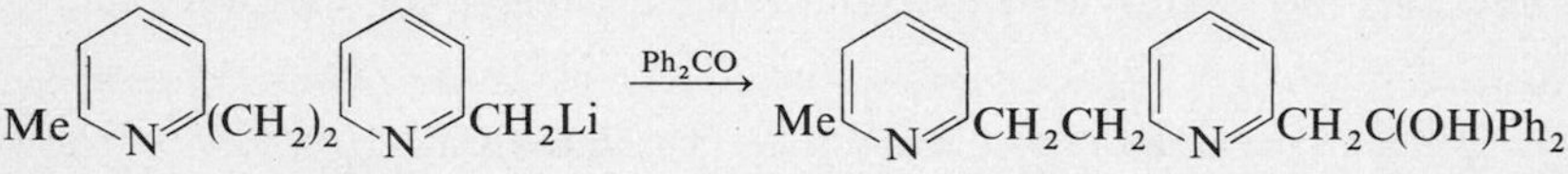

VII-36a

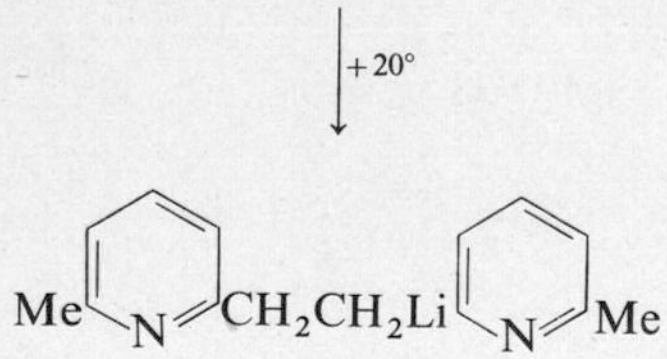

VII-36c

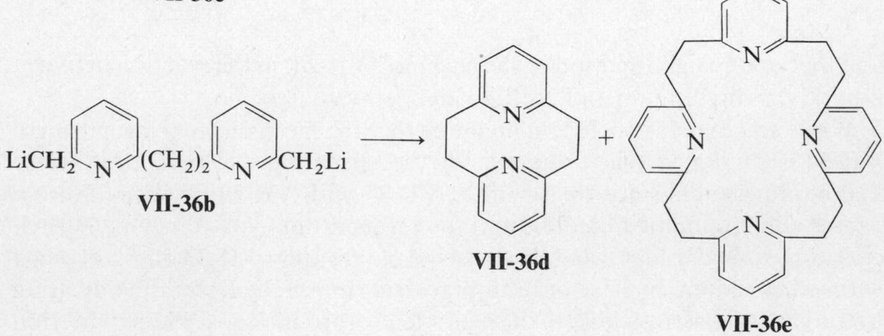

VII-36b

VII-36d

VII-36e

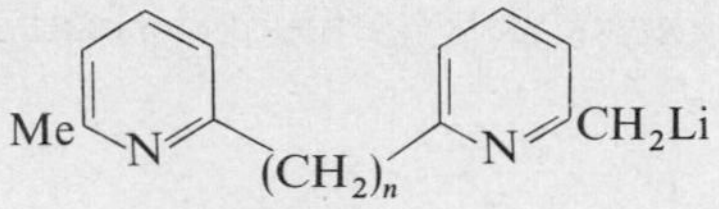

VII-36f; $n = 3, 5, 7$

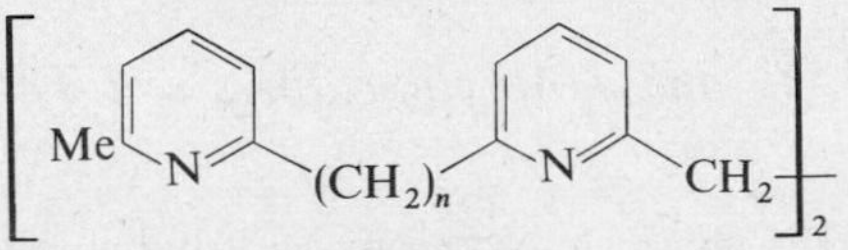

VII-36g; $n = 3, 5, 7$

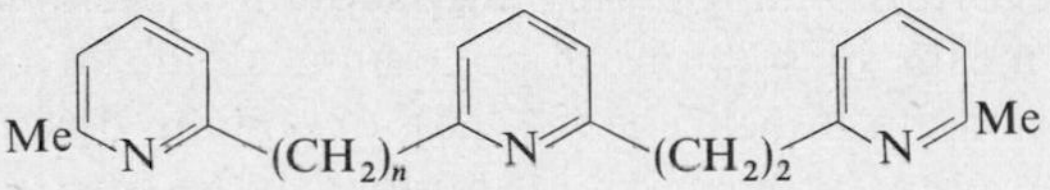

VII-36h; $n = 3, 5, 7$

As noted above with **VII-36a** and **VII-36b**, rearrangement of certain pyridylalkyllithium compounds can occur. Other examples can be found with 2-(2-chloro-1,1-diphenylethyl)pyridine and 4-(2-chloro-1,1-diphenyl-ethyl)pyridine. When each of these two compounds, in tetrahydrofuran, is treated with lithium metal at 0° rearrangement of **VII-36i** and **VII-36j** occurs, since treatment with deuterium oxide gives **VII-36k** and **VII-36l**, respectively. Measurable rearrangement occurs even at $-60°$ (203).

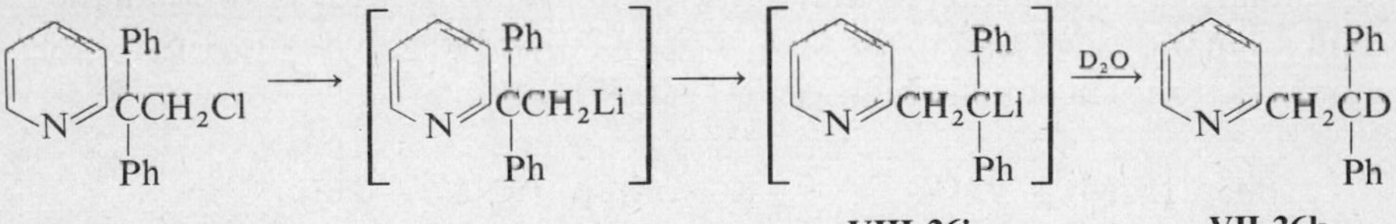

VIII-36i   **VII-36k**

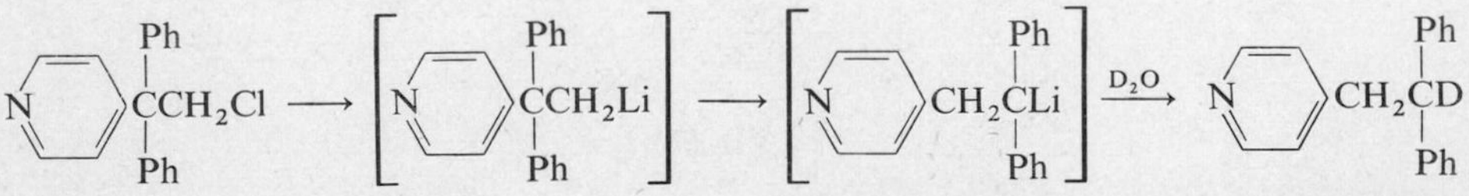

VII-36j   **VII-36l**

The slow formation of triphenylmethyllithium in the metalation of triphenylmethane by n-butyllithium is in marked contrast to the rapid formation of diphenyl-3-pyridylmethyllithium by the reaction of diphenyl-3-pyridylmethane and n-butyllithium, even at $-40°$ (45).

The reactions of the pyridylalkyllithium derivatives are listed in Table VII-6.

2. Pyridylalkylsodium and Pyridylalkylpotassium Compounds

A. From 2-, 3-, and 4-Alkylpyridines and Their 1-Oxides

The α-protons of the isomeric picolines and their higher homologs are far more acidic than the α-protons of toluene and, consequently, are more readily converted to the corresponding pyridylalkylsodium or pyridylalkylpotassium derivatives by means of sodium or potassium amide in liquid ammonia (46–54), phenylsodium in benzene (54, 55), or sodium diisopropylamide in benzene (54). The preparation in liquid ammonia is the most widely employed and the pyridylalkylsodium and potassium compounds have reacted normally, in the majority of instances, with alkyl halides, alkylene dihalides, aryl halides, and aralkyl halides to give the anticipated side chain lengthened derivatives (46–48, 52, 53, 56), and with aldehydes and ketones to give the expected secondary and tertiary carbinols, respectively (49, 51, 57). In competitive metalation experiments in liquid ammonia with sodium amide it has been demonstrated that 4-picoline is the most reactive, 3-picoline is the least reactive, and 2-picoline has an intermediate reactivity (40, 41). With the stronger bases, phenylsodium, and in particular, sodium diisopropylamide, in benzene, both 3- and 4-methylpyridines are metalated at essentially the same rate so that subsequent reactions with both aliphatic and aromatic esters lead to the formation of 3-pyridylmethyl- and 4-pyridylmethyl ketones in equally high yields (**VII-37**) (54).

$$(Me_2CH)_2NH + PhNa \xrightarrow[C_6H_6]{} (Me_2CH)_2NNa + PhH$$

VII-37

VII-37

Variable yields are obtained from the reactions of the isomeric pyridylmethylsodium and -potassium derivatives with the various alkyl halides, alkylene dihalides, aryl halides, and aralkyl halides. Presumably, the reactivity of the halogen is a primary consideration, since, in 4-chlorobutyl acetate, for example, the chlorine atom is so firmly held that it does not react with 2-pyridylmethylsodium (46).

A substantial effort has been devoted to a study of the sodium-catalyzed alkylation of 2-, 3-, and 4-picolines and their higher homologs with ethylene, propylene, isobutylene, or isoprene (58–60). At 135 to 150° in an autoclave, 2- or 4-picoline react with ethylene to give 2-n-propyl- and 2-(3-pentyl)-pyridine and 4-n-propyl- and 4-(3-pentyl)pyridine, respectively. The ethylation of the n-propylpyridine apparently occurs with greater ease than the initial ethylation of the picoline. 3-Picoline, in contrast, does not react with ethylene at 135 to 150°; at 200°, it undergoes autogenous dimerization and trimerization rather than react with the olefin. The mechanisms proposed for these reactions are shown in **VII-38** and **VII-39** (41, 56, 58–60). Under the same conditions, 3-ethylpyridine and ethylene give a variety of bicyclic structures **(VII-39a)** (210) while ω-pyridyl-1-alkenes undergo intramolecular cyclizations to give structures as shown in **VII-39b** (211).

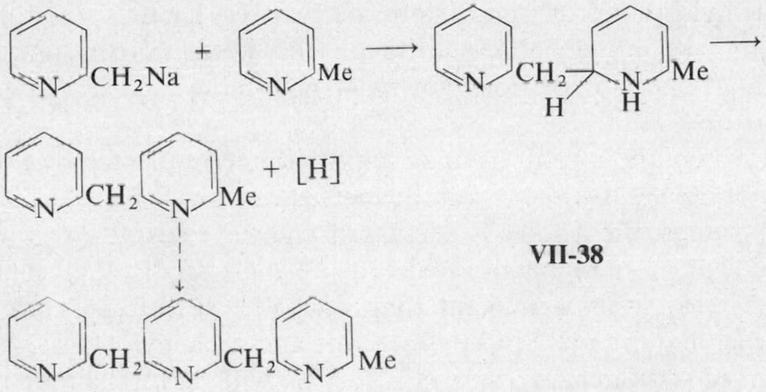

VII-38

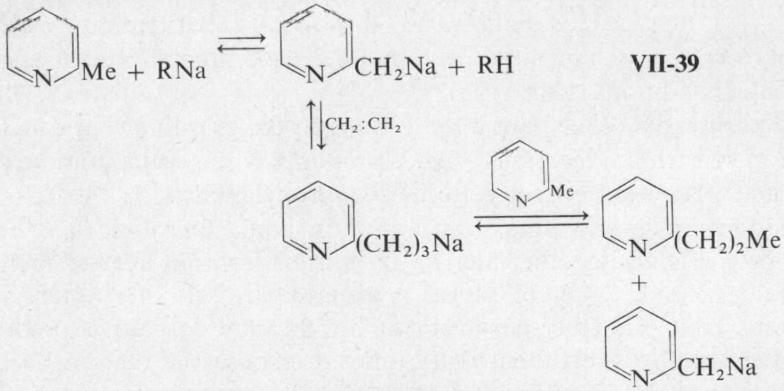

VII-39

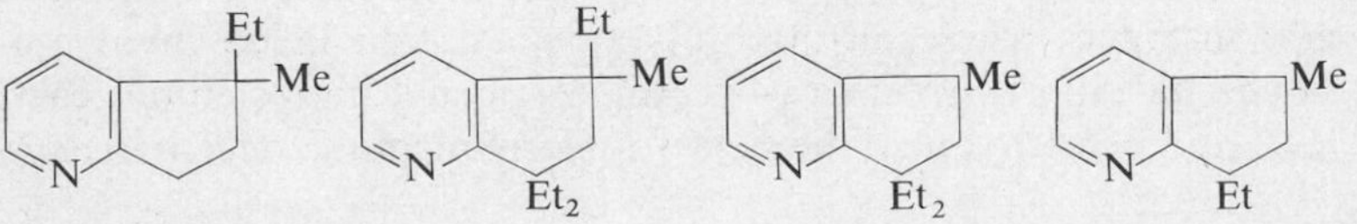

VII-39a

VII-39b

2- And 4-methylpyridine-1-oxide are metalated by either sodium amide or potassium amide in liquid ammonia; the 1-oxido-2- and 1-oxido-4-pyridylmethylsodium or -potassium derivatives participate normally in alkylations with alkyl halides although the yields of products are low. 3-Methylpyridine-1-oxide does not react with either sodium or potassium amide (46, 61).

The acidity of the α-protons in an alkylpyridine is affected by the number and kind of other substituents on the methyl carbon. When the substituents are alkyl groups, the acidity is decreased and the reactivity toward base is decreased. The more highly branched the alkyl group, the less reactive is the α-hydrogen (58). It is significant that when the substituents are phenyls, for example, in diphenyl-2-pyridylmethane and diphenyl-4-pyridylmethane, the bright red sodium salts are readily formed with sodium amide in toluene at 110°; diphenyl-3-pyridylmethane reacts slowly with sodium amide even in xylene at 135°. It should be noted, however, that triphenylmethane does not react with sodium amide or even phenylsodium under a variety of experimental conditions (62).

Substituents other than alkyl groups in the pyridine ring usually do not interfere with the reactions of alkylpyridines with sodium amide, potassium amide, phenylsodium, or sodium diisopropylamide (13, 40, 46, 61, 63–65). Sodium amide and potassium amide in liquid ammonia and lithium diisopropylamide in ether-hexane or tetrahydrofuran-hexane metalate 2,4-lutidines and 2,4,6-collidines predominantly at the 4-methyl group; n-butyllithium in ether-hexane metalates the same derivatives predominantly at the 2-methyl substituent (217). It has been observed that potassium amide in liquid ammonia may effect metalation when sodium amide is ineffective (46). Several instances where all attempts at metalation failed involve 4-methyl-(2-benzenesulfonyl)pyridine [4-methyl-2-(phenylthio)pyridine reacts readily] (61), and 6-methylpicolinic acid [5-methylpicolinic acid is readily metalated] (61, 65).

An unusual example of the ammonolysis of an amide by liquid ammonia is seen in the reactions of 2-, 3-, and 4-pyridylmethylsodium in liquid ammonia with N,N-dimethyl-11-bromoundecanamide; the products, in all instances, are the primary amides (**VII-40**). Presumably, either the pyridylmethylsodium derivative or sodium amide is a catalyst for these ammonolyses (61).

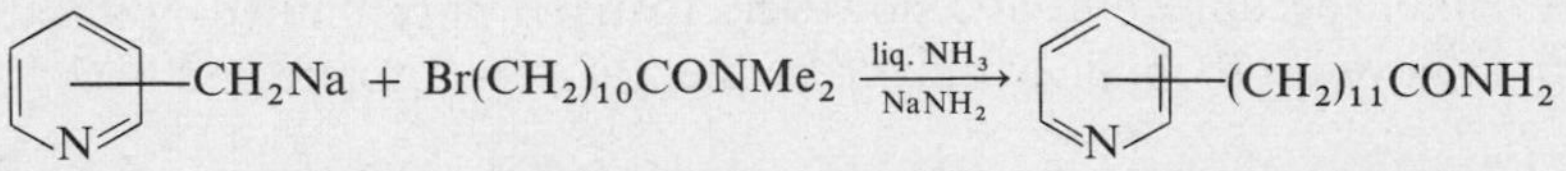

VII-40

The reactions of the pyridylalkylsodium and pyridylalkylpotassium compounds are listed in Tables VII-6, VII-7, and VII-8.

B. From the Addition of Sodium in Liquid Ammonia to Pyridylaldimines and Pyridylketimines

Pyridylaldimines and pyridylketimines are alkylated at carbon by the addition, initially, of sodium metal dissolved in liquid ammonia, followed by an organic halide; the same products are obtained by metalation of the related amine with sodium amide in liquid ammonia, followed by reaction with the halide (**VII-41**) (66).

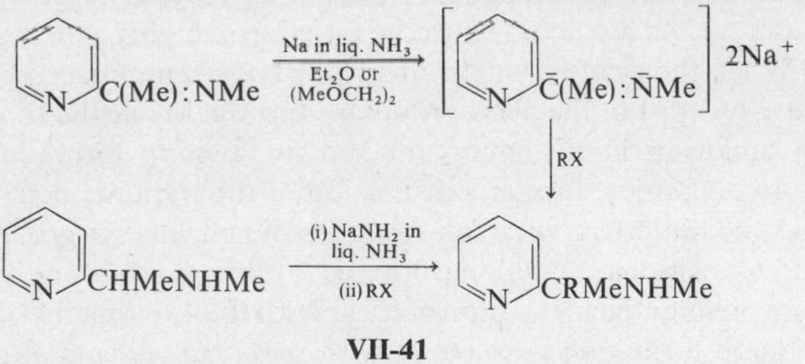

VII-41

As anticipated, the reductive alkylation gives the best yields with the 4-pyridylimines and involves only the carbon atom; with 2-pyridylimines, mixtures of products resulting from alkylation at both carbon and amino nitrogen, as well as tars, are formed; the 3-pyridylimines give only tars (66).

All of the alkylation reactions with pyridylaldimines and pyridylalkyl-ketimines are listed in Tables VII-6 and VII-8.

C. From the Addition of Sodium in Liquid Ammonia to Pyridinecarboxylic Acid Esters

The addition of methyl isonicotinate in ether to a solution of sodium in liquid ammonia gives the sodium ketyl (**VII-42**); addition of an aralkyl halide results in alkylation at carbon and the formation of ketones (**VII-43**) (66). This procedure appears, however, to have only limited application; ketones have been obtained only from methyl isonicotinate and benzyl and

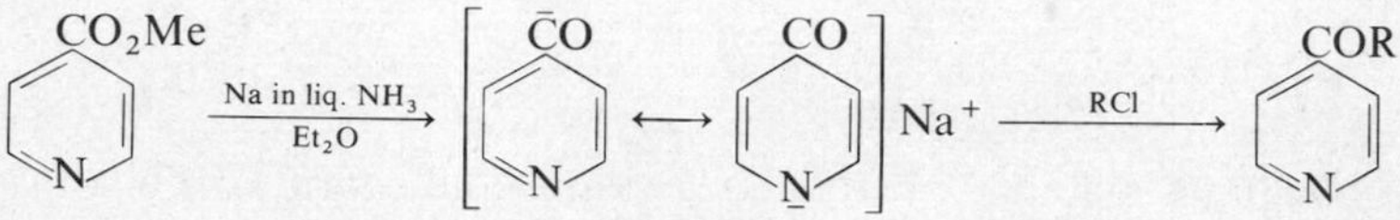

VII-42   **VII-43**

p-chlorobenzyl chlorides; that is, the reaction fails with **VII-42** and ethyl bromide, allyl chloride, and 2-dimethylaminoethyl chloride. Finally, methyl picolinate does not undergo this synthesis of pyridyl ketones.

The reactions of the sodium ketyls are listed in Table VII-6.

D. Anions Formed from Alkylpyridones, Alkylpyridinethiones, Benzoylpyridines and Acetamidopyridines

The 4-alkyl- and 6-alkyl-3-cyano-2($1H$)-pyridines are an interesting and somewhat unique class of components in that they react with two equivalents of potassium amide in liquid ammonia to form highly colored reactive dianions. When the cyano group is absent or has been replaced by carboxamide, the α-protons of the alkyl group are too weakly acidic to react with potassium amide in liquid ammonia, and no dianion formation occurs. In the cyano derivative, further substitution of the pyridine heterocycle by phenyl does not interfere with the reaction. When methyl groups occupy both 4- and 6-positions, a dianion is formed, although it has not been established where preferential metalation occurs. With the 4,6-dimethyl derivative, limited trianion formation does occur with three equivalents of potassium amide. Treatment of the dianions with slightly more than one equivalent of an active halogen compound gives, apparently, alkylation only at carbon and not at oxygen or nitrogen, and these products are obtained in high yield. Thus the dianions undergo all of the reactions of a pyridylmethylpotassium compound, and this is confirmed by the obtention of the normal products following reactions with aldehydes, ketones, and esters. With conjugated ketones, the products are those of conjugate addition, and, with diethyl oxalate, the product is the stable enol (**VII-44**), as proved both by its infrared and its ultraviolet spectra (67, 88).

Although the 4-alkyl- and 6-alkyl-2(1*H*)-pyridones do not react with potassium amide in liquid ammonia, these, as well as the 3-alkyl-2(1*H*)-pyridones and -pyridinethiones, do react with *n*-butyllithium in tetrahydrofuran-hexane to give dilithio derivatives, and the latter again function preferentially, as pyridylalkyllithium compounds in their behavior toward active halogen compounds, aldehydes, and ketones (69).

Related dilithio derivatives can be prepared from 2- and 3-acetamido-pyridines only with *n*-butyllithium in tetrahydrofuran and not with potassium amide in liquid ammonia. These derivatives, also, undergo the appropriate additions of organometallic compounds to aldehydes, ketones, and nitriles (**VII-45**) (70).

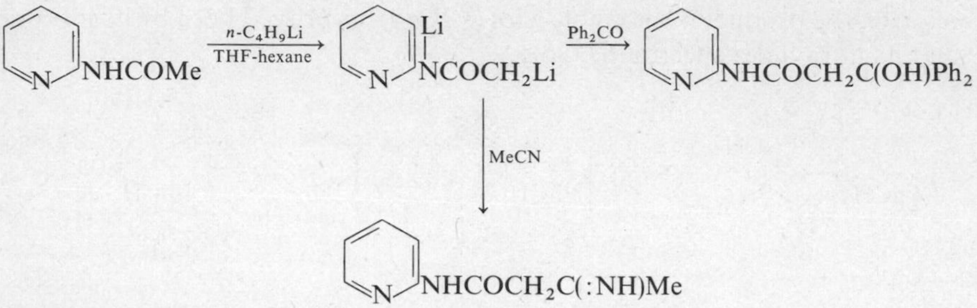

The anion generated from 3-benzoylpyridine with sodium in liquid ammonia is reported to react with several alkyl and aralkyl halides to give the tertiary carbinols **VII-45a** and with benzophenone to give **VII-45b** (228).

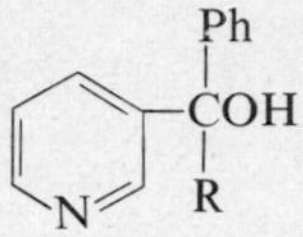

VII-45a

R = Et, isoPr, *t*-Bu, PhCH$_2$

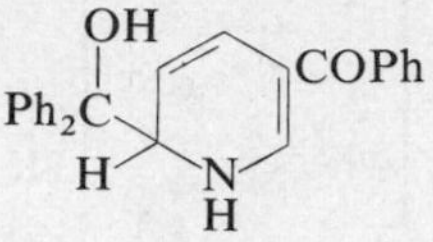

VII-45b

All of the reactions of the dianions formed from alkylpyridones, alkyl-pyridinethiones, benzoylpyridines and acetamidopyridines are listed in Tables VII-6 and VII-8.

3. Pyridylalkylmagnesium Halides

2-Pyridylmethylmagnesium bromide has been prepared by the reaction of 2-picoline and ethylmagnesium bromide in diethyl ether, while 2-pyridyl-methylmagnesium chloride has been synthesized from 2-pyridylmethyl chloride and magnesium in di-*n*-propyl ether. Both compounds give the normal products with chloromethyl methyl ether, aldehydes, and ketones (**VII-46**) (71, 72).

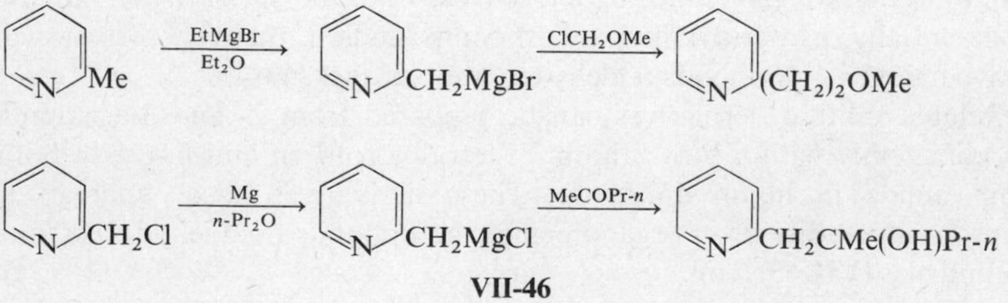

VII-46

Complex reactions have been observed when 3-(3-pyridyl)propyl-, 3-(4-pyridyl)propyl-, and 3-methyl-3-(4-pyridyl)butylmagnesium chlorides are prepared in tetrahydrofuran, (**VII-46a**) (73, 74). Along similar lines, a contrast in behavior upon hydrolysis was noted between **VII-46b** and **VII-46c**, obtained by the reaction between allylmagnesium bromide and 2- and 4-styrylpyridine (stilbazoles) in diethyl ether, respectively. The former gives 2-(2-phenyl-4-pentenyl)pyridine in 70% yield while the latter gives only trace amounts of 4-(2-phenyl-4-pentenyl)pyridine, since the major product of the hydrolysis is a polymeric material (202).

The reactions of the pyridylalkylmagnesium halides reported since the last compilation (1st ed., Part 2, pp. 423ff) are listed in Table VII-6.

V. Pyridylalkynylmetallic Compounds

A large number of 2-, 3-, and 4-pyridylalkynes have been prepared (*a*) *via* the lithium or sodium acetylide followed by alkylation with an alkyl halide or (*b*) from the pyridylalkylsodium derivative and the ω-iodoacetylene (**VII-47**) (75).

$(CH_2)_4C\vdots CH$ $\xrightarrow[\text{liq. NH}_3\text{—Et}_2\text{O}]{\text{LiNH}_2}$ $(CH_2)_4C\vdots CLi$ $\xrightarrow{n\text{-BuBr}}$ $(CH_2)_4C\vdots CBu\text{-}n$

$CH_2Na + I(CH_2)_4C\vdots C(CH_2)_4Me \longrightarrow (CH_2)_5C\vdots C(CH_2)_4Me$

VII-47

The known reactions of the pyridylalkynylmetallic compounds are listed in Table VII-9; no derivatives of this type had been known previously.

TABLE VII-9. Reactions of Pyridylalkynylmetallic Compounds (75)

Compound	Reactant	Product (yield)
2-Py(CH$_2$)$_2$C$\vdots$CLi	n-PrBr	2-Py(CH$_2$)$_2$C$\vdots$CPr-n (46%)
2-Py(CH$_2$)$_2$C$\vdots$CLi	n-C$_8$H$_{17}$Br	2-Py(CH$_2$)$_2$C$\vdots$CC$_8$H$_{17}$-n (35%)
4-Py(CH$_2$)$_4$C$\vdots$CLi	n-BuBr	4-Py(OH$_2$)$_4$C$\vdots$CBu-n (37%)
2-Py(CH$_2$)$_5$C$\vdots$CLi	n-BuBr	2-Py(CH$_2$)$_5$C$\vdots$CBu-n (54%)
2-Py(CH$_2$)$_5$C$\vdots$CLi	n-C$_5$H$_{11}$Br	2-Py(CH$_2$)$_5$C$\vdots$CC$_5$H$_{11}$-n (47%)

VI. Addition of Sodium Amide and of Organometallic Compounds to the Azomethine System of Pyridine

The addition of sodium amide to pyridine and the picolines has been reexamined. Data that suggest the catalytic role of sodium or potassium nitrate in increasing the yields have been presented (221). Sodium amide and 2-phenylpyridine give 2-amino-6-phenylpyridine (223).

Prior to the appearance of the first edition (Part 2, pp. 423ff) it had been established that the reaction of 4-picoline with phenyllithium or with sodium amide involved both lateral metalation and 1,2-addition across the azomethine system of the pyridine heterocycle (VII-48). These reactions are not limited to the picolines and a striking example is to be found in the industrial preparation of 2-aminopyridine by the addition of sodium amide to pyridine. It is significant for this discussion that 4-aminopyridine is a by-product of this synthesis, so that 1,4-addition is a minor but still competing reaction (see also Chapter IX).

About 25 papers dealing with the addition of organometallic compounds to pyridine derivatives have been published since the first edition. Several of these addition reactions lead to new organometallic compounds of pyridine;

VII-48

most do not, since the intermediate products formed possess a nitrogen metal linkage rather than a carbon metal bond. It is readily apparent, however, from a review of this literature, that only a very few of the many examples reported yield products arising from a 1,4-addition to a 1,4-unsubstituted pyridine derivative; these exceptions involve the reactions between pyridine and an alkylmagnesium bromide or benzylmagnesium chloride where very low yields of only the 4-alkyl- and 4-benzylpyridine, respectively, are reported to be the products (14) and between pyridine or 3-picoline and benzyllithium, where again, only the 4-benzyl derivatives are obtained, in each instance, but now in 56 and 41 % yields, respectively (76). It is also of interest that although phenylmagnesium bromide forms precipitates with pyridine and the three isomeric picolines, these solids regenerate only the parent pyridine base when treated with water (77). Phenylmagnesium bromide adds to pyridine-1-oxide, 2-picoline-1-oxide, and 4-picoline-1-oxide to give, following hydrolysis, the oxygen-stable, 1,2-dihydro-2-phenyl derivatives (**VII-48a–c**) (230). Quaternary pyridine derivatives undergo 1,2-addition of benzylmagnesium chlorides, but the 1,2-dihydro derivatives formed are very sensitive to oxidation (92, 93) (see p. 578, **VII-62**). With pentachloropyridine-1-oxide and methylmagnesium iodide or ethylmagnesium bromide, nucleophilic displacement of the chlorine atoms at positions-2 and -6 by methyl and ethyl groups are the major reactions and

VII-48a: R = R' = H
VII-48b: R = H, R' = Me
VII-48c: R = Me, R' = H

VII-48a

yield **VII-48d** and **VII-48e**; with phenylmagnesium bromide, the formation of the 2- and 2,6-diphenyl derivatives is accompanied by that of **VII-48f** (129).

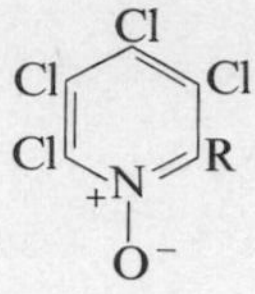

VII-48d
R = Me or Et

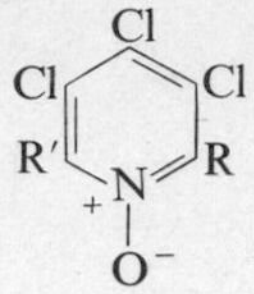

VII-48e
R = Me, Et
R' = Cl, Me, Et

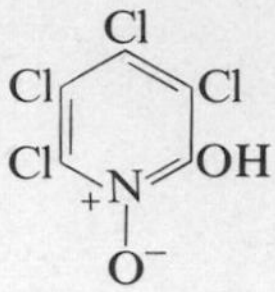

VII-48f

Triphenylsilylpotassium and diphenylmethylsodium do not add to pyridine (14), but lithium piperidide and 4-chloropyridine give 1-(4-pyridyl)-piperidine as the final product by a mechanism postulated to involve a 1,4-addition followed by elimination of lithium chloride (**VII-49**) (24, 78).

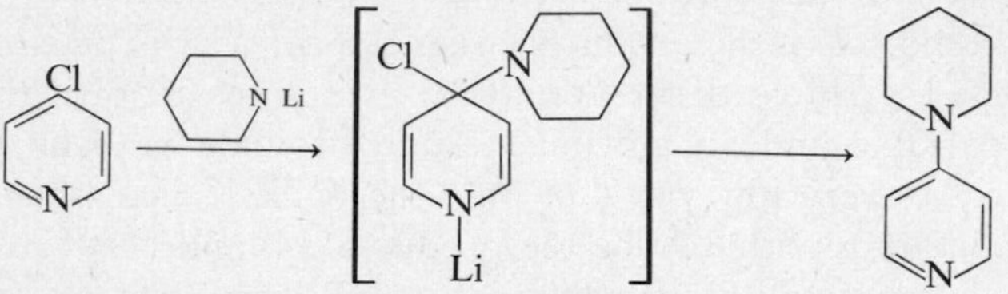

VII-49

The fourth example of a 1,4-addition involves the reaction of triphenyl-silyllithium (**VII-50**) and pyridine and this occurs *in situ* when hexaphenyl-disilane is treated in pyridine solution with lithium metal.

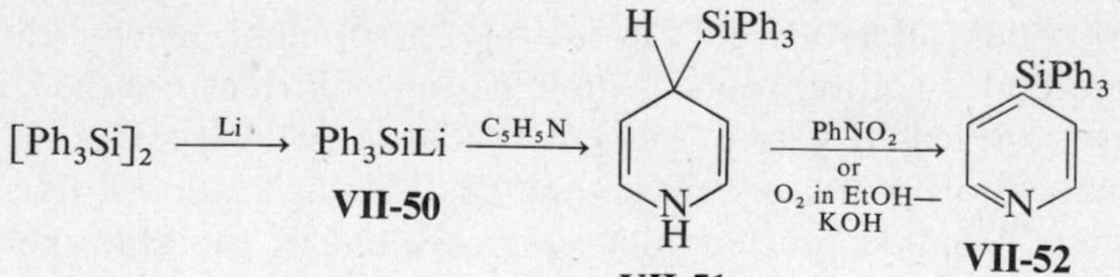

$[Ph_3Si]_2 \xrightarrow{Li} Ph_3SiLi \xrightarrow{C_5H_5N}$

VII-50

VII-51

VII-52

The structure of the 1,4-adduct (**VII-51**) follows as a consequence of its ready oxidation to **VII-52**, identical with an authentic sample of 4-pyridyl-triphenylsilane (14). It should be noted in this context that the palladium-on-carbon catalyzed addition of trimethylsilane to pyridine gives at least seven products and several of these appear to be 1,4-addition products (**VII-53**) (79, 80).

Phenyllithium is usually prepared from bromobenzene and lithium metal in diethyl ether. To synthesize phenyllithium free of lithium salts,

VII-53

diphenylmercury in diethyl ether is treated with lithium metal and the solution of phenyllithium decanted from the precipitated amalgam. The dropwise addition of pyridine to this solution, at 0°, causes a crystalline product to separate; its PMR spectrum in tetramethylethylenediamine establishes this substance to be the 1,2-adduct (Table VII-10, Adduct 1); if any of the 1,4-adduct is present, it cannot be detected from this spectrum (81). Another spectrum, that of the addition product of 3-picoline (Table VII-10, Adduct 2) does, however, have a special significance for this discussion; this spectrum does not reveal what has been established by several studies, namely, that 3-picoline and phenyllithium give 2-phenyl-3-picoline and 6-phenyl-3-picoline in a ratio of about 95:5 (82–84). Oxidation of this adduct mixture does give the product in the correct ratio. No 1,4-addition occurs within the limits of detection by glc in the phenyllithium additions. The addition of methyllithium in the presence of lithium iodide, to 3-(1-methyl-2-pyrrolidyl)pyridine (nicotine) gives three products, the major one being the 6-methyl derivative (22 to 37% yield) and the minor ones being the 4-methyl (<1 to 4%), and the 2-methyl derivative (trace) (85).

The data on these addition reactions are listed in Tables VII-12 and VII-15. There are numerous contradictions and these have been cited (82, 84, 85–87). A review of the data for phenyllithium additions, in the presence of lithium bromide, would indicate that the predominant mode of addition with 2-picoline is 1,6-; with 3-alkylpyridines, including 3-cyclohexyl (but excluding 3-t-butyl where it is mainly 1,6-), it is 1,2-; with 3-amino-, 3-bromo-, 3-N,N-diethylsulfonamido-, or 3-methoxypyridine, it is 1,2-; with 3-phenylpyridine and with nicotine, it is 1,6-. With methyllithium, in the presence of lithium iodide, and 3-picoline, the addition is mainly 1,2-; with 4-phenylpyridine it is 1,2-, and with nicotine 1,6-. The role of the solvent in the methyllithium reactions has been studied and appears to be unimportant.

TABLE VII-10. Chemical Shifts of Pyridine Ring Protons in Phenyllithium Adducts in Tetramethylethylenediamine at 100 mHz

Adduct	τ-Values					Hz			
	H_2	H_3	H_4	H_5	H_6	$J_{2,3}$	$J_{3,4}$	$J_{4,5}$	$J_{5,6}$
1	5.22	5.57	3.99	5.32	3.21	4.40	8.00	5.75	5.75
2	5.23	—	4.37	5.69	3.48	—	—	5.50	6.00
3	5.22	5.57	—	5.32	3.33	4.12	—	—	6.50
4	5.27	—	4.42	—	3.69	—	—	—	—

$1; R^1 = Ph, \quad R^2, R^3, R^4 = H$
$2; R^1 = Ph, \quad R^2 = Me, R^3, R^4 = H$
$3; R^1 = Ph, \quad R^3 = t\text{-Bu}, R^2, R^4 = H$
$4; R^1 = Ph, \quad R^2, R^4 = Me, R^3 = H$

TABLE VII-11. Lithium Compounds from Acetamidopyridines (70)

Compound	Reactant	Product (yield)
2-PyNCOCH$_2$Li (Li)	PhCHO	2-PyNHCOCH$_2$CH(OH)Ph (55%)
2-PyNCOCH$_2$Li (Li)	Ph$_2$CO	2-PyNHCOCH$_2$C(OH)Ph$_2$ (55%)
2-PyNCOCH$_2$Li (Li)	PhCO$_2$Me	2-PyNHCOCH$_2$COPh (56%)
2-PyNCOCH$_2$Li (Li)	PhCN	2-PyNHCO CH$_2$C(:NH)Ph (43%)
2-PyNCOCH$_2$Li (Li)	MeCN	2-PyNHCOCH$_2$C(:NH)Me (16%)
3-PyNCOCH$_2$Li (Li)	PhCHO	3-PyNHCOCH$_2$CH(OH)Ph (18%)
3-PyNCOCH$_2$Li (Li)	Ph$_2$CO	3-PyNHCOCH$_2$(OH)Ph$_2$ (48%)

TABLE VII-12. Addition Reactions of Organometallic Compounds to Pyridines

Compound	Organometallic compound	Product (yield if reported)[a]	Ref.
N,N-Diethyl-3-pyridine-sulfonamide	PhLi	N,N-Diethyl-2-phenyl-3-pyridine-sulfonamide (20%)	86
3-Bromopyridine	PhLi	2-Phenylpyridine (2%)	86
3-Phenylpyridine	PhLi	2,5-Diphenylpyridine (29%) 2,3-Diphenylpyridine (6%)	86, 93
N-Benzylpyridinium chloride	PhLi	N-Benzyl-2-phenylpiperidine (5%) (after hydrogenation)	86
N-Benzyl-3-picolinium chloride	PhLi	Tars	86
PyH	MeLi	2-Picoline (36%)	76
PyH	isoPr	2-Isopropylpyridine	76
PyH	BuLi	2-n-Butylpyridine	78
PyH	PhLi	2-Phenylpyridine (69%)	82, 85, 91, 182
PyH	$PhCH_2Li$	4-Benzylpyridine	76
PyH	$2\text{-}PyCH_2Na$	$2\text{-}PyCH_2Py\text{-}2$ (35%), $2\text{-}PyNH_2$ (39%)	94
2-Picoline	$2\text{-}PyCH_2Na$	2-(6-Methyl-2-pyridyl)methylpyridine	94
3-Ethylpyridine	PhLi	$\left\{\begin{array}{l}\text{3-Ethyl-2-phenylpyridine}\\\text{3-Ethyl-6-phenylpyridine}\end{array}\right\}$ 23%	82
3-(Isopropyl)pyridine	PhLi	$\left\{\begin{array}{l}\text{3-Isopropyl-2-phenylpyridine}\\\text{3-Isopropyl-6-phenylpyridine}\end{array}\right\}$ 25%	82, 182
3-t-Butylpyridine	PhLi	$\left\{\begin{array}{l}\text{3-}t\text{-Butyl-2-phenylpyridine}\\\text{3-}t\text{-Butyl-6-phenylpyridine}\end{array}\right\}$ 25%	82
Pyridine + pyridine-2-d	PhLi	2-Phenylpyridine (27%)	83
3-Picoline	MeLi	$\left\{\begin{array}{l}\text{84%, 2,3-Lutidine}\\\text{16%, 2,5-Lutidine}\end{array}\right\}$ 28%	76
3-Picoline	isoPrLi	$\left\{\begin{array}{l}\text{80%, 5-Methyl-2-isopropylpyridine}\\\text{20%, 3-Methyl-2-isopropylpyridine}\end{array}\right\}$	76
3-Picoline	PhLi	$\left\{\begin{array}{l}\text{95%, 3-Methyl-2-phenylpyridine}\\\text{5%, 5-Methyl-2-phenylpyridine}\end{array}\right\}$ 48%	82, 83, 84
3-Picoline	$o\text{-EtC}_6\text{H}_4\text{Li}$	2-o-Ethylphenyl-3-methylpyridine (53%)	183
3-Picoline	$p\text{-MeOC}_6\text{H}_4\text{Li}$	$\left\{\begin{array}{l}\text{94%, 2-}p\text{-Methoxyphenyl-3-methylpyridine}\\\text{6%, 2-}p\text{-Methoxyphenyl-5-methylpyridine}\end{array}\right\}$	183
3-Picoline	$PhCH_2Li$	4-Benzyl-3-methylpyridine (41%)	76
3-Picoline + 3-picoline-2-d	PhLi	$\left\{\begin{array}{l}\text{96%, 3-Methyl-2-phenylpyridine}\\\text{4%, 3-Methyl-6-phenylpyridine}\end{array}\right\}$ 26%[b]	83
3-Picoline + 3-picoline-2-d	PhLi	$\left\{\begin{array}{l}\text{95%, 3-Methyl-2-phenylpyridine}\\\text{5%, 3-Methyl-6-phenylpyridine}\end{array}\right\}$ 34%[c]	83
3-Ethylpyridine	PhLi	$\left\{\begin{array}{l}\text{87%, 3-Ethyl-2-phenylpyridine}\\\text{13%, 5-Ethyl-2-phenylpyridine}\end{array}\right\}$ 42%	83
3-Ethylpyridine	PhLi	$\left\{\begin{array}{l}\text{89%, 3-Ethyl-2-phenylpyridine}\\\text{11%, 5-Ethyl-2-phenylpyridine}\end{array}\right\}$ 14%[b]	83
3-Ethylpyridine	PhLi	$\left\{\begin{array}{l}\text{87%, 3-Ethyl-2-phenylpyridine}\\\text{13%, 5-Ethyl-2-phenylpyridine}\end{array}\right\}$ 23%[c]	83

Table VII-12 (*Continued*)

Compound	Organometallic compound	Product (yield if reported)[a]	Ref.
3-Isopropylpyridine	PhLi	$\left\{\begin{array}{l}76\%, \text{3-Isopropyl-2-phenylpyridine}\\24\%, \text{5-Isopropyl-2-phenylpyridine}\end{array}\right\}$ 32%	83, 87
3-Aminopyridine	PhLi	3-Amino-2-phenylpyridine (25%)[d]	184
3-Methoxypyridine	PhLi	3-Methoxy-2-phenylpyridine (21%)[e]	184
Nicotine	PhLi	$\left\{\begin{array}{l}30\%, \text{2-Phenylnicotine}\\70\%, \text{6-Phenylnicotine}\end{array}\right\}$ 34%	84, 87, 182
Nicotine	MeLi	$\left\{\begin{array}{l}96\%, \text{6-Methylnicotine}\\3\%, \text{4-Methylnicotine}\\\text{Trace 2-methylnicotine}\end{array}\right\}$	85
Nicotine	MeLi	$\left\{\begin{array}{l}\text{2-Methylnicotine (20\%)}\\\text{4-Methylnicotine (4\%)}\\\text{6-Methylnicotine (11\%)}\end{array}\right.$	185
3-Picoline	$o\text{-MeC}_6\text{H}_4\text{Li}$	$\left\{\begin{array}{l}\text{3-Methyl-2-}o\text{-tolylpyridine}\\\text{3-Methyl-5-}o\text{-tolylpyridine}\end{array}\right\}$ 96:4 3-Methyl-6-o-tolylpyridine 3-Methyl-1,2,5,6-tetrahydro-2-o-tolylpyridine	89, 183
3-Cyclohexylpyridine	PhLi	$\left\{\begin{array}{l}65\%, \text{3-Cyclohexyl-2-phenylpyridine}\\35\%, \text{5-Cyclohexyl-2-phenylpyridine}\end{array}\right.$	87, 182
Me$\underset{N}{\diagup\!\diagdown}$Me	PhLi[f]	$\left\{\begin{array}{l}39\%, \quad\text{Me}\underset{\text{Ph}}{}\overset{}{N}\text{Me}\\ \\56\%, \quad\text{Me}\overset{}{}\underset{N}{}\text{CH}_2\text{CO}_2\text{Et}\end{array}\right.$	38
2-Picoline	PhLi	2-Methyl-6-phenylpyridine	169
4-Phenylpyridine	MeLi	2-Methyl-4-phenylpyridine	169
3-Picoline	MeLi	$\left\{\begin{array}{l}\text{2,3-Lutidine (47\%)}\\\text{2,5-Lutidine (1\%)}\end{array}\right.$	76, 185
3-Picoline	MeLi	$\left\{\begin{array}{l}\text{2,3-Lutidine (30\%)}\\\text{2,5-Lutidine (3\%)}\end{array}\right.$	185
4-Ethoxypyridine	Me$_3$CLi	2-t-Butyl-4-ethoxypyridine	88
2-t-Butyl-4-ethoxypyridine	Me$_3$CLi	2,6-Di-(t-butyl)-4-ethoxypyridine	88
4-Ethylthiopyridine	Me$_3$CLi	2-t-Butyl-4-ethylthiopyridine	88
2-t-Butyl-4-ethyl-thiopyridine	Me$_3$CLi	2,6-Di-(t-butyl)-4(1H)-pyridinethione	88
PyH	PhCaI	$\left\{\begin{array}{l}\text{2-Phenylpyridine (41\%)}^f\\\text{2,5-Diphenylpyridine (6\%)}\end{array}\right.$	91
PyH	PhCaI	$\left\{\begin{array}{l}\text{2-Phenylpyridine (42\%)}^g\\\text{2,6-Diphenylpyridine (10\%)}\end{array}\right.$	91

Table VII-12 (*Continued*)

Compound	Organometallic compound	Product (yield if reported)[a]	Ref.

| | MeLi | 6,6'-Di-(*t*-butyl)-2,2'-dipyridyl | 84 |

| | PhCH$_2$MgCl | | 93 |

| | *p*-MeOC$_6$H$_4$CH$_2$MgCl | | 93 |

| | PhCH$_2$MgCl | | 92 |

| | *p*-MeOC$_6$H$_4$CH$_2$MgCl | | 92 |

4-Picoline	EtLi	2-Ethyl-4-methylpyridine	212
4-Picoline	isoPrLi	4-Methyl-2-isopropylpyridine	212
4-Picoline	*n*-BuLi	4-Methyl-2-*n*-propylpyridine	212
4-Picoline	*n*-BuLi	2-*n*-Butyl-4-methylpyridine	212
4-Picoline	*sec*-BuLi	2-*sec*-Butyl-4-methylpyridine	212
4-Picoline	*t*-BuLi	2-*t*-Butyl-4-methylpyridine	212
4-Picoline	PhLi	2-Phenyl-4-methylpyridine	212
4-Ethylpyridine	*n*-PrLi	4-Ethyl-2-*n*-propylpyridine	212
4-Ethylpyridine	PhLi	4-Ethyl-2-phenylpyridine	212
4-*n*-Propylpyridine	*n*-PrLi	2,4-di-*n*-Propylpyridine	212
4-Phenylpyridine	EtLi	2-Ethyl-4-phenylpyridine	212
4-Phenylpyridine	*n*-PrLi	4-Phenyl-2-*n*-propylpyridine	212
4-Phenylpyridine	PhLi	2,4-Diphenylpyridine	212
4-Phenylpyridine	PhCH$_2$Li	2-Benzyl-4-phenylpyridine	212
4-(*p*-Anisyl)pyridine	EtLi	4-(*p*-Anisyl)-2-ethylpyridine	212

| | CH$_2$:CHCH$_2$MgBr | 70% | 202 |

Table VII-12 (*Continued*)

Compound	Organometallic compound	Product (yield if reported)[a]	Ref.

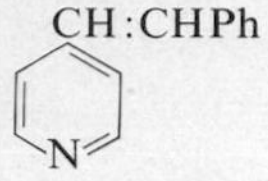

CH:CHPh (pyridine, N)

| | $CH_2:CHCH_2MgBr$ | $CH_2CHPhCH_2CH:CH_2$ (pyridine, N) trace | 202 |

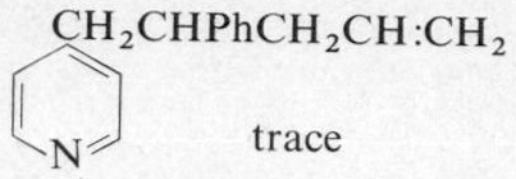

(Major product is a polymeric material)

|  pyridine N-oxide | (i) PhMgBr (ii) H_2O | 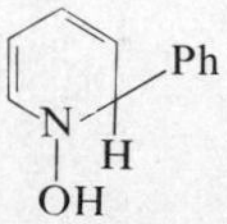 (N—Ph, H, OH structure) 60–80% | 236 |

| 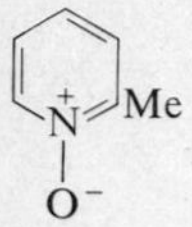 2-Me pyridine N-oxide | (i) PhMgBr (ii) H_2O | 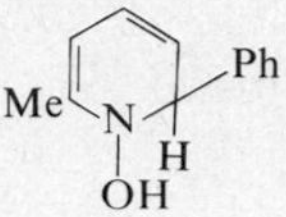 Me—N—Ph, H, OH 55% | 236 |

| 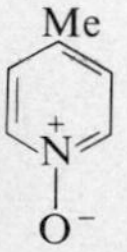 Me (pyridine N-oxide) | (i) PhMgBr (ii) H_2O | 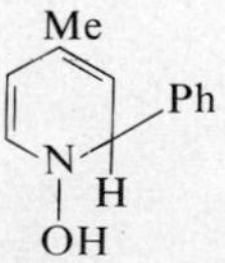 Me (N—Ph, H, OH) | 236 |

| 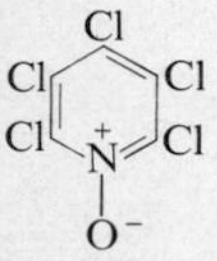 Cl Cl Cl Cl pyridine N-oxide | (i) MeMgI (ii) H_2O | 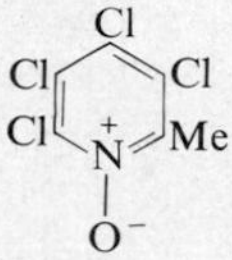 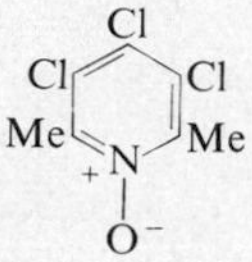 Cl Cl Cl—N(+)—Me O(−) , Cl Cl Me—N—Me O(−) (+) | 237 |

| 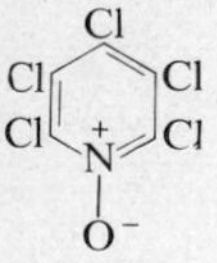 Cl Cl Cl pyridine N-oxide | (i) EtMgBr (ii) H_2O | 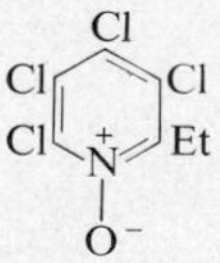 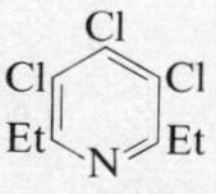 Cl Cl Cl—N(+)—Et O(−) , Cl Cl Et—N—Et O(−) | 237 |

| 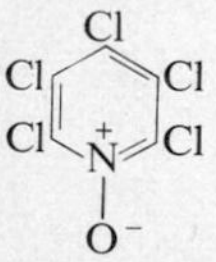 Cl Cl Cl Cl pyridine N-oxide | (i) PhMgBr (ii) H_2O | 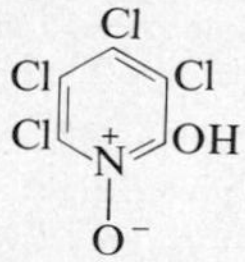 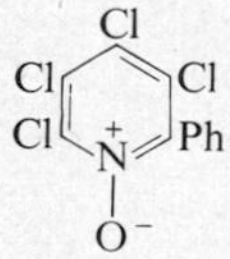 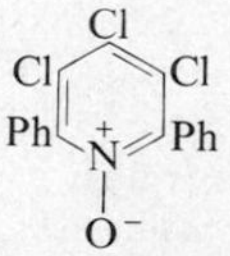 Cl Cl Cl—N(+)—OH O(−) , Cl Cl Cl—N(+)—Ph O(−) , Cl Cl Ph—N(+)—Ph O(−) | 237 |

[a] Isomer ratio (%) appears before name if determined.
[b] Phenyllithium, from PhBr + Li, plus anhydrous LiBr.
[c] Phenyllithium, from Ph_2Hg + Li.
[d] 3-Aminopyridine added to 3 molar equivalents PhLi from PhBr + Li.
[e] 3-Methoxypyridine added to 1 molar equivalent PhLi from PhBr + Li.
[f] Reaction carried out at −60°.
[g] Reaction carried out at −20°.
[h] Unstable.

All of the data summarized directly above are derived from experiments designed to determine the mode of addition of alkyl- and aryllithium compounds to pyridine bases. Several other examples of the same addition reaction have been observed as a consequence of the synthesis of several pyridine derivatives. For example, treatment of the metalation product from 2,5-lutidine and phenyllithium with ethyl chloroformate gives a 56% yield of ethyl 5-methyl-2-pyridylacetate and a 39% yield of 6-phenyl-2,5-lutidine (38). 4-Ethoxypyridine and *t*-butyllithium initially form the 1,2-adduct; loss of lithium hydride in the conventional manner gives 2-*t*-butyl-4-ethoxypyridine. The latter reacts readily with a second molecule of *t*-butyllithium to form 2,6-di-*t*-butyl-4-ethoxypyridine (**VII-54**). 4-Methyl-thiopyridine, under the same conditions and with the same reactant, gives 2-*t*-butyl-4-methylthiopyridine; but with the addition of the second molecule of *t*-butyllithium, cleavage of the sulfur-methyl carbon bond occurs, and the products isolated are **VII-55** and **VII-56** (88).

The reaction between *o*-tolyllithium and 3-picoline has yielded four products (**VII-57** to **VII-60**); **VII-59** is presumed to be formed by a modification of the mechanism proposed for the formation of 2.5-diphenylpyridine

from pyridine and phenylcalcium iodide (see below) (**VII-61**). The structure of **VII-60** is unique for this type of reaction and can arise only through a disproportionation occurring between the corresponding dihydro derivatives (89).

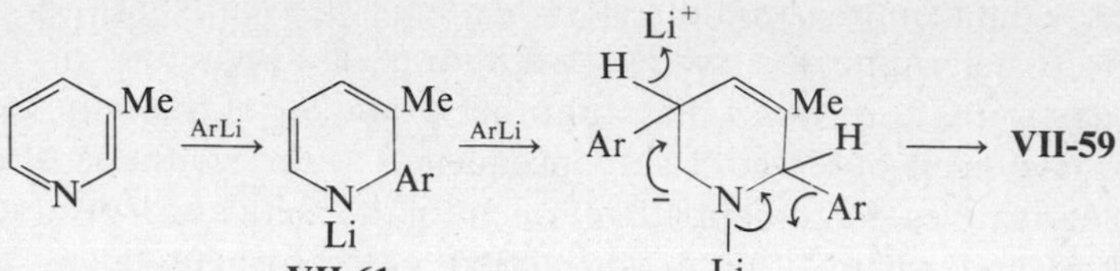

VII-61

Ar = o-tolyl

When the 1-lithio-2-phenyl-1,2-dihydropyridine adduct formed by the reaction of phenyllithium and pyridine (81) is dissolved in tetrahydrofuran and treated, at 0°, with methyl iodide, 5-methyl-2-phenylpyridine is obtained in 34–45% yield; the same adduct, with iodobenzene, benzyl chloride, or bromine gives, in unspecified yields, 2,5-diphenylpyridine, 5-benzyl-2-phenylpyridine, and 5-bromo-2-phenylpyridine. Finally, the 1-lithio-2-butyl-1,2-dihydropyridine adduct and methyl iodide give, again in unspecified yield, 2-butyl-5-methyl pyridine (90).

1-Lithio-2-phenyl-1,2-dihydropyridine can function as a reducing agent since benzophenone is converted to benzhydrol, acetophenone to α-phenylethanol, and cyclohexanol, all in low yield (164).

Phenylcalcium iodide adds readily to pyridine and the elimination of calcium hydride appears to be more rapid than with the phenyllithium adducts. Under conditions favoring the more rapid elimination of hydride ion, that is, prolonged reaction at room temperature, the products are 2-phenylpyridine (41%) and 2,5-diphenylpyridine (6%); with a shorter reaction period at room temperature the products are 2-phenylpyridine (42%) and 2,6-diphenylpyridine (10%) (91).

Quaternary pyridine derivatives undergo 1,2-addition of benzylmagnesium chlorides and these oxygen sensitive 1,2-dihydro derivatives have been utilized in the synthesis of benzomorphans (**VII-62**) (92, 93).

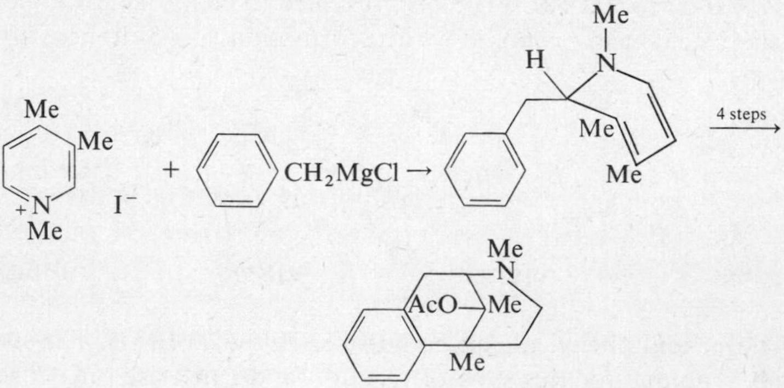

VII-62

A final example of the addition of an organometallic compound to pyridine is found in the preparation of di-(2-pyridyl)methane (**VII-63**) by the reaction of 2-pyridylmethylsodium with pyridine at an elevated temperature (94).

VII-63

VII. Pyridynes

A considerable number of reactions involving nucleophilic displacement of halogens in halopyridines proceed *via* an intermediate halogenated pyridylmetal compound (95). In this intermediate, the metal and the halogen occupy vicinal positions on the pyridine ring so that loss of metal halide generates a highly reactive species called pyridyne. As an example, the prolonged treatment of 3-bromo-2-chloropyridine in furan with lithium amalgam gives a 1.5% yield of quinoline (**VII-64**), while 3-bromo-4-chloropyridine under the same conditions forms isoquinoline in 14% yield

VII-64

(96–100). When 2-chloro-3-fluoro-4-pyridyllithium, prepared by the metalation of 2-chloro-3-fluoropyridine with *n*-butyllithium, is heated under reflux in ether with furan, the product, in 20% yield, is the *endo* oxide **VII-64a** derived from the 3,4-pyridyne (215). Under different experimental conditions, 2-fluoro-3-bromopyridyne can give either 2,3-pyridyne or 2-fluoro-3,4-pyridyne, and both species have been trapped by reaction with furan to give **VII-64b** and **VII-64c**, respectively; related derivatives are obtained with 2-methylfuran, 2-ethylfuran, 2,5-dimethylfuran or 1-methylpyrrole. By-products of these reactions are 2-fluoropyridine and 3-butyl-2-fluoropyridine (224). The reaction of 1-(2,3,5,6-tetrachloro-4-pyridyl) piperidine with *n*-butyllithium at −75° involves initially metal-halogen exchange at position 3; the interaction of this product with furan at room temperature gives the

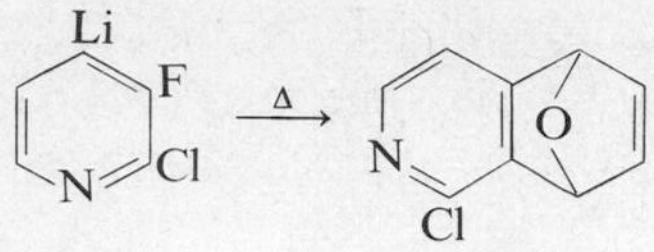

VII-64a

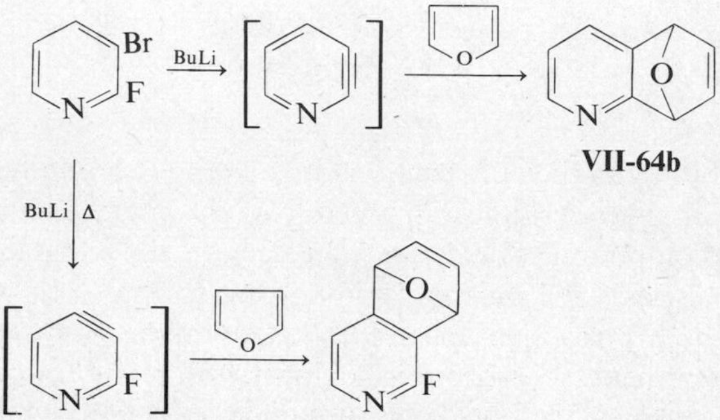

VII-64b

VII-64c

stable *endo* oxide (**VII-65**), by way of the pyridyne, in 45% yield (101). 4-Aryl-2,3,5,6-tetrachloropyridines react with *n*-butyllithium to give the 3-lithio derivatives, but pyrolysis of these in ether-furan yields only trace amounts of the *endo* oxides; the only products isolated are the 4-aryl-2,3,6-trichloropyridines (233).

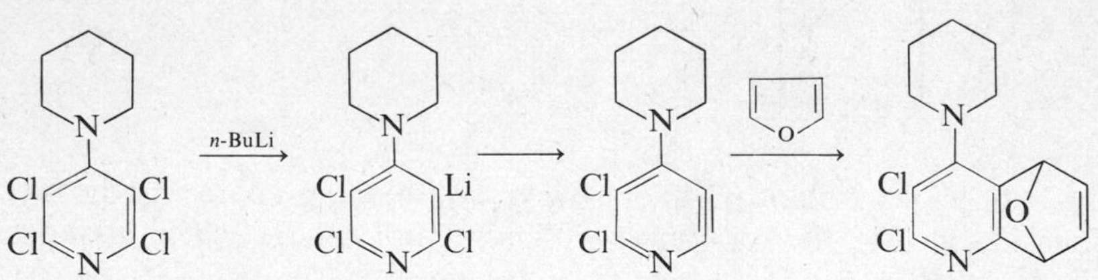

VII-65

As noted previously (p. 491), 2,5,6-trichloropyridyne, which is probably generated by the thermolysis of 2,3,5,6-tetrachloro-4-pyridyl-lithium, reacts with durene to give (**VII-4**) (6, 17, 102, 226). Unusually good yields of furan adducts, *via* 3,4-pyridynes, have been obtained with a series of 6-substituted 2,3,5-trichloro-4-pyridyllithium derivatives (**VII-65a**) (238).

An unusual demonstration of the limitations of this synthetic approach involves the reaction between 4-bromo-2,3,5,6-tetrafluoropyridine and

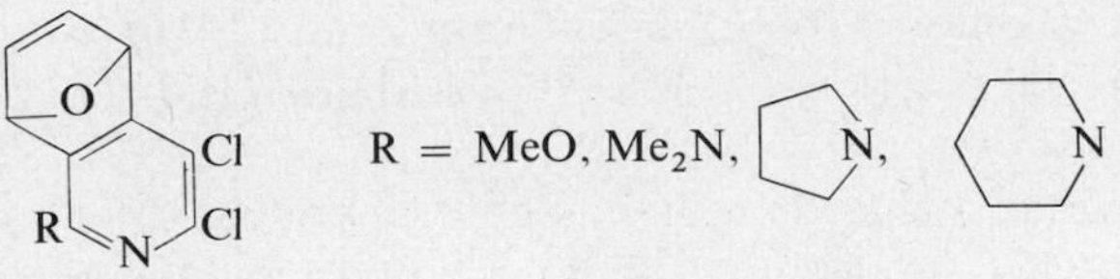

VII-65a   R = MeO, Me$_2$N, $\quad$N, $\quad$N

lithium amalgam in furan to give the diarylmercury (**VII-66**) as the principal product. No evidence could be found for the possible formation of a trifluoropyridyne from this or any other attempts to trap that intermediate (103).

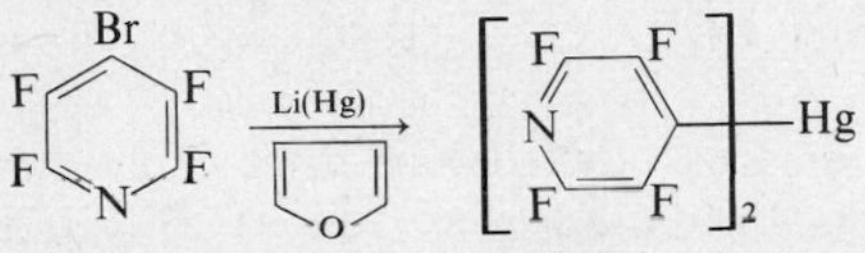

VII-66

No reaction occurs when equivalent amounts of 3-bromopyridine and sodioacetophenone are heated in toluene, under reflux. Replacing the toluene with liquid ammonia, and adding two equivalents of sodium amide has, as a consequence, the formation of 4-phenacylpyridine (13.5%), 4-aminopyridine (10%), and large amounts of amorphous nitrogenous material. The reaction can thus be visualized as involving initially the formation of pyridyne and that intermediate being trapped either by the two reagents present, or, undergoing autogenous polymerization (**VII-67**) (53) (see also Chapter VI).

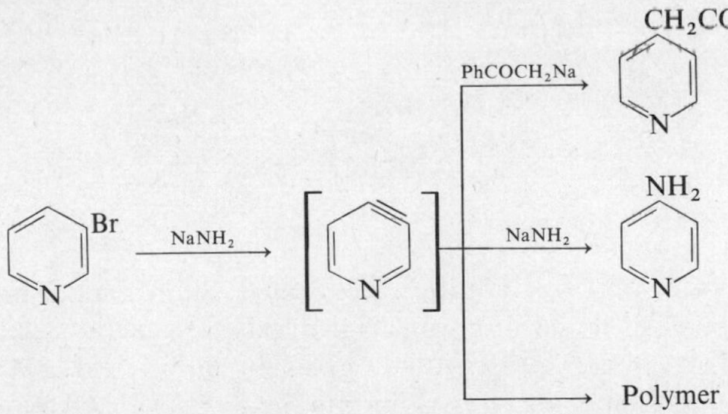

VII-67

A similar behavior has been reported with the pyridyne generated from 2-bromo-6-ethoxypyridine by the anion derived from potassium amide and 3-pentanone, in liquid ammonia. The products isolated in about 15% total yield are 2-[2'-(6'-ethoxypyridyl)]-3-pentanone and 2-[4'-(6'-ethoxypyridyl)]-3-pentanone (ratio 5:1) along with a 55% combined yield of 2- and 4-amino-6-ethoxypyridine (231).

The reactions of 3-chloropyridine, 4-chloropyridine, 3-iodopyridine, and 4-iodopyridine with potassium amide in liquid ammonia yield essentially the same mixture of products: 3-aminopyridine (25%), 4-aminopyridine (45 to 55%), and bis(4-pyridyl)amine (3 to 4%), implicating, again, the common intermediate, 3,4-pyridyne. In contrast, 2-chloropyridine and 2-iodopyridine give 75 to 80% yields of the single isomer, 2-aminopyridine (104). 3-Fluoropyridine and potassium amide behave anomalously and yield 3,3'-difluoro-4,4'-bipyridyl, 3-fluoro-4,4'-bipyridyl and either 3,3'- or 3,5'-difluoro-2,4'-bipyridyl (191). A complete discussion of these reactions is also given in Chapter VI.

Table VII-13 summarizes data from other experiments which illustrate the extent of pyridyne formation from a variety of halopyridines and potassium amide in liquid ammonia. It is apparent from several of the experiments that 1,2-addition of potassium amide can be a competing reaction during the generation of the pyridyne.

Lithium diethylamide and 2-(5-bromo-3-pyridyl)ethylamine react in ether, under reflux, to generate a 3,4-pyridyne, and this then cyclizes to 2,3-dihydro-1H-pyrrolo[3,2-c]pyridine (VII-67a) in 56% yield. Lithium piperidide and 3-benzoylamino-5-bromopyridine, in glyme under reflux, similarly generate a 3,4-pyridyne that cyclizes spontaneously to 2-phenyloxazolo[4,5-c]-pyridine (VII-67b) in 8% yield (234).

VII-67a   VII-67b

Finally, in a competitive addition between sodium amide and sodium methylmercaptide in liquid ammonia 3,4-pyridyne, generated either from 3-bromopyridine or 4-chloropyridine, gives in 60% yield a 1:1 molar mixture of 3- and 4-(methylthio)pyridine; none of the 3- or 4-aminopyridine was formed (105).

Another procedure for the formation of pyridynes involves the reaction of a 3-halopyridine with lithium piperidide. In this preparation a 3-halo-4-pyridyllithium derivative is postulated as the intermediate that generates the pyridyne *via* an elimination addition mechanism (VII-69); an

TABLE VII-13. Reactions of Halopyridines with Potassium Amide in Liquid Ammonia

Pyridine derivative	Products (yield)	Ref.
EtO, Br (pyridine)	EtO, NH₂ (pyridine), 90%	104
EtO, Br (pyridine)	EtO, NH₂ (pyridine), 95–100%	104
OEt, Br (pyridine)	OEt (pyridine), NH₂, 50–55%; OEt (pyridine), 10–15%; H₂N—(OEt, Br pyridine), 10–15%	104
OEt, Br (pyridine)	OEt, NH₂ (pyridine), 99%; OEt (pyridine), 1%	106
Br, EtO, N (pyridine)	NH₂, EtO, N (pyridine), 65–75%; EtO, N, NH₂ (pyridine), 10–15%	104
EtO, N, Br (pyridine)	EtO, N, NH₂ (pyridine), 80–85%; NH₂, EtO, N (pyridine), 10–15%	104

583

Table VII-13 (*Continued*)

Pyridine derivative	Products (yield)	Ref.
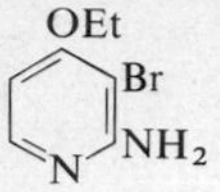 OEt, Br, NH₂ pyridine (3-OEt, 4-Br, 2-NH₂)	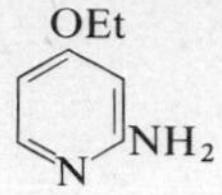 OEt, NH₂ pyridine , 5% (90% recovery of starting material)	107
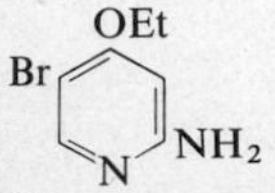 Br, OEt, NH₂ pyridine	No reaction	107
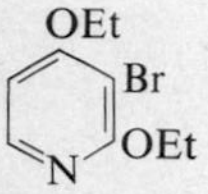 OEt, Br, OEt pyridine	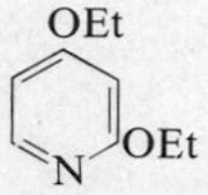 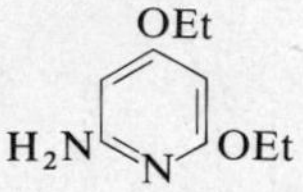 OEt, OEt pyridine , 10–15% OEt, H₂N, OEt pyridine	107
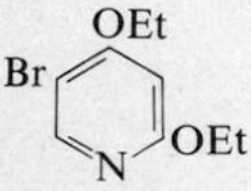 Br, OEt, OEt pyridine	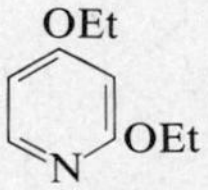 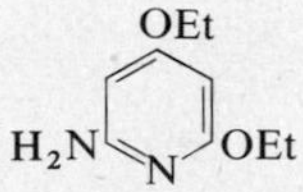 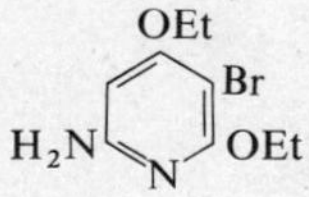 OEt, OEt pyridine OEt, H₂N, OEt pyridine OEt, Br, H₂N, OEt pyridine	107
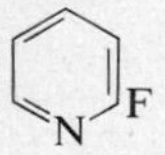 F pyridine	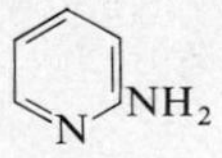 NH₂ pyridine , 75–80%	99
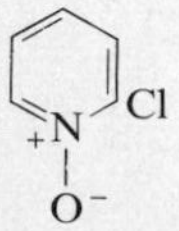 Cl pyridine N-oxide	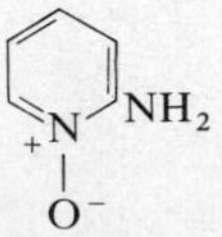 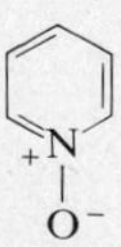 NH₂ pyridine N-oxide , 5% pyridine N-oxide , 1%	99, 108

584

Table VII-13 (*Continued*)

Pyridine derivative	Products (yield)	Ref.

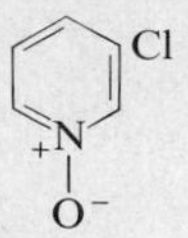

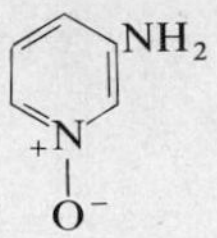

Pyridine N-oxide with Cl → NH$_2$ derivative, 80% — 99, 108

Cl-substituted pyridine N-oxide → NH$_2$ derivative, 60–70% — 99

NH$_2$ derivative, trace

No 2-amino derivative

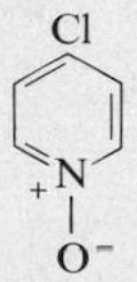

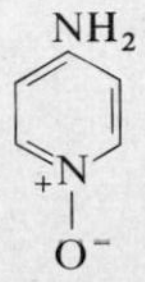

OEt/Br-substituted pyridine N-oxide → OEt, NH$_2$ derivative, 14% — 99

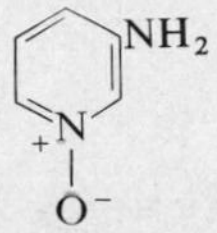

OEt, NH$_2$ derivative, 1%

OEt derivative, trace

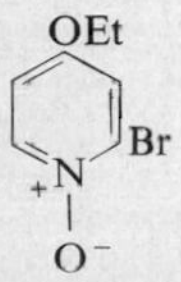

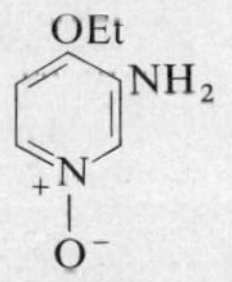

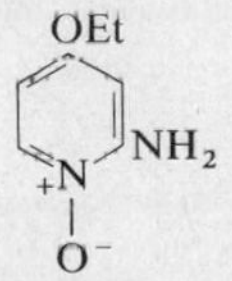

OEt/Br-substituted pyridine N-oxide → OEt, NH$_2$ derivative, 60% — 99

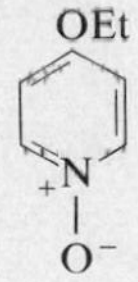

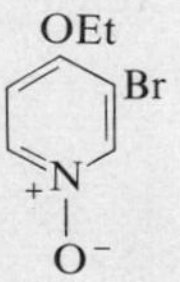

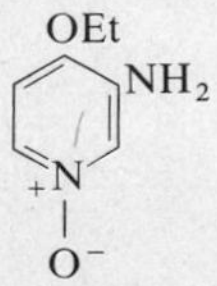

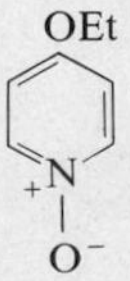

OEt derivative, trace

No 2-amino derivative

585

Table VII-13 (*Continued*)

Pyridine derivative	Products (yield)				Ref.
H₂N—[pyridine]—Br	NC—CH:CHCH₂CN, 5%				109
Me—[pyridine]—Br	Me—[pyridine]—NH₂ , 22%				109
	Me—[pyridine-NH₂] , 3%				
Ph—[pyridine]—Br	Ph—[pyridine]—NH₂ , 90%				109
	Me—[pyrimidine]—Ph , 5%				
PhO—[pyridine]—Br	PhO—[pyridine]—NH₂ 28%	PhO—[pyridine-NH₂] 9%	PhO—[pyrimidine]—Me 50%	[cyclohexadiene]—OH 5%	109
EtO—[pyridine]—Br	EtO—[pyridine]—NH₂ 72%	EtO—[pyridine-NH₂] 13%			109
Br—[pyridine]—Br	H₂N—[pyrimidine]—Me , 20%				109
O₂N—[pyridine]—Br	H₂N—[pyrimidine]—Br , 5%				109
	NC—CH:CHCH₂CN, 5%				
Me—[pyridine]—Cl	Me—[pyridine]—NH₂ 11%	Me—[pyridine-NH₂] 63%			235
Me—[pyridine]—Br	Me—[pyridine]—NH₂ 9%	Me—[pyridine-NH₂] 53%			235

Table VII-13 (*Continued*)

Pyridine derivative	Products (yield)			Ref.
Me, Cl pyridine	Me—NH₂ pyridine 14%	Me, NH₂ pyridine 56%		235
Me, Br pyridine	Me—NH₂ pyridine 14%	Me, NH₂ pyridine 76%		235
Br, Me pyridine	NH₂, Me pyridine 24%	NH₂, Me pyridine 11%		235
Cl, Me pyridine	NH₂, Me pyridine 16%	NH₂, Me pyridine 16%	H₂N, Me pyridine 3%	235
Br, Me pyridine	NH₂, Me pyridine 14%	NH₂, Me pyridine 42%	H₂N, Me pyridine 10%	235
Cl, Me pyridine	NH₂, Me pyridine 45%	H₂N, Me pyridine 15%		235
Br, Me pyridine	NH₂, Me pyridine 50%	H₂N, Me pyridine 20%		235
Br, CMe₃ pyridine	NH₂, CMe₃ pyridine 75%	H₂N, CMe₃ pyridine 15%		235

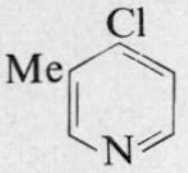

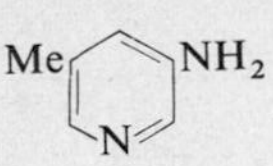

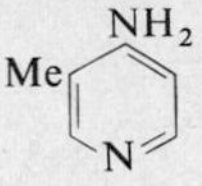

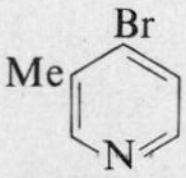

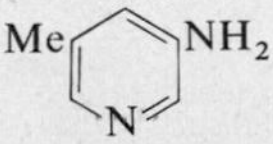

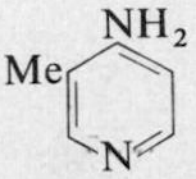

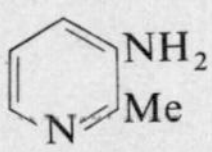

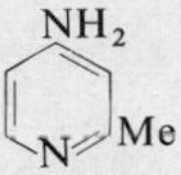

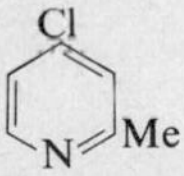

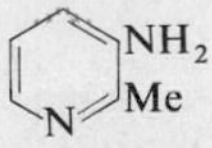

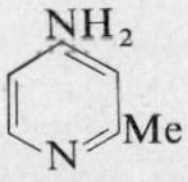

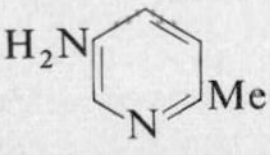

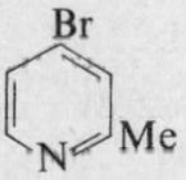

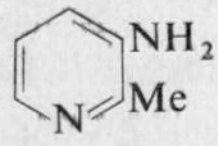

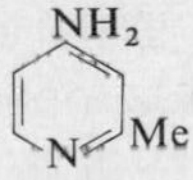

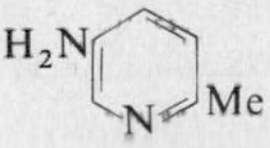

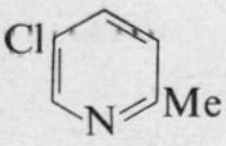

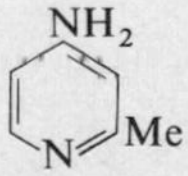

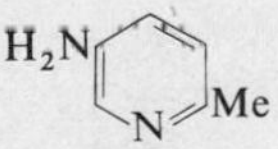

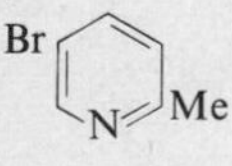

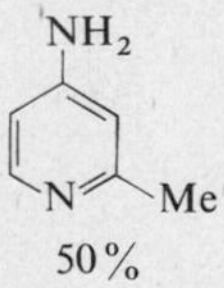

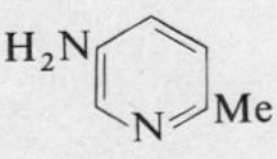

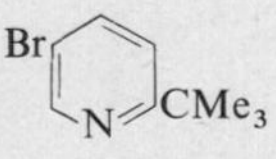

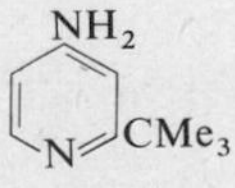

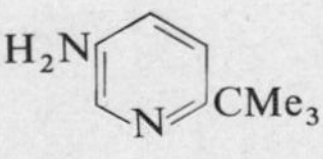

587

Table VII-13 (*Continued*)

Pyridine derivative	Products (yield)		Ref.
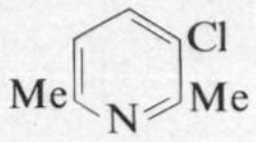 Me⟋N⟍Me, Cl	 Me⟋N⟍Me, NH₂ (30%)	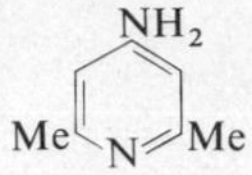 Me⟋N⟍Me, NH₂ (30%)	235
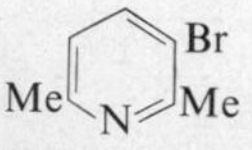 Me⟋N⟍Me, Br	 Me⟋N⟍Me, NH₂ (27%)	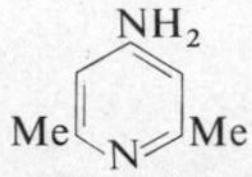 Me⟋N⟍Me, NH₂ (23%)	235
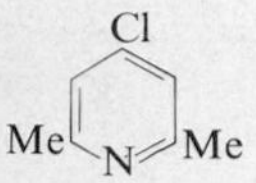 Me⟋N⟍Me, Cl	 Me⟋N⟍Me, NH₂ (14%)	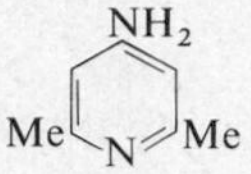 Me⟋N⟍Me, NH₂ (16%)	235
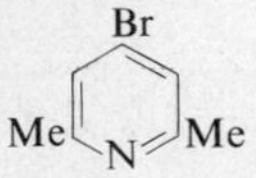 Me⟋N⟍Me, Br	 Me⟋N⟍Me, NH₂ (28%)	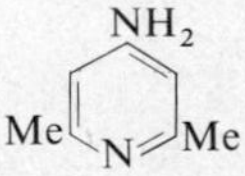 Me⟋N⟍Me, NH₂ (22%)	235

addition-elimination mechanism **VII-70** not involving a pyridyne has been proposed for the reactions of 2- or 4-halopyridines with the same reagent (24–26, 191). Table VII-14 summarizes data on isomer formation from 2-, 3-, and 4-halopyridines. The data offer adequate evidence for the formation

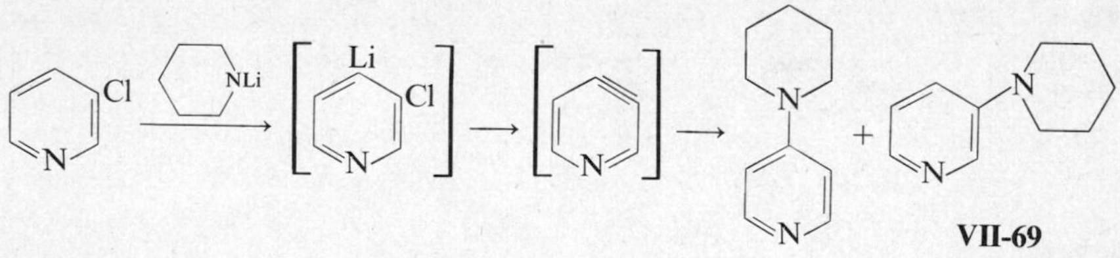

VII-69

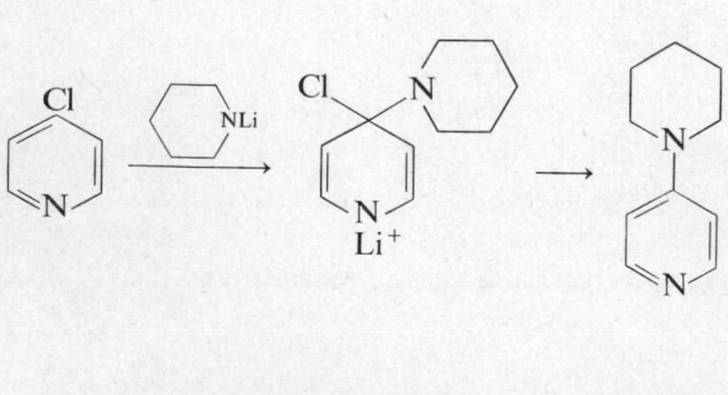

VII-70

TABLE VII-14. Reactions of Halopyridines with Lithium Piperidide

Substituent in pyridine	Total yield	Isomer Distribution in		
		2-	3-	4-
2-F	97%	100	0	0
2-Cl	<1%	Trace	0	0
2,6-Cl(Me)	75%	100	0	0
3-F	84–92%	0	100, 96	0, 4
3-Cl	66–87%	0	52, 48	48, 52
3-Br	78–88%	0	48, 48	52, 52
4-Cl	88–95%	0	0, 0.4	100, 99.6

of an intermediate pyridyne from the 3-halopyridines and, also, that a different mechanism is in operation with the 2- and 4-halopyridines.

The mechanism of pyridyne formation from halopyridines has been discussed recently (205, 206).

VIII. Mercury Compounds of Pyridine

2(1H)-Pyridone (VII-71) undergoes electrophilic substitution at positions 3 and 5 with great ease; for example, all that is required to obtain 2(1H)-pyridone-3,5-bis(mercuricacetate) (VII-72) is to warm an aqueous solution of VII-71 with mercuric acetate (110). Under the same conditions, 4(1H)-pyridone forms only the 3-mercuriacetate (100). With saturated aqueous sodium chloride, both derivatives yield the corresponding mercurichlorides. The carbon–mercury bond in the 4-pyridone derivative is significantly stronger than that in the 2-pyridone, since aqueous KI_3 at room temperature converts the former to the mercuriiodide (VII-73), but displaces both mercury atoms in the latter compound. At 100°, the reagent eliminates the mercury atom from VII-73 to give 3-iodo-4(1H)-pyridone. Alkylation of VII-73 with methyl iodide in methanolic potassium hydroxide gives 1-methyl-4(1H)-pyridone-3-mercuriiodide (110).

3-Pyridinol does not react with aqueous mercuric acetate under the conditions described above; mercuration at position 2 occurs under reflux, however, and the 2-mercuriacetate behaves normally, yielding the 2-mercurichloride with aqueous sodium chloride and 2-iodo-3-pyridinol with aqueous potassium triiodide at room temperature (110).

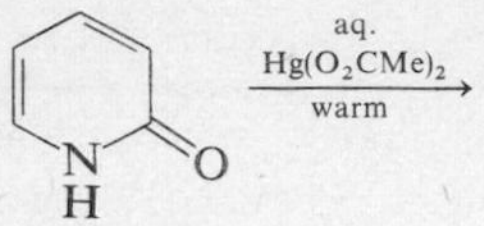

VII-71

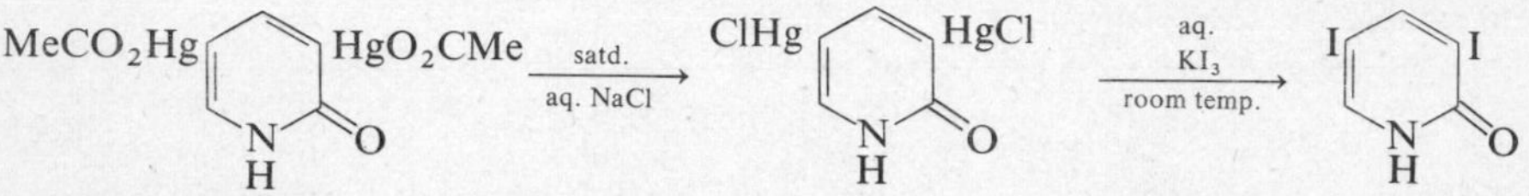

VII-72

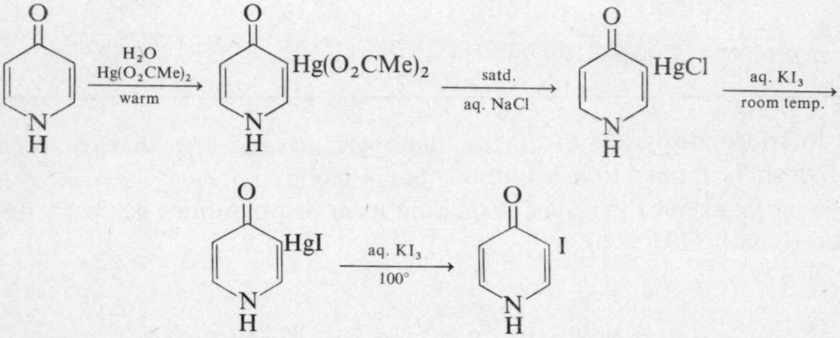

VII-73

4-Aminopyridine is more difficult to mercurate than is 4(1*H*)-pyridone. The reaction requires mercuric acetate in glacial acetic acid at 140° and yields a mixture of mono- and disubstituted mercury compounds. The mercurated 4-aminopyridine derivatives undergo the typical reactions described above as well as several additional chemical transformations (**VII-74**) (111).

With pyridine-1-oxide, mercuric acetate in 30% aqueous acetic acid at 100 to 110° gives as the major product 2-pyridylmercuric acetate-1-oxide along with a small amount of 2,6-pyridylbis(mercuriacetate)-1-oxide. This characteristic orientation of substitution by the 1-oxide group is due again to the enhanced electron densities at the 2- and 6-positions. The reactions of these derivatives are summarized below (**VII-75**) (112).

The kinetics of the addition of methoxymercuriacetate, prepared by dissolving mercuric acetate in methanol, to 5-vinyl-2-picoline have been investigated. No information is available as to the structure of the adduct, that is, either **VII-76** or **VII-77**; the addition, however, is catalyzed by acetic, nitric, and perchloric acids, and various lithium, sodium, and ammonium salts (113). The energy of activation for the reaction is estimated to be 11.4 kcal/mole.

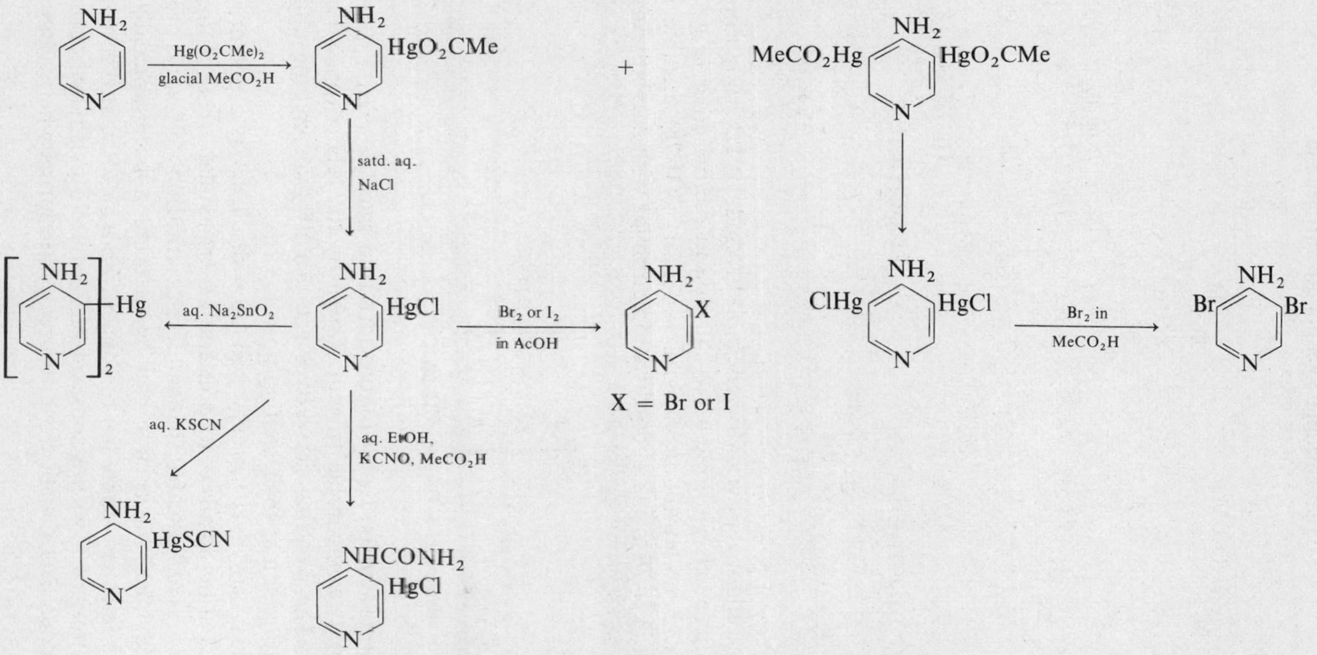

NH₂ HgO₂CMe

NH₂ HgO₂CMe
MeCO₂Hg

+

satd. aq.
NaCl

$\dfrac{\text{Hg(O}_2\text{CMe)}_2}{\text{glacial MeCO}_2\text{H}}$

NH₂

NH₂ HgCl

NH₂ HgCl
ClHg

$\dfrac{\text{Br}_2\text{ in}}{\text{MeCO}_2\text{H}}$

NH₂ Br
Br

NH₂ X

$\dfrac{\text{Br}_2\text{ or I}_2}{\text{in AcOH}}$

X = Br or I

$\dfrac{\text{aq. EtOH,}}{\text{KCNO, MeCO}_2\text{H}}$

NHCONH₂ HgCl

aq. Na₂SnO₂

aq. KSCN

NH₂ HgSCN

$\left[\begin{array}{c} \text{NH}_2 \\ \end{array} \right]_2$ Hg

VII-74

591

VII-76   **VII-77**

2,3,5,6-Tetrafluoro-4-pyridylmagnesium bromide or 2,3,5,6-tetrafluoro-4-pyridyllithium reacts with mercuric chloride in diethyl ether to give good yields of di-(2,3,5,6-tetrafluoro-4-pyridyl)mercury (**VII-78**), a stable, crystalline product; **VII-78** forms a coordination complex with 2,2'-bipyridyl (103).

VII-66

It would appear from the available data that pyridylmercury derivatives possess stability comparable to that shown by the related phenylmercury compounds. The pyridine derivatives are usually high melting and show some decomposition at their melting points.

Di-[(3-pyridyl)ethynyl] mercury had been disclosed as a potential antifungal agent but no details of its preparation are available (114–116).

A brief report has indicated that small concentrations of what is presumed to be 3-pyridylmercuric acetate can be detected polarographically (117). A method of detecting sulfite ion at concentrations as low as 0.03γ involves treating the unknown with a solution of 2-amino-5-pyridylmercuriacetate in 20% aqueous acetic acid; a white precipitate forms when sulfite ion is present.

The mercury compounds of pyridine reported since the last compilation (Part 2, pp. 462 ff) are listed in Table VII-16.

TABLE VII-15. Copper Compounds of Pyridine (44)

2-PyCH$_2$Cu[a]

[a] Postulated intermediate.

TABLE VII-16. Mercury Compounds of Pyridine

Compound	M.p. (°C)	Ref.
(?)-PyHgCl[a]	—	144
(?)-PyHgBr[a]	—	144
2-Amino-5-pyridylmercuric acetate	—	186, 187
	—	113
3-PyHgI	—	145
3-PyHgCl	—	145
3-PyHgO$_2$CMe	—	145
4-Amino-3-pyridylmercuric chloride	293–295° (dec.); hydrochloride, m.p. 191–192°	111
4-Aminopyridine-3,5-bis(mercuric chloride)	283–285° (dec.)	111
4-Amino-3-pyridylmercuric thiocyanate	195° (dec.)	111
4-Ureido-3-pyridylmercuric chloride	168–170°	111
	278° (dec.)	110
	292° (dec.)	110
	265° (dec.)	110
	310° (dec.)	110
	> 320°	110

593

Table VII-16 (*Continued*)

Compound	M.p. (°C)	Ref.
(structure: N-Me pyridone with HgI)	290°	110
(structure: pyridine with OH and HgO$_2$CMe)	84–86°	110
(structure: pyridine with OH and HgCl)	130°	110
(3-PyC⋮C)$_2$Hg	—	114–116
2-Pyridylmercuric acetate-1-oxide	194–198°	112
2-Pyridylmercuric chloride-1-oxide	—	112
(structure: ClHg–pyridine–HgCl)	—	112

[a] Position occupied by the mercury is not specified.
[b] Structure assumed by authors.

IX. Copper and Gold Compounds of Pyridine

2,3,5,6-Tetrachloropyridylmagnesium chloride, prepared in tetrahydro-furan at −10°, treated with cuprous iodide gives **VII-78**; at 0°, **VII-78** is surprisingly reactive and gives **VII-78a** and **VII-78b** in 63 and 76% yields, respectively (74). Table VII-15 lists the known copper compounds of pyridine.

CH$_2$:CHCH$_2$Br   **VII-78**   BzCl

2-Pyridylgold (I) (possibly trimeric) may be obtained in 90 to 100% yield from 2-pyridyllithium and Ph₃AsAuCl (118). On being heated to its melting point it decomposes to give gold metal and 2,2′-dipyridyl. Methylated derivatives are similarly prepared and, unlike the parent compound, are stable indefinitely to light.

X. Boron Compounds of Pyridine

The introduction of boron into the pyridine heterocycle can be affected by one of the conventional techniques, namely reaction of a pyridylmagnesium bromide or a pyridyllithium derivative with a borate ester. The dialkyl pyridineborate intermediate formed is hydrolyzed to the boronic acid (VII-79) (30).

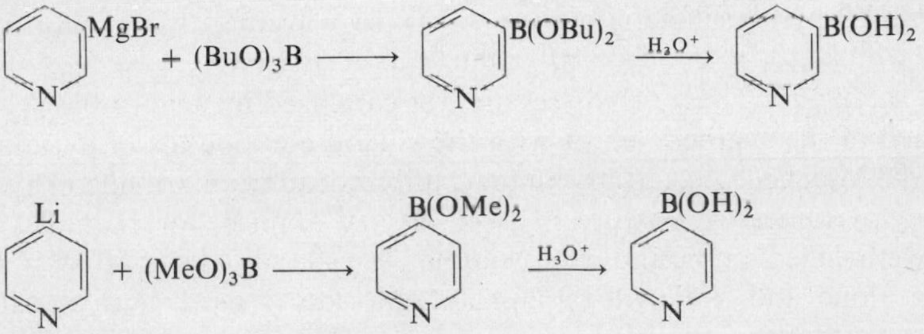

VII-79

The pyridineboronic acids are stable compounds with melting points >300°. With ethylene glycol and benzene, azeotropic distillation gives the cyclic ester (VII-80). The boron is expelled from the pyridine ring by means of hydrogen peroxide in t-butanol (VII-81) (30).

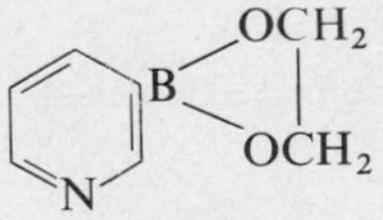

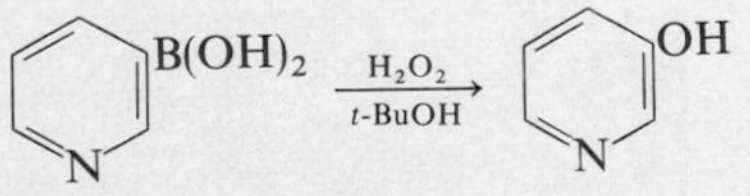

VII-80

The known boron compounds of pyridine are listed in Table VII-17; there were no compounds of this type mentioned in the earlier edition.

TABLE VII-17. Boron Compounds of
Pyridine (30)

Compound	M.p. (°C)
3-PyB(OH)$_2$	>300°
3-PyB(OCH$_2$)$_2$	185°
4-PyB(OH)$_2$	>300°
4-PyB (structure)	175°

XI. Silicon Compounds of Pyridine

Trimethyl-3- and -4-pyridylsilane are as stable as is trimethylphenyl-silane toward solvolysis by water, methanol, and ethanol. In contrast, trimethyl-2-pyridylsilane is cleaved by these solvents; the reaction is first order with $k_{MeOH} = 0.615 \times 10^{-2}$ min^{-1} at 40.15° and $k_{H_2O} = 2.42 \times 10^{-2}$ min^{-1} at 39.30°. These solvolyses are not catalyzed by acid or base (119).

Much of the synthetic effort with the silicon compounds of pyridine has involved Michael type, tertiary amine-catalyzed, additions of silicon hydrides to vinylpyridines. A mixture of two tertiary amines, for example, tetra-methylethylenediamine and tributylamine, are stated to be superior to either amine alone and, frequently, cuprous chloride is used as a co-catalyst. By this procedure, 2- and 4-vinylpyridine have been reacted successfully with dichloromethylsilane and with dimethoxymethylsilane (120). Trichloro-silane adds to 2- or 4-vinylpyridine, or to 2,6-dimethyl-4-vinylpyridine at 160° without a catalyst to give the corresponding 2-(pyridyl)trichlorosilanes (**VII-82**) (121–123). These chlorinated silanes react normally when treated

VII-82

with water, triethyl orthoformate, and a Grignard reagent (**VII-83**) (120).

2-Pyridylmethyltrimethylsilane (**VII-84**) is prepared from 2-pyridyl-methyllithium and trimethylchlorosilane in diethyl ether (124) or benzene

(123). With sulfuric acid at 0°, **VII-84** undergoes an unusual silicon-carbon cleavage involving loss of a methyl group and the formation of a disylyl

ether (**VII-85**) (124). Treatment of pentachloropyridine (**VII-86**) in tetrahydrofuran at − 70° with either methyllithium or phenyllithium followed by chlorotrimethylsilane does not give a silicon derivative; the products isolated are an unidentified heptafluorobipyridyl and 4,4′-octafluorobipyridyl (**VII-87**), respectively (see also p. 491). When, however, **VII-86** is treated with triphenylsilyllithium, a 3% yield of 2,3,5,6-tetrachloropyridyltriphenylsilane is obtained along with the major product, hexaphenyldisilane (85% yield), and a small amount of **VII-87** (33).

Although the addition of an organometallic compound to pyridine usually involves the azomethine linkage and does not yield a new organometallic product (see p. 568), an exception is noted with the silicon hydrides and with silyllithium compounds, since, with both types, new silicon derivatives of dihydro and tetrahydropyridine are obtained. Thus in the cleavage of hexaphenyldisilane in pyridine solution by means of lithium metal, the product isolated in 63% yield is 4-(triphenylsilyl)-1*H*-1,4-dihydropyridine (**VII-88**); the same product is formed from equimolar amounts of preformed triphenylsilyllithium and pyridine. Oxidation of **VII-88** in ethanolic potassium hydroxide with air or with nitrobenzene, both at reflux temperature, gives 4-pyridyltriphenylsilane (14).

$$[Ph_3Si]_2 \xrightarrow{\text{Li}} Ph_3SiLi \xrightarrow{C_5H_5N}$$

VII-88

The second example of addition involves the palladium on carbon catalyzed reaction between pyridine and trimethylsilane. At least seven products (see Table VII-18, **a** to **g**) have been isolated by vapor phase chromatography (v.p.c.). The quantity of each formed is somewhat dependent upon temperature and reaction time. Both proton magnetic resonance and infrared spectroscopy have been employed to establish the structure of these compounds, with the former being the most useful. It is implied, also, that ultraviolet absorption spectra of reference compounds and chemical reactions, especially hydrogenation and oxidation, have furnished additional confirmatory support for these structure assignments; these data, however, have not appeared (79, 80).

TABLE VII-18. Percent Yields of Trimethylsilylpyridines (**VII-89**) Under Varying Experimental Conditions.

Conditions	a	b	c	d	e	f	g	%Me₃SiH consumed
30°, 25 hr vig. stir	1.5	12	<25	35	0.8	0.2	25	95
42°, 3 days slow stir	1.0	7	1	43	9	1	30	95
60–80°, 4 hr slow stir	<1	2.4	2.2	60	17.5	4.5	12.5	60
0°, 11 days slow stir	0.6	5.8	4.7	51	8.5	1.2	28	37

Trace amounts of 1-trimethylsilylpiperidine are also formed in these reactions. A study of other catalysts showed that palladium gives the fastest rate of addition. The order found is Pd–C > PdCl₂ ⋙ Raney Ni, Pt–C, and Ru–C. Although rhodium-on-carbon gives a somewhat slower rate, it is more selective, producing only **VII-89c, d, f**. Analyses by v.p.c. during the reaction at 30° demonstrates that **VII-89d** is formed by the isomerization of **VII-89c**.

At room temperature, 2-picoline and palladium-on-carbon do not react with trimethylsilane; 4-picoline reacts at about one-quarter of the rate, and 3-picoline somewhat faster than does pyridine. 3-Picoline gives three products (VII-90 to VII-92) after 18 hr at 24° but only VII-90 after 20 hr at 40°. 4-Picoline shows an anomalous behavior in giving four products (Table VII-19 a–d), the principal one being the single example of a lateral metallation by a silicon hydride (79).

VII-90   VII-91   VII-92

Despite the variety of products obtained from these reactions there appear to be several simple mechanisms in operation. The addition of the trimethyl-silyl group to pyridine must occur on the surface of the catalyst where intermediate resonance hybrids, perhaps radicals, exist; the latter can then

TABLE VII-19. Percentage Yields of Trimethylsilylpicolines (VI-93) under Varying Experimental Conditions

Conditions	Me / N-SiMe₃ a	Me / N-SiMe₃ b	Me / N-SiMe₃ c	CH₂SiMe₃ / N d	% Me₃SiH consumed
35–40°, 40 hr or 50°, 5 hr	35	5	20	40	95
24°, 5 days	30	18	17	35	90

undergo hydrogenation, coupling, or isomerization. The formation of VII-89e, f may involve an addition-elimination mechanism while VII-93d may be formed in a variation of the coupling reaction. Steric hindrance, presumably, explains the failure of 2-picoline to react with trimethylsilane.

Since VII-89d contains the elements of a cross-conjugated dienamine, it is a reactive species; this has been demonstrated recently (125) by its reactions with ethyl azidoformate and p-tolueneisocyanate (XV-94).

Silicon derivatives of pyridine are crystalline solids or high boiling liquids, stable on storage under anhydrous conditions. Solvolysis, particularly noted with trimethyl-2-pyridylsilane (VII-94) is a general reaction, especially with the pyridylchlorosilanes, and leads to pyridylpolysiloxanes; VII-94a reacts with o- or p-chlorobenzaldehyde in alcohol to give VII-94b and that derivative in turn, is solvolyzed to VII-94c (126).

The silicon compounds reported since the last compilation (Part 2, pp. 444 ff) are listed in Table VII-20.

XII. Tin Compounds of Pyridine

A number of tin compounds of pyridine have been prepared by the addition of an organotin hydride to 4-vinylpyridine (127–129). No solvent is employed in these reactions. The 4-pyridylethyl bond to tin is far stronger than that of the *n*-propyl to tin, since bromination of **VII-95** gives di-*n*-propyl-4-(pyridylethyl) tin bromide (**VII-96**); **VII-96** is converted to the hydroxide (**VII-97**) with aqueous potassium hydroxide, and **VII-97** gives the acetate with glacial acetic (**VII-98**) (130).

2-Pyridyltrimethyltin (**VII-99**) is readily solvolyzed by water, methanol, and ethanol (119). In this behavior, **VII-99** resembles 2-pyridyltrimethylsilane (p. 596); it may be presumed, therefore, that 3- and 4-pyridyltrimethyltin would not be solvolyzed under the same conditions.

The tin compounds of pyridine described since the last compilation (Part 2, pp. 443 ff) are listed in Table VII-21.

TABLE VII-20. Silicon Compounds of Pyridine

Compound	M.p. (°C)	Ref.
2-PySiMe$_3$	—	119
3-PySiMe$_3$	—	119
4-PySiMe$_3$	—	119
2-Py(CH$_2$)$_2$SiMeCl$_2$	b.p. 100°/10 mm	120
2-Py(CH$_2$)$_2$SiMe$_3$	b.p. 117–118°/30 mm; n_D^{20} 1.4865	124
Me—[pyridine]—CH$_2$CH$_2$SiMeCl$_2$	—	120
2-PyCH$_2$CH$_2$SiPhCl$_2$	—	120
2-PyCH$_2$CH$_2$Si(C$_6$H$_4$Cl-p)Cl$_2$	—	120
2-PyCH$_2$CH$_2$SiMe(OMe)$_2$		120
2-PyCH$_2$CH$_2$SiMe(OEt)$_2$	b.p. 100–125°/1 mm	120
2-PyCH$_2$CH$_2$SiMe$_3$	b.p. 106–107°/21 mm; d^{25} 0.8982; n_D^{25} 1.4843; picrate, m.p. 89°; hydrochloride, m.p. 102–103°	123
[tetrahydropyridine]—SiMe$_3$ SiMe$_3$	b.p. 70°/2.5 mm	79
[dihydropyridine]—SiMe$_3$ SiMe$_3$	b.p. 70°/2.5 mm	79
4-PyCH$_2$SiMe$_3$	b.p. 201°	79
2-PyCH : N(CH$_2$)$_3$Si(OEt)$_3$	—	188
2-PyCH$_2$NH(CH$_2$)$_3$Si(OEt)$_3$	—	188
2-Py(CH$_2$)$_2$SiMe$_2$OSiMe$_2$(CH$_2$)$_2$Py-2	b.p. 182–183°/3 mm; n_D^{20} 1.5095; d^{20} 1.001	124
[dihydropyridine N-H]—SiPh$_3$	226–227°	14
4-PySiPh$_3$	232.5–233.5°, b.p. 430–435° (slight dec.); picrate, m.p. 198°	14
[tetrachloropyridine]—SiPh$_3$ Cl Cl Cl Cl	—	33

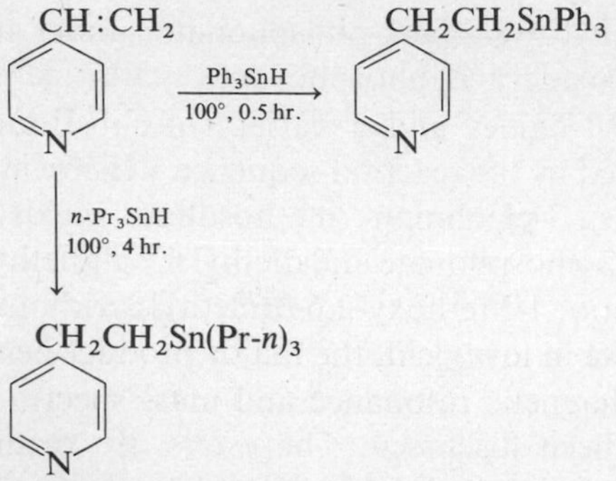

CH:CH₂

Ph_3SnH, 100°, 0.5 hr. → CH₂CH₂SnPh₃

n-Pr_3SnH, 100°, 4 hr. ↓

CH₂CH₂Sn(Pr-n)₃

VII-95

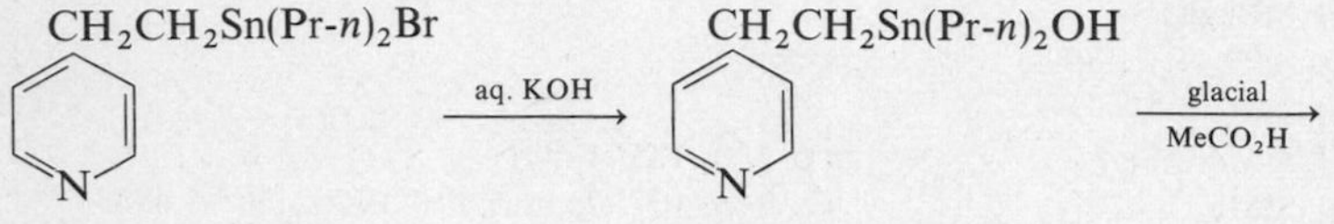

CH₂CH₂Sn(Pr-n)₂Br

VII-96

aq. KOH →

CH₂CH₂Sn(Pr-n)₂OH

VII-97

glacial MeCO₂H →

CH₂CH₂Sn(Pr-n)₂O₂CMe

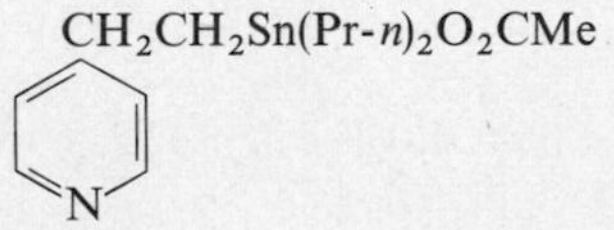

VII-98

SnMe₃

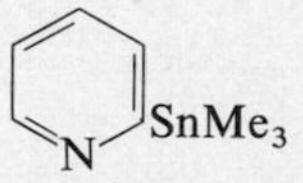

VII-99

TABLE VII-21. Tin Compounds of Pyridine

Compounds	M.p. (°C)	Ref.
4-PyCH₂CH₂SnPh₃	112–113°	127, 128, 130
2-PySnMe₃	—	119
4-PyCH₂CH₂Sn(Pr-n)₃	b.p. 121–125°/0.0009 mm; picrate, m.p. 147–148°	129
4-PyCH₂CH₂Sn(O₂CMe)(Pr-n)₂	96–97°	128

XIII. Phosphorus Compounds of Pyridine

Pyridinephosphonic acid derivatives have been prepared *via* nucleophilic displacement reactions. 3-Pyridinephosphonic acid is formed by treatment of 3-pyridinediazonium fluoroborate with phosphorous trichloride followed by treatment with water (**VII-99a**) (192). Triethyl phosphite displaces the 2-nitro group in 2-nitropyridine-1-oxide when the two are heated under reflux in acetonitrile and gives diethyl pyridine-2-phosphonate; under the same conditions, 4-nitropyridine-1-oxide and phosphorous trichloride do not react (**VII-99b**) (193). Pyridine-1-oxides and a variety of substituted pyridine-1-oxides have been converted by the reaction sequence **VII-99c** into pyridine-2-phosphonates. 1-Methoxy-3-picolinium methosulfate yields a mixture of diethyl 3-methylpyridine-2-phosphonate and diethyl 4,5-dimethyl-pyridine-2-phosphonate in a 3:1 ratio. 1-Methoxy-2,6-dimethylpyridinium methosulfate gives the 4-phosphonate in low yield, the major product being 2,6-dimethylpyridine. The proton magnetic resonance and mass spectra of these 2- and 4-phosphonates has been discussed. The esters are readily converted by heating with dilute aqueous hydrochloric acid to the corresponding phosphonic acids (194, 195).

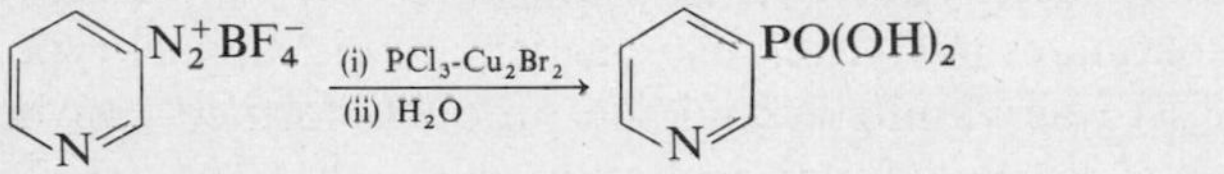

VII-99a

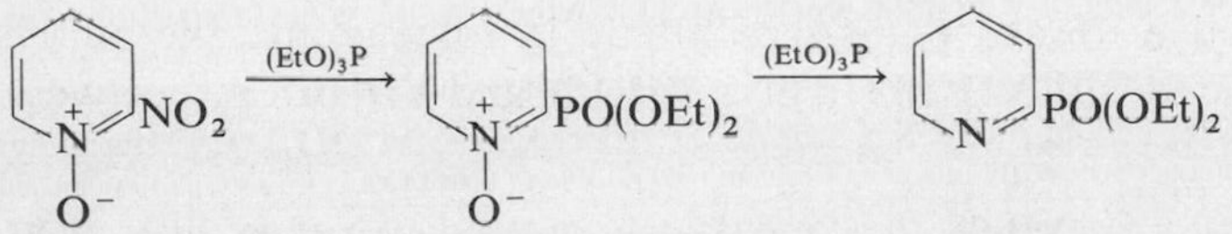

VII-99b

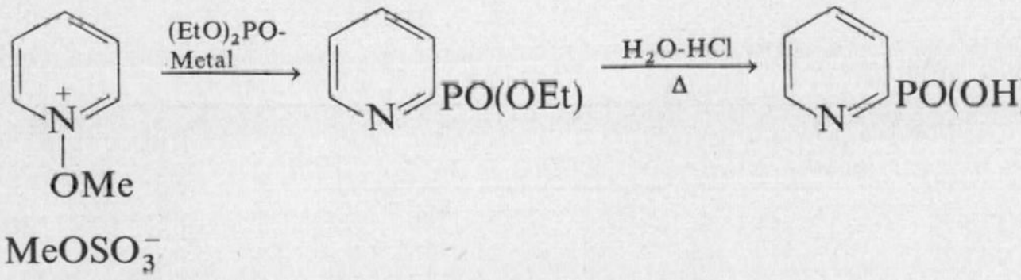

VII-99c

Dialkyl esters of phosphonic and thiophosphonic acids add to 2-vinyl-pyridine to give 2-(2-pyridyl)ethylphosphonates and 2-(2-pyridyl)ethyl-thiophosphonates, respectively. The reaction is catalyzed by base at 100°, but may be carried out without a catalyst at higher temperatures (**VII-100**) (131). This Michael-type addition has been extended to include the reactions of diethyl vinyl phosphonate with ethyl 2-pyridylacetate; again, these reactions are base-catalyzed and are carried out at 60°. The products formed undergo hydrolysis and decarboxylation in aqueous hydrochloric acid to form 3-(2-pyridyl)propylphosphonic acid (**VII-101**) (132).

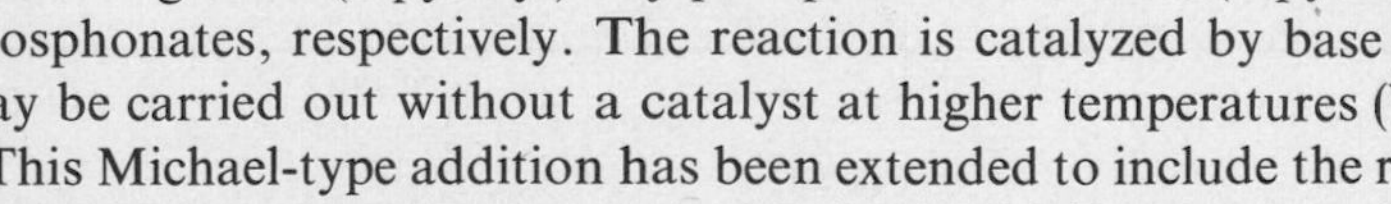

VII-100

2- And 4-pyridylmethylphosphonates (**VII-102**) are prepared by the reaction of the corresponding pyridylmethyl chloride (**VII-103**) and diethyl phosphonate; the phosphonates (**VII-102**) also participate in Michael-type, base-catalyzed additions to acrylonitrile and the products formed can be hydrolyzed to the mixed carboxylic-phosphonic acids (**VII-104**) (133). Diphenylphosphine oxide reacts with **VII-103** to give diphenyl-2-pyridyl-methylphosphine oxide (**VII-105**). The methylene protons in **VII-102** and **VII-105** are reactive toward potassium amide in toluene and the anions formed are alkylated with **VII-103** to give **VII-106** and **VII-107**, respectively.

Me$_2$ ... CHO $\xrightarrow{\text{CH}_2[\text{PO(OEt)}_2]_2}$ Me$_2$ ... CH=CHPO(OEt)$_2$ **XV-106b**

CH$_2$OH HO ... CH=CHPO(OH)$_2$ Me ... N **XV-106d** ← CH$_2$OH HO ... CH=CHPO(OEt)$_2$ Me ... N **XV-106c**

VII-106a

CH$_2$OH HO ... CH$_2$CH$_2$PO(OH)$_2$ Me ... N **XV-106e** ← CH$_2$OH HO ... CH$_2$CH$_2$PO(OEt)$_2$ Me ... N **XV-106f**

CHO HO ... CH=CHPO(OH)$_2$ Me ... N **XV-106h** ← CHO HO ... CH$_2$CH$_2$PO(OEt)$_2$ Me ... N **XV-106g**

No side reactions are observed with **VII-102**; along with **VII-106**, however, there are isolated diphenylphosphonic acid, *sym*-(2,2'-dipyridyl)ethylene, and 2-pyridylphosphonic acid (133).

The sequence of reactions shown in **VII-106a** led to a series of pyridyl-ethenyl- and pyridylethylphosphonates that possessed structural relationships to pyridoxal phosphate and pyridoxol phosphate (134). The initial reaction gave the anticipated *trans*-2-(2,2,8-trimethyl-5-(4*H*-*m*-dioxino[4, 5-*c*]pyridyl)ethenylphosphonate (**XV-106b**) (135), and this key intermediate led, by conventional procedures, to the other derivatives (**XV-106c–h**).

Tri-2-pyridylphosphine is formed from 2-pyridyllithium and phosphorus trichloride at -68 to $-58°$ (10).

The known phosphorus compounds of pyridine are listed in Table VII-22; there were no compounds of this type in the previous Edition.

XIV. Arsenic Compounds of Pyridine

Diazotization of 3-amino-2,6-dimethylpyridine in concentrated hydrochloric acid containing arsenic trichloride gives 2,6-dimethyl-3-pyridylarsine dichloride (**VII-108**); by reactions which are conventional for this class of organometallic compound (1st ed., Part IV, pp. 439 ff) **VII-108** is converted to a variety of pyridylarsenic derivatives (**VII-109**) (136).

A group of bidentate chelate compounds has been prepared from dimethyl-2-pyridylmethylarsine (**VII-110**). The arsine is readily obtained from the reaction between dimethylarsine iodide and 2-pyridylmethyllithium in diethyl ether; **VII-110** is a light-sensitive liquid with a pungent, unpleasant odor; it is readily oxidized by air through an unstable intermediate of unknown structure, to Me_2AsO_2H and 2-picoline.

The reaction of **VII-110** with cupric sulfate in aqueous alcohol results in reduction to cuprous ion, oxidation to dimethyl-2-pyridylmethylarsine oxide, and the formation of the soluble diamagnetic blue green sulfate **VII-111**; the addition of sodium perchlorate precipitates **VII-112** as the insoluble perchlorate, a salt with remarkable stability since it can be recrystallized from boiling water. Attempts to resolve **VII-112** to establish the tetrahedral disposition of the coordination covalences of the copper atom have been unsuccessful. Aeration of **VII-112** results in a slow oxidation to

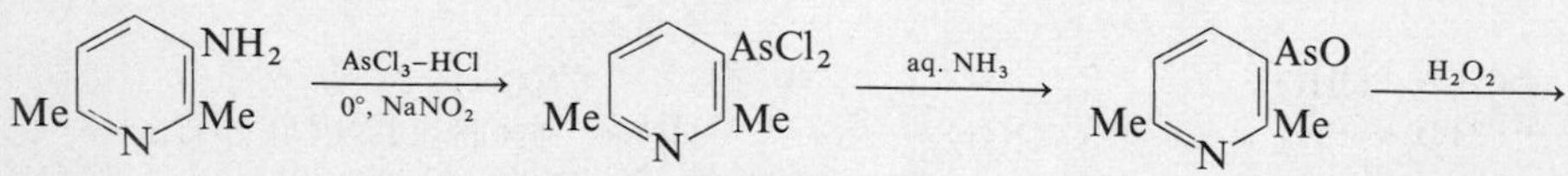

VII-108

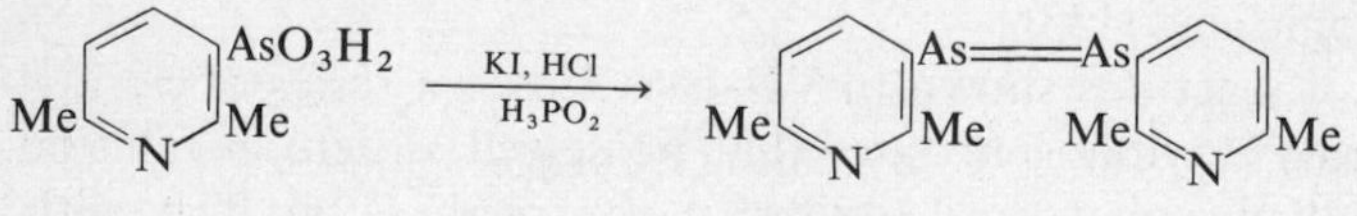

VII-109

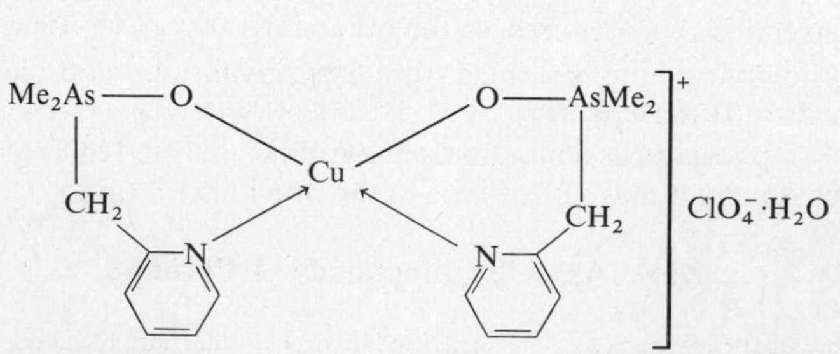

VII-112

TABLE VII-22. Phosphorus Compounds of Pyridine

Compound	M.p. (°C)	Ref.
2-PyCH(CH$_2$)$_2$PO(OEt)$_2$ 　　| 　　CN	n_D^{20} 1.4935; picrolonate, m.p. 102–103°	189
2-PyCH(CH$_2$)$_2$PO(OEt)$_2$ 　　| 　　CO$_2$Et	b.p. 122–123°/0.01 mm; n_D^{20} 1.4891	189
2-Py(CH$_2$)$_3$PO(OH)$_2$	123–124°	189
2-Py(CH$_2$)$_2$PO(OEt)$_2$	b.p. 101°/0.05 mm; n_D^{25} 1.4938, d^{25} 1.1156; picrolonate, m.p. 91–92	131
2-Py(CH$_2$)$_2$PO(OPr-n)$_2$	b.p. 105°/0.01 mm; n_D^{25} 1.4880, d^{25} 1.0725; picrolonate, m.p. 98–99°	131
2-Py(CH$_2$)$_2$PO(OPr-iso)$_2$	b.p. 99°/0.05 mm; n_D^{25} 1.4812, d^{25} 1.0691; picrolonate, m.p. 134–135° (dec.)	131
2-Py(CH$_2$)$_2$PO(OBu-n)$_2$	b.p. 132°/0.2 mm; n_D^{25} 1.4841, d^{25} 1.0460	131
2-Py(CH$_2$)$_2$PO(OEt)Ph	b.p. 143–145°/0.1 mm; n_D^{25} 1.5560, d^{25} 1.1419; picrate, m.p. 158–159°	131
2-Py(CH$_2$)$_2$PO(OCH$_2$Ph)$_2$	118°; picrate, m.p. 158–168°	131
2-Py(CH$_2$)$_2$PS(OEt)$_2$	b.p. 97°/0.05 mm; n_D^{25} 1.5101, d^{30} 1.1085	131
(2-Py)$_3$P	114°	10
4-PyCH$_2$PO(OEt)$_2$	b.p. 89/0.05 mm; n_D^{25} 1.4955; picrate, m.p. 129–130°	133
4-PyCH$_2$PO$_3$H$_2$	241–243°	133
4-PyCH$_2$POPh$_2$	220–221°; hydrochloride, m.p. 223–224°	133
2-PyCH$_2$POPh$_2$	133–134°; hydrochloride, m.p. 186–188°	133
4-PyCH$_2$PO(CH$_2$Ph)$_2$	221–222°	133
2-PyC(CH$_2$CH$_2$CN)$_2$PO(OEt)$_2$	78°, b.p. 185–190°/0.05 mm	133
2-PyC(CH$_2$CH$_2$CO$_2$H)$_2$PO$_3$H$_2$	143°	133
4-Py(CH$_2$CH$_2$CO$_2$H)$_2$PO$_3$H$_2$	220°	133
2-Py(CH$_2$CH$_2$CN)$_2$POPh$_2$	—	133
4-PyCH(CH$_2$CH$_2$CN)POPh$_2$	199–200°	133
4-PyCH(CH$_2$CH$_2$CO$_2$H)POPh$_2$	200°	133
2-PyCH(CH$_2$Py-2)PO(OEt$_2$)$_2$	b.p. 131–133°/0.01 mm; n_D^{25} 1.5350	133
2-PyCH(CH$_2$Py-2)PO$_3$H$_2$	222–224°	133
2-PyCH(CH$_2$Py-2)POPh$_2$	143.0–143.5°; dihydrochloride, m.p. 201–202°	133

Me$_2$ (structure: dioxane-fused pyridine with Me on N) —CH=CHPO(OEt)$_2$(trans) | 67–68°; hydrochloride, m.p. 147–149° | 134

(structure: HO, CH$_2$OH, Me-pyridine) —CH=CHPO(OEt)$_2$ | 97–98°; hydrochloride, m.p. 128–130° | 134

607

Table VII-22 (*Continued*)

Compound	M.p. (°C)	Ref.
CH₂OH / HO—[pyridine]—CH₂CH₂PO(OEt)₂ / Me, N	108–109°	134
CHO / HO—[pyridine]—CH₂CH₂PO(OEt)₂ / Me, N	Syrup	134
CHO / HO—[pyridine]—CH=CHPO(OEt)₂ / Me, N	Syrup	134
CH₂OH / HO—[pyridine]—CH=CHPO(OH)₂(*trans*) / Me, N	>270°	134
CH₂OH / HO—[pyridine]—CH₂CH₂PO(OH)₂ / Me, N	>270°	134
CHO / HO—[pyridine]—CH=CHPO(OH)₂ / Me, N	—	134
CHO / HO—[pyridine]—CH₂CH₂PO(OH)₂ / Me, N	Syrup	134
2-PyPO(OEt)₂	b.p. 96–97°/0.03 mm; picrate, m.p. 86–87°	194
2-PyPO(OH)₂	224–227°	194
[pyridine]—Me / N—PO(OEt)₂	b.p. 109–110°/0.07 mm	194

Table VII-22 (*Continued*)

Compound	M.p. (°C)	Ref.
(pyridine, 2-Me, 3-PO(OH)$_2$)	279–282°	194
(pyridine, 3-Me, 2-PO(OEt)$_2$)	b.p. 109–112°/0.05 mm	194
(pyridine, 3-Me, 2-PO(OH)$_2$)	272–276° (dec.)	194
(pyridine, 6-Me, 2-PO(OEt)$_2$)	b.p. 110–111°/0.1 mm	194
(pyridine, 3,4-diMe, 2-PO(OEt)$_2$)	b.p. 125–126°/0.05 mm	194
(pyridine, 4,6-diMe, 2-PO(OEt)$_2$)	b.p. 107°/0.03 mm	194
(pyridine, 4,6-diMe, 2-PO(OH)$_2$)	> 300°	194
(pyridine, 3,6-diMe, 2-PO(OEt)$_2$)	—	196
(pyridine, 3-Cl, 2-PO(OEt)$_2$)	b.p. 125–126°/0.2 mm	195
(pyridine, 3-Cl, 2-PO(OH)$_2$)	252–254°	195
(pyridine, 3-F, 2-PO(OEt)$_2$)	b.p. 124–127°/0.1 mm	195
(pyridine, 3-F, 2-PO(OH)$_2$)	220–222°	195

Table VII-22 (*Continued*)

Compound	M.p. (°C)	Ref.
Ph / pyridine-PO(OEt)$_2$	—	195
Ph / pyridine-PO(OH)$_2$	268–271°	195
CH$_2$Ph / pyridine-PO(OEt)$_2$	—	195
CH$_2$Ph / pyridine-PO(OH)$_2$	269–272°	195
pyridine-PO(OCD$_2$CD$_3$)$_2$	b.p. 93–95°/0.03 mm	195
pyridine-POCl$_2$	b.p. 88–90°/0.1 mm	195
Me, Me / pyridine-PO(OEt)$_2$	—	195
[Me, Me / pyridine-PO(O)]$_2$O·H$_2$O	296–302° disodium salt of acid, m.p. >300°	195
Me / pyridine-PO(OEt)$_2$	b.p. 110–112°/0.03 mm	195
Me / pyridine-PO(OH)$_2$	>300°	195
Me / pyridine-Et / PO(OEt)$_2$Me	—	195

Table VII-22 (*Continued*)

Compound	M.p. (°C)	Ref.
$\left[\begin{array}{c}\text{Et}\\ \text{Me} \quad \text{N} \quad \text{PO(O)}\end{array}\right]_2 \text{O·H}_2\text{O}$	278–284° disodium salt of acid m.p. > 300	195
PO(OEt)$_2$ Me N Me	b.p. 105°/0.2 mm	194
PO(OH)$_2$ Me N Me	> 300°	194

the cupric form. Similar chelates are also found with silver, palladium, and ruthenium salts (137).

Tri-(2-pyridyl)arsine is prepared from arsenic trichloride and 2-pyridyl-lithium in diethyl ether (10).

The arsenic compounds of pyridine reported since the last compilation (Part 4, pp. 439 ff) are listed in Table VII-23.

XV. Chromium Compounds of Pyridine

The reactions of benzyl and substituted benzyl anions with chromous ion leads to the formation of transient highly colored solutions. An interpretation of the data derived from the degradation products recovered from such solutions has been that organochromium compounds, for example a benzyl-chromium(III) cation, have been formed. None of the compounds has been isolated or characterized due (*a*) to their short half-life and (*b*) to their reactivity toward oxygen (138–142). Whether or not these explanations are valid in view of the fact that other highly reactive organometallic compounds have been isolated and characterized cannot now be answered. In any event, this approach has now been developed further and the related 2-, 3-, 4-, and 1-methyl-2-pyridylmethylchromium(III) cations (**VII-113** to **VII-116**) have been prepared by the addition of an excess of chromous sulfate to an aqueous solution of the bromomethylpyridininum bromide. An immediate red-brown color appears but when an excess of chromous

TABLE VII-23. Arsenic Compounds of Pyridine

Compound	M.p. (°C)	Ref.
2,6-Dimethyl-3-pyridyldichloroarsine	175–185°	136
2,6-Dimethyl-3-pyridylarsine	112°	136
2,6-Dimethyl-3-pyridinearsonic	225° (dec.)	136
3,3'-Arseno-bis(2,6-dimethylpyridine)	115° (dec.)	136
2-PyAsMe$_2$	b.p. 86–90°/2 mm; picrate, m.p. 106°	137
Bis(2-mercaptobenzothiazolyl)-2-pyridylarsine	—	190
Bis(2-mercapto-6-chlorobenzothiazolyl)-4-chloro-2-pyridylarsine	—	190
Bis(2-mercapto-6-methylbenzothiazolyl)-4-acetylamino-2-pyridylarsine	—	190

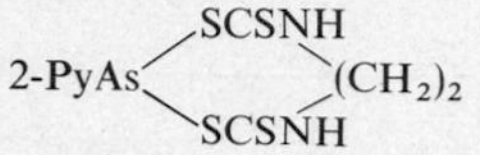

2-PyAs$\begin{smallmatrix} SCSNH \\ \\ SCSNH \end{smallmatrix}$(CH$_2$)$_2$ — 146

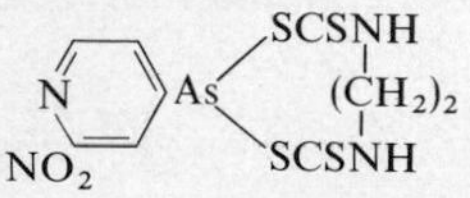

— 146

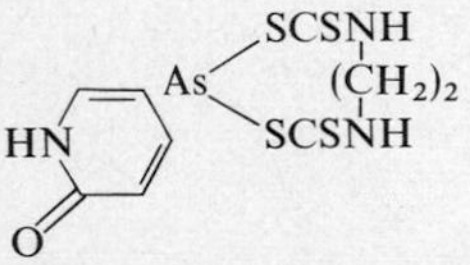

— 146

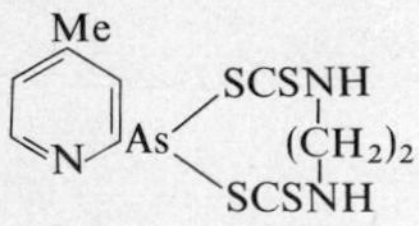

— 146

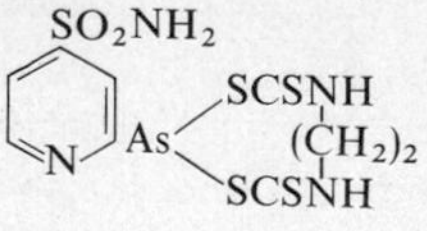

— 146

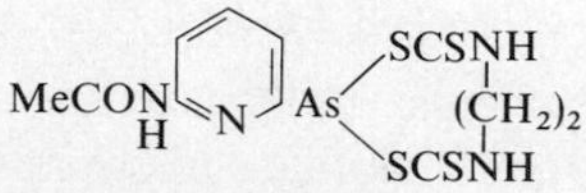

— 146

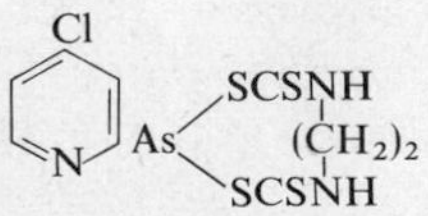

— 146

| (2-Py)$_3$As | 85° | 10 |

sulfate is present, the red-brown color changes to green. Aeration of the green solution with a source of oxygen restores the red color, presumably by selective oxidation of the inorganic chromous cation but not the pyridyl-methylchromium(III) cations. Despite the stability of these compounds in solution toward oxygen, attempts to isolate the organochromium salts have given only decomposition products.

$$R = H, \begin{cases} \textbf{VII-113; } 2\text{-CH}_2\text{Cr(H}_2\text{O)}_5{}^{2+} \\ \textbf{VII-114; } 3\text{-CH}_2\text{Cr(H}_2\text{O)}_5{}^{2+} \\ \textbf{VII-115; } 4\text{-CH}_2\text{Cr(H}_2\text{O)}_5{}^{2+} \end{cases}$$

$$R = Me, \quad \textbf{VII-116; } 2\text{-CH}_2\text{Cr(H}_2\text{O)}_5{}^{2+}$$

VII-113 to 116   VII-117

The selection of **VII-113** to **VII-116** and not **VII-117** as the structure of the cation has been based on the following considerations: (a) **VII-113** gives 4-picoline in deuterium oxide-sodium carbonate with no deuterium incorporation into the 4-methyl group or into the 2-position; (b) the ultraviolet spectra of **VII-113** to **VII-116** and of the benzylchromium(III) cation are similar (see Table VII-118), and (c) the compounds **VII-113** to **VII-116** are decomposed by perchloric acid at 55° to give the corresponding 1,2-di-(pyridyl)ethanes.

TABLE VII-118. Ultraviolet Maxima of Benzyl- and Pyridylmethylchromium Cations in Aqueous Solution

Compound	pH	$\lambda_{Max.}$ (mμ)	(ε)
VII-113	1	262, 318, 550	6,200, 10,400, 73
VII-113	5	266, 333, [a]	ε not given
VII-114	1	285, 320, 534	15,000, 9,600, 49
VII-114	5	289, 325, [a]	ε not given
VII-115	1	225, 308, 550	6,750, 15,600, 92
VII-115	5	[b] 324, [a]	ε not given
VII-116	1	266, 322, 548	8,700, 15,900, 127
VII-116	5	271, 335, [a']	ε not given
Benzylchromium cation	—	274, 299, 360	8,380, 7,920, 2,470

[a] Too weak to measure.

[b] No corresponding maximum.

The solutions containing the pyridylmethylchromium cations can be chromatographed on ZeoKarb 225SRC10 ion exchange resin, previously washed with $5M$ perchloric acid; elution with water followed by progressively higher concentrations of perchloric acid ($0.1M$ to $1.0M$) gives the purified solutions of the pyridylmethylchromium(III) cations.

The kinetics of the acid catalyzed hydrolysis of **VII-113** to **VII-116** have been investigated (138).

The known chromium compounds of pyridine are listed in Table VII-24; there were no compounds of this type in the previous edition.

TABLE VII-24. Chromium Compounds of Pyridine (138, 139)

Compound
$CH_2Cr^{2+} \cdot 5H_2O$ (pyridinium ring, $\overset{+}{N}H$ at position)
$CH_2Cr^{2+} \cdot 5H_2O$ (pyridinium ring, $\overset{+}{N}H$)
$CH_2Cr^{2+} \cdot 5H_2O$ (pyridinium ring, $\overset{+}{N}Me$)
$CH_2Cr^{2+} \cdot 5H_2O$ (pyridinium ring, $\overset{+}{N}H$)
$CH_2Cr^{2+} \cdot 5H_2O$ (pyridinium ring, $\overset{+}{N}Me$)

XVI. Potential Uses of Organometallic Compounds of Pyridine

A number of pyridine compounds of mercury have been evaluated as fungicidal, bactericidal, insecticidal, and anthelminthic agents (144–146, 111, 114–116); one pyridylarsenic derivative has been converted to polysiloxanes useful as anion exchange resins (121) and copolymerized to give silicone fluids and silicone rubbers (120, 124).

A series of 3-hydroxy-4-(hydroxymethyl)-2-methyl-5-pyridylethenyl- and 5-pyridylethylphosphonates were less potent than pyridoxol phosphate in inhibiting tyrosine decarboxylate; the same compounds were essentially equipotent with pyridoxol phosphate in inhibiting aspartate aminotransferase (134).

References

1. D. W. Adamson, P. A. Barrett, J. W. Billinghurst, and T. S. G. Jones, *J. Chem. Soc.*, 2315 (1957).
2. D. W. Adamson, P. A. Barrett, J. W. Billinghurst, and T. S. G. Jones, *J. Chem. Soc.*, 312 (1958).
3. P. A. Barrett, *J. Chem. Soc.*, 325 (1958).
4. P. A. Barrett and K. A. Chambers, *J. Chem. Soc.*, 338 (1958).
5. Ciba Ltd., Netherlands Patent 6,515,775, 1966; *Chem. Abstr.*, **65**, 15339a (1966).
6. J. D. Cook and B. J. Wakefield, *Tetrahedron Lett.*, 2535 (1967).
7. J. Finkelstein and W. Solodar, *J. Amer. Chem. Soc.*, **81**, 6508 (1959).
8. R. Lukes and O. Cervinka, *Colleot. Czeoh. Chem. Commun.*, **26**, 1893 (1961).
9. L. J. Pandya and B. D. Rilak, *J. Sci. Ind. Research*, **18B**, 371 (1959); *Chem. Abstr.*, **54**, 17391b (1960).
10. E. Plazek and R. Tyka, *Zeszyty Nauk. Politech. Wrocław., Chem.* (4) 79 (1957); *Chem. Abstr.*, **52**, 20156e (1958).
11. E. M. Roberts, G. P. Claxton, and F. G. Fallon, U.S. Patent 3,316,272, 1967; *Chem. Abstr.*, **68**, 21849c (1968).
12. Upjohn Co., Netherlands Patent 6,506,693, 0000; *Chem. Abstr.*, **64**, 15851f (1966).
13. F. J. Villani, U.S. Patent 3,188,315, 1965; *Chem. Abstr.*, **63**, 16311A (1965).
14. D. Wittenberg and H. Gilman, *Chem. Ind.* (London), 390 (1958).
15. R. L. Chambers, F. G. Drakesmith, and W. K. R. Musgrave, *J. Chem. Soc.*, 5045 (1965).
16. Société des Usines Chimiques Rhone-Poulenc, French Patent M4889, 1967; *Chem. Abstr.*, **69**, 52020a (1968).
17. J. D. Cook, B. J. Wakefield, and C. J. Clayton, *Chem. Comm.*, 150 (1967).
18. E. Ager, B. Iddon, and H. Suschitzky, *Tetrahedron Lett.*, 1507 (1969); *J. Chem. Soc., C*, 193 (1970).
19. J. D. Cook and B. J. Wakefield, *J. Chem. Soc., C*, 2376 (1969).
20. R. E. Banks, J. E. Burgess, W. M. Cheng, and R. N. Haszeldine, *J. Chem. Soc.*, 575 (1967).
21. R. D. Chambers, J. Hutchinson, and W. K. R. Musgrave, *J. Chem. Soc.*, 3736 (1964).
22. R. D. Chambers, W. K. D. Musgrave, and F. Drake-Smith, British Patent 1,134,651, 1968; *Chem. Abstr.*, **70**, 57661r (1969).
23. R. D. Chambers, C. A. Heaton, W. K. R. Musgrave, and L. Chadwick, *J. Chem. Soc., C*, 1700 (1969).
24. T. Kauffmann, *Angew. Chem.*, **4**, 543 (1965).

25. T. Kauffmann and F.-P. Boettcher, *Chem. Ber.*, **95**, 1528 (1962).
26. R. Huisgen and J. Sauer, *Angew. Chem.*, **72**, 91 (1960).
27. R. A. Abramovitch, M. Saha, E. M. Smith, and R. T. Coutts, *J. Amer. Chem. Soc.*, **89**, 1537 (1967).
28. R. A. Abramovitch and E. E. Knaus, *J. Heterocycl. Chem.*, **6**, 989 (1969).
29. R. Levine and W. M. Kadunce, *Chem. Comm.*, 921 (1970).
30. F. C. Fischer and E. Havinga, *Rec. Trav. Chim. Pays-Bas*, **84**, 439 (1965).
31. R. D. Chambers, J. Hutchinson, and W. K. R. Musgrave, Belgian Patent 660,873, July 1, 1965; *Chem. Abstr.*, **65**, 7152a (1966).
32. R. D. Chambers, J. Hutchinson, and W. K. R. Musgrave, *J. Chem. Soc.*, 5040 (1965).
33. S. S. Dua and H. Gilman, *J. Organometal. Chem.*, **12**, 299 (1968).
34. I. F. Mikhailova and V. A. Barkhash, *Zh. Obshch. Khim.*, **37**, 2792 (1967); *Chem. Abstr.*, **69**, 59056a (1968).
35. Imperial Chemical Industries Ltd., Netherlands Patent 6,512,461, 1966; *Chem. Abstr.*, **65**, 15348a (1966).
36. D. I. Davies, J. N. Done, and D. H. Hey, *J. Chem. Soc.*, C, 2019 (1969).
37. R. Bodalski, J. Michalski, and K. Studniarski, *Rocz. Chem.*, **40**, 1505 (1966); *Chem. Abstr.*, **66**, 94890c (1967).
38. J. Izdebski, *Rocz. Chem.*, **39**, 1625 (1965); *Chem. Abstr.*, **64**, 17535h (1966).
39. J. Shavel, Jr., and G. C. Morrison, U.S. Patent 3,359,273, 1967; *Chem. Abstr.*, **69**, 2872j (1968).
40. H. L. Lochte and T. H. Cheavens, *J. Amer. Chem. Soc.*, **79**, 1667 (1957).
41. H. Pines and B. Notari, *Amer. Chem. Soc. Div. Petrol. Chem. Preprints*, **4** (4) B47 (1959); *Chem. Abstr.*, **57**, 11151g (1962).
42. R. F. Shuman and E. D. Amstitz, *Rec. Trav. Chim. Pays-Bas*, **84**, 441 (1965).
43. M. E. Derieg, I. Douvan, and R. I. Fryer, *J. Org. Chem.*, **33**, 1290 (1968).
44. T. Kauffmann, G. Beissner, E. Koeppelmann, D. Kuhlmann, A. Schott, and H. Schrecken, *Angew. Chem. Int. Ed. Engl.*, **7**, 131 (1968).
45. P. P. Otto, J. P. Wibaut, and G. W. Groenendaal, *Rec. Trav. Chim. Pays-Bas*, **78**, 446 (1959).
46. D. E. Ames and J. L. Archibald, *J. Chem. Soc.*, 1475 (1962).
47. Chas. Pfizer & Co., Inc., South African Patent 67/4638(1967).
48. D. J. Berry, B. J. Wakefield, and J. D. Cook, *J. Chem. Soc.*, C, 1227 (1971).
49. D. M. Lynch and W. Cole, *J. Org. Chem.*, **31**, 3337 (1966).
50. B. M. Mikhailov and G. S. Ter-Sarkisyan, *Izv. Akad. Nauk S.S.S.R., Otdel. Khim. Nauk*, 1267 (1960); *Chem. Abstr.*, **55**, 542h (1961).
51. Abbott Laboratories, British Patent 1,110,087, 1968; *Chem. Abstr.*, **69**, 52022c (1968).
52. E. M. van Heyningen and H. M. Taylor, U.S. Patent 3,397,273, 1968; *Chem. Abstr.*, **69**, 86829m (1968).
53. W. W. Leake, A. D. Miller, and R. Levine, "Abstracts," 129th Meeting, American Chemical Society, Dallas, Tex., April 8–13, 1956, p. 25N. See, also, R. Levine and W. W. Leake, *Science*, **121**, 780 (1955).
54. S. Raynolds and R. Levine, *J. Amer. Chem. Soc.*, **82**, 472 (1960).
55. R. Grzeskowiak, G. H. Jeffery, and A. Watling, *J. Chem. Soc.*, C, 326 (1967).
56. H. Pines and B. Notari, *J. Amer. Chem. Soc.*, **82**, 2209 (1960).
57. A. D. Miller and R. Levine, *J. Org. Chem.*, **24**, 1364 (1959).
58. H. Pines and D. Wunderlich, *J. Amer. Chem. Soc.*, **81**, 2568 (1959).
59. H. Pines and S. V. Kannan, *Chem. Comm.*, 1360 (1969).
60. W. M. Stalick and H. Pines, *J. Org. Chem.*, **35**. 1712 (1970).
61. D. E. Ames and B. T. Warren, *J. Chem. Soc.*, Suppl. No. 1, 5518 (1964).

62. O. Martensson and E. Nilsson, *Acta Chem. Scand.*, **15**, 1021 (1961).
63. F. E. Cislak, C. K. McGill, and G. W. Campbell, U.S. Patent 3,317,550, 1967; *Chem. Abstr.*, **67**, 21842z (1967).
64. Merck & Co., Inc., Netherlands Patent 6,604,298, 1966; *Chem. Abstr.*, **66**, 104913d (1967).
65. K. Steiner, U. Graf, and E. Hardegger, *Helv. Chim. Acta*, **46**, 1690 (1963).
66. M. Winn, D. A. Dunnigan, and H. E. Zaugg, *J. Org. Chem.*, **33**, 2388 (1968).
67. S. Boatman, T. M. Harris, and C. R. Hauser, *J. Amer. Chem. Soc.*, **87**, 5198 (1965).
68. S. Boatman, T. M. Harris, and C. R. Hauser, *J. Org. Chem.*, **30**, 3593 (1965).
69. R. E. Smith, S. Boatman, and C. R. Hauser, *J. Org. Chem.*, **33**, 2083 (1968).
70. I. T. Barnish, C. R. Hauser, and J. F. Wolfe, *J. Org. Chem.*, **33**, 2116 (1968).
71. D. Eilhauer, U. Steinke, and G. Kurtschinski, German (East) Patent 45,082, 1967; *Chem. Abstr.*, **64**, 17552b (1966).
72. E. Profft and H. W. Linke, *Chem. Ber.*, **93**, 2591 (1960).
73. J. J. Eisch and D. A. Russo, *J. Organometal. Chem.*, **14**, 13 (1968).
74. S. S. Dua, A. E. Jukes, and H. Gilman, *Org. Prep. Proc.*, **1**, 187 (1969); *Chem. Abstr.* **71**, 70456t (1969).
75. M. Miocque, *Bull. Soc. Chim. Fr.*, **2**, 326 (1960).
76. R. A. Abramovitch and G. A. Poulton, *J. Chem. Soc.*, B, 901 (1969).
77. A. B. Lal, *J. Indian Chem. Soc.*, **38**, 257 (1961).
78. G. Fraenkel and J. C. Cooper, *Tetrahedron Lett.*, 1825 (1968).
79. N. C. Cook and J. E. Lyons, *J. Amer. Chem. Soc.*, **88**, 3396 (1966).
80. General Electric Co., U.S. Patent 3,466,270, 1969; See, also, R. A. Sulzbach, *J. Organometal. Chem.*, **24**, 307 (1970).
81. C-S. Giam and J. L. Stout, *Chem. Comm.*, 142 (1969),
82. R. A. Abramovitch and C-S. Giam, *Can. J. Chem.*, **40**, 213 (1962).
83. R. A. Abramovitch and C-S. Giam, *Can. J. Chem.*, **41**, 3127 (1963).
84. R. A. Abramovitch, C-S. Giam, and A. D. Notation, *Can. J. Chem.*, **38**, 761 (1960).
85. F. Haglid, *Acta Chem. Scand.*, **21**, 329 (1967).
86. R. A. Abramovitch, K. S. Ahmed, and C-S. Giam, *Can. J. Chem.*, **41**, 1752 (1963).
87. R. A. Abramovitch and G. A. Poulton, *J. Chem. Soc.*, B, 267 (1967).
88. H. C. van der Plas and H. J. den Hertog, *Rec. Trav. Chim. Pays-Bas*, **81**, 841 (1962).
89. R. A. Abramovitch and G. A. Poulton, *Chem. Comm.*, 274 (1967).
90. C-S. Giam and J. L. Stout, *Chem. Comm.*, 478 (1970).
91. D. Bryce-Smith and A. C. Skinner, *J. Chem. Soc.*, 577 (1963).
92. E. L. May and E. M. Fry, *J. Org. Chem.*, **22**, 1366 (1957).
93. N. B. Eddy, J. G. Murphy, and E. L. May, *J. Org. Chem.*, **22**, 1370 (1957).
94. K. Thiele, A. Gross, K. Posselt, and W. Schuler, *Chim. Ther.*, **2**, 364 (1967) *Chem. Abstr.*, **69**, 51949y (1968).
95. The subject has been reviewed by H. J. den Hertog and H. C. van der Plas, *Advan. Heterocycl. Chem.*, **4**, 121 (1965) and by R. W. Hoffmann, "Dehydrobenzene and Cycloalkynes," Academic, New York, 1967.
96. T. Kauffmann and F.-P. Boettcher, *Angew. Chem.*, **73**, 65 (1961).
97. R. J. Martens and H. J. den Hertog, *Tetrahedron Lett.*, 643 (1962).
98. T. Kauffmann and F.-P. Boettcher, *Chem. Ber.*, **95**, 61 (1962).
99. R. J. Martens and H. J. den Hertog, *Rec. Trav. Chim. Pays-Bas*, **83**, 621 (1964).
100. Société des Usines Chimiques Rhône-Poulenc, French Patent M4576, 1966; *Chem. Abstr.*, **70**, 3840e (1969).
101. J. D. Cook and B. J. Wakefield; *Chem. Comm.*, 297 (1968); J. D. Cook and B. J. Wakefield, *J. Chem. Soc.*, C, 1973 (1969).

102. J. D. Cook, B. J. Wakefield, H. Heany, and J. M. Jablonski, *J. Chem. Soc.*, *C*, 2727 (1968).
103. R. D. Chambers, F. G. Drakesmith, J. Hutchinson, and W. K. R. Musgrave, *Tetrahedron Lett.*, 1705 (1967).
104. M. J. Pieterse and H. J. den Hertog, *Rec. Trav. Chim. Pays-Bas*, **80**, 1376 (1961).
105. J. A. Zoltewicz and N. Carlo, *J. Org. Chem.*, **34**, 765 (1969).
106. H. J. den Hertog, M. J. Pieterse, and D. J. Buurman, *Rec. Trav. Chim. Pays-Bas*, **82**, 1173 (1963).
107. M. J. Pieterse and H. J. den Hertog, *Rec. Trav. Chim. Pays-Bas*, **8**, 855 (1962).
108. T. Kato, T. Niitsuma, and N. Kusaka, *Yakugaku Zasshi*, **84**, 432 (1964); *Chem. Abstr.*, **61**, 4171d (1964).
109. J. W. Streef and H. J. den Hertog, *Tetrahedron Lett.*, 5945 (1968).
110. T. Takahashi and F. Yoneda, *Chem. Pharm. Bull.* (Tokyo), **6**, 611 (1958).
111. E. Profft and K. H. Otto, *J. Prakt. Chem.*, **8**, 156 (1959).
112. M. van Ammers and H. J. den Hertog, *Rec. Trav. Chim. Pays-Bas*, **77**, 340 (1958).
113. A. P. Kreshkov and L. N. Balyatinskaya, *Zh. Obshch. Khim.*, **37**, 2211 (1967); *Chem. Abstr.*, **68**, 113734q (1968).
114. H. Sumi and K. Nakamura, *Takamine Kenkyusho Nempo*, **10**, 247 (1958); *Chem. Abstr.*, **55**, 4861b (1961).
115. H. Sumi, Y. Tanaka, and Y. Kondo, *Takamine Kenkyusho Nempo*, **10**, 240 (1958); *Chem. Abstr.*, **55**, 4861a (1961).
116. K. Tanaka, I. Iwai, Y. Yura, and K. Tomita, *Chem. Pharm. Bull.* (Tokyo), **8**, 252 (1960).
117. P. S. Casey, J. J. Carroll, and N. R. Stalica, *Proc. Penna. Acad. Sci.*, **32**, 63 (1958); *Chem. Abstr.*, **53**, 5005a (1959).
118. L. G. Vaughan, *J. Amer. Chem. Soc.*, **92**, 730 (1970).
119. D. G. Anderson, M. A. M. Bradney, B. A. Loveland, and D. E. Webster, *Chem. Ind.* (London), 505 (1964).
120. B. A. Bluestein, U.S. Patent 3,071,561, 1966; *Chem. Abstr.*, **59**, 1692a (1963).
121. J. F. Brown, U.S. Patent 2,924,601, 1960; *Chem. Abstr.*, **54**, 15406f (1960).
122. F. E. Cislak, U.S. Patent 2,854,455, 1966; *Chem. Abstr.*, **53**, 5292i (1959).
123. S. Nozakura, *Bull. Chem. Soc. Jap.*, **29**, 784 (1956); *Chem. Abstr.*, **51**, 8086d (1957).
124. Midland Silicones Ltd., British Patent 757,855 (1957); *Chem. Abstr.*, **51**, 15600e (1957).
125. E. J. Moriconi and R. E. Misner, *J. Org. Chem.*, **34**, 3672 (1969).
126. T. Ogawa, M. Yasui, M. Matsui, *Agr. Biol. Chem.* (Japan), **34**, 970 (1970); *Chem. Abstr.*, **73**, 45279w (1970).
127. G. J. van der Kerk, J. G. A. Luijten, and J. G. Noltes, *Angew Chem.*, **70**, 298 (1958).
128. G. J. M. van der Kerk and J. G. Noltes, *J. Appl. Chem.*, **9**, 106 (1959).
129. F. Binns and H. Suschitzky, *Chem. Commun.*, 750 (1970); F. Binns and H. Suschitzky, *J. Chem. Soc.*, *C*, 1223 (1971).
130. G. J. M. van der Kerk and J. G. Noltes, *J. Appl. Chem.*, **9**, 179 (1959).
131. E. Maruszewska-Wieczorkowska and J. Michalski, *Bull. Acad. Polon. Sci.*, *Ser. Sci. Chim.*, *Geol. Geograph.*, **6**, 19 (1958); *Chem. Abstr.*, **52**, 16349g (1958).
132. O. Martensson and E. Nilsson, *Acta Chem. Scand.*, **15**, 1026 (1961).
133. E. Maruszewska-Wieczorkowska and J. Michalski, *Rocz. Chem.*, **38**, 625 (1964); *Chem. Abstr.*, **61**, 10702h (1964).
134. T. L. Hullar, *J. Med. Chem.*, **12**, 58 (1969).
135. C. E. Griffith, private communication cited in Ref. 134.
136. T. Batkowski and E. Plazek, *Rocz. Chem.*, **36**, 51 (1962); *Chem. Abstr.*, **57**, 15066g (1962).
137. H. A. Goodwin and F. Lions, *J. Amer. Chem. Soc.*, **81**, 311 (1959).
138. R. G. Coombes and M. D. Johnson, *J. Chem. Soc.*, *A*, 177 (1966).

139. R. G. Coombes, M. D. Johnson, and N. Winterton, *J. Chem. Soc.*, 7029 (1965).
140. F. A. L. Anet and E. Leblane, *J. Amer. Chem. Soc.*, **79,** 2649 (1957).
141. L. H. Slaugh and J. H. Raley, *Tetrahedron*, **20,** 1005 (1964).
142. J. K. Kochi and D. D. Davis, *J. Amer. Chem. Soc.*, **86,** 5264 (1964).
143. J. K. Kochi and D. Buchanan, *J. Amer. Chem. Soc.*, **87,** 853 (1965).
144. M. Gravereaux, French Patent 1,174,137, 1959; *Chem. Abstr.*, **54,** 23169g (1960).
145. S. Miyama, *Nippon Yakurigaku Zasshi*, **54,** 658 (1958); *Chem. Abstr.*, **53,** 18310g (1959).
146. M. Nagasawa, F. Yamamoto, and Y. Imamiya, Japanese Patent 7750, 1959; *Chem. Abstr.*, **54,** 21625f (1960).
147. R. J. Mohrbacher and V. Paragamian, U.S. Patent 3,389,144, 1968; *Chem. Abstr.*, **69,** 67250z (1968).
148. E. M. Roberts, G. P. Claxton, and F. G. Fallon, South African Patent 6801,121, 1968; *Chem. Abstr.*, **70,** 77794q (1969).
149. G. E. Hardtmann and H. Ott, U.S. Patent 3,408,358, 1968; *Chem. Abstr.*, **70,** 47313q (1969).
150. F. J. Villani, C. A. Ellis, R. F. Tavares, M. Steinberg, and S. Tolksdorf, *J. Med. Chem.*, **13,** 359 (1970).
151. J. D. Cook and B. J. Wakefield, *J. Chem. Soc.*, *C*, 1973 (1969).
152. R. A. Abramovitch and M. Saha, *Can. J. Chem.*, **44,** 1765 (1966).
153. T. R. Govindachari, N. S. Narasimhan, and S. Rajadurai, *J. Chem. Soc.*, 560 (1957).
154. P. N. Gordon and W. C. Austin, South African Patent, 6704,638, 1968; *Chem. Abstr.*, **70,** 57658v (1969).
155. I. L. Kotlyareoskii, L. N. Korolenok, L. G. Stadnikova, and T. G. Shishmakova, *Izv. Akad. Nauk, SSSR. Ser. Khim.*, 1224, 1966; *Chem. Abstr.*, **65,** 16932g (1966).
156. E. M. Roberts, G. P. Claxton, and F. G. Fallon, U.S. Patent 3,306,895, 1967; *Chem. Abstr.*, **68,** 39486s (1968).
157. N. S. Prostakov, K. J. Mathew, and E. N. Sedykh, *Khim. Geterotsikl. Soedin.*, 1072 (1967); *Chem. Abstr.*, **69,** 59048z (1968).
158. V. Hněvsová and I. Ernest, *Collect. Czech. Chem. Commun.*, **25,** 748 (1960).
159. C. F. Heubner, U.S. Patent 2,947,756 (1961), *Chem. Abstr.*, **55,** 575i (1961).
160. C. F. Heubner, U.S. Patent, 3,189,612, 1965; *Chem. Abstr.*, **63,** 11514 (1965).
161. I. E. Uhlemann, *J. Prakt. Chem.*, **14,** 281 (1961).
162. I. E. Uhlemann and H. Muller, *J. Prakt. Chem.,*, **30,** 163 (1965).
163. E. Uhlemann and H. Mueller, *J. Prakt. Chem.*, **30,** 163 (1965).
164. R. A. Abramovitch and B. Vig, *Can. J. Chem.*, **41,** 1961 (1963).
165. T. Y. Shen and C. H. Shunk, U.S. Patent 3,381,015, 1968; *Chem. Abstr.*, **9,** 52025f (1968). **65,** 7153c (1966).
166. D. F. Berringer, Jr. and G. Berkelhammer, Belgian Patent 667,078 (1966); *Chem. Abstr.*, **65,** 7153c (1966).
167. H. A. DeWald, U.S. Patent 3,320,269, 1967; *Chem. Abstr.*, **67,** 90678n (1967).
168. W. T. Ely, U.S. Patent 3,306,896, 1963; *Chem. Abstr.*, **67,** 73528w (1967).
169. J. M. Bonnier, J. Court, and T. Fay, *Bull. Soc. Chim. Fr.*, 1204 (1967).
170. R. H. Wiley, C. H. Jarboe, P. X. Callahan, and J. T. Nielson, *J. Org. Chem.*, **23,** 780 (1958).
171. A. P. Gray, H. Kraus, D. E. Heitmeir, and R. H. Shiley, *J. Org. Chem.*, **33,** 3007 (1968).
172. H. B. Wright, D. A. Dunnigan, and U. Biermacher, *J. Med. Chem.*, **7,** 113 (1964).
173. F. Hoffmann La Roche & Co. A.-G., Netherlands Patent 6,511,532, 1966; *Chem. Abstr.*, **65,** 3847a (1966).
174. B. Brust, R. I. Fryer, and L. H. Sternbach, U.S. Patent 3,400,126, 1968.
175. M. E. Derieg, B. Brust, and R. I. Fryer, *J. Heterocycl. Chem.*, **3,** 165 (1966).

176. F. Hoffmann La Roche & Co. A.-G., Netherlands Patent 6,508,725, 1966; *Chem. Abstr.*, **64**, 15855f (1966).
177. B. Brust, R. I. Fryer, and L. H. Sternbach, U.S. Patent 3,400,131, 1968; *Chem. Abstr.*, **69**, 106562z (1968).
178. R. I. Fryer, B. Brust, J. V. Early, and L. W. Sternbach, *J. Org. Chem.*, **31**, 2415 (1966).
179. B. Brust, R. I. Fryer, and L. H. Sternbach, U.S. Patent 3,309,375, 1967; *Chem. Abstr.*, **68**, 12862z (1968).
180. H. Zimmer and D. K. George, *Chem. Ber.*, **89**, 2285 (1956).
181. G. Fraenkel and J. W. Cooper, *Tetrahedron Lett.*, 599 (1968).
182. R. A. Abramovitch and G. A. Poulton, *J. Chem. Soc.*, *B*, 267 (1967).
183. R. A. Abramovitch, C-S. Giam, and G. A. Poulton, *J. Chem. Soc.*, *C*, 128 (1970).
184. R. A. Abramovitch and A. D. Notation, *Can. J. Chem.*, **38**, 1445 (1960).
185. F. Haglid and J. O. Noren, *Acta Chem. Scand.*, **21**, 335 (1967).
186. I. M. Korenman and A. A. Belyakov, *Uchenye Zapiski Gor'kovsk. Gosudarst. Univ. N.I. Lobacherskogs, Ser. Khim.*, (32) 85, 1958; *Chem. Abstr.*, **54**, 5327d (1960).
187. I. M. Korenman and A. A. Belyakov, *Uchenye Zapiski Gor'kovsk. Gosudarst. Univ. N.I. Lobachevskogo, Ser. Khim.*, (32) 93, 1958; *Chem. Abstr.*, **54**, 18161e (1960).
188. R. J. Lisanke, U.S. Patent, 3,008,993, 1958; *Chem. Abstr.*, **56**, 8744 (1962).
189. E. Maruszewska-Wieczorkowska and J. Michalski, *Rocz. Chem.*, **37**, 1315 (1963); *Chem. Abstr.*, **60**, 5546a (1964).
190. M. Nagasawa, Z. Aiki, and T. Maeda, Japanese Patent 81478, 1959; *Chem. Abstr.*, **54**, 2652a (1960).
191. R. J. Martens, H. J. den Hertog, and M. van Ammers, *Tetrahedron Lett.*, 3207 (1964).
192. R. D. Bennett, A. Burger, and W. A. Volk, *J. Org. Chem.*, **23**, 940 (1958).
193. J. I. G. Cadogan, D. J. Sears, and D. M. Smith, *J. Chem. Soc.*, *C*, 1314 (1969).
194. D. Redmore, *J. Org. Chem.*, **35**, 4114 (1970).
195. D. Redmore, *J. Org. Chem.*, **38**, 1306 (1973).
196. D. Redmore, *Chem. Rev.*, **71**, 315 (1971).
197. R. A. Damico, Canadian Patent 841,632 (1970).
197a. E. E. Knaus, Ph.D. Thesis, University of Saskatchewan (1970).
198. W. G. Kofron and L. M. Baclawski, *Org. Synth.*, **52**, 75 (1972).
199. T. Kauffmann, G. Beissner, W. Sahm, and A. Woltermann, *Angew. Chem. Int. Ed. Engl.*, **9**, 808 (1970).
200. T. Kauffmann, G. Beissner, and R. Maibaum, *Angew. Chem. Int. Ed. Engl.*, **10**, 740 (1971).
201. (a) T. Kauffmann, E. Wienhöfer, and A. Woltermann, *Angew. Chem. Int. Ed. Engl.*, **10**, 741 (1971); (b) T. Kauffmann, G. Beissner, and R. Maibaum, *Angew. Chem. Int. Ed. Engl.*, **10**, 740 (1971).
202. J. J. Eisch and R. L. Harrell, Jr., *J. Organometal. Chem.*, **21**, 21 (1970).
203. J. J. Eisch and C. A. Kovacs, *J. Organometal. Chem.*, **25**, C-33 (1979).
204. German Patent 2,236,488 (1973).
205. J. A. Zoltewicz, G. Grahe, and C. L. Smith, *J. Amer. Chem. Soc.*, **91**, 550 (1969).
206. J. A. Zoltewicz and C. L. Smith, *Tetrahedron*, **25**, 4332 (1969).
207. C. S. Giam and E. E. Knaus, *Tetrahedron Lett.*, 4961 (1971).
208. J. D. Cook and B. J. Wakefield, *J. Organometal. Chem.*, **13**, 15 (1968).
209. D. J. Berry, J. D. Cook, and B. J. Wakefield, *J. C. S. Perkin I*, 2190 (1972).
210. S. V. Kannan and H. Pines, *J. Org. Chem.*, **36**, 2304 (1971).
211. H. Pines, S. V. Kannan, and W. M. Stalick, *J. Org. Chem.*, **36**, 2308 (1971).
212. P. S. Anderson, U.S. Patent 3,591,592 (1971); *Chem. Abstr.*, **75**, 76618b (1971).
213. C. G. Screttas, U.S. Patent 3,691,174; French Patent 1,585,052 (1970); *Chem. Abstr.*, **73**, 130895 (1970).
214. C. Kaiser and S. T. Ross, U.S. Patent 3,661,917 (1972).

215. F. Marsais, M. Mallet, G. Queguiner, and P. Pastour, *C. R. Acad. Sci. Paris, Ser. C*, **275**, 1535 (1972).
216. R. D. Bowden, British Patent 1,268,194 (1972); *Chem. Abstr.*, **76**, 126790v (1972).
217. E. M. Kaiser, G. J. Bartling, W. R. Thomas, S. B. Nichols, and D. R. Nash, *J. Org. Chem.*, **38**, 71 (1973).
218. E. M. Kaiser, *J. Amer. Chem. Soc.*, **89**, 3659 (1967).
219. E. M. Kaiser and G. J. Bartling, *J. Org. Chem.*, **37**, 490 (1972).
220. F. J. Villani and C. A. Ellis, *J. Med. Chem.*, **13**, 1245 (1970).
221. N. Radulescu, V. Pelloni-Tamas, and I. P. Ambrus, *Rev. Chim.* (Bucharest), **23**, 11 (1972); *Chem. Abstr.*, **77**, 5296a (1972).
222. G. N. Dorofeenko, A. V. Koblik, B. A. Tertov, and T. I. Polyakova, *Khim. Geterotsikl. Soedin.*, 1580 (1972); *Chem. Abstr.*, **78**, 58189v (1973).
223. M. Goshaev, O. S. Otroshchenko, A. S. Sadykov, and M. P. Azimova, *Khim. Getroskikl. Soedin.*, 1642 (1972); *Chem. Abstr.*, **78**, 58201t (1973).
224. M. Mallet, G. Queguiner, and P. Pastour, *C. R. Acad. Sci., Paris, Ser. C*, **274**, 719 (1972).
225. R. A. Abramovitch, E. M. Smith, E. E. Knaus, and M. Saha, *J. Org. Chem.*, **37**, 1690 (1972).
226. J. D. Cook, N. J. Foulger, and B. J. Wakefield, *J. C. S. Perkin* I, 995 (1972).
227. R. A. Abramovitch, R. T. Coutts, and E. M. Smith, *J. Org. Chem.*, **37**, 3584 (1972).
228. D. V. Ioffe and T. R. Strelets, *Khim. Geterotsikl. Soedin.*, 129 (1972); *Chem. Abstr.*, **76**, 153519v (1972).
229. E. Ager, G. E. Chivers, and H. Suschitzky, *Chem. Commun.*, 505 (1972).
230. T. Kato and H. Yamanaka, *J. Org. Chem.*, **30**, 910 (1965).
231. H. Boer and H. J. den Hertog, *Tetrahedron Lett.*, 1943 (1969).
232. R. A. Fernandez, H. Heaney, J. M. Jablonski, K. G. Mason, and T. J. Ward, *J. Chem. Soc., C.* 1908 (1969).
233. J. D. Cook and B. J. Wakefield, *J. Chem. Soc., C*, 2376 (1969).
234. T. Kauffmann and H. Fischer, *Chem. Ber.*, **106**, 220 (1973).
235. L. van der Does and H. J. den Hertog, *Rec. Trav. Chim. Pays-Bas*, **91**, 1403 (1972).

General Reference

R. A. Abramovitch and J. G. Saha, *Adv. Heterocycl. Chem.*, **6**, 280 (1966).

Index

623